Biotechnology in Agriculture, Industry and Medicine

Biotechnology and Conservation of Species from Arid Regions (2 Volume Set)

Volume 1

Biotechnology in Agriculture, Industry and Medicine

Additional books in this series can be found on Nova's website
under the Series tab.

Additional e-books in this series can be found on Nova's website
under the e-book tab.

BIOTECHNOLOGY IN AGRICULTURE, INDUSTRY AND MEDICINE

BIOTECHNOLOGY AND CONSERVATION OF SPECIES FROM ARID REGIONS (2 VOLUME SET)

VOLUME 1

SARDAR A. FAROOQ
RAEID ABED
AND
SENAN BAQIR
EDITORS

nova publishers
New York

For permission to use material from this book please contact us:
Telephone 631-231-7269; Fax 631-231-8175
Web Site: http://www.novapublishers.com

Additional color graphics may be available in the e-book version of this book.

Library of Congress Cataloging-in-Publication Data

ISBN: 978-1-63117-869-6

Published by Nova Science Publishers, Inc. † New York

CONTENTS

PREFACE

Arid regions harbor a unique variety of bacterial, plant and animal species that are very important in terms of diversity, conservation and biotechnology. Having to tolerate extreme environmental conditions, these species have developed unique adaptation strategies that can be utilized for biotechnology and their conservation poses a serious challenge. Research on plant and animal species from arid environments have received scant attention in most major biotechnology books precisely due to the difficulties in applying biotechnological approaches to their improvements. This book is designed to bring together articles from biotechnology scientists from universities, government research institutes and private sector laboratories as well as policy makers and industry leaders who are interested in exploring biotechnological application of species from arid regions. The articles in this book encompass reviews and current progress and explore potential biotech applications. The contents are divided into three main tracks; each includes topics related to microbes, animals and plants, respectively. The bacterial track focuses on the diversity and adaptation strategies of microbial species under halophilic, thermophilic and desiccation conditions and their wide applications in biotechnology including bioremediation, biofuel production, oil recovery and production of different biomolecules. The animal section includes articles on current attempts to conserve endangered species, stem cell generation, differentiation and cryobanking. The chapters will discuss advancements in human assisted reproductive technologies; emphasize latest research on animals from arid regions such as camels that include reproduction, semen characterization, freezing, IVF and ET. The ethics of using stem cells, tissues and embryos in different religions will be discussed as well. The plant track will include topics such as micropropagation and plant tissue culture, genetic transformation for salinity and stress tolerance as well as the discovery of drugs and secondary metabolites from plants .

This book will provide a platform to acquire the reviews on current state of the art of biotechnology and conservation of species in arid regions and discusses how science in this field can contribute to finding solutions to the challenges we face. Stalwarts in the concerned discipline are identified and are invited to contribute reviews on programmatic topics. The book will benefit scientists with similar interests, research students, regulatory agencies, industry representatives, and public interest groups especially those working with arid environments.

Sardar A. Farooq
Raeid M. M. Abed
Senan Baqir

BIOTECHNOLOGY AND CONSERVATION OF SPECIES FROM ARID REGIONS.

VOLUME 1

In: Biotechnology and Conservation … (2 Volume Set)
Editors: S. A. Farooq, R. Abed and S. Baqir
ISBN: 978-1-63117-869-6

Chapter 1

SCOPE OF BIOTECHNOLOGY IN OMAN

Saif N. Al Bahry[1], Nallusamy Sivakumar[1*] and Salma K. Al-Musharafi[2]

[1]Department of Biology, College of Science, Sultan Qaboos University, Muscat, Oman
[2]Department of Applied Biotechnology, Sur College of Applied Science, Oman

ABSTRACT

Biotechnology refers to the use of living organisms or their products to modify human health and the human environment. It also includes the commercial utilization of microbes and its products. Microorganisms play a crucial role in different fields such as biocontrol, antibiofouling agents, fermentation and also environmental remediation. Isolation of microorganisms with biotechnological potential from the arid region is of significant importance. In this chapter, the present status of biotechnology in Oman and the possibility of utilizing biotechnology to solve some of the problems in the environment are discussed.

Keywords: Biocontrol, biofouling, mushroom, aquaculture, waste management, MEOR

INTRODUCTION

Biotechnology is the application of technology to living systems. Biotechnology is therefore a combination of scientific and engineering principles that use biological agents to process material into valuable products and solving environmental and ecological problems. Biotechnology draws on the pure biological sciences such as genetics, microbiology, animal cell culture, molecular biology, biochemistry, cell biology etc. and in many instances it also needs knowledge and methods from outside the sphere of biology. Though various biotechnological applications have been practiced for long time, genetic engineering or

* Corresponding Author: Department of Biology, College of Science, Po.box:36, PC 123, Sultan Qaboos University, Muscat, Oman.

recombinant DNA technology is the recent focal point of biotechnology. Genetic engineering involves taking one or more genes from one organism and transferring them to another organism or putting them back into the original organism in different combinations. Biotechnology has wide applications in agriculture (e.g., improved crop production, bio-control), environment (e.g., bioremediation), medicine (e.g., therapeutics and diagnostics) and industry (e.g., production of enzymes, oil recovery). Many nations move toward this discipline of biological sciences to create new promising sources of income as well as to provide new job opportunities. It is well established globally the role of biotechnology in employment and improvement of economy of both the developed and developing nations. The biobased economy focuses on biological tools and products from renewable resources to create wealth and sustainability. The economy from biobase industry is responsible for one third of the world's economy where the USA is at front followed by Canada. Leading Chinese scientists estimated that China's biobase industry output will reach US$ 250 billion, by 2020.

In Oman, biotechnology is still in the beginning stage, though it has a lot of scope. Oman is now facing different concern including devastated agricultural lands, pollution, fear of extinction of rare indigenous species and introduced plant diseases. Biotechnology can play a significant role in providing solutions to such agricultural and environmental problems. Biotechnological researches found possibilities for creation and development of biocontrol from locally-isolated fungal species for some plant pathogens like nematodes. Another biotechnological application that can find its way in Oman is bioremediation, which is the employing of specific organisms to remove toxic chemicals, degrade biomass of the municipal waste and treat wastewater. Oil industry is a backbone of Omani economy. Oil production in some of ageing oil fields in Oman is said to drop drastically and therefore being abandoned. Recent researches have demonstrated that oil recovery can be enhanced biologically. In this chapter we will review the possible role of biotechnology in different concern faced by Oman.

BIOLOGICAL CONTROL OF PESTS

There is both public and scientific apprehension about the widespread application of synthetic chemicals as pesticides. The annual usage of chemical pesticides continuously increased throughout the world. These chemical pesticides may accumulate in soil and run off into water bodies and cause adverse effects on the environment. An attractive alternative is biocontrol using of bio-insecticides, which are either insecticide chemicals made by microorganisms or microorganisms that infect and kill insects. These insect pathogens are harmless to plants and animals. Hence, the potential danger of chemical residues could be eliminated (Copping and Menn, 2000).

Examples of microorganisms now being used to destroy insects are some strains of *Bacillus thuringiensis*. These are pathogenic for many insect larvae, such as the gypsy moth, cabbage worm and tent caterpillar. These bacteria are spore-formers and produce a protein called parasporal crystal body or BT toxin. When it is ingested it dissolves and causes destruction of the surface tissue of the larval gut, causing death of the insect. The toxin is harmless to higher animals and humans (Heimpel, 1967). *B. thuringiensis* spores and protein are applied as a dust or spray to field crops, trees, ornamental plants and home vegetable gardens. Parasporal genes of *B. thuringiensis* have been incorporated into plants to resist pests by genetic engineering. Agricultural plants are one of the common examples of genetically modified organisms (GMOs).

In Oman, most of agricultural lands have been affected by introduced diseases, drought and soil salinity. For example, lemon used to be one of the major cultivars in Oman. Oman lemon is very popular and exported to different countries. At present most of lemon plants have been seriously affected by the introduced witch broom disease as a result of importing agricultural material having a strong and advanced agricultural quarantine. Several other introduced plant diseases have also been affected other important crops such as mango, dates and banana. Some nematodes and fungal diseases have affected different varieties of plants. About 35 parasitic nematodes species that infect 35 plant hosts are found in Batina region of the Sultanate of Oman. Nematodes infect all plant parts including root, stem and leaf. They also infect the flower, buds and seeds. Plant parasitic nematodes in the Sultanate of Oman infect different types of vegetable and fruit crops.

A soil fungi feed on nematodes has been found to control the disease avoiding the chemical pesticides. About 70 genera and 160 species of fungi have been found associated with nematodes but few of them are used in biological control of nematodes because of their efficiency (Kiewnick and Sikora, 2006). Fungi can attack nematodes or their eggs and utilize them as a source of nutrient. Some are predacious fungi which form specific nematode trapping organs. So the application of these nematode eating fungi as a biocontrol would be a very good remedy to get rid of these nematode infections in Sultanate of Oman.

The other approach at the experimental stages in Oman is the development of genetically modified plants. Still studies needed to know the positive and negative impacts of the GMOs on environment and possibility of commercialization. GMO plants are grown in many countries. However, many countries prohibit the use of GMOs, genetic engineering of plants and animals has wide applications.

BIOFOULING

The impairment or degradation of underwater surfaces or equipment as a result of the growth of living organisms is called biofouling. Organisms such as bacteria, protozoans, algae, and crustaceans can accumulate in large numbers on surfaces like pipes, tanks, and ships' hulls resulting in corrosion, clogging, contamination or a decrease in the efficiency of moving parts.

Marine biofouling is an undesirable growth of micro and macroorganisms on man-made surfaces causes enormous problems such as clogging pipes and membranes of desalination plants and steam-injection plants in oil industry, decreasing speed and increasing corrosion and fuel consumption in vessels, damaging fishing nets and cages and decreasing their fishing efficiency (Armstrong et al., 2000). Countries worldwide are spending more than US$5 billion per year to fight biofouling using toxic compounds that also kill non-target organisms and accumulate in the marine environment (Schultza et al., 2011).

Biofouling is causing serious impact on all industries in the Sultanate. In comparison to other regions, there is no effective strategy to deal with biofouling and its economic and environmental impact in Oman waters. Antifouling strategies in Oman can be significantly improved by utilization of low cost and low environmental impact methods effective against biofouling organisms. Due to low rainfall and lack of fresh water for drinking, Oman is heavily depending on desalinated water. Two main processes are commonly used to desalinate sea and brackish water. Salt water is either boiled or condensed to be used for human consumption. The other method of water desalination is reverse osmosis. It is the opposite process of osmosis where water is passed through membrane filters by high pressure to remove salt. However, microbes may grow heavily on the expensive membranes cause serious fouling and damage. So the application of microbes as antibiofouling agents would reduce these problems to a greater extent.

Microbial Fermentation

Microbial fermentation products are the largest industries in the world and have significant impact on the economy of the country. Some of microbial products such as vinegar, baker yeast and yoghurt have been produced using microbial fermentation from the early history of humanity. These products in Oman were commonly produced in houses without knowledge of fermentation. Most of these products are currently imported to Oman. Recently very few companies in Sohar region of Oman started to produce yoghurt. If this simple technology has been developed and utilized properly, it will increase the economy of the country.

Mushroom Cultivation, Fungiculture

Mushrooms are heterotrophs grow well at relative humidity levels of around 95-100%, and substrate moisture levels of 50 to 75%. Mushrooms contain high contents of protein and essential amino acids. Mushrooms are rich in folic acid, which reduces the risk of anaemia and brain or spinal cord defects. Mushrooms are high in riboflavin and fibre but low in calories. They don't have fat, cholesterol and sodium and is considered as a good source of potassium (Wasser, 2002). All these factors reduce heart diseases, blood pressure and stroke which are common in Oman.

To cultivate mushrooms asexually on a large scale, actively growing mushroom culture, mycelium or spawn is placed on growth substrate. This is known as inoculation or spawning. Spores are another inoculation option, but are less developed than established mycelium. The Gulf Mushroom Products Company in Barka, Oman, cultivates more than 50 tons/year. The product is supplied throughout the country and exported to different parts of the world. In Oman, the market price of the button mushroom, *Agaricus bisporous*, is around Omani Rials 2.7 /KG.

Food and Water

Droughts in grain producing nations in addition to increased oil prices caused significant increase in food prices across the globe during 2007-2008. With the oil prices remaining high, the Food and Agricultural Organization (FAO) Food Price Index in August 2011 remained unchanged from July and was 26% higher than in August 2010. The harvested crop land in Oman decreased from 72000 ha in the year 2000 to 63606 ha in the year 2004 mainly because of drought and salinity. Water shortage in some regions, salinity increase in wells and surface irrigation are limiting factors in terms of food productivity.

Food consumption in Oman is projected to grow at 5.3% over 2011-2015. Cereals will remain the leading segment, while fruits, meat and vegetables are expected to register strong growth. Other food items, which include pulses, sugar, oil, fish, eggs and potato would also register a strong growth of 8.0%. Most of food material in Oman is imported and the value of food imports by the year 2020 is expected to cost 1.8 billion Omani Rials as a result of population growth and declining of agricultural lands. In 2010, Oman has spent about 0.8 billion Omani Rails for imported food. Some of the imported food may contain unwanted chemical and biological contaminants. The Sultanate produces nearly half of its total consumption of vegetables. Thus, food security is a prime concern and major challenge for the Sultanate.

In some areas in Oman the underground water used for drinking is very contaminated. Several groundwater pollution incidents, originating from leakage of underground petroleum storage tanks, sewage storage tanks and seepage of leachate from mine tailings have been reported. Agricultural developments utilizing inorganic fertilizers and pesticides are also a major concern of groundwater pollution. Unplanned and uncontrolled practices of development and other related activities cause groundwater pollution especially bacteriological and nitrate contamination from sewage, and agricultural activities, heavy metals and salinity from tailing dams, hydrocarbon contamination from underground storage tanks, pipeline, car wash, repair workshops and re-injection of oily water.

Generally most of diseases occur as a result of ingestion of contaminated food and water (Kosek et al., 2003). The biological contamination of food occurs due to the effects of bacteria, viruses, fungi, parasites and toxins, while chemical contamination occurs when pesticides, high levels of antibiotics, additives, cleansers, metal leaching (copper, lead, cadmium), pesticides and other non-permissible chemicals get in touch with food. Microbial and chemical food poisoning are the major causes of foodborne and waterborne diseases. Some of the biotoxins, such as botulinum toxin have immediate effect and may lead to death. However, mycotoxins such as aflatoxin, have a long term effect which may lead to cancer.

Due to huge amount of food imported to the country and threats of food and water contamination produced locally, it is crucial to have well equipped laboratories and trained personnel in this area needed to analyze the quality of huge amount of imported and local food.

MEDICAL BIOTECHNOLOGY (PHARMACEUTICALS, ANTIBIOTICS AND THERAPEUTIC DRUGS)

Medical biotechnology finds potential applications in areas such as drug production, pharmaceutical and pharmacogenomics and gene therapy. The application of biotechnology in pharmacology is often associated with the use of genetically altered microorganisms such as *Escherichia coli* or yeast for the production of substances like synthetic insulin or antibiotics (Miralles et al., 2009). It can also refer to transgenic animals and transgenic plants. Genetically altered mammalian cells, such as Chinese Hamster Ovary cells (CHO), are also used to manufacture certain pharmaceuticals. Another promising new biotechnology application is the development of plant-made pharmaceuticals.

Medical biotechnology can be used to produce existing medicines relatively easily and inexpensively. For example insulin, widely used for the treatment of diabetes was previously extracted from the pancreas of animals. In 1978 Genentech, USA, developed synthetic insulin by joining its gene with a plasmid vector inserted into *E. coli*. The resulting genetically engineered bacterium enabled the production of vast quantities of synthetic human insulin at relatively low cost.

Antibiotics are antibacterial compounds or substance that kills or slows down the growth of bacteria. The first antibiotic discovered by Alexander Fleming was penicillin produced by *Penecillium notatum*. In 1941 there was no antibiotic industry, but 10 years later, the net worldwide sales of antibiotics were 1.5 million Omani Rials per year, and antibiotics had revolutionized the practice of medicine. In addition to antibacterial agents, specific compounds are used to treat fungal and viral infections. In general, these are termed as therapeutic drugs. Many of such drugs are also produced from plant secondary metabolites. Many of the well-known medicinal plants in Oman have not yet been commercially utilized though they have anti-microbial activity (Al Saidi et al., 2012). Many microorganisms have been isolated and were found to produce antibiotics in Oman. However, so far there are very few antibiotic producing industries exist and still no industries for production of drugs from medicinal plants. This industry is promising if a gap between research findings and commercial production is filled in the future.

Applications of fungicidal products have also been used as ingredients of paints. Deterioration of painted surfaces in Oman is very common specifically during the humid seasons. These discolorations are caused by metabolic products of molds growing on the organic constituents of the paint. In Oman, Many species of molds have been isolated from mildewed or "moldy" painted surfaces. Unless paint contains an effective fungicidal ingredient, mold spotting or mildew may appear on painted surfaces.

Aquaculture

Aquaculture or aquafarming is the farming of aquatic organisms such as fish, crustaceans, molluscs, and aquatic plants and algae. Aquaculture involves cultivating freshwater and saltwater populations under controlled conditions and can be contrasted with commercial fishing. Mariculture refers to aquaculture practiced in marine environments and in underwater habitats (Boyd, 2003). The reported output from global aquaculture operations would supply one half of the fish and shellfish directly consumed by humans. Particular kinds of aquaculture include fish farming, shrimp farming, oyster farming, algaculture (such as seaweed farming), and the cultivation of ornamental fish. Particular methods include aquaponics and integrated multi-trophic aquaculture, both of which integrate fish farming and plant farming.

Aquaculture in the Sultanate of Oman is presently at an early stage of development. Since 1997, the Ministry of Fisheries Wealth is involved in aquaculture development abalone hatchery, mussel and oyster culture, shrimp farming and pilot trials on cage and pond culture of finfish. In addition site selection and feed development projects have been successfully completed (Siddeek et al., 1999).

Experience gained from Ministry-funded projects and subsequent technology transfer has stimulated entrepreneurial interest in commercial aquaculture ventures. Aquaculture in Oman began in 1986 with a production trial of the giant tiger prawn (*Penaeus monodon*) conducted in a private farm in the Al-Sharqiyah South Region, Sur. In 1994, maturation and seed production of the native abalone *Haliotis mariae*, locally known as "a'sufailah", was initiated at the Mirbat Aquaculture Facility in Dhofar. Cage farm in Quriyat Aquaculture Company produced gilthead seabream 460 tonnes in 2004. The total value of aquaculture production was about one million Omani Rials in 2004

For freshwater aquaculture, the Nile tilapia (*Oreochromis niloticus*) is now popularly farmed in many areas in Oman. This is in response to a high demand mainly from expatriates living in the country. The tilapia fry were imported from Egypt and Thailand. Small-scale Nile tilapia farming is now a widespread activity in different areas of Oman. Small-scale aquaculture production also contributes in terms of providing livelihood and income for local communities.

The majority of aquaculture production during 2003 and 2004 was exported as whole fish products to the United Arab Emirates with only very minor quantities having been consumed on the domestic market. Production from the tuna fattening project was exported directly to Japan. Marketing of cultured white shrimp is wide-spread in Oman. Currently, the annual production reaches 100 tonnes. The quantity and value of aquaculture production is still small compared to the contribution from capture fisheries. However, contribution from aquaculture towards the national economy will increase in the near future as the sector progresses.

In Oman, more than 170 thousand tons of fish catch is done annually. The Sultanate consumes less than 40 thousand tons and the rest is being exported. The Sultanate has not fully utilized the fishery wealth. In addition, new industries related to aquaculture such as fish glue, oil, emulsion or fertilizer, hydro lysate and fish meal and fish sauce can be developed.

Algae that can be eaten and used in the preparation of food is called edible seaweed. It typically contains high amounts of fiber and protein. Seaweeds are also harvested or cultivated for the extraction of alginate, agar and carrageenan, gelatinous substances collectively known as hydrocolloids or phycocolloids. Hydrocolloids have attained commercial significance, especially in food production as food additives. Most of the collected seaweeds in Oman are simply exported though it has a scope for starting processing industries.

Spirulina farming is another simple biotechnology. *Spirulina* is a microalga that can be consumed as a nutritional supplement. It is also used as a feed supplement in the aquaculture, aquarium and poultry industries (Celekli and Yavuzatmaca, 2009). Most cultivated *Spirulina* is produced in open channel raceway ponds, with paddle-wheels used to agitate the water. It is an excellent source of protein and vitamins. Dried *Spirulina* contains about 60% protein. It is a complete protein containing all essential amino acids, though with reduced amounts of methionine, cysteine and lysine when compared to the proteins of meat, eggs and milk. Phycocyanin, the blue pigment found only in *Spirulina* and cyanobacteria. Phycocyanin is a phycobiliprotein that has pharmacological properties such as antioxidant, anti-inflammatory, neuroprotective and hepatoprotective effects. The cost of 1g of purified phycocyanin is more than 9000 Omani Rials. Lot of free lands available in Oman could be utilized for *Spirulina* farming. This will increase the income of local people and will boost up the economy of the country.

BIOREMEDIATION

Several physical, chemical, biological techniques are used to remove pollutants from environment by the process of bioremediation. It is the use of microorganism's metabolism to degrade pollutants. Biodegradation is the chemical dissolution of materials by bacteria or other biological means. In nature, different materials underwent biodegradation at different rates (Sayler and Ripp, 2000). Plants are also used for bioremediation. The process is known as phytoremediation. Mycoremediation is a form of bioremediation in which fungi and fungal mycelia are used to decontaminate the area. Technologies can be generally classified as *in situ* or *ex situ*. *In situ* bioremediation involves treating the contaminated material at the site, while *ex situ* involves the removal of the contaminated material to be treated elsewhere. Some examples of bioremediation technologies are phytoremediation, bioventing, bioleaching, landfarming, bioreactor, composting, bioaugmentation, rhizofiltration and biostimulation.

There are a number of advantages to bioremediation, which can be employed in areas that are inaccessible without excavation. For example, hydrocarbon spills (specifically, petrol spills) or certain chlorinated solvents may contaminate groundwater. Heavy metals such as cadmium and lead are not readily absorbed by microorganisms. Phytoremediation is useful in these circumstances because natural or transgenic plants are able to bioaccumulate these toxins in their above-ground parts, which are then harvested for removal. The heavy metals in the harvested biomass may be further concentrated by incineration or even recycled for industrial use (Ali et al., 2013). The use of genetic engineered microorganisms specifically designed for bioremediation has great potential. The bacterium *Deinococcus radiodurans* (the most radioactive resistant organism known) has been modified to consume and digest toluene and ionic mercury from highly radioactive nuclear waste.

Waste water is another pollutant contains different type of contaminants. Studies in Oman showed that sewage water treatment does not remove heavy metals. The amount of treated waste water in Oman is estimated to be more than 70 million m^3. Without appropriate remediation, the treated waste water may become a serious pollutant. It is possible to treated polluted water by phytoremediation to obtain clean water for planting of edible crops and other uses.

One of the primary roles of fungi in the ecosystem is decomposition, which is performed by the mycelium. The mycelium secretes extracellular enzymes and acids that break down lignin and cellulose, the two main building blocks of plant fiber. These are organic compounds composed of long chains of carbon and hydrogen, structurally similar to many organic pollutants. The key to mycoremediation is determining the right fungal species to target a specific pollutant.

In one experiment, a plot of soil contaminated with diesel oil was inoculated with mycelia of oyster mushrooms. After four weeks, more than 95% of many of the PAH (polycyclic aromatic hydrocarbons) had been reduced to non-toxic components in the mycelial-inoculated plots. It appears that the natural microbial community participates with the fungi to break down contaminants, eventually into carbon dioxide and water. Wood-degrading fungi are particularly effective in breaking down aromatic pollutants (toxic components of petroleum), as well as chlorinated compounds (Sanchez, 2009). In Oman, Similar mycoremediation fungi were isolated and were found to degrade tar (byproduct of petroleum industry). Several bacterial strains were also isolated from oil contaminated soil that has high potential to treat oil spill and pollution. Diaper disposal is considered as one of the major pollutant globally. Some mushrooms have been used to break down diaper materials by the process of mycoremediation.

Bioremediation of Oil Spills

Crude oil spills is a global problem. The magnitude of oil spill internationally is overwhelming and it is very frequent in the Arabian Gulf and the Gulf of Oman (El Samra and El Deeb, 1988). The dark red circles indicate oil spills. Marine wild and terrestrial wild life is badly affected from oil pollution. Oil spills may be due to release of crude oil from tankers, offshore platforms, drilling rigs and wells, as well as spills of refined petroleum products and their by-products (Atlas and Bragg, 2009).

About 50% of world's oil is produced in the Gulf and transported internationally through Musandam and the Gulf of Oman. Due to the geographical position of Oman, its shore lines are vulnerable to oil pollution from normal tanker operations, ballast water, illegal discharges and accidental spills. When unloaded, returning tankers fill sea water as ballast to be carried in the compartments previously occupied by oil. The walls of compartments are cleaned with dinging oil by powerful seawater jets hence ballast water inevitably acquires a considerable quantity of oil, which when discharged causes unacceptable oil spill pollution. One of the major sources of oil pollution in Oman arose from operational discharges of oil from passing vessels traffic. Because of the high sea and temperatures oil evaporates at high rates and tends to arrive to the coast as heavy petroleum particulate residues called tar balls. The amount of tar ball deposition was reported to reach 1 kg/m3 indicate the severity of oil pollution in this region. Within this region, Mina AI Fahal, where all of the petroleum activities in the Sultanate of Oman exist, was found to be the heaviest polluted area by oil in the region (Elshafie et al., 2007).

In December 19, 1972 one of the worst major oil spills in history was recorded in the Gulf of Oman. The amount of spilled oil was 35.3 million gallons when the South Korean super tanker, Sea Star, collided with a Brazilian tanker, the Horta Barbosa off the coast of Oman. In August 4, 2001 pressure built up in a newly drilled well at the Zauliyah field in central Oman and blew open. Oil released uncontrollably and caused serious oil pollution.

Oman depends heavily on desalination plants for daily uses of fresh water. The largest desalination plants in Alghubra, Barka and Sur are located at the seashores. Oil spills are serious security threat to these desalination plants. Cleaning contaminated plants will render people with no water for days.

Oil spill cleanup is difficult and depends upon many factors, including the type of oil spilled, the temperature of the water and the types of shorelines and beaches involved. Spills may take weeks, months or even years to clean up. Methods for cleaning up include bioremediation using oil degrading microorganisms. These native bacteria feed on oil, turning it into carbon dioxide and harmless organic by-products (Emara, 1990).

Different types of bacteria and fungi have been isolated in Oman feeding on oil and therefore, can be used in oil spills cleanup avoiding chemical and physical means, which can lead to other environmental pollutions. The microbes were isolated from tar balls and oil contaminated soil. Some microbes were found to survive in harsh dry, salt-rich conditions of central Oman and found to be effective in bioremediation of oil spills in the desert habitats such Zawliah.

ALGAL BLOOMS AND RED TIDE IN THE GULF OF OMAN

Oman fish reserve is one of the largest in the world. Fisheries have become the second largest industry in the Sultanate after oil. During recent years, occurrences of red tide in the coast waters of Oman has become a common phenomenon and led to immense fish mortality. The red tide is a natural phenomenon occurring due to the abundance of phytoplanktons results in hydrographic and biological changes in the marine environment (Morton et al., 2002).

Right conditions of temperatures and nutrients such as organic matters forms a favourable environment for the growth of microorganisms some of which are responsible for massive mortality of marine organisms including fish. Overgrowth of algae leads to colorization of water surfaces to red, brown or green and leading to oxygen depletion (Bauman, 2010). Although not all algal blooms are toxic, some are toxic due to the secretion of some potent toxins that may enter food chain and cause food poisoning to people who consume intoxicated fish and shell. These toxins are heat stable and can cause paralysis, respiratory difficulty, and memory loss even if food is cooked. Thus, toxic blooms have raised serious concern on local consumption of fish and fish export in the Sultanate of Oman and has led a big loss of economy.

According to the Ministry of Fisheries reports, Arabian Sea has one of the thickest oxygen-depleted layers of ocean water found anywhere in the world. Poor oxygenated water at depths 100 meters in the Gulf of Oman and the northern Arabian Sea is believed to create the right conditions for growth of planktons and form a bloom that has life span of 1-3 days. The death and the decay of the sinking bloom lead to oxygen depletion in the ocean and that upwelling of the poor oxygen water may lead to massive kill of fish.

Whether mass fish mortality occurs as a result of toxins from algal bloom or depletion of oxygen, the question remains is why these events have become more frequent in the last decade. Is massive pollution in the marine habitats in the Gulf of Oman that resulted in increased human activities or as a climate change is the cause is still a question?

WASTE MANAGEMENT AND POLLUTION

Pollution is the introduction of contaminants into the natural environment that leads to detrimental effect on air, water or soil. Pollution can take the form of chemical substances or energy, such as noise, heat or light. The components of pollution are referred as pollutants, which affect organisms negatively. Some pollutants cause simple effects and others can be very serious leading to cancer in human and extinction of animals, plants and microorganisms.

SEWAGE EFFLUENT

Oman is located in arid region. An arid region is characterized by a severe lack of available water, to the extent of hindering or preventing the growth and development of plant and animal life. The FAO organization of the United Nation has classified Oman as one of the 22 countries that would meet a water barrier before 2025. The shortage of water resources in arid countries has become a major environmental issue. Due to lack of rainfall, the main source of fresh water used to be from underground water.

Rapid development and significant population increase during the last 40 years in Oman resulted in an increasing water consumption causing severe damage of habitats. Excessive use of underground water has caused desertification of many areas in Oman. Because of this, most of the farmlands have been damaged and soil has become saline. In coastal areas, such as Al-Batinah, the intrusion of sea water has increased the magnitude of soil salinity.

The severity of the arid condition and the developmental needs has forced the authorities to solve the water-demand problem. In the arid zone, many countries have practicing desalination of brackish water and seawater. The used water is recycled and used for irrigation of public parks and greeneries. Sludge is the semi-solid material left from sewage treatment processes can be used as a fertilizer. The government estimated that water usage and sewage produced is steadily increasing. It is estimated that in 2035 the quantity of treated sewage effluent will be more than 70 million m^3. In 2006, it was reported that only 11% of sewage produce is recycled and the rest is unused.

Industrial sewage water and sludge may contain heavy metals. Current methods used for recycling of sewage reduce disease causing pathogenic microbes. Studies in Oman showed that sewage water treatment does not eliminate all pathogens and some of toxic chemicals such as pesticides (Al-Bahry et al., 2011). Additionally, heavy metals in sewage remain intact. All these factors cause several environmental and public health problems as a result of accumulation of toxic compounds in soil and rapid infiltration into underground water affecting all living organisms. Heavy metals and antibiotic resistant bacteria were detected in marine turtles and fish is an indication of sewage water and terrestrial contaminants reaching marine wild life (Al-Bahry et al., 2009). This is specifically true for those sewage treatment plants located at the coastal regions of Oman. Antibiotics resistant bacteria are not supposed to be found in wild life unless polluted with contaminated effluents. Also, vast quantities of nitrogen and phosphorus fertilizers applied on crops are washed off by the rain in to water supplies and coastal regions of Oman. Together, these nutrients entering water set off a chain reaction leading to algal blooms and fish kill.

One of the effective and affordable technological solutions is the application of a process known as phytoremediation. Phytoremediation is used for *in situ* or *ex situ* treatment/removal of contaminated soils, sediments and water. The green plants degrade, assimilate, metabolize, or detoxify inorganic and organic pollutants from the environment or render them harmless (Aronsson et al., 2010). It is a cost effective green technology based on the use of specially selected metal-accumulating plants to remove toxic metals from soils and water. The detoxified water can then be safely use in agriculture. The demand of agricultural irrigation can be fulfilled if the amount of sewage treated effluent is fully utilized. This will reduce the costly importing of food materials, such as vegetables, fruits, and dairy products.

MUNICIPAL SOLID WASTE

Municipal solid waste is a type of waste consisting of everyday items that are discarded into the environment. The largest solid waste globally is plastic. It is estimated that over 1000 marine mammals die every day from lost fishing nets made of non-degradable materials such as plastic. Every year tens of thousands of turtles have also died from swallowing plastic bags or entanglement in nets. Oman has one of the largest breeding populations of turtles in the world. Unless the coastal region is protected this will decline (Taha et al., 2004).

Components of solid municipal waste in Oman consist largely of organic materials, paper and textile. Most of these wastes can be treated biotechnologically to produce energy and organic fertilizer (Slack et al., 2005). In addition, Oman produces several tons of medical wastes. Estimation of hazardous waste quantity on Gross Domestic Product (GDP) basis in 2006 was 13.8 million Omani Riyals. The international Food & Agricultural Organization (FAO) estimates that one third of world food production is lost or wasted. An average Omani family wastes about a third of all food prepared within the household. Omani household typically throws away 70 Omani Riyals worth of leftover food every month. According to a survey of ministry of economy in 2010, the average food consumption by a Omani family was 205 Omani Riyals per month. It is estimated that in 2020 the Omani population will be closer to 5 million with the estimated total waste more than 42 million tons. In 2010, the authorities have drafted law on ways to manage municipal waste from an economic and environment safety perspective. One of the important objectives of the draft law was to encourage competition to achieve public safety and to make large invests in the waste management (Al-Badi et al., 2011).

Municipal solid waste can be used to generate energy. Waste-to-energy (WtE) or energy-from-waste (EfW) is the process of creating energy in the form of electricity or heat from waste source. Most WtE processes produce electricity directly through combustion or produce a combustible fuel commodity, such as methane, methanol, ethanol or synthetic fuels. Methane, methanol, ethanol are products of microbial bio fermentation (Al-Badi et al., 2009). Some microbes isolated from Oman secrete high amount of cellulase in a short time (Sivakumar et al., 2012). These microbes can be used to utilize cellulosic biomass by degrading cellulose to sugar which can be converted to ethanol (Carere et al., 2008). Thus, cellulosic biomass is an important renewable source of energy that has not been utilized yet in Oman. Some neighbouring countries have already started to produce energy from municipal wastes.

In addition to energy production by biotechnological means, cellulase secreting microbes can be used for paper recycling. De-inking the waste paper using cellulase enzyme will avoid dangerous effect of the chemical de-inking on environment. Cellulase enzyme is also used in laundry detergent and in bipolishing of fabrics by removing excess dye from them. In food industry, cellulases are used in extraction and clarification of fruit and vegetable juices. Moreover, cellulases are used in carotenoid extraction in production of coloring agents.

Cellulases are also used for extraction of olive oil and in animal food industry by improving the nutritional quality of animal feed.

BIOTECHNOLOGY IN OIL INDUSTRY

Only 30% of total world reservoir is conventional oil. Oil and gas are the backbone of the economy, their contribution to national economy reached 81%. Oman has total proven reserves of 5.5 billion barrels of crude oil, which is the second lowest in the Arabian Peninsula after Yemen. The oil fields of south Oman contain heavier and more viscous oil than fields in north and central Oman. In Arabian Peninsula, most of the fields contain around 7 billion barrels of heavy oil. Omani fields are smaller, more widely scattered, less productive, and pose higher production costs compared to other Gulf countries. Aged oil fields and the geological composition makes production costs some of the highest in the region and its crude oil production declined in years 2000-2009. Presently, Oman is one of the leading countries, which uses the Enhanced Oil Recovery (EOR) techniques for its oil production. One of the EOR technologies is the biotechnological use of microbes to enhance oil recovery (Al-Bahry, 2013). The process is known as Microbial Enhancement Oil Recovery (MEOR).

MEOR is a biological based technology consisting of manipulating function or structure, or both, of microbial environments existing in oil reservoirs. The ultimate aim of MEOR is to improve the recovery of oil entrapped in porous media while increasing economic profits.

Crude oil or petroleum oil are formed naturally due to the decomposition of plant materials, dead maters and decaying bacteria under heat and high pressure. Normally, crude oil is driven from the pores to the surface as a result of high pressure and as fluids are produced, the pressure tends to fall off and the production will decrease. The primary stage of recovery includes the production of oil naturally by reservoir pressure. Gas and water are the most important natural reservoir mechanisms used. The secondary stage starts when the oil is depleted due to the fall in pressure and in order to enhance oil production, water and immiscible gases are injected into the wells. This stage can recover 5-50% of the remaining oil. When the amount of the recovered oil drops, the enhanced oil recovery technologies are being applied. This is known as a tertiary stage or enhanced oil recovery. In this stage, miscible gas displacement (CO_2), steam injection, chemical injection (synthetic surfactant, polymers and/or solvents), microbial technologies, and their bio-products are used. Although the non-biological technologies are easy to apply and the results are promising, they are very expensive and are not environmental friendly. Therefore, the MEOR biotechnology is an attractive method.

Oil reservoir is a subsurface pool of porous rock that occurs in several layers in deep depth. Besides water, oil, gases, plant materials and dead matters are also accumulate in the pores of the rocks. The reservoir heterogeneous ecosystem contains diverse microbial communities, which in turn are able to affect reservoir behaviour and oil mobilization. Microbial metabolites, excretion products and new cells may interact with each other or with the environment, positively or negatively. All these entities, i.e. enzymes, extracellular polymeric substances (EPS) and the cells themselves, may participate in MEOR. The technology of MEOR depends either by injection of exogenous microbes and their bio-products or by in situ generation of indigenous microbial bio-products (Al-Bahry, 2013).

Microorganisms utilize complex nutrients through catabolism converting them to simpler compounds like miscible gas (carbon dioxide, hydrogen and methane), organic acids (lactic acid, acetic acid, and butyric acid) and solvents (ethanol, acetone, butanol, and methanol) and generate ATP, which is used in anabolism to make new complex compounds like biosurfactants (lipopeptides), biopolymers (polysaccharides) etc. These bio-products are non-toxic and have been used to release oil from the wells by changing the physical and chemical properties of oil.

Many research projects have been conducted in Oman related to EOR and MEOR. Different types of microbes have been isolated and were found to have potential to enhance oil recovery and reduce heavy oil viscosity to light oil.

BIOMINING AND BIOLEACHING

Biomining is an important approach to the extraction of desired minerals from ores being explored by the mining industry in the past few years. Microorganisms are used to leach out the minerals, rather than the traditional methods of extreme heat or toxic chemicals, which have a deleterious effect on the environment including mines in Oman, such as Sohar Copper Mining.

Nature has its own slow process of breaking down the ore over many years by changing the acid chemistry of the ore. This chemical change activates natural bacteria in the ore. The result is a slow decomposition of the ore, releasing sulphuric compounds that, in turn, also release metals into the environment. This releasing of chemicals from ore is referred to as natural biological leaching.

Commercially applied bio-leaching technologies accelerate this natural process. Air temperature, oxygen, water, acidity and different bacterial populations are manipulated in a controlled environment so as to accelerate bacterial freeing of metals from the ore. Bioleaching uses billions of rock-eating bacteria, acting as catalysts, to extract iron, gold, silver, cobalt and other metals.

The potential applications of biotechnology to mining and processing are countless. Some examples of past projects in biotechnology include a biologically assisted in situ mining program. Approximately 25% of all copper mined worldwide is now obtained from leaching processes. 40 to 60% copper extraction was achieved in primary reactors and more than 90% extraction in secondary reactors with overall residence times of about 6 days.

In addition to microbial biomining, plants are also used to mine valuable heavy metal minerals from contaminated or mineralized soils. In fact, 25% of all copper is mined this way, amounting to $1 billion in revenue annually. This ranks it as one of the most important applications of biotechnology today. Bioprocessing is also being used to economically extract gold from very low grade, sulfidic gold ores, once thought to be worthless.

PLANT AND ANIMAL TISSUE CULTURE

Tissue culture has become one of the very important tools in biotechnology. The term tissue culture refers to the removal of organ, tissue or cell from plant or animal and growing it in artificial environment of liquid or solid medium.

Tissue culture is an important technique to study cell in molecular biology and biochemistry. It is used to understand the interaction of disease causing agent and cells. Also, it is used to study the effect of drugs on cells, aging process, nutritional studies, toxicity testing, cancer research, gene therapy and stem cell studies. Plant tissue culture is a technique to culture plant cells, tissues or organs in an artificial medium in aseptic conditions. Plant cells are totipotent i.e., the cell is capable of regenerating unto a whole plant. Therefore, it is possible to propagate plants vegetatively under controlled conditions. Still the development of plants by tissue culture is in the beginning stage in Oman.

BIODIVERSITY

Biodiversity is a measure of the total species richness of a given area. Extinction is the permanent loss of a species or sub-species. Extinction is natural process that occurs over large time scales as new species evolve and replace existing ones. There have been several major extinction events over the history of the Earth, possibly due to catastrophic collisions with meteors or due to volcanic activity (Collins, 2013). During the past 200 years, human activities have begun to cause another great extinction event, in which species are being lost at a much higher rate than normal. This loss of biodiversity is closely correlated with the growth in the human population.

Humans cannot live alone in the environment. We require biodiversity for food, biomaterials (e.g., many medicines are derived from plants and microorganisms) and energy. So there is an exchange of energy and nutrients among the biodiversity in the environment. Carbon, nitrogen, phosphorus and many other chemical elements are cycled through ecosystems by complex communities of bacteria and other organisms, so that these elements are available to us. Atmospheric nitrogen is converted by special type of soil bacteria into nitrate and ammonium that can be absorbed readily by plants (Binder and Polasky, 2013). Land with vegetation and soil allow water to flow slowly and to sink into the aquifers below the surface, where it can be safely stored for future use. Soil is produced from eroded rock by microorganisms adding organic material, so that plants can live. This is converted in to high quality soil by burrowing animals such as earthworms improve the soil by aeration and recycling dead plant material. Plant communities provide protection from erosion by wind and water, conserving the valuable topsoil which takes centuries to form. We would have almost no crops (other than wind pollinated grasses), if insects did not pollinate flowers to produce fruit, vegetables, etc.

There are many reasons for the loss of biodiversity. Over-hunting of wild species by humans in not a new phenomenon. 11,000 years ago when humans first colonised North America they caused a mass extinction of early mammals in a period of a few hundred years. These included 10 species of horse, Giant ground sloth, 4 species of camel, 2 species of bison, Native cow, Native antelope, Mastodon and American lion.

More recently the Passenger Pigeon was one of the most abundant birds in North America with populations estimated at 3-5 billion in the early 1800s. Merciless hunting for food decimated their population and the last specimen died in a zoo in 1924. Lake Victoria is the largest lake in Africa and was a biodiversity hot spot with over 300 species of fish belonging to the Family Cichlidae. The Nile perch (*Lates niloticus*) is a native of Africa but was not found in Lake Victoria. In an effort to establish a commercial fishery in the lake the Nile perch was introduced in the 1950s. For 20 years nothing happened, then its population exploded, causing the extinction of over 90% of the cichlid species in the lake.

Humans have caused extensive damage to the animal population in many parts of the world by introducing predators such as cats, rats and cane toads, which the local animals have no defences against. Similarly, local plants have been destroyed by introduced herbivores such as camels (in Australia) and goats.

Every year almost 80 million acres of tropical rainforest is destroyed. The causes include the clearance for existing agriculture because the land can only be farmed for 2-3 years, due to rapidly decreasing fertility, weed invasion and erosion of the soil in the high rainfall. So the farmer then has to move on and destroy more rain forests and changing to other agriculture crops. Most of the Malaysian and much of the Indonesian rain forests have been replaced by oil palm to make vegetable oil and also fuel wood harvesting destroys most of the forest ecosystem. In Africa and many other parts of developing world, wood is the major source of energy (Barbault, 2013).

THREATS TO BIODIVERSITY IN OMAN

The rapid modernization of Oman over the past 40 years has come at a cost to the environment. Most large mammals are now very rare, either because they were predators that are a threat to livestock (Arabian leopard and Arabian wolf), or else hunted for food (Arabian oryx, Arabian Tahr and Arabian gazelle). All are officially protected now.

Large scale farming using pumped artesian water from boreholes has destroyed large areas of native plants on the Batinah coast. Even more important are the huge numbers of goats in Dhofar and camels selectively eating native plants. Urbanisation is also at the expense of the native plants. Tourist resort development has extensively damaged the coastal habitats, especially by development of boat marinas. These require dredging, which stirs up mud and alters the water currents that kill most of the invertebrate animals living on the shore. The other threat to the biodiversity is by the invasive species or introduced species by humans. *Prosopis juliflora* was originally introduced from the USA as an ornamental tree, but is highly poisonous, so goats do not eat it and prevent the growth of native plants by releasing toxins in the soil. There are now estimated to be 20 million plants covering extensive areas of the Batinah.

The ring-necked parakeet is a bird introduced in to Oman as pets that later escaped. It is now extremely abundant on the Batinah as causes extensive damage to maturing date fruits. The mynah is another bird that has escaped as a pet and is extremely abundant on the Batinah. It kills young birds and has caused the decline of many native birds.

During the summer of 2002 a large number of trees, initially Ficus and later Delonix, on the campus of Sultan Qaboos University, Muscat, began to show symptoms of leaf yellowing and branch dieback. Sheets of white mycelium were seen under the bark of both species. From the trunks of affected trees, bracket-like, spongy, sessile asidiocarps appeared. These were initially rounded, pale brown above and cream-yellow below, eventually reaching a diameter of 20 cm and a thickness of 8 cm. Trees died within 6 months of symptoms first appearing. Fruiting bodies were sent to the Royal Botanic Garden, Kew, UK (RBGK) and identified as *Ganoderma collosum*. The pathogen appears to be a pantropical species on a wide variety of woody hosts. It causes selective and extensive delignification and decay of wood. It has been reported as a root and stem rot pathogen of pine, Eucalyptus and Callitris in South Africa. The species was originally described from Costa Rica and it has been reported on *Phoenix canariensis*, *Ficus carica* and *Celtics laevigata* in the USA and on *Delonix regia* trunk (with no symptoms) in Vietnam.

Due to the importance of biodiversity and the loss of different organisms as well as the fear of distinction of habitats in Oman, the authorities have decided to establish centers for genetic and environmental studies. The role of the centers is to survey and screen different types of organisms and their habitats. Further, the centers are aimed to conserve the endangered species. Several research projects are being conducted in Oman for conservation. Some of these projects involve animal and plant tissue culture to propagate endangered species. For example, the Omani Ibex (tahr) population is in serious decline. Researchers have established means to conserve this endangered species through cell bank and tissue culture. They succeeded in conserving the sperms of wild tahr to be used to increase the tahr population. This is a unique biotechnology method to conserve this animal globally (Ghazanfar, 1998).

Mangrove habitats play an important role in the marine environment and ecosystems. The forests' function is to enrich the biodiversity in marine flora and fauna (Brinkmann et al., 2009). The forest of mangrove includes several species of flowering plant that inhabit coastal environments of the inter-tidal zones in in Oman characterized by their slow growth under extreme ecological conditions. There are several species of mangroves some have hard-wood that is difficult to decompose while some are resistant to salinity. Such species consist of salt glands while others concentrate salt in their leaves. There are ten mangrove habitats in the Sultanate of Oman. Al-Qurum in Muscat and Mahout Island in Al-Wusta are the most developed mangrove forests in the Sultanate.

Historically, Omanis have used mangrove for building houses, fencing carving boats and fuel. Mangrove leaves and seeds have been used in making medicines, paper, and animal feed. However, overuse and lack of environmental awareness have led to drastic deterioration of mangrove habitats. New cities, towns and road constriction nearby the mangrove forests have further led to deterioration of the forest and fauna. As a result of the climate changes that led to Cyclone Gonu in 2007, some of the mangrove habitats were badly damaged. In 2000, the Sultanate of Oman launched a successful project in mangrove transplantation and rehabilitation. One of the objectives of this project was to raise public awareness.

Seven of the eight recognized species of marine turtles are listed as endangered or threatened and almost all populations are declining. The Omani coast is thought to be one of the last strongholds of sea turtle (*Chelonia mydas*) in Ras Al-Hadd and Olive ridley (*Caretta caretta*) in Masira Island. Masira Island is the only place in the Sultanate which hosts nesting grounds of four species of sea turtles, the Loggerhead, the Green, the Hawksbill and the Olive Ridley (Al-Bahry, 2009).

Ras Al-Hadd Reserve is one of the largest and the most successful nesting grounds in the world for the green turtles. Thousands of green sea turtles (*Chelonia mydas*) lay their eggs from September to November. Many of nesting beaches in Oman, such as Ras Al-Hadd, are ideal for nesting because of lack of human disturbance of feeding areas and unpolluted nesting beaches. These conditions make the Reserve one of the most strategic location for nesting in the world and the Arabian Peninsula.

CONCLUSION

In Oman, biotechnology is still in the beginning stage and it has a potential applications in different fields. The proper progress and utilization of biotechnology would definitely solve the problem of loss of biodiversity, waste treatment and bioremediation. The development of research and starting of industries would play a crucial role in the food and pharmaceutics.

REFERENCES

Al-Badi AH, Malik A, Gastli A. 2009. Assessment of renewable energy resources potential in Oman and identification of barrier to their significant utilization. *Renewable and Sustainable Energy Reviews*. 13: 2734-2739.

Al-Badi AH, Malik A, Gastli A. 2011. Sustainable energy usage in Oman—Opportunities and barriers. *Renewable and Sustainable Energy Reviews*,15: 3780-3788.

Al-Bahry SN, Al-Wahaibi YM, Elshafie AE, Al- Bemani AS, Joshi SJ, Al-Makhmari HS, Al-Sulaimani HS. 2013. Biosurfactant production by *Bacillus subtilis* B20 using date molasses and its possible application in enhanced oil recovery. *International Biodeterioration & Biodegradation*. 81: 141–146.

Al-Bahry SN, Elshafie AE, Al-Wahaibi YM, Al- Bemani AS, Joshi SJ, Al-Maaini RA, Al-Alawi WJ, Sugai Y, Al- Mandhari M. 2013. Microbial consortia in Oman oil fields: a possible use in enhanced oil recovery. *Journal of Microbiology and Biotechnology*. 23: 106–117.

Al-Bahry SN, Mahmoud IY, Al-Amri IS, Ba-Omar TA, Melgheit KO, Al-Kindi AY. 2009. Ultrastructural features and elemental distribution in eggshell during pre and post hatching periods in the green turtle, *Chelonia mydas* at Ras Al-Hadd, Oman. *Tissue and Cell*, 41: 214-221.

Al-Bahry SN, Mahmoud IY, Al-Belushi KIA,. Elshafie AE, Al-Harthy A, Bakheit CK. 2009. Coastal sewage discharge and its impact on fish with reference to antibiotic resistant enteric bacteria and enteric pathogens as bio-indicators of pollution. *Chemosphere*, 77: 1534-1539.

Al-Bahry SN, Mahmoud IY, Al-Zadjali M, Elshafie A, Al-Harthy A, Al-Alawi W. 2011. Antibiotic resistant bacteria as bio-indicator of polluted effluent in the green turtles, *Chelonia mydas* in Oman. *Marine Environmental Research*,71: 139-144.

Ali H, Khan E, Sajad EA. 2103. Phytoremediation of heavy metals—Concepts and applications. *Chemosphere*. 91: 869–881.

AlKindi AYA, Mahmoud I, Al-Omairi R, AL-Bahry SN, Al-Amri I, Khan T. 2008. Temperature variations in relation to sex determination and plasma progesterone and estradiol levels during embryogenesis and post hatching in the green turtle, Chelonia mydas, at Ras Al-Hadd Reserve, Oman. Abstracts/*Comparative Biochemistry and Physiology Part A: Molecular & Integrative Physiology*, 151: S46.

Al-Saidi S, Rameshkumar KB, Hisham A, Sivakumar N, Al-Kind S. 2012. Composition and Antibacterial Activity of the Essential Oils of four Commercial Grades of Omani Luban, the Oleo-Gum Resin of *Boswellia sacra*Flueck. Chemistry & Biodiversity. 9: 615-624.

Armstrong E, Boyd KG, Burgess JG. 2000. Prevention of marine biofouling using natural compounds from marine organisms. *Biotechnology Annual Review* 6: 221–241.

Aronsson P, Dahlin T, Dimitriou I. 2010. Treatment of landfill leachate by irrigation of willow coppice – Plant response and treatment efficiency. *Environmental Pollution*, 158: 795-804.

Atlas R, Bragg JR. 2009. Bioremediation of marine oil spills: when and when not - the Exxon Valdez experience. *Microbial Biotechnol*ogy. 2: 213–221.

Barbault R. 2013. Loss of Biodiversity, Overview. Encyclopedia of Biodiversity (Second Edition), 656-666.

Bauman AG, Burt JA, Feary DA, Marquis E, Usseglio P. 2010. Tropical harmful algal blooms: An emerging threat to coral reef communities? *Marine Pollution Bulletin*. 60: 2117-2122.

Binder S, Polasky S. 2013. Biodiversity, Human Well-Being, and Markets. Encyclopedia of Biodiversity (Second Edition) 435-439.

Boyd CE, 2003. Guidelines for aquaculture effluent management at the farm-level. *Aquaculture*. 226: 101– 112.

Brinkmann K, Patzelt A, Dickhoefer U, Schlecht E, Buerkert A. 2009. Vegetation patterns and diversity along an altitudinal and a grazing gradient in the Jabal al Akhdar mountain range of northern Oman. *Journal of Arid Environments*, 73: 1035-1045.

Carere CR, Sparling R, Cicek N, Levin DB. 2008. Third Generation Biofuels via Direct Cellulose Fermentation. *International Journal of Moecular Sciences* 9: 1342-1360.

Celekli A, Yavuzatmaca M. 2009. Predictive modeling of biomass production by Spirulina platensis as function of nitrate and NaCl concentrations. *Bioresource Technology*. 100: 1847–1851.

Collins MG. 2013. International Organizations and Biodiversity. Encyclopedia of Biodiversity (Second Edition), 324-331.

Copping LG, Menn JJ. 2000. Biopesticides: a review of their action, applications and efficacy. *Pest Management Science*. 56:651-676.

El Samra MI, El Deeb KZ. 1988. Horizontal and vertical distribution of oil pollution in the Arabian Gulf and the Gulf of Oman Original Research Article. *Marine Pollution Bulletin*. 19:14-18.

Elshafie A, AlKindi AY, Al-Busaidi S, Bakheit C, Albahry SN. 2007. Biodegradation of crude oil and n-alkanes by fungi isolated from Oman Original Research Article. *Marine Pollution Bulletin*, 54: 1692-1696.

Emara HI. 1990. Oil pollution in the southern Arabian Gulf and Gulf of Oman. *Marine Pollution Bulletin*. 21: 399-401.

Ghazanfar SA. 1998. Status of the flora and plant conservation in the sultanate of Oman. *Biological Conservation*, 85: 287-295.

Heimpel AM. 1967. A critical review of *Bacillus thuringiensis* var. *thuringiensis* Berliner and other crystalliferous bacteria. *Annual Review of Entomology* 12: 287–322.

Kiewnick S, Sikora RA. 2006. Biological control of the root-knot nematode Meloidogyne incognita by *Paecilomyces lilacinus* strain 251. *Biological Control* 38: 179–187.

Kosek M, Bern C, Guerrant RL 2003. The global burden of disease, as estimated from studies published between 1992 and 2000. *Bull World Health Organ*. 81:197–204.

Miralles NF, Espín JD, Corchero JL, Vázquez E, Villaverde1A. 2009. Microbial factories for recombinant pharmaceuticals. Microbial Cell Factories, 8:17.

Morton SL, Faust MA, Fairey EA, Moeller PDR. 2002. Morphology and toxicology of *Prorocentrum arabianum* sp. nov., (dinophyceae) a toxic planktonic dinoflagellate from the Gulf of Oman, Arabian Sea. *Harmful Algae*.1: 393-400.

Sanchez C, 2009. Lignocellulosic residues: Biodegradation and bioconversion by fungi. *Biotechnology* Advances. 27: 185–194.

Sayler GS and Ripp S, 2000. Field applications of genetically engineered microorganisms for bioremediation processes. *Current Opinion in Biotechnology*. 11:286–289.

Schultza MP, Bendick JA, Holmb ER, Hertelb WM. 2011. Economic impact of biofouling on a naval surface ship. *Biofouling* 27: 87–98.

Siddeek MSM, Fouda MM, Hermosa Jr GV. 1999. Demersal Fisheries of the Arabian Sea, the Gulf of Oman and the Arabian Gulf. *Estuarine, Coastal and Shelf Science*, 49: 87-97.

Sivakumar N, Al Zadjali AA, Al Bahry SN. 2012. Production of cellulase by *Bacillus licheniformis*. The 2nd Korea - Indonesia Workshop & International Symposium on Bioenergy from Biomass. Indonesia.

Slack RJ, Gronow JR, Voulvoulis N. 2005. Household hazardous waste in municipal landfills: contaminants in leachate. *Science of The Total Environment*, 337: 119-137.

Taha R, Al-Rawas A, Al-Jabri K, Al-Harthy A, Hassan H, S Al-Oraimi. 2004. An overview of waste materials recycling in the Sultanate of Oman *Resources, Conservation and Recycling*, Volume 41, Issue 4, July 2004, Pages 293-306.

Wasser S. 2002. Medicinal mushrooms as a source of antitumor and immunomodulating polysaccharides. *Applied Microbiology and Biotechnology* 60: 258-274.

In: Biotechnology and Conservation … (2 Volume Set)
Editors: S. A. Farooq, R. Abed and S. Baqir
ISBN: 978-1-63117-869-6

Chapter 2

STEM CELLS: THERAPEUTIC ALTERNATIVES OF NERVOUS TISSUE INJURIES IN THE NEUROTMESIS OF SCIATIC NERVE IN RABBITS

Karim Al-Jashamy[1,*] and Hameed Al-Timmemi[2]
[1]Faculty of Medicine/ SEGI University, Malaysia
[2]Faculty of Veterinary Medicine, Baghdad University, Iraq

ABSTRACT

Peripheral nervous system injuries include penetrating injuries, crush, traction and ischemia compression. The peripheral nervous system is able to regenerate after injury. Microsurgical suture repair remains the current gold standard in clinical practice but obvious deficiencies remain with this technique, given that surgical repair of peripheral nerves does not result in complete functional recovery. However, the availability of various nerve coaptation and other techniques for the attainment of functional nerve regeneration is still inadequate. Recent approaches have been directed towards biological factors to promote an environment conducive to growth and overcome limitations in regeneration and functional recovery. The mesenchymal stem cells (MSCs) and omental pedicle transposition have come to be the most interesting targets for the study of tissue and organ regeneration because of their plasticity. The implantation of neural stem cells, MSCs, or fibroblasts has been shown to exert a beneficial effect on peripheral nerve regeneration. Thus, cell transplantation has been proposed as a method of improving peripheral nerve regeneration. In treated experimental animals of epineural repaired nerve that was implanted with MSCs in the proximal and distal segments of the transected sciatic nerve showed the best regeneration and functional recovery. The assessment of the nerve regeneration was based on functional (motor and sensory), histological and electron microscopy.

Keywords: Alternatives therapeutic, MSCs, peripheral nervous tissue Injuries

* Corresponding author: Karim Al-Jashamy. E-mail: karim@segi.edu.my, Jashamy@yahoo.com.

INTRODUCTION

Traumatic peripheral nerve injuries are common in companion animals due to trauma, iatrogenic lesions and surgical misadventure. There are many current conventional techniques of nerve repair such as epineural suturing, perineural suturing, perineurol nerve grafting, and free vascularized nerve grafting (Alluin et al., 2008). Mesenchymal stem cells derived from bone marrow represent a source of pluripotent cells (Horwitz et al., 2002), manifest multilineage differentiation capacity, including osteogenesis chondrogenesis, and adipogenesis (Zuk, 2010). Thus, MSCs isolated and cultured from various tissues appear to be morphologically similar and have similar epitope profiles and the expression of commonly used cell surface markers (Musina et al., 2005, Liu et al., 2007). Bone marrow derived mesenchymal stem cells (MSCs) are attractive candidates for cell-based therapies. The need for a good method of repairing transected nerve was the reason for our experimental study, which presents the histopathological and ultrastructural of regenerating sciatic nerve after transection, immediate suturing of the stumps and using autologous transplanted stem cells. The objective of this study was aimed to provide an understanding the role of MSCs in peripheral nerve regeneration by utilizing an established rabbit sciatic nerve experimental model. Thirty two male adult New Zealand White rabbits weighting 2.0-2.3 kg were divided into two groups consisting of sixteen animals each. In the first group, the sciatic nerve was severed and anastomosed using epi-perineural suturing served as a control group (CG), while the treatment group (TG) was treated with MSCs post severance and anastomosed. All animals were acclimatized for 3 weeks in individual cages, fed with commercial rabbit pellets and given water *ad libitum*. Broad spectrum prophylactic antibiotics (Pencillin 20,000 IU Streptomycine 20mg/kg) and subcutaneous antihelminth injections of 0.2 mg/Kg Ivermectin was given prior to the onset of the experiment. All procedures used in this study were approved by the Faculty's Animal Care and Use Committee (08 R13/Dec 08). Four animals from each group were sacrificed at day 14, 28, 56 and 112 post operation day (POD). All animals were pre anesthetically inducted using an intramuscular injection of a mixture of 40 mg/kg Ketamine hydrochloride (Bioketan, Vetoquinol Biowet, Sp. Zo.O, France), 5mg/Kg of Xylazine hydrochloride (lium xylazil-20, Australia) and Acepromazine maleate 1 mg/kg (Calmivet. Vetoquinol. Ltd. Lure cedex, France). Maintenance was carried out by 1-2% halothane using a flow rate of 500ml/min in a non-rebearthing circuit (modified Jackson-Reed Bain) through face mask. Rabbit fur of the left hind limb was clipped from upper midline to the stifle joint. The skin was disinfected using Chlorohexidine gluconate (ibiscrub, 4% w/v Durham, UK), Isopropyl alcohol 70% (Jaya Pelita Pharma. SDN. BHD) and finally Tincture Iodine 1.8% (Jaya Pelita Pharma SDN. BHD). A 2 cm long skin incision was made from the caudo lateral side of femur bone from greater trochanter of the femoral bone to the stifle joint using scalpel blade NO. 21. Subcutanous tissue and fascia lata were separated using scalpel blade NO. 15. Biceps femoris muscle was separated cranially while the semitendenous muscle posteriorly by blunt dissection using Mayo scissors; this exposed the sciatic nerve which separated from the surrounding tissue using ophthalmic scissors. Wooden tongue depressor was placed gently under the nerve to enable its severance with scalpel blade NO. 21, followed by immediate cooptation of its new ends under the operating microscope with 6 equidistant epi and perineural simple interrupted sutures about 2 mm from the edge of transected section using 8-0 nylon (Monofilment, Ethicon).

The anastomosed nerve was replaced into the normal position. Then muscles were closed using 3-0 Vicryl (Biovek, Dynek Pty Ltd) simple continuous suturing and the skin was closed using 3-0 Vicryl using subcuticular suture pattern. The left sciatic nerve was transected and immediately anastomosed using the standard method above mentioned. The clinical assessment data were recorded daily up to 21st POD and weekly till the end of the experiment at the 112th POD (data did not show). Animals were scarified and the anastomosed sciatic nerves were obtained for further studies. Animals were anesthesed as above mentioned. The area from external angle of the ilium to the hip joint was clipped and disinfected and 1.5 ml of bone marrow was collected from ilium using syringe (5ml) with needle gauge 18. The bone marrow aspirated was immediately mixed with 3ml of the Dublecco's modified Eagle's medium (DMEM) and high glucose-DMEM was supplemented with 30% fetal bovine serum (FBS), 10U/ml pencillin G, 10U/ml streptomycin, 25mg/ml amphotericin B, 1% nonessential amino acid (Gibco) and sodium pyruvate (100 ng/ml) all supplied by GIBCO® Invitrogen Corporation. Three ml of FBS was placed in a 75 cm^2 flask for 3 minutes and then the mixed media with bone marrow was added to the flask. The flask was incubated at 37°C in 5% CO^2 in air for 3 days, and after 72 hours the non-adherent cells were removed while replacing the medium. After 12 days, the culture reached to confluence stage and the monolayer cells were washed twice with 2 ml of phosphate buffered saline (pH 7.2). Two ml trypsine was distributed on the surface of its layer for 2 minutes while checking the cells under the microscope till the regained their rounded shape and then discarded the trypsine. Dublecco's modified Eagle's medium containing 10% FBS was added to the medium and then gently tapped to detach cells for next three sub cultures. The cells were harvested by decanting the medium from the flask, washing with PBS, trypsine bathing and replacing with 10 ml of DMEM. The medium and cells were collected in sterile test tubes, centrifuged at 2000 rpm for 10 minutes and separating the precipitated pellets, 1ml of DMEM was then added to the pellets and mixed. Cells were counted using hematocytometer to assure the minimum count $1x10^6$ MSCs in 10 μl of culture medium. The transected sciatic nerve was immediately anastomosed as mentioned early and injected with 10 μl of culture medium containing $1x10^6$ MSCs into each distal and proximal stump using a glass capillary tube connected to a Hamilton syringe. Four animals from each group were euthanized on the day 14, 28, 56 and 112 day POD using intracardial injection of pentobarbitone (Dolethal 180mg/ml) 1ml/kg under ketamine/xylazine anesthesia. The anastomosed left sciatic nerve was exposed, examined grossly and then harvested for histopathological and transmission Electron Microscopy (TEM) studies. Each specimen was divided into two parts, the first part was fixed with 10% neutral buffer formalin and completed the processing for light microscope examinations, while the second part was fixed with 2.5% glutaraldehyde for 12 hours, washing by *cacodylate* buffer three times, post fixative by osmium oxide 2% for two hours, after that dehydrated with different concentrations of acetone (35%, 50%, 75%, 90%, 100%), imbedding by resin and ultra-sectioned using ultramicrotome (Leica) and stained using Lead citrate and uranyl acetate.

Histopathological findings of the CG specimens at the nerve stump sections on the 14th POD showed degeneration, vacuolation and axons infiltrated with inflammatory cell, remarkable angiogenesis, demyelination and disorientated nerve fibers. Granulation tissues were observed around the stitches with scatter inflammatory cells (Figure 1a). Comparing to the TG, there was less degeneration and vacuolation, increasing number of Schwann cells and slightly disorientation of nerve fibers were observed (Figure 1b).

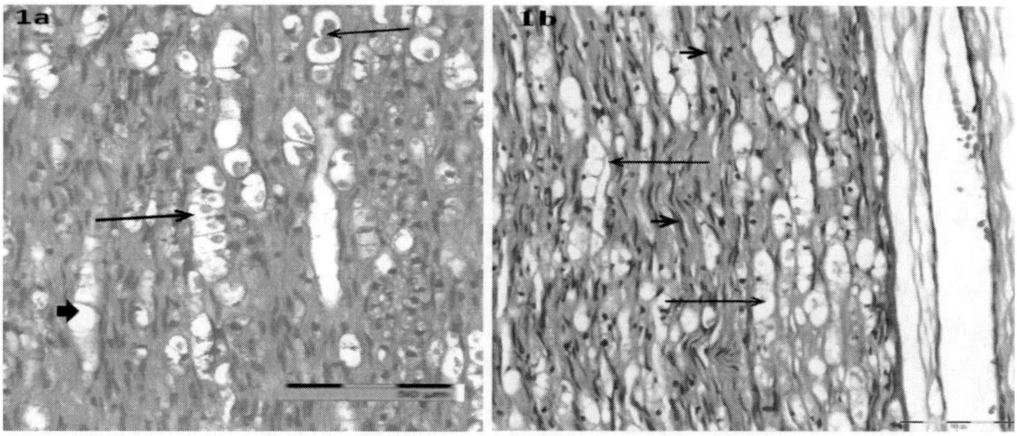

Figure 1. Photomicrographs of the sciatic nerve at 14th POD. a) in CG, demonstrated the degeneration, vacuolation (arrows head), ovoid (thin arrow) and digestive chambers of collagen (Thick arrow). b) TG showed vacuolated degenerative nerve fibers (arrow), highly activated Schwann cells (arrow head) and scar tissue at epi and perineurium H and E.

Ultrastructural examinations at 14th POD in CG showed neuronal sprouting and direction of regenerative nerve fibers was intraneural orientation, nerve fibers has extraneural orientation and Schwann cells formed bands of Bungner surrounded by remyelination of newly formed axon (Figure 2a), whereas in the TG showed the remyelinated nerve fibers, surrounding newly form of basement membrane, inner mesoaxon and external mesoaxon (Figure 2b).

Histopathological observations on the day 28 showed similar findings with slightly lower degeneration, vacuolation and increasing angiogenesis. On the 56th POD, there was a further increase in the number of Schwann cells, less degeneration, vacuolation and collagen deposition, also an increase in the angiogenesis and good parallel orientation of the nerve fibers were noted.

Sections at the 112th POD in the CG demonstrated the degeneration, vacuolation, ovoid and digestive chambers of collagen were still observed from moderate to minor changes (Figure 3a), whereas in TG showed the absent of degeneration and vacuolation, increased number Schwann cells, good orientation and scanty scar tissue in the epineurium of nerve fibers. The segments showed remarkable angiogenesis good orientation, and myelination of nerve fibers (Figure 3b). Ultrastructural examinations in CG at the 112th day POD showed thin myelinated nerve fibers and thick collagen fibers surrounding of axons (Figure 4a). In TG showed a good myelination of regenerative nerve fibers and deposited collagen fibers (Figure 4b). The macroscopic examinations of the coaptated nerve segments did not show any dehiscence and neuroma formations. It is indicated that the accurately of the surgical technique such as good alignment of the sciatic nerve. The regeneration of anastomosed sciatic nerve in the TG group showed better recovery than CG group, this finding was be in agreement with previous researches (Cuevas et al., 2002; Henery et al. 2009). This study shows an improvement in clinical signs and functions in MSCs treated nerve anastomoses. This improvement might be due to the MSCs actively acceleration of regeneration processes and decrease in the neuropathic pain generation after peripheral nerve injury.

Other researchers (Musolino et al., 2007) findings demonstrated that the BMSCs might prevent the generation of mechanical allodynia.

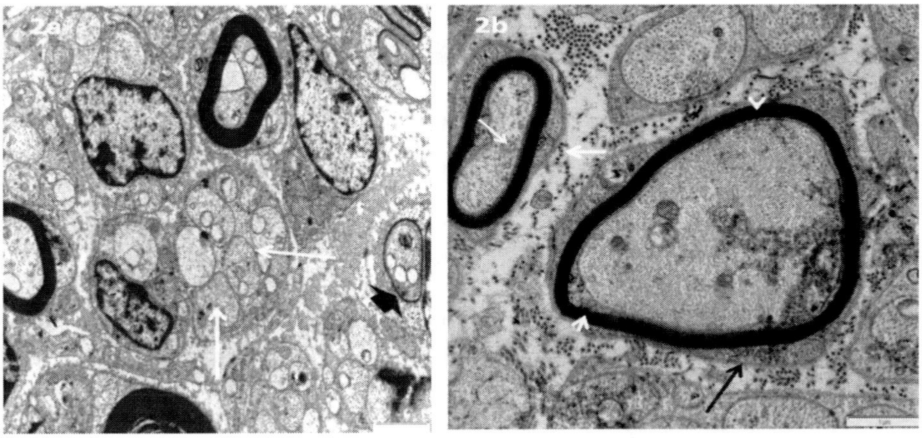

Scale bar = 2 μm.

Figure 2. Electromicrographs of sciatic nerve at 14th POD. a) CG demonstrated Schwann cells formed bands of Bungner (arrow), remyelination of newly formed axon (arrow head), Schwann cells surrounded many of axon. b) TG showed remyelinated nerve fibers (arrow head), surrounding newly form of basement membrane (black arrow), inner mesoaxon (thin arrow) and external mesoaxon (thick arrow) Uranyl acetate and lead stain.

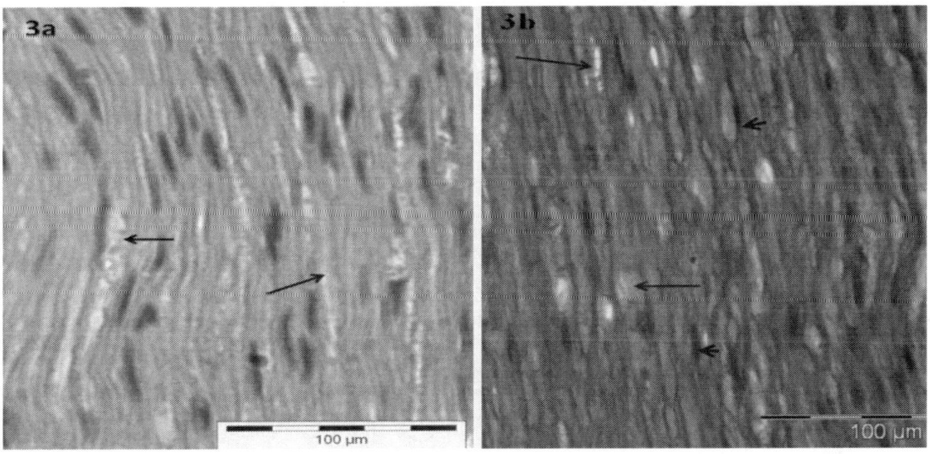

Figure 3. Photomicrographs of the sciatic nerve at 112th POD. a) CG demonstrated the degeneration, vacuolation, ovoid and digestive chambers of collagen were still observed from moderate to minor changes (arrows). b) TG demonstrated few degeneration and vacuolation of the nerve fibers (arrows), Schwann cells (arrow heads) H and E.

Moreover, MSCs constitutively secreted a diverse spectrum of interleukins (IL), growth factors, chemokines and expressed chemokine receptors (Ji et al. 2004), these cytokines are shown to play roles in survival, growth or differentiation and might modulate primary neurons response to injury and thus influence the pain conception. Histopathology results of this study demonstrated that the MSCs promoted and improved the rate of functional recovery within a short period of time compared to the control group. In addition, it was also observed that the MSCs enhanced the onset and acceleration of nerve fibers regeneration, plenty of nerve fibers regeneration such as nerve fibers density, myelin sheath thickness, number of Schwann cells and supportive tissues in the MSCs group compared with control group (Sobeski et al. 2001; Meck et al. 1999).

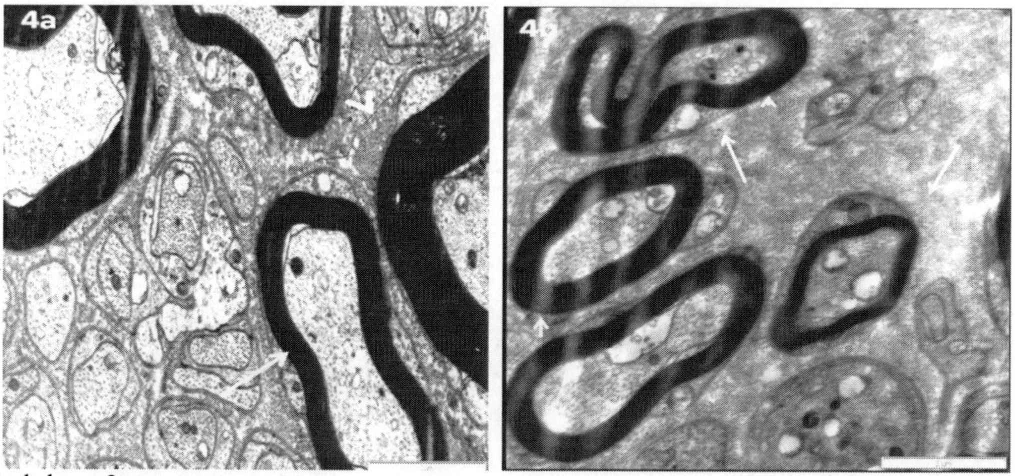

Scale bar = 2 μm.

Figure 4. Electromicrographs of the sciatic nerve at 112th POD. a) TG showed good myelination of regenerative nerve fibers (arrow head) and deposited collagen fibers (arrow). b) CG showed low, thin myelinated nerve fibers (arrow head), thick collagen fibers surrounding of axons (arrow). Uranyl acetate and lead stain.

These findings are consistent with those of Shen et al. (2006) who reported the effects of stem cell therapy on neovascularization which might be explained by multiple nonexclusive pathways. First, transplanted stem cells differentiate into endothelial cells to support new blood vessels components. Secondly, transplanted cells might release the angiogenic and/or trophic factors, which facilitated the angiogenesis. These findings further support the idea of that human bone marrow stromal cells being able to promote the angiogenesis by increasing the endogenous levels of vascular endothelial growth factor (VEGF) (Chen et al. 2003). Ultrastructural examinations showed that the MSCs demonstrated developed Schwann tube or basement membrane which is very important as a guide for axonal outgrowth due to mesenchymal stem cells differentiate to Schwann-like cell that synthesized of basement membrane and extra-cellular matrices and thick myelinated nerve fibres (Dahluin, 2008). In a study by Tohill et al. (2004) reported that undifferentiated MSCs incorporated into a nerve conduit and used to bridge a sciatic nerve defect were shown to express Schwann cell markers, indicating successful *in vivo* differentiation into Schwann cells. Schwann cells are the major supporting cell type in the peripheral nerve and play an integral role in peripheral nerve regeneration and secrete growth factors with neurotrophic and neurotropic properties including nerve growth factor (NGF), neurotrophin 4/5, brain-derived neurotrophic factor (BDNF), vascular endothelial growth factor (VEGF), insulin-like growth factor (IGF) 1 and 2 and glial cell line derived neurotrophic factor (GDNF) (Cui, 2006). The application of MSCs in peripheral nerve injuries has expanded beyond the use of bone marrow as a stem cell source. Pan et al. (2006) demonstrated improved functional outcomes with the incorporation of human amniotic-derived MSCs into a cellulose gauze to bridge a rat sciatic nerve defect. Another study used skin-derived MSCs have been successfully incorporated into artificial nerve conduits with improved axonal regeneration and functional outcomes (Marchesi et al. 2007). In conclusion; BMSCs are capable improved and strongly promoted the regeneration of neurotmesis of the sciatic nerve and *in vivo* might promote cells transplantation in peripheral nerve treatments.

The ultrastructural examinations demonstrated developed Schwann which is very important as guides for axonal outgrowth tube and thick myelinated nerve fibers.

ACKNOWLEDGMENTS

We would like to take this opportunity to thank all staff of virology, histology laboratories and electron microscopy unit Universiti Putra Malaysia for helping me to get to this point.

REFERENCES

Alluin, O., Wittmann, C., Marqueste, T., Chabas, J., Garcia, S., Lavaut, M., Guinard, D., Feron, F., and Decherchi, P. 2008. Functional recovery after peripheral nerve injury and implantation of a collagen guide. *Biomaterials* 30: 363-373.

Chen, J., Zhang, Z. G. and Li, Y. 2003. Intravenous administration of human bone marrow stromal cells induces angiogenesis in the ischemic boundary zone after in rates. *Circ. Res.* 92: 692-699.

Cuevas, P., Carceller, F., Dujovny, M., Garcia-Gomez, I., Cuervas, B., Gonzalez-Corrochano, R., Diaz-Gonzalez, D., Reimers, D. 2002. Peripheral nerve regeneration by bone marrow stromal cells. *Neurol. Res.* 24: 634-8.

Cui, Q. 2006. Actions of neurotrophic factors and their signaling pathways in neuronal survival and axonal regeneration. *Mol. Neurobiol.* 33: 155-179.

Dahluin, L. B. 2008. Techniques of peripheral nerve repair. *Scandinavian Journal of Surgery*, 97: 310-316.

Henery, F., Goyal, N., David, W., Wes, D., Bujold, K., Randolph, M., and Winogard, J. 2009. Improving electrophysiologic and histologic outcomes by photochemically sealing amnion to the peripheral nerve repair site. *Surgery*, 145: 313-322.

Horwitz, E. M., Gordon, P. L., Koo, W. K., Jeffrey C. Marx, Michael D. Neel, Rene Y. McNall, Linda Muul, Ted Hofmann 2002. Isolated allogeneic bone marrow-derived mesenchymal cells engraft and stimulate growth in children with osteogenesis imperfecta: Implications for cell therapy of bone. *Proc. Natl. Acad. Sci. US* 99: 8932–8937.

Ji, J. F., He, B. P., Dheen, S. T., Tay, S. S. 2004. Interaction of chemokines and chemokine receptors mediate the migration of mesenchymal stem cells to the impaired site in the brain after hypoglossal nerve injury. *Stem Cells*, 22: 415-427.

Liu, T. M., Martina, M., Hutmacher, D. W., Hui, J. H., Lee, E. H., Lim, B. 2007. Identification of Common Pathways Mediating Differentiation of Bone Marrow- and Adipose Tissue-Derived Human Mesenchymal Stem Cells into Three Mesenchymal Lineages. *Stem Cells,* 25: 750–760.

Marchesi, C., Pluderi, M., Colleoni, F., Belicchi, M., Meregalli, M., and Farini, A. 2007. Skin-derived stem cells transplanted into resorbable guides provide functional nerve regeneration after sciatic nerve resection. *Glia* 55: 425-438.

Meck, M. F., Dijkstra, J. R., Den Dunnen, W. F., Ijkema- Paassen, J., Schakenraad, J. M., Gramsbergen, A., and Robinson, P. H. 1999. Functional assessment of sciatic nerve reconstruction: Biodegradable poly (DLLA-epsilon-CL) nerve guides versus autologous nerve grafts. *Microsurgery,* 19: 381-388.

Musolino, P. L., Coronel, M. F., Hokfelt, T., Villar, M. 2007. Bone marrow stromal cells induce changes in pain behaviors after sciatic nerve constriction. *Neurosc.*, 418: 97-101.

Musina, R. A., Bekchanova, E. S., Sukhikh, G. T. 2005. Comparison of mesenchymal stem cells obtained from different human tissues. *Bull. Exp. Biol. Med.* 139: 504 –509.

Pan, H. C., Yang, D. Y., Chiu, Y. T., Lai, S. Z., Wang, Y. C., and Chang, M. H. 2006. Enhanced regeneration in injured sciatic nerve by human amniotic mesenchymal stem cell. *J. Clin. Neurosci.* 13: 570-575.

Shen, L., Li, Y., Chen, J., Zhang, J., Vanguri, P. 2006. Intracarotid transplantation of bone marrow stromal cells increasing axon-myelin remodeling after stroke. *Neuroscience*, 137: 393-399.

Sobeski, J. K., Kems, J. M., Safanda, J. F., Shott, S., and Gonzalez, M. H. 2001. Functional and structural effects of GM-1 ganglioside treatment on peripheral nerve grafting in the rat. *Microsurgery*, 21: 108-115.

Tohill, M., Mantovani, C., Wiberg, M., and Terenghi, G. 2004. Rat bone marrow mesenchymal stem cells express glial markers and stimulate nerve regeneration. *Neurosci. Lett.,* 362: 200-203.

Zuk, P. A. 2010. The Adipose-derived Stem Cell: Looking Back and Looking Ahead. *Mol. Biol. Cell* 1; 21(11): 1783–1787.

In: Biotechnology and Conservation … (2 Volume Set) ISBN: 978-1-63117-869-6
Editors: S. A. Farooq, R. Abed and S. Baqir

Chapter 3

COMPARISON OF OMENTUM PEDICLE AND AUTOGENOUS BONE MARROW STROMAL CELLS IMPLANTATION EFFECTIVENESS ON PERIPHERAL NERVE REGENERATION

***H. A. Al-Timmemi*[*1], *R. Ibrahim*[2], *A. Z. Zuki*[3], *T. I. Azmi*[3] *and K. Al-Jashamy*[4]**

[1]Department of Surgery and Obstatric, College of Veterinary Medicine, Baghdad University
[2]Department of Clinical Studies, Faculty of Veterinary Medicine, Universiti Putra Malaysia, Serdang, Selangor, Malaysia
[3]Department of Preclinical Studies, Faculty of Veterinary Medicine, Universiti Putra Malaysia, Serdang, Selangor, Malaysia
[4]SIGA, Medical School, Kuala Lumpur, Malaysia

ABSTRACT

The peripheral nerve injuries (PNI) are due to stretching and laceration. Therefore, the epineural repair technique is currently accepted as a standard method for peripheral nerve repair. Sixty adult male (60) rabbits were divided equally into perineural nerve sutures (ENS) as control group, omental pedicle transposition (OMPT) and bone marrow stromal cells implantation (BMSCs) as the treatment groups. The aim of this study was to investigate the effect and compare of OMPT and BMSC on the regeneration of sciatic nerve neurotmesis in rabbits. The OMPT and BMSc groups were significant improvement ($p<0.05$) of the motor and sensory functions compared to the ENS. Histopathological, morphometric criteria, including the number of myelinated nerve fibers, nerve fiber diameter, axon diameter, myelin sheath thickness, g ratio and relative gastrocnemius muscle weight examinations showed that the OMPT group had the best regeneration and functional recovery compared to BMSCs and ENS

[*] Corresponding author. E-mail: hameed_altmeme@yahoo.com.

Keyword: Omentum pedicle, Bone marrow stromal cells (BMSCs), histopathology, morphometric, peripheral nerve regeneration

INTRODUCTION

The peripheral nervous system is able to regenerate after injury. Etiologies of injuries include penetrating injury, crush, traction, ischemia compression (Robinson, 2004).

There was research focused on a biological approach to healing, looking at omentum and mesenchymal stem cells implantation has become the focus of attention. Omentum was secreted and repository of many factors to promote an environment conducive to growth and overcome limitations in regeneration and functional recovery. The mesenchymal stem cells (MSCs) was become one of the most interesting targets for the study of tissue and organ regeneration because of their plasticity (Prokop, 1997). The implantation of neural stem cells, bone marrow stromal cells (BMSCs), or fibroblasts has been shown to exert a beneficial effect on peripheral nerve regeneration (Mimura et al., 2004). The MSCs was expression of many cytokines and cellular factors (Bhagavati and Xu, 2004). This experimental was aimed to compare the effectiveness of omentum pedicle transposition and bone marrow stromal cells implantation on peripheral nerve regeneration in a rabbit model. Sixty male New Zealand white rabbits (3-5 months old) weighing between 2.0 to 2.3 kg (animals unit, UKM, Malaysia). The animals were kept in their respective in cages and gave broad-spectrum antibiotics and antihelmintic. All procedures used in this study were approved by the Faculty of Veterinary Medicine, Universiti Putra Malaysia Animal Care and Use Committee (08 R13/Dec 08). Rabbits were randomly divided into three groups (n=20). The three groups comprised of animals coaptated transected sciatic nerve with epineurial nerve suture (ENS) as control group, omental pedicle transposition (OMPT) and bone marrow stromal cells implantation groups. Five animals were euthanized from each group on days 14, 28, 56 and 112 post operations. BMSCs were prepared from ilium bone of adult New Zealand white male rabbits was described preparation, identification and differentiation previously by (Al-timmemi et al., 2011). The motor and sensory sciatic nerve clinical reflexes were evaluated at 14th, 28th, 56th and 112th POD. Animals were monitored daily for onset and the ability to walk to the 112th POD. The animals examined for type of walking including crouching, crawling on heel, and normal, as well as knuckling, which was classified into severe, moderate, mild and normal. The muscle contraction force was graded from weak, moderate to strong, and muscle mass atrophy graded as severe, moderate, mild or normal.

Sensory functions of the coaptated sciatic nerves were tested daily to the end of experiment. Toe spreading reflex, lateral aspect leg sensation, toe pinch and toe prick were evaluated as either present or absent. In addition, the foot withdrawal and vocalization tests of lateral aspect leg sensation, toe pinch and toe prick were recorded as positive responses indicating recovery and improved function. Following sacrifice of the rabbit the gastrocnemius muscles were harvested and immediately weighed. The contralateral muscle was also harvested as control for weigh variation between individual rabbit. Each muscle was then weighed separately using 0.0001g weight (Sartorius Analytic Balance Model 2603, Munich, Germany) to calculate the percentage reduction in muscle mass (denervated muscle weight vs. contra lateral muscle weight).

The muscles weight data were expressed as a ratio of the operated limb (left) to un-operated right limb as a negative control to calculate the RGMW. The anastomosed left sciatic nerve was exposed and harvested from each animal three samples 1-cm long were collected from the proximal, middle (coaptate site) and distal segments of the *coaptated* sciatic nerve. The samples were trimmed of any excess length and divided into two parts 5 mm-long from proximal segment each for electron microscopy studies and the other for light microscopy to prepare longitudinal section. 1-cm-long nerve sample was obtained from the intermediate segment, which corresponded to the lesion site for longitudinal sections for light microscopy and the distal segment was prepared as for the proximal segment. Fixed with the 10% neutral buffered formalin, dehydrated in a graded ethanol series, cleared in xylene, embedded in paraffin and cut into 5-μm thick sections and stained with hematoxyline and eosin and Meyer's modified trichrome stain. The specimens for semithin section were fixed with 4% glutaraldehyde overnight at 4°C. The specimens were dehydration, infiltrated with resin and polymerized. Following polymerization the samples were sectioned using an ultra microtome (Leica). Semithin section were stained with 1% toluidine blue and some slide stained using Thionine and Acridine Orang stains (T&AO) and examined using Olympus image analysis (BX 51 TF attachment of CC 12 camera). The number of myelin nerve fibers, total fiber and axon diameters were determined using image software (Abramoff et al., 2004). The myelin thickness was derived from the differences between the fiber and axon diameter. In addition, the g ratio of each fiber was calculated as the axon diameter to the fiber diameter ratio. Right sciatic nerve (normal right hind limb) specimen was also collected and prepared as negative control. All data were expressed as a means and standard deviations (M± SD). Statistical comparisons between groups were performed using Statistical Package for the Social Sciences (SPSS) 16.0 software (non parametric tests), Kruskal Wallis and Mann-Witney tests for clinical observation and for morphometric analysis by one way analysis of variance test (ANOVA), followed by Tukey and Duncan post test. P value ≤0.05 was considered significant.

The onset and ability to walk on day 14 PO the OMPT and BMSCs groups was significant (p≤0.05) compared to the ENS group. The type of gait, disappearance of crouching in the all animals in the OMPT group were significant (p ≤0.05) compared to the ENS and BMSCs groups. Crawl in the OMPT and BMSC groups was significant (p ≤0.05) compared to the ENS group on day 28 PO. On day 56 PO, the type of gait became normal in the OMPT and BMSCs groups significant (p ≤0.05) compared to the ENS. On day 112 PO, the knuckling disappeared in the OMPT and BMSCs groups, significance difference (p ≤0.05) compared to ENS. On day 28 PO, muscle force contraction became stronger in the OMPT and BMSC groups with significant difference (p ≤0.05) compared to the ENS groups. When the animals of OMPT group were used the operated limb, the muscle mass was regaining (recovered) to mild in OMPT group with significant difference (p ≤0.05) compared to the ENS and BMSCs groups on day 112 PO. Sensory clinical signs included toe spread, lateral leg sensation; toe pinch and toe prick did not show sensory reflexes on days' 14, 28 and 56 PO. The interested results of the animals in groups OMPT and BMSCs animals showed sensory reflexes with significance (p ≤0.05) compared to ENS group on day 112 PO. Histopathology findings of the ENS group on day 112 PO coaptated site showed presence of vacuolated and degenerative nerve fibers, granulomatous tissue surrounding the stitches and intra-neural scar tissue (Fig 1a), The distal longitudinal segment showed mild adhered with surrounding tissue seen, low vacuolated degenerative nerve fibers, and presence collagen fibers (scar) at peri and epineurium (Fig 2a). The distal cross section appeared high thickness of fibrous tissue of proliferation of fibroblast at perineurium and epineurium and attached with surrounding muscle (Fig 3a). Semithin transverse section of the distal segment showed thickness of internal perineurium and extraneural nerve fibers at perineurium (Fig 4a).

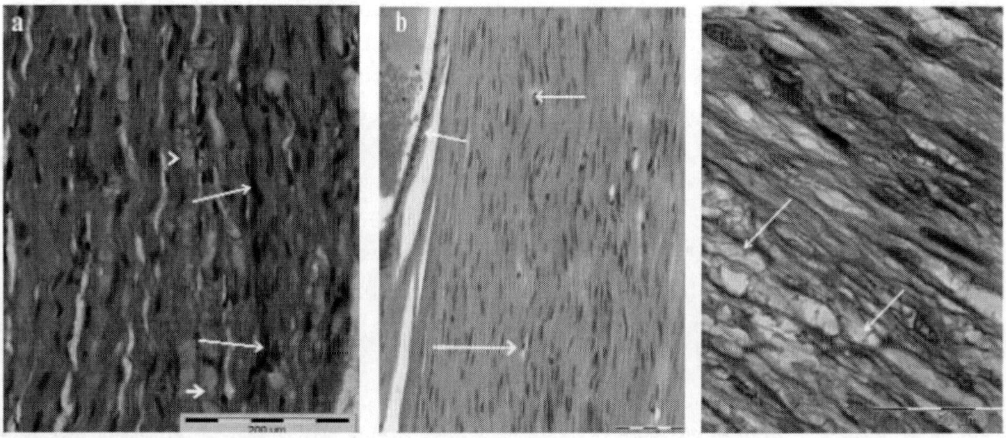

Figure 1. Light micrograph of coaptation site, (a) Mid-segment of ENS group was presence of vacuolated degenerative nerve fibers (arrow heads), granulomatous tissue surrounded the stitches, fibroblast cells (arrows) X 100. (b) Mid-segment of the OMPT group showed high number of Schwann cells, good parallel arrangement nerve fibers and good angiogenesis (thin arrows) and low degenerative nerve fibers (thick arrow). (c) Mid-segment of the BMSCs group showed arrangement of nerve fibers, vacuolated degenerative nerve fibers (arrow) and good angiogenesis H&E X 200.

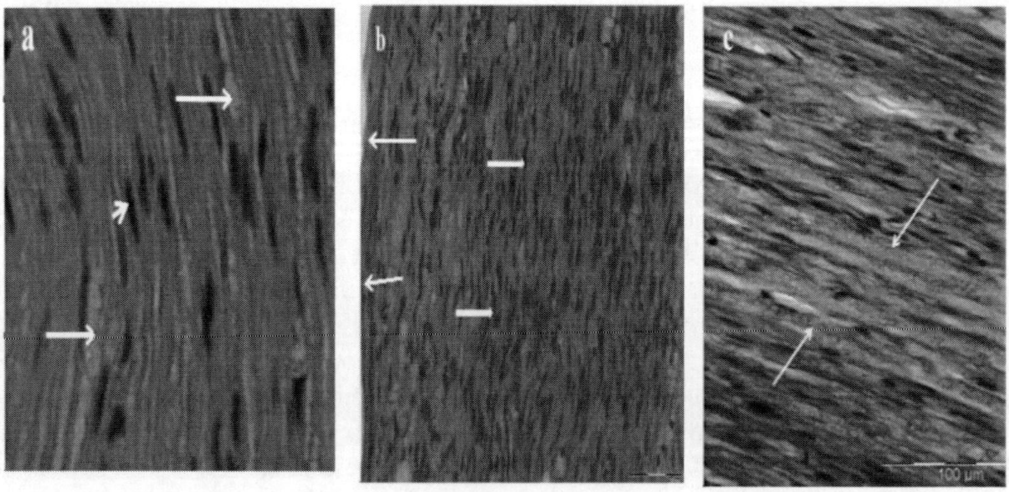

Figure 2. Light micrograph of the distal segment of the sciatic nerve (a) Longitudinal section of the ENS group showed vaculated degenerative nerve fibers, scar tissue in the epineurium and low numbers of Schwann cells. H & E X 200. (b) Longitudinal section of the OMPT group showed normal parallel arrangement of the nerve fibers, few collagen fibers (thin arrows) and high density of Schwann cells (thick arrow). H&E X100. (c) Longitudinal section of the BMSCs group showed vacuolated degenerative nerve fibers, less scar tissue and good myelination (arrows). H&E X100.

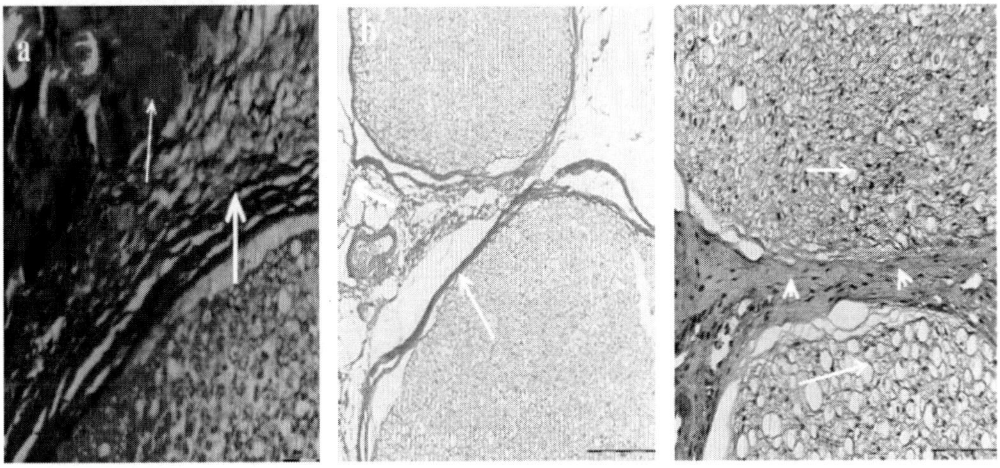

Figure 3. Light micrograph of the transverse section of distal segment of sciatic nerve (a) The section of the ENS group showed thickness of fibrous tissue at perineurium and epineurium (thick arrow) attached with surrounding muscles (thin arrow). MMTS X 200. (b) Cross section of distal segment of sciatic nerve of the OMPT group showed the thin layer of peri and epineurium connective tissue (arrow). MMTS X100. (c) Distal segment of sciatic nerve in the BMSCs group showed deposition of fibrous tissue at perineurium (arrow heads) and endoneurium (arrows). MMTS X100.

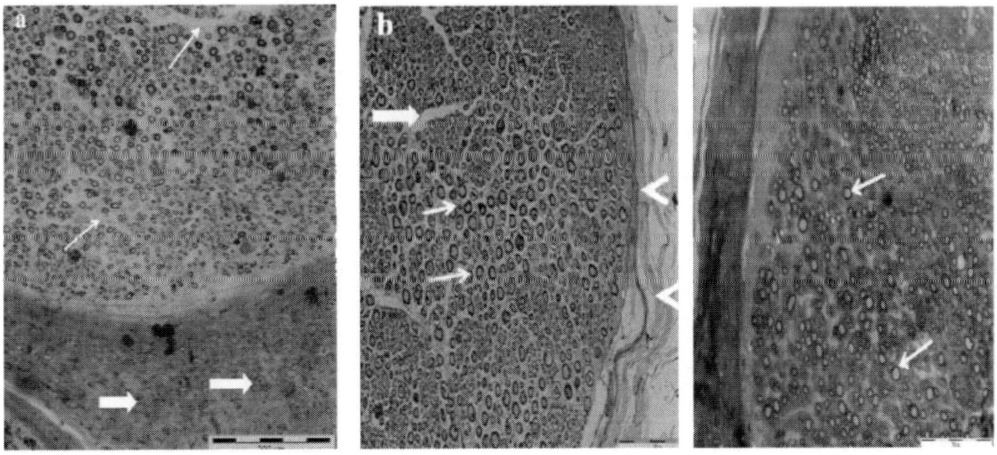

Figure 4. Light micrograph of semithin section of the distal segment of sciatic nerve (a) 5-mm transverse section of distal coaptated site of the ENS group illustrated thickness of internal perinerium (thin arrow), extraneurial nerve fibers (thick arrows) and thick collagen fibers deposit in the epineurium 1μm section. Toluidine blue X100. (b) 5-mm transverse section distal of coaptated site of the OMPT group showed normal myelination and well distribution of nerve fibers (thin arrows), fasciculation with normal perineurium (thick arrow) and normal epineurium (arrow heads). (c) 5-mm transverse section distal of coaptated site of the BMSCs group showed myelinated nerve fibers (thin arrows) 1 μm section. Toluidine blue X100.

Histopathological finding of the OMPT group on day 112 PO anastomosed segment showed a few degenerated nerve fibers, increased presence of Schwann cell and good parallel orientation nerve fibers (Fig 1 b), The distal segment showed normal parallel orientation of the nerve fibers, few collagen fibers and increased vasculature (Fig 2 b). Transverse section showed thin epi-perineurium sheath and very low scar tissues (Fig 3 b). Semithin section of the distal segment showed different sizes of myelinated fibers, with good fasciculation (Fig 4 b). Histopathological examination of the BMSCs on day 112 PO of the middle segment there were vacuolated degenerative nerve fibers in the presence of good angiogenesis and foreign body giant cells surrounded the stitches (Fig 1 c). The distal segment show vacuolated degenerated nerve fibers, minimum scar tissue and good myelination (Fig 2 c). Transverse section of distal sciatic nerve showed thickening of fibrous tissue at perineurium and endoneurium (Fig 3 c). Semithin section of the distal segment showed thin myelinated nerve fibers (Fig 4 c).

ANALYSIS ON DAY 112 POST OPERATION

Histomorphometric Analysis

Results of the negative control showed mean values of myelinated nerve fiber number, diameter fiber, myelin sheath thickness, axon diameter, and g ratio were 13292, 10.35 μm, 3.47 μm, 6.86 μm and 0.66) respectively.

The statistical analysis of the proximal and distal segments of sciatic nerve showed in (Table 2).

Table 1. Statistically analysis of mean values of the number of fibers, fiber diameter, myelin thickness, axon diameter and g ratio of the negative control, ENS, OMPT and BMSC Groups on day 112 PO

sample	Negative Control	ENS Group	OMPT Group	BMSCs Group
Proximal				
Number of fiber	13292±1091 ^{a}	8435±406^{b}	12730±692^{a}	10841±537^{d}
Diameter of the fiber	10.352±0.73^{a}	6.14±1.90^{b}	7.70±1.39^{ab}	6.25±2.4^{b}
Thickness of the sheath	3.472±0.41^{a}	2.28±0.85^{b}	2.60±0.54^{ab}	2.22±0.66^{b}
Diameter of the axon	6.86±0.28^{a}	3.86±1.69^{b}	5.10±0.91^{ab}	4.03±1.32^{b}
g ratio	0.66±0.07^{a}	0.62±0.08^{a}	0.66±0.03^{a}	0.63±0.04^{a}
Distal				
Number of fiber	13292±1091^{a}	8072±52^{b}	10923±570^{a}	10355±698^{cd}
Diameter of the fiber	10.352±0.73^{a}	4.56±1.94^{b}	6.226±1.64^{b}	5.42±2.13^{b}
Thickness of the sheath	3.472±0.41^{b}	1.73±0.73^{a}	1.95±0.73^{a}	1.72±0.54^{a}
Diameter of the axon	6.86±0.28^{a}	3.03±1.30^{b}	4.27±1.09^{b}	3.70±1.66^{b}
g ratio	0.66±0.07^{a}	0.63±0.02^{a}	0.68±0.06^{a}	0.67±0.07^{a}

$^{a, b, c}$ Value bearing similar superscript in the same row not significant at ($p \leq 0.05$)compared to control negative

The transection of the sciatic nerve produced the loss of neural innervation of the gastrocnemius muscle, which led to a decrease in gastrocnemius muscle mass. The statistical analysis of RGMW showed in (Table 2).

Table 2. Statistically analysis of mean values of the relative gastrocnemius muscle weight measurement in the ENS, OMPT and BMSCs Groups

Time	ENS Group	OMPT Group	BMSCs Group
14 days	0.55±0.01[a]	0.80±0.02[b]	0.64±0.03[c]
28 days	0.46±0.1[a]	0.46±0.09[a]	0.48±0 08[a]
56 days	0.50±0.06[a]	0.63±0.04[a]	0.49±0.13[a]
112 days	0.57±0.01[a]	0.81±0.04[b]	0.67±0.06[c]

[a, b, c] Value bearing similar superscript in the same row not significant at ($p<0.05$)

Results of this study showed that the motor function of the BMSCs and OMPT groups were short time (significance difference $p \leq 0.05$) compared to the ENS group. The type of motor clinical sign which included the ability to walk in its operated hind feet were determined based on the severity of pain was classified into neuropathic and inflammatory pain The omentum in the present study might secrete the analgesic substance such as opioids, neurotransmitter including gamma aminobuytric acid (GABA), norepinephrine and other monoamines in their roles in the modulation of pain. This result was in agreement with that Agner et al., (2001) which reported the role of analgesic substance in the mechanism of modulation of pain. The roles of the BMSCs in this respect could be to the diverse spectrum of interleukins (IL), growth factors and chemokines which promote the recovery function of transected sciatic nerve. These conclusions are in agreement with previous works (Liu et al., 2005) which reported that cytokines acted as survival, growth or differentiation factors and may modulate primary sensory neurons response to injury, and thus influence pain behavior.

Knuckling disappeared in all animals in the OMPT and BMSCs groups due to functional recovery of transected sciatic nerve on day of 112 PO, which innervate extensor and flexor muscle that control of normal locomotion of the limb. Result in the animals group treated with omental pedicle transposition was in agreement with Agner et al., 2001 who reported that the omentum are rich in macrophages and mononuclear cells and that the macrophage are essential for successful nerve regeneration by release growth factors and cytokines that stimulate Schwann cell to proliferation (Harman-Boehm et al., 2007). The effects of BMSCs in peripheral nerve injuries are theorized to be supportive in nature and are of two folds. BMSCs are believed to act as Schwann cells in that they function to prevent neuronal cell death and promote directional axonal growth. In addition to its role phagocytosis, BMSC proliferate in response to macrophage stimulation and absence of axonal contact. As they proliferate to fill endoneurial sheaths, they form longitudinal columns commonly known as bands of Büngner. This finding in the current investigation is consistent with previous study which showed that macrophage elaborate Schwann cell mitogens and release of neurotrophin (Baichwal et al., 1988). This muscle force contraction and muscle mass atrophied were related with muscular denervation and muscle disuse and it was increased muscle mass was indicated to the progress of the motor function of sciatic nerve. The degree of sensory reflex included spreading of toe increases in from day to day, gradually regaining of function involving the second, the third and fourth toes which was agreement with the report of (Schmitz and Beer, 2001) who described the gradual return of peroneal nerve function. This index might be applicable to the monitoring of recovery in an animal to express differences in the final degree of recovery. Therefore, in a study of the recovery of the sciatic nerve in rabbits, the reliability and desensitivity of the toe-spreading reflex was evaluated as a sign of functional recovery alongside concurrent muscle weights.

In the OMPT group, the omentum released chemotactic substances (Logmans et al., 1996), vascular endothelial growth factor and basic fibroblast growth factor (bFGF) (Bikfalvi et al., 1990) which might play a role of nerve regeneration which VEGF stimulates axonal outgrowth from dorsal root neuron and promotes survival of neurons and satellite cells. Progress of functional recovery of transected sciatic nerve in BMSC group was because the cultured fibroblast-like BMSCs constitutively expressed trophic factors and supporting substances, including nerve growth factor, brain-derived neurotropic factor, glial cell line-derived neurotrophic factor, collagen, fibronectin and laminin this agreement with (Chen et al., 2007). The histomorphometric analysis showed the numbers of myelinated nerve fibers in the proximal segment of sciatic nerve no significant difference ($p \leq 0.05$) to OMPT group compared with negative control (normal nerve). Omentum rich of macrophages which strongly expressed FGF-2 and FGFR 1–3 (Grothe and Nikkhah, 2001). The omentum pedicle was secreted the growth factors, which are play role of decrease of scar tissue and fibrosis (Castaneda and Kinne, 2002) at the site of coaptation between proximal and distal segment which facilitate of the regenerative nerve fibers to navigate to connect at the end-organ (MacMillan and Stanffer, 1991). Al-Timmemi et al. (2010) who reported that the omentum pedicle might promote of sciatic nerve regeneration due to decreased scar tissue, reduced fibrosis and improved nerve regeneration. The diameter of myelinated nerve fibers of the proximal segment in the OMPT did not show significant difference ($p \leq 0.05$) with compared with the negative control. There are many factors effect on nerve fibers diameter such as the connection with target organs to re-establish correctly. The growth factor secreted from target organs to active the sensory and motor neurons, therefore that the fiber diameter is determined by the size of the connective tissue tubes (Schwann tube) remaining in the distal stump after degeneration. In addition, another study detected that when added the BMSCs increased the number and diameter of mylinated axons in the regenerated nerve (Choi et al., 2005). The omentum pedicle is comprised of adipocytes secreted angiopoietin-1 (Ang-1) which control neurite growth of sensory neurons through induced neurogenesis involves transactivation of the high-affinity nerve growth factor (NGF) receptor tyrosine kinase A (trkA) (Kosacka et al., 2006). Another study showed angiopoietin-1 as a growth factor mediate neurite outgrowth (Chen et al., 2009). The ENS group showed increased collagen deposited around the distal stump endoneurium, which reduced the caliber of endoneurium correlate with smaller fibril diameters. These results was in agreement with those of previous studies (Roytta and Salonen, 1988), which reported that after nerves injury increased the amount of collagen in the distal stump endoneurium reduced their caliber. The diameter of myelinated nerve fibers in the distal stump showed smaller than proximal and normal diameter nerve fibers so this result contrary of the result found by (Sanders, 1948) that the diameter of distal nerve stump similar to those enclosing axons of the same diameter in normal nerve. The OMPT group was appeared the diameter of myelinated nerve fibers improved due to omental pedicle might reduce fibrosis and scar tissue formation, which induced of the nerve fibers maturation. The BMSCs group showed the role of the BMSCs in cell replacement, trophic factors production, extracellular matrix molecule synthesis and remyelination. This physiological function of stem cell improved of diameter myelinated fibers, this result consistent with Mimura et al. (2004) which reported the pluripotency of BMSCs elicited neuroprotection has been explored in peripherial nerve injury. The previous researches are illustrated the diameter of axon and myelin sheath recover fully when the transected axons regenerate and reinnervate denervated sensory and muscle targets (Titmus and Faber, 1990). Another study found that connection with peripheral end-organs is essential to maintain the normal size of peripheral nerve fibers.

In the current study, the distal segment showed decreased in diameter of axon and thickness of myelin sheath compared with negative control which was contrary with the result of (Sanders, 1948) which found that the thickness of myelin sheaths not change and small axons compared with normal fibers at 100 days after sciatic nerve transaction. It means that the OMPT and BMSCs groups provided growth factors, which were principle of synthesis of basement membrane. The basement membrane is considered the main factor that effect on axon diameter and myelin thickness more developed in the OMPT and BMSCs groups and are rich of laminin as extra cellular matrix (ECM) that has numerous biological activities promotion of cell adhesion, migration, growth and differentiation (Yurchenco, 2010). The g ratio in the proximal and distal segments of regenerative sciatic nerve did not show significant difference between ENS and OMPT, BMSCs groups and negative control. The thickness of myelin sheath and axon diameter in proximal segments on day 112 PO were greater than distal segment in all groups. These results are indicated that the maturation process of the regeneration through the increasing of diameter of the axon and decreasing the thickness of myelin sheath and this result in agreement with Perrot et al. (2007) who found in the initial period of the regeneration showed hyper remyelination and increased diameters which then eventually revert to the normal g ratio. On the OMPT and BMSC groups, the myelin sheath was thinner while in the ENS group was thicker than normal nerve fibers of corresponding axon diameter. Moreover, the axon showed larger diameter in the OMPT and BMSCs groups more than in the ENS group. On day 112 PO, the regeneration process was rapid in the OMPT and BMSCs groups than ENS group and the time was influence of the regeneration of peripheral nerve. The comparative relative gastrocnemius muscle weight in the OMPT and BMSCs groups showed significant improvement compared to the epineural sutures, which considered current golden method (ENS groups), it means well-reinnervated target organ and regeneration. The temporary prevention of denervation-induced biochemical changes using OMPT and BMSCs treated may be prompted by atrophic signal for increased synthesis of creatine kinase (CK). Thus preserving a reservoir of high-energy phosphate available for quick resynthesis of adenosine triphosphate (ATP), this decrease of the function of denervated muscles can be restored this congruent with found by Goldspink (1976) who mentioned that the increased synthetic rates denervated tissues were reflected as proportional increases in both myofibrillar and soluble proteins. It maintains functional activity of the injured nerve for a long period might decreases scar tissue formation at the injury site, decreases degeneration in corresponding motor neurons of the spinal cord. There was significantly increasing of the axonal growth and myelination

ACKNOWLEDGMENTS

Authors would like to thank the tissue culture laboratory staff for collaboration with us and are grateful to the Histology Laboratory staff.

References

Abramoff MD, Magelhaes PJ. and Ram SJ. 2004. Image processing with Image. *Journal Biophoton International* 11: 36-42.

Agner C, Yeomans D. and Dujovny M. 2001. The neurochemical basis for the application of the greater omentum in neurosurgery. *Neurolog Research* 23: 7-15.

Al-Timmemi H, Ibrahim R, Al-Jashamy K, Zuki A, Azmi T. and Ramasamy R. 2011. Neurobiological observation of bone mesenchymal stem cells in vitr and in vivo of injured sciatic nerve in rabbit. *Journal of Animals and Veterinary Advance* 10: 686-691.

Al-Timmemi HA, Ibrahim R, Zuki AZ. and Azmi T.I. 2010. Effects of Omental Pedicle Transposition on Regeneration of Neurotmesis Sciatic Nerve in Rabbit. *Pertanika Journal of Tropical Agriculture Sciences* 33: 145–150.

Baichwal RR, Bigbeet Jw. and Devries GH. 1988. Macrophage-mediated myelin-related mitogenic factor for cultured Schwann cells. *Proceedings of the National Academy of Sciences of the U.S.A* 85: 1701-1705.

Bhagavati S, Xu W, 2004. Isolation and enrichment of skeletal muscle progenitor cells from mouse bone marrow. *Biochem. Biophys. Res. Commun.* 318:119–124.

Bikfalvi A, Alterio J, Inyang AL, Dupuy E, Laurent M, Hartmann MP, Vigny L, Raulais D, Courtois Y. and Tobelem G. 1990. Basic fibroblast growth factor expression in human omental microvascular endothelial cells and the effect of phorbolester. *Journal of Cell Physiology* 144: 151–158.

Castaneda F. and Kinne R. 2002. Omental graft improve functional recovery of transected peripheral nerve. *Muscle and Nerve* 26: 527-532.

Chen J, Ou Y, Liao S, Chen W, Chen S, Wu C, Wang C, Wang W, Huang Y. and Hsu S. 2007. Transplantation of bone marrow stromal cells for peripheral nerve repair. *Experimental Neurology* 204: 443–453.

Chen X, Fu W, Tung C. and Ward N. 2009. Angiopoietin-1 induces neurite outgrowth of PC12 cells in a Tie2-independent, β1-integrin dependent manner. *Neuroscience Research* 64: 348-354.

Choi BH, Zhu SJ, Kim BY, Huh JY, Lee SH. and Jung JH. 2005. Transplantation of cultured bone marrow stromal cells to improve peripheral nerve injury. *International Journal of Oral and Maxillary Surgery* 34: 537-542.

Goldspink DF. 1976. The effects of denervation on protein turnover of rat skeletal muscle. *Biochemistry Journal* 156: 71–80.

Grothe C, Heese K, Meisinger C, Wewetzer K, Kunz D, Cattini P. and Otten, U. 2001. Expression of interleukin-6 and its receptor in the sciatic nerve and cultured Schwann cells: relation to 18kD fibroblast growth factor-2. *Brain Research* 885: 172–181.

Harman-Boehm I, Blüher M, Redel H, Sion-Vardy N, Ovadia S, Avinoach E, Shai I, Klöting N, Stumvoll M, Bashan N. and Rudich A. 2007. Macrophage infiltration into omental versus subcutaneous fat across different populations: effect of regional adiposity and the co morbidities of obesity. *Journal of Clinical Endocrinology and Metabolism* 92: 2240–2247.

Kosacka J, Nowicki M, Kacza J, Borlak J, Engele J. and Spanel-Borowski K. 2006. dipocyte-derived angiopoietin-1 supports neurite outgrowth and synaptogenesis of sensory neurons. *Journal Neuroscience Research* 83: 1160-1169.

Liu LH, Sun Z, Sun QY, Huang YJ, Man QH, Guo M, Zhao CH. and Ai HS. 2005. Study on biological characteristics of cultured rhesus mesenchymal stem cells. *Zhongguo Shi Yan Zue Ye Xue Za Zhi* 13:417-421.

Logmans A, Schoenmakers CH, Haensel SM, Koolhoven I, Trimbos JB, van Lent M. and van Ingen, H.E. 1996. High tissue factor concentration in the omentum, a possible cause of its hemostatic properties. *European Journal of Clinical Investigation* 26: 82–83.

MacMillan M. and Stauffer ES. 1991. The effect of omental pedicle graft transfer on spinal microcirculation and laminectory membrane formation. *Spine* 16: 176-80.

Mimura T, Dezawa M, Kanno H, Sawada H. and Yamamoto, I. 2004. Peripheral nerve regeneration by transplantation of bone marrow stromal cell-derived Schwann cells in adult rats. *J Neurosurg* 101: 806-812.

Perrot R, Lonchampt P, Peterson AC. and Eyer J. 2007. Axonal neurofilaments control multiple fiber properties but do not influence structure or spacing of nodes of Ranvier. *Journal of Neuroscience* 27: 9573–9584.

Prokop DJ. 1997. Marrow stromal cells as stem cells for nonhematopoitic tissue. Science 276: 71-76.

Robinson LR. 2004. Traumatic injury to peripheral nerves. *Supplement Clinical Neurophysiology*.57: 173–86.

Roytta M. and Salonen V. 1988. Long-term endoneurial changes after nerve transection. *Acta Neuropathology* 76: 35-45.

Sanders FK. 1948. The Thickness of the Myelin Sheaths of Normal and Regenerating Peripheral Nerve Fibres. *Proceedings of the Royal Society of London* 135: 323-357.

Schmitzl H and Beer G. 2001. The toe-spreading reflex of the rabbit revisited functional evaluation of complete peroneal nerve lesions. *Laboratory Animals* 35: 340–345.

Titmus MJ and Faber D S. 1990. Axotomy-induced alterations in electrophysiological characteristics of neurons. *Progress in Neurobiology* 35: 1–51.

Yurchenco PD. 2010. Basement membranes: cell scaffoldings and signaling platforms. Cold Spring Harbor Perspectives in Biology. *www.cshperspectives.cshlp.org.*

In: Biotechnology and Conservation … (2 Volume Set)
Editors: S. A. Farooq, R. Abed and S. Baqir
ISBN: 978-1-63117-869-6

Chapter 4

IMMUNOLOCALIZATION OF ANDROGEN RECEPTOR IN THE OVARY OF THE BEDOUIN GOAT (*CAPRA HIRCUS*) IN NON-BREEDING SEASON

***S. Kassouri*[1], *N. Boukenaoui*[1,2], *S. Charallah*[1], *Z. Amirat*[1] *and F. Khammar*[1]**

[1]University of Sciences and Technology Houari Boumediene, Faculty of Biological Sciences, Laboratory of Research on Arid Zones, BP 44, Alger Gare, Algiers, Algeria
[2]University Saad Dahlab Blida, Faculty of Agro-Veterinary Sciences, Department of Veterinary Sciences, BP 270, Road Soumaa Blida, Algeria

ABSTRACT

The role of androgens on the growth and follicular development (by their direct action via their specific receptors, or indirectly by the estrogens), is the subject of many studies in mammals. However, very few works treat androgen localization in ovaries of ruminants living in arid regions. The aim of this study was to determine the androgen receptor localization, by immunohistochemical technique, in the ovaries of the bedouin goat *(Capra hircus)* in the non-breeding season. Ovaries obtained from the mature female goat during the sexual quiescence period (spring), were removed and then immediately fixed in Bouin's solution. After paraffin embedding, sections were realized and used for immunohistochemistry analysis using a polyclonal antibody androgen receptor. At non breeding season, the results obtained showed that androgen receptors: 1/ were detected in different compartments of the ovary and in various stages of follicles; 2/ their localization is essentially nuclear in granulosa, theca and stroma cells; their cytoplasm also exhibited a weak positive immunostaining; 3/ extern theca cells were completely deprived of immunoreactivity; 4/ the immunostaining was more intense at preovulatory stage and 5/ in atretic follicle, the theca intern cells lose the expression of androgen receptor. These findings suggested that in the non-breeding season androgens presented important activities in the regulation of growth and follicular development as it was reported in other species of mammals like the rat, the squirrel, the woman and primates. The present study contributes to a better knowledge of the physiology of the reproduction of this species, thus allowing in the longer term to better plan its reproduction, to increase its productivity and to put conservation strategy for this race living in arid lands.

Keywords: Androgen receptor, ovaries, bedouin goat (*Capra hircus*)

INTRODUCTION

Androgens are involved in the development and the physiological function of female re productive organs and more precisely they antagonize follicular development and may induce apoptosis in granulosa cells in ovaries (Duda et al., 2004). The androgen action is mediated by the androgen receptor (AR) (Myamoto et al., 2007), which belongs to the superfamily of ligand-responsive transcription regulators (Carson-Jurica et al., 1990), little was known about the actions of androgen and AR in female reproduction, although AR expression in growing follicles have beendescribed (Shiina et al., 2006). For better understanding the role of androgens in the growth and follicular development, it is important to define the hormone site(s) action. The aim of this study was to determine the AR localization, by immunohistochemical technique, in ovaries of bedouin goat (*Capra hircus*) during non breeding season.

MATERIALS AND METHODS

The experiment was conducted at Beni-Abbes station (30° 7' N., 2° 10' W.). Two bedouin goat two years aged and living in nomadic herding were used in this study. After slaughter of the goats, right and left ovaries were removed and then sectioned along the axis of the hilum. Immediately after, samples were fixed in Bouin's solution, dehydrated in increasing concentrations of ethanol, cleared in toluene and embedded in paraffin. Sections of paraffin at 3μm were stored at room temperature. Androgen receptor immunohistochemical studies were performed using the avidin–biotin complex method (ABC), using the VectastainElite ABC kit (Vector Laboratories, Burlingame, CA, USA).

Paraffin sections were deparaffinized, hydrated through a graded ethanol series (100%, 95% and 70%), and washed in PBS. Immunohistochemistry was performed on deparaffinized sections with heat-induced antigen retrieval in citrate buffer (pH 6.0) using a pressurecooker, as previously described in the prospectus for the kit "Vector Antigen Unmasking Solutions" (Vector Laboratories, CA, H3300). This was followed by endogenous peroxidase blocking 3% H_2O_2 in PBS for 5 minutes at room temperature. All washes between antibody or reagent incubations were rinsed 2 × 5 min at room temperature in PBS, and all incubations were carried out in a wet chamber. Tissue sections were first submitted to the appropriate serum in order to block nonspecific binding sites. After that, sections processed for AR labelling were incubated with normal horse serum at room temperature for 5 min, and then with both avidin and biotin sites subsequently blocked (Vector Laboratories, CA, SP-2001). All sections were incubated over-night at 4°C with the primary antibody, which was diluted at 1:200 in PBS. Bound antibodies were visualized by incubating the sections with biotinylated secondary antibody (Vectastain Elite ABC kit-Vector Laboratories, CA, #PK-6200) for 30 min. Labelling of AR was visualized with3,3'-diaminobenzidine-tetra-hydrochloride chromogenic substrate (SK-4100, DAB substrate kit for peroxidase;Vector Laboratories) and were monitored microscopically. Sections were counterstained with hematoxylin (Hematoxylin

QS, H-3404; Vector Laboratories, Burlingame, CA, USA). These sections were dehydrated and mounted. Images were captured using a digital camera Canon Power Shot A640 integrated with a Zeiss light microscope. Sections incubated with normal horse serum instead of primary antibody were used as negative controls.

RESULTS

At non breeding season, the results obtained showed that androgen receptors were detected in different compartments (granulosa, theca, stroma) of the ovary and in various stages of follicles (Figure 1). In primary follicle the positivity for AR was found in granulosa and interstitial nuclei cells (Figure 1a).

In preovulatory (Figure 1b) and antral (Figure 1c) follicles the labelling is localized in granulosa cells, theca and stroma nuclei with an intense labelling at preovulatory stage. However, the cytoplasm of granulosa and theca cells showed also a weak positive immunostaining. The extern theca cells were completely deprived of immune reactivity (Figure 1c). In atretic follicle, the intern theca cells lose the expression of androgen receptor (Figure 1d). No immunostaining was detected in control sections when normal horse serum was substituted for the primary antibody (Figures 1b, 1c and 1d).

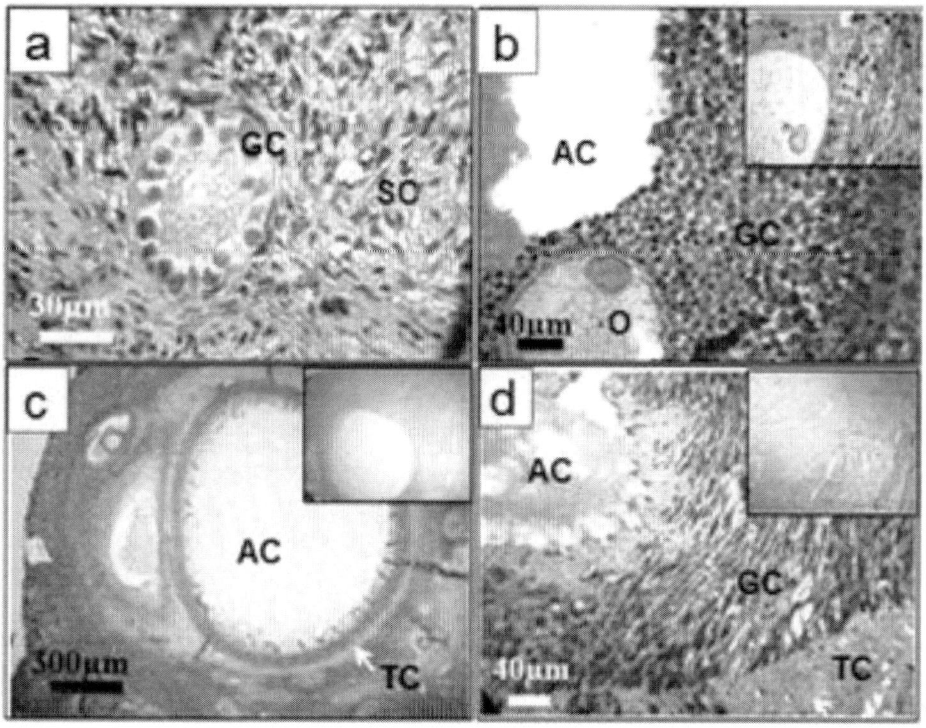

Figure 1. Immunolocalization of androgen receptor in the ovary of the bedouin goat during non breeding season. Primary follicle (a), ovulatory follicle (b), large antral follicle (c) and atretic follicles (d). Inserts in (b), (c), and (d) show negative controls. AC: Antral cavity; GC: Granulosa Cell; SC: Stroma Cell; O: Oocyte; TC: Theca Cell.

DISCUSSION

These findings suggested that in non breeding season, androgens present important activities in the regulation of growth and follicular development as it was reported in rat (Szoltys et al., 2000), squirrel (Li et al., 2012), woman (Ahonen et al., 2000) and primates (Weil et al., 1998).

In conclusion, bedouin goat *(Capra hircus)* might offer a good model to study the regulation pathway of androgen during the process of follicular development in the ovary of mammals living in arid region. Folliculogenesis in this species is a physiological process influenced probably by the season; Although the follicular growth remains similar to those of other Mammals.

ACKNOWLEDGMENTS

We would like to thank all the personnel of Beni-Abbes station in the Algerian sahara. This study was supported by the Accord program Algéro-Français 09MDU 756 CMEP.

REFERENCES

Ahonen MH, Zhuang YH, Aine R, Ylikomi T, Tuohimaa P. 2000. Androgen receptor and vitamin D receptor in human ovarian cancer: growth stimulation and inhibition by ligands. *Int. J. Canc.* 86: 40-46.

Carson-Jurica MA, Schrader WT, O'Malley BW. 1990. Steroid receptor family: structure and functions. *Endocr. Rev.* 11: 201–220.

Duda M, Burek M, Galas J, Kozioł K, Koziorowski M, Słomczyńska M. 2004. Immunohistochemical localization of androgen receptor and aromatase in the ovary of the pregnant pig. *Biol. Reprod.* 4,3: 289–298

Li X, Zhang H, Sheng X, Li B, Zhou J, Xu M, Weng Q, Watanabe G, Taya K. 2012. Immunoreactivities of androgen receptor, estrogen receptors, p450arom, p450c17 proteins in wild ground squirrels ovaries during the non breeding and breeding seasons. *J. Ovarian Res.* 5(1), 26.

Miyamoto J, Matsumoto T, Shiina H, Inoue K, Takada I, Ito S, Itoh J, Minematsu T, Sato T, Yanase T, Nawata H, OsamuraYR, Kato S. 2007. The Pituitary Function of Androgen Receptor Constitutes a Glucocorticoid Production Circuit. *Mol. Cell Biol.* 27: 4807–4814.

Shiina H, Matsumoto T, Sato T, Igarashi K, Miyamoto J, Takemasa S, Sakari M, Takada I, akamura T, Metzger D, Chambon P, Kanno J, Yoshikawa H, Kato S. 2006. Premature ovarian failure in androgen receptor-deficient mice. *Proc. Natl. Acad. Sci.* 103: 224–229.

Szołtys M, Słomczyn´ska M. 2000. Changes in distribution of androgen receptor during maturation of rat ovarian follicles. *Exp. Clin. Endocr. Diab.* 108: 228–34.

Weil SJ, Vendola K, Zhou J, Adesanya OO, Wang J, Okafor J, Bondy CA. 1998. Androgen receptor gene expression in the primate ovary: cellular localization, regulation, and functional correlations. *J. Clin. Endoc. Met.* 83: 2479–2485.

In: Biotechnology and Conservation … (2 Volume Set) ISBN: 978-1-63117-869-6
Editors: S. A. Farooq, R. Abed and S. Baqir

Chapter 5

Reference Values for Haematological, Biochemical and Physiological Parameters in Arabian Oryx

***Mansoor H. AlJahdhami*[1*], *J. Anne Brown*[2], *Abulaziz Y. A. Alkindi*[3], *Khalid AlRasbi*[4] and *Elias Nikolakopoulos*[5]**

[1]Office for Conservation of the Environment, Diwan of Royal Court, Sultanate of Oman
[2]Biosciences, College of Life and Environmental Sciences, Geoffrey Pope, University of Exeter, Stocker Road, Exeter, UK
[3]University of Nizwa, Nizwa, Sultanate of Oman
[4]Omani Mammals Breeding Centre, Directorate of Veterinary Services, Royal Court Affairs, AlSeeb, Sultanate of Oman
[5]Directorate of Veterinary Services, Royal Court Affairs, Sultanate of Oman

Abstract

The vulnerable Arabian oryx, *Oryx leucoryx* faces a wide range of issues that potentially have adverse effects on their welfare while they are free-ranging in their natural habitat, housed in captivity for conservation-breeding or when they are translocated from the wild to captivity or vice versa. Furthermore, the global increase in the number of captive Arabian oryx gives rise to particular concern for their welfare and health within captive conditions. Thorough assessment of the welfare of animals involves physiological and behavioural measures. For assessment and monitoring of health of animals, one of the important things is to establish reference or expected range of values for various parameters. Published reference values for haematological, biochemical, hormonal and clinical parameters for Arabian oryx are limited, with little information for non-immobilised and non-tranquillised oryx or consideration of possible age and sex differences. Therefore, reference values and inter-percentile ranges (2.5 and 97.5 percentiles) were established for 32 parameters, in separate groups of male and female adult oryx, without using immobilising or tranquillising chemicals during capture. The

[*] Corresponding author: mhjahdhami@gmail.com.

haematological parameters investigated were white blood cell count and differentiation (%) of cell types (neutrophils, lymphocytes, monocytes, eosinophils, basophils), number of platelets, red blood cell count, haemoglobin concentration and haematocrit, erythrocyte cell volume, erythrocyte haemoglobin content and concentration, serum osmolality and ions (sodium, potassium, chloride, calcium, magnesium and phosphorus). Biochemical parameters investigated were serum urea, glucose, total protein, albumin and plasma lactate concentrations. Clinical parameters investigated were body temperature, heart and respiratory rates. Hormonal parameters measured were cortisol, free-thyroxine, free-triiodothyronine and insulin concentrations. Near basal values for serum cortisol were measured in Arabian oryx sampled within 2 min, while values were significantly higher in oryx sampled within 5-10 min. The reference values established in these studies are considered valuable tools for diagnosis of disease and physiological alterations in Arabian oryx. The establishment of reference values for Arabian oryx that considers the differences between males and females has importance for future monitoring of the well-being of this endangered species.

Keywords: Biochemistry, Biotechnology, Arabian oryx, Oryx leucoryx

INTRODUCTION

Maintaining good health and freedom from disease are essential elements of good animal welfare (Botreau et al., 2007). One of the 'Five freedoms' for animal welfare is Freedom from pain, injury and disease by prevention and rapid diagnosis and treatment (FAWC, 1993). The role of veterinary medicine is important for conservation of wildlife through assessment and monitoring of health in the wild or under captive conditions (Karesh & Cook, 1995; Kirkwood, 1993). Maintaining good health of wildlife particularly under captive conditions is of high importance. Examples of health problems are injuries, endo and ecto-parasites, lameness, infectious diseases, diarrhoea and chronic stress. For assessment and monitoring of health of animals, it is important to establish reference or expected range of values for various parameters which include data for blood parameters such as complete blood count, serum biochemistry profile, ions, hormones and clinical parameters e.g., heart rate, respiratory rate and body temperature (Deem et al., 2001; Karesh et al., 1997). These reference values aid in the diagnosis of certain pathologies (Deem et al., 2001; Kaneko et al., 1997). For example, haematological data such as haemoglobin and neutrophil count are used for the diagnosis of pathologies such as anaemia and bacterial infections, respectively (Hawkey, 1991; Junqueria & Carnerio, 2005). The values used for diagnosis or screening of diseases are often called "normal values" (Walton, 2001). Because these values might vary according to the sex, age or in captive or wild individuals of the same species, the term "reference values" should be used to avoid confusion (Solberg, 2006; Walton, 2001). Reference values are often given as the range between minimum and maximum values, which might include extreme outliers. Therefore, the International Federation of Clinical Chemistry (IFCC) recommended the use of the central 95 % reference ranges defined as the 2.5 and 97.5 percentiles (Dimauro et al., 2008; Harris & Boyd, 1995; Horn & Pesce, 2003; Lumsden, 1998; Solberg, 2006; Walton, 2001). Haematological and biochemical reference values have been established for a wide range of animals. For example, reference values for haematological and/or biochemical parameters were established for chital deer (*Axis axis*) (Chapple et al., 1991), southern chamois (*Rupicapra pyrenaica*) (Lopez-Olvera et al., 2006a), Spanish ibex (*Capra pyrenaica*) (Perez et al., 2003), black-faced impala (*Aepyceros melampus petersi*) (Karesh et al., 1997), working horses (Pritchard et al., 2009) and some species of wild ruminants in captivity (Peinado et al., 1999). Most Arabian oryx around the world (more than 95 %) live in some form of captivity nowadays. The total number of Arabian oryx in Arabia is currently estimated at about 8000 oryx (Strauss, 2008). Holding of Arabian oryx in captivity requires close monitoring of their health and well-being. The availability of accepted reference values for haematology, biochemistry, ions and hormones and clinical parameters will help in this monitoring and in the diagnosis of diseases within the population. The reference values of some haematological, biochemical, ions and other physiological parameters such as heart and respiratory rates vary according to the sex and age of the same species (Bush et al., 1983; Jain, 1986; Lopez-Olvera et al., 2006a). The variation might also depend on the method of capture and whether immobilising or tranquilising chemicals have been used or whether animals have been captured with or without using chemicals, as reported in red deer (*Cervus elaphus*) (Marco & Lavin, 1999) and koalas (*Phascolarctos cinereus*) (Hajduk et al., 1992). Some haematological and biochemical parameters, and data for body temperature, heart and respiratory rates were reported for neonates of Arabian oryx (Bounous-Dalton & Hood, 1980; Ferrell et al., 2001). There are clear differences in some haematological parameters between some of the seven age-groups reported for scimitar-horned oryx (*Oryx dammah*) (Hawkey & Hart, 1984). Ferrell et al. (2001) demonstrated important differences in biomedical parameters between neonates of closely related species of hippotragini: Arabian oryx (*Oryx leucoryx*), Addax (*Addax nasomaculatus*), scimitar-horned oryx (*Oryx dammah*) and sable antelope (*Hippotragus niger*). Therefore, the reference values established for neonates of one species cannot be applied in neonates of other species. Reference values of adult animals are also not recommended as a reference for neonates of the same species or other closely related species (Ferrell et al., 2001). Available information for haematological and biochemical parameters, serum ions and osmolality, hormones and clinical parameters (e.g., body temperature, heart rate and respiratory rate) in adult Arabian oryx have all been obtained either from immobilised animals (with data combined for both sexes) (Ancrenaz et al., 1996; Greth et al., 1993; Vassart & Greth, 1991) or from a single female that experienced capture myopathy (Vassart et al., 1992). In 1986, when there was an outbreak of tuberculosis in a recently reintroduced herd of Arabian oryx in Saudi Arabia (Vassart & Greth, 1991) the first haematological and serum biochemical values for Arabian oryx were established during a campaign for eradication of the tuberculosis. The oryx that were used for establishment of the reference values were captured by immobilisation and tranquillisation. The idea of establishing reference ranges was excellent; however, as samples were collected from animals that might have tuberculosis and that were captured after chemical injection, the values reported need to be considered with care. Other available values for Arabian oryx of more than 50 parameters measured by 17 different institutions have been collected in a database (International Species Information System, 2002), however, the data combine data for both sexes and all age groups. In this database, the number of samples per parameter varied from a single sample for some parameters to 265 samples for others, with variable numbers of data points per animal. Because of mixing data from all age groups and sexes and using variable numbers of samples per animal, these values should also be considered with care. One study that was looking particularly at the blood-gas and acid-base parameters also investigated a limited number of haematological parameters and serum ions from non-immobilised and non-tranquillised Arabian oryx (Kilgallon et al., 2008). However, this study combined the data for 14 males and 5 females. Ostrowski et al. (2006) also reported some biochemical parameters for Arabian oryx, but did not specify the method of capture and whether it was with or without immobilisation. So all available information for haematological, biochemical or other physiological data for adult Arabian oryx were obtained from either immobilised oryx in a mixed sex group, or after unspecified methods of capture or analytical methods. The use of chemical restraint e.g., immobilisation and tranquillisation, has been reported to depress some blood parameters in white-tailed deer (*Odocoileus virginianus*) compared to restraint without using chemicals (Kocan et al., 1981; Presidente et al., 1973; Seal et al., 1972; Wesson III et al., 1979). In this study, we present data for some haematological and biochemical parameters, serum ions and hormones, respiratory and heart rates and body temperature for male and female adult Arabian oryx that were captured physically without chemical immobilisation or tranquillisation. The differences between sexes are examined. Thirty six adult Arabian oryx (24 males and 12 females) were used for the establishment of reference values. All reference values presented in this study were obtained from oryx that were captured physically without using chemical immobilisation or tranquillisation. The total number of oryx at the Omani Mammals Breeding Centre was 146 by the end of 2008, which means that about 25 % of the oryx at the centre were included in the studies. The sampled oryx represented about 8 % of Oman's total population of captive and wild oryx (430 individuals, at the end of 2008). Arabian oryx at the Omani Mammals Breeding Centre are held in two enclosures of areas 37,331 m^2 and 12,674 m^2. Adult Arabian oryx were randomly selected based on inclusion and exclusion criteria from the paddocks and captured by immobilisation or sometimes physically, without using chemicals, and transferred to small holding pens. The inclusion criteria included adults of age between 1 and 5 years, apparently healthy and from both sexes. Exclusion criteria included calves, old oryx, dominant males, pregnant and lactating females, and apparently sick animals. Food and water were provided for the oryx in the morning and evening. They are fed with hay, fresh alfalfa (lucerne), concentrate animal feed pellets (Barakat, Oman Flour Mills, Muscat, Oman), and sometimes with some cabbage and lettuce. The oryx were blind-folded immediately after capture and blood samples were collected by a qualified veterinarian. Blood samples were collected from the jugular vein using a disposable 50 ml syringe with a 18 G x 1.5" needle (Becton Dickinson) and then transferred into vacutainer (Becton Dickinson, 7 ml) tubes containing anticoagulant (EDTA) for haematological analysis, anticoagulant (fluoride oxalate) for analysis of plasma lactate and tubes without anticoagulant for analysis of serum ions, osmolality, hormones and biochemistry, by the passive force of the vacuum. This method was preferable to direct collection into vacutainers which in early trials was found to be unreliable for collection of required amount of blood in the various types of vacutainer tubes as well as taking longer overall. The inner diameter (ID) of the needle employed in the present studies 18G (ID: 0.838 mm) was wider than the 21G (ID: 0.514 mm) commonly used for blood sampling (Lopez-Olvera et al., 2006b), which minimised cell damage and erythrocyte haemolysis. The collected samples were labelled and placed in a cool-box containing ice until processing, within 3-6 h after collection. Rectal body temperature was measured by using a digital thermometer (Thermoval, Hartmann, Heidenheim, Germany). Heart and respiratory rates were measured using a stethoscope and a stopwatch.

Blood Samples Analyses

Haematological analysis was done within about 3 h after blood collection, using an automated blood analyser (Cell-DYN 4000, Abbott Diagnostics Santa Clara, CA, USA) at the Sultan Qaboos University Hospital, Muscat, Sultanate of Oman. The inter- and intra-variations of data for haematological parameters were measured for 8 males using duplicate samples (Table 1) as recommended by (Murray et al., 1993).

Table 1. Inter- and intra-variations (%) of haematological parameters calculated for 8 male Arabian oryx after measurement by Cell-Dyn 4000

Parameters	Intra-variation	Inter-variation
WBC	1.77	19.48
Neutrophils	1.08	4.92
Lymphocytes	3.36	12.10
Monocytes	126.20	247.09
Eosinophils	20.61	28.45
Basophils	21.83	44.90
RBC	1.34	4.69
Haemoglobin	0.80	6.82
Haematocrit	1.29	7.58
MCV	0.23	6.66
MCH	1.58	6.38
MCHC	1.64	1.40
Platelets	36.31	74.24

Blood samples for the analysis of hormones, ions, osmolality and biochemical parameters were centrifuged at 2500 g for 15 min. Serum and plasma were transferred into polypropylene micro-centrifuge tubes, which were stored at - 80 °C until analysis. Serum ions (calcium, phosphorus, magnesium) and a range of biochemical parameters (serum glucose, urea, albumin, total protein and plasma lactate) were analysed with an auto-analyser (Cobas Integra 800, Roche, Switzerland), at the Department of Biochemisty, Sultan Qaboos University Hospital using absorbance photometry (enzymes and substrates). This instrument was also used in the potentiometric mode (measuring electrical potential) to determine serum concentrations of sodium, potassium and chloride. Serum hormones where analysed by immuno-assays (Access® immuno-assay system, Beckman Coulter Inc.). The studied parameters were cortisol, free thyroxine (T_4), free triiodothyronine (T_3), and insulin. The instruments that measured blood ions, biochemistry and hormones (Cobas Integra 800 and Access, respectively) were calibrated as a routine practice. As a further check on the validity of measurements for Arabian oryx blood samples, a blood sample was subjected to dilution with normal saline (sodium chloride (0.9 %, w/v)) by preparation of different ratios of serum to saline to give: 0 % 25 % 50 % 75 % and 100 % serum. The dilution of serum with saline resulted in the expected decline in the concentration of selected representative parameters. Interassay and intraassay variability in the concentrations of ions, biochemical parameters and hormones calculated from 4 duplicated samples are shown in Table 2.

Serum osmolality was determined by freezing point depression using 20 µl of serum in a micro-osmometer (Advanced Micro-osmometer 3300, Advanced Instruments, Inc., Norwood, MA, USA).

Table 2. Inter- and intra-variations (%) of serum ions, biochemistry and hormones calculated from 4 duplicated samples

Parameters	Intra-variation	Inter-variation
Sodium	13.80	15.73
Potassium	12.92	11.56
Chloride	13.39	13.80
Urea	12.32	26.32
Calcium	14.38	6.32
Total protein	13.37	15.28
Glucose	13.45	16.54
Albumin	14.48	8.91
Phosphorus	14.15	11.70
Cortisol	7.66	22.70
Free T_4	7.73	9.02
Free T_3	21.79	12.86
Insulin	30.30	56.84

The statistical differences between sexes (male versus female) were compared by one way ANOVAs followed by multiple comparison tests (Holm-Sidak method). Where data could not be tested by one way ANOVAs because of missing values or failure to achieve normality, data were analysed by t-tests or Mann-Whitney tests, as appropriate. Data are presented as means, standard deviation of mean (SD), standard error of mean (SEM) and 2.5 and 97.5 percentiles. Within the discussion section of this study, the data from the present studies are presented as mean and standard error of mean, unless stated otherwise.

The haematological data for male and female Arabian oryx are presented in (Table 3). All reported haematological parameters other than white blood cells and the percentages of neutrophils, lymphocytes and basophils, did not differ between males and females (Table 3). The white blood cell count ($P = 0.018$) and the percentage of neutrophils ($P < 0.001$) were significantly higher in males than females while the percentages of lymphocytes ($P < 0.001$) and basophils ($P = 0.047$) were significantly lower (Table 3).

The values for biochemical and clinical parameters, ions, osmolality and hormones for male and female oryx are shown in Table 4. Most parameters did not differ significantly between males and females. The only parameters that differed significantly between males and females were sodium ($P = 0.027$), urea ($P < 0.001$), glucose ($P < 0.001$), cortisol ($P < 0.001$) and free T_3 ($P = 0.016$). Males have significantly higher values of sodium, urea, glucose and free T_3 and significantly lower cortisol than females (Tables 4).

The main aim of this study was to establish reference values for haematological, biochemical hormonal and clinical parameters. These parameters are the main parameters assessed in analytical laboratories and made available to veterinarians to assess health and abnormalities in animals. In the present study, all animal used in assessing reference ranges were from the Omani Mammals Breeding Center within easy reach of the analytical laboratories at Sultan Qaboos University. No samples from the other captive population in Oman, at Jaaluni (the AlWusta Wildlife Reserve, known previously as the Arabian Oryx Sanctuary) were included because of the complexities of dealing with samples collected 630 km from the laboratory. Capturing and collecting blood samples from a few individuals per day and travelling the very long distance for analysis of unfrozen samples was not feasible for this study. However, ideally to obtain reference values for the population of Arabian oryx in the Oman, values for samples obtained from the Jaaluni population and free-ranging animals should be incorporated in later studies.

Table 3. Reference values of haematological parameters for male and female oryx. n, number of animals; SD, standard deviation of mean; SEM, standard error of mean; 2.5 and 97.5 percentiles; WBC, white blood cells (leukocytes); RBC, red blood cells (erythrocytes) MCV, mean corpuscular volume; MCH, mean corpuscular haemoglobin; and MCHC, mean corpuscular haemoglobin concentration

Parameter	Sex	Unit	n	Mean	SD	SEM	2.5 %	97.5 %
WBC	Males	$x10^9/L$	24	8.03	1.60	0.33	5.11	11.20
	Females	$x10^9/L$	12	6.93	1.48	0.43	4.26	9.64
Neutrophils	Males	%	24	69.66	5.90	1.21	57.10	78.40
	Females	%	12	58.52	6.60	1.91	45.60	67.20
Lymphocytes	Males	%	24	25.50	5.13	1.05	17.10	36.10
	Females	%	12	34.20	7.50	2.16	24.60	49.10
Monocytes	Males	%	24	0.69	1.27	0.26	0.00	4.05
	Females	%	12	0.98	1.04	0.30	0.00	3.02
Eosinophils	Males	%	24	2.15	1.30	0.27	0.41	5.48
	Females	%	12	3.07	1.64	0.47	0.40	6.02
Basophils	Males	%	24	2.01	1.35	0.28	0.00	4.21
	Females	%	12	3.23	0.90	0.26	1.80	4.99
RBC	Males	$x10^{12}/L$	24	12.22	0.75	0.15	10.70	13.70
	Females	$x10^{12}/L$	12	12.17	0.59	0.17	11.20	13.10
Haemoglobin	Males	g/dL	24	18.29	1.26	0.26	15.50	20.60
	Females	g/dL	12	18.05	0.98	0.28	15.90	19.30
Haematocrit	Males	%	24	54.80	4.97	1.01	47.10	70.30
	Females	%	12	52.60	2.88	0.83	47.50	56.60
MCV	Males	fL	24	44.80	2.25	0.46	41.60	51.20
	Females	fL	12	43.23	1.71	0.49	40.50	45.90
MCH	Males	pg/cell	24	14.98	0.74	0.15	13.10	16.70
	Females	pg/cell	12	14.86	0.70	0.20	13.60	15.90
MCHC	Males	g/dL	24	33.47	1.59	0.32	29.00	35.30
	Females	g/dL	12	34.33	0.70	0.20	33.60	36.00
Platelets	Males	$x10^9/L$	24	215.64	109.51	22.35	44.90	429.00
	Females	$x10^9/L$	12	157.97	177.78	51.32	38.30	663.00

The International Federation of Clinical Chemistry (IFCC) recommends a sample size of at least 120 to establish reference values (International Federation of Clinical Chemistry, 1987; Solberg, 2006). However, sample size should take account of the total number of available animals, particularly for endangered species. The total number of Arabian oryx in Oman is (2010, time of study) less than 500 individuals so IFCC recommendations would have required sampling of more than 24 % of the then current Omani population of oryx. As an endangered species, Arabian oryx must be handled with maximum care in a way that keeps them in good welfare and reduces stress such as capture stress. For example, capture myopathy that has been experienced by some captured Arabian oryx (Greth & Vassart, 1989; Vassart et al., 1992). Although the sample size in the present study (36) was smaller than the IFCC recommend, it is in line with the 'three R's' relating to welfare that encourage reduction of the number of animals used to obtain information (Russell & Burch, 1959).

Table 4. Blood ions, osmolality, biochemistry, hormonal and clinical parameters for male and female Arabian oryx. n, number of animals; SD, standard deviation of mean; SEM, standard error of mean; 2.5 and 97.5 percentiles; T_3, triiodothyronine, T_4, thyroxine; bpm, breaths per min (for respiratory rate) or beats per min (for heart rate)

	Parameter	Sex	Unit	n	Mean	SD	SEM	2.5%	97.5%
Ions and osmolality	Sodium	Males	mmol/L	24	151.00	13.17	2.69	137.00	188.00
		Females	mmol/L	12	140.83	5.36	1.55	132.00	148.00
	Potassium	Males	mmol/L	24	6.28	1.38	0.28	4.20	10.10
		Females	mmol/L	12	6.03	0.93	0.27	4.50	7.50
	Chloride	Males	mmol/L	24	107.00	9.83	2.01	97.00	135.00
		Females	mmol/L	12	101.25	4.09	1.18	95.00	107.00
	Calcium	Males	mmol/L	24	2.30	0.25	0.05	1.95	2.93
		Females	mmol/L	12	2.26	0.14	0.04	2.05	2.48
	Magnesium	Males	mmol/L	16	1.20	0.14	0.04	0.98	1.39
		Females	mmol/L	4	1.18	0.08	0.04	1.09	1.27
	Phosphorus	Males	mmol/L	24	2.71	0.52	0.11	1.94	4.26
		Females	mmol/L	12	2.38	0.52	0.15	1.82	3.58
	Osmolality	Males	mosmol/Kg	16	309.53	10.53	2.63	286.50	325.00
		Females	mosmol/Kg	4	298.13	1.97	0.99	296.50	301.00
Biochemistry	Urea	Males	mmol/L	24	6.47	1.80	0.37	3.80	10.70
		Females	mmol/L	12	5.02	1.84	0.53	3.50	8.90
	Glucose	Males	mmol/L	24	6.08	1.94	0.40	3.85	10.76
		Females	mmol/L	12	5.14	2.21	0.64	3.46	11.48
	Lactate	Males	mmol/L	8	9.91	3.89	1.38	5.03	18.07
		Females	mmol/L	8	12.00	4.52	1.60	7.13	21.80
	Total protein	Males	g/L	24	71.67	8.99	1.84	56.00	98.00
		Females	g/L	12	63.42	3.83	1.10	57.00	68.00
	Albumin	Males	g/L	24	52.29	8.11	1.65	41.00	77.00
		Females	g/L	12	46.00	4.55	1.31	40.00	54.00
Hormones	Cortisol	Males	nmol/L	21	102.71	54.33	11.86	4.00	213.00
		Females	nmol/L	12	123.17	42.81	12.36	69.00	208.00
	Free T_4	Males	pmol/L	17	9.26	1.73	0.42	7.00	12.70
		Females	pmol/L	12	10.07	1.96	0.57	6.00	12.80
	Free T_3	Males	pmol/L	16	4.41	1.49	0.37	2.41	8.69
		Females	pmol/L	12	4.04	1.31	0.38	0.73	5.77
	Insulin	Males	mIU/L	17	0.61	0.27	0.07	0.40	1.50
		Females	mIU/L	12	0.73	0.31	0.09	0.20	1.30
Clinical	Body temperature	Males	°C	8	38.29	0.62	0.22	37.50	39.30
		Females	°C	8	38.49	0.56	0.20	37.30	39.20
	Heart rate	Males	bpm	8	101.00	23.62	8.35	80.00	150.00
		Females	bpm	8	113.00	21.19	7.49	90.00	142.00
	Respiratory rate	Males	bpm	8	34.00	9.91	3.51	20.00	48.00
		Females	bpm	8	43.50	9.67	3.42	30.00	54.00

The samples in the present study were exclusively from Arabian oryx housed at the Omani Mammals Breeding Centre, which is a Royal Property of H M Sultan Qaboos bin Said. Thirty-six individuals were made available for the present study out of about 150 individuals present at the centre at the time of the study, which represents a relatively high proportion, about a quarter (24 %) of the herd at the centre. Application of the exclusion criteria (calves, old, lactating and pregnant females and sick animals) would in any case not have provided the 120 individuals that the IFCC recommend. In future studies, the inclusion of animals from Jaaluni would provide data for a total number of oryx that is closer to the recommended 120 for establishing reference values as well as incorporating a different sub-population of Omani Arabian oryx. Applying exclusion and inclusion criteria focused the study on apparently healthy adult male and female Arabian oryx. This helps in the establishment of reference values for adult Arabian oryx of specified age group that are not affected by social status, like dominant males, or by the reproductive status, for example of pregnant and lactating females.

One of the challenges of comparing the data from the present study with previous studies that report haematological, biochemical or other physiological parameters for Arabian oryx or other closely related species, is that most if not all the previous studies used chemical restraint (e.g., immobilisation, sedation, and tranquillisation). There is a high possibility that reported values were affected by restraint chemicals (Cross et al., 1988; Dehghani et al., 1991; Marco & Lavin, 1999) and therefore, the values reported by those studies might differ from the reference values reported in present studies.

Immobilisation has significant effects on many haematological and biochemical parameters, as found in Arabian oryx using xylazine (AlJahdhami 2010). This agrees with the effects of chemical capture by immobilisation that caused significant changes in haematological and biochemical parameters compared to three physical capture methods in bighorn sheep (*Ovis canadensis*) (Kock et al., 1987). Therefore, the literature that report haematological and biochemical parameters from immobilised oryx are not compared to those obtained in the present study. Future researchers who measure blood parameters of Arabian oryx should consider the method of capture (i.e. with or without immobilisation) before comparing their results with the reference values obtained in the present studies.

The reference values obtained in the present study are compared with those reported in the few previous studies that did not use restraint chemicals. It might be argued that if Arabian oryx and the closely related species are mostly captured by chemical means, why not obtain reference values from those animals? The difficulty is that many types of restraint chemicals are used and each has a different mechanism of action and therefore effects on the haematological, biochemical or other physiological parameters vary according to the type of chemical. For this reason, obtaining reference values from animals captured without chemical means provides the most representative reference values that are closest to "normal".

Haematology Studies

Bush et al. (1983) evaluated haematological and some serum chemistry values of neonate (less than a month old), juvenile (less than a year old) and adult (more than a year old) scimitar-horned oryx, and found significant differences between neonates and adults in 15 out of 29 parameters. In the present study, only adult Arabian oryx were investigated. Haematological values of neonate Arabian oryx have also been measured by Bounous-Dalton & Hood (1980) and Ferrell et al., (2001). Comparing their haematological values for neonatal Arabian oryx, the values of red blood cell counts, haematocrit and haemoglobin concentrations were clearly lower in neonates than in the combined values of male and female adults included in the present study (Table 5 and Table 6).

The mean leukocyte counts in the present study are comparable to those of neonates reported by Bounous-Dalton and Hood (1980) (Table 5). However, Ferrell et al. (2001) reported a lower median WBC count than that obtained in the present study (Table 6). The values of MCV, MCH and MCHC are comparable between neonates and adult Arabian oryx (Bounous-Dalton & Hood, 1980) (Table 5). This idea is supported by the evidence of a lower haematocrit (27 %) in neonates than adults (Ferrell et al., 2001) with a similar hematocrit in neonate oryx to those found by Bounous-Dalton and Hood (1980). However, Ferrell et al. (2001) did not measure red blood cell count and haemoglobin of neonates to make further comparison.

A study by Kilgallon et al. (2008) is the only investigation of haematological parameters of adult Arabian oryx without using any immobilising or tranquilising chemicals. Comparison of the values reported by Kilgallon et al. (2008) and combined values of males and females in the present study is presented in Table 7. This comparison indicates that the white blood cell data are very close to each other, but other parameters vary between the two studies. The mean percentages of neutrophils and monocytes are higher and the mean haematocrit, percentages of lymphocytes, eosinophils and basophils are lower in the study of Kilgallon et al. (2008) than in the present study. The neutrophilia and lymphopenia reported by Kilgallon et al. (2008) might be due to more stress during capture in that study than the present study.

Kilgallon et al. (2008) moved oryx from paddocks through corridors and into a handling chute, which involved more manipulation. Stress of handling results in excessive secretion of glucocorticoids, that cause neutrophilia and lymphopenia (Burton et al., 1995; Iseki et al., 1991; Sapolsky et al., 2000). The main difference between the present study and that of Kilgallon et al., (2008) is that the oryx that were used in the present studies were housed for a prolonged period in holding pens and exposed to a series of handling events, while those used by Kilgallon et al. (2008) had not been exposed to previous handling. Familiarisation of animals by frequent exposure to handling is likely to play a role in reducing the physiological response to handling over time and allow derivation of more valid reference data (Broom & Johnson, 1993). Cortisol and catecholamine-mediated distortions of haematology and blood biochemistry have been reported in captured impala and red deer (Marco & Lavin, 1999).

The mean haematocrit reported by Kilgallon et al. (2008) (41.79 %, Table 7) and ISIS (2002) (44.00 %, Table 8) are lower than those measured in the present study (54.10 %). The intra-variation between samples for haematocrit in the present study within duplicate samples was 1.29 % and inter-variation between animals was 7.58 % (Table 1) which are below the satisfactory accepted percentage of variation (10 %) suggested by Murray et al. (1993). Arabian oryx in the holding pens were provided with water twice a day (morning and evening) in a water container, but could have been dehydrated by the time of handling, as water was provided after handling. Serum osmolality is another indicator of hydration status but urine osmolality is a better indicator of hydration status than serum osmolality or haematocrit (Shirreffs, 2003). In the present study, the urine osmolality was not examined and therefore, further investigations are recommended to look at the correlation between hydration status, haematocrit, serum osmolality and urine osmolality.

Stress is known to increase haematocrit by causing splenic contraction and therefore increasing the number of red blood cells in circulation (Stewart & McKenzie, 2002), and consequently causing an increase in haematocrit. Probably the stress of capture caused an increase in haematocrit in Arabian oryx in the present study. However, the previous studies such as those of Kilgallon et al. (2008) probably used more stressful approaches of capture than the present study but they reported much lower haematocrit. Further investigations are needed to explain this apparent discrepancy.

Table 5. Comparison of haematological values for neonates as reported by Bounous-Dalton and Hood (1980) and for combined values of male and female adults as reported in the present study. n, number of animals; SD, standard deviation of mean; SEM, standard error of mean; Abrreviations as in Table 3

Parameter	Bounous-Dalton and Hood, 1980					The present study			
	Unit	n	Means	SD	SEM	n	Mean	SD	SEM
WBC	x10^9/L	17	7.3	1.87	0.45	36	7.67	1.63	0.27
Neutrophils	%	17	61.4	9.61	2.33	36	65.95	8.06	1.34
Lymphocytes	%	17	37.2	9.15	2.22	36	28.40	7.23	1.21
Monocytes	%	17	0.74	0.64	0.16	36	0.78	1.19	0.20
Eosinophils	%	17	0.57	0.82	0.20	36	2.46	1.47	0.24
Basophils	%	17	0.0		0.00	36	2.42	1.34	0.22
RBC	x101^2/L	19	6.6	0.8	0.18	36	12.20	0.70	0.12
Haemoglobin	g/dL	17	11.0	1.37	0.33	36	18.21	1.17	0.19
Haematocrit	%	17	33.9	4.24	1.03	36	54.10	4.46	0.74
MCV	fL	19	50.1	4.56	1.05	36	44.28	2.19	0.37
MCH	pg/cell	19	16.5	3.72	0.85	36	14.94	0.72	0.12
MCHC	g/dL	19	32.7	2.64	0.61	36	33.76	1.41	0.24

Table 6. Comparison of haematological values for neonates as reported by Ferrell et al. (2001) and for adults as reported in the present study. WBC, white blood cells; min, minimum; max, maximum and n, number of animals

Parameter	Ferrell et al., (2001)				The present study			
	Median	Min	Max	n	Median	Min	Max	n
WBC (x10^9/L)	5.73	4.18	9.25	10	7.45	4.26	11.20	36.00
Neutrophils (%)	65.97	43.80	111.34	9	66.60	45.60	78.40	36.00
Lymphocytes (%)	23.56	20.24	49.39	9	27.10	17.10	49.10	36.00
Monocytes (%)	2.48	0.73	2.74	9	0.10	0.00	4.05	36.00
Eosinophils (%)	0.00	0.00	1.08	9	2.08	0.40	6.02	36.00
Basophils (%)	0.00	0.00	0.00	9	2.69	0.00	4.99	36.00
Haematocrit (%)	38.5	30.0	42.0	10	53.10	47.10	70.30	36.00

Table 7. Comparison between some haematological parameters reported by Kilgallon et al., (2008) and the combined values for males and females in the present study. SD, standard deviation of mean, SEM, standard error of mean, n, number of animals and WBC, white blood cells

Parameter	Kilgallon et al., (2008)				The present study			
	n	Mean	SD	SEM	n	Mean	SD	SEM
WBC ($x10^9$/L)	19	7.31	1.99	0.46	36	7.67	1.63	0.27
Neutrophils (%)	19	76.29	9.02	2.07	36	65.95	8.06	1.34
Lymphocytes (%)	19	19.73	7.85	1.80	36	28.40	7.23	1.21
Monocytes (%)	19	3.08	1.28	0.29	36	0.78	1.19	0.20
Eosinophils (%)	19	0.33	0.27	0.06	36	2.46	1.47	0.24
Basophils (%)	19	0.57		0.00	36	2.42	1.34	0.22
Haematocrit (%)	19	41.79	1.84	0.42	36	54.10	4.46	0.74

Table 8. Comparison between some haematological parameters reported by the database of ISIS (2002) and the combined values for males and females in the present study. SEM, standard error of mean, n, number of animals, WBC, white blood cells; RBC, red blood cells (erythrocytes) MCV, mean corpuscular volume; MCH, mean corpuscular haemoglobin; and MCHC, mean corpuscular haemoglobin concentration

		ISIS, 2002				The present study		
Parameter	Unit	Mean	SEM	Samples	Animals	Mean	SEM	N
WBC	$x10^9$/L	6.87	0.15	250	137	7.67	0.27	36
Neutrophils	%	70.69	0.14	214	112	65.95	1.34	36
Lymphocytes	%	23.24	0.05	214	112	28.40	1.21	36
Monocytes	%	2.58	0.01	166	97	0.78	0.20	36
Eosinophils	%	2.74	0.02	152	89	2.46	0.24	36
Basophils	%	1.12	0.01	43	34	2.42	0.22	36
RBC	$x10^{12}$/L	10.14	0.19	159	95	12.20	0.12	36
Haemoglobin	g/dL	15.70	0.22	158	98	18.21	0.19	36
Haematocrit	%	44.00	0.01	265	145	54.10	0.74	36
MCV	fL	45.20	0.59	152	93	44.28	0.37	36
MCH	pg/cell	15.50	0.19	135	79	14.94	0.12	36
MCHC	g/dL	34.40	0.28	152	96	33.76	0.24	36
Platelets	$x10^9$/L	330.00	30.00	35	26	196.42	22.72	36

Acute stress has been found to cause large, rapid and reversible changes in the distribution of leukocytes in peripheral blood, such as increase in neutrophils and decrease of lymphocytes as found in rats (Dhabhar et al., 1996).

In the present study significant differences between male and female Arabian oryx occurred in the white blood cell count, percentages of neutrophils, lymphocytes and basophils. Therefore, these parameters should be reported and presented for each sex separately. No previous studies have reported haematological parameters for male and female Arabian oryx separately. Therefore, the present study is the first to look into the differences between sexes and the first to present separate reference values for haematological parameters that differ significantly between sexes. Some sex differences in haematological parameters were found between males and females of southern chamois (*Rupicapra pyrenaica*) such as a higher neutrophil count in males (Lopez-Olvera et al., 2006a), as seen in oryx in the present study. However, in some species there are no sex differences in haematological parameters exposed to capture. For example, Rispat et al. (1993) found no significant difference between 36 males and 35 females of Yucatan micropigs in all haematological parameters except in platelets count. Thus, sex differences are species dependant and it seems that each species has to be studied to elucidate these differences.

BIOCHEMISTRY, IONS, OSMOLALITY, HORMONES AND CLINICAL PARAMETERS

The reference values for potassium, chloride, calcium, magnesium, lactate, insulin, body temperature, heart rate and respiratory rate showed no significant differences between male and female Arabian oryx.

Kilgallon et al. (2008) reported the concentrations of potassium and calcium in serum for a mixed group male and female Arabian oryx captured by non-chemical means as 4.75 ± 0.19 mmol/L and 1.16 ± 0.01 mmol/L, respectively (n = 19, 14 males and 5 females). These values are slightly lower but comparable to those obtained in the present study for mixed groups of both sexes (6.19 ± 1.24 mmol/L and 2.29 ± 0.04 mmol/L respectively) and values did not differ significantly in males and females (Table 4).

The concentrations of chloride, magnesium, lactate, insulin, body temperature, heart rate and respiratory rate have not previously reported for oryx captured without chemical restraint. However, the values for chloride, magnesium and body temperature in the ISIS database (International Species Information System, 2002), after unknown methods of capture, and for unknown sex and age groups are comparable to those in the present study (Table 9).

Ideally, to get more accurate measurements for the baseline of body temperature, heart rate and respiratory rate, animals should be fitted with telemetric recording devices that measure these parameters at set intervals for long periods of time (Lopez-Olvera et al., 2006b). Some measurements of body temperature have now been acquired for Arabian oryx and show clear seasonal patterns (Hetem et al., 2010), but telemetric devices were not available for the present study. The collection of respiratory and heart rates by auscultation is a common and widely used method by veterinarians (Gonzalez et al., 2008) but suffers from the immediate effects of proximity and contact with the animals. In future studies, telemetry devices should be utilised for Arabian oryx to obtain measurements close to baseline for comparison to published values.

In the present study, there was a significant difference between the serum glucose concentrations of male and female oryx (6.08 ± 1.94 mmol/L in males; 5.14 ± 2.21 mmol/L in females). A number of studies have reported the concentration of glucose in Arabian oryx, but for mixed sex groups. Ostrowski et al. (2006) reported a lower glucose concentration of 3.70 ± 0.30 mmol/L (n = 7), while Kilgallon et al. (2008) and Vassart and Greth (1991) reported higher glucose concentrations of 10.24 ± 2.27 mmol/L (n = 19), and 10.82 ± 0.83 mmol/L (n = 73, immobilised) respectively. Finally, the ISIS database gives a value of 7.99 ± 3.11 mmol/L (n = 163 samples from 98 animals). The concentration of glucose for merged sexes in the present study was 5.76 ± 2.05 mmol/L.

Table 9. Comparison between the concentration of chloride, magnesium and body temperature reported by ISIS, 2002 data base and the present study

		ISIS, 2002				The present study		
Parameter	Units	Mean	SEM	Samples	Animals	Mean	SEM	n
Chloride	mmol/L	104	0.368	118	69	105.08	1.46	36
Magnesium	mmol/L	0.971	0.057	14	12	1.69	0.50	20
Body temperature:	°C	38.8	0.197	37	17	38.39	0.15	16

The concentration of circulating glucose is influenced by acute stress, and secretion of glucocorticoids and catecholamines increase the glucose in the blood (Steffens and de Boer, 1999). The animals in the present study and in the study of Ostrowski et al. (2006) were exposed to many occasions of handling and capture, before collection of the samples used in calculating reference values. In contrast, other studies (Vassart and Greth, 1991; Kilgallon et al., 2008) used naïve animals, captured for the first time and this is likely to explain the higher concentrations of circulating glucose. Familiarisation of animals to handling might play a role in reducing the level of stress in oryx, as implied by the lower concentration of glucose in familiarised animals compared to unfamiliar ones. Creatine kinase was not measured in the present study but would be a useful parameter for inclusion in future studies as an indicator of muscle damage and capture myopathy (Vassart et al., 1992). Additional parameters that might be useful in future studies are aspartate aminotransferase (AST) and alanine aminotransferase (ALT) for monitoring liver function (Kramer & Hoffmann, 1997). This study presents a broad range of valuable parameters that are useful for diagnosis of abnormalities in Arabian oryx and aid the conservation programmes to keep the oryx populations healthy. Using no chemical immobilisation makes the presented reference values closer to normality. The establishment of reference values for Arabian oryx that considers the differences between males and females has importance for future monitoring of the well-being of this endangered species.

ACKNOWLEDGMENTS

The study was supported by Diwan of Royal Court, Sultante of Oman. Analyses of samples were carried out at Sultan Qaboos University. Special thanks are due to Sultan Qaboos University Hospital (Departments of Haematology and Biochemistry). At the Royal Court Affairs, we would like to thank Eng. Mahmood AlAbri, the Director General of Veterinary Services and all staff who helped in the study.

REFERENCES

AlJahdhami, MH 2010. Physiological monitoring of well-being for conservation of Arabian oryx, *Oryx leucoryx*. PhD Thesis, *University of Exeter*, UK.

Ancrenaz, M, S Ostrowski, S Anagariyah and A Delhomme 1996. Long-duration anesthesia in Arabian oryx (*Oryx leucoryx*) using a medetomidine-etorphine combination. *Journal of Zoo and Wildlife Medicine*. 27, (2): 209-216.

Botreau, R, I Veissier, A Butterworth, M Bracke and L Keeling 2007. Definition of criteria for overall assessment of animal welfare. *Animal Welfare*. 16, (2): 225-228.

Bounous-Dalton, D and HB Hood 1980. Hematologic values for neonatal Arabian oryx (*Oryx leucoryx*). *Journal of Zoo Animal Medicine*. 11, (4): 118-119.

Broom, D and K Johnson 1993. *Approaching questions of stress and welfare. Stress and Animal Welfare*. Kluwer Academic Publishers. 1-7.

Burton, JL, ME Kehrli, Jr., S Kapil and RL Horst 1995. Regulation of L-selectin and CD18 on bovine neutrophils by glucocorticoids: effects of cortisol and dexamethasone. *Journal of Leukocyte Biology*. 57, (2): 317-25.

Bush, M, RS Custer, JC Whitla and RJ Montali 1983. Hematologic and serum chemistry values of captive scimitar-horned oryx (*Oryx tao*): variations with age and sex. *Journal of Zoo Animal Medicine*. 14, (51): 55.

Chapple, R, A English, R Mulley and E Lepherd 1991. Haematology and serum biochemistry of captive unsedated chital deer (*Axis axis*) in Australia. *Journal of Wildlife Diseases*. 27, (3): 396.

Cross, JP, CG Mackintosh and JF Griffin 1988. Effect of physical restraint and xylazine sedation on haematological values in red deer (*Cervus elaphus*). *Research in Veterinary Science*. 45, (3). 281-6.

Deem, SL, WB Karesh and W Weisman 2001. Putting theory into practice: wildlife health in conservation. Conservation Biology. 15, (5): 1224-1233.

Dehghani, S, N Sharifnia, M Yahyaei and A Souri 1991. Clinical, haematological and biochemical effects of xylazine, ketamine and their combination in caprine and feline. *Veterinary Anaesthesia and Analgesia*. 18, (s1): 129-133.

Dhabhar, F, A Miller, B McEwen and R Spencer 1996. Stress-induced changes in blood leukocyte distribution. Role of adrenal steroid hormones. *The Journal of Immunology*. 157, (4): 1638-1644.

Dimauro, C, P Bonelli, P Nicolussi, S Rassu, A Cappio-Borlino and G Pulina 2008. Estimating clinical chemistry reference values based on an existing data set of unselected animals. *The Veterinary Journal*. 178, (2): 278-281.

FAWC 1993. Second report on priorities for research and development in farm animal welfare. FAWC, *Ministry of Agriculture, Fisheries and Food*, Tolworth, UK (Now DEFRA, London).

Ferrell, ST, RW Radcliffe, R Marsh, CB Thurman, CM Cartwright, TWJ De Maar, ES Blumer, E Spevak and SA Osofsky 2001. Comparisons among selected neonatal biomedical parameters of four species of semi-free ranging hippotragini: Addax (*Addax nasomaculatus*), scimitar-horned oryx (*Oryx dammah*), Arabian oryx (*Oryx leucoryx*), and sable antelope (*Hippotragus niger*). *Zoo Biology*. 20, (1): 47-54.

Gonzalez, R, F Berlinguer, G Espeso, F Ariu, A del Olmo, JJ Garde, M Gomendio, S Ledda and ERS Roldan 2008. Use of a neuroleptic in assisted reproduction of the critically endangered Mohor gazelle (*Gazella dama mhorr*). *Theriogenology*. 70, (6): 909-922.

Greth, A and M Vassart 1989. Capture myopathy in Arabian oryx *Oryx leucoryx*, what is the best translocation technique? In, unpublished manuscript.

Greth, A, M Vassart and S Anagariyah 1993. Evaluation of medetomidine-induced immobilization in Arabian oryx (*Oryx leucoryx*) - clinical, hematologic and biochemical effects. *Journal of Zoo and Wildlife Medicine*. 24, (4): 445-453.

Hajduk, P, MD Copland and DA Schultz 1992. Effects of capture on hematological values and plasma cortisol levels of free-range koalas (*Phascolarctos cinereus*). *Journal of Wildlife Diseases*. 28, (3): 502-506.

Harris, E and J Boyd, 1995. Statistical Bases of Reference Values in Laboratory Medicine. CRC Press.

Hawkey, C and M Hart 1984. Age-related changes in the blood count of the scimitar-horned oryx (*Oryx tao*). *The Journal of Zoo Animal Medicine*. 157-160.

Hawkey, CM 1991. The value of comparative haematological studies. Comparative Haematology International. 1, (1): 1-9.

Hetem, R, W Strauss, L Fick, S Maloney, L Meyer, M Shobrak, A Fuller and D Mitchell 2010. Variation in the daily rhythm of body temperature of free-living Arabian oryx (*Oryx leucoryx*): does water limitation drive heterothermy? *Journal of Comparative Physiology B: Biochemical, Systemic, and Environmental Physiology*. 1-9.

Horn, P and A Pesce 2003. Reference intervals: an update. *Clinica Chimica Acta*. 334, (1-2): 5-23.

International Federation of Clinical Chemistry 1987. Approved recommendation (1987) on the theory of reference values. Part 5. Statistical treatment of collected reference values. Determination of reference limits. *Journal of Clinical Chemistry and Clinical Biochemistry*. 25, 645-656.

International Species Information System 2002. Reference ranges for physiological data values, Arabian oryx (*Oryx leucoryx*). *ISIS*, www.isis.org.

Iseki, R, M Mukai and M Iwata 1991. Regulation of T lymphocyte apoptosis: Signals for the antagonism between activation and glucocorticoids induced death. *Journal of Immunology*. 147, 4286-4292.

Jain, N, 1986. *Schalm's Veterinary Hematology*. Lea & Febiger, Philadelphia.

Junqueria, LC and J Carnerio, 2005. *Basic Histology*. Prentice Hall International, London.

Kaneko, JJ, JW Harvey and ML Bruss, 1997. *Clinical Biochemistry of Domestic Animals*. Academic Press Inc, San Diego, USA.

Karesh, W and R Cook 1995. Applications of veterinary medicine to in situ conservation efforts. *Oryx*. 29, (4): 244-252.

Karesh, W, A Rothstein, W Green, H Reuter, W Braselton, A Torres and R Cook 1997. Health evaluation of black-faced impala (*Aepyceros melampus petersi*) using blood chemistry and serology. *Journal of Zoo and Wildlife Medicine*. 361-367.

Kilgallon, C, T Bailey, B Arca-Ruibal, M Misheff and D O'Donovan 2008. Blood-gas and acid-base parameters in nontranquilized Arabian oryx (*Oryx leucoryx*) in the United Arab Emirates. *Journal of Zoo and Wildlife Medicine*. 39, (1): 6-12.

Kirkwood, J 1993. Interventions for wildlife health, conservation and welfare. *The Veterinary Record*. 132, (10): 235-238.

Kocan, A, B Glenn, T Thedford, R Doyle, K Waldrup, G Kubat and M Shaw 1981. Effects of chemical immobilization on hematologic and serum chemical values in captive white-tailed deer. *Journal of the American Veterinary Medical Association*. 179, (11): 1153.

Kock, M, D Jessup, R Clark and C Franti 1987. Effects of capture on biological parameters in free-ranging bighorn sheep (*Ovis canadensis*): evaluation of drop-net, drive-net, chemical immobilization and the net-gun. *Journal of Wildlife Diseases*. 23, (4): 641-651.

Kramer, JW and WE Hoffmann 1997. Clinical Enzymology. In, Clinical Biochemistry of Domestic Animals. Academic Press Inc, San Diego, USA, pp. 303-325.

Lopez-Olvera, JR, I Marco, J Montane and S Lavin 2006a. Haematological and serum biochemical values of southern chamois (*Rupicapra pyrenaica*). *Veterinary Record*. 158, (14): 479-484.

Lopez-Olvera, JR, I Marco, J Montane and S Lavin 2006b. Transport stress in Southern chamois (*Rupicapra pyrenaica*) and its modulation by acepromazine. *Veterinary Journal*. 172, (2): 347-355.

Lumsden, J 1998. "Normal" or reference values: questions and comments. *Veterinary Clinical Pathology*. 27, (4): 102.

Marco, I and S Lavin 1999. Effect of the method of capture on the haematology and blood chemistry of red deer (*Cervus elaphus*). *Research in Veterinary Science*. 66, (2): 81-84.

Murray, W, A Peter and R Teclaw 1993. The clinical relevance of assay validation. *The Compendium on Continuing Education for the Practicing Veterinarian* (USA). 15, (12): 1665-1675.

Ostrowski, S, JB Williams, P Mesochina and H Sauerwein 2006. Physiological acclimation of a desert antelope, Arabian oryx (*Oryx leucoryx*), to long-term food and water restriction. *Journal of Comparative Physiology B: Biochemical, Systemic, and Environmental Physiology*. 176, (3): 191-201.

Peinado, V, J CeldrÚn and J Palomeque 1999. Blood biochemistry values in some wild ruminants in captivity. *Comparative Clinical Pathology*. 9, (4): 175-181.

Perez, JM, FJ Gonzalez, JE Granados, MC Perez, P Fandos, RC Soriguer and E Serrano 2003. Hematologic and biochemical reference intervals for Spanish ibex. *Journal of Wildlife Diseases*. 39, (1): 209-215.

Presidente, P, J Lumsden, K Presnell, W Rapley and B McCraw 1973. Combination of etorphine and xylazine in captive white-tailed deer. II. Effects on hematologic, serum biochemical and blood gas values. *Journal of Wildlife Diseases*. 9, (4): 342.

Pritchard, JC, CC Burn, ARS Barr and HR Whay 2009. Haematological and serum biochemical reference values for apparently healthy working horses in Pakistan. *Research in Veterinary Science*. 87, (3): 389-395.

Rispat, G, M Slaoui, D Weber, P Salemink, C Berthoux and R Shrivastava 1993. Haematological and plasma biochemical values for healthy Yucatan micropigs. *Laboratory Animals*. 27, (4): 368-373.

Russell, W and R Burch, 1959. The Principles of Humane Experimental Technique. Methuen, London, UK.

Sapolsky, RM, LM Romero and AU Munck 2000. How do glucocorticoids influence stress responses? Integrating permissive, suppressive, stimulatory, and preparative actions. *Endocrine Reviews*. 21, (1): 55-89.

Seal, U, J Ozoga, A Erickson and L Verme 1972. Effects of immobilization on blood analyses of white-tailed deer. *The Journal of Wildlife Management*. 1034-1040.

Shirreffs, S 2003. Markers of hydration status. *European Journal of Clinical Nutrition*. 57, S6-S9.

Solberg, HE 2006. Establishment and use of reference values. *In*, Tietz Textbook of Clinical Chemistry and Molecular Diagnostics. Saunders, St. Louis, pp. 425-48.

Steffens, AB and SF de Boer 1999. Impact of stress on animal intermediate metabolism. In, Stress Physiology in Animals. Sheffield Acadimic Press, Sheffield, pp. 108-129.

Stewart, IB and DC McKenzie 2002. The human spleen during physiological stress. *Sports Medicine*. 32, 361-369.

Strauss, WM 2008. Where are we in terms of Arabian oryx conservation? *Wildlife Middle East News*. 2, (4): 3-4.

Vassart, M and A Greth 1991. Hematological and serum chemistry values for Arabian oryx (Oryx leucoryx). *Journal of Wildlife Diseases*. 27, (3): 506-8.

Vassart, M, A Greth, S Anagariyah and F Mollet 1992. Biochemical parameters following capture myopathy in one Arabian oryx (*Oryx leucoryx*). *Journal of Veterinary Medical Science*. 54, (6): 1233-1235.

Walton, RN 2001. Establishing reference intervals: Health as a relative concept. *Seminars in Avian and Exotic Pet Medicine*. 10, (2): 66-71.

Wesson III, J, P Scanlon, R Kirkpatrick and H Mosby 1979. Influence of chemical immobilization and physical restraint on packed cell volume, total protein, glucose, and blood urea nitrogen in blood of white-tailed deer. *Canadian Journal of Zoology*. 57, (4): 756-767.

In: Biotechnology and Conservation … (2 Volume Set) ISBN: 978-1-63117-869-6
Editors: S. A. Farooq, R. Abed and S. Baqir

Chapter 6

TRIALS ON TREATMENT OF CRYPTOSPORIDIOSIS IN BALB/C MICE. SHEDDING OF OOCYSTS

Nabeel E. Salih[1] and Seena A. A. Al-Jarjary[2]
[1]Department of Biology, Education College, Mosul University, Iraq
[2]Department of Biology, Science College, Mosul University, Iraq

ABSTRACT

A mixture of two drugs, Azithromycin and Spiramycin, was administered, orally, with and without vitamin E, to mice, experimentally infected with cryptosporidiosis, at different concentrations, for the first time. Oocyst shedding was the criterion taken into consideration. The shortest period at which oocysts stopped shedding in faeces was 48 hours, as oocysts disappeared in the faeces at day 3 post-treatment, at certain concentration. It is concluded that the mixture used, with vitamin E, may well be considered a promising drug against human cryptosporidiosis.

Keywords: Cryptosporidiosis, treatment, experimental infection, Azithromycin, Spiramycin

INTRODUCTION

Cryptosporidiosis is one of zoonotic parasitic diseases of world wide distribution, causing severe diarrhoea. The disease is self-limited in immunocompetent patients but is severe enough to cause death in immunocompromised (immunodeficient) persons. The disease is caused by an apicomplexan parasite of the genus *Cryptosporidium* (Current, 1985).

Because of the seriousness of the disease, especially in children, it's treatment has attracted the attention of many research workers who commenced trials on many chemotherapeutic agents, such as Azithromycin, Paromomycin, Roxithromycin, Nitazoxanide, Rifabutin, Letrazuril, Spiramycin, Somatostatin, Diclazuril, Halofuginone and Highly Active Antiretroviral Therapy (HAART). However, more than a 100 chemotherapeutic agents have been tested, *in vitro*, and only 40 of them showed some activity (Woods et al., 1996; Maggi et al., 2000. Research workers pointed out that until 2003, no chemotherapeutic agent seems to have shown convincing results (Hommer et al., 2003). This motivated some workers to initiate another line of research in this respect, that is admixing of more than one drug together in the course of their fight against the disease such as Nitazoxanide with Azithromycin and Rifabutin and also cataionic peptides with inhibitors of ion transport system (ITS) and found that mixing gave better results (Giacometti et al., 2000a,b). The present study is, therefore, a continuation in this line of research in which the activity of Azithromycin and Spiramycin, mixed together, emulcified in distilled water (DW) and in phosphate buffer saline (PBS), with and without the antioxidant, vitamin E, is tested against cryptosporidiosis in experimentally infected BALB/c mice. Oocyst shedding is the criterion taken into consideration.

*Sub*100 samples of stool were collected from children, 20 days-5 years old, of both sexes (54 males and 46 females) complaining of gastroenteritis and diarrhea in two hospitals in Mosul city. Samples were preserved in 2.5% Potassium dichromate solution and left in refrigerator until use. An infection dose of $3x10^4$ oocysts, prepared according to Freire-Santos et al. (1999) was used to infect mice, orally, using a stomach tube (Woods et al., 1996; Lindsay et al., 2000). 351 mice (3 –veC-not infected, not treated; 12 +ve C, infected, not treated; 336 infected and treated). 0.5 ml of emulsion, containing the above mentioned number of cysts, was introduced into each mouse. Stool of the infected mice was examined daily, starting at two days post-infection, for detection of oocysts in the stool (Viu et al., 2000; Khalifa et al., 2000). A mixture of the drugs Azithromycin and SpiramycinC:\spiramycin, emulcified in DW and in PBS, with and without vitamin E, was administered once the oocysts appeared in the stool. Samples preserved in 2.5 % Potassium dichromate were washed three times in D. water, examined for oocysts, and filtered through four layers of mesh, centrifuged (using a Gallenkamp centrifuge-England) for 10 min. at 1000 rpm. Residues were collected, washed with D. water, centrifuged, and using sugar flotation method, at 700 rpm for 20 min. The filtrate was collected in each tube, washed with PBS (pH 7.2) few times and a drop was examined to ensure purity of the oocysts. Infection dose was prepared using a hemocytometer (Superior- Germany). Results of the present study showed that the effect of the two drugs was obvious from the second or the third day post treatment. For Azithromycin. Figure (1) shows that administration of the drug, emulcified in DW and PBS, at different concentrations, with or without vitamin E, caused a significant decrease in the number of oocysts, shed daily. The oocysts disappeared in faeces at different times, ranging between day 5 and day 10 post treatment. Under similar conditions, Spiramycin showed, more or less, similar results, although it had a little bit faster effect as the oocysts disappeared in faeces at days ranging between day 5 and day 8 post treatment. Figure (2). The interesting and non-expected result was obtained when the two drugs were mixed together, along with vitamin E, forming the formula ASE, as the oocysts started decreasing in faeces at the second day post-treatment at different concentrations. At the concentration 4mg/20gm body weight for Azithromycin and the concentration 1.5mg/20gm body weight for Spiramycin, the oocysts disappeared in faeces at day 3 post-treatment Figure (3).

Due to the failier of many antibiotics in treating cryptosporidiosis properly because of the inadequate knowledge of the mechanism of action of the parasite (Armson et al.,2002; Kadappu et al.,2002; Hommer et al.,2003), in addition to the assurance of Hoepeiman and Schneider (1995) and Hoepeiman (1996) on the necessity of production of drugs or modification of some antiviral drugs to be used in treating this disease, we thought of admixing two drugs, Azithromycin and Spiramycin, at different concentrations, with and without vitamin E, as a first trial to investigate the possibility of improving the action of these drugs. In the present work, the number of oocysts shed decreased when the two drugs, Azithromycin and Spiramycin, were used, each on it's own. Azithromycin has been shown to be a good inhibitor for the parasite's growth (Rehg, 1999; Kimata et al., 1991 and Dunne, 1996). Similar results were obtained with Spiramycin by previous workers (Saez-Liorens et al., 1989; Mantovani et al., 1995).

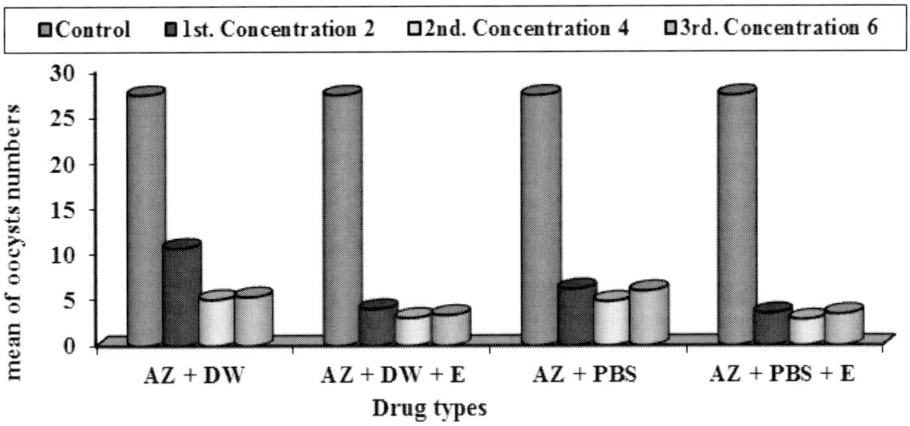

AZ +DW: Azithromycin emulcified in DW.
AZ + DW + E: Azithromycin emulcified in DW with Vitamin E.
AZ + PBS: Azithromycin emulcified in PBS.
AZ + PBS + E:: Azithromycin emulcified in PBS with Vitamin E.

Figure 1. Numbers of oocysts of *C.parvum* in various periods (days) after Azithromycin administration with different concentration.

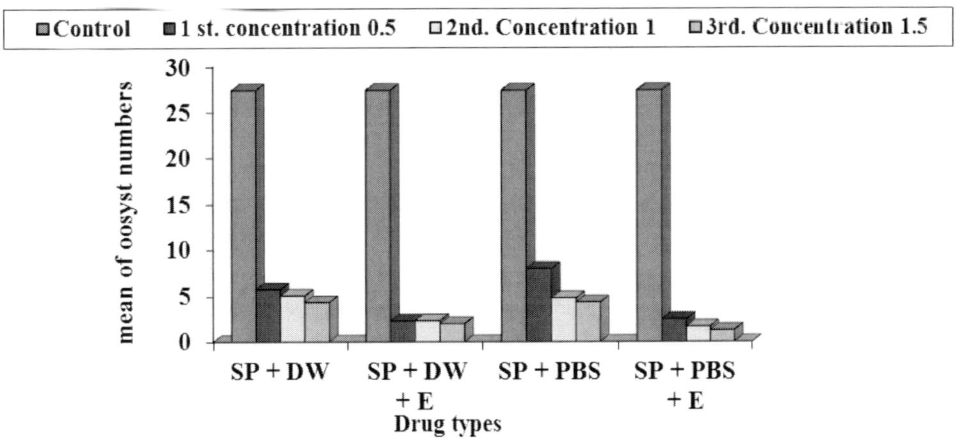

SP + DW: Sprimycin emulcified in DW.
SP + DW + E: Sprimycin emulcified in DWwith Vitamin E.
SP + PBS: Sprimycin emulcified in PBS.
SP + PBS + E: Sprimycin emulcified in PBS with Vitamin E.

Figure 2. Numbers of oocysts of *C.parvum* in various periods (days) after Sprimycin administration with different concentration.

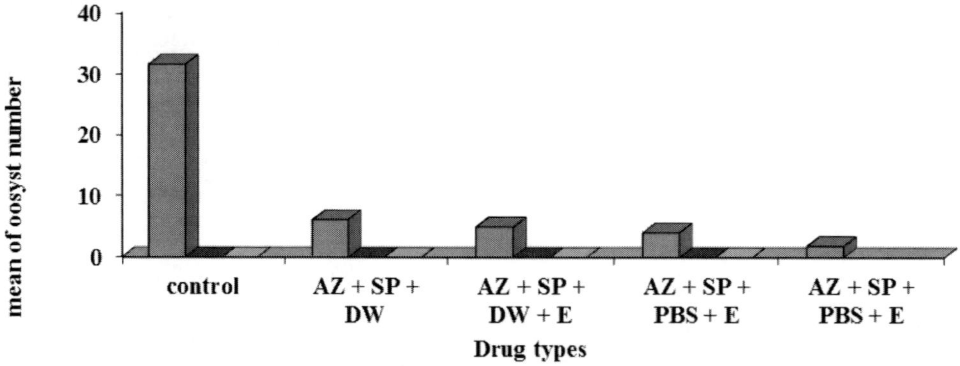

Figure 3. Numbers of oocysts of *C.parvum* in various period (days) after mixture of the drug Azithromycin and Spiramycin administration with different concentration.

However, much better results were obtained when the two drugs were admixed with each other as the oocysts shedding stopped on the third day post-treatment, at certain concentration. Admixing more than one drug together has been pointed out by some authors to be more effective on oocyst shedding, such as Woods et al. (1996) on Azithromycin and Clindamycin; Giacometti et al.(2000) on Nitazoxanide, Azithromycin and Rifabutin; and Maggi et al.(2000) on different antiviral drugs but did not reach results of the present study. The present results support the findings of Clark (1999) who reported that admixing the two drugs Azithromycin and paromomycin led to a better clinical picture. On the other hand, our results disagree with those of Theodos et al. (1998) who found that admixing the two drugs Nitazoxanide and paromomycin did not show any effect better than that of Nitazoxanide on it's own. The only available explanation for this controversy is that an antagonistic action could have occurred between the two drugs whereas, in the present study, there could have been a synergistic action between the two drugs used, a fact which is very well known to the research workers, in general. In the present study, decreasing of oocyst shedding started at 2 days post- treatment with Azithromycin whereas it was not shown in the first 3 days post-treatment with Spiramycin. This might be due to a recurrence after treatment as a result of resistance of the parasites when the dose was inadequate, leading to their activation. In the present study, the decrease in oocyst shedding was inversely proportional to the period of treatment. Previous workers, however, reported different periods of oocyst disappearance when given different doses of different drugs (Giacometti et al., 1999; Armson et al., 2002; Kadappu et al., 2002). However, the shortest period for disappearance of oocysts from faeces was obtained in the present study. In the present study, using the two drugs, Azithromycin and Spiramycin, each on it's own and mixed together, in the form of emulshion in PBS, gave better results than when emulsified in DW. PBS may have worked as a better dissolving agent and helped in a better absorption by the intestine. This result is in accordance with that of Hommer et al. (2003) Who pointed out that, using different solvents with drugs, such as ethanol dimethyl sulphpxide (DMSO) and PBS showed that ethanol inhibited the excystment of the sporozoites at concentrations less than 2%. Armson et al. (2002) ensured that the activity of initroaniline oryzalin, *in vitro* and *in vivo*, requires more solubility and more absorption. In the present study, addition of vitamin E to the mixture led to disappearance of the oocysts in a shorter period. The only available explanation, at this time, is that, since it is an antioxidant (Brigelius –Flohe et al., 2002; Kolleck et al., 2002), it may have decreased, or prevented, the oxidation of the drugs in the intestine by free radicals, in case oxidation occurs, similar to what occurs to the drug Albendazole, used to treat hydatid cysts, as it is oxidized in the intestine to Albendazole sulphoxide by the free radicals with further oxidation to Sulphone which can not be absorped by the intestine, leading to an inactive form of the drug (see Sida,2005). From the above mentioned results, it may well be concluded that the mixture of Azithromycin and Spiramycin, given with vitamin E, administered orally, at a certain concentration, could be considered a promising drug against cryptosporidiosis in human beings.

REFERENCES

Armson A, menon K, O'Hara A, Macdonld LM, Reads CM, Sargent K, Thompson RCA and Reynoldson JA. 2002. Efficacy of oryzalin and association of histological changes in *Cryptosporidium* infected neonatal rats. *Parasitology*, 125: 113 – 117.

Clark DP. 1999. New insights into human cryptosporidiosis. *Clin. Microbiol. Rev.*, 2 (4): 554 – 563.

Current WL.1985. Cryptosporidiosis. *JAMA*, 187 (12): 1334 - 1338.

Dunne MW. 1993. Open label azithromycin in the treatment of Cryptosporidiosis. *Ninth Inter. Conf. AIDS, Berlin, abstract PO-B*10-1500.

Freire-Santos F, Oteiza-Lopez AM, Vergara-castiblanco, CA and Area-Mazas ME. 1999. Effects of salinity, temperature storage time on mouse experimental infection by *Cryptosporidium parvum*. *Vet. Parasitol.*, 87: 1 – 7.

Giacometti A, Burzacchini F, Cirioni O, Barchiesi F, Dini M and Scalis G. 1999. Efficacy of treatment with paromomycin, Azithromycin and nitazoxanide in a patient with disseminated cryptosporidiosis. *Europ. J. Clin. Microbiol. Infect. Dis.*, 18: 885 – 889.

Giacometti A, Cirioni O, Barchiesi F, Ancarani F and Scalis G. 2000a. Activity of nitazoxanide alone and in combination with azithromycin and rifabutin against *Cryptosporidium parvum* in cell culture. *J. Antimicrob. Chemother.*, 45: 453 – 456.

Giacometti A, Cirioni O, Barchiesi F, Ancarani F and Scalis G. 2000b. *In vitro* anti-cryptosporidial activity of cationic peptides alone and in combination with inhibitors of ion transport systems. *J. Antimicrob. Chemother.*, 45: 651 – 654.

Hoepeiman IM. 1996. Human cryptosporidiosis. *Int. J. St. D. AIDS*, 7 (1): 28 – 33.

Hoepeiman IM and Schneider MME.1995. Azithromycin the first of tissue-selective azalided. *Int. J. Antimicrob. Agent.*, 5: 145 – 167.

Hommer V, Eichholz J and Petry F. 2003. Effect of antiretroviral protease inhibitors alone and in combination with paromomycin, on the excystation, invasion and in vitro development of *Cryptosporidium parvum*. *J. Antimicrob. Chemoth.*, 52: 359 – 364.

Kadappu KK, Nagaraja MV, Rao, PV and Shastry BA. 2002. Azithromycin as treatment for cryptosporidiosis in human immunodeficiency virus disease. *J. Postgrad. Med.*, 48 (3): 179 – 181.

Khalifa AM, Ibrahim IR and EL-Kerdany ED. 2000. Coccidial infection in immunosuppressed mice: prophylaxis and treatment with dehydroepiandrosterone. *East. Medit. Helth. J.*, 6 (5): 908 – 918.

Kimata I, Uni S and Iseki M. 1991. Chemotherapeutic effect of azithromycin and lasalocid on *Cryptosporidium parvum* infection in mice. *J. Protozool.*, 38: 232 – 233.

Lindsay DS, Woods KM, Upton SJ and Blagburn BL.2000. Activity of decoquinate against *Cryptosporidium parvum* in cell culture and neonatal mice. *Vet. Parasitol.*, 89: 307 – 311.

Magg P, Larocca A, Quarto M, Serio G, Brandonisio O, Angarano G and Pastore G. 2000. Effect of antiretroviral therapy on cryptosporidiosis and microsporidiosis in patients infected with human immunodeficiency virus type 1. *Europ. J. Clin. Microbiol. Infect. Dis.*, 19: 213 – 217.

Mantovani PM, Martino DL, Dettori G, Vajro P, Scotti S, Ditullio MT and Guandalini S. 1995. Asymptomatic carriage of intestinal *Cryptosporidium* in immunocompetent and immunodeficient children. *Pediatr. Infect. Dis. J.*, 14 (12): 1042 – 1027.

Rehg JE. 1991. Activity of azithromycin against *Cryptosporidia* in immunosuppressed rats. *J. Infect. Dis.*, 163: 1293 – 1296.

Saez-Liorens X, Odio CM, Umana MA and Morales MV. 1989. Spiramycin, placebo for treatment of acut diarrhea caused by *Cryptosporidium. Pediatr infect. Dis. J.*, 8: 136 – 140.

Sida LAA. 2005. Trials in treatment of secondary hydatid disease in laboratory mice (BALB/c). Ph.D. *Thesis, University of Baghdad.*

Theodos CM, Griffiths JK, Donfro J, Fairfield A and Tzipori S. 1998. Efficacy of nitazoxanide against *Cryptosporidium parvum* in cell culture and in animal models. *Antimicrob. Agen. Chemother.*, 42 (8): 1959 – 1965.

Viu M, Quilez J, Sanchez-Acedo C, Delcacho E and Lopez-Bernad F. 2000. Field trial on the therapeutic efficacy of paromomycin on natural *Cryptosporidium parvum* infections in lambs. *Vet. Parasitol.*, 90: 163 170.

Woods KM, Nesterenko MN and Upton SJ. 1996. Efficacy of 101 antimicrobials and other agents on the development of *Cryptosporidium parvum in vitro. Ann. Trop. Med. Parasitol.*, 90 (6): 603 – 615.

In: Biotechnology and Conservation … (2 Volume Set) ISBN: 978-1-63117-869-6
Editors: S. A. Farooq, R. Abed and S. Baqir

Chapter 7

A MICROSATELLITE POPULATION GENETIC ANALYSIS OF THE SMALL-SCALED GROUPER, *EPINEPHELUS POLYLEPIS*, IN THE GULF OF OMAN AND ARABIAN SEA

W. N. Al-Amry and W. F. Hutchinson
Department of Biological Science,
University of Hull, Hull, UK

ABSTRACT

The coastal waters off Oman contain a rich diversity of fishes, which support a series of economically and socially important commercial fisheries. *Epinephelus polylepis* is considered to be one of the most important commercial fish species in Oman; however it is potentially at risk of a serious population decline due to overfishing. Despite its commercial importance, very little is known about the biology of the species or the degree of structuring within the populations along the Oman coast. Such information is essential for developing a viable management plan that ensures that it is exploited sustainably. This project was undertaken to estimate the levels of genetic structuring and biodiversity of six populations of *E. polylepis* collected from two geographic regions along the Omani coastline: the Gulf of Oman (2 site; Dibba and Muttrah) and the Arabian Sea (3 sites; Shuwamiah, Duqum and Salalah). A total of 293 samples were collected during 2004-2006, and analysed using nine microsatellite molecular genetic markers to assess genetic differentiation between the populations. Tests for linkage disequilibrium verified that the chosen microsatellite loci could be considered as independent markers and evidence of Hardy-Weinberg equilibrium indicated that there was no significant sub-structuring was present within the regional samples. Tests of genic and genotypic differentiation, and *F*-statistic estimates (θ F_{ST} and ρ R_{ST}), were used to determine if significant genetic divergence existed between the populations. The genetic distance measure, Dc, was calculated and plotted against logged geographic distance to test for Isolation by Distance. Of these various statistics, a very weak divergence could be seen between Duqum and Muttrah (Genic: P = 0.0047; Genotypic: P = 0.005; θ Fst; P = 0.029), although only the genic value remained significant after a highly conservative Sequential Bonferroni correction (α = 0.005). Estimates of Dc showed no significant correlation between genetic and geographic distance (P = 0.607), which indicated that there is no evidence of isolation by physical distance. This present study shows very little evidence of stock structuring in this species between the Gulf of Oman and the Arabian Sea, with only minor population sub-structuring between the two sites of Muttrah and Duqum. Interestingly, these results suggest that the Rass Al-Hadd Jet does not act as a physical barrier between the Gulf of Oman and the Arabian Sea for *E. polylepis*.

Keywords: Microsatellite, *E. polylepis*, Small-scalled grouper, Gulf of Oman, Arabian Sea

INTRODUCTION

The Sultanate of Oman is located in the northwest Indian Ocean, with its northeastern coastline dominated by the Gulf of Oman, and its southern coastline dominated by the Arabian Sea, both of which are part of the Indian Ocean. Fishing is the most important occupation of the Omani people because the Omani diet is based heavily on sea food, therefore, the fishing sector in Oman contributes significantly to the overall economic growth of the country. The Gulf of Oman and the Arabian Sea have a very rich variety of fish and other seafood. One of the most important fish groups in Oman are the Grouper fishes (family Serranidae), known locally as Hamoor, and one of these species, *Epinephelus polylepis,* the small-scale grouper, was the focus of this study. There are many reasons necessitating the study of this species, one of which is insuring the sustainability of this stock from any risks, threats or population declines in the future. This species was described by Randall and Heemstra (1991); and is a third species in this group described by them (Heemstra & Randall 1993). The fish has a wide geographic distribution, extending from the northwest of the Indian Ocean, including the Gulf of Oman and the Arabian Sea, to the West coast of India (Heemstra & Randall 1993). Fish is very important locally to the Omani people as a food source, as well as having a high market value, making it very important commercially and economically (Sluka et al., 2001) for export purposes. Other popular types of fish with high market values, such as kingfish and tuna, have experienced stocks declines in recent years. Very little work been done on the biology of *E. polylepis* and it has never been a target for aquaculture (Randall 1995). Furthermore, due to potential separation of the Arabian Sea and the Gulf of Oman it was thought that populations from the two regions are genetically different.

In this study, genetic DNA analyses were carried out, because such information can be used to improve fisheries management, maintain sustainable levels of fish exploitation, and resolve geographic variations (Ihssen et al., 1981). In addition, it is used to identify the stock structure, and estimate the population size. We attempted to study the population genetics of *E. polylepis* using microsatellite markers, which will examine the Omani population, size, structure and diversity. This type of marker has been selected due to its selective neutrality and high mutation rates (Hutchinson et al., 2001). The study focuses on the population stock structure in the Gulf of Oman and Arabian Sea. The importance of this study area has to do with the hypothesis of the physical water dynamics, such as the cyclonic eddy in the the Gulf of Oman, and the anticyclonic heeddy in the Arabian Sea (Kindle 2001; Kindle & Anone 2001). Hence this is a very dynamic area in terms of thesurface current during the monsoon, that changes with the seasonal wind patterns. The Gulf of Oman and the Arabian Sea are considered as two separate seas from each other, because of the dynamic activity of the Ras Al Hadd jet stream separating these geographic areas at approximately the midpoint of the easterly facing coast of Oman. For at least 4-5 months of the year from the south east Monsoon, the Ras Al Hadd jet stream is active (see Figure 1.1), which diverts surface water eastwards into the Indian Ocean and away from the coast. This could act as a physical barrier, so we suspect that during this time, larval fish cannot cross northwards into the Gulf of Oman. This might also be due to the water dynamics. If this is the case, then it may be reflected in the genetic structures, giving rise to two distinct populations of benthic grouper species like *Epinephelus*. Hence, there is a need to test this hypothesis of the physical water dynamics which might affect the genetic exchange between the fish populations in the Gulf of Oman and the Arabian Sea, suggesting the presence of two genetically diverce populations. Additional results should show the presence of one or more populations.

Threats to the Marine Environment

There are various none-genetic threats to marine species in the ecosystem. Overfishing of commercial species can cause significant declines in stock abundance (Molony et al., 2003). Pollution of the marine environment can also lead to high mortality in fishes and other species (Taylor 1991; Wilson et al., 2002).

There are also genetic impacts of intensive fishing and evidence of genetic changes in exploited population. Erosion of genetic diversity in small population; genetic drift, inbreeding and effective population size. Selective removal from the population; Fishing is often a highly selective form of exploitation, typically leading to size selection (Hutchinson 2001), and may reduce the genetic composition of a population. The genetic diversity may be lost by three major types of selection in populations, directional selection, disruptive selection, and stabilising selection (Smith 1994).

Loss of locally adapted population; The loss of locally adapted populations or stocks (Extirpation) attracts more consideration in fishery management and conservation biology. Genetic diversity is very important for the conservation of individual species. Local adaptation is important for the survival of different population, and the species as a whole. Genetic variation between individuals or between populations is crucial for the survival of species because it allows populations to respond to environmental change (Carvalho 1993). Without genetic variation, depleted populations may become more vulnerable to disease or environmental fluctuation because they may lack individuals with the genetic potential to adapt (Carvalho & Pitcher 1995). However, in marine species the depletion of populations to levels where inbreeding becomes detrimental is unlikely to occur because fisheries will become commercially non-viable long before reaching this point.

Fisheries Management Objectives and Challenges

Generally, there are several objectives that managers want to achieve. However, the main objective is to maintain the marine fish stocks at a sustainable level and to avoid population crashes. If a fishery consists of more than one fishing ground, as most do, then the question is whether the fishery consists of more than one stock. If so, separate assessments of sustainable fishing levels should be made for each stock (Ward 2000). Ecosystem-based management has been proposed as a more sustainable option and is inherently much more complex. Ultimately, management is responsible for the functioning of a particular ecosystem, and this requires a determination of stock performance and the identification of sustainable targets within the ecosystem or fishery, and should also take account of such aspects as the recovery of endangered species, and sharing catch of commercial species between boundaries or countries. Ecosystem-based management requires the ability to determine credible performance targets, growth ranges, mortality rates and recruitment of the fishery stocks (Molony et al., 2003). The major problem that the managers might encounter has to do with obtaining adequate and correct information about fish stock to prevent over-exploitation. Stock assessment depends on reliable data on fish stocks as it is upon this that fisheries management is based.

Furthermore, fisheries managers may need to take into account economic, social and sometimes political factors in determining the controls that need to be practised in any particular fishery (Ward 2000). From an economic perspective, if the fishing industry harvests mixed fish species, this can cause a problem of overexploitation of common or commercial species. If the industrial fisheries are not regulated by law, then there could be a total collapse of the fishery, resulting in significant economic losses.

Socially, the relationship between the fisheries management and the fisherman is a very difficult relationship. They need to understand and help each other. In order to gather information, cooperation is necessary between the fisherman, with their knowledge gained from working experience, and the scientists. This can improve the quality of the scientific information and add value to the data gathered. From the policy side, a country needs to focus on developing a modern fishery sector, increasing fish exportation thus enhancing the economy, and developing the traditional fishing methods by training the fisherman and improving the industrial fisheries using modern equipment. However, it is very difficult to monitor the success of new policies where waters are shared by multiple countries. They might be targeting the same fish stocks, which may also migrate across international boundaries for feeding or spawning.

TECHNIQUES USED TO ESTIMATE THE LEVEL OF STOCK STRUCTURE

Artificial tags such as tagging. Tagging is a very important method used to identify fish stock separation and migration, and can provide useful information to estimate the parameters of fish growth, mortality or population size (Ricker 1975; Pawson & Jennings 1996).

Natural markers. There are two types of natural marker techniques used to identify the stock structure: biological and molecular genetics. Biological markers (e.g., morphological, meristic, scale and otolits) and molecular genetics. Previously used (immunonogenetics and allozymes) and recently used (sequencing of nuclear and mitochondrial DNA, RAPDs, AFLPs and microsatellites).

Oceanographic Conditions of the Gulf of Oman and the Arabian Sea

As the Rass al Hadd jet collapses, the jet curves around the anticyclonic eddy, both the cyclonic and anticyclonic eddies drift to the south. Further to the east, in the central Arabian Sea, circulation patterns formed by the dynamic effects of the sustained monsoon winds and ocean currents seem to marbelize the ocean surface (Figure 1.1). This region is considered as a mid-point between the Gulf of Oman and the Arabian Sea. The Ras al Hadd jet is very variable, sometimes extended out of the coast to the east and sometimes extended to the northeast or southeast. The Ras al Hadd Front represents the seasonal boundary between the northern Arabian Sea and the Gulf of Oman. During the SW monsoon, the Ras al Hadd transport is at least 10Sv (Elloit & Savidge 1990). In 1995 it was observed that the Ras al Hadd jet intensified in August following a reverse in direction of the current along NE the northeastern coast from northward to southward increasing the flow along the Ras al Hadd Front (Kindle & Anone 2001).

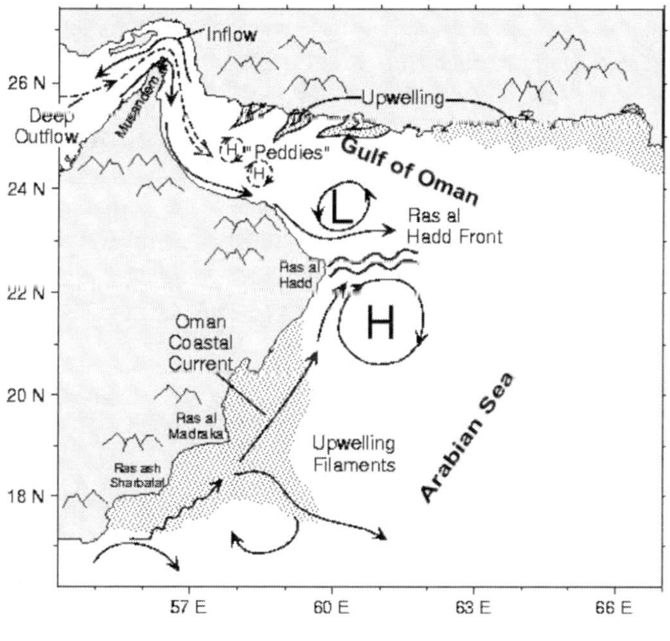

Figure 1.1. The map of the geographic dynamics of low and high pressure variabilities causes the cyclonic and anticyclonic eddies in the Gulf of Oman and the Arabian Sea (Kindle & Anone 2001). This map show hows the conditions in half the year during the SW monsoon (June to September) cause the cyclonic and anticyclonic eddies in the Gulf of Oman and the Arabian Sea (Johns et al., 2000).

Role of Marine Science and Fisheries in the Gulf of Oman and Arabian Sea (MSFC)

Marine Science and Fisheries Centre (MSFC) of the Directorate General of Fisheries, which established in 1986, carry out fisheries research in Oman. The research programmes of

MSFC designed to provide data and information necessary for decision making regarding development and management of marine resources in Oman. Such information should serve also the private sector companies involved in fishery production, processing, export, aquaculture and other fisheries activities.

EXPLOITATION AND MANAGEMENT OF E. POLYLEPIS IN THE GULF OF OMAN AND ARABIAN SEA

In this study will investigate the potential genetic structure of the population of Smallscaled grouper, *E. polylepis*, from Oman using a neutral microsatellite marker. This research focuses on the stock structure of this species in the Gulf of Oman and Arabian Sea. As this species was recently described by Heemstra and Randall (1993) little is known of its life-history traits and it has never been a target of aquaculture (Randall 1995). It is the third grouper species recently described from this area, the other two being the Brownspotted grouper, *E. chlorostigma* and the Multispotted grouper *E. gabriellae*. *E. chlorostigma* has puzzling gaps in its worldwide distribution, from the Red Sea and east coast of Africa to the western Pacific (from southern Japan to New Caledonia), Gulf of Aden, Kenya, Tanzania, Seychelles, Chagos Islands, Maldives, Lak-shadweep Island, Andaman Islands, Nicobar Islands, Viet Nam, China, Hong Kong, Korea, Papua New Guinea, New Ireland, Caroline Islands, American Samoa, and Fiji (Randall & Heemstra 1991). The Multispotted grouper *E. gabriellae* is distributed in the North Indian Ocean from Somalia to Oman (Randall & Heemstra 1991; Heemstra & Randall 1993). The latter is sympatric with *E. polylepis* in the Gulf of Aden and along Oman's southern coast, and *E. chlorostigma,* although very similar looking to *E. polylepis* is alopatric (Figure 1.2), but to date has not been recorded from Oman.

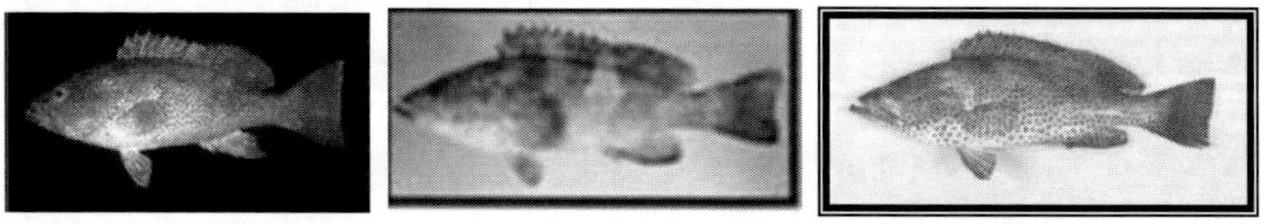

(a) (b) (c)

Figure 1.2. Three species photographs of grouper, which ocur in the NW Indian Ocean, (a) the Brownspotted grouper *E. chlorostigma* Picture (Epchl_u0.jpg) by Randall, J.E. and (b) Multispotted grouper *E. gabriellae* Picture (Epgab_u0.jpg) by Randall, J.E. which, both look similar to the (c) Smallscaled grouper *E. polylepis* Picture (Eppol_u6.jpg) by Randall, J.E. (Heemstra & Randall 1993). These pitures from (fishbase.org website).

E. polylepis is found only in the North west Indian Ocean, including the Gulf of Aden, Gulf of Oman, Arabian Gulf, Pakistan, and western India (Randall & Heemstra 1991; Heemstra & Randall 1993; Randall 1995) see Figure 1.3.

Because *E. polylepis* is largely demersal and found at depths of 33-100 metres, they are heavily targeted by fishermen using a variety of fishing gears such as baited fish traps, hand lines and bottom gill nets (Al-Abdessalam 1995). In Oman, they are caught predominately by the artisanal fishermen (Randall 1995). The Smallscaled grouper is a preferred table fish, and owing to its high market price is considered one of the most valuable fish groups in the artisan fisheries of Oman (Al-Abdessalam 1995).

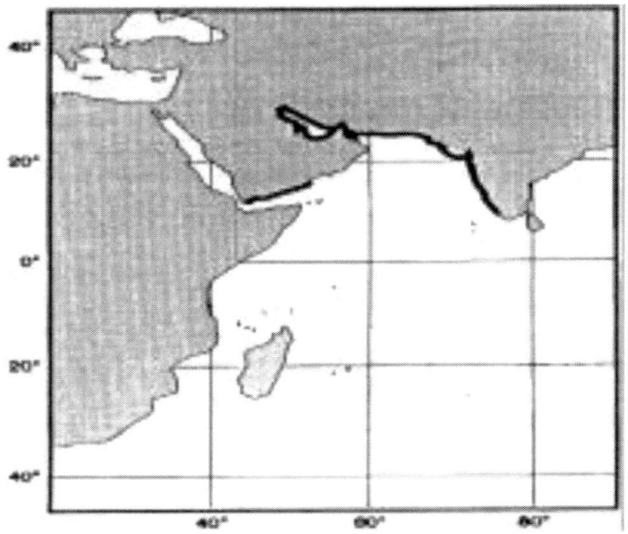

Figure 1.3. The distribution of *E. polylepis* (Heemstra & Randall 1993).

EXPLOITATION OF *E. POLYLEPIS*

The combination of the slow growth rates and the tendency towards site specificity makes grouper vulnerable to overexploitation by fishermen (Heemstra & Randall 1993). Most groupers are heavily targeted by fishermen using a variety of fishing gears, such as hand lines and fishing traps. In Oman some species are caught by trawlers (Heemstra & Randall 1993) and heavily targeted by fishermen using a variety of fishing gears such as baited fish traps, hand lines and bottom gill nets. In 1993-1997, the industrial fishery sector accounted for, on average, only 20% of the total grouper landings for Oman (Anon 2001). Artisanal fishermen, operating out of small fibreglass boats and traditional dhows, accounted for the remaining 80% or approximately 3000 tonnes of groupers.

AIM AND OBJECTIVES

In this project, it was thought that populations from the two regions of the Gulf of Oman and the Arabian Sea are genetically different due to potential separation of the Arabian Sea and the Gulf of Oman. Therefore, this may effect on the population diversity of the *E. polylepis*. This study aims to investigate the level of genetic stock structure and population diversity of the small-scaled grouper, *E. polylepis*, along the coast of the Sultanate of Oman, within the Gulf of Oman and the Arabian Sea. A range of microsatellite primers from related grouper species will be tested for cross-amplification in *E. polylepis,* and a subset of ten molecular markers will be selected to genotype samples collected from along the Omani coast. The aim is to determine whether significant genetic structuring exists in this region, and to what extent eddies and seasonally variable currents, in particular the Ras al Hadd jet, influence the presence or lack of such differentiation. Potential strategies for sustainable exploitation and the conservation of inter and intra-population genetic diversity may then be considered.

This study was conducted using *Epinephelus polylepis* grouper samples from various landing sites along the Omani coast collected over a period of two years from 2004 to 2006 (Tables 1.1). The Five sampling sites were located in the Gulf of Oman (Dibba from Musandam and Muttrah from Muscat) and the Arabian Sea (Duqum from Al-Wusta, Shawarmiah, and Salalah from Dhofar) along the coast of Oman (from the north of the Gulf of Oman to the south of the Arabian Sea). Tissue plugs and pelvic fin clips were taken from freshly caught fish obtained from local fishermen, and stored in small plastic screw-cap tubes filled with 96% ethanol. Samples were collected by two technicians (Nader Al-abri and Nasser Al-gardani) from Sultan Qaboos University (SQU) under the supervision of Dr Jenny McIlwain Sample sizes ranged from 36 for Shuwaimiah to 73 for Muttrah, and were dependant upon the availability of fish, which varied seasonally and geographically.

Table 1.1. Number of samples collected from different landing sites in the Gulf of Oman and Arabian Sea in the Sultanate of Oman

Landing site region (ID)	Region	Number of fish Samples
Dibba from Musandam	Gulf of Oman	53
Muttrah from Muscat	Gulf of Oman	73
Duqum from Al-Wusta	Arabian Sea	59
Shuwaimiah from Dhofar	Arabian Sea	36
Salalah from Dhofar	Arabian Sea	72
Total		293

The ethanol was removed from the tissue, prior to the DNA extraction by squeezing the excess ethanol from the sample and allowing the sample to air dry on clean tissue paper. The DNA was extracted from the fin clip tissues and muscles using the phenol chloroform method (Taggart et al., 1992). Electrophoresis of the eluted DNA in a 1% agarose gel was used to test for yield and purity. The DNA was stained with Ethidium bromide and visualised under UV light (figure 3.1). Finally, the DNA concentration was measured using a GeneQuant spectrophotometer (Pharmacia Biotech), and the DNA diluted using double distilled water to give a final concentration of 5ng/µl. A total of 18 microsatellite primers selection (forward & reversed primer) isolated from three different species of grouper *Mycteroperca bonaci, Epinephelus quernus, Mycteroperca microlepis)*, five microsatellite loci (Forward & Reversed primer) isolated from black grouper, *M. bonaci* (Zatcoff et al., 2002), seven microsatellite loci (forward & reversed primer) isolated from Hawaiian grouper, *E. quernus* (Rivera et al., 2004), and six microsatellite loci (forward & reverse primer) isolated from Gag grouper, *M. microlepis* (Chapman et al., 1999; De Innocentiis et al., 2001), were tested with the DNA extracted from the *Epinphelus polylepis* using a range of PCR annealing temperatures and magnesium chloride concentrations. PCR programme used a touch-down (60-50°C) PCR cycling programme used to test the cross-amplification of 18 pairs of microsatellite primers. In the first 10 cycles, the annealing temperature was reduced by 2°C every other cycle. Of the original 18 microsatellite primer pairs, only 12 pairs were successfully optimised to produce consistently strong PCR products with the small-scaled grouper *E. polylepis* DNA.

These loci (Mbp029, Mbo061, Mbo066, Mbo088, CA-2, CA3, CA-6, GAG007, GAG010, GAG023, GAG045, and GAG049) were thus chosen for screening the market samples using the conditions. The 12 microsatellite loci PCR products were used for Beckman screening for microsatellite analysis. The statistical analysis used Test for genotyping errors due to null alleles, stuttering and large allele dropout, test for Linkage disequilibrium, Calculation of observed and expected heterozygosity, and tests for deviation from Hardy-Weinberg expectations (HWE), Estimation of allelic richness, test for genic and genotypic differentiation, test for genetic differentiation with F_{ST} and R_{ST} and Calculation of Cavalli-Sforza and Edward's Dc (1967) and test for isolation by distance.

The DNA quantification measurements revealed a high quantity of DNA concentration with low protein concentration (Figure 3.1). The DNA dilution 5ng/100ul was used for the PCR reactions. Figure 3.1 shows the agarose gel of 11 DNA extractions of *E. polylepis* muscles. using the phenol/chloroform method (Taggart et al., 1992), indicating good quality, high molecular weight DNA.

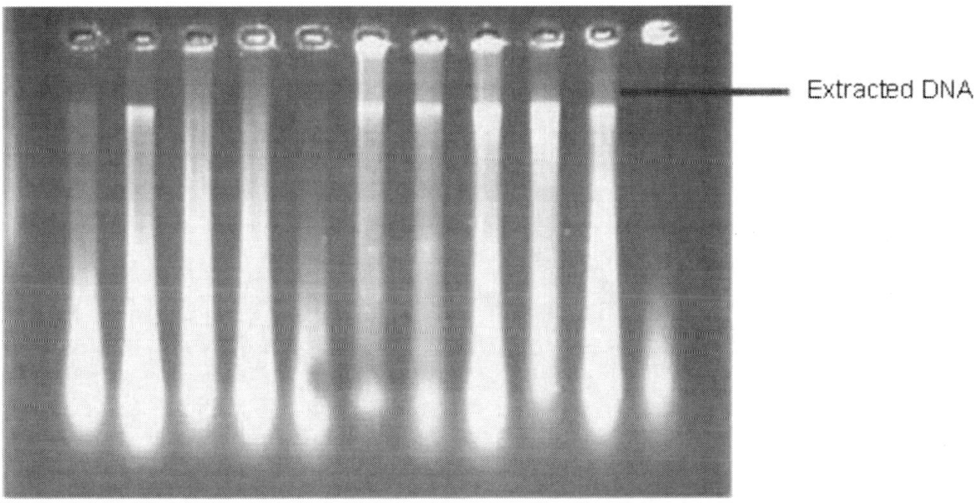

Figure 3.1. Verification of the agarose gel of 11 DNA extractions of *E. polylepis* muscles.

Only 12 primers (microsatellite loci) were employed successfully with 309 *E. polylepis* DNA extracted samples in the PCR reactions. These successful primers were Mbo029, Mbo061, Mbo066, Mbo088, CA-2, CA-3, CA-6, GAG007, GAG010, GAG023, GAG045, and GAG049 while the unsuccessful primers were Mbo48, CA-1, CA-4, CA-7, GA-1, and Mbo38.

Overall, five populations were genotyped with eleven microsatellite loci. A total of 293 samples from five populations. Screening with Microchecker yielded evidence of null alleles at the Mbo29 locus in the Dibba and Shuwaimiah samples, and at the Mbo66 locus in the Muttrah sample. Significant heterozygote deficiencies were also observed at these loci. Re-scoring the chromatograms failed to alleviate this genotyping error and hence these two loci were removed from the data. The remaining five samples were genotyped using nine microsatellite loci GAG007, CA2, CA6, GAG010, GAG045, Mbo88, GAG023, GAG049, and Mbo61.

CALCULATION OF OBSERVED AND EXPECTED HETEROZYGOSITY, AND TESTS FOR DEVIATION FROM HARDY-WEINBERG EXPECTATIONS (HWE)

Mean observed heterozygosities across loci within samples ranged from 0.89 to 0.91 and the expected heterozygosities ranged from 0.90 to 0.91 (Table 3.1).

Table 3.1. Expected (He) and Observed (Ho) heterozygosity for 5 populations of grouper *E. polylepis* analysed at 9 microsatellite loci

Locus/ Population		Dibba	Duqum	Muttrah	Salalah	Shuwaimiah	Mean
GAG007	*HetExp*	0.86	0.87	0.88	0.87	0.87	0.87
	HetObs	0.89	0.90	0.84	0.86	0.83	0.86
Locus CA2	*HetExp*	0.90	0.89	0.89	0.91	0.92	0.90
	HetObs	0.89	0.92	0.81	0.89	0.94	0.89
Locus CA6	*HetExp*	0.88	0.85	0.87	0.84	0.88	0.87
	HetObs	0.87	0.92	0.86	0.85	0.83	0.87
GAG010	*HetExp*	0.94	0.95	0.95	0.94	0.95	0.95
	HetObs	0.89	0.90	0.96	0.92	0.92	0.92
GAG045	*HetExp*	0.96	0.95	0.95	0.94	0.95	0.95
	HetObs	0.96	0.93	0.99	0.94	0.92	0.95
Mbo88	*HetExp*	0.91	0.88	0.93	0.91	0.92	0.91
	HetObs	0.91	0.78	0.96	0.94	0.94	0.91
GAG023	*HetExp*	0.94	0.92	0.93	0.94	0.92	0.93
	HetObs	0.96	0.92	0.96	0.93	0.89	0.93
GAG049	*HetExp*	0.87	0.90	0.88	0.87	0.86	0.88
	HetObs	0.79	0.86	0.85	0.85	0.81	0.83
Mbo61	*HetExp*	0.92	0.92	0.88	0.90	0.93	0.91
	HetObs	0.87	0.92	0.92	0.94	0.97	0.92
Mean	*HetExp*	0.91	0.90	0.91	0.90	0.91	0.91
Mean	*HetObs*	0.89	0.89	0.91	0.90	0.89	0.90

HETEREOZYGOTE DEFICIENCY AND EXCESS

Prior to the removal of the Mbo29 and Mbo66 loci, significant heterozygote deficiencies were observed in seven of the 55 tests (eleven loci in five populations). Removal of these two loci from the data due to the potential presence of null alleles (*Microchecker*) left three significant results (CA2 in Muttrah, Mbo88 in Duqum and Mbo61 in Dibba). Testing of these significant results using Kinnison et al. (2002) binomial likelihood method indicated that these results were significant by chance (i.e. Type I error) and thus could be rejected ($\alpha = 0.05$, $L = 0.299$ for 1 significant test across nine replicate loci; $L = 0.204$ for 1 significant test across five replicate populations). Therefore, we can assume that no population or locus exhibits a significant heterozygote deficiency. Tests for heterozygote excess were not significant at any locus or population ($P<0.05$) (Table 3.2).

Table 3.2. Significant deviations from Hardy-Weinberg Expectation due to heterozygote deficiencies, with standard errors (S.E)

Sample	Loci	F_{IS} P-values	S.E.
Dibba	Mbo29	0.0160	0.0022
Shuwaimiah	Mbo29	0.0022	0.0005
Muttrah	Mb066	0.0000	0.0000
Salalah	Mb066	0.0298	0.0088
Muttrah	CA2	0.0257	0.0013
Duqum	Mb088	0.0213	0.0017
Dibba	Mb061	0.0271	0.0017

All pair-wise tests for genotypic linkage disequilibrium were non-significant, and therefore allelic variation at the nine remaining loci could be treated as independent. The Shuwaimiah population had the highest (19.0) mean allelic richness across loci (Table 3.3) and the Salalah population had the lowest (18.0). The loci GAG045 and CA2 exhibited the highest (24.2) and lowest (13.9) mean allelic richness across the five populations.

Table 3.3. Allelic diversity of 5 populations of Omani Grouper (*E. polylepis*) analysed at 9 microsatellite loci

Locus/ Population	Dibba	Duqum	Muttrah	Salalah	Shuwaimiah	Mean across Samples
GAG007	16.8	17.4	19.0	18.7	20.0	18.4
CA2	13.9	12.5	14.0	13.1	16.0	13.9
CA6	18.7	17.3	16.8	16.1	16.0	17.0
GAG010	19.4	22.0	22.1	20.3	22.0	21.3
GAG045	25.5	24.7	25.4	23.6	23.0	24.2
Mbo88	19.2	17.9	20.9	19.2	21.0	19.2
GAG023	20.9	20.2	19.9	21.3	20.0	20.5
GAG049	13.6	14.4	13.5	13.5	11.0	14.0
Mbo61	18.0	19.3	17.2	16.4	22.0	17.8
Mean across Loci	18.4	18.4	18.8	18.0	19.0	18.5

Pair-wise exact tests of genic and genotypic differentiation between all five populations were non-significant with the exception of the Duqum and Muttrah pair-wise test (Genic test: P = 0.0047; Genotypic test: P = 0.0051). However, a conservative sequential Bonferroni adjustment (adjusted alpha value = 0.005) rendered the genotypic value non-significant (Table 3.4).

Tests of differentiation between the five populations using Weir & Cockerham's θ F_{ST} and Goodman's (1997) ρ R_{ST} yielded very weak (≤ 0.0011) or often negative values (Table 3.5). Only one pair-wise test, Duqum versus Muttrah, was significant (θ $F_{ST} = 0.0022$, P = 0.029), but became non-significant after a conservative Sequential Bonferroni correction for ten pair-wise tests ($\alpha = 0.005$).

Table 3.4. Genic and genotypic differentiation of *E. polylepis* in the Gulf of Oman and Arabian Sea. The exact p-value of genic differentiation are presented in the upper triangle, and genotypic differentiation in the lower triangle. Significant values at P<0.01 are identified with two stars (**). Values that remained significant following sequential Bonferroni correction (α = 0.005) are highlighted in bold

	Dibba	Duqum	Muttrah	Salalah	Shuwaimiah
Dibba		0.691	0.121	0.720	0.918
Duqum	0.006		0.0047**	0.739	0.598
Muttrah	0.145	0.0051**		0.132	0.833
Salalah	0.769	0.850	0.121		0.853
Shuwamia	0.942	0.619	0.867	0.871	

Table 3.5. Population pair-wise θ F_{ST} and ρ R_{ST} estimates of genetic differentiation for samples from Dibba, Duqum, Muttrah, Salalah, and Shuwaimiah. Values shown in the upper triangle are F_{ST} and lower triangle is R_{ST}. Values significant at P<0.05 are identified with a star (*). No values remained significant following sequential Bonferroni correction (α = 0.005)

	Dibba	Duqum	Muttrah	Salalah	Shuwaimiah
Dibba		-0.0001	0.0011	-0.0005	-0.0025
Duqum	-0.00349		0.0022*	0.0005	-0.0007
Muttrah	-0.00388	0.00108		0.0005	-0.0018
Salalah	-0.00676	-0.00001	-0.0047		-0.0024
Shuwaimiah	-0.00429	-0.00739	-0.00394	-0.00489	

Table 3.6. The upper values are Dc values and the lower values are P-values

	Dibba	Duqum	Muttrah	Salalah	Shuwaimiah
Dibba		0.013	0.011	0.010	0.014
Duqum	0.128		0.012	0.009	0.013
Muttrah	0.283	0.033		0.009	0.012
Salalah	0.758	0.837	0.300		0.012
Shuwaimiah	0.887	0.814	0.889	0.934	

Of the ten pair-wise tests of genetic distance (Dc) between the five populations, only the relationship between Duqum and Muttrah (Dc = 0.012, P = 0.033) was significant at the P < 0.05 level (Table 3.6). However, this value became non-significant following sequential

Bonferroni (α = 0.005). A test of the relationship between the geographic and genetic distance using the Mantel test (Figure 3.2) indicated that there was no significant isolation by distance (P = 0.607).

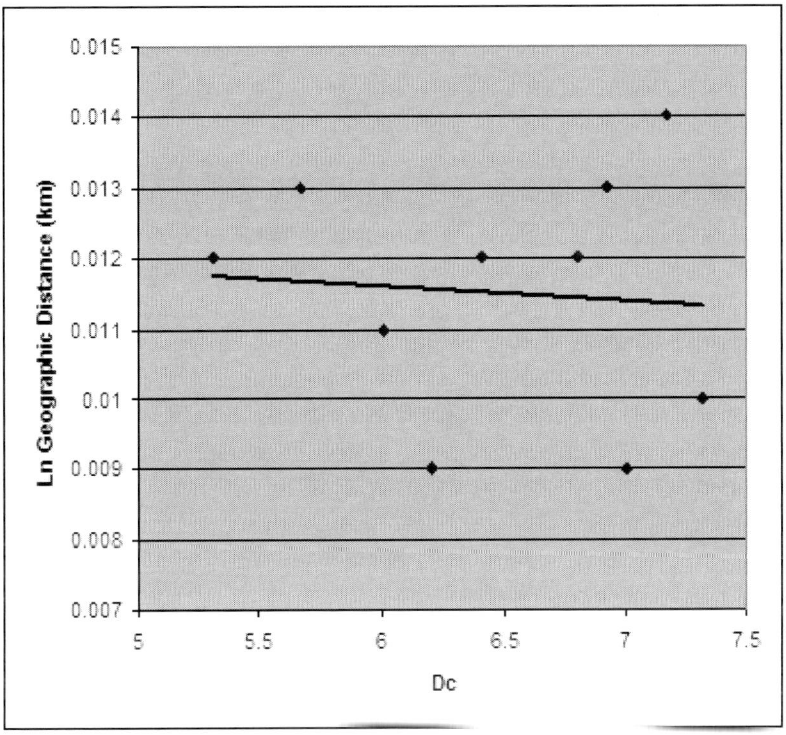

Figure 3.2. Plot of log-transformed geographic distance (km) against Cavalli-Sforza & Edwards (1967's) genetic distance Dc for each of the ten pair-wise comparisons.

In this study, the population structure of *E. polylepis* in the Gulf of Oman and the Arabian Sea was investigated using nine microsatellite molecular genetic markers (Mbo061, Mbo088, CA-2, CA-6, GAG007, GAG010, GAG023, GAG045, and GAG049). The samples were distributed along ~1800 km of coastline, within two regions (five sites) separated by the Ras al Hadd jet, an important oceanographic frontal zone between the Gulf of Oman and the Arabian Sea. Although the Ras al Hadd jet is affected seasonally by the SW and NE monsoons, it forms a geographical boundary zone that may prevent transport of eggs, larvae and adults between the Arabian Sea and the Gulf of Oman, and may therefore be instrumental in the formation of genetically distinct fish stocks. Indeed, the distributions of several fish species, such as *E. gabriellae, E. Latifasciatus, E. malabaricus, E. multinotatus, E. radiatus, E. stoliczkae, E. tukula* and *E. undulosus* are restricted to either the Arabian Sea or the Gulf of Oman (Randall, 1995). There are at least 40 types of Serranidae families restricted to specific parts of Omani waters (Gulf of Oman and the Arabian Sea) that show a similar distribution to *E. polylepis*, such as Areolated grouper *E. areolatus*, Brown-spotted grouper *E. chlorostigma*, Spinycheek grouper *E. dicanthus*, Striped grouper *E. latifasciatus*, Longteeth grouper *E. moara*, and Oblique-banded grouper *E. radiatus* (Al-Abdulssalam 1995). Although many factors affect fish distribution, this distinct boundary zone remains noteworthy. Therefore, it was hypothesized that the Ras al Hadd jet may restrict the movement of *E. polylepis* between the two water masses, giving rise to two or more genetically distinct populations of the species. The *E. polylepis* stocks within the Arabian Sea and the Gulf of Oman are currently managed as a single stock, however if multiple distinct stocks exist within this region, changes to the management policies are required to ensure that the stocks are exploited sustainable.

The results of the microsatellite analysis of fish from five sites showed that:

- There is significant differentiation between samples collected from Duqum and Muttrah (Genic test, P = 0.0047; Genotypic test, P = 0.0051; F_{ST}, P = 0.029), which lie either side of the Ras Al Hadd jet, although the latter two tests became non-significant after a highly conservative sequential Bonferroni ($\alpha = 0.005$), suggesting very weak structuring. However, pair-wise tests between more distant samples separated by the front were non-significant, making this data difficult to interpret.
- There is no significant correlation between genetic distance and geographic distance (P = 0.607), indicating no evidence of isolation by distance.
- The evidence of weak structuring between Duqum and Mutrah suggest that the populations in the Arabian Sea and the Gulf of Oman are not panmictic. Although the Bonferoni test turns these results non-significant, this test is very stringent and we might be in a danger of rejecting a hypothesis that is actually significant. However, F_{ST} tests for samples collected from regions separated by longer distances such as Dibba (North of the Sultanate) and Salalah (South of the Sultanate) were not significant, when we would expect them to be if the front was causing differentiation.

The observed heterozygosities at the five sites varied between 0.89 and 0.91 (Table 3.1). A study of the dusky grouper *E. marginatus* in the Mediterranean Sea used five of the microsatellite loci used in the present study (GAG007, GAG010, GAG023, GAG045, and GAG049) and revealed high mean observed heterozygosity used the same loci. Other studies found relatively lower observed heterozgosity for microsatellites in *E. marginatus* (0.61 mean observed heterozygosity) (De Innocentiis et al., 2001), and in five other *Epinephelus* species (0.48-0.66; Nugroho et al., 1998; De Innocentiis et al., 2001) with the same loci (GAG007, GAG010, GAG023, GAG045 and GAG049). The Shuwaimiah population had the highest (19.0) mean allelic richness across loci and the Salalah population had the lowest (18.0). The loci GAG045 and CA2 exhibited the highest (24.2) and lowest (13.9) mean allelic richness across the five populations. Based on these results no significant difference in allelic richness or heterozygosity was apparent between the five populations of Omani Grouper *E. polylepis*, suggesting similar evolutionary histories and effective population sizes.

Suitability of the Chosen Molecular Markers and Sampling for Detecting Genetic Structuring

The general lack of detectable genetic structuring may have been influenced by the methods used in this study, and several factors could have reduced the statistical power of the techniques utilised (i.e. sample size, number of loci, statistical power of tests used, and the composition of samples). Initially, we investigated a total of 18 microsatellite primers isolated from three different species of groupers, including the black grouper *Mycteroperca bonaci* (Zatcoff et al., 2002), the Hawaiian grouper *Epinephelus quernus* (Rivera et al., 2004), and the Gag grouper *Mycteroperca microlepis* (Chapman et al., 1999). Of these, six primer pairs (Mbo048, CA-1, CA-4, CA-7, GA-1, and GAG038) were excluded as they failed to amplify the *E. polylepis* DNA, or yielded non-specific bands. Two more (Mbo029 and Mbo066) were subsequently omitted due to evidence of null alleles in the genotype data. Microchecker estimated the proportion of null alleles at the Mbo029 and Mbo048 loci to be 37% and 13%, respectively. To increase the chance of a successful amplification of all alleles, the stringency of the PCR could have been reduced or primers re-designed (DeSalle & Shearwater 1998), however due to the availability of other amplifiable loci, Mbo029 and Mbo048 were eliminated without further optimisation. A third marker (CA-3) was removed from the data set due to severe stuttering causing unreliable genotyping. There remained nine microsatellite loci that produced polymorphic and reproducibly scorable products, with no evidence of Hardy Weinberg or linkage disequilibrium, and thus it is unlikely that any weak structuring could have been lost through errors in the data caused by inaccurate genotyping. It is generally accepted that the greater the number of loci used the more reliable the estimate of structuring. This is particularly the case in marine fish where weak structuring is predicted due to large population sizes and a high potential for gene flow. However, significant structuring has been detected with very few loci. For example, in a study of Atlantic herring *Clupea harengus*, only four microsatellite loci showed highly informative spatial structuring (Shaw et al., 1999). Another population study used only five microsatellite loci to detect fine-scale spatial structuring in cod (*Gadus*) *morhua* within the North Sea (Hutchinson et al., 2001). Therefore, the nine loci used in this study offered greater potential power for detecting structuring than other studies that detected significant divergence on a comparable scale with fewer loci.

Ruzzante (1998) showed that for microsatellite analysis, equal sample sizes of at least 50 per site are required, but a maximum of 100 per site would be optimum. Sample sizes of this magnitude generally lead to more precise and accurate estimates of structuring, and avoid major statistical bias. The numbers of *E. polylepis* sampled per site in this study ranged from 36 to 73 individuals, with only one site, Shwamiah, with less than fifty. The power of a study depends on the number of loci, number of alleles at each locus, and range in allele size, and this interacts with sample size to affect bias and sampling variance (Ruzzante 1998). Kalinowski (2005) suggested that if the coefficient of variation of the genetic distance estimator F_{ST} is >0.05 then a sample of 20 individuals per population is reasonable, but when Fst <0.01 then 100 individuals per population were recommended. In this study of *E. polylepis* the F_{ST} estimates fell well below 0.01 (maximum significant value = 0.002) and hence larger sample sizes would have been preferable had the fish been available. Nevertheless, there was no evidence that comparisons between the larger samples (e.g., Duqum and Salalah, Dibba and Salalah) yielded higher F_{ST} values. There are several reasons why the selected loci could reduce statistical power, e.g., too few alleles, low heterozygosities, or the allele frequency distribution may be dominated by a single allele, which is common in some populations. Should a small sample size be combined with large numbers of alleles, it would potentially reduce statistical power (De Innocentiis et al., 2001). Statistical power may also be reduced if the allele frequency distribution is dominated by a single allele that is common in all the sampled populations. In our study, different size ranges of alleles were observed i.e. GAG045 was 70bp, GAG049 was 38bp, Mbo029 was 32bp, Mbo066 was 66bp, Mbo088 was 54bp, GAG010 was 70bp, GAG023 was 68bp, Mbo061 was 62bp, GAG007 was 110bp, CA-2 was 42bp, CA-3 was 300bp and CA-6 was 64bp. Furthermore, allelic distributions were bimodal or multimodal, with the dominant allele varying between samples. Most studies of population structuring assume that their chosen genetic marker is not under selection and that any detected genetic differences are mainly due to genetic drift and mutation. Although there is some recent evidence of selection acting directly on microsatellite loci in tilapia in high-salinity environments (Kornfield 2004) and evidence of hitch-hiking selection in Atlantic cod *Gadus morhua* at the microsatellite GMO132 locus in cod (Nielsen et al., 2006), it is unlikely to have affected the distribution of alleles in the present study, not least because the observed structuring is rather weak. One of the prerequisites for detecting significant divergence in weakly structured populations is the specific targeting of spawning fish on the spawning grounds, thus ensuring that the samples reflect the stocks at their genetically most distinct. In the present study, fish were scarce and thus samples had to be collected throughout the year, and since spawning grounds are largely unknown, they could not be specifically targeted. Furthermore, since the samples were collected from fish markets their exact origin was difficult to ascertain and they could have been composed of mixtures from outside the intended sampling region. This could potentially explain the weak structuring observed between Muttrah and Duqum, not least because both Duqum and Muttrah are very important landing sites. However, this appears unlikely as the majority of the fishing boats are small inshore vessels with a limited range and fish are seldom moved between markets by truck or larger boats. Furthermore, there was no evidence of a sample-specific heterozygote deficiency in any of the samples, which might have been observed if multiple stocks had been present in individual samples due to the sampling of mixed feeding aggregations or mixed catches. Although larger sample sizes would have provided greater power for investigating weak structuring, the selection of nine suitably variable loci combined with accurate genotyping should have been sufficient to detect any significant genetic differentiation across such an extensive geographical range (i.e. 1800 km of coast). However, it is reasonable to suggest that the very weak structuring seen in this study is a real result and not an artefact of the data or analysis methods. The question now remains as to whether environmental and species-specific biological factors may have contributed to the observed weak structuring.

Influence of Life History and Oceanography on Genetic Structuring in *E. polylepis* in the Arabian Sea and the Gulf of Oman

Where extensive migration to and from feeding or spawning grounds occurs, the occasional straying of migrants can potentially prevent significant divergence evolving between geographically isolated populations. There are many species of pelagic fish, which migrate over long distances (e.g., hundreds to thousands of kilometres) for spawning. Some adult fish may spawn in a different stock each year, or recruit to a non-natal stock and then always return to that stock for spawning, either of which will result in gene flow. However, the demersal small-scaled grouper described in the present study would probably migrate no more than a few hundred kilometres to spawn, if migration takes place at all. In general, groupers tend to have a suite of identifiable life history traits which are favourable to population structuring, such as highly specific habitat preferences, male territorial behaviour, and stable spawning aggregations (Heemstra & Randall 1993; Morris et al., 2000), (Rivera et al., 2004). Therefore, we might expect our species to be more sedentary and there be a lesser potential for gene flow than in pelagic fishes.

Larvae and eggs can be carried over large distances by currents, although gyre-like circulation patterns may act as localised transportation and retention mechanisms (Ruzzante et al., 1998; Smith & Morse, 1984; Lough & Bolz, 1989). Our study showed a possible mixing mechanism for the eggs and an oceanographic feature can separate larvae across two bodies of water, which for a large part of the year. Recent evidence from both population studies of marine organisms and simulations of larval transport in oceanic currents highlight the potential for local retention within limited geographical ranges as an important component of the population dynamics of marine recruitment (Swearer et al., 1999; James et al., 2002). Both long distance dispersal and local retention of larvae can be heavily influenced by a number of key factors, including timing and duration of spawning, planktonic duration, adult life expectancy and larval behaviour (Largier 2003). In this study, the key environmental factors are the timing of the monsoons and associated change in direction of the water current, which occur at the same time every year (SW in the summer, NE in autumn), along with the associated atmospheric forcing and air-sea fluxes (William et al., 1999). Although gyres develop along the Omani coast during the monsoons, which could help to retain eggs and larvae, the lack of genetic structuring suggests that they and the Rass al Hadd jet are insufficient to counteract the mixing effects of the seasonally reversible Oman Coastal Current, which would tend to homogenise coastal waters. Furthermore, due to a lack of biological data it is difficult to determine whether the egg and larval phase actually coincides with the development of these gyres, and thus a temporal mismatch may equally render this retention mechanism ineffective. Rhodes et al. (2003), found genetic homogeny within open-ocean environments in camouflage grouper, *E. polyhekadion*, in the western central Pacific due to a lack of apparent barriers to gene flow (Rosenblatt & Waples 1986). However, several physical and biological retention mechanisms were thought to operate with populations distributed around the islands, and it was thought that features such as eddies, wind-driven and topographically induced fronts, and reef-oriented swimming behaviour in *E. polyhekadion* may enhance population structuring (Leis & Carson-Ewart 1998; Swearer et al., 1999). In a study of the Hawaiian grouper, *E. quernus* (Serranidae) using mitochondrial DNA, Rivera et al. (2004) found that genetic differentiation across the Hawaiian archipelago was often low, although breaks in gene flow could be detected, which were potentially associated with the patterns of the oceanic currents. Lack of differentiation of neutral genetic markers across scales of several thousands kilometres may reflect recent colonization histories coupled with large populations experiencing low levels of genetic drift, and need not invoke high gene flow per-se (McPherson et al., 2003). However, since the Arabian Sea is thought to have formed over 50 million years ago, as the Indian subcontinent collided with Asia, it is unlikely that the colonisation of Gulf of Oman and Arabian Sea by the grouper would have been too recent for genetic differentiation to have time to develop.

Implications for Management and Further Research

The results from this study show weak genetic stock structuring of *E. polylepis* between the two regions of Muttrah and Duqum in the Gulf of Oman and the Arabian Sea. However, there was no evidence of structuring on a larger scale, as might be expected if the identified divergence was due to a process of isolation by distance. Nevertheless, although insufficient genetic evidence currently exists to treat the Gulf of Oman and the Arabian Sea as separate stocks, a cautionary approach would be advisable, and more data should be collected before any long-term management plans are developed. *E. polylepis* is potentially particularly vulnerable to intensive fishing due to their protogynous reproduction (age based sex reversal from female to male), particularly if stocks vary regionally in their resilience to exploitation. Furthermore, gear selectivity may reduce the reproductive capacity of these populations by removing the larger male fish in the older year classes and therefore decreasing the probability of successfully fertilisation (McIlawain et al., 2006). This study was weakened by a lack of critical biological data (e.g., spawning times and locations, age at maturity, catch rates and sex ratios), which made it difficult to design the sampling to maximise the statistical power for detecting structuring. This was further exacerbated by the problems of obtaining reasonable samples (e.g., 100) of adult fish, since the fork length of the sampled fish ranged from 215-755cm, which no doubt included both immature and mature fish. The presence of immature pre-recruit fish in the samples is likely to have further concealed any significant structuring in the adult stocks. As such, further genetic analysis based on better sampling is required before firm conclusions can be drawn about levels of genetic structuring. Furthermore, samples need to be collected from a larger geographical range, including neighbouring countries such as Yemen, Iran, and Pakistan, which share the same water mass (Varghese & Somvanshi 2001). This will help to determine the geographic scale at which significant genetic structuring occurs, and verify if the selected loci are sufficiently sensitive to detect it. A variety of other non-genetic tools can be applied to examine population structuring in fishes, including such approaches as the analysis of life-history traits and parasite loads, morphometry, artificial tagging, and micro-chemical analysis of hard parts. All these approaches can provide valuable complementary information on population structuring, which may reflect different degrees of stock separation, whether it be daily, seasonal, developmental, life-long or over many generations. Additional estimates of structuring based on a range of such markers are required for this species, combined with an in-depth survey of its biology. Indeed, in a similar large-scale study of fish on the Great Barrier Reef fish species by Dudgenon et al. (2000), such data has highlighted age demographic differences in scarid fishes between populations across the continental shelf where genetics has failed to detect structuring.

Future Perspectives

1. The results of the present study on *E. polylepis* suggested that populations in the Arabian Sea and the Gulf of Oman should be managed as a single stock.
2. Data were collected over a longer time-period than would have been optimal. Samples were also collected from fish markets, and consequently there may be some

doubt as to the exact locations were sample fish were captured. It is thus suggested that parts of the study be repeated, particularly for the Muttrah and Duqum sites, because these are important landing areas for grouper.

3. Analysis for mitochondrial DNA should be attempted, as alternative markers to nuclear DNA.
4. There are at least 34 species of the subfamily Epinephelinae in Oman (Randall, 1995), and some of these have been wrongly assigned to species level. A genetic comparison among species would be useful from a taxonomic perspective.
5. Morphometric and meristic studies are suggested to support genetic evidence.
6. This study was the first of its kind in Oman, and did not succeed in answering all the key questions. Rather, it has raised further questions, which may form the basis of continued genetic studies in the region.

ACKNOWLEDGMENTS

I thank my supervisor Dr. Bill Hutchinson and my other supervisors, Dr. Ian Cowx and Dr. Bearnd Heanfling at Hull University, and Dr. Jennifer Mcllawain from Sultan Qaboos University.

REFERENCES

Al-Abdessalam, T. Z. (1995) Marine species and Fisheries Centre of the Sultanate of Oman. An identification Guid, Ministry of Agriculture and Fisheries, Sultanate of Oman., Publication No. 46195.

Carvalho, G. R. & Pitcher, T. J. (1995) Molecular genetics in fisheries London, Chaman & Hall.

Chapman, R. W., Sedberry, G. R., Koenig, C. C. & Eleby, B. M. (1999) Stock Identification of Gag, *Mycteroperca microlepis*, Along the Southeast Coast of the United States. --> 1: 137-146. De Innocentiis et al., 2001.

DeSalle, R. & Shearwater, B. (1998) Molecular approach to ecology and evolution.

Dudgeon, C. L., Gust, N. & Blair, D. (2000) No apparent genetic basis to demographic differences in scarid fishes across the continantal shelf of the Great Barrier Reef. *Marine Biology* 137: 1050-1066.

Elloit, A. J. & Savidge, G. (1990) Some features of the upwelling of Oman. *J. Mar. Res.,* 48: 319-333.

Heemstra, P. C. & Randall, J. E. (1993) FAO species catalogue, Grouper of the world (Family Serranidae, Subfamily Epinephelinae). *An Annotated and Illustrated Catalogue of the grouper, Rockcod, Hind, Coral Grouper and Lyretail Species Known to Date.*, FAO Fisheries Synopsis.

Hutchinson, W. F. (2001) PhD Thesis: Temporal and spatial genetic diversity in the North Sea cod (*Gadus morhua*). University of Hull.

Hutchinson, W. F., Carvalho, G. R. & Rogers, S. I. (2001) Marked genetic structuring in localised spawning populations of cod *Gadus morhua* in the North Sea and adjoining waters as revealed by microsatellites. *Marine Ecology Progress Series* 223: 251-260.

James, M.K., Armsworth, P.R., Mason, L.B. & BL. (2002) The structure of reef fish metapopulations: Modelling larval dispersal and retention patterns. *Proceedings of the Royal Society of London, Series B* 269: 2079–2086.

Johns, W. E., Jacobs, G. A., Kindle, J. C., Murray, S. P. & Carron, M. (2000) Arabian Marginal Seas and Gulfs (AMSG). Report of a workshop held at stennis space center, Mississippi, p. 43. Sponsored by the office of naval research, University of Miami RSMAS.

Kalinowski, S. T. (2005) Do polymorphic loci require large sample sizes to estimate genetic distances? *Heredity* 94: 33-36.

Kindle, J. C. & Anone, R. A. (2001) Areview of the surface circulation of the Northern Arabian Sea. *In Claereboudt, M., Goddard, S., Al-Oufi, H., and Mcliwain, J. eds. Proc. 1st International Conference on Fisheries, Aquaculture and Environment in the NW Indian Ocean, Sultan Qaboos University, Muscat, Sulatante of Oman,*: 113-122.

Kindle, J. C. (2001) The surface circulation of the Northern Arabian Sea. *Naval research lab stennis space center MS Oceanography div.*

Kinnison, M. T., Bentzen, P., Unwin, M.J. & Quinn, T.P. (2002) Reconstructing recent divergence: evaluating nonequilidrium population structure in New Zeland chinook salmone. 11. *Mol. Ecol.* 11: 739-754.

Kornfield, I. (2004) Genetic differentiation among Atlantic cod (Gadus morhua) from Browns Bank, Georges Bank, and Nantucket Shoals.

Largier, J. L. (2003) Considerations in estimating larval dispersal distances from oceanographic data. *Ecological Applications* 13: S71–S89.

Leis, J.M. & Carson-Ewart, B.M. (1998) Complex behaviour by coralreef larvae in open-water and near-reef pelagic environments. *Environ Biol Fishes* 53: 259–266.

Lough, R. G. & Bolz, G. R. (1989) The movement of cod and haddock larvae onto the shoals of Georges Bank. *Journal of Fish Biology* 35(Suppl. A): 71-79.

McIlawain, J., Hermosa, G. V., M., C., Al-Oufi, H.S. & Al-Awi, M. (2006) Spawning and reproductive patterns of six exploited finfish species from the Arabian Sea,

McPherson, A. A., O'Reilly, P. T. & Taggart, C. T. (2003) Genetic differentiation, temporal stability, and the absence of isolation by distance among atlantic herring populations. *Trans. Amer. Fish. Soc.* 133: 434-446.

Molony, B. W., Lenanton, R., Jackson, G. & Norriss, J. (2003) Stock inhancment as a fisheries management tool. *Review in fish biology and fisheries* 13: 409-432.

Morris, A. V., Roberts, C. M. & Hawkins, J. P. (2000) The threatened status of grouper (Epinephelinae). --> 9: 919-942.

Nielsen, E. E., Hansen, M. M. & Meldrup, D. (2006) Evidence of microsatellite hitch-hiking selection in Atlantic cod (Gadus morhua L.): Implications for inferring population structure in nonmodel organisms. *Molecular Ecology* 15 (11): 3219-3229.

Nugroho, E., Takagi, M., Sugama, K. & Taniguchi, N. (1998) Detection of GT repeats microsatellite loci and their polymorphism for grouper of the genus *Epinephelus*. --> 64: 836-837.

Randall, J. E. (1995) Coastal fishs of Oman, University of Hawaii Press.

Randall, J. E. & Heemstra, P. C. (1991) Revision of the Indo-Pacific groupers (Perciformes: Serranidae: Epinephelinae), with description of five new species. --> 20: 1-296.

Rhodes, K. L., Lewis, R.I., Chapman, R.W. & Sadovy, Y. (2003) Genetic structure of camouflage grouper, *Epinephelus polyphekadion* (Pisces: Serranidae), in the western central Pacific. *Marine Biology* 142: 771-776.

Ricker, W. E. (1975) Computation and interpretation of biological statistics of fish populations. *Ball. Fish. Res. Bd. Can.* 191, 382.

Pawson, M.G. & Jennings, S. (1996) A critique of methods for stock identification in marine capture fisheries. *Fisheries research* 25: 203-217.

Rivera, M. A. J., Kelley, C. D. & Roderick, G. K. (2004) Subtle population genetic structure in the Hawaiian grouper, *Epinephelus quernus* (Serranidea) as revealed by mitichondrial DNA analyses. *Biological journal of the linnean society* 81: 449-468.

Rosenblatt, R. H. & Waples, R. S. (1986) A genetic comparison of allopatric populations of shore fish species from the eastern and central Pacific Ocean: dispersal or vicariance? *Copeia* 1986.

Ruzzante, D. E. (1998) A comparison of several measures of genetic distance and population structure with microsatellite data: bias and sampling variance. *Canadian Journal of Fisheries and Aquatic Sciences* 55: 1-14.

Ruzzante, D. E., Taggart, C. T. & Cook, D. (1998) A nuclear DNA basis for shelf- and bank-scalc population structure in northwest Atlantic cod (*Gadus morhua*): Labrador to Georges Bank. *Molecular Ecology* 7: 1663-1680.

Shaw, P. W., Pierce, G. J. & Boyle, P. R. (1999a) Subtle population structuring within a highly vagile marine invertebrate, the veined squid *Loligo forbesi*, demonstrated with microsatellite DNA markers. *Molecular Ecology* 8: 407-417.

Shaw, P. W., Turan, C., Wright, J.M., O'Connel, M. & Carvalho, G. R. (1999b) Microsatellite DNA analysis of population structure in Atlantic herring (*Clupea harengus*), with direct comparison to allozyme and mtDNA RFLP analyses. *Heredity* 83: 490-499.

Sluka, R. D., Chiappone, M. & Sullivan, S. K. M. (2001) Influence of habitat on grouper abandance in the Florida Keys, U.S.A. *Jurnal of Fish Biology* 58: 682-700.

Smith, P. J. (1994) Genetic diversity of marine fisheries resources: Possible impacts of fishing. Fisheries Research Center, Ministry of Agriculture and Fisheries, Wellington, New Zealand., FAO Fisheries Technical Paper.

Smith, W. & Morse, W. W. (1984) Retention of larval haddock Melanogrammus aeglefinus in the Georges Bank region, a gyre-influenced spawning area. *Mar. Ecol. Prog. Set.* 24: 1-13.

Swearer, S. E., Caselle, J. E., Lea, D. W. & Warner, R. R. (1999) Larval retention and recruitemnt in an island population of a coral-reef fish. *Nature* 402: 799-802.

Taggart, J. B., Haynes, R. A., Prodöhl, P. A. & Ferguson, A. (1992) A simplified protocol for routine total DNA isolation from salmonid fishes. *Journal of Fish Biology.* 40: 963-965.

Taylor, J. (1991) Gulf war impact on marine environment and species.

Ward, R. D. (2000) Genetics in fisheries management. *Hydrobiologia* 420: 191-201.

Williams, J. G. K., Kubelike, A. R., Livak, K.J., Rafalski, J. A. & Tingey, S. (1990) DNA polymorphisms by arbitrary primers are useful as genetic markers. *Nucleic Acids Research* 18: 6531-6535

Zatcoff, M.S., Ball, A.O. & Chapman, R.W. (2002) Characterization of polymorphic microsatellite loci from black grouper, *Mycteroperca bonaci* (Teleostei: Serranidae). *Molecular Ecology* 2: 217-219.

In: Biotechnology and Conservation … (2 Volume Set) ISBN: 978-1-63117-869-6
Editors: S. A. Farooq, R. Abed and S. Baqir

Chapter 8

Histopathological Effects of Variable Doses of Hydroxyurea in Male Swiss Rats

Fatima. Q. M. Al-Hayyali[1], Hamad. J. Al- Jumaily[2] and Hafidh. I. Al-Sadi[3]

[1,2]Dept of Biology, Science College, [3]Pathology Department, College of Veterinary Medicine, Mosul University, Mosul. Iraq

Abstract

The purpose of this study was to investigate the histopathological changes induced by variable doses of hydroxyurea (Hu) in male albino Swiss rats. Giving Hu in oral doses of 1, 5, 10, 15, 20, 25, 50, 100, 150, 200, 250, 500, 750 and 1000 mg/kg.b.w. for 4 weeks lead to the occurrence of degenerative and necrotic lesions in the liver, proliferative lesions in the lungs, degenerative and necrotic lesions in the kidneys, lymphoid depletion in the spleen and degenerative lesion in the testes. These lesions were more apparent with the dose 200 mg/kg. b.w. and they were more severe with the doses 250, 500, 750 and 1000 mg/kg. b.w. Furthermore, the lesions were more severe when the same doses were used for a longer period of time (8 weeks). From results of this study, it was concluded that the administration of Hu in doses higher than the normal dose induces histopathological changes in the liver, lungs, kidneys, spleen and testes, and that the severity of these changes -increased with the increase of the dose and time of exposure.

Keywords: Hydroxyurea, histopathological changes, pathological changes, male swiss rats, histological changes

Introduction

Hydroxyurea is an inhibitor of the enzyme ribonucleoside diphosphate reductase which catalyzes the reductive conversion of ribonucleotides to deoxyribonucleotides (Calabresi and Chabner, 2001). It acts specifically on the S phase of the cell cycle, in which concentrations of the target reductase are maximal, and it causes cells to arrest at the Gl-S interface (Schilsky et al., 1992). Therefore, the drug is used on wide base to treat malignant and nonmalignant diseases. Clinically, hydroxyurea has been used commonly as a myelosuppressive agent in the myeloproliferative syndromes particularly chronic granulocytic leukemia, polycythemiavera, and essential thrombocytosis (Sweetman, 2005; Kumer et al., 2012). It has been also used in corporation with irradiation, as it is able to synchronize cells into a radiation sensitive phase of the cell cycle. Hydroxyurea could also be useful in the treatment of adult patients with sickle cell disease (Charache et al., 1995; Almeida et al., 2012). The purpose of this study was to investigate the histopathological effects of variable doses of hydroxyurea in male Swiss rats. One hundred thirty five, 2.5-3 months old, male Swiss rats, weighting 200 - 250 gm were used. They were kept in 30X40X20 cm plastic cages in room in the Animal House, College of Veterinary Medicine, University of Mosul. The room temperature was 20-25C^0 and there was a light circle of 14 hours lightening and 10 hours darkness and the rats were given a standard concentrated standard food provided by the Animal House. Food and water were given to the rats *adlibitum*. In the first experiment, 75 rats were randomly allocated into 15 groups of 5 rats each. Group (1) was used as control. The other 14 groups were given hydroxyurea orally at daily doses of 1, 5, 10, 15, 20, 25, 50, 100, 150, 200, 250, 750 and 1000 mg/kg. b.w. for four weeks. In the second experiment, 60 rats were allocated randomly into 12 groups. Group (1) served as control and the other 11 groups were given hydroxyurea orally at daily doses of 1, 5, 10, 15, 20, 25, 50, 100, 150, 200 and 250 mg/kg. b.w. for 8 weeks. At the end of each experiment, the rats were euthanatized using ether, necropsied, and tissue specimens were collected from the liver, lungs, spleen, heart, kidney, testes, skin and the brain. The tissue specimens were fixed in 10% neutral formalin solution for histopathological study. After 48-72 hours, these tissue pieces were washed overnight in running tap water, dehydrated in ascending grades of alcohol, cleared in xylol and embedded in paraffin wax (60-62 C° melting point). Sections of 4-6 mm thickness were cut and stained with hematoxylin and eosin and trichrome stains as per the standard procedures recommended by (Kiernan, 1999).

The morphologic changes in rat's organs which treated with hydroxyurea at (200, 250, 500, 750, 1000mg/kg b.w.) showed pallor in liver and kidneys. While didn't appear any changes in hydroxyurea at (1, 5, 10, 15, 20, 25, 50, 100, 150 mg/kg b.w.). The clinical changes were include on alopecia, kyphosis, heamorrahge and enophthalmus, anorexia, nausea, disorders in motility, swelling and leg ulceration in rats which treated with hydroxyurea at (750,1000 mg/kg b.w.) while other doses didn't appear any changes of clinical changes (Figure 1).

Microscopic examination of sections prepared from the specimens of rats treated with the variable doses of hydroxyurea for 4 weeks revealed the occurrence of vacuolar degeneration and necrosis of the hepatocytes and the accumulation of mononuclear cells (particularly the lymphocytes) around the blood vessels (perivascular cuffing) (Figure 2). These lesions increased in severity with increase of the dose of hydroxyurea. The lungs showed emphysema, hyperemia, and the infiltration of mixed inflammatory cells in the interalveolar septa (Figure 3). In the kidneys, there were necrosis of the epithelial cells lining the proximal convoluted tubules and a moderate degree of glomerulonephritis (Figure 4). The spleen exhibited depletion of lymphocytes in the white pulp and in the testes there were edema in the intertubular tissue and only few spermatozoa. In the group that was given the dose 750 mg/kg b.w. for 4 weeks, skin ulcers were seen (Figure 5).

Figure 1. Morphological changes of rat's male which treated with hydroxyurea (750 mg/kg, b.w.). Showing heamorrahge and enophthalmus, anorexia, nausea, disorders in motility, swelling and leg ulceration.

In groups of rats that were treated with the variable doses of hydroxyurea for 8 weeks, there were severe vacuolar degeneration and necrosis of hepatocytes, disarrangement of the hepatic cords, and perivascular cuffing with lymphocytes. In the lungs there were emphysema, thickening of the interalveolar septa with focal mononuclear inflammatory cells, peribronchiolar accumulation of mononuclear inflammatory cells, and fibrous thickening of the interalveolar septa and walls of the bronchioles (Figure 6). The kidneys showed coagulative necrosis of epithelial lining of the proximal and distal convoluted tubules. Sections prepared from the spleen exhibited depletion of lymphocytes from the white pulp. In the testes there were edema in the interstitium and hyperplasia of the spermatocytes. In some cases (dose of 150 and 200 mg/kg b.w.) severe vascular degeneration and necrosis of the spermatocytes and desquamation of these cells into the lumen of the seminiferous tubules were visible (Figure 7). Sperms were not observed in these cases. Other histopathological changes included foci of necrosis in the pancreas, gliosis and vacuolar degeneration of the neurons in the brain, and focal areas of coagulative necrosis in the myocardium.

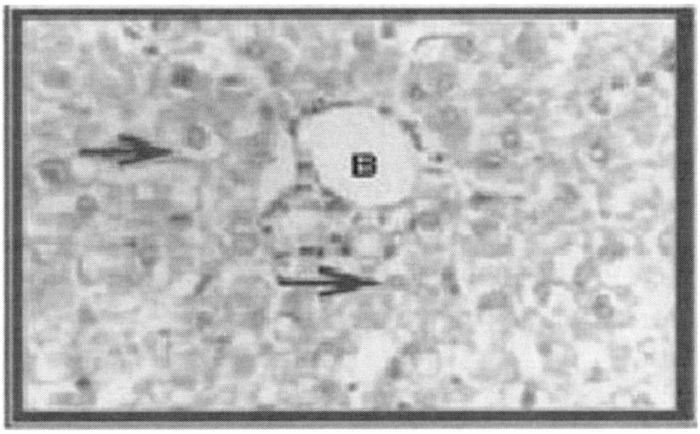

Figure 2. Photomicrograph of rat's liver section which treated with hydroxyurea (25 mg/kg, b.w. for 4 weeks). Showing vacuolar degeneration in hepatocytes cells (arrows) with a few of inflammatory cells (lymph cells) around blood vessels (BV). 400X. H&E,a).

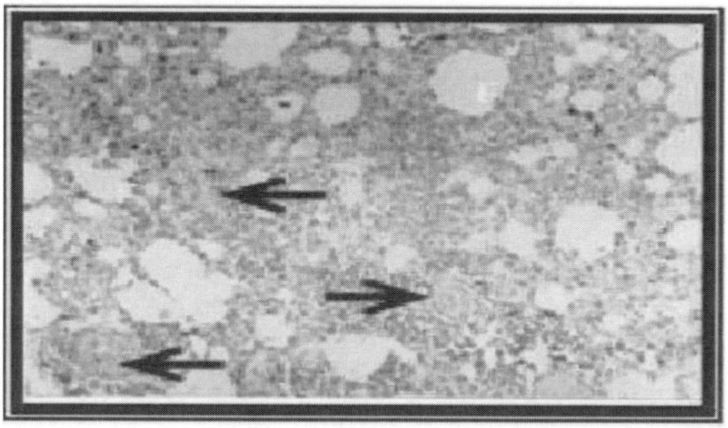

Figure 3. Photomicrograph of rat's lung section which treated with hydroxyurea (25 mg/kg, b.w. for 4 weeks). Showing congestion of blood vessels (arrows) Emphysema (E), interstitial pneumonitis with lymph, plasma and macrophage cells. 100X. H&E, b).

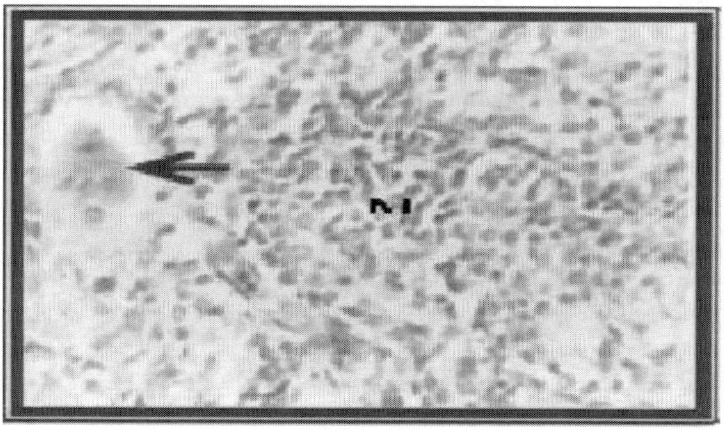

Figure 4. Photomicrograph of rat's kidney section which treated with hydroxyurea(50mg/kg, b.w. for (4weeks), showing necrosis in epithelium of proximal convoluted tubules and aggregation of epithelial cells inside some tudules (arrows), interstitial nephritis and glomerulitis (N). 400X, H&E,c).

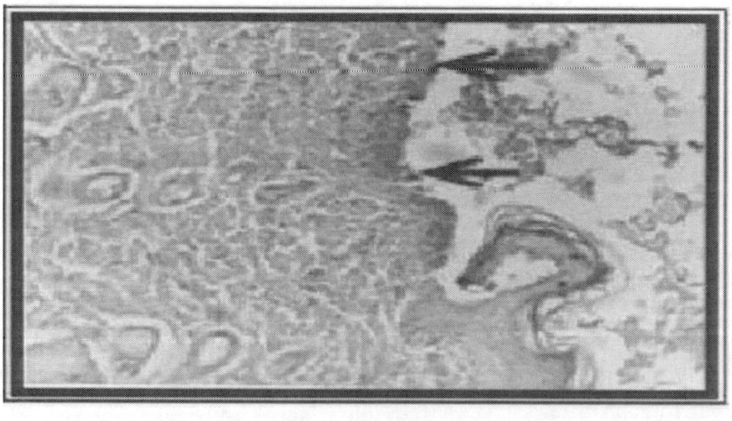

Figure 5. Photomicrograph of rat's skin section which treated with hydroxyurea (750 mg/kg, b.w. for 13 days, showing skin ulcer and necrosis in subepidermis tissue, 100X,H&E, d).

Figure 6. Photomicrograph of rat's lung section which treated with hydroxyurea (150 mg/kg, b.w. for 8 weeks). Showing thinking of trabeculi between pulmonary alveoli with inflammatory discharge and increasing of trabeculithikness and walls of respiratory bronchioles with collagenous fibers (arrows) Masson's Trichrometichneque, 100X,e).

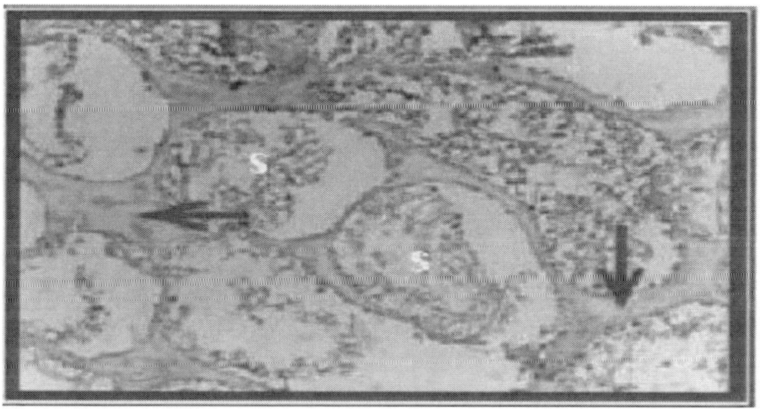

Figure 7. Photomicrograph of rat's testis which treated with hydroxyurea (150 mg/kg, b.w. for 8 weeks). showing degeneration and necrosis and desquamation of spermatocytes cells (S), apsence of sperm and oedema between the seminiferous tubules (arrows) 100X, H&E, f).

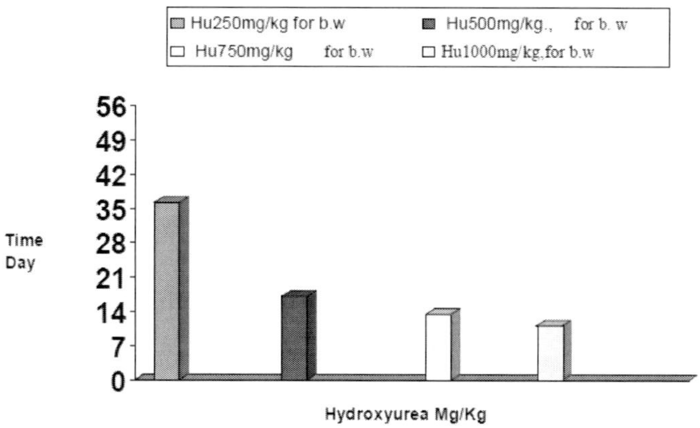

Figure 8. Ulstrating the effect of different doses from hydroxyurea on percentage of rat's death through 8 weeks. g).

The differences in the distribution and severity of lesions in the two groups of rats (4 and 8 weeks treatment) are presented in (Figure 8). From data presented in this table it could be observed that the lesions became, more extensive and more severe when the dose and time of exposure were increased.

In this study, low and high doses of hydroxyurea were used which administered to the rat's in the current investication. This study revealed that the doses of hydroxyurea (5, 10, 20, 50, 100, 150, 200mg/kg b.w) for 4weeks caused histopathological changes in liver, lung, kidney, testis, spleen and skin. These changes were increases with the increase of drug dose to 250,500,750 and 1000mg/kg b.w) for 4 weeks. Also the severity of these changes increases with the increase of the period of drug administration (for 8 weeks). These results are correlated with the findings of other investigators (IARG, 2000; Anonymous, 2010). The pathological changes of hydroxyurea in this research results from the side effects of this drug. The side effects of hydroxyureahave included bone marrow suppresion, gastrointestinal disturbaces, impairment of renal function, pulmonary oedema, pulmonary fibrosis, skin ulceration, dermatomysitis, neurologiclreacti (Best et al., 1998; Sweetman, 2005; Dissemond et al., 2006; Hoff et al., 2009) and inhibition of spermatogensis (Masood et al., 2007). In human medicine, two dosage schedules for hydroxyureahave been followed, (Data Sheet, 1999) intermittent therapy with 80 mg/kg administered orally as a single dose every third day, or continuous therapy with 20-30 mg/kg administered as a single daily dose (Data Sheet, 1999; Calabresi and Chabner, 2001). When hydroxyurea is used in combination with irradiation, the drug is given at a dose of 80 mg/kg b.w. as a single dose every third day (Data Sheet, 1999; Abrams and Goldsmith, 2001). In cases of resistant chronic myelogenousleukemia, hydroxyurea is used as a single oral dose of 20 - 30 mg/kg b.w. (Haskell and Berek, 2001). It has been found in human beings that the single dose of hydroxyurea that causes acute toxicity (mild nausea and vomition) and chronic toxicity (hyperpigmentation and bone marrow suppression) is 500-1500 mg/kg orally daily (Tomiyama et al., 1997; Tierney et al., 2004). In rats, the results of studies onsubacute or chronic toxicity of hydroxyurea indicated that giving hydroxyureaat a dose of 140- 240 mg/kg b.w. (or 140 - 1260 mg/kg b.w.) per weekand for a period of 12 weeks or a dose of 400- 800 mg/kg b.w. daily for7 - 15 days, or a dose of 1260 mg/kg b.w. daily for 37 days, or a dose of 2520 mg/kg b.w. daily for 40 days, it will lead to the appearance of specific pathological changes whose severity increases with increase of the dose (Chabner et al., 2001).

REFERENCES

Abrams, A.C. and Goldsmith, T.h. (2001). Clinical Drug Therapy. *Rationales for Nursing Practice*. 6th ed. Lippincott Williams and Wilkins, Company. Philadelphia. New York. P: 949.

Almeida, C. B; Scheiemann, C and Jang, J.E.(2012). Hydroxyurea and acGMP-amplifying agent have immediate benefits on acute vaso-occlusive events in sickle cell disease mice. *Blood*. 120(14): 2879-2888.

Anonymous, (2010). Report problems to the food and drug administration (Hydroxyurea). *Inc. Version*: 402. FDA Med watch.

Best, P.J. ;Daoud, M.S.; Pittelkow, M.R. and Petitt, R.M. (1998). Hydroxyurea induced leg ulceration in 14patient. *Ann. Intern. Med.*, 128(1):29-32.

Calabresi, p. and Chabner, B.A. (2001). Chemotherapy of Neoplastic Diseases In: *The Pharmacological Basis therapeutics.* Hardman, J.G.;Limbird, L.E. and Gilman, A.G. 4th ed. McGraw- Hill, New York. p: 1381.

Chabner, B.A.; D.P.; Paz- Ares, L.;Carbonero, R.G. and Calabresi, p. (2001). Antineoplastic agents In. *The Phamacology Basis of Therapeutic.* Hardman, J.G.; Limbird, L.E. and Gilman, A.G. (eds). McGraw- Hill, New York, pp: 13989.

Charache, S.;Terrin, M.h.; Moore, R.D.; Dover, G.J. McMahan, R.P.; Barton, F.B. and Eckert, S.V. (1995). The effect of hydroxyurea on the frequency of painful crises in sickle cell anemia. *N. Eng. J. Med.*, 332: 1317-1322.

Data sheet, (1999), *Hydrea, Hydroxyurea (hydroxyl-Carbamide)* file://A:Data%20Sheet.htm.

Dissemond, J. ;Hoft,D. and Knab,J.(2006)Leg ulcer in apatient associated with hydroxyurea therapy. *Int. J. Dermatol.* 2:58-60.

Haskell, C. and Berek, J.S. (2001). Cancer Treatment, 5th ed. W.B. Saunders. Company. Philadelphia. London. New York. p: 159- 327.

Hoff, N.P.; Diesel, A.J; Pippirs, U.;Schutte, K.W. and Hanneken,S. (2009) Kutane Nebenwirkugeneiner Hydroxyurea-Therapiebei Polycythamiavera. *Der Hautarzt* 60:783-787.

IARC (2000). Monographs on the evaluation of the carcinogenic risk of chemicals to man. Genova: World Health Organization, *International Agency For Research on Cancer.* (Multivolume Work). 76(364).

Kiernan, J.A. (1999). Histological and Histochemical Methods: theory and practice.3rd.ed., Oxford: Butter worth/ Heinemann, pp:111- 112.

Kumer, S.K.; LeeJ. H. and Lahucrta, J.J. (2012). Risk of progression and survival in myeloma relapsed after therapy with Lmids and botezomib:amulticenter international myeloma group study. *Leukemia.*26:149-157.

Masood, J.; Hafeez, A.; Hughes, A and Barua, J.M. (2007). Hydroxyurea therapy:arare cause of reversible azoospermia. *Int. Urol. Nephrol.*39:905-907.

Schilsky, R.L.; Ratain, M.J.; Vokes, E.E.; Vogelsang, N.J.; Anderson, J. and Peterson, B.A. (1992). Haboratory and clinical studies of biochemical by hydroxyurea modulating semin. *Oncol.* 19: 84.

Sweetman, S. C. (2005). Martindale. *The complete srug reference.* 33rd ed. London: The pharmaceutical press. PP: 544-545.

Tierney, L.M.; Mcphee, S.J. and Papadakis, M.A. (2004). *Current medical diagnosis and treatment.* 43rd ed. McGraw- Hill companies. Ins. pp 486- 610.

Tomiyama J.; Yano, K.; Uchino, S.; Ito, T.; Kudo, H.; Irimajiri, T.; Ohkubo, H.; Kanazawa, A. and Kawahara, Y. (1997). Bilateral leg ulcer with pathological evidence of small vessel vasculitis by skin biopsy during hydroxyurea therapy of chronic myelogenous Lenkemia. *Rinsho Ketsueki*, 38(3): 231-233.

In: Biotechnology and Conservation … (2 Volume Set) ISBN: 978-1-63117-869-6
Editors: S. A. Farooq, R. Abed and S. Baqir

Chapter 9

Relationship between Age, Breed, Body Weight, Gestational Status and Parity on the Body Condition Score of Omani Breeding Camels

S. Mathan Kumar [1*]**, *E. H. Johnson*** [1] ***and D. A. El-Badry*** [2]
[1]Department of Animal and Veterinary Sciences, College of Agricultural and Marine Sciences, Sultan Qaboos University, Muscat, Sultanate of Oman
[2]Assayel-Al Fairuz Farm, Al- Suwaiq, Sultanate of Oman

Abstract

Body condition score (BCS) is a qualitative assessment of the quantity of available energy from an animal's fat storage that can be used for reproduction, lactation and is an important herd management tool. There is an association of lower BCS of female camelids to reduced ovarian activity resulting in reduced reproductive efficiency. Therefore, the factors influencing BCS of breeding camels are to be elucidated. The aim of the present study was to observe the influence of age, breed, body weight, gestational status and parity on BCS of Omani breeding camels. Unlike results reported for cattle that documented that BCS is influenced by breed, age and body weight, in the present study our preliminary results indicate that in camels BCS is influence primarily by body weight. To the best of our knowledge this is the first study in relation to this subject in Oman and elsewhere on breeding camels managed on unified husbandry management.

Keywords: BCS, body condition score, camelid reproduction, dromedary camels, Omani camels, Omani camel breeds, reproduction

[*] Corresponding author: email: mathan@squ.ed.om.

Reproduction in dromedary camels is complicated by the typicality of short breeding season, length of follicular wave, coitus induced ovulation, longer gestation period and longer weaning period of the camel calves resulting in reduced reproductive efficiency (Tibary, A, and Anouassi, A. 2000). Nutritional influence on reproduction is well documented in all production animals and is been considered as the most controllable factor of reproduction. There is a great body of scientific literature is available for dairy cattle in relation to the factors that influence BCS and the bench mark status of desired BCS at different reproductive and productive phases.

However this information is not available for camels and it cannot be extrapolated. Therefore, the aim of the present study was to observe the influence of age, breed, body weight, gestational status and parity on BCS of Omani breeding camels. BCS of 87 she camels was recorded by a single operator on a scale of 1 to 5. Score of 1 = emaciated and score of 5 = extremely fat with increments of 0.25 as adopted by Sghiri, A and Driancourt MA (1999). The independent variables were grouped into categories. There were five age categories (years) (≤ 5,6-10,11-15,16-20 and 21-25 years), four body weight (BW) categories(kg) (≤ 400, 401-500, 501- 599 and ≤ 600kg), gestational status as five categories (months) (≤ 3,3-6,6-9,9-13 months and non-pregnant), parity as five categories (heifer, first, second, third and above 3) and four BCS categories (≤2.5, 3.0, 3.25 and ≥3.5). The initial results were analysed in Microsoft Excel^{R}2010, using general tools as filtering. Ages varied from≤5 years (n=9), between 6-10 years (n=31), between 11-15 years (n=32), between16-20 years (n=13) and between 21-25 years (n=2). There were thirteen different breed types observed. Arja (n=4), Musaiha (n=14), Buwadah (n=5), Al-Bahree (n=5), Zabeia (n=2), Shabbar (n=2), Samha (n=18), Al-Kawara (n=9), Gazella (n=7), Al Derehiah (n=2), Farhan (n=1) Salalah (n=16) and Al-Sudani (n=2).

The number of camels in the respective categories was: BW ≤400 kg (n=12), 400-500 kg (n= 46), between500-600 kg (n=26) and BW≥600 (n=3). Gestational status as ≤ 3 months of pregnancy (n=28), 3-6 months (n=5), 6-9 months (n=5), ≥ 9months (n=8) and non-pregnant animals (n=41). There were 10 heifers, on first parity (n=9), second parity (n=13), third parity (n=27) and above third parity (n=28). The observed BCS were: ≤2.5 (n=22), 3.0 (n=28), 3.25 (n=4) and ≥3.5 (n=33). ANOVA was carried out to determine the influence of all the independent variables on the BCS. Our preliminary results report that, there is a significant influence of BW on BCS.

Gestational status should hypothetically influence greater than the other variables such as age, breed, body weight and parity on BCS as the additional energy requirements to the growing foetus are to be met by increased dietary intake during pregnancy. The herd of the present study follows a unified husbandry system (housing, zero grazing) and nutritional supplements are adjusted against their reproductive status (from just prior to breeding of she camels and until the time of calf weaning) such as provision of wheat bran, barley mixture and dates in addition to the fresh alfalfa and Rhodes hay. Further the data has to be analysed in detail to see the effect of class of variables on BCS by statistical tests such Kruskal-Wallis test and linear regression.

Figure a. Camel with BCS of 4.0.

Figure b. Camel with a BCS of 1.75.

REFERENCES

Sghiri, A. & Driancourt, M. A. (1999). Seasonal effects on fertility and ovarian follicular growth and maturation in camels Camelus dromedarius. *Anim Reprod Sci*, *55*, 223-237.

Tibary, A. & Anouassi, A. (2000). Reproductive Disorders in the Female Camelid In: Recent Advances in Camelid Reproduction, Skidmore L. and Adams G.P. (Eds.). International Veterinary Information Service, Ithaca NY (www.ivis.org)

In: Biotechnology and Conservation … (2 Volume Set) ISBN: 978-1-63117-869-6
Editors: S. A. Farooq, R. Abed and S. Baqir

Chapter 10

First Report about Growth, Partial Record Egg Production and Morphological Characters of a Newly Characterized Native Saudi Chicken Lines Hajar 1 and Hajar 2

A. S. Ahmed[1,2] *and A. H. Al Abbad*[1]
[1]Department of Animal and Fish Production,
College of agriculture and food science
King Faisal University, Al Ahsaa, Saudi Arabia
[2]Animal production department, College of agriculture,
Cairo University, Egypt

Abstract

This report is the first of its kind to describe the productive and phonotypic characteristics of first-hatch birds in two recently established Saudi locally-owned chicken lines, Hajar 1 and Hajar 2. As such, the report forms part of a large sustainable conservation plan for locally-owned chicken production. The majority of the locally-owned chicken gene pool in KSA is represented by the parents of those chickens collected for the current study during a previous project in the late nineties. Three hundred and sixty-four dam and 26 sired chickens were divided equally for use as parents in this study. After the hatch, the yields of Hajar 1 and Hajar 2 were 732 and 789 chicks respectively. Weekly body weight, biweekly weight gain, weight and age at first egg, and number of eggs per 90 days of production were all recorded for each chicken. Phenotypic characterization of feather color, shank color, and comb ecotype was performed. Results indicated that body weight at six weeks of age, body weight at first egg, and age at first egg where similar between the two lines. However, Hajar 1 birds were significantly superior in egg production and recorded a significantly lower total weight gain. Black feather color was dominant (96.5%) in Hajar 1 birds, while brown feather color was dominant (90.71%) in Hajar 2 birds. Also, 60.13% of Hajar 1 birds showed single comb, compared to 41.35 % for birds in Hajar 2. However, buttercup comb made up 30.76% of birds in Hajar 1 and 46.41% in Hajar 2. V-shaped comb was the lowest in both lines. Hajar 1 birds distinguished by their black shank accounted for 88.11%, against 45.99 %

for those of Hajar 2. The study suggested that the two locally-owned chicken lines have high genetic potential. They are mutually diverse in certain production indicators, which might indicate their validity for both conservation and improvement as locally-owned lines of production.

Keywords: Local Saudi Chicken, characterization, conservation, production

INTRODUCTION

Conservation of indigenous livestock breeds possessing genetic variation specific to their particular environment is essential for sustainable development. In spite of their relatively small number, locally-owned chickens, in addition to a high adaptability to their natural environment, are also an integral part of local life (Eltayeb, 2009). Unlike pure breeds and hybrids, which have a limited, selected number of economically important genes, the gene pools of locally-owned chicken populations probably contain all possible genes in Gallus gallus domesticus (Msoffe et al., 1998; Mwalusanya et al., 2001). Rural chicken ecotypes which have evolved in and adapted to stressful environments are likely to carry valuable genes and gene combinations controlling specific behavioral, physiological, and disease and parasite resistance traits (Giovambattista et al., 2001).

A survey conducted by Al-Yousef (2007) indicated that about 23% of the owners of large Saudi chicken farms raised chickens as their primary income source. Seventy-seven percent of owners considered raising chickens a secondary economic activity (8% were farmers, 23% were governmental employees, and 46% were merchants). Less attention is currently paid to the genetic characterization and conservation of locally-owned Saudi chickens. For instance, due to the reduction in the number of locally-owned Saudi chickens, a statistic provided by the Saudi Ministry of Agriculture in 2010 reported the total poultry wealth in Saudi Arabia as 444,680,000 birds, without mentioning the number of locally-owned chickens (Ministry of Agriculture, KSA, 2010). In 2008, the Animal Production and Health division issued a national poultry sector review for Saudi Arabia. Regarding village or backyard production, the author stated that "this information has not yet been sourced" (FAO, 2008).

Various studies have been conducted using locally-owned Saudi chickens. These include assessing the effect of water restriction (Alamer and Ahmed, 2011), describing immunological and productive performance (Ahmed, 2010), attempting to enhance performance through feed modulation (Ahmed et al., 2009), phenotypic characterization of locally-owned birds in different areas of the kingdom and analysis of their potential improvement (Al-Yousef, 2007; Al-Yousef et al., 1999), evaluating egg production performance (Al-Aqil, 1998; Najib, 1994), investigating the effects of nutrition on modulate fertility and hatchability (Alsobayel 1992), studying rearing regimes and their effect on egg quality parameters (Attia et al., 1991), and evaluating egg weight (AlSobayl, 1986). The research has provided variable measurements with regard to performance and characterization. The difference in results is thought to be due to such factors as location, genetic background of the birds, experiment design, and different environmental conditions. The current report relates to first-generation chickens from a base population of adult, locally-owned chickens used for a valuable piece of research undertaken in the late nineties (Al-Yousef et al., 1999). In 1999, the chicken population collected by Al-Yousef et al.

represented the majority of the locally-owned chicken gene pool in KSA. Since the completion of this research, the population has been retained at an experimental station in King Faisal University (KFU). For the purposes of this research, two plumage phenotypes within this population – the black feather line (Hajar1) and the brown feather line (Hajar2) - have been used to produce the current experimental generation. This report is the first report to assess first hatch growth, partial record egg production, and morphological characteristics for a base population of Saudi locally-owned chicken lines seeking genetic conservation.

2. Materials and Methods

2.1. Parent Population

Adult, locally-owned (Baladi) chickens at the KFU experimental station - the harvest of previous research (Al- Yousef et al., 1999) – were obtained. Two plumage phenotypes were characterized: black feather with black shank color (line 1), and brown feather with yellowish shank color (line 2).

The base population consisted of 182 25-week-old dams and 13 sires for each line. Birds in both lines were wing-banded and distributed equally and at random to 13 breeding floor pens (14 dams and one sire per bin) in an open-house system provided with feeders, drinkers, wood shavings for litter, and trap nests. The birds were exposed to 17 hours/day of light and seven hours/day of darkness.

Feed and water were available *ad libitum*. The birds were fed a commercial layer diet with 16% protein and a ME of 2700 Kcal/kg feed. The regular experimental station vaccination system was implemented with slight modifications to its schedule.

2.2. Offspring

The first hatch of the current offspring population was used in this research. Fertile eggs were produced and collected using trap nests. Each egg was then labeled by sire and dam numbers for pedigree. A total number of 1,690 eggs from both lines were sprayed with a sanitizer and stored in a cold room at 18°C for 10 days. After storage, the eggs of each pullet were set separately in a forced air incubator. At hatching, chicks were wing-banded and placed in a cage system for optimal brooding conditions. The birds were vaccinated against prevailing diseases in the area. Female birds were moved to production cages for individual egg production at age 18 weeks.

During their first five weeks of life all chicks were fed a commercial starter layer ration (20% crude protein and 2800 kcal ME/kg). From age six weeks to 17 weeks of age, they were fed a commercial growth ration (14% crude protein and 2700 kcal ME/kg). Thereafter, the birds received a commercial layer diet with 16% protein and ME of 2700 Kcal/kg feed. Feed and water were provided *ad libitum*.

2.3. Parameters and Data Collection

The hatches from line 1 (Hajar 1) and line 2 (Hajar 2) were 732 and 789 chicks respectively. Day- old body weights were recorded for all chicks after hatching and wing-banding. Body weights were recorded at 2, 4, 6, 8, 10, 12, 14, 16, and 18 weeks of age. Biweekly weight gains were then calculated. Aged 12 weeks, the chicks were phenotypically characterized through physical examination in order to determine sex, plumage color, shank length, and comb ecotype. At sexual maturity, body weight at first egg, age at first egg, and number of eggs per 90 days of production were recorded for each hen.

2.4. Statistical Analysis

The data from this study were subjected to a two-way analysis of variance in the effects of line, sex, and the interaction of these factors. Data were analyzed using the general linear model procedure of SAS software (SAS 2000). Descriptive statistics were applied for phenotypic characterization in the current study. A P value of ≤ 0.05 was considered statistically significant.

RESULTS

Mean body weights indicated that there were no significant differences between males and females in both lines, starting from the 6th week of age (Figure 1).

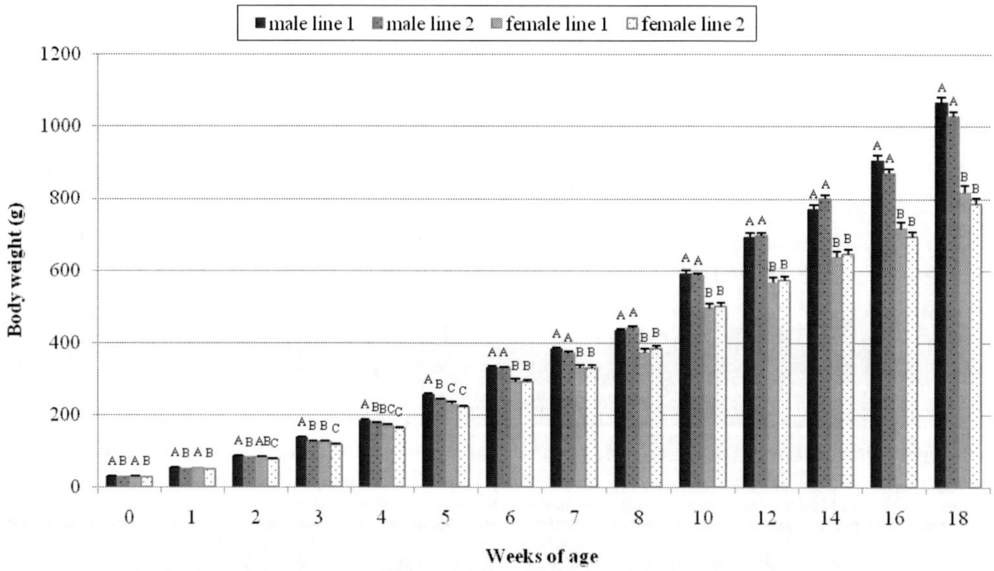

[a,b,c] columns within a week with different superscript differ significantly (P ≤0.05).

Figure 1. Weekly body weight of the Saudi chicken lines Hajar 1 and Hajar 2.

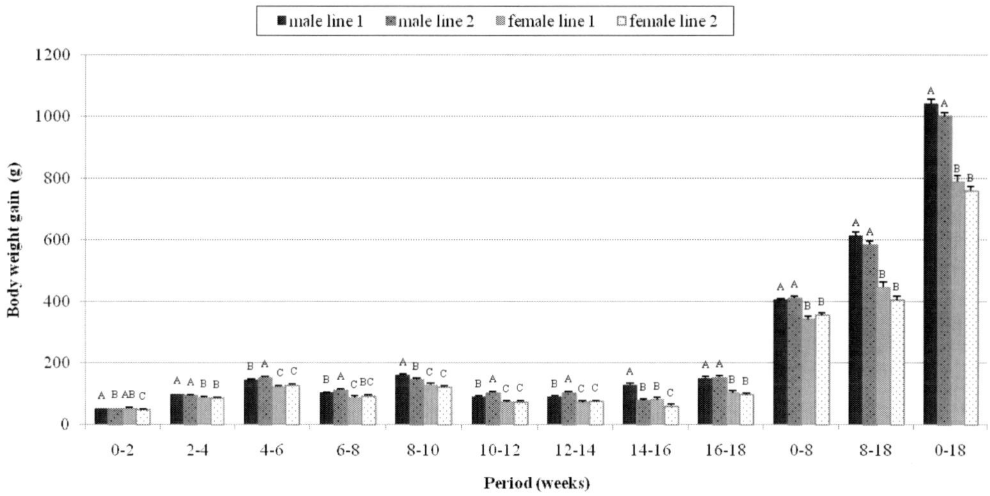

a,b,c columns within a period with different superscript differ significantly ($P \leq 0.05$).

Figure 2. Bi-weekly weight gain of the Saudi chicken lines Hajar 1 and Hajar 2.

From the third week of age, males in both lines were heavier than females in the same line as themselves. There were no differences in body weight between male and female chicks in both lines during the first week of age. Until the end of the first week of age, line 1 male and female chicks were heavier than those of line 2 (Figure 1). Weight gain results demonstrated that the differences in biweekly weight gain varied from one period to the next (Figure 2). A clearer picture can be achieved by looking at 0-8, 8-18, and 0-18 week weight gain statistics. These indicate that the male birds in both lines weighed more than the female birds.

Total weight gain from day-old to first egg showed significant ($P < 0.05$) differences between the two lines, with Hajar 2 recording the higher figure of the two (Table 1).

Table 1. Egg production parameters of the Saudi chicken lines Hajar 1 and Hajar 2

Trait	Hajar 1 (Line 1)	Hajar 2 (Line 2)
Age at first egg (AFE)	161.49 ± 9.01^{A}	160.64 ± 6.52^{A}
Body weight at first egg (BWFE)	1151.54 ± 18.85^{A}	1195 ± 13.84^{A}
Egg production (eggs)[1]	59.51 ± 1.61^{A}	54.56 ± 1.19^{B}
Total Weight gain[2]	1122.41 ± 18.72^{B}	1170.37 ± 13.86^{A}

a,b values within a parameter with different superscript differ significantly ($P \leq 0.05$).

[1] Egg production calculated as number of eggs produced in 90 days partial record.

[2] Total weight gains calculated as female weight gain from day old till body weight at first egg.

There were no differences between the lines for age at first egg and body weight at first egg (Table 1). However, Hajar 1 hens produced a significantly ($P < 0.05$) higher number of eggs within the first 90 days of production than those in Hajar 2 (Table 1). Black (96.50%) and brown (90.71%) were the dominant feather colors in Hajar1 and Hajar 2 respectively (Table 2) (Figure 3). Of the Hajar 1 birds, 60.13% had a single comb shape (Figure 3), 32.09% had a buttercup comb shape, and 9.87% showed V-shaped comb. In Hajar 2, 46.41%

of birds had a buttercup comb shape, compared to 41.35% for single comb, and 9.87% for V-shaped comb (Table 2). The black shank was dominant in Hajar 1 birds, especially for female birds (98.38%), while Hajar 2 birds showed similar percentages of black and white shank colors (Table 2).

Table 2. Some morphological characterizations of the Saudi chicken lines Hajar 1 and Hajar 2

Trait	Description	Hajar 1 (Line 1)			Hajar 2 (Line 2)		
		Male	Female	Combined sexes	Male	Female	Combined sexes
Feather colour	Black	93.82%	100%	96.50%	0%	6.40%	3.37%
	Brown	4.93%	0%	2.79%	91.96%	89.60%	90.71%
	White	0%	0%	0%	2.67%	0%	1.26%
	Mixed	1.23%	0%	0.69%	5.35%	4.00%	4.64%
Comb type	Single	58.02%	62.90%	60.13%	35.71%	46.4%	41.35%
	Buttercup	32.09%	29.03%	30.76%	50.00%	43.20%	46.41%
	V-Shape	9.87%	8.06%	9.09%	14.28%	10.40%	12.23%
Shank colour	Black	80.24%	98.38%	88.11%	38.39%	52.80%	45.99%
	White	17.28%	1.61%	10.48%	50.89%	39.20%	44.72%
	Yellow	2.46%	0%	1.39%	10.71%	8.00%	9.28%

(a) (b) (c)

Figure 3. (a) Male and female Saudi chicken line Hajar 1; (b) Male and female Saudi chicken line Hajar 2.

DISCUSSION

Despite the introduction of exotics breeds, genetically unimproved, locally-owned chickens remain predominant in rural areas. This is thought to be due to their natural ability to tolerate the harsh nature of their environment. Research into the characterization of locally-owned Saudi chickens is relatively scarce. Those few papers to have been published mainly describe locally-owned chickens according to their geographical location, and the last such paper was in 2007 (Al-Yousef, 2007).

International research on the Saudi poultry sector mentions that there is no existing data concerning locally-owned Saudi chickens (FAO, 2008). This situation brought to our attention the importance of conserving locally-owned Saudi chicken lines, especially in light of the fact that these chickens represent the majority of the kingdom's total chicken gene pool (Al- Yousef et al., 1999).

The current project was focused on the general characteristics of two main indigenous ecotypes that have been recently detected in the locally-owned Saudi chicken population. The significantly higher egg production of the Hajar 1 line, coupled with its lower total weight gain, indicates a potential genetic diversity between the two lines. The average egg production of the chickens in this study falls into the same range as that of locally-owned Jordanian chickens (Abdelqader et al., 2007). There were some similarities between birds of the same sex, regardless of the line. For instance, there were similarities in body weight from the sixth week of age.

Similarities in body weight gain from the 16th week of age were also apparent, and may be understood to come about due to the absence of the maternal effect. Weight gain at this stage of the birds' life could equally be explained as a consequence of hormonal changes. Generally, the average body weight in both lines was lower than the reported average body weight of locally-owned chickens in Ethiopia and South Africa (Halima et al., 2007; Marle-Köster, 2008) and similar to that of locally owned chickens in Nigeria and Asia (Ajaya, 2010; Abdelqader et al., 2008).

Diversity in production traits was accompanied by feather, comb type, and shank color phenotypic diversity. Significant morphological variations between the two lines showed great discriminatory potential - one assumed to reflect the high genetic potential previously described (Abdelqader et al., 2008). The next step of our research project - genetic assessment of locally-owned breeds – is currently being undertaken. It is important that the locally-owned Saudi chicken population and its valuable gene pool are conserved. Such an approach is consistent with worldwide attitudes towards future breed improvement. This trend could be made more efficient by increasing the use of molecular genetics applications in poultry production (Fulton, 2008).

In conclusion, the primary function of this study was to demonstrate the production and morphological traits of two locally-owned Saudi chicken lines. The results indicated the potential use of current lines for further investigation, in addition to their potential use in genetic conservation and improvement. The reported diversity between the two lines could be considered advantageous for further studies in this field.

ACKNOWLEDGMENTS

The authors gratefully acknowledge the Deanship of Scientific Research at King Faisal University (KFU) for its financial support. The authors also acknowledge the efforts of Dr. Yousef Al-Yousef, Dr. Huthail Najeeb and their research team - the founders of the base population used for this research. Thanks are also extended to Dr. Fahad Al Hur of the Agriculture Research Station for his support in conducting this research. Also, the Animal and Fish Production department at the College of Agriculture and Food Science, KFU, for encouraging our decision to adopt this approach. And finally, thanks to Abdullah Alnazr, Hassan Bogrein, Jaafar Al Hamada, Zafer Hakami and all the staff at the poultry research unit for their technical assistance.

REFERENCES

Abdelqader A., C.B.A. Wollny and M. Gauly 2007. Characterization of local chicken production systems and their potential under different levels of management practice in Jordan. *Trop. Anim Health Prod.* 39: 155-164.

Abdelqader A., C.B.A. Wollny and M. Gauly 2008. On-farm investigation of local chicken biodiversity and performance potentials in rural areas of Jordan. *Animal Genetic Resources Information* 43: 49-57.

Ahmed A. S. 2010. Immunological and productive performance of local saudi chicken differing in their primary response to SRBC antigen. *Proc. 3rd scientific conference of animal wealth research in middle east and north Africa*. pp123-134.

Ahmed A. S., Y.M. Al-Yousif and H. Najib 2009 Effect Of Dietary High Levels Of Heat Treated Locally Produced Flaxseeds On Body Weight Gain And Immune Response Of Pre-Layer Local Chicken. *Egyptian journal of poultry science*. 29: 821-835.

Ahmed A.S. and M. A. Alamer 2011 Effect of short-term water restriction on body weight, egg production, and immune response of local and commercial layers in the late phase of production. Asian Australasian journal of animal science Asian-Aust. *J. Anim. Sci.* 6: 825 – 833.

Ajaya F.O. 2010. Nigerian indigenous chicken: A valuable genetic resource for meet and egg production. *Asian Journal of Poult. sci.*4: 164-172.

Al- Yousef Y.M., H. Najib, F. Al Mulhim, and M.H. Sadek 1999. studying the productivity of native chickens and the possibilities of their genetic improvement in the Kingdom of Saudi Arabia. Final report, project No AR -12-72 King Abdul Aziz city for science and technology (KACST).

Al- Yousef, Y.M. 2007. A Survey Study on the Distribution of Saudi Baladi Chickens and Their Characteristics. *International Journal of Poultry Science* 6 (4): 289-292.

Al-Aqil, A. M. 1998. Biological and physiological responses of Baladi and exotic layers to seasonal changes in temperature. A thesis submitted in partial fulfillment of the requirements for Master's degree. Dept. of Animal sciences, King Faisal University.

Al-Sobayel, A.A. 1986. some egg quality characteristics of Baladi chaicken kept under two housing systems in Saudi Arabia. *J. College. Agric.*, King Saud University, Vol 8, No. 1, pp79-87.

Alsobayel, A.A. 1992. Effect of protein rearing diet and age on fertility and hatchability parameters of Saudi Arabian Baladi chickens. *J. King Saud Univ.,* Vol. 4, Agri. Sci., 1: 47-54.

Attia, F.M., A.A. Alsobayel, M.S. Bayoumi and M.A. El-Badry 1991. Effect of rearing regimen and age of bird on egg weight and weight of the component parts of Saudi Arabian Baladi chicken eggs. *J. King Saud Univ.*, Vol. 3, Agri. Sci., 1: 43-55.

El Tayeb N.M.N. 2009. Phenotypic characterization of native chicken of Sudan. M.Sc. thesis. Faculty of Animal Production university of Khartoum, Al Sudan.

FAO 2008. FAO animal production and health division emergency center for transboundary animal diseases socio economics, production and biodiversity unit, "*Poultry sector country review- Saudi Arabia*" September 2008.

Filton J.E. 2008. Molecular genetics in a modern day breeding organization. *World's Poult. Sci. J.* 64: 171-176.

Giovambattista, G., M.V. Ripoli, P. Peral-Garcia, and J.L. Bouzat 2001. Indigenous domestic breeds as reservoirs of genetic diversity: the Argentinean Creole cattle. *Animal Genetics* 32: 240 – 247.

Halima H., F.W.C. Neser, E. van Marle-Koster and A. D. Kock 2007. Phenotypic variation of native chicken population in northwest Ethiopia. *Trop. Anim Health Prod.* 39: 507-513.

Marle-Köster E. van, C.A. Hefer, L.H. Nel and M.A.M. Groenen 2008. Genetic diversity and population structure of locally adapted South African chicken lines: Implications for conservation. *South African Journal of Animal Science* 38:271-281.

Ministry of Agriculture, KSA. 2010. Indicators for agriculture Kingdom of Saudi Arabia, twenty fourth edition, Department of studies planning and statistic. pp 5.

Msoffe, P.L.M., M.M.A., Mtambo, U.M., Minga, P.S., Gwakisa, A.M., Katule, S.K., Mutayoba, S.K., Mdegela, R.H., Mwalusanya, M.G.S., Yongolo, and J.E. OLSEN, 1998. Scavenging local chickens in Tanzania: Is there room for improvement. Proceedings of the Tanzania Veterinary Association Scientific Conference, *Tanzania Veterinary Journal* 18 (Suppl. 4): 117 – 126.

Mwalusanya, N.A., A.M., Katule, S.K., Mutayoba, M.M.A., Mtambo, J.E. Olsen, and U.M. Minga. 2001. Productivity of local chickens under village management conditions. *Tropical Animal Health and Production* 34: 405 – 416.

Najib, H. 1994. Effect of feeding different levels of protein on production traits of Saudi Baladi chickens raised in confinement. *J. Appl. Anim. Res.* 6:177-181.

SAS Institute 2000. *SAS/STAT User's Guide*, statistics SAS institute Inc. Cary, NC27513, USA.

In: Biotechnology and Conservation … (2 Volume Set) ISBN: 978-1-63117-869-6
Editors: S. A. Farooq, R. Abed and S. Baqir

Chapter 11

LIZARDS BIODIVERSITY IN ARID REGIONS OF IRAQ

Zohair I. F. Rahemo* and Serbaz I. Mohamad
Kurdistan Natural History Museum, College of Science,
University of Salahaddin, Erbil, Kurdistan, Iraq

ABSTRACT

Arid regions in Iraq are extensive and the lizards have been adapted to different substrates. Certain Iraqi lizards are adopted for life in or on eolian sand dunes or sandy soils and are restricted to such zones. The lizards include species of *Acanthodactylus*, *Scincus*, and *Phrynocephaluss*. Members of genus *Scincus* and to lesser extent species of *Phrynocephalus*, are adapted for burrowing in soft sand and sand running. Furthermore, the species of *Laudakia*, such as *L. nupta* appear to be restricted to areas such as limestone outcrops and rocky cliff faces, where basking surfaces and deep crevices for retreat are provided. Species of *Trapelus*, such as *T. agilis, T. persicus, T. ruderatus* occur on plains, valleys, and alluvial fans, on sandy, loam, clay, and gravel soils. These lizards are usually found in the vicinity of small rock piles, such as those erected by local inhabitants to mark the boundaries of grain fields. They retreat into these piles for shelter. Species of *Phrynocephalus* show distinct preference for particular soil types; some species prefer open clay and gravel plains, while others are usually found on sandy plains and steps. Rock-inhibiting species are able to negotiate the rough vertical surfaces of rock outcrops and mountain habitats. These include species of *Laudakia* and several species of *Lacerta* which utilize the many angles and shadows of this environment for temperature regulation by basking and make use of the crevices for retreat from predators and temperature extremes. The gecko, *Cyrtopodion* usually are found on rocky slopes and cliff faces, in crevices and caverns; and in human habitation particularly *C. scrabum*. *Asaccus elisae* is found in caverns in gypsum deposits and limestone. Species of *Uromastix* are confined in their local distribution to well-drained alluvial soils wherein they are able to excavate their burrows, for instance *U. lorcatus* prefers silty-clay soils. *Varanus sp.* inhabits the continuously distributed substrate. From the biodiversity of lizard habitat, it can be concluded that the evolution of lizards is correlated with their specific affinity for a substrate type.

* Corresponding Author: zohair_rahemo@yahoo.com.

Keywords: Lizards, substrate, Iraq

INTRODUCTION

Iraq is a rich geographic area for many animals, especially for lizards, due to its wide range of deserts and diversified habitat. The desert plateau comprises the largest part of Iraq (57% of the total land area). The reptiles of Iraq have been a subject of investigation for a long time, starting from Werner (1895), then Khalaf, (1959) who wrote a book on his observations, without giving any locality of the species recorded and identified 22 species of lizards. Mahdi and George (1969) gave a comprehensive list of vertebrates of Iraq including lizards, they listed a total of 48 species of lizards without giving any locality in Iraq or any unique character in specimens collected. Later on, Dixon and Anderson (1973) described a new genus and species of Gecko collected from Iran and Iraq. Afrasiab and Mohamad (2009) described a new species of gecko from Saffin mountain, Kurdistan, Iraq. Recently, Afrasiab et al. (2013) published a valid review of the five species of Lacertini from Iraq. Nevertheless, the information on lizards is still incomplete due to the difficulty in catching and their classification needs expertise. Ctyroky (1972) described the ecological and geographical habitats of northern Iraq, and stated that mountains ranges mostly built by limestone of different age, the intra mountainous valleys originated in softer clay or sandy strata, the slopes of ranges are covered mainly by low xerothermous annual vegetation, oak bushes and 5 different species of junipers and pines. In valleys along streams the vegetation is livelier with low grown oak woods, popular woods and single willow trees. Greater or smaller streams are running in most of valleys and few wells on mountainous slopes. Climatically this area has a dry and warm summer from June to September; winter is cold with regular snow fall from November to February. Rainfall periods are mainly in November and from March to April. The present investigation aimed to collect ecological information about distribution of lizards in different regions of Iraq especially from mountains and arid regions as very little information is available.

COLLECTION, EXAMINATION AND CLASSIFICATION

Specimens were collected using a hand net or directly by hand using sticks or steel bars. After being captured, the specimens were brought to the laboratory of Kurdistan Natural history museum to record some of the external features such as color, and preserved in 10% formalin in order to examine the details of their morphology under a microscope. From other parts of Iraq, especially from arid regions, scattered information was collected and analyzed and compared with museum records or personal consultations. Classifications were based on available literature such as those published in Iraq (Khalaf, 1959; 1960), Afrasiab and Mohamad (2009), Afasiab et al. (2013), or the comprehensive book on lizards (Anderson 1999), and Sandaco (2006) and others.

After carrying out several collection trips in mountainous regions and other regions a total of 13 species of lizards were identified, belonging to the order (order Squamata), and 5 families, Gekkonidae (4 species), Scinidae (2 species), Lacertidae (2 species), Agamidae (3 species) and Anguinidae (2 species).

(1) ***Asaccus elisae*** (Werner, 1895)
Phyllodactylus elisae Werner, 1895
Leaf-toed gecko
Remarks: Szczerbak and Golubev (1996) listing of gecko of USSR not reporting any *Asaccus* species. Khalaf (1959) mentioned that this species is present in Iraq without giving any location, later on (1960) gave its name as Iraqi lizards collected from both Baghdad and Mosul, Ana, Arbil, while Mahdi and Georg (1969) listed in the geckos of Iraq depending on observation of Werner (1895). Anderson (1999) reported this species from the ruins of Niniveh, near Mosul, Iraq, and it is also present in Iran (Fars province) in caves, foothills of Khuzestan, southern Turkey (Birecik) and eastern Syria.

(2) ***Asaccus griseontus*** (Dixon and Anderson, 1973)
Gray-spotted leaf-toad gecko.
Remarks: This species is neither reported by Khalaf (1959;1960) nor by Mahdi and Georg (1969). Type locality of this species is Shahabad, Kermanshahan, Iran as reported by Anderson (1999), he gave the name of Palegawra cave, Sulaimaniyah province, Iraq which is the same locality from which our specimens were collected, while Hammad ju na-garrawa near Dokan dam represent a new locality in Iraq for this species.

(3) ***Asaccus saffinae*** (Afrasiab and Mohamad, 2009)
Remarks: This species differ in shape and morphology from all other members of the genus. It is somewhat close to *A. kermanshahensis* (Rastegar-Pouyani, 1996), but differs from it by having a faint dark ring on the tail and the diameter of the tubercle; which is not seen in *A. kermanshahensis*. As such this species is recently designated as a new species by Afrasiab and Mohamad (2009).

(4) ***Cyrtopodion scabrum*** (Heyden, 1827)
Gymnodactylus scaber
Remarks: Khalaf (1959) reported its presence in Iraq without giving the exact location, but in 1960 he reported it to have been collected from Baghdad, Hindiya, Mousaiab, Ana, Rawa, Arbil, Basrah and Mahdi and Georg(1969) giving its name in the list of geckonids in Iraq. The type locality of this species is Tor Sinai (Egypt) as reported by Anderson (1999), it is also present in southwest Asia- southern Anatolia, eastern Syria and Iraq.

(5) ***Mabya aurata*** (Linnaeus, 1758): Golden grass skink
Remarks: According to the key proposed by Khalaf (1959; 1960) it belongs to *Mabuya aurata septemtaeniata*, and its location is Shaklawa, Baghdad, Mosul, and Hindya. However, Anderson (1999) reported its presence in the lowlands of Iraq beside Iran, but in the present work these specimens were collected from the mountainous areas or foothills of Kurdistan where rock crevices provide a retreat as reported by Anderson (1963).

(6) ***Mabuya vittata*** (Olivier, 1804). Bridled skink
Remarks: It is worthy of note that anal opening is not evident in the three specimens, may be because that they have been collected during hibernation or possibly other reason which is

very interesting to investigate. Khalaf (1959) reported this species as part of Iraqi fauna but did not mention the site of collection. Anderson (1999) pointed out that this species inhabit oases and not clay walls. He cited mediterrenean coasts of North Africa, Algeria, Egypt, Cyprus, Rhodes, Turkey, Lebanon, Syria, Jordan, Western Iran, and Iraq.

(7) ***Lacerta media media*** Lantz and Cyren, 1920 (syn *Lacerta viridis* Laurenti, 1768)

Three-lined lizards, giant green lizards

Remarks: Anderson (1999) reported this species from eastern Turkey, northeastern Iraq, Caucasus and Western Iran; 1000-1500 m elevation in Iran. He identified the Iranian subspecies as *L. m. media-zagrosform.*

(8) ***Ophisopis elegans persicus*** Boulenger

Remarks: On the basis of the materials examined, Anderson (1999), concluded that the applicable subspecies names in various publications of this species do not help to clarify systematic and zoogeographical problems. Haas and Werner (1969) pointed out that characters usually employed in the diagnosis of subspecies may prove to vary. *O.e. blanfordi* will prove a valid taxon of the Mesopotamian plain, intergrading with one or more subspecies; *O. e. elegans,* and *O.e. persicus* are exiting names over a broad zone in the Zagaros foothills. Khalaf (1960) reported this lizard from Sadi, Haj Umran, Sakran, and Makhmoor. Further site collections and subspecies identification are useful for revealing the information about the populations of different subspecies in the Kurdistan region of Iraq.

(9) ***Laudakia*** (Agama) ***nupta*** (De Fillippi, 1843)

Large scaled rocks Agama

Remarks: Khalaf (1959) believed that *Agama nupta fusca* Blanford, 1876 might be a subspecies found in Iraq. It has a darker color than the typical subspecies; dark gray, with greatest development of spines on head and neck.

(10) ***Trapelus ruderatus*** (Olivier, 1807)

Agama ruteratus

Horny scaled agama, Olivers agama

Remarks: This species is reported to be present in Rawa (Khalaf, 1960). Comparing to the Iranian specimens reported by Anderson (1999) our specimens seem to be larger.

(11) ***Agama caucasic*** (Eichwald)

Caucasian agama, northern rock agama

Remarks: These lizards were collected from Sedakan series of mountains. According to Anderson (1999) these are mountain and upland lizards, invariably associated with rock outcrops of sandstone, limestone, and basalt, tree-covered slopes, or with boulders in river beds. In cultivated areas they are seen on brick or rock fences and on abandoned buildings. They are discovered in the mountains of northeastern Iraq and Sedakan which is situated in northeast Iraq.

(12) ***Ophisaurus apodus*** (Pallas, 1775)

Glass lizard

Remarks: Khalaf (1959) did not report this lizard as an Iraqi fauna, but Mahdi and George (1969) were listed as Iraqi lizards. Anderson (1999) did not mention Iraq as site of its distribution but Sandaco (2006) in his checklist mentioned Balkanas, Ionian and Aegean island, Mediterranean and eastern Anatolia, Syria, Lebanon, Jordan as its range of distribution. It is reported for the first time in Kurdistan, Iraq.

(13) ***Ophisaurus apodus*** (Pallas, 1775)
Glass lizard
Remarks: Sandaco (2006) reported its presence in Balkanas, Ionian and Aegean islands; Mediterranean and eastern(arrarat area), Anatolia, Syria, Lebanon, Jordan but not from Iraq. (Anderson 1999) did not report it in Iraq but he believes the subspecies *O. a. thrracius* Obst (1978) is found in western Anatolia. Further study needed to verify the subspecies found in Iraq whether it is similar to Iranian or Anatolian subspecies.

Species listed by Khalaf (1960) with their geographical distribution are mentioned below:

(14) *Hemidactylus persicus*, Baghdad; Mousaiab
(15) *Hemidactylus flaviviridis* Baghdad
(16) *Ptyodactylus hasselquisti;* Haditha.
(17) *Agama stellio,* Sheikhan region.
(18) *Uromastix microlepis;* Tharthar.
(19) *Varanus griseus;* Bagdad.
(20) *Diplometopon zarudnyi*; Hindia
(21) *Acanthodactylus boskianus asper*; Ana, Rawa
(22) *Ophisops blanfordi*; Bagdad,
(23) *Ablepharus kitaibelli;* Bagdad,
(24) *Eumees schneiderii princeps*; Arbil
(25) *Scincus scincus*; Hindiya, Najaf.
(26) *Typhlos vermicularis;* Ana.
(27) *Leptotyphos macrorhynchus;* Bagdad, Mosul, Shaklawa.
(28) *Eryx jaculus*; Baqouba, Hindiya.
(29) *Diplometopon shueaibi* after Niazi (1979)

Niazi (1979) described a new species *Diplometopon shueaibi* from Iraq. The study included the ecological observations, habitat, factors affecting their distribution such as humidity, temperature and type of food. Ethological observations were obtained concerning the mode of locomotion, tunnel construction, reaction to light, effect of temperature on depth occupied, seasonal and diurnal vertical shifts. Specimens were collected from Karballa city, and later expanded westward. The specimens were collected from the Karballa desert, south and southwest of Lake Rhazaza and few from Anbar southeast of Lake Habbanya. Anderson (1999) devoted a chapter on (Ecological and behavioral influences on Lizard distribution) of Iran. Based on this, lizards from Iraq are included in the same sequence, as follows:

Eolian Sand Dunes

Certain lizard species are adapted for life in or on eolian sand dunes, or on sandy soils and are restricted to such zones, among these are species of *Acanthodactylus, Scincus* and species of *Diplometopon* which are sand dwellers. Imprints of different species of lizards were noticed on the sandy dunes at the town Senia and north of Tikrit city.

Surface of the Sand

Some lizards live beneath the surface of the sand, and their limbs have been greatly reduced, and this adaptation facilitates different types of subsurface locomotion employed by these animals. Some of the species have the digits equipped with comb like fringes of scales an adaptation which has arisen independently in many groups in various sandy deserts. In Najaf and Hindia deserts *Scincus* (Khalaf, 1960) is adapted both for burrowing in soft sand and for sand running. *Ptyodactylus puiseuxi, P. hasselquistii* were collected from sandy areas of Rawa in Anbar province (Afrasiab and Mohamad, 2009).

Limestone Outcrops

Laudakia (Agama) is restricted to areas such as limestone outcrops and rocky cliffs where surfaces and deep crevices for retreat are provided. Species in this group include *L. nupta* in Iraqi soils and rocks. These lizards are usually found in the vicinity of small rock piles, such as those erected by local inhabitants to mark the boundaries of grain fields. Such rock piles provide basking areas upon which the lizards are able to orient themselves to sunlight for temperature control. They retreat into these piles for shelter. Some prefer open clay and gravel plains, while others usually found on sandy plains and steppes; some prefer mountain slopes, and others are found on both sandy and clay surfaces of the flatlands. *L. nupta* were collected from the sandy area mixed with limestone rocks in a village known as Al-Bakra south of Mosul city.

Crevices

Local distribution of many lacertid species may be determined by the availability of cracks and holes in clay and gravel soils or burrows in plant-stabilized sandy soil. These crevices provide a retreat from predators and from temperature extremes. The *Uromastix* sp observed in the Khozar river passing through Mosul city is a typical example of such habitat.

Rock-Inhabiting

A number of species are rock-inhabiting, able to negotiate the rough vertical surfaces of rock outcrops and mountain habitats. These include the species of *Laudakia* These lizards utilize the many angles and shadows of this environment for temperature regulation by basking and make use of many crevices for retreat from predators and temperature extremes.

Among the geckos, sand-dwelling species have been recorded. Various species of *Stenodactylus* occur on sand, but to what extent they occupy other substrates is not clear. *Cyrtopodium* is usually found on rocky slopes and cliff faces, in crevices and in caverns and in about places of human habitation, particularly true of *Cyrtopodium scabrum*. This lizard is abundant in Mosul houses.

Asacus elisae is found in caverns in gypsum deposits and limestone, and occasionally as a house gecko. *Hemidactylus* is similarly adapted to life on vertical surfaces. These large geckos are dependent upon the cavernous areas in gypsum where water persists throughout the year and a high relative humidity is maintained. Afrosaib and Mohammad (2009) reported similar observations while describing a new cave-dwelling gecko from Saffine mountain (Iraq). It is a nocturnal species found inside the Shera Swar cave, between oak trees after sunset. Furthermore *Asaccus griseonotus* was also collected from different caves, (Palegawra cave, Hazarmerd cave; Sargat Daray Mar caves by Afrasiab and Mohamad, (2009). *Cyrtopodiom heterocercum* collected from Shera swar cave, Saffine mountain (Afrasiab and Mohamad, 2009). *Cyrtopodion scabrum* is a common house gecko but also found in Sadam cave, south of Mosul, in the same cave with *A. elisae* (Afrasiab and Mohamad, 2009).

Mud-brick Habit

Structures usually built of mud-brick provide additional habitats not only for the geckos mentioned above, but also for rock-dwelling species of *Laudakia* as well. These lizards are often quite numerous on walls, houses, and monuments. There is usually an abundance of insect prey in such situations, attracted by the human inhabitants and their domestic animals and cultivated plants.

Alluvial Soil

The Iraqian species of *Uromastix* are confined in to well-drained alluvial soils wherein they are able to excavate their burrows. Many *Uromastix* sp. were observed in burrows at Al-Khossar River passing through left-bank of Mosul city. Furthermore, *Uromastix* are hit by cars some times when crossing roads in Hamama-Alil southern of Mosul city. Species occupying the greatest range of substrates such as *Trapelus sp.*, *Mabuya sp* or those inhabiting the most continuously distributed substrates. According to Anderson (1999), evolution of the various lizard groups in southwest Asia (including Iraq and Iran) may be significantly correlated with this specific affinity for substrate type and the discontinuous distribution of these substrates. Noteworthy, that specimens of *Mabuyae vittata* have been reported from sandy areas near north east of Mosul city with their parasites (Rahemo et al., 2002).

Vegetation and Lizards Distribution

Certain types of shrubs, for instance, stabilize dune sands and provide suitable sites for burrow excavation in the roots. These burrows may be constructed by rodents, or large arthropods and the lizards inhabit them. So depend upon the presence of these animals as well as on plants. e.g., *Uromastix* which live in alluvial excavation between roots of licorice at banks of Khosser and Tigris river in Mosul city. *Uromastix aegypticu* is also reported near vegetation (Kevork and Ul-Uthman 1972).

In the arid areas of sparse vegetation, many lizards may depend for sustenance upon the insect species attracted to the vegetation and consequently their local distribution depends upon the frequency and the occurrence of certain plants. Some species such as *Trapelus* climb into the branches of low steppe vegetation to orient themselves relative to the sun rays and to escape the hot soil surface for temperature control. Some prefer heavy vegetation such as *Lacerta*, skinks such as *Mabuya* show riparian habitats.

Effect of Temperature on Lizards Distribution

The animal's body temperature at any given time is a product of its relation to ambient temperature, and reflected radiation from substrata. Temperature may be most critical to developmental stages as lizard must be able to place its eggs where they will be protected from lethal extremes as well as exposed to temperatures sufficiently high for development to proceed. Oviparous species development is depend on their ability to endure the inevitable fluctuations and inconsistences in temperature and moisture characteristic of arid regions.

Sand Surface

By moving onto the sand surface, or near the surface, during the warm hours, they come quickly to activity temperature, while both high and low extremes are readily avoided by burrowing a few centimeters below the surface. The small lacertids are able to extend their activity periods into the hottest hours of the day by utilizing the small areas of shade provided by rock or bush, making brief forays into the sunlight to catch insects. The agamids position themselves relative to the incident sunlight so that the maximum surface is exposed during basking the minimum during the hottest period.

Nocturnal Species

One of the striking aspects of the lizard fauna of this desert region is the diversity of gecko species. These animals are able to circumvent the problem of light time, temperatures through exploitation of nocturnal activity. During the hottest season, when diurnal lizards are restricted to brief activity periods, nocturnal air temperatures remain high, due to the re-radiation from the heated ground surface. The activity of many insects and other arthropods is also largely confined to the night hours. A few geckos apparently have become secondarily diurnal, or partially so. Such behavior is indicated in *Cyrtopodium* all of which have darkly pigmented peritoneum. *Asaccus saffinae* is a nocturnal species collected from caves in Kurdistan region (Afrasiab and Mohammad 2009).

Mountains and Plateau

With the general increase of continental temperatures and increased aridity, some plateau species may find refuge in the higher elevations of the mountain masses, while in response to climatic cooling and increased precipitation, species isolated in such mountain areas may descend to the plateau to become more widely distributed. In Iraq many species were collected from mountainous regions of Kurdistan such as *Asaccus saffinae*, *Cyrtopodiom heterocercum* collected from Shera swar cave, Saffine mountain and *Cyrtopodion scabrum* (Afrasiab and Mohamad, 2009).

References

A Directory of Wetlands in the Middle East(Iraq)-1994.

Afrasiab SR and Mohamad SI 2009. A study on cave-dwelling geckos in Iraq, with the description of a new species from *Saffine Mountain. Zool.* Mid. East 47:49-56.

Anderson S 1999. The Lizards of Iran. Society for the study of Amphibians and Reptiles.

Ctyroky P 1972. White-throated robin *Iraniagutturallis*(Guerin) breeding in Northern Iraq. *Bull. Iraq nat. Hist. Mus.* Vol. v(3): 1-10.

Dixon JR and Anderson SC 1973.A new genus and species of Gecko (Sauria: Geckonidae) from Iran and Iraq. *Bulletin of the southern California Academy of Science*, 72(3): 155-160.

KevorkO and Al-Uthman HS 1972. Ecological observations on the Eqyptian spiny-tailed lizard, *Uromastixaegypticus*: 1_on Food and Feeding Habits, with Notes on the Climate and vegetation of the study area. *Bull. Iraq nat. Hist. Mus.* vol.v(2): 26-44.

Khalaf KT1959. *Reptiles of Iraq with Notes on the Amphibias.* Published by grant from the ministry of education of Iraq.

Khalaf K T1960. Notes on a collectoion of lizards and snakes from Iraq. *Iraq nat. Hist. Mus.* Publ. No.18, Feb. 1960.

Mahdi N and Georg PV 1969.A systematic list of Vertebrate of Iraq. *Iraq Natural History Museum* pub. no. 26.

Mahdi N and Georg PV 1969.A systematic list of the Vertebrates of Iraq. *Iraqi natural history Publication* no.26.

Niazi A 1979. The Arabian worm lizard: its ecology, morphology and phylogeny and Phyllogeny with a description of a new species. *Irq. J. Biol. Sc.* 7(1): 3-31.

Rahemo ZIF 1972. Unpublished data.

Rahemo ZIF 1975. Smashing of Uromastyx at roads-unpublished data.

Rahemo ZIF; Mohammad MSA and Al-Zako SS 2002. *Pneumonemamabuyae* sp. nov.(Rictulariidae: Nematatoda) from the Golden Skink *Mabuyavittata* from *Iraq. Raf. J. Sci.* 13(2): 8-12.

Rastegar-Pouyani N 1996. A new species of Asaccus (Sauria-Geckkonidae) from Zagros mountains, Kermanshahanprovince, western Iran.-*Russian journal of herpetology* 3: 11-17.

SandacoR 2006. Annotated checklist of the reptiles of the Mediterranean countries, with keys to Asiatic and African species. Part 1-Turtles, crocodiles, amphisbaenians and Lizards (Reptilia). Annals de museuoCivico di StoriaNaturale"G. Doria", *Genovapp* 85-191.

Szcerbak N and Golubev M 1996. Gecko Fauna of the USSR and Contiguous regions. Published by the aid of a grant from Atherton SiedellEndowoment fund by *Society for the study of Amphibians and Reptiles.*

Werner F 1895. Uber eineSammlung von ReptilienausPersien, Mesopotamien, und Arabien. *Verh. Zool.-Bot. Ges.*Wien, 45.

In: Biotechnology and Conservation … (2 Volume Set) ISBN: 978-1-63117-869-6
Editors: S. A. Farooq, R. Abed and S. Baqir

Chapter 12

ENVIRONMENTAL CHANGES AND THEIR EFFECTS ON THE FATE OF SEA TURTLE REPRODUCTIVE POTENTIAL AND CONSERVATION

I. Y. Mahmoud[1], S. K. Al-Musharafi[2], S. N. Al-Bahry[3] and I. S. Al-Amri[1,*]

[1]Department of Biological Science and Chemistry, University of Nizwa, Nizwa, Oman
[2]Department of Biotechnology, Sur College of Applied Sciences, Sur, Oman
[3]Department of Biology, College of Science, Sultan Qaboos University, Muscat, Oman

ABSTRACT

This study tends to address the major environmental problems which threaten the nesting beaches and feeding grounds of the sea turtles throughout the world. Today, sea turtles are exposed to a contaminated environment throughout their migratory routes. The destruction of natural nesting beaches is escalating due to development of coastal industries, tourism, residential areas, excessive commercial fishing, water recreational activities, artificial lighting near the nesting areas, and renourishment of beaches. These conditions discourage and frequently prevent the turtles from nesting. Hatchlings and nesting turtles are confused by artificial lighting and frequently take a wrong direction toward the light rather than to the sea. Also, the turtles are exposed to heavy metals and antibiotic- resistant bacteria in sewage effluent which reaches the nesting beaches via urban sewage discharge. In addition, worldwide nesting beaches are subjected to drastic changes in sand temperature due to global warming and global dimming. Sex determination in sea turtles and in some other reptiles is determined by temperature during the incubation period. Eggs incubated at a temperature of 30-31 °C produce mostly females, while those incubated at 26-27 °C produce mostly males. Fluctuation in

* Corresponding Author: Ibrahim.younis@unizwa.edu.om.

the global temperature may change sex ratio which consequently threatens their survival. It is increasingly urgent that these issues must be addressed internationally by biologists and conservationists.

Keywords: Sea turtles, environmental changes, pollution, reproductive potential conservation

CLIMATIC AND ENVIRONMENTAL CHANGES

At the end of Permian period, 251 million years ago there was complete destruction of habitats, which almost destroyed life on earth. As a result coral reefs, fishes, shellfish, trilobites, plankton, and many other groups disappeared from the marine environment. Terrestrial reptiles and others were extinct from the Earth and only 5% of species survived for the next 500 thousand years. The full ecosystem complexity took 50 million years to recover (Benton, 2003).

The oldest turtle fossils from the late Triassic period, 220-million-years ago, were found in south-western China. These turtles did not have full carapaces (Owen, 2008). Aquentin (2010) reported that the first turtles were terrestrials, which appeared 210 million years ago in the Triassic period. Some became aquatic forms 180 to 160 million years ago in the Jurassic period (Owen, 2008).

There are two theories about climatic changes on Earth in the Triassic periods which lead to massive extinction of more than 90% of living organisms. The earth possibly collided with an asteroid or a comet. The other theory suggests that there were massive and prolonged violent eruptions (National Geography, 2013). Another massive extinction of life on Earth was during the Jurassic period 65 million years ago. The meteorite impacted in present-day Mexico. Plants and many other groups of animals such as dinosaurs were wiped out completely. The impact caused shock waves of dust and rocks over the entire Earth (Benton, 2003). Five to eight glacial periods were reported (Augustin et al., 2004, Eldredge and Biek, 2013). The earliest glacial period was about 2 billion years ago. The latest glacial period began 3 million years ago at the beginning of the Pleistocene period, as a result of prolonged precipitation of ice, specifically in the northern hemisphere (Eldredge and Biek, 2013; Gradstein et al., 2004). Abnormal winds took place, with changes in the sea level, causing deposition of material and major erosion of the continents. Many river routes were modified and new lakes formed. The entire biological communities were affected and many habitats were destroyed (Gradstein et al., 2004). An annual depression of temperature is about 8.4°C (Kuhle, 2011). The maximum temperature during the last glacial event ranged between -10 to 4°C (Petit et al., 1999).

In the last fifty years the global warming and global dimming have been on a rise caused by human made pollution. Both are said to have major impacts on global biodiversity and can cause major climatic disaster.

Global dimming is a gradual reduction in temperature due a steady increase in global irradiance at earth atmosphere as a result of an increase of particulates such as sulphate aerosols affecting aquatic and terrestrial habitats. Global dimming can cause a cooling effect on the Earth blocking effect of the green house global warming. The pollutants can block incoming sunlight and reflected back into space resulted in gradual in gradual drooping in global atmospheric temperature (Stanhill and Cohen, 2001).

Global warming, on the other hand, is an opposing environmental force causing a global rising of temperature. Both global dimming and global warming may eventually lead to drastic changes in earth habitats leading to extinction of wildlife on earth (Petit et al., 1999).

MODERN CIVILIZATION THREATS

The availability of nesting beaches to sea turtles is in rapid decline in the tropical and semi-tropical regions of the world. There are physical factors that lead to the destruction of the nesting beaches. For example, changes in coastal habitats, which mainly involve establishment of resorts, residential areas and hotels, are major factors destroying nesting beaches. Other man-made structures, such as beach armouring and rock revetment can dramatically change the natural nesting beaches. Also, beach nourishment, removing or depositing sand, alters the natural nesting beaches. Compacted sand prevents turtles from nesting. In Oman, beach driving is another factor that contributes to the destruction of nesting beaches and should be banned immediately by law.

At Jupiter Island Florida, a study was conducted on the effects of renourished beaches on sea-turtle nesting activities. Adjoining natural sand beaches were used for controls. Females made fewer nests on the renourished beaches than on the natural ones; however, the percentage of hatchlings proportionally did not differ. Less nesting occurred, overall, on renourished beaches over the seven- year study period, due to the formation of berms near the surf line, the fill used, and the sand hardness, making these beaches unsuitable for nesting (Steinitz et al., 1998).

Approximately ninety-percent of the loggerhead turtles (*Caretta caretta* L.) found in the western Atlantic Ocean nest along Florida's oceanic shores (Erhart 1989). Before the beaches were developed for residential, business, and tourist sites, the effect of erosion due to storms, currents, and sea-level changes was minimal, with sand being naturally replaced from adjacent locations. However, after coastal development these conditions changed drastically, with the disruption of normal patterns of sand transport and the building of jetties and inlets (Pilkey, 1991). This, in turn, brought about the strategy of renourishment of beaches to prevent sand loss which posed a threat to the sea turtle nesting sites (National Marine Fisheries Service and US Fish and Wild life Service, 1991). With this procedure; however, other hazards occurred, such as berm formation at the surf line, limiting the turtles to nesting at a lower site making the nests vulnerable to wave action (Nelson et al., 1987). Also, sand transported from other areas was frequently too dry, too moist, and less permeable to oxygen, consequently affecting the development of the embryo during incubation (Grain et al., 1995).

It was unclear why some females preferred some sites over others; however, the preferred beaches were easily accessible with fewer predators (Erhart and Raymond, 1983) and frequently females stopped nesting when there was artificial lighting (Salmon et al., 1995; Witherington, 1992).

At Ras Al-Hadd, Oman, the nesting beaches were kept in natural condition for the nesting green turtles, which resulted in fluctuations in nesting population over the years, but without significant decline in population (Al-Kindi et al., 2003). The annual monsoon which disrupts the nesting shore lines has no effect on the nesting turtles. Sand texture and lack of human activities are the main reason for successful nesting from successful nesting.

Polluted beaches are affecting the natural conditions of coastal ecosystems such as lagoons, salt marshes and rivers carrying pollutants and emptying into the coastal water, thus affecting the marine habitats and turtle feeding-grounds. The pollutants include a variety of toxic materials such as heavy metals, pesticides, fertilizers, oil spills and sewage discharge. Industrial sewage water and sludge may contain heavy metals. Current methods used for recycling of sewage reduce disease-causing microbes (pathogens) (Al-Bahry et al., 2011; Al-Musharafi et al., 2012; 2013; *Han* et al., 2003). Studies in Oman showed that sewage water treatment does not eliminate all pathogens and some of the toxic chemicals such as pesticides (Al-Bahry 1999; Al-Bahry et al., 2006; 2007; 2009a; 2009b; 2009c; 2010; 2012). Additionally, heavy metals remain intact in sewage (Al-Bahry et al., 2011; Al-Musharafi et al., 2012; 2013; *Han* et al., 2003). All these factors cause environmental damage as a result of accumulation of toxic compounds deposited into marine habitats. Heavy metals and antibiotic-resistant bacteria detected in marine turtles and fish is a good indication of sewage water and terrestrial contaminates reaching marine wild life (Al-Bahry et al., 2009a; 2009b; 2009c; 2010; 2012). This is specifically true for those sewage-treatment plants located in the coastal regions of Oman (Al-Bahry et al., 2009c). Antibiotic-resistant bacteria are not usually found in wild life unless contaminated effluent pollution exists (Al-Bahry et al., 2010; 2012, Kumerrer, 2009a).

Antibiotic-resistant bacteria are rapidly invading various habitats, both terrestrial and aquatic. Antibiotics are used widely throughout the world as growth promoters in agriculture and aquaculture (Al-Bahry et al., 2007). They are also used for treatment of infectious diseases in humans and animals. The wide spread of multiple-antibiotic-resistant bacteria (MARB) in various terrestrial and aquatic habitats is related to the extensive usage of pharmaceutical products in medicine, agriculture and farm animals (Guardabassi et al., 1988; Kumerrer, 2009a, Al-Bahry et al., 2009a; 2010, 2012). The exposure of normal flora and pathogenic bacteria to antibiotics will cause the emergence of MARBs (Al-Bahry et al., 1999; 2006). The extensive usage of antibiotics throughout the world has resulted in a rapid spread of MARBs contaminating terrestrial and marine life (Al-Bahry et al., 2009a and 2009b; Foti et al., 2009; Kumerer, 2009a and 2009b; Miranda and Zemelman 2001; Schmidt et al., 2000). Antibiotic-resistant bacteria were found in pelagic fish (Miranda and Zemelman 2001). Al-Bahry et al., 2009b and 2009c, reported that there is evidence that infection by MARBs of fish is related to sewage effluent, even after treatment, which was consequently disseminated into the marine environment. Moreover, most of the antibiotic compounds used in medicine are released through urine and feces into the sewage system. Recently, Al-Bahry et al. (2011) reported that oviductal fluid secreted periodically on the eggs during oviposition by the nesting green turtle was heavily contaminated with MARBs, an indication that the turtles were subjected to contaminate effluents reaching the sea, and consequently infecting feeding grounds and nesting shorelines. The presence of MARBs was used as bioindicators and biomonitoring of marine pollution (Al-Bahry et al., 2009a, 2012; Foti et al., 2009). The green turtles of Oman migrate from different geographical locations, such as the Arabian Gulf, Arabian Sea, Red Sea and the northern sector of the Indian Ocean, and are exposed to various polluted effluents. In addition, some of the feeding-grounds near the nesting beaches are also heavily contaminated by the effluent run-off (Al-Bahry et al., 2012). Using the sea turtle for the presence of MARBs and other contaminates in the survey of the regions of marine habitats is a novel idea. MARBS were also isolated from juveniles of loggerhead sea turtles in North Carolina, USA (Harms, et al., 2006).

After ovulation, the eggs remain in the reproductive tract for approximately two weeks for the process of shelling which involves the formation of eggshell and eggshell membrane. During this period the secretory glands in the oviduct are very active (Alkindi et al., 2006) and there is a possibility of contamination. Another possibility of contamination is by feces as the egg passes the cloacal chamber during the process of oviposition (Al-Bahry et al., 2009a).

About 25-30% of the world's oil is produced in the Gulf and transported internationally through Musandam and the Gulf of Oman (The Geography of Transport, 2013). Due to the geographical position of Oman its shore lines are vulnerable to oil pollution from normal tanker operations which include ballast water, illegal discharges and accidental spills. Passing vessel traffic is considered one of the major sources of oil pollution in Oman (Vaidya, 2013). On December 19, 1972 one of the worst major oil spills in history was recorded in the Gulf of Oman. The amount of spilled oil was 35.3 million gallons of crude oil (Moss, 2010). Some of the most protected habitats in Oman are very sensitive to oil pollution. These include Daymaniyat Islands, Musandam, Khawr Kalba, Ras Al Hadd & Ras Al Jinz, Barr Al Hickman which have nesting beaches for sea turtles (Ministry of Environment, Oman, 2012).

During recent years, occurrences of red tide in the coastal waters of Oman have become a common phenomenon leading to great fish mortality, as well as affecting the environment of sea turtles. Data related to red tide affecting turtles are not available (Benny, 2011).

EFFECT OF INCUBATION TEMPERATURE ON SEX DETERMINATION

Sex determination in reptiles, including turtles, is controlled by temperature during the incubation period. All eggs incubated at 30 °C produce 100% females. Eggs incubated at 26-27 °C produce 100% males. A middle temperature 28.5 °C produce a mixed sex ratio but the majority will be female (about 85%). The sexual differentiation of the gonads takes place during the middle third of incubation period (20-30 days) (Godfrey et al., 2003).

Oviparous reptiles have a temperature dependent sexual differentiation (TSD) mechanism (O'steen and Janzen, 1999). The thermo-sensitive period (TSP) is about 10 days (day 20-30) in both male and female-promoting temperatures. The rate of the gonadal development was faster in the female-promoting temperature than in the male-promoting temperature. Formation of the genital ridge began at day 20–22 and histological differentiation of the gonads occurred after that until 30 days of incubation (Mahmoud et al., 2011a; Mahmoud et al., 2013).

On day 18 of incubation the gonads were still undifferentiated. There was no progesterone receptors (PRs) expression in the gonads.

At day 20 the gonads started to differentiate. There were PRs expressions in the gonads. Also, follicles cells were well developed in the cortex and poorly developed in the medulla. However, in the male, the reverse condition takes place where the medulla is well developed with seminiferous tubules (Mrosovsky, 1988).

The role of progesterone (P) has been more extensively studied in the female reproductive organs (ovary, reproductive tract, mammary gland) and the brain, in which it is an important regulator and modulator in conjunction with estradiol (E). In nonmammalian species, little research has been done on P metabolites involved in ovulation (Gregory and Schmid, 2001). Moreover, P induces the expression of egg-white, decreases myometrial contractility, and facilitates egg processing, eggshell formation, and deposition of egg-white proteins. Actions of P may be synergistic with, or antagonist to the actions of E, depending on hormone ratios, timing of exposure, and physiological state. The effects of P are mediated through progesterone receptor isoform A (PRA), a general transcription inhibitor of P targets genes, and isoform B (PRB), a specific transcriptional stimulator of some reproductive tract genes (Custodia-Lora and Callard, 2002). The expression of intracellular PR in the ovary has been reported in mammalian species, including humans (Suzuki et al., 1994), and in nonmammalian species, including the chick (Pasanen et al., 1997). P action is critical for ovulation in nonmammalian species, such as the frog (Schuetz and Lessman, 1982). In rats, P controls the timing of ovulation by modulating the expression of adenylate cyclase activity in the granulosa cells (Ko and Park-Sarge, 2000).

The detection of progesterone receptors during different stages of embryogenesis may give us some information on how temperature influences the gonadal activities, which reflect on the endocrinological mechanisms that consequently determine the sex of the embryo. It appears that once sex differentiation occurred during TSP, the protein levels remain stable until further differentiation after hatching. The protein levels remain the same during the thermosensitive period (TSP), which extends between day 20-30. Once the ovarian protein receptors appeared during day 20 they remained stable with a slight increase for the rest of TSP. Such study may provide valuable information on hatching success and sex determination for the green turtles.

Moreover, understanding the reproductive mechanism during developmental stages will add more information for the overall understanding of reproductive potential of this species.

The mechanism of temperature-dependent sexual differentiation (TSD) has been found in all species of sea turtles (Godfrey et al., 1999; O'Steen and Janzen, 1999; Pieau et al., 1999). The eggs of the green turtles are buried deep under the sand surface, approximately one meter (Alkindi et al., 2003), thus the temperature in the nesting chamber does not fluctuate significantly. The study of sex determination under the influence of temperatures would be of value to the overall understanding of reproductive physiology during the embryological development. At Ras Al-Hadd, the temperature variations under natural conditions indicate that the incubated eggs in the nesting chamber are usually under constant temperature, with very little temperature fluctuations. Because of this condition, the majority of the embryos in the eggs have the same embryonic staging during the period of incubation, and they usually hatch out within a few hours of each other, (Mahmoud et al., 2005).

In this study, subjecting the eggs to different fixed temperatures in the incubators would be the best way to monitor the effect of temperatures on sex determination. Moreover, the steriodogenic activities in the gonads will be examined, such as sex hormone levels and development of male or female gonad.

The duration of incubation depends on the temperature of the nest chamber where sexual differentiation takes place during the middle third of incubation. Lower temperatures produce males while higher temperatures produce females (Lance, 1994; Godfrey et al., 1999, and 2003).

CONSERVATION

The Archie Carr Refuge represents the nation's most significant land-acquisition effort to protect the Florida's populations of marine turtles. The Refuge is an example of how coastal habitats can be protected. Unfortunately, rapid coastal development in Brevard and Indian River Counties threatens the future effectiveness of the Refuge. Supporters of the Refuge are literally in a race against time to acquire the best remaining parcels of undeveloped land (The Archie Carr National Wildlife Refuge, 2013). The Archie Carr National Wildlife Refuge was designated by Congress in 1990 to protect sea turtles, now threatened with extinction throughout the world. Leading sea turtle researchers and concerned citizens have watched with increasing concern as turtle populations worldwide have plummeted due to over-exploitation and destruction of nesting habitats. The Carr Refuge offers hope for saving one of the most important sea turtle nesting sites in the world. (The Archie Carr National Wildlife Refuge, 2013).

In Oman, the nesting population at Ras Al-Hadd Reserve represents one of the largest populations in the world. Thousands of green turtles lay their eggs on the nesting beaches of Ras Al-Hadd. The peak of nesting season activities extends from May to October and then there is a gradual decline in nesting (Alkindi et al., 2003). Up to the present time the nesting grounds in Oman have been well protected; however, tourism and fishing activities are still interfering with the nesting process. It is suggested that tourists should not be allowed to enter the nesting beaches and turtle-watch should be limited to platforms. Hopefully, such platforms will be built in the near future, limiting access and damage to the turtle nests. During the last 250 million years, the turtles have survived drastic changes in the climatic conditions. However, current changes in the environment, such as major destruction of their feeding and nesting grounds, beach pollution and hunting have driven these animals close to extinction. Unless global conservation programs are put into effect, extinction of these animals is evident.

The nesting beaches, worldwide, are in decline due to pollution, residential development areas, tourism, recreation, industries, artificial lighting, renourishment of the beaches, and oil spills.

It is incumbent on authorities to bring about international agreements and enforcement regarding the cessation of rearmament of beaches, shore-front development, and the oil-spill problems. The protection of natural beaches is of utmost urgency or the sea turtle population will continue to decline, consequently leading to extinction.

REFERENCES

Al-Bahry SN, Al-Mashani BM, Elshafie AE, Pathare N, Al-Harthy AH. 2006. Plasmid profile of antibiotic resistant *E. coli* isolated from chicken intestines. *Alabama Journal of Academy Sciences*. 77: 152-159.

Al-Bahry SN, Al-Zadjali MA, Mahmoud IY, Elshafie AE. 2012. Biomonitoring marine habitats in reference to antibiotic resistant bacteria and ampicillin resistance determinants from oviductal fluid of the nesting green sea turtle, *Chelonia mydas*. *Chemosphere*. 87: 1308–1315.

Al-Bahry SN, Elshafie AE, Al-Busaidi S, Al-Hinai, JA, Al-Shidi I. 2007. Antibiotic resistant *Salmonella* spp from human and none-human resources in the Sultanate of Oman. *Eastern Mediterranean Health Journal*. 13: 49-55.

Al-Bahry SN, Mahmoud I, Al-Rawahi S and Paulson J. 2011. Egg contamination as an indicator of environmental health. *In*: Impact of Egg Contamination on Environmental Health. Nova Science Publisher Inc. New York.

Al-Bahry SN, Mahmoud IY, Al-Belushi KI, Elshafie AE, Al-Harthy A, Bakheit CK. 2009c. Coastal sewage discharge and its impact on fish with reference to antibiotic resistant enteric bacteria and enteric pathogens as bio-indicators of pollution. *Chemosphere*. 77: 1534-1539.

Al-Bahry SN, Mahmoud IY, Al-Khaifi A, Elshafie AE, Al-Harthy A. 2009a. Viability of multiple antibiotic resistant bacteria in distribution lines of treated sewage effluent used for irrigation. *Water Science & Technology*. 60: 2939-2948.

Al-Bahry SN, Mahmoud IY, Al-Zadjali M, Elshafie A, Al-Harthy A, Al-Alawi W. 2010. Antibiotic resistant bacteria as bio-indicator of polluted effluent in the green turtles, *Chelonia mydas* in Oman. *Marine Environmental Research*. 71: 139-144.

Al-Bahry SN, Mahmoud IY, Elshafie AE, Al-Harthy A, Al-Ghafri S, Al-Amri I, Alkindi AY. 2009b. Bacterial flora and antibiotic resistance from eggs of green turtles *Chelonia mydas*: an indication of polluted effluents. *Marine Pollution Bulletin*. 41: 214-221.

Al-Bahry SN. 1999. Antibiotic resistance of Salmonella isolated from Muscat, Oman. *Pakistan Journal of Biological Sciences*. 2: 523-528.

Al-Kindi AY, Mahmoud IY, Al-Habsi AA., Al-Bahry SN, Al-Ghailani HM, Bkheit CS. 2005. The effect of physical and human factors on nest selectivity in the green turtles (*Chelonia mydas*) at Ras Al-Hadd Reserve in Oman. *Journal of Chelonian Conservation and Biology*. 5: 289-294.

Alkindi AY, Mahmoud IY, Woller MJ, Plude JL. 2006. Oviductal morphology in relation to hormonal levels in the snapping turtle, *Chelydra serpentine*. *Tissue and Cell*. 38: 19-33.

Alkindi AYA, Mahmoud IY, Al-Gheilani HM, Bakheit CS, Al-Habsi A, Al-Kiyumi AA. 2003. Comparative study of the nesting behavior of green turtle, *Chelonia mydas*, during high and low population density periods at Ras Al-Hadd Reserve, Oman. *Chelonian Conservation and Biology*. 4: 603-611.

Al-Musharafi SK, Mahmoud IY, Al-Bahry SN. 2012. Heavy metal contamination from treated sewage effluents. 11th International Conference on Modelling, Monitoring and Management of Water Pollution. New Forest, UK. www.witpress.com, ISSN 1743-3541 (on-line). WIT Transactions on Ecology and The Environment, Vol. 164, © 2012 WIT Press. UK. Page 381-389.

Al-Musharafi SK, Mahmoud IY, Al-Bahry SN. 2013. Heavy metal pollution from treated sewage effluent. Fresh water habitat pollution by treated sewage effluent in relation to multiple-antibiotic-resistant bacteria. ICESD2013. APCBEES Procedia (*In press*).

Al-Rawahy SH, Alkindi AY, Elshafie AE, Mahmoud IY, Al-Bahry SN, Al-siyabi S, Al Mansori M, Al Kiyumi A. 2007. Accumulation of metals in the egg yolk and liver of hatchling of green turtles *Chelonia mydas* at Ras Al Hadd, Sultanate of Oman. *Journal of Biological Sciences*. 7: 925-930.

Anquetin J. 2010. The anatomy of the basal turtle *Eileanchelys waldmani* from the Middle Jurassic of the Isle of Skye, Scotland. Earth and Environmental Science Transactions of the Royal Society of Edinburgh 101: 67–96.

Augustin L, Barbante C, Barnes PR, Barnola JM, Bigler M, Castellano E, Cattani O, Chappellaz J, Dahl-Jensen D, Delmonte B, Dreyfus G, Durand G, Falourd S, Fischer H, Flückiger J, Hansson ME, Huybrechts P, Jugie G, Johnsen SJ, Jouzel J, Kaufmann P, Kipfstuhl J, Lambert F, Lipenkov VY, Littot GC, Longinelli A, Lorrain R, Maggi V, Masson-Delmotte V, Miller H, Mulvaney R, Oerlemans J, Oerter H, Orombelli G, Parrenin F, Peel DA, Petit JR, Raynaud D, Ritz C, Ruth U, Schwander J, Siegenthaler U, Souchez R, Stauffer B, Steffensen JP, Stenni B, Stocker TF, Tabacco IE, Udisti R, Van De Wal RS, Van Den Broeke M, Weiss J, Wilhelms F, Winther JG, Wolff EW, Zucchelli M; EPICA community members. 2004. Eight glacial cycles from an Antarctic ice core. *Nature.* 429 (6992): 623-628.

Benny, J. 2011. Red tide hits Omani waters yet again. Muscat Daily. January 08.

Benton M. 2003. Wipeout: when life nearly died. *New Scientist.* 178 (2392-26): 38-41.

Custodia-Lora N, Callard IP. 2002 Progesterone and progesterone receptors in reptiles. *General and Comparative Endocrinology.* 127: 1-7.

Davies E. (2012-05-07). "BBC Nature-Dinosaur gases 'warmed the Earth'". Bbc.co.uk. http://www.bbc.co.uk/nature/17953792. Retrieved 2012-08-07. (Accessed February, 2013).

Eldredge S, Biek B. 2013. Ice Ages-What are they and what causes them? Utah Gelological Survey. http://geology.utah.gov/surveynotes/gladasked/gladice_ages.htm___(Accessed February, 2013).

Erhart LM, Raymond PW. 1983. The effect of beach restoration on marine turtles nesting in south Brevard County, Florida. Progress Report to the U.S. Army Copr of Engineers, Jacksonville, Florida, pp. 1-47.

Erhart LM. 1989. Status report on the loggerhed turtle. *In*: Ogren, L. (ed.). Proceedings of the Second Western Antlantic Turtle Symposium, NOAA Technical memorandum NMFS SEFC-226. pp. 122-144

Fleming A, Crews D. 2001. Estradiol and incubation temperature modulate regulation of steroidogenic factor 1 in the developing gonad of the red-eared slider turtle. *Endocrinology.* 142: 1403-11.

Foti M, Giacopello C, Bottari T, Fisichella V, Rinaldo D, Mamminac C, 2009. Antibiotic resistance of Gram negatives isolates from loggerhead sea turtles (*Caretta caretta*) in the central Mediterranean Sea. *Marine Pollution Bulletin.* 58: 1363-1366.

Godfrey MH, D'Amato AF, Marcovaldi MÂ, Mrosovsky N. 1999. Pivotal temperature and predicted sex ratios for hatchling hawksbill turtles from Brazil. *Canadian Journal of Zoology.* 77: 1465-1473.

Godfrey MH, Delmas V, Girondot M. 2003. Assessment of patterns of temperature-dependent sex determination using maximum likelihood model selection. *Ecoscience.* 10: 265-272.

Gradstein F, Gradstein FM, Ogg JG, Smith AG. 2004. A geologic time scale. 2004. New York: Cambridge University Press. Page 412.

Grain DA, Bolten AB, Bjorndal KA. 1995. Effects of beach nourishment on sea turtles: review and research initiatives. *Restoration Ecology.* 3: 95-104.

Gregory LF, Schmid JR. 2001. Stress responses and sexing of wild Kemp's ridley sea turtles (*Lepidochelys kempii*) in the northeastern Gulf of Mexico. *General and Comparative Endocrinology.* 124: 66-74.

Guardabassi L, Petersen A, Olsen JE, Dalsgaard A., 1998. Antibiotic resistance in *Acinetobacter* spp. isolated from sewers receiving waste effluent from a hospital and a pharmaceutical plant. *Applied and Environmental Microbiology*. 64: 3499-3502.

Han FX, Banin A, Kingery WL, Triplett GB, Zhou LX, Zheng SJ, Ding WX. 2003. New approach to studies of heavy metal redistribution in soil. *Advances in Environmental Research*. 8: 113- 120.

Harms CA, Mihnovets AN., Braun-McNeill J, Kelly TR, Avens L, Goodman MA, Goshe LR, Godfrey MH, Hohn AA. 2006. Cloacal bacterial isolates and antimicrobial resistance patterns in juvenile loggerhead turtles in North Carolina, USA. *In*: Proceedings of Annual Symposium on Sea Turtle Conservation and Biology. Page 58.

Ko C, Park-Sarge OK. 2000. Progesterone receptor activation mediates LH-induced type-I pituitary adenylate cyclase activating polypeptide receptor (PAC1) gene expression in rat granulosa cells. *Biochemical and Biophysical Research Communications*. 277: 270-279.

Kuhle M. 2011. "Ch 53: The high-glacial (Last Glacial Maximum). Glacier cover of the Aconcagua group and adjacent massifs in the Mendoza Andes (South America) with a closer look at further empirical evidence". *In*: Ehlers J, Gibbard PL, Hughes PD. Quaternary glaciations-extent and chronology: a closer look. Development in Quaternary Science. Amsterdam: Elsevier. Page 735-8.

Kümmerer K, 2009a. Antibiotics in the aquatic environment. A review. Part I. *Chemosphere*. 75: 417-434.

Kümmerer K, 2009b. Antibiotics in the aquatic environment. A review. Part II. *Chemosphere*. 75: 435-441.

Lance VA. 1994. Life in the slow lane: hormones, stress, and the immune system in reptiles. Perspectives in comparative endocrinology. Ottawa: National Research Council of Canada. Page 529-534.

Li C, Wu XC, Rieppel O, Wang LT, Zhao LJ. 2008. An ancestral turtle from the Late Triassic of southwestern China. *Nature*. 456 (7221):497-50.

Mahmoud IY, Alawi M, Yaish MW, Al-Bahry SN. 2011b. Sex determination in the green turtle Chelonia mydas during embryogenesis in reference to progesterone receptor. World Congress on Reproductive Biology & Society for Reproductive Biology, SRB/WCRB. Cairns, Australia.

Mahmoud IY, Alawi M, Yaish MW, Al-Bahry SN. 2013. Embryonic gonadal differentiation of progesterone receptors in the green turtle Chelonia mydas, Ras Al-Hadd, Oman. 33rd Sea International Symposium for Sea Turtles, I.S.T. Society. 2013: Baltimore, USA.

Mahmoud IY, AlKindi AY, Khan T. Al-Bahry SN. 2011a. Detection of low plasma estradiol concentrations in nesting green turtles (*Chelonia mydas*) by HPLCMs-Ms. *Journal of Experimental Zoology Part A: Ecological Genetics and Physiology*. 315: 170-174.

Mahmoud IY, Alkindi AYA, BaOmar TA, Al-Siyabi S, Al-Bahry SN, Elshafie AE, Bkheit CS. 2005. Emergence pattern of the green turtle, *Chelonia mydas*, hatchlings under laboratory and natural conditions. *Zoology in the Middle East*. 35: 19-27.

Ministry of Environment and Climate Affairs, Oman. 2012. Sea Turtle Reserves. Ministry of Environment and Climate Affairs.

Miranda CD, Zemelman R. 2001. Antibiotic resistant bacteria in fish from the Concepcion Bay, Chile. *Marine Pollution Bulletin*. 42: 1096-1102.

Moss, L. 2010. The 13 largest oil spills in history. Mother Nature Network. http://www.mnn.com/earth-matters/wilderness. (Accessed February, 2013).

Mrosovsky N. 1988. Pivotal temperatures for loggerhead turtles (*Caretta caretta*) from northern and southern nesting beaches. *Canadian Journal of Zoology*. 66: 661–669.

National Geography. 2013. http://science. (Accessed February, 2013).

National Marine Fisheries Service and U.S. Fish and Wildlife Service. 1991. Recovery plan for the U.S. population of loggerhead turtle. National Marine Fisharies Service. Washington, DC. Page 1-52.

Nelson DA, Mauck K, Fletmeyer J. 1987. Pysical effects of beach renourishment on sea turtle nesting, Delray Beach, Florida. Technical Report EL-87-15. US Department of Army, U.S. Army Coprs of Engineers, Jacksonville, FL. Page 1-56.

O'steen S, Janzen FJ. 1999. Embryonic temperature affects metabolic compensation and thyroid hormones in hatching snapping turtles. *Physiological and Biochemical Zoology*. 72: 520-533.

Owen J. 2008. Earliest swimming turtle fossils found - new species. (http://news.nationalgeographic.com/news/2008/11/081119-aquatic-turtle-fossil-missions). (Accessed February, 2013).

Pasanen S, Ylikomi T, Syvala H, Tuohimaa P. 1997. Distribution of progesterone receptor in chicken: novel target organs for progesterone and estrogen action. *Molecular and Cellular Endocrinology*. 135: 79–91.

Petit, Jean-Robert; Jouzel, Jean; Raynaud, Dominique; Barkov, N I; Barnola, Jean-Marc; Basile, I; Bender, Michael L, Chappellaz, Jérôme A, Davis, John C, Delaygue, G, Delmotte, M; Kotlyakov, Vladimir; Legrand, Michel R; Lipenkov, Vladimir Y; Lorius, C; Pepin, L; Ritz, Catherine; Saltzman, Eric S; Stiévenard, Michel. 1999: Climate and atmospheric history of the past 420,000 years from the Vostok ice core, Antarctica. *Nature*. 399 (6735): 429-436.

Pieau C, Dorizzi M, Richard-Mercier N. 1999. Temperature-dependent sex determination and gonadal differentiation in reptiles. *Cellular and Molecular Life Sciences*. 55: 887–900.

Pilkey O. 1991. Coastal erosion. *Episodes*. 14: 56-51.

Salmon M, Reiners R, Lavin C, Wyneken J. 1995. Behavior of loggerhead sea turtle on an beach. I. correlates of nest placement. *Journal of Herpatology*. 29: 560-567.

Schmidt A, Bruun MS, Dalsgaard I, Pedersen K, Larsen JL. 2000. Occurrence of antimicrobial resistance in fish pathogens and environmental bacteria associated with four danish rainbow trout farms. *Applied and Environmental Microbiology*. 66: 4908-4915.

Schuetz AW, Lessman C. 1982. Evidence for follicle wall involvement in ovulation and progesterone production by frog (*Rana pipiens*) follicles *in vitro*. *Differentiation*. 22: 79-84.

Schulman A, Milton S, Lutz P. 1994. Aragonite sand as a substrate and its effect on *Cretta caretta* nests. *In*: Bjordal K., Bolten A., Johnson D, Ellazar P. editors. Proceedings of the furteenth Annual Symposium on Sea Turtles Biology and Concervation. NOAA Technical Memorandum NMFS-SEFC-351. Miami. Florida. Page 134.

Stanhill G, Cohen S. 2001. Global dimming: a review of the evidence for a widespread and significant reduction in global radiation with discussion of its probable causes and possible agricultural consequences. *Agricultural and Forest Meteorology*. 107: 255-278.

Steinitz MJ, Salmon M, Wyneken J. 1998. Beach renourishment and loggerhead turtle reporduction: a seven year study at Jupiter Island, Florida. *Journal of Coastal Research*. 14: 1000-1013.

Suzuki T, Sasano H, Kimura N, Tamura M, Fukaya T, Yajima A and Nagura H. 1994. Immunohistochemical distribution of progesterone, androgen and oestrogen receptors in the human ovary during the menstrual cycle: relationship to expression of steroidogenic enzymes. *Human Reproduction.* 9: 1589-1595.

The Archie Carr National Wildlife Refuge - A Refuge for Sea Turtles. 2013. Sea Turtle Conservancy. http://www.conserveturtles.org/habitats. (Accessed February, 2013).

The Geography of Transport. 2013. Share of OPEC and the Persian Gulf in the World Crude Oil Production, 1960-2008. http://people.hofstra.edu/geotrans/eng/ch5en/appl5en/shareopec.html.

Vaidya SK. 2013. Silent oil spill poses risk to Arabian Sea. Gulf News. March 9. 17: 46.

Witherington BA. 1992. Behavioral reponses of nesting sea turtles to artificial lighting. *Herpatology*. 48: 31-39.

In: Biotechnology and Conservation … (2 Volume Set) ISBN: 978-1-63117-869-6
Editors: S. A. Farooq, R. Abed and S. Baqir

Chapter 13

ANALYSIS OF HEAVY METAL IN EGGSHELLS OF GREEN TURTLES, *CHELONIA MYDAS*, BY SCANNING ELECTRON MICROSCOPY AND X-RAY MICROANALYSIS

S. K. Al-Musharafi
Department of Applied Biotechnology,
Sur College of Applied Sciences, Sur, Oman

ABSTRACT

Ras Al-Hadd, Oman is considered to be one of the largest nesting grounds in the world for the green turtle. Fresh eggs were collected from different nests immediately after oviposition and examined for heavy metal contamination in the eggshell. Eggshells in green turtles are made up of three major layers, an outside loose calcareous layer, a middle compact layer with multistrata and an inner fibrous eggshell membrane. The three layers from each egg were separated and analyzed for heavy metal detection using Oxford energy dispersive X-ray spectrometer (EDS). The heavy metal spectrum x-microanalysis as well as other elemental spectra were analyzed. The most common detected heavy metals were Aluminum (Al), Copper (Cu) and Zinc (Zn). The less common detected were Tin (Sn), Bromine (Br), Tellurium (Te) and Zirconium (Zr). Heavy metals contamination in the egg layers indicate environmental pollution and may have an impact on the embryonic development. The presence of heavy metals in green turtle eggs may have harmful developmental effect on turtle embryo.

Keywords: Green turtle, heavy metals, contamination, eggshells

INTRODUCTION

The toxicity of heavy metals in organisms is derived from their ability to bind to biomolecules. Some of these metals can be mutagens or carcinogens (He et al., 2005). The sources of heavy metals include sewage water reclaimed for irrigation, land-applied wastewater, sludge, municipal and industrial water reuse (Al-Bahry et al., 2011a, 2014; Al-Musharafi et al., 2012, 2014a; 2014b; Shahidul and Tanaka, 2004; Han et al., 2003). Heavy metals such as Cu, Zn, Pb and Cd have been often reported to cause contamination of soil, water, and food chains. On the other hand, heavy metals such as Co, Cu, Mn and Zn are essential elements at certain concentrations for the enzymatic activities. However, when they are present at higher concentrations they become toxic. An accumulation of nonessential heavy metals in tissues, which are not involved in any metabolic process, are considered to be indicators of environmental pollution (Kojadinovic et al., 2007). Different types of heavy metals accumulation have been reported from terrestrial and marine wild life in Oman (Shahidul and Tanaka, 2004; Al-Bahry et al., 2011a; Al-Musharafi et al., 2013a; 2013b; 2014a; 2014b; Al-Rawahy et al., 2007).

Sewage is composed of different microbial and chemical pollutants as well as other waste products discharged into the coastal regions. Some of microbes resistant to antibiotics contaminate the aquatic environment (Al-Bahry et al., 2009a; 2009b; 2009c; Kümmerer, 2009; Kim and Aga, 2007; Mahmoud et al., 2013). Antibiotic resistant bacteria remain viable even after tertiary treatment and were found to infect fish and marine turtles (Al-Bahry et al., 2009a, 2009b; 2011a and 2012).

Al-Bahry et al. (2009e) conducted an analysis of trace elements in TSErevealed that Fe and Cu concentrations declined fromsewage treatment plants in 3sites, 1.5 km apart from each other. In addition, the declining of trace elements were probably used as micronutrient for microbial growth.

Akinbowale et al. (2007) and Bass et al. (1999) reported that the mechanisms of heavy metal tolerance may be linked to antimicrobial resistance mechanism and that heavy metals may select antibiotic resistant bacteria. Hölzelet al. (2012) they also reported that the presence of some heavy metals such as copper and zinc in pig manure increased microbialresistance to antibiotics. They reported that Zinc in manure increased bacterial resistance to piperacillin and doxycycline antibiotics. Al-Musharafiet al. (2014b) reported that heavy metal contaminated sludge is dumped into the environment. Al-Musharafi et al. (2012; 2013a; 2013b) also reported that many treated sewage effluents from industrial sources consisted of several heavy metals, including Zn and Cu, of which some end up in the marine environment. Therefore, determination of heavy metals contamination in terrestrial and aquatic habitats may be used for bio-monitoring the environment.

Between May and November, the nesting beaches at Ras Al-Hadd, Oman attract one of the largest nesting population of green turtles in the world. These turtles migrate predominantly to the nesting beaches of Ras Al-Hadd from the Arabian Sea, Red Sea and East Africa (Al-Bahry et al., 2009a; 2012).

Edwards et al. (2001) reported the transfer of heavy metals from turtles to their eggs. Such transfer was found to take place during eggshell formation (Burger and Coachfeld, 1991). In another study, an increased accumulation of Co, Hg, Pb and Sn in turtle eggs at Ras Al-Had Reserve over a 10-year period was reported (Al-Bahry et al., 2011a). Al-Rawahyet al., (2007) reported that the source of Cr, Co, Cu, Pb, Hg, Ni and Se accumulation in green turtle egg yolk was from the sand nest. However, the sand cannot be the only source of heavy metal pollution. Contaminated effluents from industrial sources play an important role in contamination of turtle feeding areas. Heavy metal contamination in fish from a nearby treated sewage effluent dumping site, was reported (Al-Muasharafi et al., 2013b). In the current study, freshly laid eggs contaminated with heavy metals from nesting beaches at Ras Al-Hadd, is reported using x-microanalysis technique of scanning electron microscopy.

SCANNING ELECTRON MICROSCOPY AND MICROANALYSIS

Scanning electron microscopy is used to analyze the topography and composition of biological materials at high resolution. Some scanning electron microscopies are equipped with x-ray microanalysis tools. Using wavelength dispersive or energy-dispersive x-ray spectroscopy, elements, such as heavy metals contamination, can be detected at micro-scales levels and its distribution can be mapped in samples. One of the techniques used to study heavy metal contamination in eggshells is the collection of turtle eggs immediately after oviposition to avoid structural changes in the eggshells. The eggshells are rinsed with distilled water to remove sand particles and then separated from albumen and yolk (Al-Bahry et al., 2009e; 2011b). The back scattered electron detector at 20 kV is used to view and analyze the eggshells for elemental composition and topographical structures using Oxford energy dispersive X-ray spectrometer (EDS), at 20 kV with working distanceof 20mm. The EDS is calibrated with a pure cobalt sample operated under the above conditions. The spectrum of elemental peaks are cross examined and data (weight and atomic percentages) are normalized (Al-Bahry et al., 2009e; 2011b). The Oxford Inca EDS system, is used in this analysis.

Heavy Metals in Turtle Eggshells

Al-Bahry et al. (2009e; 2011b) reported the presence of three eggshell layers (outer, middle and inner). The outer calcareous layer consists of loose nodules, the middle layer is composed of compacted strata while the inner layer has loose spicules attached to the inner membrane (Figure 1).

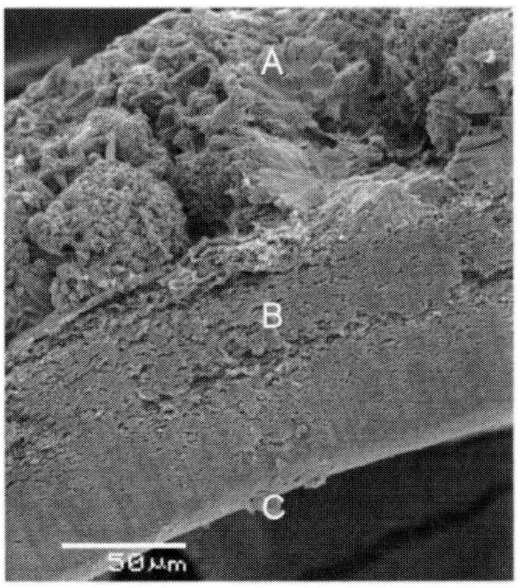

Figure 1. Eggshell layers. A = calcareous layer with nodular units; B = middle layer with tightly compacted strata; C = innershell membrane from the green turtle at Ras Al-Hadd, Oman.

In the current study, Aluminum (Al), bromine (Br), copper (Cu), iron (Fe), magnesium (Mg), phosphorus (P), sulfur (S), silica (Si), tin (Sn), tellurium (Te), yttrium (Y), ytterbium (Yb), zinc (Zn) and zirconium (Zr) were detected in eggshells. The x-ray micoroanalysis indicated that Al and S were the most frequent followed by Mg and Si, while the rest were at low levels (Figure 2). However, the weight percentages of the heavy metal indicated that S, Si and P were the most dominant, followed by the rest metals (Figure 3). Only some spectra of heavy metals are shown in Figure 4.

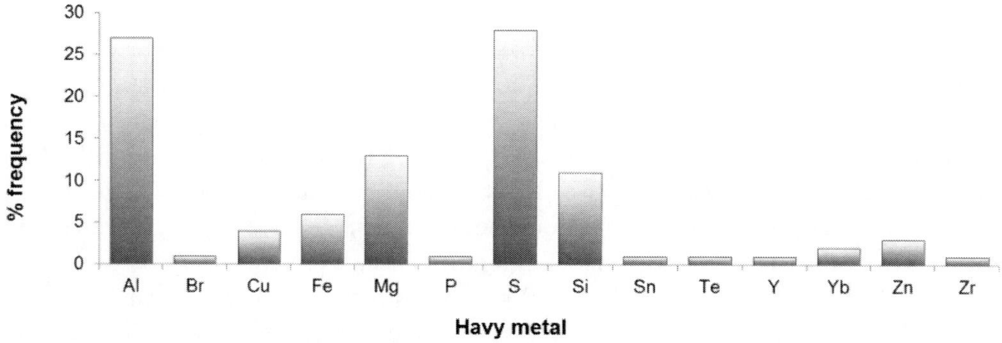

Figure 2. Percentage frequency of heavy metals detection in eggshell of the green turtle at Ras Al-Hadd, Oman.

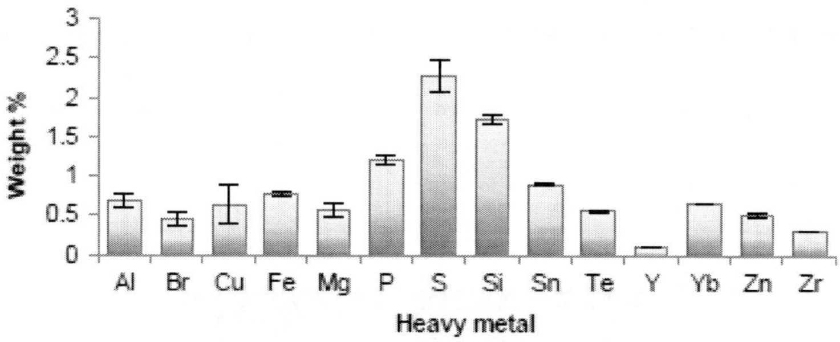

Figure 3. Percentage weight of heavy metals in the green turtle at Ras Al-Hadd, Oman.

Several heavy metals were detected in green turtle eggshells. Some of those metals exist naturally and they are crucial for the bioactivities. However, some can be toxic at high concentration, and some of the detected heavy metals are non-essential elements. Their presence indicates an environmental contamination (Al-Bahry et al., 2011a; Al-Musharafi et al., 2012; 2013a; 2013b; 2014a; 2014b; Kojadinovic et al., 2007). The industrial source of some of those metals reveals the role of industrial pollution in marine life.

It is well documented that heavy metals are not eliminated by the process of sewage treatment and the contaminants could be released into the terrestrial and aquatic habitats (Al-Bahry et al., 2009a; 2009b; 2011a and 2012;Al-Musharafi et al., 2012; 2013a; 2013b; 2014a; 2014b).

Detection of multiple antibiotics resistance in the green turtle eggs in Oman, suggests that the green turtles are exposed to contaminated effluents in their migratory routes feeding areas (Al-Bahry et al., 2009a; 2012). Antibiotics and antibiotic-resistant bacteria are released via sewage effluent into the aquatic environment (Weigel, 2003). The discharge of sewage effluent in seawater is considered to be one of the routes were antibiotic resistant bacteria reach marine habitats. Heavy metals contaminated effluents may also reach the feeding areas exposing turtles to pollutants.

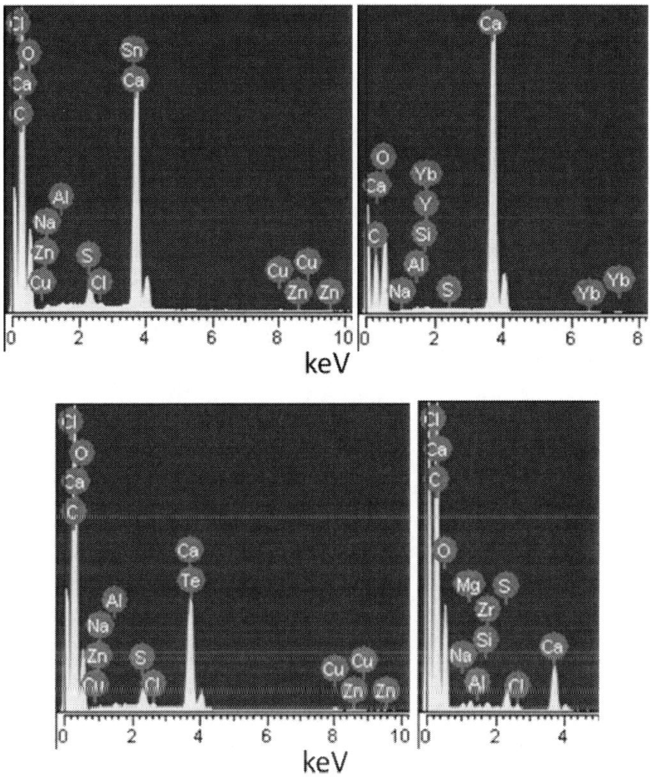

Figure 4. Spectrum of x-ray microanalysis of heavy metals from the green turtles.

Fish and marine turtles were exposed to polluted effluents containing pharmaceutical residues, antibiotic resistant bacteria and heavy metals. This occurred throughout their habitats and migratory routes (Miranda and Zemelman, 2001; Al-Rawahy et al., 2007; Al-Bahry et al., 2009a, 2009b, 2011b; Foti et al., 2009; Kümmerer, 2009). It was suggested that green turtles may use spatially distinct feeding habitats along the coastal areas (Talavera-Saenz et al., 2007) and that tissue metal profiles could be used as means to determine these specific feeding areas (Franzellitti et al., 2004). Also, eggs taken from green turtles during oviposition were heavily contaminated which is harmful to the sea turtle population (Al-Rawahy et al., 2007; Al-Bahry et al., 2009a; 2009b; 2009c).

The decline of various marine animals has been attributed to heavy metal intoxication (Storelli et al., 2008; Harper et al., 2007). Continuous contamination of living organisms to heavy metals can lead to a gradual accumulation in various tissues and ultimately fatal (Fent et al., 1996). A total of 12 different heavy metals in eggs and hatchlings of green turtles in Oman were detected (Al-Rawahi et al., 2007). Such contamination could be as a result of contaminated effluent (Al-Bahry et al., 2009a; 2009b; 2009c). Some of these detected metals applied to industrial applications. For instance, Y is used for manufacturing color TV's, computers' screens and radars. Yb is known to be used in metallurgical and chemical experiments. Te is used for coloring of glass, ceramics and thermoelectric devices manufacturing while Zr is known to be for nuclear applications (ECHM, 2013).

The Ministry of Regional Municipalities, Environment and Water Resources (MRMEWR, 2006) conducted the tracking of nesting sea turtles. The data showed a decline in sea turtle population in Oman. A continuing trend will have a serious detrimental effect on the already endangered species of green turtles at Ras Al-Hadd, as well as other marine turtles around the world.

References

Akinbowale, O.L., Peng, H., Grant, P., Barton, M.D., 2007. Antibiotic and heavy metal resistance in motile aeromonads and pseudomonads from rainbow trout (*Oncorhynchus mykiss*) farmsinAustralia. *International Journal of Antimicrobial Agents.* 30: 177–182.

Al-Bahry, S.N., Mahmoud, I.Y., Elshafie, A.E., Al-Harthy, A., Al-Ghafri, S., Al-Amri,I., Alkindi, A.Y. (2009a). Bacterial flora and antibiotic resistance from eggs of green turtles *Chelonia mydas*: an indication of polluted effluents. *Marine Pollution Bulletin.* 41: 214-221.

Al-Bahry, S.N., Mahmoud, I.Y., Al-amri, I.S., Ba-Omar, T.A., Melgheit, K.O., Al-Kindi, A.Y. (2009b). Ultrastructural features and elemental distribution in eggshell during pre and post hatching periods in the green turtle, *Chelonia mydas* at Ras Al-Hadd, Oman. *Tissue and Cell.* 58: 720-725.

Al-Bahry, S.N., Mahmoud, I.Y., Al-Belushi, K.I., Elshafie, A.E., Al-Harthy, A., Bakheit, C.K. (2009c). Coastal sewage discharge and its impact on fish with reference to antibiotic resistant enteric bacteria and enteric pathogens as bio-indicators of pollution. *Chemosphere.* 77: 1534-1539.

Al-Bahry, S.N., Mahmoud, I.Y., Al-Khaifi, A., Elshafie, A.E., Al-Harthy, A. (2009d). Viability of multiple antibiotic resistant bacteria in distribution lines of treated sewage effluent used for irrigation. *Water Science & Technology.* 60: 2939-2948.

Al-Bahry, S.N., Mahmoud, I.Y., Al-amri, I.S., Ba-Omar, T.A., Melgheit, K.O., Al-Kindi, A.Y. (2009e). Ultrastructural features and elemental distribution in eggshell during pre and post hatching periods in the green turtle, *Chelonia mydas* at Ras Al-Hadd, Oman. *Tissue and Cell.* 58: 720-725.

Al-Bahry SN, MahmoudI, Al-Rawahi S and Paulson J. 2011a. Egg contamination as an indicator of environmental health. *In*: Impact of Egg Contamination on Environmental Health. Nova Science Publisher Inc. New York.

Al-Bahry SN, Mahmoud IY, Melghit K, Al-Amri I. 2011b. Analysis of Elemental Composition of the Eggshell before and after Incubation in the Loggerhead Turtle *Caretta caretta* in Oman. *Microscopy and Microanalysis.* 17: 1-9.

Al-Bahry SN, Mahmoud IY, Al-Musharafi SK. 2014. The effect of physical factors on fecal coliform viability rate in sewage sludge. *Journal of Geoscience and Environment Protection.* (In press)

Al-Bahry SN, Al-Zadjali MA, Mahmoud IY, Elshafie AE. 2012. Biomonitoring marine habitats in reference to antibiotic resistant bacteria and ampicillin resistance determinants from oviductal fluid of the nesting green sea turtle, *Chelonia mydas*. *Chemosphere*. 87: 1308-1315.

Al-Musharafi SK, Mahmoud IY, Al-Bahry SN 2012. Heavy Metal Contamination From Treated Sewage Effluents. 11th International Conference on Modelling, Monitoring and Management of Water Pollution. New Forest, UK. WIT Transactions on Ecology and the Environment, Vol. 164. WIT Press. UK. 2012.

Al-Musharafi SK, Mahmoud IY, Al-Bahry SN. 2013a. Heavy metal pollution from treated sewage effluent. *Procedia APCBEES*. 5: 344-348.

Al-Musharafi SK, Mahmoud IY, Al-Bahry SN. 2013b. Contamination of marine fish by heavy metals from coastal sewage treated effluent runoff. The Third Asian Conference on Sustainability, Energy and the Environment. acsee2013. Osaka, Japan. The International Academic Forum (IAFOR). Pages 100-107.

Al-Musharafi SK, Mahmoud IY, Al-Bahry S.N. 2014a. Environmental contamination by industrial effluents and sludge relative to heavy metals. *Journal of Geoscience and Environment Protection*. (in press).

Al-Musharafi SK, Mahmoud IY, Al-Bahry SN 2014b. Environmental hazards and pollution from liquid waste lagoons. *IPCBEE* (In press).

Al-Rawahy SH, Al Kindi AY, Elshafie A, Mahmoud IY, Al-Bahry SN, Al-Siyabi SS, Mansour MH, Al-Kiyumi AA. 2007. Accumulation of metals in the egg yolk and liver of hatchling of green turtles *Chelonia mydas* at Ras Al Hadd, Sultanate of Oman. ANSI: *Journal of Biological Sciences*. 7: 925-930.

Bass L, Liebert CA, Lee MD, Summers AO, White DG, Thayer SG, Maurer JJ. 1999. Incidence and characterization of integrons, genetic elements mediating multiple-drugresistance, in avian *Escherichia coli*. *Antimicrobial Agents and Chemotherapy*. 43: 2925–2929.

Burger J, Coachfeld M.1991. Cadmium and lead in common terns (Aves: *Sterna hirundo*): relationship between leveles in parents and eggs. *Envirnmental Monitoring and Assesmennt*. 16: 253-258.

ECHM (Environmental Chemistry and Hazardous Materials). Periodic table of elements. http://environmentalchemistry.com/yogi/periodic. Acessed March, 31.2013.

Edwards JW, Edyvane KS, Boxall VA, Hamann M, Soole KL. 2001. Metal lelvels in Seston and marine fish flesh near industrial and metroplitan centers in South Australia. *Marine Pollution Bulletin*. 42: 389-396.

Foti M., Giacopello C., Bottari T., Fisichella V., Rinaldo D., Mamminac C. 2009. Antibiotic resistance of Gram negatives isolates from loggerhead sea turtles (Caretta caretta) in the central Mediterranean Sea. *Marine Pollution Bulletin*. 58: 1363-1366.

Fent K. Ecotoxicology of organotin compounds. 1996. *Critical Reviews in Toxicology*. 26: 1-117.

Franzellitti S, Locatelli C, Gerosa G, Vallini C, Fabbri E. 2004. Heavymetals in tissues of loggerhead turtles (*Caretta caretta*) from the northwestern Adriatic Sea. *Comparative Biochemistry and Physiology* C. 138:187–94.

Harper ER, St Leger JA, Westberg JA, Mazzaro L, Schmitt T, Reidarson TH, Tucker M, Cross DH, Puschner B. 2007. Tissue heavy metal concentrations of stranded California sea lions (*Zalophus californianus*) in Southern California. *Environmental Pollution*. 147:677-682,.

Han FX, Banin A, Kingery WL, Triplett GB, Zhou LX, Zheng SJ, Ding WX. 2003. New approach to studies of heavy metal redistribution in soil. *Advances in Environmental Research*. 8: 113- 120.

He ZL, Yang XE, Stoffella PJ. 2005. Trace elements in agroecosystems and impacts on the environment. *Journal of Trace Elements in Medicine and Biology*. 19: 125-140.

Hölzel CS, Müller C, Harms KS, Mikolajewski S, Schäfer S, Schwaiger K, Bauer J. 2012. Heavy metals in liquid pig manure in light of bacterial antimicrobial resistance. *Environmental Research* 113:21–27.

Kim S, Aga DS. 2007. Potential ecological and human health impacts of antibiotics and antibiotic-resistant bacteria from wastewater treatment plants. *Journal of Toxicology Environment and HealthB*. 10: 559–573.

Kojadinovic J, Potier M, Le CM, Cosson RP, Bustamante P. 2007. Bioaccumulation of trace elements in pelagic fish from the Western Indian Ocean. *Environmental Pollution*. 146: 548-566.

Kümmerer K., 2009. Antibiotics in the aquatic environment. A review. Part II. *Chemosphere* 75, 435-441.

Mahmoud IY, Al-Bahry SN, Al-Musharafi SK. 2013. Fresh water habitat pollution by treated sewage effluent in relation to multiple-antibiotic-resistant bacteria. *APCBEES Procedia*. 5: 363-367.

Miranda, C.D., Zemelman, R., 2001. Antibiotic resistant bacteria in fish from Concepcion Bay, Chile. *Marine Pollution Bulletin*. 42: 1096–1102.

MRMEWR, 2006. Marine Pollution and Coastal Zone Management Section: Indian Ocean South-East Asian Marine Turtle Memorandum of Understanding. Ministry of Regional Municipalities and Environment and Water Resources (MRMEWR). Sultanate of Oman.

Shahidul, I. and Tanaka, M., Impacts of pollution on coastal and marine ecosystems including coastal and marine fisheries and approach for management: a review and synthesis. Marine Pollution Bulletin, 48, pp. 624-649, 2004.

Storelli MM, Barone G, Storelli A, Marcotrigiano GO. 2008. Total and subcellular distribution of trace elements (Cd, Cu and Zn) in the liver and kidney of green turtles (*Chelonia mydas*) from the Mediterranean Sea. *Chemosphere*. 70: 908-913.

Talavera-Saenz A, Gardner SC, Riosmena Rodriquez R, Acosta Vargas B. 2007. Metal profiles used as environmental markers of green turtle (*Chelonia mydas*) foraging resources. *Science of the Total Environment*. 373: 94-102.

Weigel S. 2003. Occurrence, Distribution and fate of pharmaceuticals and further polar contamination in the marine environment. PhD Dissertation, University of Hamburg, Hamburg.

In: Biotechnology and Conservation … (2 Volume Set)
Editors: S. A. Farooq, R. Abed and S. Baqir
ISBN: 978-1-63117-869-6

Chapter 14

DISTRIBUTION OF RH BLOOD GROUP IN TABUK, KINGDOM OF SAUDI ARABIA

***Amal Ibrahim Zair*[2], *Ahlaam Rajaa Al Shahrani*[2], *Shafiq Hussain*[2] *and S. A. Farooq*[1]**
[1]Department of Biology, Sultan Qaboos University, Al Khoud, PC123, Oman
[2]Department of Biology, College of Science for Women, Mahrjan, University of Tabuk, Tabuk, KSA

ABSTRACT

A study was conducted to find the frequency of the Rhesus blood group in Saudi female students at Tabuk. The results were compared with the hospital data and other studies from the Kingdom of Saudi Arabia. Samples were collected from apparently healthy Saudi female students studying at College of Science for Women, Mahrjan, Tabuk. Data from King Khalid Hospital was used for comparison. Percentages of Rh- and Rh+ were calculated in each group. Rhesus positive college students were 91.7% and Rhesus negative were 8.3%. The percentage of Rh+ and Rh- in male hospital cases was 93.3% and 6.6 %, while in females it was 90.7% and 9.2% respectively. The overall frequency of Rhesus positive in this group was 91.8 % and that of Rh negative was 8.18 %. The information of frequencies of Rh blood group in Saudi Arabia is very important because of its role in blood transfusions and haemolytic disease in various populations.

Keywords: Frequency, Rhesus blood group, haemolytic disease, Tabuk, Saudi Arabia

INTRODUCTION

The Rh blood group is the best known system after the ABO in clinical practice. The Rh antigen is named after the rhesus monkey, *Macaca mulatta* where it was initially detected. The mode of inheritance of the antigen is complex and is effectively used in modern medicine, genetic research, anthropology and in tracing ancestral relations in humans. The Rh system is highly polymorphic, employing 44 distinct antigens, but clinically the most significant one is the presence or absence of the Rh (D) antigen on red blood cells. The Rh antigens are carried on three nonglycosylated trans-membrane proteins that are encoded by two genes, RHD and RHCE (Arce et al., 1993; Avent et al., 1991; Simse et al., 1994). Alternative mRNA splicing is responsible for the production of two distinct polypeptides from the single RHCE gene. The lack of D antigen expression is usually due to the absence of the entire RHD gene in the genome of Rh (D) negative individuals (Colin et al., 1991). The ABO gene is located on the long arm of the ninth human chromosome (9q34.1) (Al-Arrayed et al., 2001) while the Rh(D) gene encoding the Rh protein is located on chromosome 1p34–p36 (Cartron, 1994). The Rh blood system is important because the Rh antibody causes severe hemolytic disease of the newborn (HDN) (Van der Schoot et al., 2003; Wagle and Deshpande, 2010). The population studies all over the world indicate varied ratios for ABO and Rh blood group systems. The gene cloning studies of the Rh system has tremendously advanced our understanding of the Rh alleles and further studies are still in progress as new alleles begin to be elucidated (Chou, 2010). The information on the gene frequencies of the ABO and Rh groups is limited from the Middle Eastern countries especially from Tabuk, a province in Kingdom of Saudi Arabia. An attempt is made to provide information on the distribution pattern of the phenotypes and genotypes of the genetic variants in people living Tabuk, KSA.

People are positive if they have a certain Rh antigen (the D antigen) on the surface of their erythrocytes, and people are Rh – negative if they do not have this Rh antigen. Rh incompatibility can pose a major problem in some pregnancies when the mother is Rh – negative and the foetus is Rh – positive (Avent, 1999). If foetal blood leaks through the placenta and mixes with the mother's blood, the mother becomes sensitized to the Rh antigen. The mother produces Rh antibodies that cross the placenta and cause agglutination and haemolysis of foetal erythrocytes. This disorder is called Haemolytic disease of the newborn (HDN) or erythroblastosis foetalis, and it may be fatal to the foetus (Dennis et al., 1998). Rh-D distribution varies in different parts of the world. Rh-D negative blood group is documented as 5.5% in south India, 5% in Nairobi, 4.8% in Nigeria, 7.3% in Lahore, 7.7% in Rawalpindi (Mwangi, 1999; Omotade et al., 1999; Bhatti and Amin, 1996). About 95% of African- Americans are Rh-positive whereas indigenous Africans are virtually 100% Rh-positive. As regards to A, B, AB and O blood groups in Saudi Arabia, the frequencies are 0.24, 0.17, 0.04, and 0.52 respectively (Khattak et al., 2008). Rhesus (Rh) blood group antigens are useful in population genetic studies in resolving medico-legal issues and more importantly in compatibility test in blood transfusion practice.

The frequencies of Rh-D blood groups vary from one population to another. The main objective in this type of study is to determine the frequencies of the Rhesus D blood groups and to assess the preparedness of the hospital based patient targeted blood donor unit in the provision of blood and blood products and also to determine the blood groups of patients attending the antenatal clinic over the same period. The ABO blood groups and Rhesus (Rh) blood group antigens are the most frequently studied genetic markers in a large group of people (Worlledge et al., 1974) The knowledge of the distribution of ABO and Rh blood groups is essential for the effective management of blood bank inventories, be it a facility of a smaller local transfusion service or a regional or national transfusion service. Apart from their importance in blood transfusion practice, the ABO and Rh blood groups are useful in population genetic studies, researching population migration patterns, as well as resolving certain medico-legal issues, particularly of disputed parentage. It is, therefore, imperative to have information on the distribution of these blood groups in any population group.

Of the thirty blood group systems known so far, the Rh (Rhesus) factor is the most researched blood group after ABO. It consists of 50 defined blood-group *antigens,* among which the 5 antigens D, C, c, E, and e are the most important. These are complex oligosaccharides present on red cell surface. The commonly-used terms Rh factor, Rh positive and Rh negative refer to the D antigen only. If the D antigen is present, the individual is Rh positive. If it is absent, he is Rh negative. Apart from its role in *blood transfusion*, the knowledge of Rh blood group system is used to determine the risk of *hemolytic disease of the newborn* or *erythroblastosis foetalis* and to prevent it. In contrast to the ABO blood group, antibodies against Rh are formed only during *blood transfusion* or during pregnancy.

Blood was collected by finger-prick from a total of 96 apparently healthy Saudi female students at College of Science for Women, Mahrjan, Tabuk They were evaluated for the Rh factor. Presence of the Rh D antigen was determined by testing the red cells with anti-D reagent by standard tube method (Walker et al., 1993). Red cells carrying D antigen are directly agglutinated by anti-D reagent. The samples showing agglutination with antisera were classified as Rh+ and the others as Rh-. Samples were compared with standard controls. For comparison, data of 110 individuals from King Khalid Hospital, Tabuk was used. This data consisted of Saudi males and females referred to the hospital for various medical ailments. As regards the Rhesus blood group system, it was noticed that 91.7% of the sampled population were Rh (D)+ve while 8.3% were Rh (d)-ve (Table 1). There was higher proportion of Rh (D)+ve individuals than the Rh-ve in the studied population. The observed frequency for Rh positive was 91.0%.

The percentage of incidence of Rh positive in college students was 91.7% and Rh negative was 8.3%. In the hospital female cases, the percentage of Rh+ was 90.8% and Rh- was 9.2 %; whereas the males showed an incidence of 93.3% and 6.7 % respectively. The percentage of Rh- was slightly lower in students. Overall percentage of Rh- was 8.18% in the hospital data which was close to that observed in the student sample.

These studies did not show any remarkable deviation in the frequencies of Rh blood group. This observation was attributed to population drift and mixing of genes. The Rh factor is of particular importance to females at or below childbearing age, because any subsequent pregnancy may be affected by the *Rhesus D hemolytic disease of the newborn* if the baby is Rh positive. Such a situation needs to be suitably monitored. The values in the present report are in conformity with the observations of Ozsoylu and Alhejaily (1987) who found a frequency of Rh- in Al Madina Munawara and Tabuk samples to be 10 %, and the studies of Sarhan et al., (2009) where it was reported to be 7.2%.

Table 1. The distribution of Rh factor among college students and patients in hospital

Type	Number of individuals	Rh+ n (%)	Rh- n (%)
Female students	96	88 (91.67)	8 (8.33)
Total patients	110	101 (92.6)	9 (8.1)
Male patients	45	42 (93.3)	3 (6.7)
Female patients	65	59 (90.8)	6 (9.2)

Rh-D distribution also varies within any group of the human population. In overall, the total percentage of RhD positive was 91% and that of RhD negative was found to be 9%. Over the years, the Rh blood group system has been distributed among any population and maintained a low frequency of RhD negative because the clinical situations could arise through Rh blood incompatibility. Similar pattern of distribution is also observed in the following studies. The Rh-D negative blood group is documented as 5.5% in South India, 5% in Nairobi, 4.5% in Nigeria, 7.3% in Lahore, 7.7% in Ralwalpindi (Das et al., 2001; Mwangi, 1999; Omotade et al., 1999; Majeed and Hayee, 2002; Bhalti and Amin, 1996).

This study correlates well with the other studies conducted in the Kingdom of Saudi Arabia, that the incidence of Rh- is approximately 8 – 9% (Bashawri et al. 2001), which is much less as compared to the Western countries (Vengelen-Tyler,1996). Thus the number of cases of hemolytic disease of the newborn (HDN) are expected to be much lower in Saudi Arabia. According to the findings of Ohlsson and Badawi (1985), hemolytic disease of the newborn runs a moderate to severe course in the Kingdom.

Our study presents the incidence of Rh factor in Tabuk and compares it to different areas of Saudi Arabia. Such findings may help in formulating a national figure for future reference (Bashawri et al., 2001). Research on blood groups can never be over-emphasized as the information is useful not only for blood transfusion, disease association but also for the study of evolutionary mechanisms acting on populations.

CONCLUSION

The frequency of Rh phenotypes in Tabuk appears to be similar to other published data. The results show that the Rh-negative frequency is 8.33% in the female students. This study has a significant implication in the management of blood banks and transfusion services in this area. Knowledge of blood group phenotype distribution is important for clinical studies (for example disease association), as well as for population studies. It is necessary to conduct similar well designed studies in other Middle Eastern countries to determine the blood group frequencies. The data generated in the present study is useful for health planners to face the future health challenges in the region. The generation of a simple database of blood groups not only provides data on the availability of human blood in cases of regional calamities, but also serves to detect the possibilities of future burdens of disease.

REFERENCES

Al-Arrayed S., Shome D. K., Hafadh N., Amin S., Al Mukhareq H., Al Mulla M., (2001). ABO blood group and Rhd phenotypes in Bahrain: results of screening school children and blood donors. *Bahrain Med. Bull.*, 2001; 23.

Arce, M. A., Thompson, E. S., Wagner, S., Coyne, K. E., Ferdman, B. A. and Lublin, D. M. (1993) Molecular cloning of RhD cDNA derived from a gene present in RhD-positive, but not D-negative individuals. *Blood,* 82: 651-655.

Avent N. D. (1999). The rhesus blood group system: insights from recent advances in molecular biology. *Transfusion Med. Rev.*, 13: 245-266.

Avent, N. D., Ridgewell, K., Tanner, M. J. A. and Anstee, D. J. (1991) cDNA cloning of a 30 kDa erythrocyte membrane protein associated with Rh (Rhesus) blood group expression. *Biochem. J.*, 271: 821–825.

Bashawri L. A., AlMulhim A. A., Ahmed M. S., Ahmed M. A. (2001) Frequency of ABO blood groups in the Eastern region of Saudi Arabia. *Saudi Med. J.*, 22:1008-1012.

Bhalti F. A., Amin (1996). Spectrum of ABO and D blood groups of donors at Rawalpindi/Islamabad. *Pakistan J. Pathol.*, 7:26-28.

Cartron J. P. (1994) Defining the Rh blood group antigens: biochemistry and molecular genetics. *Blood Rev.*, 8:199–212.

Chou S. T., Westhoff C. M. The Rh and RhAG blood group systems. (2010) *Immunohematology,* 26:178–86.

Dennis, Y. M., N. M. Hylem, C. Fidler, I. L. Sargent and M. F. Murphy et al., 1998. Prenatal diagnosis of fetal RhD status by molecular analysis of maternal plasma. *New Engl. J. Med.*, 337: 1734-1738. PMID: 9845707.

Khattak I. D., Khan T. M., Syed P., Shah A. M., Khattak S. T., Ali A. (2008) Frequency of ABO and Rhesus blood groups in district Swat, Pakistan. *J. Ayub. Med. Coll. Abbottabad.*, 20:127–9.

Mawuagi J. (1999). Blood group distribution in an urban population of patient targeted blood donors. *East Afr. Med. J.,* 76:615 – 618.

Ohlsson A., Badawi B. (1985) Moderate to severe HDNB in a referral center in Saudi Arabia. *Annals of Saudi Medicine*, 5: 207-211.

Omotade O. O., Adeyemo A. A., Kayode C. M., Falade S. L., Ikpeme S. (1999). Gene frequencies of ABO and Rh (D) blood group alleles in a healthy infant population in Ibadan, *Nig. West Afr. J. Med.*, 18:291–297.

Ozsoylu S., Alhejaily M. (1987) The distribution of ABO and Rhblood groups in the Tabuk regions and Madinah Munnawwara, Saudi Arabia. *Turk. J. Pediatr.*, 29: 239-241.

Sarhan M. A., Saleh K. A., Bin-Dajem S. M. (2009)Distribution of ABO blood groups and Rhesus factor in Southwest Saudi Arabia. *Saudi Med. J.,* 30:116-119.

Simse, S., de Jong, C. A. M., Cuijpers, H. T. M., Westers, T. M., Overbeeke, M. A. M., Goldschmeding, R., van der Schoos, C. E. and von dem Borne, A. E. G. K. (1994) Sequence analysis of cDNA derived from reticulocyte mRNAs coding for Rh polypeptides and demonstration of E/e and C/c polymorphisms. *Vox. Sang.*, 67: 203–209.

Van der Schoot C. E., Tax G. H., Rijnders R. J. (2003) Prenatal typing of Rh and Kell blood group system antigens: The edge of a watershed. *Transfus. Med. Rev.*, 17:31–44.

Vengelen-Tyler V. (1996) ABO, H and Le blood groups. In: American Association of Blood Banks Technical Manual. 12th ed. Maryland USA: 229-254.

Wagle S., Deshpande P. G. Hemolytic disease of newborn (2010). In: Itani Oussama, editor. eMedicine.

Walker R. H., Hoppe P. A., Judd W. J. (1993) Technical manual. American Association of Blood Banks, Arlington(VA).

Worlledge S., Ogiemudia S. E., Thomas C. O., Ikoku B. N., Luzzutto L. (1974) Blood group antigens and antibodies in Nigeria. *Ann. Trop. Med. Parasitol.*, 68:249–64.

In: Biotechnology and Conservation … (2 Volume Set) ISBN: 978-1-63117-869-6
Editors: S. A. Farooq, R. Abed and S. Baqir

Chapter 15

A Comparison between Representative Types of Platyhelminthes in Their Nervous Components

Zohair I. F. Rahemo*
Department of Zoology and Comparative Physiology,
University of Birmingham, UK

Abstract

A trial was made to explore the nervous systems in some representatives of the main Platyhelminthes groups, turbellarians, digeneans, monogeneans and cestodes. *Mesostoma ehrenbergi* and *Udonella caligorum* were selected as the representatives of turbellarians; *Temnocephala fulva, Phyllodistomum folium* and *Gorgodera vitelliloba* as digeneas; *Khunia scombri* as monogenean; a cyclophyllid *Hymenolepis diminuta*; a tetraphyllid and a cestode *Phyllobothrium pirieri*. The nerves distribution, cerebral ganglia and their anterior connections and posterior nerve trunks of each species are given using the enzymatic technique esterase acetylthiochole iodide.

Keywords: Helminthes, nervous system, enzymatic technique

Introduction

Using simple techniques, the nervous structure of platyhelminth groups are studied [1, 2]. Turbellarian nervous systems are reported by investigators [3-7]. By employing advanced techniques, amide-related peptides were detected in turbelarians [8]. The nervous systems of digeneans have been studied using histological techniques in adults [9-12]; while studies on the nervous system of cercariae are reported using enzymatic techniques [13-19].

* Present address: Department of Biology, College of Science, University of Salahaddin, Erbil, Iraq. E-mail: Zohair_rahemo@yahoo.com.

A study was carried out on the nervous system and chaetotaxy of *Cercaria paludinae* using both enzymatic and silver nitrate impregnation [20]. Digeneans obtained from frogs were studied for their nervous structures [21]. Work was done on cholinergic components of the nervous system of *Haplorchoides* cahirinus and *Acanthostomum absconditum* (digenea). The gross anatomy of the muscle and systems nervous system of metacercaria *Apatemon cobiridis proterohini* is reported as visualized by confocal microscopy [23].

Researches on monogenean nervous system were initiated by [24 and 25]. Noteworthy studies of individual monogeneans have been published [26] on *Diplozoon paradoxum* [27], *Polystomoides malayai* [28] and *Diclidophora* merlangi using both conventional and enzymatic techniques. A detailed description of the nervous system of *Polystoma intregerrimum* was given in Ref. [29]. Biogenic amines were detected in the nervous system of *Eudiplozoon nippoicum* [30].

Furthermore [31] studied the nervous system and chaetotaxy of *Macrogyrodactylus clarii* and *M. congolensis* with a note on argentophilic elements in the nervous system. Using immunomicroscopical techniques [32] traced the nervous system of *Eudiplozoon nipponicum.* The cholinergic components of the nervous enzymatic techniques have been investigated in *P. anguillae* from the eel *Anguilla anguilla* in Nile Delta waters [33], while [34] studied the nervous system of the monogenean fish parasite, *Macrogyrodactylus clarii.* Recently [35] published research on the nervous system of four peculiarly clamped monogeneans namely *Gastrocotyle trachuri*, *Microcotyle donovani*, *Axine belones* and *Pseudaxine trachuri*; and it has been concluded that the haptor innervations cope with the modification of haptor with its armature. The most important work on cestodes gross morphology using classical histological collected from birds were also investigated for their nervous system using enzymatic techniques [43]. Furthermore, the nervous system of *Raillietina echinobothrida* was studied using acetylthiocholine activity and anthelmintic efficacy of certain plant extracts [44].

It is well known that the nerves of Platyhelminthes are unmylinated so they are very difficult to distinguish from surrounding mesenchyma so enzymatic techniques were developed. The aim of the present study is to make a comparative study on the nervous system using enzymatic techniques of selected types of Platyhelminthes groups namely, *Mesostoma* ehrenbergi and *Udenella caligorum* representing turbellarian, and Temnocephala *fulva*, *Phyllodistomum folium* and *Gorgodera vitelliloba* representing digeneans, *Kuhnia scombri* representing monogenean, and *Phyllobothrium pireri* and *Hymenolepis nana* representing cestodes.

Specimens of *M. ehrenbergi* were taken from laboratory of genetics, University of Birmingham, UK. *U. caligorum* is found on the parasitic copepod, *Caligus minimum* which infects the buccal cavity of the bass *Dicentrarchus labrax* was obtained from marine biological station at Plymouth. *T. fulva* lives on the exoskeleton of the Tasmanian crayfish, *Parasitacoides tasmanicus*; *K. scombri* from Markel; *Scomber scombrus* were obtained from marine biological station at Plymouth, UK. *P. folium* was recovered from urinary bladder of three-spine stickleback and *G. vitelliloba* was collected from the urinary bladder of the toad. *H. nana* was collected from the intestines of white laboratory rats. Specimens were fixed in 10% formalin, then washed in water, and incubated in the working solution of acetylthiocholine iodide for esterases [45].

Temnocephala fulva (figure 4) consists of two prominent cerebral ganglia connected to a pair of eyes which lie just dorsal to these ganglia. Two large nerves arise from these ganglia to supply the anterio-lateral part of the worm while 2 less prominent nerves pass in the posterior quarter of the body. The nervous system of *M. ehrenbergi* has a characteristic coarse-meshed plexus making a superficial cover all over the body. This plexus is connected to the anterior and posterior nerve cords and becomes denser at the anterior end of the nerves, possibly to connect nerve endings, since this area is an exploratory region of the animal to detect its food and surroundings. The presence of coarse-meshed plexus that forms a superficial cover all over the body is similar to that found in other flatworms [4, 36- 38] using ultra structural studies [39- 40] enzymatic techniques [41- 42]. The cestode *U. caligorum* is similar in its basic pattern to that of monogeneans, but the ventral nerve cord becomes stout before entering the posterior sucker and gives off fine nerves. The supply of the posterior sucker is similar to innervation supply in monogeneans [35] but no such supply was found in digeneans [13, 21] which may indicate close relation to the monogeneans than to the digeneans.

The main character of the nervous system of *T. fulva* is the presence of the three nerve trunks namely the ventral, lateral and the dorsal nerve cords. In addition, there are 2 other nerve cords situated close to the lateral margins of the animal. All these cords are connected together by many transverse commissures to give a network appearance. The other interesting feature of *T. fulva* is the prominent innervation of the anterior tentacles, "anterior lobe" and the posterior sucker. The posterior sucker has characteristic radial nerves. Innervation of the anterior lobes "tentacles" are located in the exploratory organelles of the animal to detect its surroundings and are similar to the innervation observed in *M. ehrenbergi*. The high number of nerve cords and transverse commissures detected in this worm is similar to the high number (more than 60) of nerve cords present in cestodes [1].

The nervous system of *Phyllodistomum folium* (Figure 5) consists of 2 cerebral ganglia which give off 3 pairs of small anterior nerves to supply the oral sucker and 3 pairs of prominent nerves anteriorly, viz. ventral, lateral and dorsal, to supply to the main body parts. From the ventral nerve cords two ventral commissures arise near the ventral sucker and surround it on all sides and then send fine nerves to the sucker. The sucker itself has an intrinsic nerve ring which sends a number of radiating nerves anteriorly. In addition, a few multipolar cells are present near the inner border of the ventral sucker. It is worthy of note that *P. folium* has numerous ganglia just beneath the body surface with some commissures connecting them (commissural ganglia). A similar pattern of innervations were recovered in other digeneans such as *Ceylonocotyle scoliocoelium* [15] *Fasciola hepatica* and *Dicrocoelim dendriticum* [17] and others. The nervous system of *Gorgodera vitelliloba* is very similar to that of *P. folium*, including the network of commissural ganglia near the surface of the body. The innervation recovered in this study is very similar to the *P. folium* parasitizing frog in Iraq [21]. In general, the digeneans examined in this investigation are similar in their nervous system to the generalized pattern of the trematode nervous system reported by [2].

MONOGENEA

The nervous system of *Kuhnia scombri* (Figure 6) consists of two cerebral ganglia giving off 2 -3 pairs of anterior nerves to supply the anterior feeding or attachment organs and 3 pairs of prominent posterior nerves. The three pairs of posterior nerve trunks fuse together forming two pre-haptoral ganglia from which 2 main nerves arise to supply the main attachment organ, the posterior haptor, an organ corresponding in function to the scolex of cestodes and the ventral sucker of digeneans. The nervous system of monogeneans is modified according to the type of haptor and its attachment organs. The haptor of *K. scombri* possesses four pairs of clamps and the posteriormost part of the haptor (the languette) has one pair of large hamuli and 2 pairs of haptors. 2 main nerves arise which run along each side of the haptor. These nerves, after supplying each of the 4 pairs of clamps, run posteriorly and supply to each hamulus with 2 nerves; one of them is more prominent and supplies the gurd (spur) muscles of the hamuli and form a large irregular ring; while the other, which is less prominent, supplies the proximal end of the hamulus by fusing with the tendons of the posterior muscles present near the proximal region of the hamulus. Similar results were obtained in four peculiarly-clamped monogeneans, such as *Gastrocotyle trachuri*, *Microcotyle donovani*, *Axines belones Pseudoaxine trachuri* [35] and in *Macrogyrodactylus clarii* a gill parasite of Nile catfish [34].

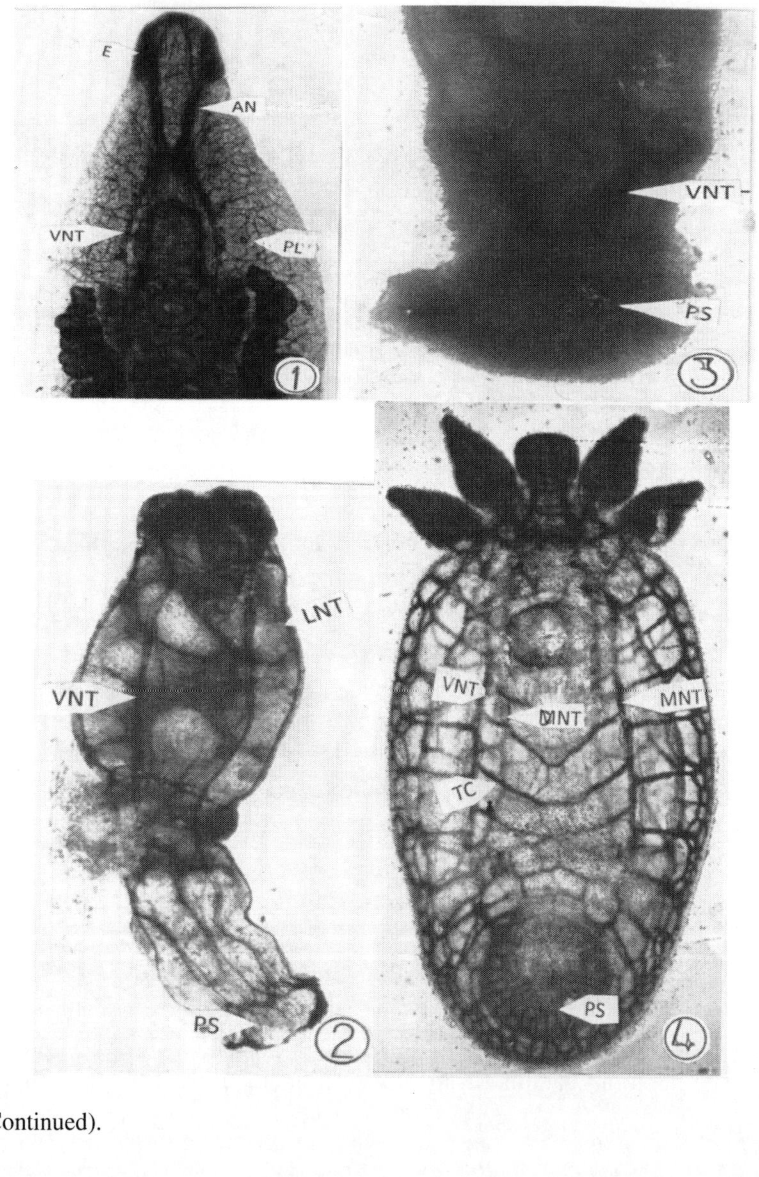

Figure (1): (Continued).

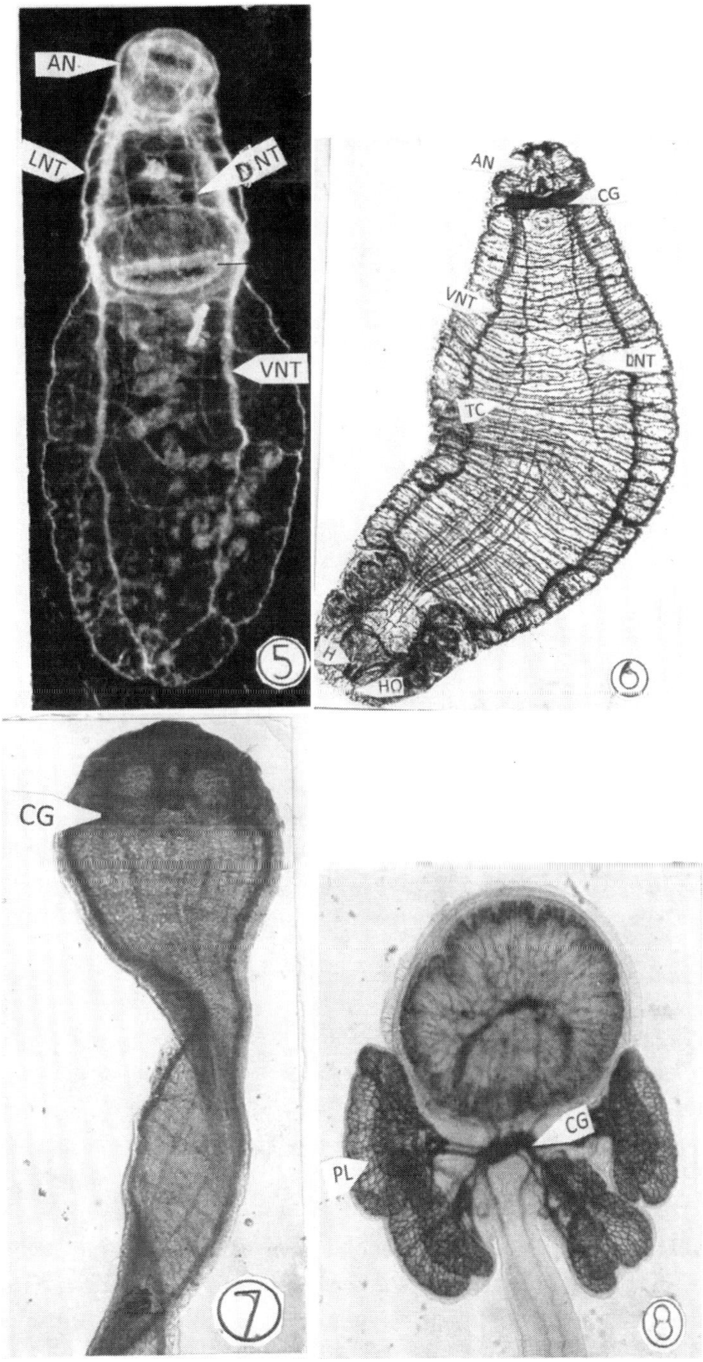

Abbreviations: AN: anterior nerve; CG: cerebral ganglia; DNT: dorsal nerve trunk; E: eye; H: hamuli; HO; hooks; LNT: lateral nerve trunks; MNT: median nerve trunk; PL: plexus; PS: posterior sucker; PHG: pre-haptoral ganglia; TC: transverse commissure; VNT: posterior nerve trunk.

Figure (1): Photomicrograph (transmitted light) of *M. ehrenbergi* showing its nervous system as revealed by acetylthiocholine iodide technique (AcThT); Figure (2): *U. caligorum*. Figure (3): The posterior end of *U. caligorum*. Figure (4): *T. fulva*. Figure (5): Epi-illumination of P. folium. Figure (6): K. scombri. Figure (7): anterior end of *H. diminuta*. Figure (8): scolex of *P. pirieri*.

CESTODA

In the two species studied, *Hymenolepis diminuta* and *Phyllobothrium pirieri* the main concentration of the nervous system is in the scolex (Figures 7, 8). The scolex itself exhibits different forms in different groups of cestodes as observed in the cyclophyllid and tetraphyllid studied in the present investigation.

From the two cerebral ganglia present in the scolex of *H. diminuta,* many nerves arise to supply the four suckers. In the tetraphyllid, the cerebral ganglia are connected together by a short thick transverse commissure. From each ganglion 2 pairs of nerves arise to supply the two bothridia on each side. The terminal sucker receives about 5 fine nerves. It is worthy of note that the bothria have a nerve network resembling the venation of plant leaves. In each segment of the tetraphyllid and cyclophyllid, there are four pairs of nerves: the main nerve trunk is the lateral and 3 pairs of slender median nerve cords are present. At the posterior end of each segment the lateral nerve trunk swells into a small ganglion.

Many fine commissures are present connecting the main nerve trunk. The details of scolex innervations of *P. pireri* is comparable to those reported by [2] in *Acanthobothrium coronatum*. Similar innervation of scolex and the presence of numerous longitudinal nerve cords and transverse commissures were also seen in *Oochritica sigmoides* and *Raillietina tetragona* [42] and in fowl tapeworm, *Raillietina echinobothria* [44].

REFERENCES

[1] T. H. Bullock, and G. A. Horridge,"*Structure and function in the nervous system in Invertebrates",* Vol.1: W.H.Freeman, 1965.

[2] L. R. Roberts and J. Janovy, "Gerad D. Schmids and Larry S. Roberts. *Foundations of Parasitology*", 8th ed. McGraw Hill-Higher Education, ch. 29, 2009.

[3] V. Hickman."Tasmanian Temnocephalidea" The *Paper and Proceedings of the Royal Society of Tasmania*, vol, 101. pp. 227-250,1967.

[4] H. Koopowitz, P. Chien," Ultrastructure of the nerve plexus in Flatworms: 1.Peripheral organization," *Cell and Tissue Res*, vol.15.pp.337-351, 1974.

[5] J. Moraczewski, A. Czuba, J. Kwiatowska,"Localization of neurosecretion in the nervous system of Catenulida (Turbellaria)", *Zoomorphologie.* Vol, 87, pp. 97-102, 1977.

[6] J. Faisst, C. L. Keenman, and H. Koopowit, "Neuronal repair and avoidance behavior in the flatworm, *Notoplana acticola*", *J.Neurobiol.* vol. 11, pp.483-496.1980.

[7] C.Keenan, R.Coss, H.Koopowitz, "Cytoarchitecture of Primitive Brains: Golgi Studies in Flatworms", *The J.Comp. Neuro.vol.*195, pp. 697-716, 1981.

[8] A. G. Maule, C. Shaw D. W. Halton, W. J. Curry, L. Thim, "PYIR-Famide: a turbellarian FMRF amide related peptide (FaRP)", *Reg.Pep.* vol.50, pp. 37-43.1994.

[9] E. J. Kolmogorova,"Structure of the central parts of the nervous system in *Ophisthorchis felineus* ", (in Russian), *Zool. Zhrn*, vol.38, pp.1627-1633, 1959.

[10] K. Rohde,"Das nervensystem der cercarie von *Catatropis indica* Srivastava, 1935(Digenea: Notocotylidae) under der geschlechtsreifen from von *Diaschistorchis*

multitesticulars Rohde, 1962 (Digenea: Pronocephalidae", *Zeit. Morph. Tiere*. Vol. 62, pp.77-102, 1968, a.

[11] S. K. Lee."A light microscopical study of the nervous system of three species of amphistomes (Trematoda: Paramphistomidae)", *Zool. Anz.* vol.187, pp. 1-25, 1971.

[12] J. B. Jennings, W. B. LeFlore,"The Histochemical demonstration of certain aspects of cercarial morphology", *Trans. Amer. Microsco.Soc.* vol. 91, pp.56-62, 1972.

[13] K. Shyamsundari, K. H. Rao."The structure and cytochemistry of the neurosecretory cells of *Fasciola gigantica* Cobbold and *Fasciola hepatica* L."Zeitsch.Parasitenk. vol.47, pp.103-109, 1975.

[14] Z. Zdarska."Morphology and Histochemisrty of the sensory endings in the metacercaria of the trematode *Echinoparyphium acaniatum*", *Folia Parasit.*vol. 22, pp, 25-31, 1975.

[15] A. K. Bhatnagar, A. N. Gupta, R. S. Srivastav. "Histology and Cytochemistry of the neuroendocrine components of *Ceylonocotyle scoliocoelium* (Digenea: Trematoda)" *Zeitsch. Parasitenk.* Vol.64, pp.77-84, 1980.

[16] D. Halton, "Histochemical studies of carboxylic esterase activity in *Fasciola hepatica*" *The J. Parasit.* Vol.53, pp. 1210-1216, 1967.

[17] A. Ramisz, Z. Szankowska"Studies on the Nervous system of *Fasciola hepatica* and *Dicrocoelim dendriticum* by means of histochemical method for active acetylcholinesterase. *Acta Parasit. Pol.* vol.15, pp.217-223, 1970.

[18] W. B. LeFlore "Histochemical observations on hydrolytic enzymes in cercariae of *Plagiorchis elegans*, with notes on morphology of nervous system". *Trans. Amer. Microsc. Soc.* vol.98, pp. 225-232, 1979.

[19] B. Grabda-Kazubska, T. Moczon. "The nervous system and Chaetotaxy in the cercaria of *Haplometra cylindricea* (Zeder, 1800) (Digenea: Plagiorchiidae)" *Zeitsch. Parasitenk.* Vol. 65, pp.53-61, 1981.

[20] Z. I. F. Rahemo"The nervous system and Chaetotaxy in *Cercaria paludinae* Fillippi, 1854 (Digenea: Monorchidae)", *Mesopotamia J. Agricul.* vol.22, no.2, pp.17-27, 1990.

[21] M. S. A. Mohammad, Z. I. F. Rahemo "The nervous structures of two trematode parasites in frog". Mu"tah Lil-Buhooth Wa Al-Dirasat. Vol.8, no.3, pp. 9-20 December, 1993 (in Arabic).

[22] S. Z. Arafa, E. S. Reda, M. M. EL-Naggar," Cholinergic components of the nervous system of the digenean parasites, *Haplorchoides cahirinus* and *Acanthostomum absconditum* from catfish *Bagrus bayad* in Egypt". *Acta Parasit.*vol.47, pp. 272-279, 2002.

[23] T. M. Stewart., A. Mousley, B. Koubkova, N. J. Marks, D. W. Halton, " Gross anatomy of the muscle systems and associated innervation of *Apatemon* cobiridis *proterorhini* metacercaria (Trematoda: Strigeidea) as visculised by confocal microscopy", *Parasitology* vol. 126, pp. 273-282, 2002.

[24] S. Goto, "studies on the ectoparasitic trematodes of Japan", *J. Coll. Sci.Tokyo.* vol. 8. pp. 1-273, 1894.

[25] J. Andre," Zur morphologie des nerven systems von *Polystoma integerrimum*" *Froel.Zeitsch. Parasitenk.* Vol. 95, pp. 191- 202, 1910.

[26] D. W. Halton, J. B. Jennings, "Demonstration of the nervous system in the monogenetic trematode *Diplozoon paradoxum* Nordmann by the indoxyl acetate method for esterases" *Nature,* vol. 202, pp. 510-511, 1964.

[27] K. Rohde. "Des nervensystem der Gattung *Polystomoides* Ward, 1917 (Monogenea) "The nervous System of the genus Polystomoides Ward, 1917 (Monogenea)". Zeitsch. Morpholo. *Tiere.* Vol. 62, pp.58-76, 1968 b.

[28] D. W. Halton. and G.P. Morris, "Occurrence of cholinesterases and ciliated sensory structures in a fish gill-fluke, *Diclidophora merlangi* (Trmatoda: Monogenea)", *Zeitschrift fur Parasitenkunde*.vol.33, pp.21-30,1969.

[29] Z. I. F. Rahemo, Z. F. F. and N. S. Gorgees, "Studies on the nervous system of *Polystoma integerrimum* as revealed by acetylthiocholine activity" *Parasitol. Res.vol.* 73pp. 234-239, 1987.

[30] L. M. Lyukshina, B. H. Shinov "Biogenic amines in the nervous system of *Eudiplozoon nipponicum* (Monogena)"(in Russian), In: Sakharov D.A.(ed.) simple nervous system". Nauka, Moscow, pp.173-176, 1988.

[31] M. M. El-Naggar, S. Z. Arafa, S. A. El-Abassy,G. C. Kearn,"Chaetotaxy of the monogeneans *Macrogyrodactylus clarii* and *M. congolensis* from the gills and skin of the catfish *Clarias gariepinus* in Egypt, with a note on argentophilic elemnts in the nervous system", *Folia Parasitol.* Vol.48, pp.201-208, 2001.

[32] T. H. Zurawski, A. Mousley; G. P. Mair, G. P. Brennan, A. G. Maule, M. Gelnar and D.W. Halton, "immunomicroscopical observations on the nervous system of adult *Eudiplozoon nipponicum* (Monogenea: Diplozoidae)", *Inter. J. Parasit.* Vol.31, pp. 783-792, 2001.

[33] E. S. Red., S. Z. Arafa." Cholinergic components of the nervous system of monogenean gill parasites, *Pseudodactylogyrus bini* and *P. anguillae* from the eel *Anguilla anguilla* in Nile Delta waters", *Egyptian J. Zool.* vol. 38, pp. 41-54, 2002.

[34] M. M. El-Naggar, S. Z. Arafa, S. A. El-Abbasy, M. T. Stewart.and D.W.Halton, "Neuromusculature of *Macrogyrodactylus clarii*, a monogenean gill parasite of the Nile catfish *Clarias gariepinus* in Egypt." *Parasitol. Res.* Pp.1-15. (online publication).,16/08/2005.2004.

[35] Z. I. F. Rahemo, "Distribution of the Nervous Elements in The Haptor of Four Peculiarly-clamped Monogenean Fish Parasites to Match with its Morphology", *Trends in Parasit. Res.* vol. 1, no.1 pp.,10-14, 2012.

[36] G. Rees. "A contribution to the morphology and life history of *Proteocephalus niloticus* (L.) in China". *Parasitology.* Vo. 53, pp.201-115, 1963.

[37] G. Rees" Cestodes from Bermuda Fishes and an account of *Acompsocephalum tortum* (Linton, 1905) gen.nov.from the lizard fish *Syndus intermedius* (Agassiz). *Parasitology.* vol. 59, pp. 519-548, 1969.

[38] G. Rees, H. H. Williams. "The functional morphology of the scolex and genitalia of *Acanthobothrium coronatum* (Rud...) (Cestoda: Tetraphyllidea)". *Parasitology.* vol.55, pp. 617-651, 1965.

[39] J. Schardein, H. Waitz. "Histochemical studies of esterases in the cuticle and nerve cords of four cyclophyllidean cestodes. *The J. Parasit.* Vol. 51, no. 356-363, 1965.

[40] V.C. Wilson, E.L. Schiller."The neuroanatomy of *Hymenolepis diminuta* and *H. nana. J. Parasit.* vol.55, pp.261-270, 1969.

[41] W. B. LeFlore, B. F. Smith." The histochemical localization of esterases in whole mounts of *Cysticercus fasciolaris*, *Trans. Amer. Microsco.Soc.* vol. 95, pp. 73-79, 1976.

[42] K. Krishna, S. S. Simha. "A sstudy on the nerve arrangement and cholineesterases distribution in *Oochristica sigmoides* and *Raillietina tetragona* (Cestoda). *J. Anim. Morph. Physiol.* Vol. 27, pp. 10-13, 1980.

[43] Z. I. F. Rahemo "The nervous system of Two avian Cyclophyllid cestodes as revealed by histochemical demonstration. *Turk.Parasit. Derg*. Vol. 17, pp. 64-70, 1993.

[44] P. Pal, V. Tandon "Anthehelmintic efficacy of Flemingia vestita (Fabaceae): Genistein-induced alternations in the esterase activity in the cestode, *Raillietina echinobothria*" *J. Biosci.* Vol. 23, no. 1, pp. 25-31. 1998.

[45] G. Gomori, "*Microscopic histochemistry*" University of Chicago Press, Chicago, 1952. 273 pp.

In: Biotechnology and Conservation … (2 Volume Set)
Editors: S. A. Farooq, R. Abed and S. Baqir
ISBN: 978-1-63117-869-6

Chapter 16

EFFECT OF THE NATURE OF THE FEEDING RESOURCE ON *IN VITRO* GAS PRODUCTION KINETICS USING RUMEN FLUID OF SLAUGHTERED DROMEDARY

R. Arhab[1], M. Aggoun[3] and H. Zitouni[2]
[1]Département des Sciences de la Nature et de la Vie,
Faculté des Sciences Exactes et des Sciences de la Nature et de la Vie,
Université Larbi Ben Mhidi, Oum El Bouaghi, Algérie
[2]Département de Biochimie et Microbiologie,
Faculté des Sciences de la Nature et de la Vie,
Université Mentouri, Constantine, Algérie
[3]Unité de Recherches sur les Herbivores, INRA,
Saint-Genès Champanelle, France

ABSTRACT

Degradation aspects in terms of kinetics of the tested samples namely dates oranges and olives residues by the dromedary ruminatemicrobiota is comparatively studied with vetch-oat hay as a standard. The results indicate greater hydrolytic activity of the dromedary ruminatemicrobiota towards dates and orange residues than olives residues and hay. Fermentation of dates and oranges residues reaches their stationary phase after 24h and olives residues after 48h of incubation. Besides, fermentation process was marked by two phases; the degradation of soluble fraction and that of cellulosic one. The results showed also that nature of substrates is a determining factor for in vitro gas production. In fact, substrates rich in cellular content (dates and oranges residues) is characterised by a fast fermentation that moves towards carbon dioxide production, and it is marked by a long latency phase. On contrary, the fibrous substrates degradation (olives residues and hay) is tributary of less long latency phase and generates mainly methane. The degradation level observed indicates that the dates and oranges residues might represent an acceptable source of energy for dromedary. However, the olives residues, despite its high organic matter content, cannot be used for dromedary feeding.

Keywords: Dromedary, rumen, microbiota, agro-industrial by-products, in vitro gas production

INTRODUCTION

The lignocellulosic biomass, which is composed by harvest residues and agro-industrial by-products, represent considerable quantities. It remains unexploited or used at little scale because it is generally considered as weakly degradable and therefore without real commercial value. However, such biomass provides a potential source to ruminant feeding (Jayasuriya, 1993; Pham et al., 2001), notably in developing countries such as Algeria. Among herbivores, camels are considered particularly able to convert any type of biomass into energy due to their presumably specific ruminalemicrobiota activity and their ability to adapt to severe environmental conditions (Engelhardt et al., 1987). The available data deals mainly with the physiological properties of the dromedary, such as their resistance to heat and thirst. Its digestive aspects and physiology pattern have been illustrated only during the last decade (Kayouli et al., 1991; Kayouli et al., 1995; Jouany et al., 1995; Dulphy et al., 1995).

The use of non-conventional substrate as feeding substance has not yet been subjected to significant studies. For this reason, the present research aimed to study the fermentation capacity of dromedary ruminatemicrobiota towards some agro-industrial by-products selected for their availability in our country. The effect of the nature feeding resource on in vitro gas production kinetics was also examined. Substrates used in this experiment were dates, oranges and olives residues. Vetch-oat hay was taken as substrate standard. These substrates had a known chemical composition (table 1). Samples were taken from an industrial firm of transformation and conservation of dates (relegated dates). Orange residues were obtained from an industrial factory for jam and juice (pulps and seeds). Olives waste was obtained from a traditional olive oil unite (crushed olive). Dates and oranges residues were dried at 45°C (in order to avoid the Maillard reaction) and olive residues and vetch-oat hay at 105°C until constant weight. Samples were ground to pass a 1-mm sieve.

Table 1. Chemical composition of the experimental substrates and standard (g/100 g DM)

Substrates	Abrev.	DM	Ash	TS	CP	EE	CF
Dates residues	RD	91.1	2.40	82.6	2.85	0.54	2.93
Oranges residues	RC	19.5	4.37	25.9	5.57	2.34	11.9
Olives residues	RO	68.2	1.61	5.14	0.97	15.6	40.9
Vetch-oat hay	H	90.1	5.60	2.90	6.10	1.30	51.3
S.E.M.		0.53	0.25	1.76	0.41	0.62	2.70

DM, dry matter; CP, crude protein; TS, total sugars; EE, ether extract; CF, crude fibre; S.E.M., standard error of means.

The substrates were incubated with rumen fluid in 100 ml calibrated glass syringes following the procedure described by Menke et al. (1979) and Menke and Steingass (1988). The syringes were incubated at 39°C in an electrically heated isothermal oven equipped with a rotor, which rolled continuously at 9 rpm for 72 hours of incubation. Rumen fluid was collected for each trail from three healthy dromedaries immediately after slaughter and stored in preheated Thermos flasks saturated with CO_2. For each substrate, 200 mg of dried sample plus 30 ml of artificial saliva (10 of rumen fluid and 20 ml of buffer solution) were incubated in triplicate. Under the same conditions, controls; vetch-oat hay (standard substrate) plus blank syringe (only artificial saliva) were also incubated in triplicate. Gas production was recorded at 2, 4, 6, 10, 24, 48 and 72 hours. The quantitative analysis of gas production was done by direct reading of the level of piston displacement in the syringe and the qualitative ones is carried out using the procedure of Jouany (1982). Net gas volume at each incubation period was calculated by subtracting the mean gas volume of the blank from the gas volume in syringes with samples. The gas volume was not corrected according to a standard substrate. Data of gas production were fitted to the exponential model proposed by Orskov and McDonalds (1979) and adapted for gas production by Blümmel and Orskov (1993): $p = a + b\,(1-e^{-c*t})$, where p represents the net gas production at time t; $(a+b)$ potential gas production and c the rate of gas production. Software Neway developed by Chen (1997) was used to calculate the data. The data were analysed by one factor variance analysis (substrate effect) using STAT-ITCF program version 5. The kinetics of gas production is showed in figure 1. It follows an ascending pattern for the different substrates. The fermentation is relatively intensive during the first 24h of incubation, after which it reaches a stationary phase. However, with certain substrates, it already started to decline. The kinetic of gas production appears to be determinate by two different phases; the first one corresponds to the degradation of the soluble fraction and the second to the insoluble but potentially fermentable fraction. The examination of the specific fermentation curves shows that the by-products of dates and oranges were more fermented than vetch-oat hay and olives residues ($P < 0.05$) and their degradation occurs mainly during the 10h of incubation. Whereas in the case of vetch-oat hay and olives residues, the fermentation is tributary of a latency phase.

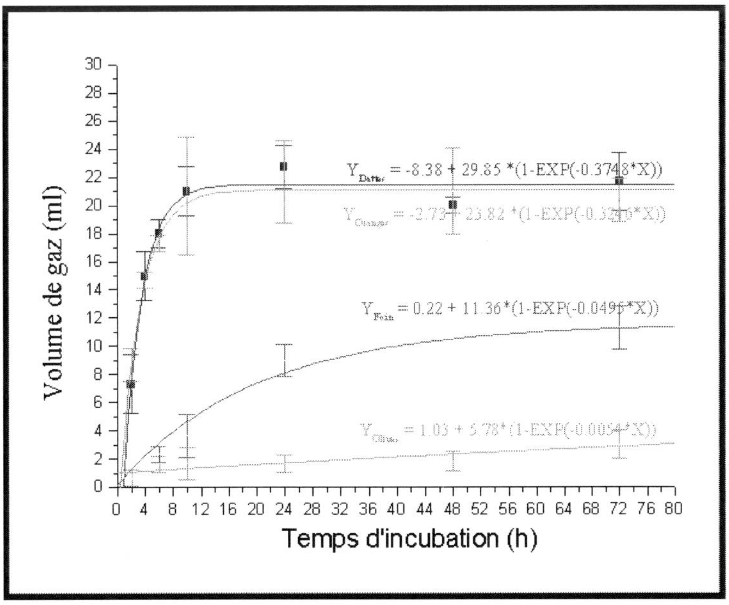

Figure 1. *In vitro* fermentation kinetics of the different substrates described by the exponential model.

The difference in the kinetics aspects of gas production between the dates and oranges residues and that of control is certainly a function of their chemical composition (tab.1) which indicates that the dates and oranges residues are rich in soluble sugars. In addition, their cell walls are less lignified compared to the control which is rich in cellulose (Gihad et al., 1989). The weak gas production observed with olives residues has also been mentioned by Theriez and Boule (1970).

This result could be explained by the fact that olives residue contain phenolic and tannic substances which are characterized by the unsolubilisation of protein and the inhibition of microbial activity. Their effect is mainly important during the first hours of incubation (Leinmüller et al., 1991). The pressed olives residues contain a part of pulp and 40% of nucleus which are rich in fatty acids. These later are converted into calcic salts in presence of calcium and magnesium (compounds of the buffer solution). These ions are primordial for the adhesion of cellulolytic bacteria to the cellulose (Tamming and Doreau, 1991). This situation could also be an explanation for the weak gas production of olive residues.

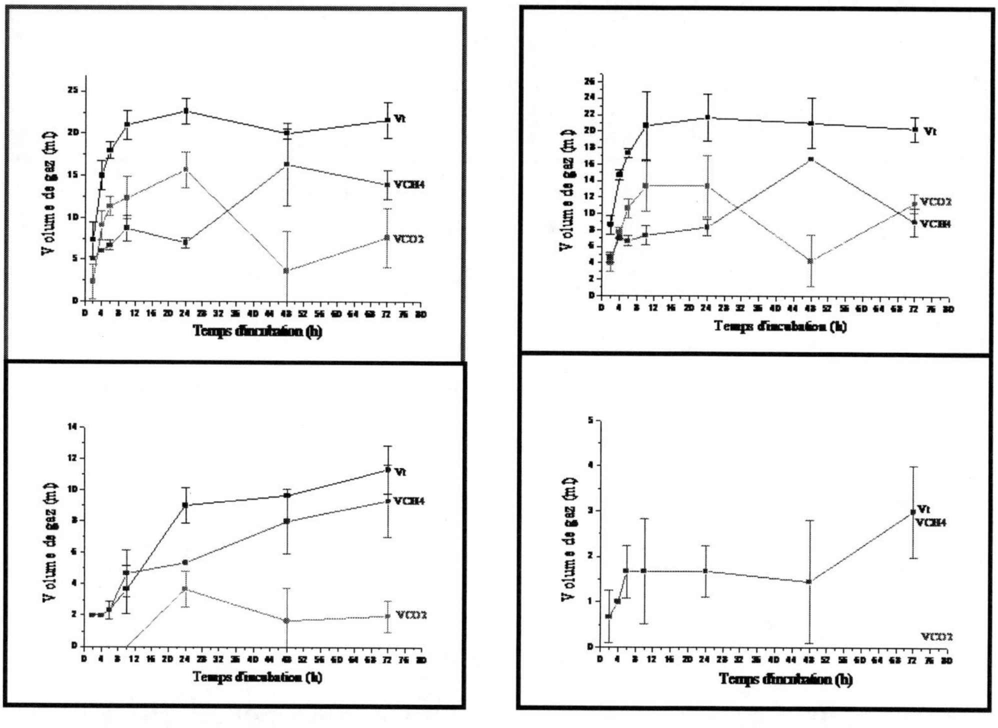

Figure 2. Effect of substrate nature on *in vitro* gas production pattern of dates (a), orange residues (b), vetch-oat hay (c) and olive residues (d).

The qualitative analysis of gas production is illustrated by figure 2. It reveals that the degradation pattern of dates and orange residues are similar. In the first hours of incubation, the dominant gas released is carbon dioxide (CO_2) and beyond 24h of incubation an inverse tendency takes place and methane (CH_4) becomes dominant. Concerning vetch-oat hay, both CO_2 and CH_4 are produced with a little disequilibrium in favour of CH_4. However, the degradation of olive residues produces exclusively CH_4. In the same way, it was found that CO_2 and CH_4 production, observed in vitro for dates and orange residues, evolves the other way around during fermentation. This result agrees with those mentioned in vitro by Vermorel (1995).

The production of gas during fermentation is correlated with qualitative production of volatile acids (Orskov and Ryle, 1990). Numerous authors suggest that the degradation of substrate rich in starch and soluble sugars favours the production of propionic and butyric acids (Orskov et al., 1988). These end-products are related to CO_2 production. Otherwise, the fermentation of fibrous substrate produces acetic acid itself being associated with an important production of hydrogen (H_2), which induces an increased production of gas in the form of CH_4. This leads us to deduce that the degradation of dates and orange residues which are rich in soluble sugars might favour the production of propionic and butyric acids but that of fibrous substrate (olive residues and hay) favours the production of acetic acid.

Table 2 indicates that the values of soluble fraction (a), deduced from the exponential model, are positive as well as negative. The negative values have also been reported by other authors working under the same conditions or *in sacco* (Orskov and Ryle, 1990; Blümmel and Orskov, 1993). They are associated to latency phase and they could be explained by the necessary time to ruminalmicrobiota to degrade soluble fraction and then to adhere to cellulose fraction of the substrate. Furthermore, the dates and orange residues are characterised by a fast fermentation than hay and olive residues (P < 0.05).

Table 2. Cumulative *in vitro* gas production (ml) after 72 hours of incubation and substrate fermentation characteristics defined by the exponential equation

Substrates	Gas production after		Exponential parameters				
	48h	72h	*b* (ml)	*a* + *b* (ml)	*c* (%/h)	Lag time (h)	R.S.D.
RD	20.0	21.6^a	30.7^a	21.4^a	38.65^a	0.86	1.73
RC	21.0	20.3^a	25.2^a	21.0^a	35.26^a	0.50	1.98
RO	2.1	3.0^c	2.5^c	2.20^c	9.23^b	0.35	0.46
H	9.67	11.3^b	12.4^b	12.5^b	5.90^b	0.16	1.02
S.E.M.	1.08	1.53	3.45	1.5	3.91		

[a, b, c] Means in the same column without letter in common differ significantly (P < 0.05). R.S.D., residual standard of deviation, for abbreviates see table 1.

The results of the present study complement the important studies made on dromedary by physiologists in the last decade and indicate the greater hydrolytic activity of the dromedary ruminatemicrobiota against substrates rich both in soluble fraction and cell wall compounds.

The results show also that the substrate nature is a determining factor for in vitro gas production. In fact, the substrate rich in cellular content is characterised by a fast fermentation that moves towards CO_2 production, and it is marked by a long latency phase. On contrary, the fibrous substrate degradation is also tributary of less long latency period and generates CH_4.

Concerning the nutritive value of the studied substrates, the results indicate that dates and orange residues might represent an acceptable source of energy for dromedary. Whereas, the olive residues, in spite of being rich in organic matter, cannot be used in animal feeding. However, their use as a constituent of feeding ration can be considered perhaps after a treatment aimed to eliminate the inhibitory factor of the ruminatemicrobiota and increase the solubility of its protein content.

REFERENCES

Blümmel M and Orskov ER 1993. Comparison of in vitro gas production and nylon bag degradability of roughages in predicting feed intake in cattle. *Anim. Feed Sci. Technol.* 40: 109-119.

Chen XB 1997. NewayExel: a utility for processing data of feed degradability and in vitro gas production (version 5.0). Rowett Research Institute, Aberdeen, UK.

Dulphy JP, Martin-Rosset W andJouany JP 1995. Ingestion et digestion comparée des fourrages chez les différentes espèces d'herbivores. *INRA Prod. Anim.*8: 293-307.

Engelhardt WV, Lechner-doll M, Heller R, Schwartz HJ and Ernest WC 1987. Utilization of agricultural waste products in animal nutrition. *Anim. Res. Dev.*26: 56-69.

Gihad EA, El-Gallad TT, Saoud A E, Abou El Nasr HM and, Farid MFA 1989. Feed and water intake, digestibility and nitrogen utilisation by camels compared to sheep and goats fed low protein desert by-products. *Options Méditerranéennes.*2: 75-80.

Jayasuriya MM 1993. Use of crop residues and agro-industrial by-products in ruminant production systems in developing countries. *British. Soc. Anim. Prod.*16: 47-55.

Jouany J P 1982. Volatile fatty acids and alcohols determination in digestive content, silage juice bacterial culture and anaerobic fermentation content. *Science des Aliments.*8: 293-307.

Kayouli C, Jouany J P and Ben Amor J 1991. Comparison of microbial activity in the forestomachs of the dromedary and the sheep measured in vitro and in sacco on Mediterranean roughages. *Anim. Feed Sci. Technol.*33: 237-245.

Kayouli C, Jouany JP, Dardillat C and Tisserand JL 1995. Particularités physiologiques du dromadaire: conséquences sur son alimentation. *Options Méditerranéennes.*13: 143-155.

Leinmüller E, Steingass H and Menke KH 1991. Tannins in ruminant feedstuffs. *Anim. Res. Dev.* 33: 33-53.

Menke KH andSteingass H1988. Estimation of the energetic feed value obtained from chemical analysis and in vitro gas production using rumen fluid. *Anim. Res. Dev.*28: 47-55.

Menke KH, Raab L, Salewski A, Steingass H, Fritz D andShneider W1979. The estimation of the digestibility and metabolizable energy content of ruminant feedingstuffs from the gas production when they are incubated with rumen liquor in vitro. *J. Agric. Sci. Camb.*92: 217-222.

Orskov ER and McDonald I 1979. The estimation of protein degradability in the rumen from incubation measurements weighted according to rate of passage. *J. Agric. Sci. Camb.*92: 499-503.

Orskov ER and Ryle M 1990. Energy, nutrition in ruminants. *Elsevier Applied Science* (Ed.) Amsterdam, pp. 1-10.

Pham Kim Coung, Vu Chi Cuong, Le Viet Ly and Jan Berg 2001. Straw yield, in sacco degradability and in vitro gas production of several rice varieties in the red river delta in

Vietnam. *Proceeding Workshop on Improved Utilisation of By-Products for Animal Feeding in Vietnam. NUFU-Project*-3/2001.

Tamminga S and Doreau M1991. Lipids and rumen digestion. In: Rumen microbial metabolism and ruminant digestion, *Jouany JP* (ed.), *INRA* (Paris), pp. 151-163.

Theriez M and Boule G 1970. Valeur alimentaire du tourteau d'olive. *Annales de Zootechnie*. 19 (2): 143-157.

Vermorel M 1995. Production gazeuses et thermiques résultant des fermentations digestives. In: Nutrition des ruminants domestiques (ingestion et digestion). Jarrige R, Ruckebusch Y, Demarquilly C, Farce MH and Journet M (eds.), *INRA,* (Paris), France, pp. 92-102.

In: Biotechnology and Conservation ... (2 Volume Set)
Editors: S. A. Farooq, R. Abed and S. Baqir
ISBN: 978-1-63117-869-6

Chapter 17

DIETARY MANIPULATION OF THE FATTY ACID CONTENT OF FARMED RED HYBRID TILAPIA, *OREOCHROMIS* SP.

Ahmed Al-Souti*
Department of Marine Science and Fisheries,
College of Agricultural and Marine Sciences,
Sultan Qaboos University, Al-Khod, Oman

ABSTRACT

Red hybrid tilapia were fed diets with fish oil content ranging from 0-12%. After 10 weeks no significant differences were seen in weight gain, specific growth rate, feed conversion ratio, whole body or fillet proximate composition. GC analysis revealed significant differences in the free fatty acid content of dorsal muscle. The n-3/n-6 ratio increased from 0.2 (fish fed 0% cod liver oil) to 1.44 (fish fed 12% cod liver oil). There was evidence of selective retention of 22:6n-3 (DHA) in the dorsal muscle, since DHA levels exceeded dietary levels. The retention of EPA was low in all dietary groups. The results demonstrate the potential to modify favorably n-3/n-6 fatty acid ratios of fatty acids in tilapia fillets by increasing dietary fish oil.n-3 enrichment enhances the beneficial health effects of fish, which is increasingly important to the health- conscious consumer.

Keywords: Fish feeds, n-3 fatty acids, red hybrid tilapia

INTRODUCTION

Polyunsaturated fatty acids (PUFA) are essential for normal growth, development and reproduction in all vertebrates. Diets should contain 18:2n-6 (Linoleic acid) and 18:3n-3

* Corresponding author: A. Al-Souti, Department of Marine Science and Fisheries, College of Agricultural and Marine Sciences, Sultan Qaboos University, PO Box 34, Al-Khod 123, Muscat, Sultanate of Oman. E-mail: souti@squ.edu.om Tel.: +968 24141211Fax.: +968 24413418.

(alpha-Linolenic acid) whichare subsequently converted to the metabolically essential C20 and C22 homologues, 20:4n-6 (Arichidonic acid), 20:5n-3 (Eicosapentaenoic acid, EPA) and 22:6n-3 (Docosahexaenoic acid, DHA. In general, freshwater fish have higher capacity to convert C18 PUFA to the longer C20 and C22 homologues than marine fish although there are exceptions (Steffens 1997). Tilapias are known to possess some ability to elongate and de-saturate alpha-Linolenic acid to DHA and EPA (Olsen, Henderson & McAndrew 1990; Tocher, Agaba, Hastings, Bell, Dick & Teale 2002).

Early studies on the dietary requirements ofNile tilapia, *Oreochromis niloticus*, reported essential requirements of 0.5-1 % linoleic acid (Takeuchi, Satoh & Watanabe, 1983) and for *Tilapia zilli* 1% linoleic acid or 1% arichidonic acid (Kanazawa, Teshima, Sakamoto & Awal 1980. More recently reports have suggested that hybrid tilapia require both n-3 and n-6 fatty acids and it has been proposed that diets for farmed tilapia should contain 0.5-1.0% of both n-3 and n-6 PUFA (Ng, 2005).

There is some ambiguity concerning the optimal gross lipid requirements of tilapia. A lipid content of 18% lipid in the diets of red hybrid tilapia was shown to spare protein for growth, and a diet containing 30% protein and 18% lipid was recommended by De Silva, Gunasekera, & Shim (1991). Jauncey (1998) however, reviewing existing information, recommended lipid levels between 6-12% as optimal for farmed tilapia up to 25 g weight and lipid levels of 6-8% in diets for larger fish. Commercial tilapia feeds typically contain 4-5% oil (Orachunwon, Thammasarat, & Lohawatanakul 2001).

In addition to providing the optimal dietary requirements of essential fatty acids to support maximum growth and feed conversion in farmed fish it is equally important to consider the final fatty acid content of tilapia at harvest, in particular the fatty acid content of the consumable fillets. Recent reports have suggested that the beneficial PUFA content in farmed tilapia may be reduced in fish farmed intensively and fed diets containing plant oils rich in n-6 fatty acids (Karapanogiotidis, Bell, Little, Yakupitiyage, & Rakshit 2006; Weaver, Iveter, Chilton, Wilson, Pandey & Chilton 2008). In contrast fish grown extensively or semi-intensivelyin green water systems, rich in phytoplankton, have fillet fatty acid content comparable with that of wild fish and contain higher levels of DHA and EPA than intensively farmed fish (Karapanagiotidis et al. 2006).

Considerable research has focused on the substitution of fish oils with cheaper plant oils (Ng, 2005). The use of n-6 rich plant oils, such as corn orsunflower oil, has been shown to reduce the n-3/n-6 fatty acid ratios in intensively-reared reared fish, which do not have access to natural plankton. It has been shown however that feeding diets containing linseed oil, which is rich in 18:3n-3, can increase the EPA/DHA content of Nile tilapia fillets when substituted against sunflower oil, which is rich in 18:2n-6 (Justi, Hayashi, Visentainer, de Souza, & Matsushita 2003; Visentainer, De Souza, Makota, Hayashi & Franco 2005. It is therefore recommended to include n-3 rich plant oils or fish oils in commercial tilapia feeds formulated for use in intensive systems (Karapanagiotidis et al. 2006). Farmed tilapia with elevated levels of n-3 PUFA may convey significant health benefits to consumers, in view of the extensive evidence of their apparent beneficial effects on cardiovascular (Russo 2009; Lecerf 2009), autoimmune (Ruxton, Reed, Simpson & Millington 2007) and inflammatory disorders (Calder 2006).

It has previously been reported that feeding diets rich in fish oil (10% cod liver oil) slightly depressed growth and significantly reduced feed efficiency in red hybrid tilapia (Ng, Lim, & Boey 2003. The present study was carried out in order to assess the effects of different n-3/n-6 fatty acid ratios on the growth and feed efficiency of juvenile red hybrid tilapia and on the n-3 content of red hybrid tilapia fillets.

Table 1. Ingredients and proximate composition of the experimental diets

Diet ingredients (g kg^{-1} dry diet)	Diet			
	0FO	4FO	8FO	12FO
Fishmeal[1]	200	200	200	200
Soybean meal[1]	300	300	300	300
Wheat flour	125	125	125	125
Wheat bran[1]	150	150	150	150
Corn Starch	100	100	100	100
Corn oil	120	80	40	-
Cod liver oil	-	40	80	120
Vitamin and mineral mix[2]	5	5	5	5
Proximate composition(g kg^{-1} dry diet)				
Dry matter (g kg^{-1} as fed)	928	930	921	931
Protein	329	318	317	321
Lipid	130	137	135	134
Ash	95	94	95	97
Gross energy (kJg^{-1} dry matter)	17.6	18.1	17.1	17.5

[1] Supplied by Oman Flour Mills, Muscat, Oman.
[2] Goddard and McLean (2001).

Analytical Procedures

The proximate composition of the experimental diets and initial and final fish carcasses was determined using standard methods (AOAC, 2000). The gross energy content of the feeds was determined using a ballistic bomb calorimeter (GALLENKAMP, Model CBB 330, Loughborough, Leics., UK) calibrated with benzoic acid. All analyses were carried out in duplicate.

Freeze- dried samples of food and dorsal muscle for fatty acid analysis were processed as described by Ulberth and Henninger (1992), except that henicosanoic acid (C21:0) was used as internal standard (Ulberth and Henninger 1995). The GC analysis of the methyl esters was performed on an AGILENT 6890NGC (Bellefonte, PA, USA) equipped with a 30 mx0.25 mm SUPELCO SP-2380 (SUPELCO INC., Santa Clara, CA, USA) fused-capillary column attached with flame ionization detector. Helium was used as carrier gas at a velocity of 20 cm/sec with electronic pressure control and split ratio 100: 1. The temperature was programmed from 50–250°C rising 4°C/min. The injector temperature was 250 °C, the detector 260°C. One µl toluene extract, containing the methyl esters of fatty acids, was injected using an Agilent 7683 series injector. Individual fatty acids were identified by comparing their retention time with commercially available mixtures of 37 components FAME MIX Standard (SUPELCO INC., Santa Clara, CA, USA). The fatty acids were quantified (mg.g^{-1} dry weight) using heneicosanoic acid (C21:0) as an internal standard (Table 2).

Table 2. Fatty acid composition of the experimental diets containing 0% (0FO), 4% (4FO) 8% (FO) and 12% (12FO) fish oil

Fatty acid % of total fatty acids	Diet			
	0FO	4FO	8FO	12FO
C14:0	3.8	6.1	2.1	10.1
C16:0	21.8	19.7	14.3	6.9
C16:1	1.2	9.8	10.2	12.4
C18:0	9.5	6.4	5.5	2.9
C18:1*n*-9	2.4	16.8	25.0	11.4
C18:2*n*-6	17.0	16.9	16.3	9.9
C18:3*n*-3	Tr	Tr	Tr	1.2
C20:0	2.5	0.1	Tr	Tr
C20:1	0.4	0.2	0.5	6.7
C20:4*n*-6	2.0	1.5	2.1	1.5
C20:5*n*-3	Tr	4.7	5.5	6.8
C22:1*n*-9	0.5	Tr	4.5	Tr
C22:6*n*-3	3.0	6.2	8.5	10.7
Total saturates[1]	40.5	35.4	21.7	19.9
Total PUFA	27.0	31.7	32.5	35.8
Total *n*-3	3.2	11.3	14.6	22.2
Total *n*-6	23.2	20.4	19.7	11.5
n-3/*n*-6	0.1	0.5	0.7	1.9

[1] Total saturates include fatty acids not listed, C6:0, C8:0, C10:0, C11:0, C12.0; C13:0, C15:0, C17:0, C21:0, C23:0, C24:0.

Tr = trace (<0.1 $g.100g^{-1}$ fatty acids).

Feed Preparation

Four isonitrogenous (32% crude protein) isoenergetic (18 $kJ.g^{-1}$) feeds were formulated containing different levels of cod liver oil substituted against corn oil (Table 1) The ingredients were ground, blended in a food mixer, extruded through a 4mm die, dried at room temperature and stored at -20^0C.

Red hybrid tilapia, *Oreochromis sp.* were supplied by Nam Sai Farms Co. Ltd., Thailand as 1g fry and grown in the Tilapia Unit at the Agricultural Experiment Station at Sultan Qaboos University. A total of 240 fish (mean weight ± SD, 15.8± 0.3 g) were randomly assigned to 12, 120 L outdoor circular tanks. Tanks were individually aerated and supplied with water from a header tank at a rate of 5 L h^{-1}. Mean ambient water temperature during the feeding experiments was 27 +/- 2.2°C. Fish were fed to apparent satiation twice a day at 0700 and 1800 by slowly introducing feed until the feeding response ceased. Each of the four diets was fed to three replicate groups, each of 20 fish and records of feed consumption were kept for each treatment group. Weight gain, specific growth rate (SGR), and food conversion ratio (FCR), were calculated as parameters of growth and feed utilization (Table 3).

Table 3. Growth and feed utilization in red hybrid tilapia fed test diets containing 0% (0FO), 4% (4FO), 8% (8FO) and 12% (12FO) fish oil for 10 weeks[4]

	Diet				
	0FO	4FO	8FO	12FO	SEM[1]
Initial fish weight (g $fish^{-1}$)	15.5	14.9	16.6	16.3	2.0
Final fish weight (g $fish^{-1}$)	61.7	58.4	59.6	58.8	4.9
Weight gain (g $fish^{-1}$)	46.1	43.5	42.9	42.5	3.4
Specific growth rate[2]	1.97	1.96	1.85	1.84	0.1
Feed intake (g $tank^{-1}$)	1422	1467	1411	1376	33.4
Food conversion ratio[3]	1.56	1.70	1.64	1.65	0.12
Survival (%)	100	97	100	100	100

[1] Pooled standard error of mean.
[2] Specific growth rate, SGR = (ln initial weight – ln final weight)/t.
[3] Food conversion ratio, FCR = g dry food fed/g wet weight gain.
[4] No significant differences observed at P<0.05.

One-way ANOVA was used to compare the growth and feeding data, whole body proximate composition and liver and fillet fatty acid composition. Homogeneity of variances was tested using Levene's test and multiple comparisons among treatments were performed using a Tukey HSD *post-hoc* test. Treatment effects were considered significant at P<0.05. The statistical package SPSS (v.12.0 for Windows) was used for all statistical analyses.

Growth and Feed Efficiency

The growth rates and feed conversion ratios of red hybrid tilapia fingerlings fed diets 0, 4, 8 and 12% cod liver oil, substituted against corn oil, with n-3/n-6 ratios of 0.1, 0.2, 0.6 and 1.0 respectively showed no significant differences (P>0.05) (Table 3). At the end of the 10-week feeding trial the range of weight gain fish $^{-1}$ was 42.4-46.1g and the feed conversion ratio 1.56-1.70. It was observed however that specific growth rate and overall weight gain decreased slightly as dietary fish oil increased. A single mortality occurred during the feeding trial.

Proximate Composition

Dietary lipid composition did not significantly affect either the whole body or dorsal muscle proximate composition (moisture, ash, protein and lipid) of fingerling red hybrid tilapia (P>0.05) (Table 4). In the whole body samples, moisture and ash ranged from 71.9-73.2% and 3.7-4.2% respectively, crude protein from 14.3-15.1% and total lipid from 7.0-7.8%. In the dorsal muscle samples, moisture and ash ranged from 77.7-78.4%, crude protein from 17.7-18.1% and total lipid from 1.2-1.3%. The total lipid contents of both whole body and dorsal muscle samples were similar from each treatment group. It was observed that the whole body protein levels of the fish at the start of the feeding trial were lower than the values from any of the treatment groups and that the lipid levels were lower in the dorsal muscle samples from the initial fish than from the treatment groups at the end of the feeding trial (Table 4)

Table 4. Proximate compositionof whole body tissue and dorsal muscle of red hybrid tilapia fed experimental diets containing 0% (0FO), 4% (4FO), 8% (8FO) and 12% (12FO) fish oil for 10 weeks[1]

Initial	Final					
		0FO	4FO	8FO	12FO	SEM[2]
Whole body tissue %						
Moisture	73.1	72.5	72.8	73.2	71.9	1.06
Crude protein	14.0	14.3	14.5	14.4	15.1	1.80
Lipid	7.0	7.1	7.8	7.5	7.5	1.03
Ash	3.5	3.8	3.7	3.9	4.1	0.78
Dorsal muscle						
Moisture	78.1	78.0	78.4	78.3	77.7	1.06
Crude protein	17.8	18.0	17.7	17.7	18.1	1.1
Lipid	1.0	1.2	1.2	1.3	1.3	0.93
Ash	1.2	1.4	1.3	1.2	1.4	0.33

[1] Values are the means of 9 fish from each treatment. No significant differences were found between treatment means, at $P<0.05$.

[2] Pooled standard error of the mean.

Fatty Acid Composition of Dorsal Muscle and Liver

In general, the deposition of fatty acids in tilapia dorsal muscle reflected the fatty acid content of the experimental diets.

The saturated fatty acid, palmitic acid, the monoene, oleic acid and the n-3 PUFA, DHAshowed the highest values in each class in both diets and fish. Total n-3 content in the dorsal muscle increased significantly with increased dietary n-3 levels. Of the n-3 fatty acids, DHA levels in dorsal muscle increased significantly with increased levels of fish oil and exceeded dietary levels (Table 5, Figure 1.1). Both EPA and DHA levels were highest in dorsal muscle samples from fish fed Diet 12FO, the diet containing the largest amount of cod liver oil. EPA levels in the dorsal muscle samples were much lower than DHA levels and were also lower than in the corresponding diets. The n-3/n-6 ratios increased from 0.2-1.44 in dorsal muscle samples from fish fed increasing amounts of cod liver oil (Table 5).

The fatty acids measured in fish livers also reflected the lipid content of the diets (Figure 1.2). The highest levels of total n-3 and DHA were observed in lipid samples from fish fed diets containing 12% cod liver oil. The changes in fatty acid content in liver samples were generally less pronounced than observed in dorsal muscle samples and liver EPA levels were either low or non-detectable.

Table 5. Fatty acid composition (% of total fatty acids) of the initial and final dorsal muscle of red hybrid tilapia fed diets containing 0% (0FO), 4% (4FO), 8% (8FO), and 12 % (12FO) fish oil for 10 weeks

Fatty acid	Initial fish	Diet				
		0FO	4FO	8FO	12FO	SEM
C14:0	2.0	1.8^{a}	3.2^{b}	4.5^{c}	5.1^{c}	0.60
C16:0	0.0	16.2^{a}	16.6^{a}	21.3^{b}	31.7^{c}	1.94
C16:1	3.0	0.6^{a}	1.5^{ab}	5.5^{bc}	6.2^{c}	1.12
C18:0	7.8	12.2	10.3	7.2	6.5	3.16
C18:1*n*-9	3.2	2.9^{a}	6.2^{b}	3.9^{ab}	4.0^{ab}	0.86
C18:2*n*-6	12.3	35.0^{a}	25.6^{b}	21.0^{c}	12.2^{d}	1.06
C18:3*n*-3	0.05	0.0	0.0	0.2	0.3	0.21
C20:0	Tr	0.7	0.3	0.7	0.1	0.47
C20:1	0.1	0.2^{a}	0.5^{a}	4.1^{b}	4.8^{b}	1.29
C20:4*n*-6	Tr	Tr	Tr	0.2	0.1	0.04
C20:5*n*-3	Tr	0.4	0.5	0.6	0.9	0.47
C22:1*n*-9	Tr	1.5	Tr	1.8	1.2	0.85
C22:6*n*-3	10.3	5.8^{d}	11.3^{c}	14.4^{bc}	16.1^{a}	1.03
Total saturates	12.5	18.7	35.5	33.5	41.8	5.67
Total PUFA	24.6	52.7	43.1	41.2	33.5	3.46
Total *n*-3	10.3	7.6^{a}	11.8^{b}	16.0^{c}	18.6^{c}	0.82
Total *n*-6	12.6	37.5^{a}	27.9^{b}	22.5^{c}	12.9^{d}	1.27
n-3/*n*-6	0.81	0.20^{a}	0.42^{ab}	0.71^{c}	1.44^{d}	0.08

[a] Values are the means of triplicate groups of 6 fish.

Tr = trace (<0.1 g100g^{-1} fatty acids).

Mean values in rows with different superscripts were significantly different ($P<0.05$).

Total saturates include fatty acids not listed, C6:0, C8:0, C10:0, C11:0, C12:0, C13:0, C15:0, C17:0, C21:0, C23:0, C24:0.

It has previously been reported that high levels of n-3 PUFA in *Tilapia zillii* diets (Kanazawa et al., 1980) and hybrid tilapia (*O.niloticus* x *O. aureus*) diets (Huang, Huang & Hou 1998) depressed growth and feed efficiency. Ng, Lim & Sidek (2001) also reported slight depression of growth and feed efficiency in red hybrid tilapia fed 10% fish oil. Similar growth depression has also been seen in African catfish, *Clarius gariepinus* (Ng et al., 2003) and channel catfish (Stickney, McGreachin, Lewis & Marks 1983). In the present study no significant differences were seen in the growth and feed conversion ratios of fish fed diets containing up to 12% fish oil, although weight gain and specific growth rate were observed to decrease slightly as dietary fish oil increased. Previous studies by Ng et al. (2001) with red hybrid tilapia showed that high fish oil diets (10% cod liver oil) significantly reduced feed efficiency when compared with diets containing 10% crude palm oil. The same diet had no significant effects on growth, although slightly reduced weight gains were observed. Similar overall observations in the present study may indicate that the high levels of fish oil, resulting in an imbalanced n-3/n-6 ratio, may have exceeded the physiological tolerance of red hybrid tilapia for n-3 fatty acids. This may have resulted from competition between n-3 and n-6 fatty acids for the same desaturase enzymes involved in lipid metabolism.

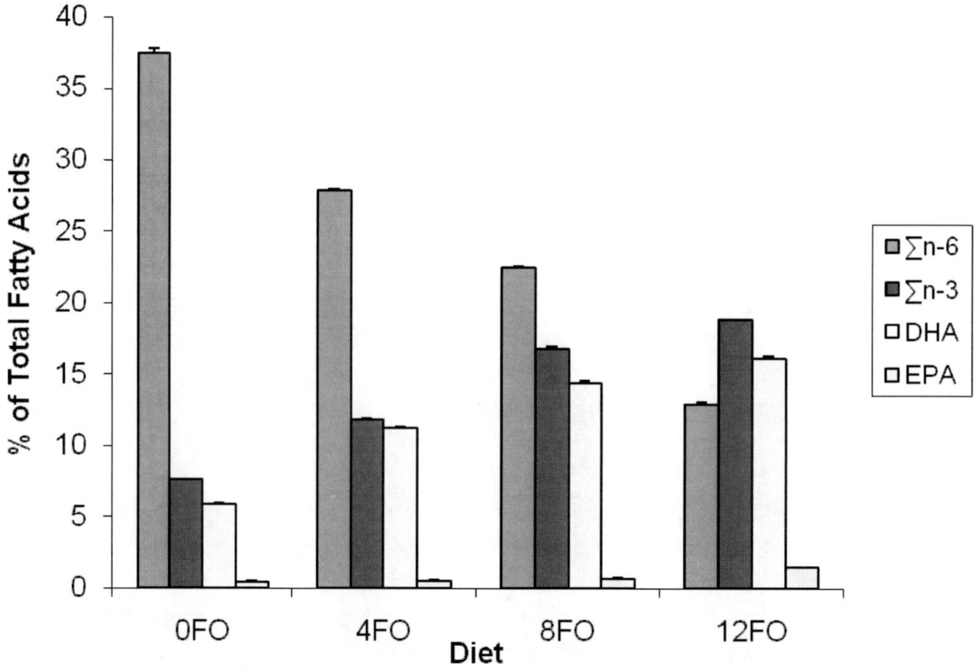

Figure 1.1. Total n-3 and n-6 fatty acids, DHA and EPA in dorsal muscle samples from hybrid red tilapia fed diets containing 0% (0FO), 4% (4FO), 8% (8FO) and 12% (12FO) fish oil.

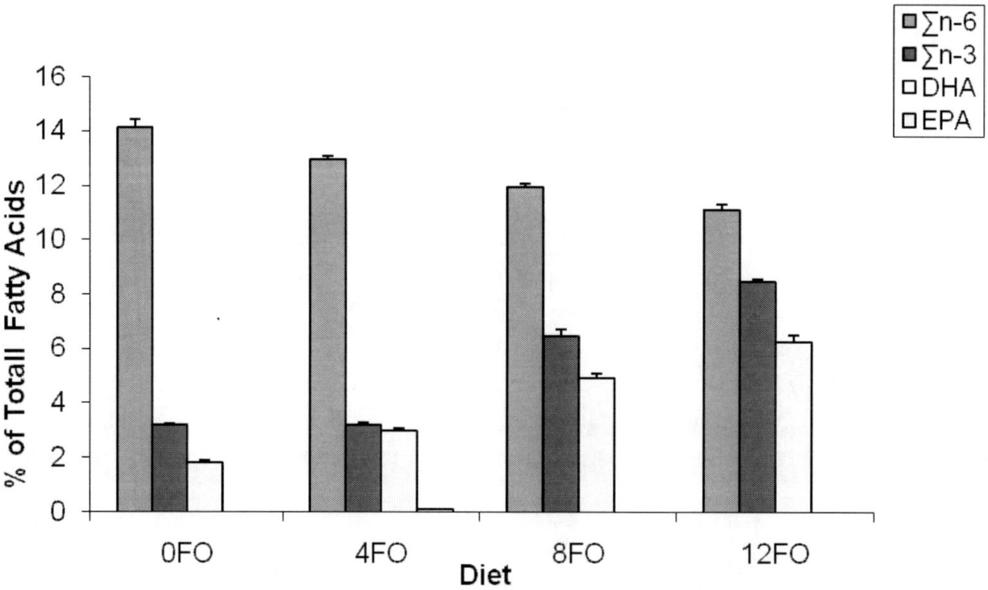

Figure 1.2. Total n-3 and n-6 fatty acids, DHA and EPA in liver samples from hybrid red tilapia fed diets containing 0% (0FO), 4% (4FO), 8% (8FO) and 12% (12FO) fish oil.

This study showedhowever that corn oil can totally replace fish oil without any significant effects on growth. Survival rates were very high and consistent between treatments, while values of SGR were normal for juvenile red hybrid tilapia and in general agreement with published values (Bahurmiz and Ng, 2007). The similar growth rates for all treatment groups of red tilapia indicate that essential fatty acids were not deficient in any of the diets.

No significant differences were observed in the proximate composition of either whole fish or dorsal muscle samples between the treated groups. The observed increase in the lipid content of the whole fish and dorsal muscle samples from all treated groups, compared with the lipid content from samples taken at the start of the feeding trial have been reported elsewhere for other species, including the humpback grouper, *Cromileptes altivelis* (Shapawi, Mustafa & Ng, 2008).

Dietary lipids affect the fatty acid composition of tilapia and other fish. Previous studies have shown that increasing dietary levels of PUFA modified the fatty acid composition of the dorsal muscle in tilapia (Ng et al., 2001; Visentainer et al., 2005; Hsieh, Hu, Hsu & Hsieh 2007).

The incorporation of dietary fatty acids into tissues is also affected by various metabolic factors, including preferential incorporation, β-oxidation, lipogenic activity, chain lengthening and desaturation (Sargent, Tocher & Bell 2002). Incorporation of fatty acids may also be influenced by various environmental factors and the age or size of the fish. In the present study there were significant decreases in linoleic acid in dorsal muscle samples as dietary corn oil was reduced and increases in total n-3 and DHA content as dietary cod liver oil increased. The deposition of EPA was low. Small, non-significant increases were seen in the EPA content of the dorsal muscle samples. This confirms previous observations from hybrid tilapia fed 5% cod liver oil (Chou, Shiau, & Hung 2001) and supports the supposition that EPA is selectively used as a substrate for β-oxidation in tilapia (Karapanagiotodis et al., 2007).

Therewas no evidence from this study of accumulation of arachidonic acid (20.4n-6, ARA) in the muscle tissue of the treated fish. The levels of ARA, the main product of 18:2n-6 desaturation, were higher in the diets than in the dorsal muscle of treated fish, where only very low or trace amounts were found (Tables 4.2, 4.5). This observation confirms previous reports that tropical wild fish, from both freshwater and marine environments, include considerable amounts of ARA in their lipids in contrast to intensively cultured fish (Karapanagiotidis et al., 2006). Similar results for ARA were reported by El-Sayed, Mansourand Ezzat (2005), following studies on Nile tilapia brood stock, where ARA was significantly lower in both dorsal muscle and in eggs than in corresponding diets. In contrast, however Ng et al. (2001) showed ARA levels of 2.3g/100g fatty acid) in muscle tissue from red hybrid tilapia fed diets containing 10% cod liver oil. Dietary ARA levels were not reported from this study however.

In their review of the role of ARA in aquaculture feeds Bell and Sargent (2003) stressed the importance of ARA as the primary eicosanoid precursor, with many roles including resistance to stress. The increased stressors experienced by fish in intensive culture systems may account for the reduced levels of ARA, in contrast to those reported from wild or semi-intensively cultured fish. The muscle levels of ARA found in this study were low or found only in trace amounts. This is in marked contrast to the observations of Weaver et al. (2008) who recorded high levels of ARA and high ratios of ARA/EPA in fillet samples of farmed tilapia. Based on their results, from fish purchased in US markets, these authors raised questions concerning potential disease risk factors associated with tilapia consumption. These risk factors included the potential production, from excessive ARA of undesirable levels of eicosanoids, such as prostoglandins, thromboxanes and leukotrienes.

Increasing the fish oil content in tilapia feeds was shown in the present study to increase significantly the n-3/n-6 ratio in the dorsal muscle of farmed tilapia. Similar results have been obtained using diets enriched with n-3 rich plant oils. Justi et al. (2003) and Visentainer et al. (2005) reported that feeding Nile tilapia diets containing linseed oil, which is rich in 18:3n-3, can increase the linolenic acid, EPA and DHA content of fillets when substituted against sunflower oil, which is rich in 18:2n-6. It may therefore be beneficial to include n-3 rich plant oils, or fish oils, in commercial tilapia feeds formulated for use in intensive systems where fish have little or no access to phytoplankton (Karapanagiotidis et al., 2007). Recent studies with adult Nile tilapia have shown that 45 days is the shortest time required for the inclusion of flaxseed oil to increase significantly the total n-3 and DHA content of tilapia muscle (Tonial, Stevanato, Matsushita, De Souza, Furuya & Visentainer 2009). To reduce the additional costs associated with the use of fish oil, additional feeding strategies have been proposed. These include feeding tilapia on diets containing flaxseed oil through most of their farmed production and then switching to a fish oil-based diet as the fish approach harvestable size (Karapanagiotidis et al., 2007).

The present study, carried out with juvenile tilapia, showed that a relatively short period of feeding (10 weeks), with a fish oil-based diet, is sufficient to modify significantly the fatty acid profile of tilapia fillets. Numerous beneficial health effects have been attributed to n-3 fatty acids. Among these, lower blood pressure, decrease in triglicerides and improved learning ability in children. Fish species, like salmon, herring and sardines are known to be naturally rich in n-3 fatty acids, while tilapias normally contain low levels of n-3. Further research, using 'finishing diets' containing fish oil on market-sized tilapia, is necessary to examine the potential to culture tilapia, as an n-3 enriched functional food. It may be possible in the future to offset the additional feeding costs, associated with the use of fish oil, by sales of n-3 enriched tilapia fillets into specialized markets, where sales strategies are based on the health promoting levels of long chain, n-3 fatty acids.

ACKNOWLEDGMENTS

The work was partially supported by a grant from HM Sultan Qaboos Fund for Strategic Research to Bassam Soussi. The author is pleased to acknowledge Stephen Goddard and the technical assistance of Nasr Al-Jardani.

REFERENCES

AOAC (Association of Official Analytical Chemists 2000. *Official Methods of Analysis*, 17th edn. AOAC, Arlington, VA, USA.

Bahurmiz, O.M. & Ng, W.K. 2007. Effects of dietary palm oil on growth, tissue fatty acid composition and nutrient digestibility of red hybrid tilapia, *Oreochromis* sp., raised from stocking to marketable size. *Aquaculture, 262*, 382-392.

Bell, J.G. & Sargent, J.R. 2003. Arachidonic acid in aquaculture feeds: current status and future opportunities. *Aquaculture,218*, 491-499.

Calder, P.C. (2006).n-3 Polyunsaturated fatty acids, inflammation and inflammatory diseases. *American Journal of Clinical Nutrition, 83*, 1505-1519.

De Silva, S.S., Gunasekera, R.M. & Shim, K.F. 1991. Interactions of varying dietary protein and lipid levels in young red tilapia: Evidence of protein sparing. *Aquaculture, 95*, 305-308.

El-Sayed, A.F-M, Mansour, C.R. & Ezzat, A.A. 2005. Effects of dietary lipid source on spawning performance of Nile tilapia (*Oreochromis niloticus*) broodstock reared at different water salinities. *Aquaculture,248*, 187-196.

Glencross, B.D., Hawkins, W.E. & Curnow, J.G. 2003. Restoration of the fatty acid composition of red seabream (*Pagrus auratus*) using a fish oil finishing diet after grow-out on plant oil based diets. *Aquaculture Nutrition ,9*, 409-418.

Goddard J.S. & McLean E. 2001. Acid-insoluble ash as an inert reference material for digestibility studies in tilapia, *Oreochromis aureus. Aquaculture, 194*,93-98.

Hsieh, S-L., Hu, C-Y., Hsu, Y-T & Hsieh, T-J. 2007. Influence of dietary lipids on the fatty acid composition and Stearoyl-CoA desaturase expression in hybrid tilapia (*Oreochromis niloticus* x O. *aureus*) under cold shock. *Comparative Biochemistry and Physiology, Part B, 147*, 438-444.

Huang, C.H., Huang, M.C. & Hou, P.C. 1998. Effect of dietary lipids on fatty acid composition and lipid peroxidation in sarcoplasmic reticulum of hybrid tilapia, *Oreochromis niloticus x O. aureus. Comparative Biochemistry and Physiology Part A, 120*, 331-336.

Jauncey, K. (1998) *Tilapia feeds and feeding*. Stirling: Pisces Press.

Justi, K.C., Hayashi, C., Visentainer, J.V., de Souza, N.E. & Matsushita, M. 2003. The influence of feed supply on the fatty acid profile of Nile tilapia (*Oreochromis niloticus*) fed on a diet enriched with n-3 fatty acids. *Food Chemistry,80*, 489-493.

Kanazawa, A., Teshima, S., Sakamoto, M. & Awal, A. 1980. Requirements of *Tilapia zilli* for essentialfatty acids. *Bulletin of the Japanese Society of Scientific Fisheries,46*, 1353-1356.

Karapanogiotidis, I.T., Bell, M.V., Little, D.C., Yakupitiyage, A. & Rakshit, S.K. 2006. Polyunsaturated fatty acid content of wild and farmed tilapias in Thailand: Effect of aquaculture practices and implications for human nutrition. *Journal of Agriculture and Food Chemistry,* 54, 4304-4310.

Karapanogiotidis, I.T., Bell, M.V., Little, D.C. & Yakupitiyage, A. 2007. Replacement of dietary fish oils by alpha-linolenic acid-rich oils lowers omega-3 content in tilapia flesh. *Lipids*, 42, 547-559.

Lecerf, J-M. 2009. Fatty acids and cardiovascular disease. *Nutrition Reviews,67 (5)*, 273-283.

Ng, W.K., Lim,P.K. & Sidek, H. 2001. The influence of a dietary lipid source on growth, muscle fatty acid composition and erythrocyte osmotic fragility of hybrid tilapia. *Fish Physiology and Biochemistry,*25,301-310.

Ng W.K., Lim, P.K. & Boey, P.L. 2003. Dietary lipid and palm oil source affects growth, fatty acid composition and muscle alpha-tocopherol concentration of African catfish, *Clarias gariepinus. Aquaculture,* 215, 229-243.

Ng, W.K. 2005. Lipid nutrition of farmed tilapia. *Global Aquaculture Advocate.* October, 60-61.

Olsen, R.E., Henderson, R.J. & McAndrew, B.J. 1990. The conversion of linoleic acid and linolenic acid to longer chain polyunsaturated fatty acids by tilapia (*Oreochromis niloticus*) *in vivo. Fish Physiology and Biochemistry,8*,261-270.

Orachunwon, C, Thammasarat, S. & Lohawatanakul, C. 2001. Recent developments in tilapia feeds. In S. Subasinghe & T. Singh (Eds.)*Proceedings of the Tilapia 2001 International Technical and Trade Conference on Tilapia*(pp.113-122) Kuala Lumpur: Infofish.

Russo, G.L. 2009. Dietary n-6 and n-3 polyunsaturated fatty acids: From biochemistry to clinical implications in cardiovascular prevention. *Biochemical Pharmacology77*,937-946.

Ruxton, C.H.S., Reed, S.C., Simpson, M.J.A. & Millington, K.J. 2007. The health benefits of omega-3 polyunsaturated fatty acids: a review of the evidence. *Journal of Human Nutrition and Dietetics,*20,275-285.

Sargent, J.R., Tocher, D.R. & Bell, J.G. 2002. The lipids. J. E. Halver and R. W. Hardy (Eds.), *Fish Nutrition* (pp. 181-257). New York: USA.

Shapawi, R., Mustafa, S. & Ng, W.K. 2008. Effects of dietary fish oil replacement with vegetable oils on growth and tissue fatty acid composition of humpback grouper, *Cromileptes altivelis* (Valenciennes). *Aquaculture Research, 39*, 315-213.

Steffens, W. (1997). Effects of variation in essential fatty acids in fish feeds on nutritive value of freshwater fish for humans. *Aquaculture, 151*, 97-119.

Stickney, R.R., McGreachin, R.B., Lewis, D.H. and Marks, J. 1983. Response of young channel catfish to diets containing purified fatty acids. *Transactions of the American Fisheries Society,112*, 665-669.

Takeuchi, T., Satoh, S. & Watanabe, T. 1983. Requirements of *Tilapia nilotica* for essential fatty acids. *Bulletin of the Japanese Society of Scientific Fisheries,49*,1127-1134.

Tocher, D.R., Agaba, M., Hastings, N., Bell, J.G., Dick, J.R. & Teale, A.J. 2002. Nutritional regulation of hepatocyte fatty acid desaturation and polyunsaturated fatty acid composition in zebra fish (*Danio rerio*) and tilapia (*Oreochromis niloticus*). *Fish Physiology and Biochemistry,24*, 309-320.

Tonial, I.B., Stevanato, F.B., Matsushita, M., De Souza, M.E., Furuya, W.M. & Visentainer, J.V. 2009. Optimisation of flaxseed oil feeding time length in adult Nile Nutrition tilapia (*Oreochromis niloticus*) as a function of muscle omega-3 fatty acids composition. *Aquaculture,15*, 564-568.

Ulberth, F. & Henninger, M. 1992. One- step extraction/methylation method for determining the fatty acid composition of processed foods. *Journal of American Oil Chemistry Society,* 69, 174-177.

Ulberth, F. & Henninger, M. 1995, Determination of the fatty acid profile of fish by a one-step extraction/methylation method. *Fat Science and Technology,97*, 77-80.

Visentainer, J.V., De Souza, N.E., Makota, M., Hayashi, C. & Franco, M.R.B. 2005. Influence of diets enriched with flaxseed oil on the alpha-linolenic, eicosapentaenoic and docosahexaenoic fatty acid in Nile tilapia (*Oreochromis niloticus*). *Food Chemistry90*,557-560.

Weaver, K.L., Iveter, P., Chilton, J.A., Wilson, M.D. Pandey, P. & Chilton, F.H. 2008. The content of favorable and unfavorable fatty acids found in commonly eaten fish. *Journal of the American Dietetic Association, 108*, 1178-1185.

In: Biotechnology and Conservation … (2 Volume Set)
Editors: S. A. Farooq, R. Abed and S. Baqir
ISBN: 978-1-63117-869-6

Chapter 18

DETERMINATION OF THE BOTANICAL AND GEOGRAPHICAL ORIGINS OF HONEY IN NORTHERN OMAN

Alia Sajwani and Sardar A. Farooq*
Department of Biology, College of Science,
Sultan Qaboos University, Oman

ABSTRACT

Knowledge about botanical and georaphical sources of honey are very important for beekeepers while it indicates adequate and abundant supply soures of nectar and pollen for bees, thus contributing toward enhanced productivity. Microscopic analysis of the pollen in honey is used to determine its flora and locality origin. This knowledge is very important for Omani beekeepers as they are producing high-quality honey with its distinctive flavour and aroma. The present study reports the most botanical and geographical sources in northern Oman where the honey is harvested twice a year, i.e. in summer and winter. The study shows that *Ziziphus spina -christi, Acacia tortilis Prosopis cineraria, Citrus* sp., *Maerua crassifolia,* and *Prosopis juliflora* constitute the chief nectar and pollen sources for honeybees in this area. Each plant source yields honey with specific biochemical characteristics. The identified pollen/nectar sources reflect a wide range of foraging plant species for honeybees and encompass sufficient potential for sustaining beekeeping ventures in Northern Oman.

INTRODUCTION

Melissopalynology is the study of pollen in honey by morphology (Jones and Bryant, 1992). This study shows that honeybees regularly exploit a restricted number of trees and bushes that have fragrant, rigid and nectar containing blooms, also it relates to characteristics of plants exhibited by floral architecture and ultraviolet reflectance. As a result it plays an

* Corresponding author. e-mail: sfarook@squ.edu.om.

important role in ascertaining the botanical and geographical origins of honey by studying the pollen contained in the honey (Song et al., 2012). Furthermore, pollen analysis of honey samples sets the baseline information for the formulation of national standards for honey products to discourage honey adulteration and outright sale of fake honeys (Ramanujam et al., 1992; Tilde and Payawal, 1992).

The importance of a taxon as a chief nectar source for the honeybees can be assessed by taking into consideration the frequency representation of its pollen in the pollen spectra of individual honeys. On the other hand, its importance as geographical or regional indicator is a function of the frequency of its occurrence in the total contingent of honey samples (Ramanujam and Kalpana, 1995). Following these parameters led to the determination of the botanical and geographical origins of Omani honey (Sajwani et al., 2007a). In Oman there are many regions with different agro-climatic conditions and flora that have potential to provide different types of honeys. Omani honey is well known due to its specific aroma and taste. It is evaluated by physical tests, which often fail to detect the quality of the honey accurately unless melissopalynological studies are carried out. With these objectives, identification of bee flora and their propagation help in improving the bees forage wealth and concomitantly the efficacy of beekeeping industry and commercial honey production. At the same time, the mapping of such vegetational units of bee plants is important for the apiary industry with new possibilities for the palynological standardization of honeys with regard to their geographical origin and the plant species represented by pollen in the honeys. For the reason that Omani honey has not been studied well and no melissopalynological studies of the native honey have been published as well as no investigations have been carried out on the analysis of its biochemical properties, Sajwani et al., (2007a) focused on the analysis of the Omani honey from hives in 14 locations of Muscat and the Al Batinah regions to determine the botanical and geographical origin of Omani honey.

BEEKEEPING IN OMAN

Honey is an important natural product in the Omani society. It is commonly used in ethnomedical treatment of various ailments (e.g. cough, chest pain, indigestion, constipation, sprains, burns and minor open injuries). The Omanis have a long experience in beekeeping and honey gathering. Oman apiculture has been practiced traditionally by small producers (professional and semi-professional). Honey production takes place in northern Oman such as at Nizwa, Ar Rustaq and northern Coastal province of the Al Batinah (Figure 1) which is an important region as the centre of farming in Oman. It is also one of the major honey production region (Anderson, 1990; Ministry of Information, 1995 and 2000; Black & White, 2010). Production areas in southern Oman include Dhofar Mountains (Jabals). These regions are important constituent of the vegetation (Anderson, 1990). The current number of honeybee colonies in Oman is small and commercial beekeeping is relatively underdeveloped therefore the production of Omani honey is quite low (Oman Daily, Observer, 2010). At the present time, Omani honey is not marketed abroad as local demand exceeds production and the market is supplied with imported honeys from different countries. Some of which are sold as native products. Nevertheless given a chance, honey production in Oman will increase if potential honey producing areas are developed. Proper marketing of honey will be necessary in order to maintain value, consumption and price and gain access to overseas markets.

Two species of honeybee, *Apis mellifera* and *Apis florea* are present in Oman. *A. mellifera* belongs to the *A. mellifera yemenitica* race that existed here for centuries (Dutton et al., 1981). The Ministry of Agriculture and Fisheries has imposed a strict ban on mixing Omani and imported strains of bees (Ministry of information, 2002). The bee species are identified by their nests with several combs in the mountain cliffs. These bee species are also kept in man-made hives. *A. florea* "Little Honeybee" are identified by their nests with a single comb between the tree branches, caves and buildings. These wild bee species cannot live in hives, and cannot stay long in one place and they all migrate once when are disturbed. In Oman they are called (*Abu tuwake)* due to an orange-red abdomen tipped by distinctive black and white bands (or *tokes*) around their abdomens (Whitcombe, 1984). Both bee species are kept in the northern regions for production of honey using traditional methods.

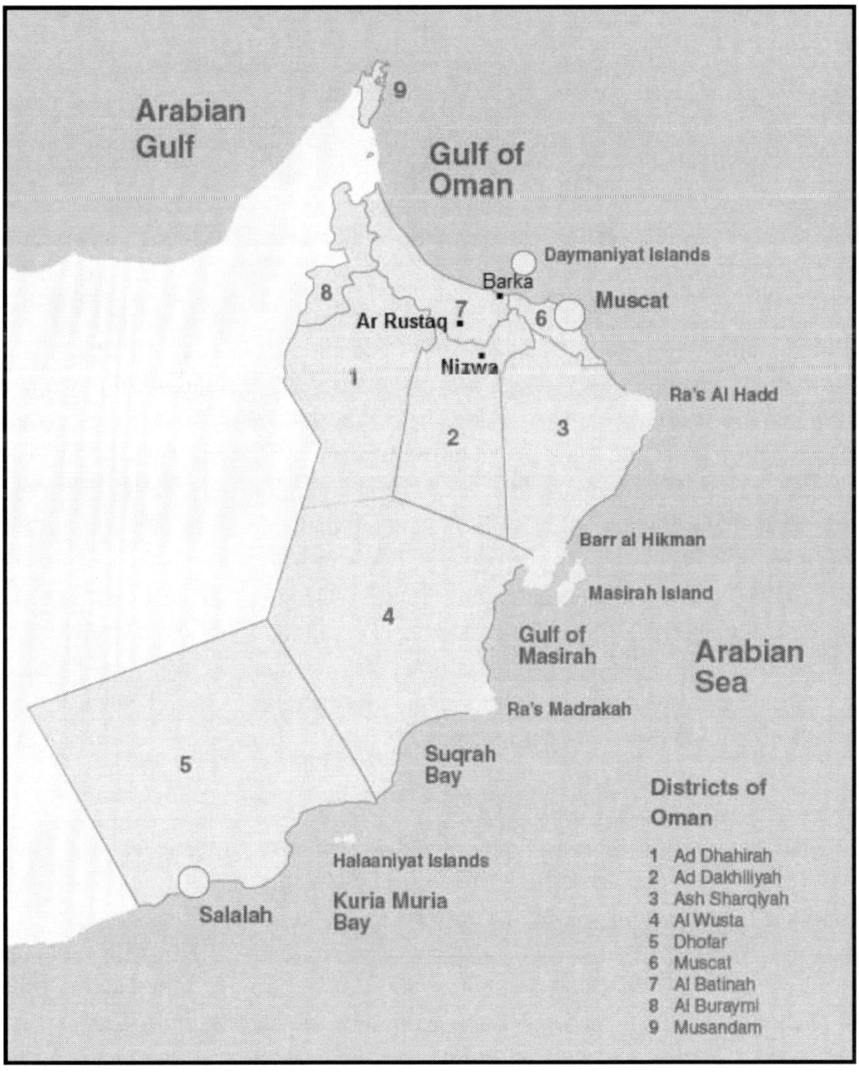

Figure 1. Sultanate of Oman map showing the districts.

Figure 2. Traditional beekeeping of *Apis mellifera* in *tubl* (hollowed out date palm logs) in Ar Rustaq.

One of the unique traditional methods of beekeeping in Oman is hiving *A. mellifera* in *tubl* (hollowed out date palm logs) (Figure 2) by covering their cavities with slim layer of date paste to encourage the bees to live inside them. This old custom is now gradually being replaced by the modern and more productive method of hiving colonies in Langstroth's movable-frame hives. The specialist keepers of *A. florea* in Oman are the only ones known in the world. They have developed a skillful way of managing the bees by cutting away a section of brood comb and clamping it between two split sticks to make a second colony with a new queen once the honey has been taken (Figure 3) (Dutton et al., 1982; Crane, 1990).

Some Omani beekeepers prefer migratory beekeeping to get most benefits from the flora to maintain their bee colonies' survive and to harvest reasonable crop. Other beekeepers feed their bees with date paste or another sources of carbohydrates during the dearth of nectar and pollen from June to August to protect the bees' life. Still others refuse feeding sugar solutions because they think that this type of feeding produces adulterated honey. Therefore, they sell their honey expensively for about 40 R.O. /kg.

Honey harvest depends on season and climatic conditions. It is collected in Oman twice a year. The first is in summer from May to June. The summer honey is mostly *Acacia* honey. The second is in winter from November to December, in which *Ziziphus* honey is produced (Hussein, 1992; Ministry of Agriculture and Fisheries resources, 1996; Sajwani et al., 2007a).

Most Omanis believe that only *Ziziphus* and *Acacia* honey is produced in Oman. It was observed in northern Oman that bees visit wild flowers such as all species of *Acacia* sp., especially *Acacia tortilis,* as well as cultivated flowers such as mangoes, limes, alfalfa, watermelons and vegetables. They gather pollen from Australian pine, datepalm, banana and corn and collect pollen and nectar from *Ziziphus* sp., and Basil (Dutton et al., 1982; Hussein,

1992). The most important naturally occurring bee plants are *Ziziphus spina christi, Acacia tortilis, Pteropyrum scoparium, Maerua crassifolia, Acacia ehrenbergiana* (Manley, 1983, Sajwani et al., 2007a).

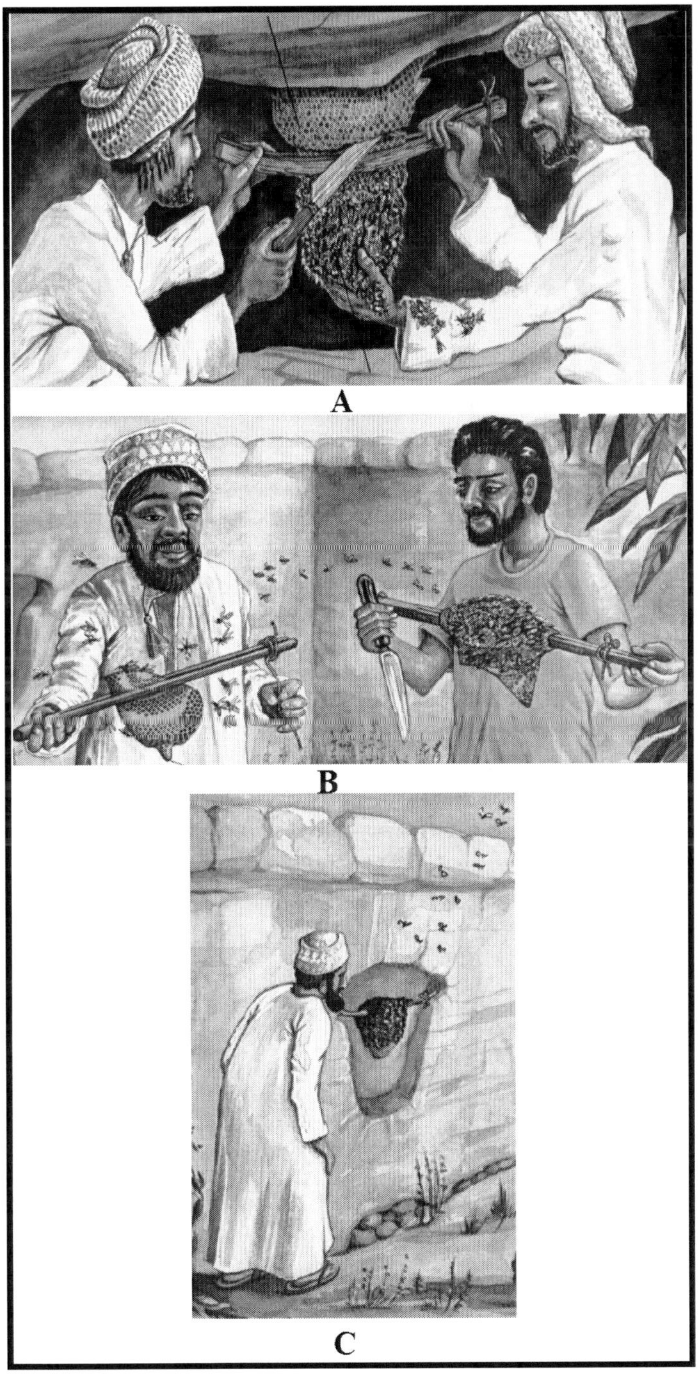

Figure 3. A-C: The Omani beekeepers are managing *Apis florea* nest by cutting away a section of brood comb and clamping it between two split sticks to make a second colony with a new queen once the honey has been taken.

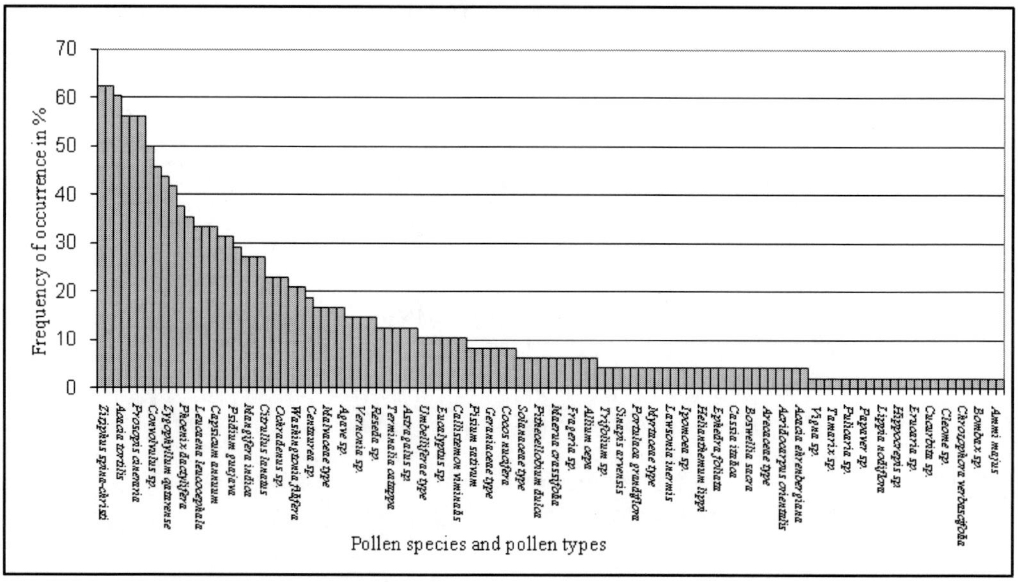

Figure 4. Frequency of melliferous pollen types in the series of Omani honey.

However, Sajwani et al., (2007a) shows other major sources of honey in this country. The predominance and high frequency of the following pollen taxa: *Ziziphus spina-christi, Acacia tortilis, Prosopis cineraria*, *Prospis juliflora,* and *Citrus* sp. in the honey samples classify these plants as the chief sources of nectar and pollen for the honey of Northern Oman. Their high frequency could be explained by the following: These plants have several features in common. They have exposed attractive flowers with protruding stamens. These flowers provide large amount of pollen and nectar. The most important fact is that these trees are large, isolated in the landscape, visible against the horizon and abundant in Omani localities (Mandaville, 1978; Miller and Morris, 1988). Therefore they are important foraging plants and their pollen are good indicators of the potentials of an area for beekeeping (Figure 4). Due to the dry nature of Oman, a relatively low number of chief nectar sources are available in some areas. This result is in agreement with the technical report on the beekeeping in Oman (Arab Organization for Agricultural Development, 1998). The pollen composition of the honey samples reflects not only the main vegetation types of the area but also the type of agrarian system.

MOST COMMON HONEY TYPES IN NORTHERN OMAN

Ziziphus spina-christi (L.) Desf. Honey

This type of honey is locally called Sidr. It is the most common as well as the most famous one in the country which characterised by pleasant taste and aroma. It is produced by both bee species *A. mellifera* and *A. florea.* Considering *Ziziphus spina-christi* plant as the major source of honeys (Sajwani et al., 2007a) is due to its prevalence throughout Oman along wadis and plains, frequently cultivated, reaching relatively great stature and has been used by local honey producers since a long time (Mandaville, 1978; Dutton et al., 1982). This tree species is native to the Arabia. It is well suited to this region's agroclimatic conditions, being drought and it is also highly salt tolerant and able to grow on alkaline soils.

Its green-yellow flowers are small with well-exposed stamens and anthers and though visually insignificant, have a pungent sweet fermenting smell, which is highly attractive to bees (Love, 1995). These flowers produce both nectar and abundant pollen, which are the important food source for honeybees (Miller and Morris, 1988). Therefore it was observed that the bees visit these flowers frequently (Whitcombe, 1984). This plant is flowering and fruiting from September to November (Ghazanfar, 1997). Therefore most winter honeys contain *Ziziphus* pollen (Sajwani et al., 2007a).

Generally, *Ziziphus* honey is of clear amber colour with moderate sugars (Sajwani et al., 2007b) and keeps as liquid for several years. Unfortunately, of late pesticides are applied to cultivated *Ziziphus* sp., which may affect the honey. To avoid this effect, native *Ziziphus* honey crop can be produced by practicing 'migratory beekeeping' in the wild *Ziziphus* areas. This type of honey is found abundantly in Arabia, Pakistan and Iran (Abu-Tarboush et al., 1993; Mossadegh and Bandpay, 1993).

Acacia tortilis (Forsskal) Hayne Honey

Acacia tortilis is one of the most common native trees throughout the country, distributed on the coastal regions, foothills, and plains. It's flowering and fruiting period is from April to June (Ghazanfar, 1997). It is therefore an important summer honey source in Oman. Honey hunters in the country search for the bee combs in the mountains during the flowering season of *Acacia* (Miller and Morris, 1988). *Acacia* honey is amber colour with quite high amount of sugars (Sajwani et al., 2007b).

This type of honey which is called locally Sumr is valuable because pesticides are not applied on *Acacia* groves. Extending migratory beekeeping could increase *Acacia* honey crop. Reports from Iran and Pakistan also mentioned about *Acacia* sp. as honey plants (Mossadegh and Bandpay, 1993; Noor et al., 2009). In Europe *Acacia* honey is produced from *Robinia pseudoacaia,* which is actually false acacia (Howes, 1979) since *Acacia tortilis* is distributed in Africa and Arabia.

Prosopis cineraria (L.) Druce Honey

Prosopis cineraria is distributed throughout Oman along wadi channels from coastal plain inland as far as the edge of the Rub al Kali. It has pale yellow flowers with well exposed anthers. These flowers produce high amount of nectar and pollen and therefore are visited frequently by honeybees (Whitcombe, 1984). This is also indicated from the high frequency of its pollen in the honey (Sajwani et al., 2007a). The flowering occurs twice a year. In autumn, from September to October and in spring, from March to May (Mandaville, 1978; Ghazanfar, 1997). This explains the presence of *Prosopis cineraria* pollen in both summer and winter honeys. This honey type has a light amber colour, moderate sugars and no ability for crystallization (Sajwani et al., 2007a).

Prosopis juliflora (Swartz) DC. Honey

This species is distributed in Oman, along the streets, gardens and roads. It is a fairly common cultivated plant, which is able to grow fast where nothing else will. It is produces catkins from November to March (El-Ghazaly, 1990; Love, 1995). This explains the presence of its pollen in both, summer and winter honeys. *Prosopis* honeys are produced by both bee species *A. florea* and *A. Mellifera* (Sajwani et al., 2007a). It is characterized by clear light amber colour, non-crystallized form, moderate amount of sugars and very low amount of sucrose (Sajwani et al., 2007b).

Kalpana and Ramanujam (1990) highlighted the importance of this plant in India as a major and reliable nectar and pollen source for the honeybees in winter. In spite of the value of this plant in beekeeping, it causes several disadvantages for other vegetation and soils (Eisikowitch and Dafni, 1988). Therefore elimination of this tree is necessary to maintain the Omani vegetation environment and will not affect honey production because other important plant sources are available for honeybees in the same season such as *Ziziphus spina-christi* and *Prosopis cineraria* (Sajwani et al., 2007a).

Citrus sp. Honey

Large-scale cultivation of *Citrus* sp. is practiced in the area. This indicates a high potential for production of unifloral honey from *Citrus*. The most common species is *Citrus aurantifolia*, which is cultivated throughout Oman and is best suited to the soil and climate of northern region (Kiew, 1993; Ghazanfar, 1997). Several other species of *Citrus* are cultivated in farms throughout Oman that include grapefruit, lemons and oranges. They are seasonal evergreen shrubs with fragrant, whitish tube-shaped blossoms, which attract honeybees. Many studies established the importance of *Citrus* as a nectar source for bees (Roubik, 1995; Lakshmi and Suryanarayana, 1997; Noor et al., 2009).

Citrus honey is not very popular in Oman, in spite of its production in large amount. This could be because people may not have enough knowledge about it and seem to prefer other native honeys such as *Acacia* and *Ziziphus*. This honey has good characteristics such as clear light colour, moderate amount of sugars and does not crystallize (Sajwani et al., 2007b).

Maerua crassifolia Forsskal (Sarah) Honey

Melissopalynological analysis revealed *Maerua crassifolia* to be a major nectar and pollen source in this honey (Sajwani et al., 2007a) since this plant is common in rock foothills and ravines of northern mountains. It produces greenish-white fragrant flowers with long conspicuous stamens. During the flowering period from March to April the flowers attract bees and butterflies (Ghazanfar, 1997). This honey is well known only in Oman and is expensive due to its low production in limited localities such as wadies and foothills. It is characterized by amber colour, has 80% sugars with low amount of sucrose 0.84% and has ability to crystallize (Sajwani et al., 2007b).

Multifloral Honeys

Multifloral honeys are produced commonly in Northern Oman. They contain various melliferous and non-melliferous pollen from a number of plants. When the plant taxa do not offer enough sources of food to the honeybees due to many reasons, for instance dry climate, low number of flowers, end of the flowering season (Seijo et al., 1993), the bees are forced to forage on several plants to gather enough nectar and pollen. The honey bees are foraging on many plants such as *Citrullus lanatus, Cucumis* sp., *Vicia* sp., *Pisium sativum, Coriandrum sativum, Allium cepa, Capsicum annum, Psidium guajava, Mangifera indica, Terminalia catappa, Cocos nucifera, Ocimum* sp., *Ochradenus* sp., *Launaea* sp., *Convolvulus* sp., extra (Figure 4) in addition to the major foraging plants.

Bee Botanical Flora

Sajwani et al., 2007a highlighted a number of plants that serve as a source of nectar and pollen for the honeybees of Muscat and Al Batinah regions (Figure 4). Both wild and cultivated vegetation are incorporating in the Omani bee flora. Wild plants are represented in the following examples: *Ochradenus* sp., *Tribulus* sp., *Blepharis ciliaris, Asphodelus fistulosus, Maerua crassifolia, Acridocarpus orientalis, Moringa peregrina* and *Olea europaea*. Cultivated plants are represented by the following: *Psidium guajava, Mangifera indica, Azadirachta indica, Terminalia catappa, Leucaena leucocephala, Citrullus lanatus, Ocimum* sp., and *Capsicum annuum* (Sajwani et al., 2007a)

Geographical and Botanical Origins of Omani Honeys

As mentioned by Bryant and Jones (2001), one of the reasons why pollen analysis in honey is required, is to determine the honey's geographical origin. Oman has a varied landscape represented in its dried-up river beds, hills, plains and deserts in which flourish the plants and trees which provide the honey bee with the nutrients it requires. Omani honey's quality depends on the geographical features and the types of crops where the bees exist. Thus, the beekeeping activities can be found in different geographical regions in Oman such as:

Plains

Plain honeys are produced in flat lands such as Barka in the Al Batinah region (Figure 1). Mostly honeys from these areas are recognise by the presence of pollen of *Citrullus lanatus, Medicago sativa, Trifolium* sp., *Thymus vulgaris, Abutilon* sp., *Capparis* sp., *Blepharis ciliaris* and *Grewia* sp.

Mountains

Mountain honeys are collected from apiaries located in the foothills of Al Hajar Mountains. Their pollen sources are from plants which grow on rock slopes such as *Acridocarpus orientalis, Olea europaea, Salvia* sp., *Helianthemum lippi.* The summer flowers of *Olea europaea*, in the proximity of collection sites, attract honeybees and yield a pale, clear honey which is requested by consumers (Miller and Morris, 1988; Ghazanfar, 1997).

Wadis

Wadi honeys are produced in the valleys, the dry areas that contains water only during times of heavy rain such as Wadi Bani Uffe and Wadi Sahtan. In these honeys the pollen from wild plants are more than the pollen of cultivated plants, such as *Ziziphus spina-christi, Acacia tortilis, Moringa peregrina, Maerua crassifolia,* and *Lychium shawii.* The use of native plants for production of honey is of particular interest to Omani people since they believe that these honeys are free from additive sugars and pesticides.

Oases

Oases honeys are collected from an apiaries located near date palm irrigation channels in Dhahira to the northwest and the Sharqiya to the east, where there are many oases irrigated by *aflaj* (aqueducts carrying water below and above ground from distant springs) (Whitcombe, 1984). In this honey the pollen belonged to plants growing in irrigated sandy soils in shade such as *Launaea* sp., *Ammi majus,* and *Lippia nodiflora.*

Coastlines

Coastal honeys are collected from apiaries located about 3 km from the sea (Oman Gulf). Their pollen came from plants, which show high salt tolerance and good drought resistance such as *Acacia tortilis, Zygophyllum qatarense* and *Heliotropium* sp.

Farms

The presence of large amount of pollen from cultivated plant species in honeys such as *Albizia lebbeck, Citrus* sp., *Terminalia cattapa, Mangifera indica, Ocimum* sp., and *Psidium guajava* points out that their origin is from agricultural plains, which are distributed all around Al Batinah region. Presence of cultivated species in Omani honeys indicates that adoption of agriculture by farmers in the area enrich the bee forage potential.

Urban

Urban honey shows small numbers of pollen types because they are collected from habitant towns where there is low cultivated vegetation such as *Ocimum* sp., *Citrullus lanatus* and *Capsicum annuum.*

Pollen analysis of honey is also required to determine the honey's botanical origin. In the light of this fact, it was observed that pollen from some plants such as *Acacia senegal* and *Boswellia sacra,* which are native to Dhofar, southern region of Oman, were present in northern honey samples. This signifies that these plants are introduced to the Al Batinah region. The presence of *Bombax* sp. pollen in honeys points that this African or South Asian, silk-cotton plant is introduced in some areas. *Helianthus annuus* (Sunflower) is very important honeybee forage plant for pollen and nectar in some countries and unifloral honeys can be obtained from its nectar (Howes, 1979). In Oman this plant is not an important honey source because it is grown in small numbers as an ornamental shrub.

It is possible to conclude that most of the pollen types found in honey are used by bee species but to a different degree because foraging of honeybees on specific sources depends upon the quality of the food, competition among sources flowering synchronously and competition among the insect visitors (Roubik, 1989). Beekeepers will have to be accordingly advised to undertake migration to locations with abundance of bee plants or to introduce suitable bee plants around their hives to increase the diversity of nectar sources in Oman and to augment the honey production.

REFERENCES

Abu-Tarboush, H., H. Al-Kahtani, and M. El-Sarrage. 1993. Floral-type identification and Quality evaluation of some honey types. *Food Chemistry*. 46: 13-17.

Anderson, D. 1990. A consultancy on beekeeping in the Sultanate of Oman. Report commissioned by FAO and UN from 25/10/1989 to 8/11/1989. Black Mountain Laboratories. Csiro.

Arab Organization for Agricultural Development. 1998. Technical report of the beekeeping in Sultanate of Oman. Leagu of Arab States. Khartoum.

Black & White. March 15/2010. Honeymen' in Feequin. http://blackandwhiteoman.blogspot.com/2010/07/honeymen-in-feequin.html. Muscat.

Bryant, V., and Jones, G. 2001. The R-values of honey: pollen coefficients. *Palynology*. 25: 11-28.

Crane, E. 1990. *Bees and beekeeping science, practice and world resources*. Heinemann Newness. London.

Dutton, R., A. Mjeni, and R. Whitcombe. 1982. *Honeybees in Oman*. Holmes McDougall Ltd. Edinburgh.

Dutton, R., F. Ruttner, A. Berkeley, and D. Manley. 1981. Observations on the morphology, relationships and ecology of *Apis mellifera* of Oman. *Journal of Apicultural Research*. 20: 201-214.

Eisikowitch, D., and A. Dafni. 1988. Thc usc and abusc of introducing honcy plants. *Bee World*. 69: 12-14.

El-Ghazaly, G. 1990. Pollen flora of Qatar. Scientific and applied research centre University of Qatar. Al-Doha.

Ghazanfar, S. 1997. Trees of Oman, an illustrated guide to the native trees of Oman. Ministry of Regional Municipalities and Environment. Muscat.

Howes, F. 1979. Plants and beekeeping. Faber and Faber Limited. London.

Hussein, M. 1992. Beekeeping in Sultanate of Oman. Ministry of Agriculture and Fisheries Resources. Oman.

Jones, G., and V. Bryant. 1992. Melissopalynology in the United States: a review and critique. *Palynology*. 16: 63-71.

Kalpana, T., and C. Ramanujam. 1990. Melittopalynology of *Prosopis juliflora* honeys from Ranga Reddy District (A. P.) and its significance. *Indian Bee Journal*. 52: 44-46.

Kiew, R. 1993. Characteristics of Malaysian flowers visited by *Apis cerana*. *Asian honeybees and bee mites*. 361-366.

Lakshmi, K., and Suryanarayana. 1997. Microscopical analysis of rockbee honeys from Nallamalai forest of Andhra Pradesh, India. *Journal of Palynology*. 33: 263-272.

Love, A. 1995. *Gardening in Oman and the Gulf*. Apex Publishing. Muscat.

Mandaville, J. 1978. *Wild flowers of northern Oman.* John Bartholomew and Son Ltd. London.

Manley, M. J. 1983. Technical aspect of extension work in northern Oman with reference to importation and distribution of movable frame hives. *In: Second international conference on apiculture in tropical climates.* Indian Agricultural Research Institute. New Delhi.

Miller, A., and M. Morris. 1988. *Plants of Dhofar, the southern region of Oman, traditional, economic and medicinal uses.* The Office of the Adviser for Conservation of the Environment, Diwan of Royal Court, Sultanate of Oman.

Ministry of Agriculture and Fisheries resources. 1996. Beekeeping. Sultanate of Oman. Muscat.

Ministry of Information. 1995. Oman 95. Sultanate of Oman. Muscat.

Ministry of Information. 2000. Oman 2000. Sultanate of Oman. Muscat.

Ministry of Information. 2002. Oman 2002. Sultanate of Oman. Muscat.

Mossadegh, M., and V. Bandpay. 1993. Plants visited by *Apis florea* F. in Iran. *Asian honeybees and bee mites.* 369-371.

Noor, M., M. Khan, and E. Camphor. 2009. Palynological analysis of pollen loads from pollen sources of honeybees in Islamabad, Pakistan. *Pakistan Journal of Botany.* 41: 495-501.

Oman Daily, Observer. 13 July 2010. Oman's honey production at 10,000 kg. Sultanate of Oman. Muscat.

Ramanujam, C., and T. Kalpana. 1995. Microscopic analysis of honeys from a coastal district of Andhra Pradesh, India. *Review of Palaeobotany and Palynology.* 89: 469- 480.

Ramanujam, C., T. Kalpana, and Khatija Fatima. 1992. Melittopalynology and recognition of major nectar and pollen sources for honey bees in some districts, Andhra Pradesh. *Geophytology.* 22: 261-271.

Roubik, D. 1989. *Ecology and natural history of tropical bees.* Cambridge Univ. Press. Cambridge.

Roubik, D. 1995. *Pollination of cultivated plants in the tropics.* Food and Agriculture Organization of the United Nations. Rome.

Sajwani, A., S. Farooq, A. Patzelt, E. Eltayeb, and V. Bryant. 2007a. Melissopalynological studies from Oman. *Palynology.* 31: 63–79.

Sajwani, A., E. Eltayeb, S. Farooq and A. Patzelt. 2007b. Sugar and protein profiles of Omani honey from Muscat and Batinah regions of Oman. *International Journal of Food Properties.* 10: 675–690.

Seijo, M., M. Aira, M. Iglesias and M. Jato. 1993. Palynological characterization of honeys from La Coruna province (Northwest Spain). *Journal of Apicultural Research.* 31: 149-155.

Song, Y., F. Yao and D. Yan. 2012. Pollen Analysis of Natural Honeys from the Central Region of Shanxi, North China. *PLoS ONE.* 7(11): e49545. doi:10.1371/journal.pone.0049545.

Tilde, A., and P. Payawal. 1992. Commercial honey in the Philippines 1. Pollen grains analysis. *The Philippine Agriculturist.* 75: 81-87.

Whitcombe, R. 1984. Aspects of the biology and management of *Apis florea* in Oman. *Proceedings of 3rd International Conference on Apiculture of Tropical Climates, Nairobi.* P. 96-103.

In: Biotechnology and Conservation … (2 Volume Set) ISBN: 978-1-63117-869-6
Editors: S. A. Farooq, R. Abed and S. Baqir

Chapter 19

ESTABLISHMENT OF SHORT AND LONG-TERM STORAGE PROTOCOL FOR OIL PALM (*ELAEIS GUINEENSIS* JACQ.) POLYEMBRYOIDS

Uma Rani Sinniah*[1]*, Sharrmila R. Palanyandy*[1]*,
***Maheran A. Aziz*[1] *and Sreeramanan Subramaniam*[2]**
[1]Department of Crop Science, Faculty of Agriculture,
Universiti Putra Malaysia, Serdang Selangor, Malaysia
[2]School of Biological Sciences, Universiti Sains
Malaysia, Minden Heights, Penang, Malaysia

ABSTRACT

This study focuses on establishment of short-term and long-term storage protocol for oil palm polyembryoids. In order to establish a successful storage technique, it is essential to identify the right stage prior to plantlet formation which will aid in simultaneous plantlet production whereby it can be stored adequately and recommenced for growth only when needed.

In the first experiment, the micromorphological changes during conversion of suspension culture into polyembryoids were observed. The morphological observation revealed that, polyembryoids with the presence of globular structures, haustorium and secondary somatic embryo (SSE) which was termed as matured polyembryoids as the right stage for rapid conversion into plantlet. Meanwhile, in the second experiment, determination of suitable gelling agent was conducted using six types of culture medium (Agar Type 900 at 8-12 gL^{-1} or Gelrite® 1.5-3.5 gL^{-1}) for better conversion of selected polyembryoids. Gelrite® with 3.5 gL^{-1} concentration was chosen as the effective gelling agent for higher conversion of polyembryoids into plantlet. Prior to storage, the samples were pretreated with sucrose to condition the polyembryoids for better adaptation to storage condition. Thus, in the third experiment, effect of encapsulation and gradual sucrose preculture on oil palm polyembryoids was studied. It showed that exposure of naked polyembryoids to sucrose preculture lead to lethal damage and in contrast, the encapsulated polyembryoids were able to withstand the treatment and gave enhanced survival. Hence, for both short-term and long-term storage study, encapsulated and subsequently sucrose-precultured polyembryoids were employed.

In the fourth experiment, method for short-term storage was established by manipulating temperature whereby the samples were stored at 5°C, 10°C and 25°C. In this study, the highest survival 73.3% was obtained at 5°C after 60 days of storage. In the fifth experiment, the encapsulated polyembryoids were subjected to long-term storage *via* cryopreservation after gradual sucrose preculture and Laminar Air Flow (LAF) desiccation using encapsulation dehydration technique. In this technique, encapsulated materials were subjected to cryopreservation after desiccation in order to avoid freezing injury upon liquid nitrogen (LN) exposure. Based on the results obtained, 73.3% survival was obtained with 23.3% water content for encapsulated polyembryoids, subjected to gradual sucrose preculture for 1 week, and desiccated under LAF prior to LN exposure. The experiments performed, showed that oil palm polyembryoids acquired sufficient tolerance for successful storage either short-term or long-term whereby good survival rate was obtained after their growth recommenced.

Keywords: Lamina air flow (LAF), Liquid nitrogen (LN), Secondary somatic embryo (SSE)

INTRODUCTION

Oil palm (*Elaeis guineensis* Jacq.) is oil–producing plant that economically most important crop in the world which play major role in international market. Production of palm oil per ha can achieve six ton per year (Sastrosayono, 2003). It is an economically important source of vegetable oil, the most traded vegetable oil in the international market, and increasingly used in the food industry (Corley and Tinker, 2003). Currently for those involved in the oil palm industry, increasing the oil yield is the main agenda to manage such a high demand for palm oil. It has been leaded to improve the performance of the crop due to increasing numbers of research done in plant genetics and breeding. However, with conventional breeding and seed production methods the maximum potential of the selected hybrid genotypes may still not be realized. Therefore, vegetative propagation of the oil palm via tissue culture seemed to be an attractive alternative which is currently widely used. Conversely, it is still not possible to propagate any palm in order. The longer time scale and non-uniformity in growth and development of oil palm cell suspension culture into plantlet renders the successful propagation of oil palm for larger scale production. In order to achieve successful propagation of oil palm a specific conversion stage must be identified by detailed micromorphological study and hence, can be stored until needed. This stage can be exploited to reduce the time scale to obtain complete plantlet formation by modifying media. In addition synchronized batches of specific conversion stage can be obtained by optimization of storage method using artificial seed so that it can be subjected to recommence when needed in order to obtain simultaneous development. The study was conducted in the Cryopreservation Laboratory in the Department of Crop Science, Universiti Putra Malaysia, Serdang, Selangor, Malaysia. Cell suspension cultures of *Elaeis guineensis* Jacq., obtained from Felda Biotechnology, Seremban, were used in this experiment. In vitro embryogenesis was achieved from embryogenic callus induced from young leaf which maintained on agar solidified MS (Murashige and Skoog, 1962) media containing 0.1 mgL^{-1} of 2, 4-dichlorophenoxyacetic acid (2, 4–D) at 25 ºC. In the first experiment, cell suspension culture was established by transferring 0.5g (fresh weight) of embryogenic callus to 125 ml Erlenmeyer flasks containing 30 ml of liquid media supplemented 30 gL^{-1} sucrose and 0.1 mgL^{-1} 2, 4 –D.

Then, embryo maturation was accomplished using MS semisolid medium consisted of 30 gL^{-1} sucrose, vitamins and 3.5 gL^{-1} gel rite at 25 °C. SEM observations were done for each morphologically distinguishes stage and specific conversion stage chosen to be used for upcoming experiments. In the second experiment, effect of different gelling agent in media for the enhancement of growth and development of chosen specific stage into oil palm plantlets was studied. The samples were cultured on MS medium with Gel Rite® (1.5 gL^{-1}, 2.5 gL^{-1} and 3.5g L^{-1}) and Agar Type 900 (8 gL^{-1}, 10gL^{-1} and 12 gL^{-1}). The percentage of viability and survival was recorded. For the detailed evaluation, morphological characteristics such as the number of shoot, root, secondary somatic embryo, callus and haustorium were measured. The medium with best gelling agent for growing oil palm embryos was selected from this experiment. The third experiment was conducted to study the effect of sucrose preculture and desiccation on naked and encapsulated embryos. The polyembryoids of identified specific stage were isolated and subsequently encapsulated as described by Grout (1995). The encapsulation solutions—0.1 M $CaCl_2$ and 3% (w/v) sodium alginate solution—were prepared in standard MS medium. The sample were drenched in the 3% (w/v) sodium alginate solution and then aliquotes were dropped in 0.1 M $CaCl_2$ solution to form beads with a diameter of 4–5 mm (each bead contained one sample). The beads were recovered and incubated for 7 days in sucrose-enriched MS liquid medium using gradual dehydration method (0.1M, 0.3M, 0.5M, 0.75M and 1M). Naked samples were also subjected to gradual sucrose dehydration. Then, both naked and encapsulated samples were desiccated in open petri dishes placed in the airflow of a laminar flow bench for 4 h. The moisture content, viability, survival and morphological characteristics were observed.

In the fourth experiment, encapsulated and sucrose precultured sample were subjected to cryopreservation following desiccation until 10 h. At every 1 h of desiccation, the beads were recovered and placed in cryotubes (5 per vial), frozen in liquid nitrogen (direct immersion) and stored for 24 h.

Samples were thawed by directly immersing the vials in a water bath (38 °C for 2 min after which no ice crystals were visible). Viability and survival of both frozen (+LN) and non-frozen (-LN) encapsulated and desiccated samples was determined following culture on recovery media. The encapsulated and sucrose precultured samples from third experiment were subjected for short-term storage in the fifth experiment. Encapsulated beads were stored in dark at three different storage temperature which was 5°C, 10°C and 25°C. Every 10 days interval samples were recovered and cultured in media until 70 days of storage. The data on viability, survival and morphological characteristics were observed.

Data from all the experiments were subjected to analysis of variance. Means were separated with Duncan's multiple range tests (DMRT) at the 0.05 level. The experiment was laid out in Randomized Complete Block Design (RCBD) with three replications per treatment, with each treatment having five polyembryoids.

In the first experiment a specific growing stage encountered using the micromorphological study during the developmental stages at which the embryos turns into an aggregation of self-potential cells. It is termed as polyembryoids. It is the immediate stage prior to plantlet formation. The polyembryoids selected and used in all the other experiments due to its self-potential and manipulative nature. In the second experiment the polyembryoids subjected to enhanced plantlet development using two types of gelling agent in MS media with different concentration as mentioned earlier.

The MS media with 3.5 gL^{-1} Gel Rite® resulted in 100% survival during 30 days of culture and it facilitated in enhanced plantlet formation which was clearly observed in morphological characteristics. No death or necrosis was observed during the 90th day in culture whereas all the other media gave 10-20% of necrosis except Gel Rite® 2.5 gL^{-1}. Hence, polyembryoids cultured in MS media with 3.5 gL^{-1} Gel Rite® gave significantly enhanced plant characteristics such as formation of multiple shoot and root.

In the third experiment, both encapsulated and naked polyembryoids which were with or without sucrose preculture displayed significantly different results at $p<0.05$. The encapsulated and sucrose precultured polyembryoids showed more than 50% survival even after 4 h of desiccation in LAF.

It also gave enhanced number of shoots (6 per plantlet) and roots (3.7 per plantlet) in comparison with other treatments. The naked and sucrose precultured poly-embryoids gave only 6.7% of survival after 4 h of desiccation which might be due to direct exposure of polyembryoids to gradual increasing of sucrose concentration. It might cause toxicity effect and death of cells.

Hence, the encapsulated and sucrose precultured polyembryoids were used for cryopreservation and short-term storage.

The encapsulated and sucrose precultured polyembryoids were desiccated in LAF and their moisture content decreased with increasing desiccation time. The fresh poly-embryoids had an initial moisture content of 67.8% and the final moisture content after 10 h of desiccation was about 21.3%. The polyembryoids with moisture content above 25.9% did not survive in LN treatment. Water removal plays a central role in preventing freezing injury and in maintaining post thaw viability (Gonzalez-Arnao et al., 2008). Loss in survival owing to high moisture content upon exposure to liquid nitrogen is a common phenomenon. This is due to the formation of ice crystals during the freezing and subsequent thawing which caused cell death.

Generally, the ice crystals are formed due to the presence of free water. Most of the freezable water is removed during dehydration and subsequently cooling is performed rapidly by direct immersion into the LN in order to obtain optimum survival. Polyembryoids with approximately 23.3% moisture content (9 h desiccation in LAF) showed the highest survival (73.3%) upon LN exposure. It was noticed that, there was a decline in the percentage of survival when the embryos were desiccated lower than 23.3% moisture content. The results showed that optimum survival of cryopreserved polyembryoids can be achieved upon desiccation to moisture content around 23.3%.

In the fifth experiment, the encapsulated and sucrose precultured polyembryoids stored at three different storage temperature which was 5°C, 10°C and 25°C showed significantly different results. Highest survival 73.3% was obtained for the polyembryoids that stored at 5°C for 60 days after two months in culture.

Meanwhile, the 53.3% of dead beads was obtained after two months of culture for the polyembryoids that stored at 25°C for 60 days. It showed that increase in temperature increase the metabolic rate and cause the death of polyembryoids in storage itself.

Hence, the ranking among favourable storage temperatures was 5°C > 10°C > 25°C for encapsulated oil palm polyembryoids. It showed that viability decreased when temperature increase. Saiprasad et al. (2003) also that reported the ranking among storage temperatures was 4°C > 20°C > 25°C > 0°C in *Dendrobium* 'Sonia' encapsulated PLBs.

ACKNOWLEDGMENTS

We would like to thank FELDA Biotechnology Center Sdn Bhd, Seremban, Malaysia for supplying us with the cell suspension culture and Yayasan Felda for the research grant.

REFERENCES

Gonzalez-Arnao, M. T., A. Panta, W. M. Roca, R. H. Escobar and F. Engelmann, 2008. Development and large scale application of cryopreservation techniques for shoot and somatic embryo cultures of tropical crops. *Plant Cell Tissue Org. Cult.*, 92: 1-13.

Grout, B. W. W. 1995. Introduction to the in vitro preservation of plant cells, tissues and organs. In: *Genetic Preservation of Plant Cells In Vitro*. Ed. B. W. W. Grout. Springer-Verlag. Heidelberg, pp. 1–20.

Murashige, T. and F. Skoog. 1962. A revised medium for rapid growth and bio assays with tobacco tissue cultures. *Physiol. Plant.* 15: 473–497.

Saiprasad G. V. S. and R. Polisetty, 2003. Propagation of three orchid genera suing encapsulated protocorm-like bodies. *In Vitro Cell. Dev. Biol. Plant*, 39: 42–44.

Sastrosayono, S. 2003. *Budidaya Kelapa Sawit*. AgroMedia Pustaka: Jakarta.

The Oil Palm. 4th Edition. Edited by R. H. V. Corley and P. B. Tinker. Oxford: Blackwell Publishing (2003), pp. 284. ISBN 0-632-05212-0.

In: Biotechnology and Conservation … (2 Volume Set) ISBN: 978-1-63117-869-6
Editors: S. A. Farooq, R. Abed and S. Baqir

Chapter 20

ISSR PHYLOGENETIC DIVERSITY OF *SORGHUM BICOLOR* (L.) MOENCH ACCESSIONS FROM DIFFERENT REGIONS IN SUDAN

***Haitham K. A. EL-Amin*[1] *and Nada B. Hamza*[2,*]**
[1]Sudan Academy of Sciences, Ministry of Science & Technology, Khartoum, Sudan
[2]Departmentof Molecular Biology, Commission for Biotechnology and Genetic Engineering, National Centre for Research, Khartoum, Sudan

ABSTRACT

Sorghum (*Sorghum bicolor* (L.)) was domesticated for the first time within the geographical range of Sudan, so studying the genetic variation of sorghum genotypes collections attracts special interest. In this study, 50 sorghum accessions with important agronomic traits, representing 11 states in Sudan and Republic of South Sudan were assayed for polymorphism using Inter-simple sequence repeat (ISSR). Seven primers out of 10 tested (807, 808, 810, 814, 848, 872 and 879) showed high polymorphism among the accessions.

The results indicated 75 polymorphic bands out of 78 bands with percentage of polymorphic bands of 97%. UPGMA result showed ISSR distance matrix were ranged between (0.04-0.47) which reflected high genetic diversity. The ISSR UPGMA dendrogram showed high molecular variance within regions. ISSR technique given different results compared to the RAPD technique which it used in earlier study and that due to the ISSR technique amplify different parts of the genome to covering wider area in the genome. Based on the results of this study ISSR technique showed the differences among closed related accessions of sorghum. Also it proved to be a useful technique to study genetic variation among the Sudanese sorghum accessions.

Keywords: Sorghum, ISSR markers, genetic variation, Cluster, UPGMA, Sudan

* Dr. Nada B. Hamza, Email: nada.hamza@gmail.com.

INTRODUCTION

Sorghum seeds are important because they provide the main meal for millions of poor people (Mehmoodet al., 2008). They also provide forage, fiber, dextrose syrup, biofuels, alcohol, and other products (Iqbal et al., 2010). Recently, Sorghum was considered a key plant species for the comparative analysis of *Poaceae* genomes for the agricultural field. Sorghum genome is relatively small with 750 Mbp (Arumuganathan and Earle, 1991). The Sorghum genetic recombination map of 2512 loci spaced at average 0.4 cm (ca. 300 kb) which open opportunities to help advance genomics in the larger genomes of many other *Poaceae* members (Bowers et al., 2003). The DNA molecular markers become the choice for the crop genetic diversity study, it has become routine, to revolutionize plant biotechnology (Kumar et al., 2009). There are many DNA markers available, like randomly amplified polymorphic DNA (RAPD) (Welsh and McCleland, 1990), inter simple sequence repeats (ISSR) (Zietkiewicz et. al, 1994), and amplified fragment length polymorphism (AFLP) (Voset al., 1995). The different types of DNA markers have the potential to differentiate and detect differences among genotypes, in relation to cost, easiness of use, consistency and repeatability of the results (Schlötterer, 2004). Inter simple sequence repeat (ISSR) is a technique based on Polymerase Chain Reaction (PCR), which includes amplification of the DNA (Reddy et al., 2002). It's considered as a dominant markers (Wang et al., 1998). Due to the long ISSR primers (16–25 mers) produce high reproducibility possibly as compared to the RAPD primers (10 mers) which permits the subsequent use of high annealing temperature (45– 60 ◦C), which it leading to higher stringency (Reddy et al., 2002). It proved the genetic variation among very closely related individuals and in crop accessions classification better than most other DNA markers (Fang and Roose, 1997). The main objective of our study was to investigate the degree of polymorphism detected using 7 ISSR (Inter-simple sequence repeat) markers in 50 Sorghum accessions from 11 different regions in Sudan and Republic of South Sudan.

SEED MATERIAL

The seeds of the 50 sorghum accessions used were provided by the Germplasm Bank of the Genetic Resources Unit (the Agriculture Research Corporation, Wad Madani). They were collected from ten different States of Sudan namly River Nile, West Darfur, North Kordofan, Sinnar, Kassala, Blue Nile, South Kordofan, White Nile, Red Sea, North Darfur, and Bahr EL Jabel State of South Sudan (Table 1). Two millet accessions were included in the study as control. Sorghum seeds were sown in pots containing equal volumes of sand and clay (1:1). Each pot contained about 20 seeds.

DNA EXTRACTION

DNA was extracted from fresh leaf tissue of *Sorghum bicolor*accessions using modified CTAB method (Porebski et al., 1997). The modification was made in intention to improve the DNA quantity and the quality. In this method the fine powdered plant materials were immediately transferred into 15 ml Falcon tubes containing 5 ml of pre-warmed lysis solution. Tubes containing the samples were thenincubated in a water bath at 60ºC with gentle shaking for 30 minand left to cool at room temperature for 10 min. Chloroform: Isoamylalcoholmixture (24:1) was added to each tube and the phaseswere mixed gently for 10 min at room temperature to make a homogenousmixture. The cell debris was removed by centrifugation at 4000 rpm for 15 min and the resulted clear aqueous phases (containing DNA) were transferred to new sterile tubes.

The step of the chloroform: isoamyl alcohol extraction was repeated twice. Thenucleic acids in the aqueous phase were precipitated by addingequal volume of deep cooled Isopropanol. The contents were mixedgently and collected by centrifugation at 4000 rpm for 10 min. Theformed DNA pellet was washed twice with 70% Ethanol and theEthanol was discarded. Theremained ethanol was removed by leaving the pellet to dry at roomtemperature. The pellet was dissolved in TE buffer (10 mM Tris, 1 mM EDTA, pH 8) and stored at -20ºC for further use.

Table 1.The names and regions of sorghum accessions used in the study

No.	Accessions Name	Regions	No.	Accessions Name	Regions
1	HSD 2790	River Nile	27	HSD 5194	Kassala
2	HSD 2791	River Nile	28	HSD 5640	Blue Nile
3	HSD 2792	River Nile	29	HSD 5641	Blue Nile
4	HSD 2793	River Nile	30	HSD 5642	Blue Nile
5	HSD 2795	River Nile	31	HSD 5643	Blue Nile
6	HSD 2939	Bahr EL Jabel	32	HSD 5650	Blue Nile
7	HSD 2941	Bahr EL Jabel	33	HSD 6001	South Kordofan
8	HSD 2945	Bahr EL Jabel	34	HSD 6002	South Kordofan
9	HSD 3220	West Darfur	35	HSD 6003	South Kordofan
10	HSD 3221	West Darfur	36	HSD 6006	South Kordofan
11	HSD 3222	West Darfur	37	HSD 6007	South Kordofan
12	HSD 3223	West Darfur	38	HSD 6541	White Nile
13	HSD 3226	West Darfur	39	HSD 6542	White Nile
14	HSD 3444	North Kordofan	40	HSD 6543	White Nile
15	HSD 3445	North Kordofan	41	HSD 6544	White Nile
16	HSD 3447	North Kordofan	42	HSD 6545	White Nile
17	HSD 3449	North Kordofan	43	HSD 6974	Red Sea
18	HSD 3901	Sinnar	44	HSD 6975	Red Sea
19	HSD 3903	Sinnar	45	HSD 6977	Red Sea
20	HSD 3905	Sinnar	46	HSD 6991	Red Sea
21	HSD 3906	Sinnar	47	HSD 7115	North Darfur
22	HSD 3907	Sinnar	48	HSD 7116	North Darfur
23	HSD 5190	Kassala	49	HSD 7117	North Darfur
24	HSD 5191	Kassala	50	HSD 7125	North Darfur
25	HSD 5192	Kassala	A	HSD 2369	South Kordofan
26	HSD 5193	Kassala	B	HSD 5564	Blue Nile

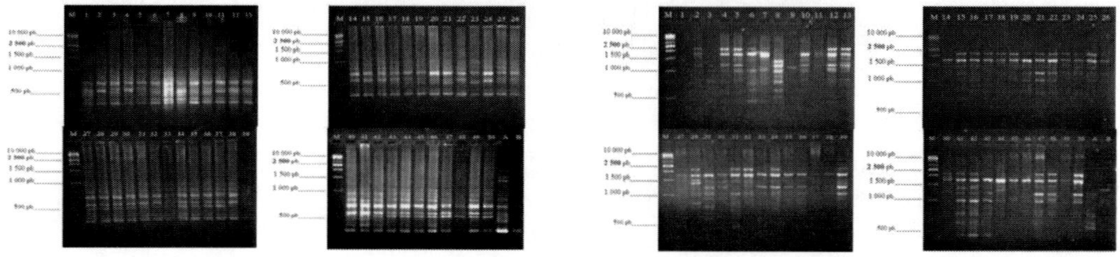

807 879

Figure 1. ISSR amplification patterns with primers 807 and 879 (M: DNA ladder (1 kbp), 1: HSD 2790, 2: HSD 2791, 3: HSD 2792, 4: HSD 2793, 5: HSD 2795, 6: HSD 2939, 7: HSD 2941, 8: HSD 2945, 9: HSD 3220, 10: HSD 3221, 11: HSD 3222, 12: HSD 3223, 13: HSD 3226, 14: HSD 3444, 15: HSD 3445, 16: HSD 3447, 17: HSD 3449, 18: HSD 3901, 19: HSD 3903, 20: HSD 3905, 21: HSD 3906, 22: HSD 3907, 23: HSD 5190, 24: HSD 5191, 25: HSD 5192, 26: HSD 5193, 27: HSD 5194, 28: HSD 5640, 29: HSD 5641, 30: HSD 5642, 31: HSD 5643, 32: HSD 5650, 33: HSD 6001, 34: HSD 6002, 35: HSD 6003, 36: HSD 6006, 37: HSD 6007, 38: HSD 6541, 39: HSD 6542, 40: HSD 6543, 41: HSD 6544, 42: HSD 6545, 43: HSD 6974, 44: HSD 6975, 45: HSD 6977, 46: HSD 6991, 47: HSD 7115, 48: HSD 7116, 49: HSD 7117, 50: HSD 7125).

DNA Quality and Quantity

The extractedDNA samples were observed under UV illumination after stainingwith Ethidium Bromide and agarose gel electrophoresis. The purityand the concentrations of the DNA were then Spectrophotometrically assessed following Sambrook et al., (1989) method.

PCR of the ISSR Technique

Seven ISSR primers were used in the polymerase chain reaction (PCR) in a final volume of 25 μL containing 1.0 μL DNA diluted, 0.5 μLTaqpolymerase, 2.5 μL 10X buffer, 2.5 μL (2mM/μl) dNTPs, 1.5 μL (50 mM) $MgCl_2$, 2.0 μL (10 pmol/ μl) ISSR primer and 15 μL d_dH_2O.

The amplifications were performed in a thermal cycler following the program: 94°C for 5 min, 40 cycles (1 min at 94°C, 1 min at 43°C and 1 min at 72°C) and final elongation of 7 min at 72°C.4 μl of PCR product were mixed with 2 μl of loading dye and 1.4 μl of 1 Kbp DNA ladder, then were electrophoresed using 2% agarose gel at 80 Volts followed by staining with Ethidium Bromide then the separated fragments were visualized with an ultraviolet (UV) transilluminator (Figure 1).

ISSR Data Analysis

The number of polymorphic and monomorphic bands were determined for each primer. Genotypes were scored (1) for present band, and (0) for absent band and then entered into a data matrix. Percentage of polymorphism was calculated as the following equation: (polymorphic bands / total number of bands x 100).

Table 2. Polymorphism detected by the use of seven polymorphic ISSR primers on 50 sorghum accessions

Primer name	Sequence (5'-3')	Total number of bands	Number of polymorphic bands	Number of monomorphic bands	Percentage of polymorphic bands
807	(AG)8 T	19	17	2	89.5%
808	(AG)8 C	10	9	1	90%
810	(GA)8 T	9	9	0	100%
814	(CT)8 A	9	9	0	100%
848	(CA)8 RG	9	9	0	100%
872	(GATA)4	14	14	0	100%
879	(CTTCA)3	8	8	0	100%
Total		78	75	3	
Average		11.1	10.7	0.4	97%

The tree diagram was produced by clustering the similarity data with the UPGMA method using STATISTCA- SPSS software Ver. 9 following the method as used by El-Amin and Hamza, (2012).

To isolate height quality of DNA, the CTAB-based procedure optimized in the present study, yielded good quality DNA free of phenols, which may inhibit the activity of Taq polymerase. Ten primers were tested on the 50 accessions (*Sorghum bicolor* L.) and the results indicated that seven primers out of the ten tested showed a high polymorphic band percentage (97%).

The seven informative primers were selected and used to evaluate the degree of polymorphism and genetic relationships among the genotypes under study. Total of 78 amplified fragments were distinguished across the selected primers and the statistical analysis showed 75 polymorphic bands among the accessions. The maximum number of bands were produced by the primer 807 (19 bands) with 89.5% polymorphism respectively while the minimum numbers of fragments were produced by the primer 879 (8 bands) with 100%, polymorphism respectively. ISSR fragments pattern produced by the seven primers is shown in Table (2).

The distance matrix table of ISSR markers the highest similarity were between accessions (43, 45), (44, 46) and (45, 46) with 0.04, and the lowest similarity was between (1, 7) with 0.47 (Table 3).

The tree diagram of ISSR markers analysis showed accessions (1) and (39), as genetically distant from all other sorghum accessions and appeared as outgroup. Accession (22) was also distant. Cluster *A* had only three accessions (2), (4) from River Nile beside accession (24) from Kassala State. Cluster *B* regrouped the States accessions together and distributed them to three groups. Several small sub-groups of the accessions appeared according to their States of sampling as shown in Figure (2).

The genetic variability was investigated for 50 sorghum accessions from Sudan and Republic of South Sudan using seven ISSR primers were used to evaluate the degree of polymorphism and genetic relationship within and among all accessions under study. ISSR primers gave good amplification and were found polymorphic with the 50 sorghum accessions.

Table 3. The Distance Matrix of sorghum accessions used in the study

	1	2	3	4	5	6	7	8	9	10	11	12	13	14	15	16	17	18	19	20	21	22	23	24	25	26	27	28	29	30	31	32	33	34	35	36	37	38	39	40	41	42	43	44	45	46	47	48	49	50	A	B
1	0.00																																																		
2	0.30	0.00																																																	
3	0.28	0.16	0.00																																																
4	0.42	0.17	0.20	0.00																																															
5	0.34	0.22	0.14	0.13	0.00																																														
6	0.38	0.24	0.24	0.25	0.14	0.00																																													
7	0.47	0.30	0.25	0.24	0.18	0.17	0.00																																												
8	0.42	0.28	0.22	0.21	0.13	0.17	0.16	0.00																																											
9	0.34	0.22	0.09	0.18	0.13	0.20	0.16	0.18	0.00																																										
10	0.34	0.20	0.14	0.16	0.08	0.17	0.13	0.18	0.11	0.00																																									
11	0.36	0.21	0.16	0.14	0.12	0.21	0.14	0.20	0.12	0.07	0.00																																								
12	0.36	0.21	0.24	0.25	0.17	0.16	0.14	0.25	0.20	0.12	0.13	0.00																																							
13	0.33	0.26	0.21	0.20	0.14	0.24	0.17	0.20	0.17	0.12	0.11	0.16	0.00																																						
14	0.38	0.18	0.16	0.17	0.09	0.18	0.20	0.17	0.12	0.12	0.16	0.21	0.21	0.00																																					
15	0.41	0.21	0.24	0.17	0.12	0.16	0.20	0.20	0.17	0.12	0.13	0.18	0.18	0.11	0.00																																				
16	0.38	0.18	0.18	0.12	0.07	0.16	0.17	0.14	0.17	0.12	0.13	0.18	0.18	0.13	0.13	0.00																																			
17	0.42	0.22	0.20	0.18	0.08	0.17	0.18	0.18	0.16	0.11	0.14	0.17	0.20	0.12	0.09	0.09	0.00																																		
18	0.42	0.20	0.22	0.21	0.16	0.20	0.16	0.18	0.16	0.13	0.12	0.17	0.20	0.12	0.12	0.14	0.11	0.00																																	
19	0.38	0.26	0.24	0.20	0.09	0.18	0.17	0.12	0.20	0.14	0.16	0.21	0.18	0.13	0.13	0.08	0.12	0.14	0.00																																
20	0.43	0.26	0.24	0.22	0.14	0.24	0.20	0.14	0.20	0.20	0.18	0.24	0.24	0.13	0.16	0.13	0.14	0.12	0.11	0.00																															
21	0.39	0.20	0.25	0.26	0.21	0.20	0.18	0.21	0.21	0.18	0.17	0.14	0.25	0.17	0.20	0.22	0.18	0.11	0.22	0.17	0.00																														
22	0.34	0.30	0.30	0.29	0.21	0.25	0.26	0.26	0.26	0.24	0.28	0.30	0.33	0.17	0.17	0.22	0.16	0.18	0.22	0.22	0.26	0.00																													
23	0.39	0.20	0.17	0.18	0.13	0.22	0.16	0.21	0.11	0.08	0.09	0.14	0.14	0.09	0.12	0.14	0.13	0.08	0.17	0.12	0.16	0.24	0.00																												
24	0.39	0.17	0.20	0.13	0.13	0.25	0.24	0.18	0.16	0.18	0.20	0.28	0.20	0.12	0.20	0.09	0.18	0.18	0.14	0.17	0.26	0.26	0.16	0.00																											
25	0.43	0.21	0.18	0.20	0.14	0.24	0.17	0.20	0.14	0.14	0.16	0.18	0.21	0.13	0.13	0.13	0.09	0.09	0.18	0.13	0.20	0.20	0.07	0.17	0.00																										
26	0.45	0.22	0.17	0.18	0.13	0.22	0.18	0.21	0.13	0.16	0.20	0.25	0.25	0.12	0.14	0.17	0.11	0.13	0.20	0.17	0.24	0.18	0.11	0.18	0.07	0.00																									
27	0.39	0.22	0.17	0.21	0.11	0.14	0.16	0.16	0.16	0.13	0.12	0.14	0.20	0.12	0.12	0.09	0.11	0.11	0.09	0.12	0.18	0.21	0.13	0.18	0.14	0.16	0.00																								
28	0.38	0.18	0.16	0.14	0.09	0.18	0.14	0.14	0.09	0.09	0.11	0.18	0.18	0.05	0.13	0.08	0.12	0.09	0.11	0.11	0.17	0.20	0.07	0.09	0.11	0.12	0.09	0.00																							
29	0.42	0.25	0.22	0.21	0.11	0.17	0.13	0.13	0.16	0.13	0.14	0.17	0.20	0.09	0.12	0.09	0.11	0.11	0.09	0.12	0.16	0.18	0.13	0.16	0.14	0.16	0.08	0.07	0.00																						
30	0.37	0.22	0.20	0.21	0.16	0.20	0.18	0.18	0.16	0.18	0.17	0.20	0.22	0.17	0.17	0.12	0.18	0.16	0.14	0.17	0.21	0.26	0.18	0.18	0.20	0.21	0.13	0.12	0.13	0.00																					
31	0.41	0.21	0.24	0.22	0.20	0.26	0.17	0.22	0.17	0.17	0.18	0.21	0.18	0.18	0.21	0.18	0.20	0.17	0.21	0.21	0.20	0.30	0.12	0.14	0.13	0.14	0.22	0.13	0.20	0.25	0.00																				
32	0.43	0.21	0.21	0.17	0.14	0.26	0.12	0.20	0.14	0.12	0.13	0.18	0.16	0.13	0.18	0.13	0.14	0.12	0.18	0.18	0.20	0.25	0.09	0.14	0.11	0.12	0.17	0.08	0.12	0.17	0.08	0.00																			
33	0.41	0.24	0.18	0.20	0.17	0.21	0.14	0.25	0.12	0.12	0.16	0.18	0.16	0.13	0.18	0.18	0.20	0.17	0.24	0.21	0.20	0.30	0.09	0.20	0.16	0.14	0.20	0.13	0.17	0.20	0.16	0.11	0.00																		
34	0.36	0.21	0.13	0.20	0.12	0.18	0.17	0.20	0.12	0.12	0.13	0.16	0.13	0.11	0.16	0.13	0.14	0.14	0.16	0.16	0.22	0.25	0.09	0.17	0.13	0.14	0.09	0.08	0.12	0.12	0.18	0.11	0.11	0.00																	
35	0.37	0.25	0.20	0.21	0.13	0.25	0.24	0.18	0.18	0.18	0.14	0.22	0.17	0.14	0.17	0.14	0.16	0.16	0.14	0.09	0.21	0.26	0.13	0.18	0.17	0.18	0.13	0.12	0.13	0.13	0.20	0.14	0.17	0.09	0.00																
36	0.39	0.20	0.28	0.29	0.29	0.25	0.26	0.34	0.24	0.24	0.20	0.17	0.25	0.28	0.25	0.25	0.29	0.21	0.28	0.28	0.24	0.39	0.18	0.26	0.25	0.26	0.21	0.22	0.26	0.18	0.22	0.20	0.17	0.17	0.18	0.00															
37	0.46	0.29	0.24	0.28	0.22	0.29	0.20	0.28	0.17	0.17	0.18	0.24	0.24	0.18	0.21	0.24	0.20	0.17	0.24	0.18	0.22	0.30	0.14	0.25	0.21	0.17	0.17	0.16	0.17	0.20	0.18	0.13	0.13	0.16	0.12	0.17	0.00														
38	0.39	0.22	0.22	0.29	0.24	0.28	0.18	0.26	0.18	0.18	0.17	0.20	0.22	0.20	0.22	0.25	0.24	0.16	0.25	0.20	0.21	0.32	0.13	0.26	0.17	0.16	0.18	0.17	0.18	0.18	0.17	0.12	0.14	0.14	0.13	0.13	0.09	0.00													
39	0.32	0.36	0.41	0.39	0.34	0.38	0.34	0.42	0.37	0.34	0.33	0.33	0.36	0.33	0.33	0.36	0.32	0.26	0.38	0.30	0.26	0.29	0.26	0.37	0.30	0.32	0.34	0.30	0.32	0.37	0.25	0.25	0.30	0.30	0.26	0.32	0.28	0.29	0.00												
40	0.45	0.22	0.25	0.21	0.16	0.17	0.13	0.24	0.18	0.16	0.14	0.17	0.20	0.17	0.12	0.14	0.13	0.11	0.17	0.20	0.16	0.24	0.16	0.21	0.17	0.16	0.13	0.14	0.11	0.16	0.17	0.12	0.17	0.17	0.21	0.21	0.17	0.18	0.26	0.00											
41	0.41	0.21	0.29	0.30	0.22	0.18	0.22	0.25	0.22	0.17	0.24	0.16	0.29	0.18	0.16	0.24	0.17	0.14	0.24	0.18	0.09	0.25	0.14	0.28	0.18	0.20	0.22	0.18	0.17	0.22	0.18	0.21	0.21	0.24	0.22	0.22	0.18	0.17	0.28	0.17	0.00										
42	0.43	0.26	0.32	0.30	0.20	0.18	0.20	0.25	0.22	0.17	0.21	0.16	0.26	0.18	0.13	0.21	0.14	0.14	0.21	0.18	0.09	0.22	0.14	0.28	0.18	0.20	0.20	0.18	0.14	0.22	0.18	0.21	0.21	0.24	0.22	0.25	0.18	0.22	0.28	0.14	0.05	0.00									
43	0.41	0.29	0.24	0.28	0.14	0.16	0.20	0.22	0.17	0.17	0.18	0.21	0.26	0.13	0.11	0.16	0.12	0.14	0.16	0.13	0.17	0.17	0.14	0.22	0.16	0.17	0.14	0.13	0.12	0.17	0.21	0.21	0.21	0.18	0.14	0.28	0.16	0.22	0.30	0.17	0.13	0.08	0.00								
44	0.37	0.25	0.28	0.32	0.18	0.17	0.24	0.26	0.21	0.21	0.22	0.17	0.28	0.17	0.14	0.20	0.16	0.18	0.20	0.17	0.18	0.21	0.18	0.26	0.20	0.21	0.18	0.17	0.13	0.16	0.25	0.22	0.22	0.20	0.16	0.21	0.17	0.18	0.32	0.16	0.12	0.09	0.07	0.00							
45	0.37	0.28	0.22	0.29	0.16	0.20	0.24	0.24	0.18	0.18	0.17	0.22	0.22	0.14	0.12	0.17	0.13	0.13	0.17	0.12	0.18	0.18	0.13	0.24	0.14	0.18	0.13	0.14	0.13	0.18	0.22	0.20	0.20	0.14	0.11	0.24	0.14	0.18	0.29	0.18	0.17	0.12	0.04	0.08	0.00						
46	0.33	0.24	0.26	0.33	0.20	0.18	0.28	0.28	0.22	0.22	0.21	0.18	0.26	0.18	0.16	0.21	0.17	0.17	0.21	0.16	0.17	0.22	0.17	0.28	0.18	0.22	0.17	0.18	0.17	0.20	0.24	0.24	0.24	0.18	0.14	0.20	0.18	0.17	0.30	0.20	0.13	0.11	0.08	0.04	0.04	0.00					
47	0.36	0.24	0.26	0.28	0.17	0.16	0.28	0.20	0.25	0.20	0.26	0.21	0.26	0.18	0.16	0.18	0.17	0.20	0.18	0.16	0.17	0.25	0.20	0.25	0.21	0.22	0.20	0.18	0.17	0.20	0.21	0.24	0.24	0.18	0.14	0.25	0.21	0.22	0.30	0.22	0.08	0.13	0.13	0.14	0.14	0.13	0.00				
48	0.32	0.20	0.22	0.24	0.16	0.17	0.26	0.21	0.21	0.16	0.22	0.20	0.22	0.17	0.14	0.14	0.18	0.18	0.17	0.17	0.24	0.26	0.16	0.21	0.17	0.21	0.18	0.14	0.18	0.16	0.20	0.20	0.20	0.14	0.16	0.21	0.22	0.21	0.32	0.21	0.14	0.20	0.17	0.16	0.16	0.14	0.07	0.00			
49	0.33	0.18	0.21	0.28	0.17	0.13	0.22	0.22	0.20	0.17	0.24	0.18	0.26	0.11	0.16	0.18	0.17	0.14	0.18	0.18	0.14	0.22	0.17	0.22	0.18	0.20	0.17	0.13	0.14	0.20	0.18	0.18	0.18	0.16	0.20	0.25	0.21	0.20	0.30	0.17	0.11	0.16	0.13	0.14	0.14	0.13	0.08	0.09	0.00		
50	0.34	0.17	0.25	0.26	0.21	0.17	0.21	0.24	0.18	0.16	0.17	0.12	0.20	0.17	0.17	0.20	0.18	0.11	0.22	0.20	0.13	0.26	0.11	0.24	0.14	0.21	0.18	0.14	0.18	0.18	0.17	0.14	0.17	0.14	0.21	0.16	0.22	0.16	0.29	0.16	0.12	0.14	0.20	0.16	0.16	0.12	0.17	0.13	0.12	0.00	
A	0.57	0.47	0.47	0.51	0.57	0.55	0.46	0.49	0.43	0.49	0.47	0.50	0.55	0.50	0.53	0.50	0.54	0.46	0.53	0.50	0.49	0.59	0.46	0.49	0.47	0.51	0.46	0.47	0.49	0.51	0.47	0.47	0.50	0.50	0.54	0.49	0.47	0.46	0.51	0.49	0.50	0.53	0.55	0.57	0.54	0.55	0.55	0.51	0.53	0.49	0.00		
B	0.38	0.45	0.50	0.51	0.57	0.55	0.49	0.51	0.46	0.54	0.53	0.50	0.55	0.53	0.53	0.50	0.57	0.51	0.53	0.47	0.54	0.49	0.49	0.49	0.47	0.51	0.51	0.50	0.51	0.46	0.50	0.50	0.50	0.50	0.49	0.46	0.50	0.41	0.38	0.51	0.50	0.55	0.55	0.49	0.54	0.50	0.53	0.46	0.53	0.49	0.29	0.00

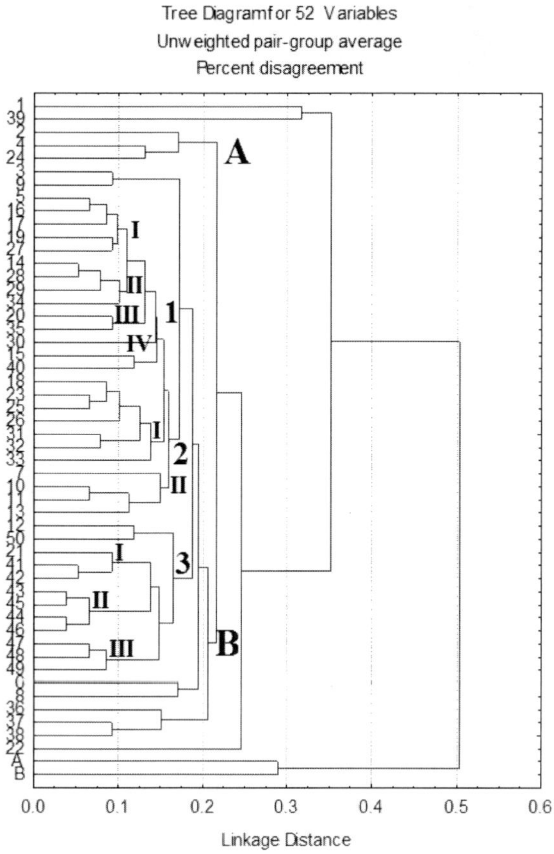

Figure 2. Tree Diagram resulting from the analysis of seven ISSR primers reflecting the relationships among the 50 sorghum accessions with two control samples.

The percentage of polymorphic bands showed that ISSR primers had high polymorphic percentage (97%). The average number of polymorphic bands produced per ISSR primer was 10.7, which is higher than the findings in earlier reports such as Iqbal et al., (2010) who reported an average of 7; Amrapaliet al., (2008) with an average of 6.5; Mohamed et al., (2008) with an average of 5.8.

The ISSR UPGMA Dendrogram showed higher distribution among sorghum accessions than the RAPD UPGMA Dendrogram. RAPD UPGMA Dendrogram grouped the accessions according to their States of sampling. Therefore, we can consider that the ISSR gave different results compared to the RAPD technique and that might be due to the fact that the ISSR primers amplify different parts on the genome compared to those amplified by the RAPD primers. Thus using both markers will cover wider area in the genome. In this study, ISSR produce more reliable and reproducible bands because ISSRs have high reproducibility and this was possibly due to the use of longer primers (16–25 mers) when compared to shorter RAPD primers (10- mers). The longer primers permits the subsequent use of high annealing temperatures (45– 60◦C) leading to higher stringency (Singh et al., 2011). Therefore, ISSR marker technique overcomes most of these limitations and it is rapidly being used by the breeders and researchers community in various fields of Sudanese sorghum improvement, because ISSR marker is a simple, quick, and efficient technique and it has high reproducibility (Reddy et al., 2002).

Acknowledgments

Appreciations are extended to the Germplasm Bank of the Agriculture Research Corporation (ARC) for cooperation for providing me with the sorghum accessions.

References

Amrapali AA, Sakhare SB, Kulwal PL, Dhumale DB, et al. 2008. RAPD profile studies in sorghum for identification of hybrids and their parents. *Inter. J. Integ. Bio.*, 3: 18-24.

Arumuganathan E and Earle ED. 1991. Nuclear DNA content of some important plant species. *Plant Mol. Biol. Rep.*, 9:208-218.

Bowers JE, Abbey C, Anderson S, Chang C, Draye X, Lattu AH, Jessup R, Lemke C, Lennington J, Li Z, Lin Y, Liu S, Luo L, Marler BS, Ming R, Mitchell SE, Qiang D, Reischmann K, Schulze SR, Skinner DN, Wang Y, Kresovich S, Schertz KF and Paterson AH. 2003. A High-density Genetic Recombination Map of Sequence-tagged Sites for *Sorghum* for Comparative Structural and Evolutionary Genomics of Tropical Grains and Grasses. *Genetics*, 165:367-386.

EL-Amin HKA and Hamza NB. 2012. Phylogenetic diversity of *Sorghum bicolor* (L.) Moench accessions from different regions in Sudan. *American Journal of biochemistry and molecular biology*, 2150-4210.

Fang DQ, Roose ML. 1997. Identification of closely relative Citrus cultivars with inter simple sequence repeat markers. *Theor. Appl. Genet.*, 95:408-417.

Iqbal A, Sadia B, Khan AI, Awan FS, KainthRAand SadaqatHA 2010. Biodiversity in the sorghum (Sorghum bicolor L. Moench) germplasm of Pakistan,l online Journal, Genetics and Molecular Research, 1676 5680, 9 (2): 756-764.

Kumar P, Gupta VK, Misra AK, Modi DR and Pandey BK. 2009. Potential of Molecular Markers in Plant Biotechnology. Plant Omics Journal *Southern Cross Journals*, 1836-3644/ 2(4):141-162.

Mehmood S, Bashir A, Ahmad A, Akram Z, Nyla Jabeen, Gulfraz M. 2008. Molecular Characterization of Regional *Sorghum Bicolor*Varieties From Pakistan. *Pak. J. Bot.*, 40(5): 2015-2021.

Porebski S, Bailey LG, and Baum R. 1997. Modification of a CTAB DNA extraction protocol for plants containing high polysaccharide and polyphenol components. *Plant Mol. Biol. Reporter*, 15(1): 8-15.

Reddy MP, Sarla N, Siddiq EA. 2002. Inter simple sequence repeat (ISSR) polymorphism and its application in plant breeding. *Euphytica,* 128: 9–17.

Sambrook J, Fritsch E, Maniatis, T. 1989. Molecular cloning. A laboratory manual. Second edition. Cold Spring Harbor Laboratory Press, Cold Spring Harbor, New York, USA.

Singh DR, Srivastava AK, Srivastava A, Srivastava RC. 2011. Genetic diversity among three morinda species using RAPD and ISSR markers. *Indian journal of biotechnology*, 10: 285-293.

Schlötterer C. 2004. The evolution of molecular markers—just a matter of fashion. *Nat. Rev. Genet.*, 5: 63–69.

Vos P, Hogers R, Bleeker M, Reijans M, Lee van de T, Hornes M, Frijters A, Pot J, Peleman J, Kuiper M, Zabeau M. 1995. AFLP: a new technique for DNA fingerprinting. *Nucl. Acids Res.*, 23: 4407–4414.

Wang G, Mahalingan R and Knap HT. 1998. (C-A) and (GA) anchored simple sequence repeats (ASSRs) generated polymorphism in soybean, *Glycine max* (L.) Merr. *Theor. Appl Genet.*, 96: 1086–1096.

Welsh Jand McClelland M. 1990. Genomic fingerprinting with AP-PCR using pairwais combinations of primer: Application to genetic mapping of the mouse. *Nucleic Acids Res.,* 19: 5275- 5279.

Zietkiewicz E, Rafalski A, Labuda D. 1994. Genome fingerprinting by simple sequence repeat (SSR) - anchored polymerase chain reaction amplification. *Genomics*, 20: 176–183.

In: Biotechnology and Conservation … (2 Volume Set)
ISBN: 978-1-63117-869-6
Editors: S. A. Farooq, R. Abed and S. Baqir

Chapter 21

GENETIC VARIABILITY AND MOLECULAR FINGERPRINTING OF SOME WILD EGYPTIAN GRAMINEAE PLANTS

Hadia A. Heikal*[1, 2*]*, Effat Badr*[3]*, H. F. Elwakil*[4]*,
Y. Mabrouk*[3]*, O. M. Badawy*[5]*, M. El-Seehy*[3]*,
***Y. Moustafa*[3] *and Sarah Aggag*[3]**

[1]Dept. Molecular Biology, Genetic Engineering & Biotechnology Research Institute, Sadat City, Menofia University, Egypt
[2]Dept. Biology, Fac. Education, Wadi Aldwaser, Salman bin Abdulaziz University, Kingdom of Saudi Arabia
[3]Dept. Genetics, Fac. Agriculture, Alexandria University, Alex., Egypt
[4]Fac. Agriculture, Saba-Basha, Alexandria University, Alex. Egypt
[5]Dept. Breeding and Genetics, Agric. Research Station, Sabbahia, Sugar Crops Res. Inst., Agric. Res. Center, Alex., Egypt

ABSTRACT

Fourteen Gramineae landraces were collected from North and North West of Delta, Egypt, and characterized using 12 RAPD and 13 ISSR primers. The polymorphic RAPD and ISSR bands were 394 and 441, whith 115 and 102 bands being useful as specific markers, respectively. Both RAPD and ISSR showed that primers Z-08 and S6 resulted in the highest number of polymorphic bands (57 and 53), respectively. The results showed extensive diversity between the *Gramineae* landraces. Similarity Index varies from 40% to 77% for RAPD and 37% to 88% for ISSR markers.

The dendrogram using RAPD and ISSR analysis showed the genotypes which were grouped together, based on ISSR analysis were more related than those based on RAPD analysis. These results demonestrated RAPD and ISSR analyses to be useful in generating specific markers of Gramineae flora in North and North West Coast of Egypt.

* Corresponding author email address: heikalhadia@gmail.com.

The information provides database that can later be used for the introduction of abiotic stress-tolerance genes into economic plants; thus the cultivated area in Egypt would be extended. In addition, not only, the results provides database for genetic biodiversity, but also is it a necessity for the protection of the Egyptian landraces againest illegal bioprospecting.

Keywords: Gramineae landraces, genetic diversity, molecular markers, random amplified polymorphic DNA (RAPD), Inter-simple sequence repeats (ISSR), Similarity Index, phylogenetic relationships

INTRODUCTION

Gramineae is one of the most important families. It includes several economic crops as well as wild species, which represent a valuable source of biodiversity. Molecular markers have proved to be valuable tools in the characterization and evaluation of genetic diversity within and between species and populations; different markers might reveal different classes of variations (Hadia et al., 2007).

The relationship between genetic diversity and eco-geographical parameters in wild Gramineae, suggests that it may be a source of useful genes related to adaptation and stress responses which can be used in breeding programs.

The PCR-based random amplified polymorphic DNA (RAPD) markers are extensively used in genetic mapping (Chalmers et al., 2001) and for the identification of markers linked with useful traits (Bai et al., 2003). Due to its technical simplicity and speed, RAPD methodology has been used for diversity analyses in several crops (Zenglu and Randall, 2001). RAPDs proved to be useful as genetic markers in the case of self pollinating species with a relatively low level of intraspecific polymorphism, such as hexaploid wheat (Joshi et al., 1993) and cultivated barley (Tinker et al., 1993 and Hadia et al., 2007).

Inter-simple sequence repeats (ISSR) markers, permit detection of polymorphisms in inter-microsatellite loci, using a primer designed from dinucleotide or trinucleotide simple sequence repeats.

It can be used to assess genetic diversity (Lalhruaitluanga et al., 2009). The main advantages of ISSR are: no need for DNA sequence information prior to amplification, low cost, simple operation, high stability, and abundance of genomic information. (Shen et al., 2006).

A study was carried out at the Genetics Department, Faculty of Agriculture, Alexandria University and Genetic Engineering & Biotechnology Research Institute, Sadat City, Menofia University. During this study, 14 Gramineae landraces were collected from North and North West of Delta, Egypt. Samples were identified and classified by the Horticulture Institute, Flora and Phyto taxonomy Researches, The Herbarium, Agricultural Research Center, Ministry of Agriculture, Dokky, Cairo, Egypt (Table 1 and Figure 1).

The AxyPrep Multisource Genomic DNA Miniprep Kit (Biosciences) was used to extract total genomic DNA from fresh plant tissues.

Twelve RAPD primers from twenty 10-mer oligonucleotides with arbitrary sequence were screened using three DNA samples and were found to be suitable for the study of all samples. Thirteen ISSR primers based on dinucleotide, tetranucleotide or pentanucleotide repeats, which produced clear and reproducible bands, were selected for the amplification of all DNA samples (Table 2).

The PCR reaction mixture consisted of 20ng genomic DNA, 5X PCR buffer, 25mM/l $MgCl_2$, 100μM/l of each dNTP (Promega, USA), 66ng/μl Primer and 5 U/μl *Taq* polymerase in a 25μl. The reaction amplification protocol included 94 °C for 5min, followed by 5 cycles of 92 °C for 30 Sec, 35 °C for 2min and 72 °C for 90 Sec, followed by 35 cycles of 92 °C for 30 Sec, 40 °C for 30 Sec and 72 °C for 90 Sec, with a final extension at 72 °C for 5 min, and eventually storage at 4 °C (Hadia et al., 2007).

The amplified products were electrophoresed in 1% agarose gel with 0.5x TBE buffer, after the gel was stained with ethidium bromide, and band patterns were visualized with a UV transilluminator. RAPD and ISSR data were scored for presence (1), or absence (0) using the Phoretix 1D image analysis system (Phoretix International, London) to integrate the data obtained. A similarity dendrogram among the 14 Gramineae landraces was produced using SPSS statistics (IBM, Version 17).

The goal of this study is to identify and molecularly characterize the genetic diversity of 14 wild gramineae genotypes collected from the North Coast of Egypt using Two types of molecular markers; RAPD and ISSR.

Genetic Polymorphism

The selected 12 useful RAPD primers gave a total of 509 amplified detected bands with an average of about 42.4 bands per primer, ranging from 2817-138 bp (Table 3 and Figure 2). Primer Z-08 had the highest total number of bands (69 bands), while primer Z-05 showed the lowest total number of bands (22 bands). However, the results showed that there were 394 polymorphic bands out of the total amplified bands, revealing 77.4% polymorphism. The Primer No. Z19 had the highest percentage of polymorphic bands (93.2%), wherease primer No.Z12 showed the lowest percentage of polymorphic bands (33.3%), respectively (Table 3). Table (4) showed that overall 115 specific markers were generated among the 14 genotypes. The highest total number of specific bands was 16 bands using primer Z-12, while primer Z-19 showed the lowest specific bands (3 bands).

The selected 13 useful ISSR primers gave a total of 543 amplified detected bands with an average of about 41.8 bands per primer. Primer S6 had the highest total number of bands (59 bands), while primer S13 showed the lowest total number of bands (19 bands) (Table 5 and Figure 3). The results revealed that 441 bands (81.2%) were polymorphic. Primer S14 had the highest percentage of polymorphic bands (90.6%), while primer S13 showed the lowest percentage of polymorphic bands (36.8%) (Table 5). Table (6) shows a total of 102 specific bands among the 14 genotypes used in this study. Primers S10 and S13 gave the highest number of specific bands (12 bands), whereas the lowest number of specific bands were produced with primers S4, S12 and S14 (5 bands).

PHYLOGENETIC RELATIONSHIP

RAPD data were used to estimate the genetic similarities and the phylogentic relationships among the 14 gramineae landraces. The highest similarity Index (77%) was observed between the landraces Ad and Ss, while the lowest similarity Index (40%) was recorded between landraces Af and Ic as shown in Table, 7. The dendrogram for the phylogentic relationships based on RAPD showed that landraces Ad and Ss were out of the group (Figure 4).

The remaining 12 landraces positioned in two groups. The first group included two landraces (Sv and Lm), whereas the second group included the remaining 10 landraces, which were divided into three sub-groups. Landraces Cf, Sc and Sp, distinct separately from the seven landraces in the first sub-group. The second sub-group contained two landraces (Pr and Pa), while the third sub-group combined the reminder five landraces, three of them (Hl, Bu and Bs) isolated separately, whereas the remaining two landraces (Ic and Af) were grouped together.

Similarity indices and consensus tree were developed on the basis of banding batterns of 14 gramineae landraces using 13 ISSR primers as shown in Table (8) and Figure (5). The most closely related landraces were between landraces Ad and Cf and between landraces Ad and Ss with the highest Similarity Index (88%). On the other hand, the lowest Similarity Index (37%) was recorded between landraces Bs and Sp, and between landraces Pa and Sp. The dendrogram for phylogentic relationships based on ISSR markers showed that landrace Ad was distantly related to the group. The remaining 13 landraces appeared to be closely related and were positioned in two groups. The first group included four landraces (Pr, Sv, Sc and Lm), whereas the second group included nine landraces, which were differentiated into two sub-groups.

The first sub-group contains four landraces (Ic, Af, Hl and Bu), while the other five landraces grouped together into the second sub-group (Ss, Cf, Bs, Sp and Pa).

Table 1. Code and Latin names of the collected Gramineae species

No.	Code	Latin name	No.	Code	Latin name
1	Ad	*Arundo donax* L.	8	Lm	*Lolium multiforum* Lam.
2	Af	*Avena fatua* L.	9	Pa	*Phragmatis australis* (Cav.) Trin.
3	Bs	*Bromus scoparius L.*	10	Pr	*Panicum repens* L.
4	Bu	*Bromus unioloides Kunth in H. B.*	11	Sc	*Stipa capensis* Thunb
5	Cf	*Centropodia forsskaolii (Vahl)*	12	Sp	*Stipa parviflora* Desf.
6	Hl	*Hordeum leporinum Link*	13	Ss	*Saspalidium seminatum*
7	Ic	*Imperata cylindrical* L. Beauv.	14	Sv	*Sorghum virgatum* (Hack). Stapf

Table 2. Code and sequence of RAPD and ISSR primers

RAPD primers			ISSR primers			
No.	Oligo Name	SEQUENCE	No.	Oligo Name	Code	SEQUENCE
1	Z-03	5'-CAG CAC CGC A-3'	1	ISSR 844A	S2	5'-CTC TCT CTC TCT CTC TAC-3'
2	Z-04	5'-ACG CTG TGC T-3'	2	ISSR 844B	S3	5'-CTC TCT CTC TCT CTC TGC-3'
3	Z-05	5'-TCC CAT GCT G-3'	3	ISSR 17898A	S4	5'- CAC ACA CAC ACA AC -3'
4	Z-06	5'-GTC CCG TTC A-3'	4	ISSR 17898B	S5	5'- CAC ACA CAC ACA GT -3'
5	Z-07	5'-CCA GGA GGA C-3'	5	ISSR 17899A	S6	5'- CAC ACA CAC ACA AG-3'
6	Z-08	5'-GGG TGG GTA A-3'	6	ISSR 17899B	S7	5'- CAC ACA CAC ACA GG-3'
7	Z-12	5'-TCA ACG GGA C-3'	7	ISSR HB-8	S8	5'- GAG AGA GAG AGA GG -3'
8	Z-13	5'-GAC TAA GCC C-3'	8	ISSR HB-10	S10	5'- GAG AGA GAG AGA CC -3'
9	Z-17	5'-CCT TCC CAC T-3'	9	ISSR HB-11	S11	5'- GTG TGT GTG TGT CC -3'
10	Z-18	5'-AGG GTC TGT G-3'	10	ISSR HB-12	S12	5'- CAC CAC CAC GC -3'
11	Z-19	5'-GTG CGA GCA A-3'	11	ISSR HB-13	S13	5'- GAG GAG GAG GC -3'
12	Z-20	5'-ACT TTG GCG G-3'	12	ISSR HB-14	S14	5'- CTC CTC CTC GC -3'
			13	ISSR HB-15	S15	5'- GTG GTG GTG GC -3'

Table 3. Total number of bands generated and percentage of polymorphism as revealed by RAPD among the fourteen Gramineae landraces

Primers	Total number of bands	Total number of polymorphic bands	% of polymorphism
Z-03	49	42	85.7
Z-04	40	32	80
Z-05	22	10	45.5
Z-06	54	47	87
Z-07	28	21	75
Z-08	69	57	82.6
Z-12	24	8	33.3
Z-13	27	15	55.6
Z-17	41	28	68.3
Z-18	64	52	81.3
Z-19	44	41	93.2
Z-20	47	41	87.2
Total	509	394	77.4

Table 4. Molecular Weight and total number of specific bands in 14 Gramineae landraces, using 12 RAPD primers

Primers	Molecular Weight of specific bands														Total specific bands
	Ad	Af	Bs	Bu	Cf	Hl	Ic	Lm	Pa	Pr	Sc	Sp	Ss	Sv	
Z-03	-	-	-	-	-	1615	-	1885 942	-	216	-	763 547 256	-	-	7
Z-04	1600 503	-	-	1199 400	-	935 662	-	-	-	1750	-	2250	-	-	8
Z-05	-	1264	497	-	989 606.5 439	-	696.5 369	-	909.5 539 400 304	-	-	-	1742	-	12
Z-06	294.5	-	429	-	-	577 266	-	-	-	-	2183	-	-	2817 2450	7
Z-07	-	-	1144 1016	-	-	-	1850 1084	-	406	703.5	-	-	1950	-	7
Z-08	458 338	535 360	-	-	710	-	-	2391 1828 1508	564 317	-	2641	-	-	1750	12
Z-12	2016 1677 1258 1058 510.5 402.5 354 253	700	-	-	-	-	-	-	-	1127 303.5	1742 867	661	-	1806 1565	16
Z-13	1192 916	-	1254 649 371	-	838	-	-	-	-	-	407 273 196 138	-	526	218	12

Primers	Molecular Weight of specific bands														Total specific bands
	Ad	Af	Bs	Bu	Cf	Hl	Ic	Lm	Pa	Pr	Sc	Sp	Ss	Sv	
Z-17	-	-	-	-	-	-	-	805 564 330	1305	-	1656 940 370	424 349	1734 1515. 5 1237	1173	13
Z-18	790	-	-	-	1906 1578	1844	-	339	-	703.5 593	726 427	1797	2172 1672	-	12
Z-19	-	-	-	1267	-	-	443	-	-	-	-	-	1485	-	3
Z-20	-	-	-	494.5	671	2025	700	-	-	-	733	-	-	1707	6
Total	16	4	7	4	8	7	6	9	8	7	14	8	9	8	115

Table 5. Total number of bands generated and percentage of polymorphism as revealed by ISSR among the fourteen Gramineae landraces

Primers	Total number of bands	Total number of polymorphic bands	% of polymorphism
S2	43	35	81.4
S3	44	35	79.6
S4	36	31	86.1
S5	43	33	76.7
S6	59	53	89.8
S7	55	45	81.8
S8	29	23	79.3
S10	48	36	75
S11	40	33	82.5
S12	26	21	80.8
S13	19	7	36.8
S14	53	48	90.6
S15	48	41	85.4
Total	543	441	81.2

Table 6. Molecular Weight and total number of specific bands in 14 Gramineae landraces, using 13 ISSR primers

Primers	Molecular Weight of specific bands														Total specific bands
	Ad	Af	Bs	Bu	Cf	Hl	Ic	Lm	Pa	Pr	Sc	Sp	Ss	Sv	
S2	-	-	-	-	571	-	1396 1278	-	-	-	316 163	417	-	263 225	8
S3	-	1049	-	-	-	-	-	198	-	-	496.5 230	269	-	959 748 662 398	9
S4	300	471	-	-	980	-	-	-	-	262	281	-	-	-	5
S5	251 143	-	-	-	-	-	-	363	-	205.5	305	-	1694 1010.5 628	1423 496.5	10
S6	161	-	1392 994	-	-	203.5	-	-	-	-	-	1155	1219	-	6
S7	220	-	1983	-	1867	256	-	538	-	164	696.5 236 92	-	1600	-	10
S8	-	-	-	690.5	1352 659	-	-	-	1629 600	-	-	1742	-	-	6
S10	-	239	1818 948	-	1652	198 144	-	743	-	1231 809 300	-	894.5	1356	-	12
S11	-	413	-	-	563	-	431	383	-	-	497 398	-	1413	-	7
S12	255	-	-	-	-	221	461 367	-	-	104	-	-	-	-	5
S13	1147 911.5 733 519 331	800 468 355	-	547 373	-	201.5	494	-	-	-	-	-	-	-	12
S14	395.5 266	703.5	-	-	1304	739	-	-	-	-	-	-	-	-	5
S15	360 300	1524	-	-	1919	-	-	-	-	404.5 342 265	-	-	-	-	7
Total	15	9	5	3	9	7	6	5	2	10	11	5	7	8	102

Table 7. Percentage of similarity between 14 different Gramineae landraces based on data obtained from RAPD-PCR

Genotypes	Ad	Af	Bs	Bu	Cf	Hl	Ic	Lm	Pa	Pr	Sc	Sp	Ss
Af	.57												
Bs	.61	.42											
Bu	.57	.42	.44										
Cf	.76	.57	.55	.51									
Hl	.70	.45	.49	.45	.58								
Ic	.65	.40	.44	.46	.57	.53							
Lm	.75	.54	.66	.58	.67	.61	.60						
Pa	.62	.53	.51	.47	.70	.62	.57	.59					
Pr	.53	.46	.48	.50	.67	.57	.48	.66	.45				
Sc	.74	.45	.57	.61	.72	.60	.55	.63	.66	.61			
Sp	.63	.52	.54	.52	.63	.61	.54	.66	.53	.52	.59		
Ss	.77	.62	.64	.64	.63	.61	.66	.70	.67	.64	.73	.66	
Sv	.74	.57	.59	.61	.70	.56	.65	.57	.66	.63	.58	.63	.57

Table 8. Percentage of similarity between 14 different Gramineae landraces based on data obtained from ISSR-PCR

Genotypes	Ad	Af	Bs	Bu	Cf	Hl	Ic	Lm	Pa	Pr	Sc	Sp	Ss
Af	.80												
Bs	.84	.56											
Bu	.78	.58	.46										
Cf	.88	.74	.42	.62									
Hl	.76	.56	.56	.42	.68								
Ic	.82	.44	.60	.54	.70	.54							
Lm	.76	.60	.54	.60	.64	.64	.64						
Pa	.74	.60	.46	.52	.42	.60	.60	.60					
Pr	.67	.65	.63	.61	.75	.57	.69	.59	.67				
Sc	.85	.75	.65	.61	.75	.61	.75	.49	.67	.64			
Sp	.77	.53	.37	.47	.47	.51	.55	.47	.37	.60	.62		
Ss	.88	.68	.44	.58	.38	.64	.68	.56	.46	.69	.67	.41	
Sv	.84	.66	.64	.70	.72	.70	.72	.62	.70	.65	.61	.65	.70

Figure 1. Types of wild Gramineae plants collected from North and North West of Delta, Egypt (Ad: *Arundo donax* LA; Af: *Avena fatua* L; Bs: *Bromus scoparius* L; Bu: *Bromus unioloides* Kunth in H. B.; Cf: *Centropodia forsskaolii* (Vahl); Hl: *Hordeum leporinum* Link; Ic: *Imperata cylindrical* (L.) Beauv.; Lm: *Lolium multiforum* Lam; Pa: *Phragmatis australis* (Cav.) Trin.; Pr: *Panicum repens* L.; Sc: *Stipa capensis* Thunb; Sp: *Stipa parviflora* Desf.; Ss: *Saspalidium seminatum*; Sv: *Sorghum virgatum* (Hack). Stapf.

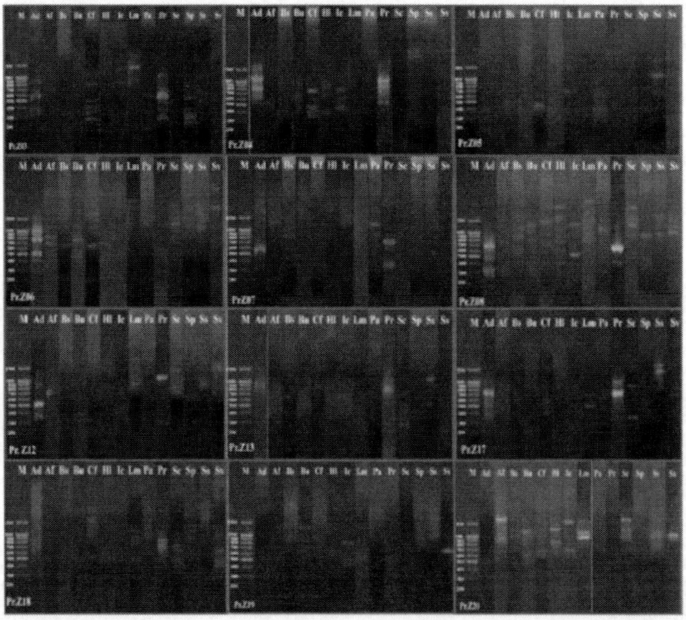

Figure 2. RAPD-PCR banding patterns of DNA for 14 Gramineae landraces using 12 RAPD primers. M refers to Molecular Weight Marker.

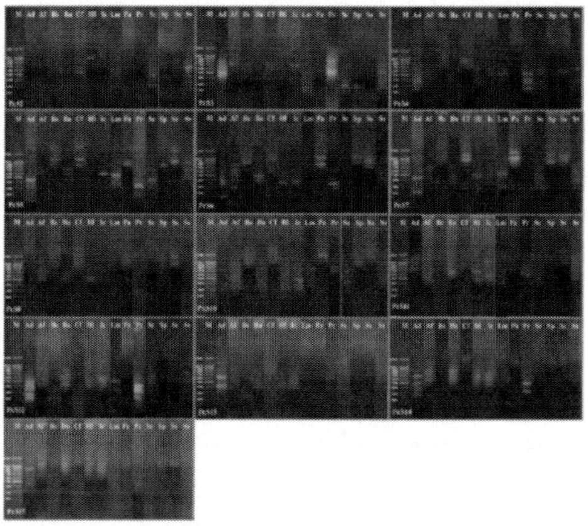

Figure 3. ISSR-PCR banding patterns of DNA for 14 Gramineae landraces using 13 ISSR primers. M refers to Molecular Weight Marker.

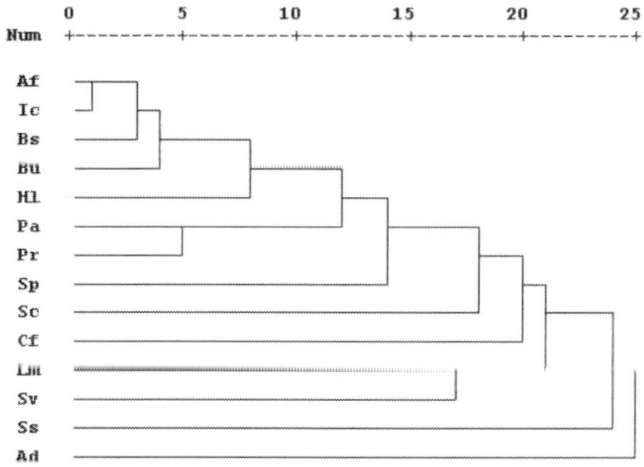

Figure 4. Dendrogram for phylogentic relationships among 14 Gramineae landraces based on RAPD data.

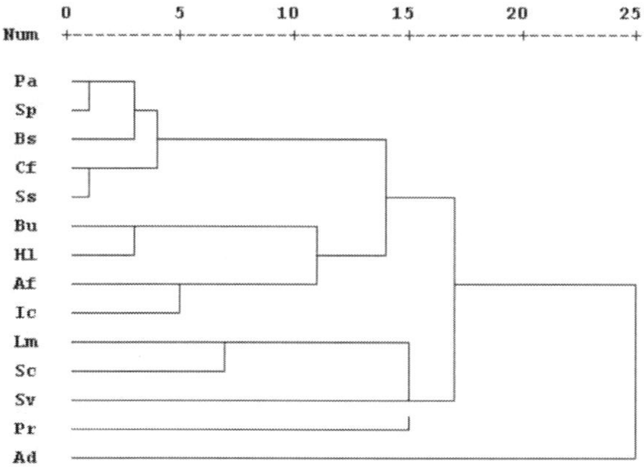

Figure 5. Dendrogram for phylogentic relationships among 14 Gramineae landraces based on ISSR data.

DISCUSSION

This study showed that RAPD and ISSR markers worked effectively and gave high polymorphism (77.4% and 81.2%), respectively. These results indicated that ISSR markers more effective and gave higher polymorphism, which meet the conclusion of Kojima et al. (1998) and Semagn et al. (2006) that ISSR markers usually show high polymorphism, although the level of polymorphism has been shown to vary with the detection method used.

Similarity Index based on RAPD and ISSR analysis among 14 Gramineae landraces detected that the highest Similarity Index were 77% and 88%, respectively, whereas the lowest Similarity Index were 40% and 37%, respectively. The dendrogram for phylogenetic relationships using RAPD and ISSR were indicated that the landraces, which were grouped together based on ISSR analysis were more related than those based on RAPD analysis. Similar results were obtained in several studies using the same markers (Farsani et al., 2012 and Etemadi et al., 2006).

Many of researchers concluded that, RAPD and ISSR markers proved to be efficient and inexpensive ways to provide molecular data. They have been used successfully in determining genetic relationship and used for DNA fingerprinting (Blair et al., 1999; Gilbert et al., 1999). This study demonstrates that molecular markers are needed to identify plant species. Roodt et al. (2002) identified different cultivars of *Cynodon dactylon* (L.) Pers. using RAPD analyses of some well known cultivars used in South Africa, as well as 10 potential new cultivars.

Only five primers were needed to obtain a specific fragment pattern for each cultivar. ISSR have been utilized in Gramineae plants to identify markers associated with seed size in wheat (Ammiraju et al., 2001) and fingerprinting in rice (Blair et al., 1999) as well as in other crops, e.g., in fingerprinting cashew (Archak et al., 2003). Both RAPD and ISSR markers have been used to assess somaclonal variation in maize and were found to be highly efficient (Osipova et al., 2003). Moreover, Kuznetsova et al. (2005) showed their reliability in analyzing DNA polymorphisms generated by long-term culture and subsequent regeneration in pea.

In this study, we have shown that RAPD and ISSR markers are essential for species identification and provide database for genetic biodiversity, as well as being useful for generating candidate- specific markers of Gramineae flora in North and North West coast of Egypt. This can be used later to identify genetic loci related to abiotic stress which will be very useful in wheat breeding programs in order to expand the area of cultivated wheat in Egypt. Molecular identification of Egyptian wild Gramineae species will be a reference for conserving intellectual property rights for these plants. In addition, the results provides database for genetic biodiversity to protect the Egyptian landraces againest illegal bioprospecting.

ACKNOWLEDGMENTS

This work was supported by the project entitled "Genetic Characterization for the Flora of Gramineae Plants Grown in North and North West of Delta and Siwa Oasis", National Strategy for Biotechnology and Genetic Engineering program, Science and Technology

Center, Academy of Scientific Research and Technology, Ministry of Scientific Research, Egypt.

REFERENCES

Ammiraju J S S, Dholakia B B, Santra D K, Singh H, Lagu M D, Tamhankar S A, Dhaliwal H S, Rao V S, Gupta V S and Ranjekar P K. 2001. Identification of inter simple sequence repeat (ISSR) markers associated with seed size in wheat. *Theor. Appl. Genet.* 102: 726-732.

Archak S, Gaikwad A B, Gautam D, Rao E V V B, Swamy K R M and Karihaloo J L. 2003. Comparative assessment of DNA fingerprinting techniques (RAPD, ISSR, and AFLP) for genetic analysis of cashew (*Anacardium occidentale* L.) accessions of India. *Genome*. 46: 362-369.

Bai G, Guo P and Kolb F L. 2003. Genetic relationships among head blight resistant cultivars of wheat assessed on the basis of molecular markers. *Crop Sci*. 43: 498-507.

Blair M W, Panaud O and Mccouch S R. 1999. Inter- simple sequence repeat (ISSR) amplification for analysis of microsatellite motif frequency and fingerprinting in rice (*Oryza sativa* L.). *Theor. Appl. Genet*. 98: 780-792.

Chalmers J K, Campbell A W, Krestschmer J and Karakousis A. 2001. Construction of three linkage maps in bread wheat, *Triticum aestivum* L. *Aust. J. Agric. Res*. 52: 1089-1119.

Etemadi N, Sayed-tabatabaei B E, Zamani Z, Razmjoo K H, Khalighi A and Lessani H. 2006. Evaluation of diversity among *Cynodon dactylon* (L.) Pers. using RAPD markers. *Agric. Biol.* 8: 198-202.

Farsani T M, Etemadi N, Sayed-Tabatabaei B E and Talebi M. 2012. Assessment of genetic diversity of bermudagrass (*Cynodon dactylon*) using ISSR markers. *In. J. Mol. Sci.* 13: 383-392.

Gilbert J E, Lewis R V, Wilkinson M J and Caligari P D S. 1999. Developing an appropriate strategy to assess genetic variability in plant germplasm collection. *Theor. Appl. Genet.* 98: 1125-1131.

Hadia A Heikal, Mabrouk Y, Badawy O M, El-Shehawy A and Badr E A. 2007. Fingerprinting Egyptian gramineae species using random amplified polymorphic DNA (RAPD) and inter-simple sepuence repeat (ISSR) markers. *Res. J. Cell and Molecular Biology* (RJCMB). 1(1): 15-22.

Joshi C P and Nguyen H T. 1993. RAPD (Random amplified polymorphic DNA) analysis based on intervarietal genetic relationships among hexaploid wheats. *Plant Sci*., 93: 95-103.

Kojima T, Nagaoka T, Noda N, Ogihara Y. 1998. Genetic linkage map of ISSR and RAPD markers in Einkorn wheat in relation to that of RFLP markers. *Theor. Appl. Genet.* 96: 37-45.

Kuznetsova O L, Ash O A, Hartina G A and Gostimskii S A. 2005. RAPD and ISSR analyses of regenerated pea *Pisum sativum* L. plants. *Rus J. Genet*. 41: 60–65.

Lalhruaitluanga H and Prasad M N V. 2009. Comparative results of RAPD and ISSR markers for genetic diversity assessment in *Melocanna baccifera Roxb*. growing in Mizoram State of India. *Afri. J. Biotech*. 8 (22): 6053-6062.

Osipova E S, Koveza O V, Troitskii A V, Dolgikh Y I, Shamina Z B and Gostimskii S A. 2003. Analysis of specific RAPD and ISSR fragments in maize (*Zea mays* L.) somaclones and development of SCAR markers on their basis. *Rus. J. Genet.* 39: 1412–1419.

Roodt R, Spies J J and Burger T H. 2002. Preliminary DNA fingerprinting of the turf grass Cynodon dactylon (Poaceae: Chloridoideae). *Bothalia.* 32(1): 117-122.

Semagn K, Bjørnstad Å and Ndjiondjop M N. 2006. An overview of molecular marker methods for plants. *Afri. J. Biotech.* 5 (25): 2540-2568.

Shen J, Ding X Y, Liu D Y, Ding G, He J, Li X X, Tang F, and Chu BH. 2006. Inter-simple sequence repeats (ISSR) molecular fingerprinting markers for authenticating populations of *Dendrobium officinale* kimura et Migo. *Biol. Pharm. Bull.* 29(3): 420-422.

Tinker N A, Fortin M G and Mather D E. 1993. Random amplified polymorphic DNA and pedigree relationships in spring barley. *Theor. Appl. Genet.* 85(8):976-984.

Zenglu L and Randall L N. 2001. Genetic diversity among soybean accessions from three countries measured by RAPDs. *Crop Sci.* 41: 1337-1347.

In: Biotechnology and Conservation … (2 Volume Set) ISBN: 978-1-63117-869-6
Editors: S. A. Farooq, R. Abed and S. Baqir

Chapter 22

Genetic Distance and Its Association with F1 Interspecific Hybrids Performance and Heterosis for Yield Components and Fiber Properties in Cotton

Yanal A. Alkuddsi**, *S. S. Patil, S. M. Manjula, H. L. Nadaf and B. C. Patil

Agricultural Research Station, University of Agricultural Sciences, Dharwad, Karnataka, India

Abstract

Knowledge of genetic diversity and relationships among breeding materials has a significant impact on crop improvement. Association between parental divergence and progeny performance has not been well documented in cotton (*Gossypium hirsutum*L.). The objectives of this study were to estimate genetic diversity among selected cotton genotypes on the basis of simple sequence repeat (SSR) markers, and to investigate the relationship between genetic diversity and F_1 population performance and heterosis. The present study was conducted to investigate the relationship between parents molecular marker diversity and interspecific hybrids of cotton to evaluate the hybrid performance and heterosis using molecular markers. Twenty eight F_4 lines of (*Gossypium barbadense* L.) were crossed with the four common diverse testers (*Gossypium hirsutum* L.) *viz.*, DH 98-27 (T_1), ZCH 8 (T_2), 178-24 (T_3) and DH 18-31 (T4) to produce 112 F_1 interspecific hybrids during 2010. These 112 F_1 hybrids, their F_5parents lines with 4 testers and ruling commercial check (MRC 6918 and DCH 32) were evaluated for yield and fiber quality traits and sown during *kharif*2011 at University of Agricultural Sciences, Dharwad, India. Genetic distances (GD) among the parents were calculated from 40 microsatellite marker data, and their correlation with hybrid performance and heterosis were analysed. The dendrogram constructed from the pooled data revealed three distinct clusters. One cluster involved testers and other clusters showed all lines were placed which are already having proven record in giving good hybrids. The similarity coefficients involved in the line x

* Correspondence to: Yanal Ahmad Alkuddsi, E-mail: y.alkuddsi@hotmail.com.

tester study ranged from 57% to 96 %, with an average of 81%. Among the parental lines, the lines [DB 533 x DB 534 F4 IPS 8] and [DB 533 x DB 534 F4 IPS 1] showed highest (96%) similarity coefficient value. While, the lines [DB 533 x DB 534 F4 IPS 48] and [DB 533 x DB 534 F4 IPS 16] exhibited lowest (57%) similarity coefficient value. Genetic distance (GD) ranged from 0.041 to 0.429, with an average of 0.183. The result implied that each cluster dendrogram substantially reflected its own genetic relationship among parents. The correlation between (number of bolls per plant, mean boll weight, seed index, fiber strength, fiber uniformity ratio and fiber elongation) and genetic distance (GD) were not significant. Highly significant positive correlation were found between genetic distance (GD) and ginning outturn for F_1 performance (0.277) and heterosis over MRC 6918 and DCH 32 (0.279), between genetic distance (GD) and seed cotton yield for F_1performance (0.359) and heterosis over MRC 6918 (0.336) and over DCH 32 (0.362) and between genetic distance (GD) and fiber micronaire value for mid parent heterosis (0.266). In conclusion, the relationship between SSR marker heterozygosity hybrid performance can be used to predict fiber micronaire value, ginning outturn and seed cotton yield during interspecific hybrid cotton breeding.

Keywords: *Gossypium hirsutum*, *Gossypium barbadense*, SSR marker, Genetic distance (GD)

INTRODUCTION

Cotton (*Gossypium* L.) is a leading fiber crop in the world today. Although the genus *Gossypium* L. has approximately 50 species, only four of them are cultivated, which include two diploids (2n = 2x = 26): *G. arboreum* L. (A_2A_2) and *G. herbaceum* L. (A_1A_1), as well as two allotetraploids (2n = 4x = 52): *G. hirsutum* L. (AADD) and G. *barbadense* L. (AADD). It was believed that tetraploid cotton was originated from an interspecific hybridization of an old world diploid species that was closely related with G. *arboreum* or G. *herbaceum* (A genome donor) and a new world diploid species relative to G. *raimondii Ulbrich* or G. *gossipioides Standley* (D genome donor), which occurred about 1–2 million years ago (Beasley 1940).

Use of heterosis in cotton production might be one of the key approaches to increase seed cotton yield. Heterosis for yield in F_1 hybrids cotton has been extensively analyzed in the past decades. Useful heterosis for yield in F_1 hybrids during 1947 and 1972 ranged from 7 to 50% in interspecific hybrids and from 10 to 138% in intraspecific hybrids`(Davis, 1978). In any hybrid programme, a large number of crosses need to be made, while only few of the hybrids will show good performance over the standard check. This process is extremely labour intensive, time- consuming and tedious. Molecular markers increasingly detect locus differences among genotypes and represent apowerful tool for the assessment of genetic diversity in plant species (Tanksley, 1983). Selection of desirable parents is an important task to initiate a hybrid – breeding programme. Because heterosis is associated with the interaction of different alleles at a locus (Jones, 1945), it has been suggested that molecular marker diversity may be used to select parents for hybridization.

In cotton, a number of efforts have been made to investigate the relationship between DNA marker- based genotype variation of the parents to be used in a hybrid- breeding programme and heterosis with varying results. For example, Diers et al. (1996) reported that marker- based genetic distance was not consistently correlated with heterosis for inbred diallels and for cultivar diallels in rape seed. Sheng et al. (2002) reported significant correlation between genetic distance and seed yield but the determinative coefficient was very low (0.1024). However, Riaz et al. (2001) found that the genetic distance of sequence – related amplified polymorphism (SRAP) in *American B.napus* inbred lines was significantly correlated with hybrid yield performance and heterosis.

Meredith and Brown (1998) studied the relationship between genetic distance estimated by restriction fragment length polymorphic (RFLP) markers among 15 cultivars and one strain from the USA and yield heterosis of 120 F2 hybrids produced by a half _ diallel genetic design and found that the correlation were very low (r=0.08). Wu et al. (2002) studied the correlation between genetic distance measured by random amplified polymorphic DNAs (RAPD), inter simple sequence repeat (ISSR) and simple sequence repeat (SSR) markers among six domestic and two exotic cultivars and interspecific F1 and F2 hybrids, and found the correlation between these was low.

Gutierrez et al. (2002) used five US, four Australian cultivars and two day- neutral converted lines of *G.hirsutum* to analysis the association between genetic distance based on SSR markers and performance of agronomic and fiber traits of F2-bulk populations and deduced that significant correlations ranged from negative to positive depending on the traits, genetic background and environment.

Zhang et al. (2007) studied the relationship between parental molecular marker diversity and hybrid performance in both intra and interspecific hybrids of cotton to evaluate the feasibility of predicting hybrid performance using molecular markers. Three cytoplasmic male sterile (CMS) lines were crossed with 10 restorer lines to produce 22 F, hybrids during 2003. Of 22 F(1)s, 14 hybrids were intraspecific (*G.hirsutum x G. hirsutum*) and eight interspecific (*G. hirsutum x G. barbadense*). These 22 F, hybrids and their parents were evaluated for yield and fiber quality traits at Zhejiang University, Hangzhou, China during 2004 and 2005. Genetic distances (GD) among the parents were calculated from 56 random-amplified polymorphic DNAs (RAPD) and 66 simple sequence repeat (SSR) marker data, and their correlation with hybrid performance and heterosis were analysed.

Mohammadi et al. (2008) investigated the correlation between the potential of molecular markers and hybrids performance in Maize. Significant correlation was found between GD value of parental lines and hybrid performance for the testcross and diallel data. In diallel analysis significant correlation was observed between total grain yield per ear (TGW) and genetic distance based on SM coefficient, whereas the correlation of GD and specific combining ability of hybrids for this trait was not. Through the stepwise multiple regression analysis a total of 19 informative SSR markers distributed over all chromosomes, except chromosomes 7 and 8, were detected. GD values based on informative markers in general were grater compared to that of based on all markers and significant improvement was observed in the correlations between GD estimates based on informative markers and TGW as well as SCA.

The objectives of the present study were 1- To investigate the relationship of genetic distance, based on SSR markers, with hybrid performance and heterosis and to determine whether these markers would be useful for predicting hybrid performance and heterosis in cotton. 2- To improve the yield and fiber quality using interspecific hybridization (*G.hirsutum*x*G. barbadense*) in cotton.

Plant Materials and Field Evaluation

During 2010 the twenty eight F4 lines of (*Gossypium barbadense* L.) (Table 1) cross (DB 533× DB 534) depending on the highest of fiber strength, are proposed to be crossed with the four common diverse (*Gossypium hirsutum* L.) viz., DH 98-27 (T1), ZCH8 (T2), 178-24 (T3) and DH 18-31 (T4) selected based on the earlier study. The crossing programme was taken up during 2010. The F4 lines and four common testers were sown on staggered dates. To obtain derived F1s seed, the flower buds of the proper size from testers (used as female) were hand emasculated in the evening between 3.00 to 6.00 pm. The emasculated flowers were covered by butter paper packets for avoiding out crossing as well as ensuring their easy identification at the time of crossing. The emasculated flowers were pollinated during the next day morning between 9.30 am to 11.30 am by brushing the pollen from one of the F4 lines (used as male) on the stigmatic surface. The pedicel of each pollinated flower was tied with price label bearing date and lines number for identification of crossed bolls. In this manner derived F1s seeds were obtained. Simultaneously, two populations of F4 lines were selfed and material was advanced to F_5 generation during the same season.

Table 1. List of F_4 *barbadense* line parents involved in the study

SL.No	Abbreviation	Lines
1	L1	(DB 533 x DB 534 F5 IPS 44)
2	L2	(DB 533 x DB 534 F5 IPS 62)
3	L3	(DB 533 x DB 534 F5 IPS 105)
4	L4	(DB 533 x DB 534 F5 IPS 26)
5	L5	(DB 533 x DB 534 F5 IPS 71)
6	L6	(DB 533 x DB 534 F5 IPS 30)
7	L7	(DB 533 x DB 534 F5 IPS 25)
8	L8	(DB 533 x DB 534 F5 IPS 49)
9	L9	(DB 533 x DB 534 F5 IPS 23)
10	L10	(DB 533 x DB 534 F5 IPS 36)
11	L11	(DB 533 x DB 534 F5 IPS 15)
12	L12	(DB 533 x DB 534 F5 IPS 1)
13	L13	(DB 533 x DB 534 F5 IPS 33)
14	L14	(DB 533 X DB 534 F5 IPS 24)
15	L15	(DB 533 x DB 534 F5 IPS 16)
16	L16	(DB 533 x DB 534 F5 IPS 52)
17	L17	(DB 533 x DB 534 F5 IPS 12)
18	L18	(DB 534 x DB 533 F5 IPS 22)
19	L19	(DB 533 x DB 534 F5 IPS 14)
20	L20	(DB 533 x DB 534 F5 IPS 34)
21	L21	(DB 533 x DB 534 F5 IPS 55)
22	L22	(DB 533 x DB 534 F5 IPS 17)
23	L23	(DB 533 x DB 534 F5 IPS 32)
24	L24	(DB 533 x DB 534 F5 IPS 38)
25	L25	(DB 533 x DB 534 F5 IPS 48)
26	L26	(DB 533 x DB 534 F5 IPS 13)
27	L27	(DB 533 x DB 534 F5 IPS 6)
28	L28	(DB 533 x DB 534 F5 IPS 8)

The experimental material was planted on a medium black soil at College of Agriculture, Dharwad under irrigated condition. The F_5 lines, derived F_1s of two populations along with the straight crosses and ruling commercial check (MRC 6918 and DCH 32) were sown during *kharif* 2011 in all a randomized block design with two replications and a spacing of 90 cm between rows and 60 cm between the plants within a row. Recommended fertilizer doses were applied and other cultural practices were carried out at regular interval. Plant protection measures were taken at appropriate time to control pests and diseases. Each set of 28 F_4 lines thus was involved in 112 crosses (refer to as derived F_1s), which were subjected to L × T analysis. The observations were recorded for number of bolls (no/plant), mean boll weight (g), seed index (g), ginning outturn (%), lint index (g), seed cotton yield (kg/ha), fiber length (mm), fiber strength (g/t), fiber micronaire value (µg/inch), fiber uniformity ratio (%), fiber maturity ratio and fiber elongation (%). Fiber quality traits were measured with the High-Volume Instrument.

SSR Molecular Marker Analysis

Leaf tissue of each parents was harvested and total genomic DNA was extracted from young leaves using the hexadecyl - trimethyl ammonium bromide (CTAB) method described by Saghai-Maroof et al., (1984). SSR assays were performed using 40 oligonucleotide primers from Sigma Aldrich Chemicals Pvt.Ltd., Co. Amplification reactions were carried out in 20 µl volumes containing 2.0 µl 10X assay buffer, 2.0 µl dNTP mix (2.5 mM each), 0.5µl forward primer (5 pM/µl), 0.5 µl Reverse (5 pM/µl), 0.5 µl Taq DNA polymerase (3U/µl), 2.0 µl Template DNA (15 ng/µl) and 7.5 µl Sterile double distill water. The amplification programmed for 5 min at 94 ^{0}C Denaturation (initial) of genomic DNA by one cycle followed by 25 cycles of 1 min at 94 ^{0}C, 1 min at 48 $\pm$ 5^{0}C and 1 min at 72^{0}C. This was followed by a final extension at 72 ^{0}C for 5 min. Amplification products were analysed by Non-Denaturing gel electrophoresis (PAGE) and viewed by silver staining.

Scoring the Amplified Fragments

The amplification of DNA profiles for all the primers were compared with each other and the bands of DNA at each amplification level of every primer were scored as present (1) or absent (0) thus generating the 0, 1 matrix.

$$\text{Per cent polymorphism (\%)} = \frac{\text{Total No. of polymorphic bands}}{\text{Total No. of bands generated by 40 primers}} \times 100$$

Analysis of SSR Profiles

Pair similarity coefficients were calculated for all pairwise combinations of the parental lines according to the method developed by Nei and Li (1979): $Sij= 2Nij / (Ni+ Nj)$, where

*Sij*is the similarity between parents *i* and *j; Nij* is the number of bands present in both parents; *Ni* is the number of bands present only in parent I; Nj is the number of bands present only in parent j. GD (genetic distance) was calculated as GD= 1- *Sij*. The similarity matrix from SSR markers, which were computed using NTSYS-PC version 2.1 (Rohlf 2001)were used to construct dendrograms based on UPGMA (the unweighted pair- group method with arithmetic means). Using the same NTSYS software, a cophenetic value matrix was calculated to test the goodness of fit for the cluster analysis to the original distance matrix.

For studying the relationship between SSR molecular maker and hybrids performance and heterosis, the mid parent heterosis (MPH) was computed using the formula 100 x (F1-MP)/ MP, where F_1 is the hybrid performance and MP is the mid- parent mean. Per cent heterosis in F_1 over commercial check (CC) was computed using the formula 100 x (F_1-CC)/CC, where CC commercial check mean.

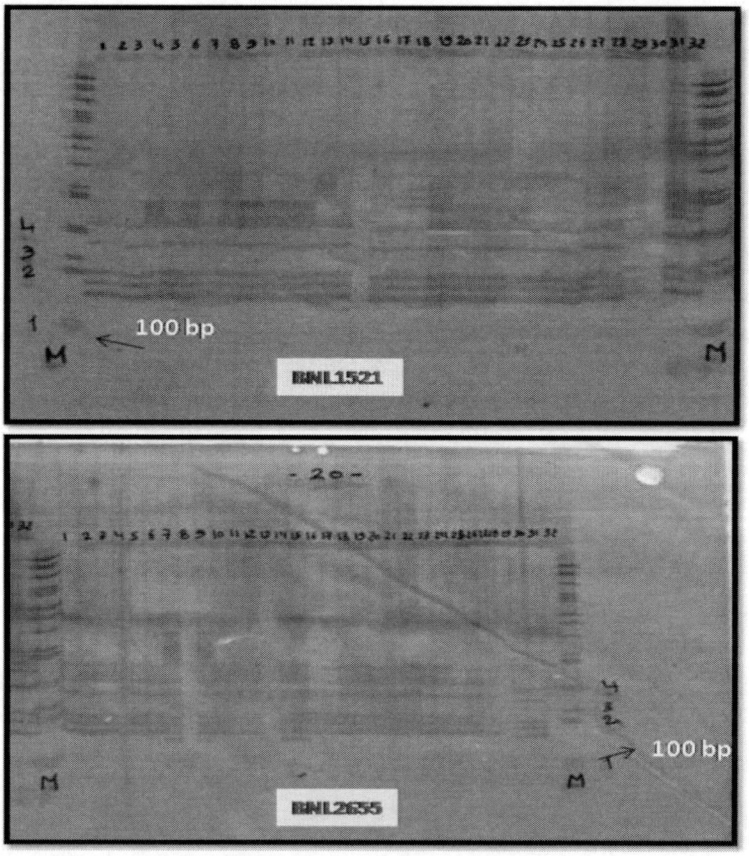

Plate 1. DNA amplification pattern of 32 parents genotype.

Marker Polymorphism

Analysis of microsatellites (SSR's) in 32 parents using 40 primers. Of these 40 SSR primers, 23 primers revealed a high DNA polymorphism among parents, these 23 primers produced a total of 134 amplified profiles (Table 2). Among these, 93 were polymorphic with an average of 68.65 per cent polymorphism. Primers viz., BNL3871, BNL3867 and BNL1611, gave highest (100%) polymorphism. The number of bands ranged from two (primers BNL3871, BNL1034, BNL1227 and BNL3867) to ten (BNL2655, BNL3145, BNL1440, BNL3171 and BNL3994) with an average of 3.35 bands per primer. The primers viz., BNL1034, BNL1227, BNL1059 and CIR246 (50%) showed the least polymorphism. DNA amplification pattern of 32 parents is shown in the (Plate 1).

Table 2. Analysis of SSR patterns generated using 40 primers in cotton genotypes

Sl. No.	SSR Name	No of bands Total	Monomorphic	Polymorphic	Polymorphism
1	BNL3627	0	0	0	0
2	BNL3147	0	0	0	0
3	BNL2921	0	0	0	0
4	BNL4082	0	0	0	0
5	BNL3871	2	0	2	100
6	BNL1034	2	1	1	50
7	BNL1227	2	1	1	50
8	BNL341	0	0	0	0
9	BNL1231	0	0	0	0
10	BNL1878	0	0	0	0
11	BNL3867	2	0	2	100
12	BNL116	4	1	3	75
13	BNL3511	8	2	6	75
14	BNL3031	0	0	0	0
15	BNL3085	0	0	0	0
16	BNL3569	0	0	0	0
17	BNL1421	7	2	5	71
18	BNL1495	5	2	3	60
19	BNL1521	7	3	4	57
20	BNL2655	10	3	7	70
21	BNL3145	10	2	8	80
22	BNL580	0	0	0	0
23	BNL542	0	0	0	0
24	BNL686	0	0	0	0
25	BNL3383	0	0	0	0
26	BNL1611	6	0	6	100
27	BNL1531	7	3	4	57
28	BNL2920	0	0	0	0
29	BNL2882	3	1	2	67
30	BNL1059	4	2	2	50
31	BNL3418	0	0	0	0
32	BNL3259	5	2	3	60
33	BNL1440	10	3	7	70
34	BNL3171	10	2	8	80
35	BNL3408	5	2	3	60
36	BNL3994	10	3	7	70
37	CIR246	4	2	2	50
38	CIR381	6	2	4	67
39	CIR070	0	0	0	0
40	CIR100	5	2	3	60
		134		93	68.65

Table 3. Similarity coefficients for the 32 parents computed from SSR molecular marker data

	L1	L2	L3	L4	L5	L6	L7	L8	L9	L10	L11	L12	L13	L14	L15	L16	L17	L18	L19	L20	L21	L22	L23	L24	L25	L26	L27	L28	T1	T2	T3	T4
L1	1	0.88	0.90	0.78	0.87	0.82	0.77	0.80	0.82	0.84	0.84	0.93	0.77	0.78	0.64	0.85	0.82	0.78	0.79	0.76	0.82	0.77	0.82	0.78	0.75	0.82	0.82	0.88	0.63	0.79	0.74	0.68
L2		1	0.90	0.84	0.84	0.85	0.79	0.77	0.79	0.81	0.84	0.90	0.82	0.81	0.68	0.88	0.82	0.75	0.82	0.82	0.82	0.74	0.85	0.78	0.78	0.85	0.85	0.85	0.63	0.79	0.77	0.68
L3			1	0.87	0.91	0.90	0.85	0.80	0.85	0.80	0.86	0.89	0.82	0.83	0.68	0.93	0.85	0.81	0.84	0.87	0.90	0.82	0.84	0.84	0.81	0.87	0.90	0.87	0.73	0.87	0.82	0.74
L4				1	0.84	0.85	0.79	0.80	0.77	0.78	0.84	0.87	0.82	0.81	0.68	0.88	0.82	0.75	0.79	0.79	0.82	0.74	0.85	0.81	0.78	0.82	0.88	0.88	0.70	0.85	0.77	0.80
L5					1	0.93	0.90	0.85	0.85	0.83	0.92	0.86	0.87	0.86	0.68	0.90	0.79	0.81	0.87	0.87	0.90	0.82	0.87	0.84	0.81	0.87	0.85	0.85	0.73	0.87	0.82	0.71
L6						1	0.86	0.89	0.83	0.90	0.90	0.88	0.86	0.87	0.67	0.89	0.81	0.85	0.85	0.89	0.92	0.84	0.89	0.85	0.85	0.92	0.89	0.86	0.72	0.85	0.81	0.73
L7							1	0.87	0.83	0.79	0.88	0.82	0.86	0.85	0.77	0.86	0.75	0.80	0.85	0.83	0.86	0.81	0.83	0.79	0.79	0.83	0.81	0.81	0.75	0.83	0.81	0.67
L8								1	0.84	0.90	0.91	0.88	0.84	0.82	0.61	0.81	0.84	0.88	0.88	0.81	0.87	0.82	0.86	0.83	0.89	0.87	0.87	0.89	0.67	0.83	0.78	0.76
L9									1	0.85	0.88	0.88	0.81	0.87	0.67	0.83	0.89	0.88	0.91	0.89	0.89	0.87	0.86	0.85	0.85	0.89	0.86	0.86	0.72	0.83	0.81	0.73
L10										1	0.92	0.89	0.79	0.83	0.64	0.82	0.82	0.86	0.84	0.81	0.87	0.82	0.84	0.84	0.84	0.90	0.85	0.87	0.60	0.81	0.79	0.74
L11											1	0.92	0.88	0.83	0.69	0.88	0.85	0.87	0.92	0.85	0.88	0.86	0.87	0.84	0.87	0.88	0.85	0.90	0.68	0.87	0.85	0.77
L12												1	0.85	0.83	0.66	0.90	0.90	0.87	0.87	0.82	0.88	0.83	0.90	0.84	0.84	0.88	0.90	0.96	0.68	0.84	0.80	0.77
L13													1	0.87	0.77	0.89	0.81	0.85	0.91	0.86	0.83	0.81	0.89	0.79	0.77	0.83	0.81	0.81	0.79	0.85	0.78	0.75
L14														1	0.75	0.90	0.79	0.86	0.84	0.93	0.90	0.85	0.84	0.84	0.78	0.87	0.85	0.82	0.73	0.84	0.79	0.74
L15															1	0.73	0.60	0.66	0.67	0.69	0.67	0.67	0.69	0.68	0.57	0.67	0.60	0.60	0.78	0.70	0.67	0.63
L16																1	0.86	0.85	0.85	0.91	0.89	0.87	0.86	0.82	0.79	0.86	0.89	0.89	0.75	0.88	0.83	0.78
L17																	1	0.88	0.88	0.83	0.83	0.84	0.86	0.82	0.88	0.86	0.92	0.92	0.72	0.83	0.78	0.81
L18																		1	0.90	0.90	0.88	0.91	0.82	0.84	0.87	0.90	0.88	0.85	0.74	0.82	0.77	0.80
L19																			1	0.88	0.85	0.86	0.85	0.79	0.85	0.85	0.83	0.85	0.72	0.87	0.85	0.75
L20																				1	0.91	0.93	0.82	0.88	0.85	0.91	0.89	0.80	0.78	0.85	0.83	0.75
L21																					1	0.90	0.89	0.91	0.88	0.94	0.92	0.89	0.72	0.85	0.81	0.73
L22																						1	0.78	0.83	0.89	0.87	0.84	0.84	0.76	0.83	0.81	0.73
L23																							1	0.88	0.82	0.89	0.89	0.89	0.75	0.80	0.74	0.72
L24																								1	0.84	0.91	0.91	0.82	0.77	0.82	0.77	0.77
L25																									1	0.88	0.88	0.88	0.70	0.79	0.74	0.71
L26																										1	0.94	0.86	0.72	0.83	0.78	0.75
L27																											1	0.92	0.72	0.83	0.78	0.81
L28																												1	0.66	0.83	0.78	0.78
T1																													1	0.75	0.66	0.69
T2																														1	0.91	0.86
T3																															1	0.84
T4																																1

MOLECULAR MARKER DIVERSITY AMONG THE PARENTS

The similarity coefficients (Table 3) involved in the line x tester study ranged from 57% to 96 %, with an average of 81%. Among the parental lines, the lines [DB 533 x DB 534 F4 IPS 8] and [DB 533 x DB 534 F4 IPS 1] showed highest (96%) similarity coefficient value. While, the lines [DB 533 x DB 534 F4 IPS 55] and [DB 533 x DB 534 F4 IPS 16] exhibited lowest (57%) similarity coefficient value. All the 32 genotypes showed diversity among themselves indicating that there is considerable amount of variation, which can be exploited through appropriate breeding programme.

The dendrogram constructed from the pooled data (Figure 1) revealed three distinct clusters. One cluster involved testers and in other clusters all F_4lines were placed which are already having proven record in giving good hybrids.

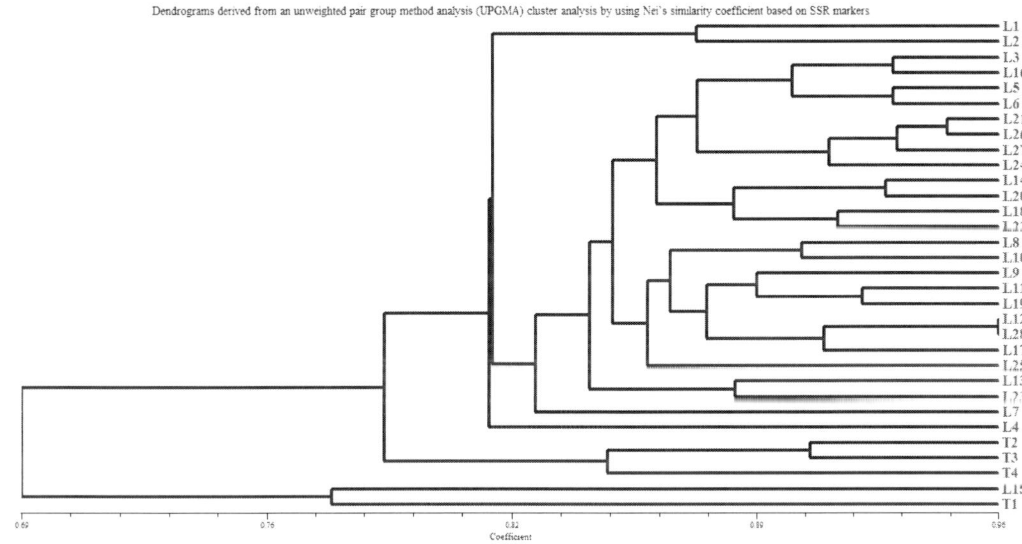

Figure 1.Dendrograms derived from an unweighted pair group method analysis (UPGMA) cluster analysis by using Nei's similarity coefficient based on SSR markers.

The similarity coefficient values between the line [DB 533 x DB 534 F4 IPS 49] and the tester [DH 98-27] showed (67%). It revealed that [DB 533 x DB 534 F4 IPS 49] was closely related to [DH 98-27] with 67 % similarity between parents. The hybrid between [DB 533 x DB 534 F4 IPS 49] and [DH 98-27] exhibited the highest yield of 2884.26 (kg/ha). Highest similarity coefficient (88%) value between lines and testers showed between the line [DB 533 x DB 534 F4 IPS 52] and the tester [ZCH8], the hybrid between these recorded an yield of 2040.757 (kg/ha). Lowest similarity coefficient value was noticed between line [DB 533 x DB 534 F4 IPS 16] and tester [DH 98-27] which revealed that they are far distinct from each other. This combination exhibited an yield 2384.62 (kg/ha).

Table 4. Genetic distance for the 32 parents computed from SSR molecular marker data

	L1	L2	L3	L4	L5	L6	L7	L8	L9	L10	L11	L12	L13	L14	L15	L16	L17	L18	L19	L20	L21	L22	L23	L24	L25	L26	L27	L28	T1	T2	T3	T4
L1	1	0.13	0.10	0.22	0.13	0.18	0.24	0.20	0.18	0.16	0.16	0.07	0.24	0.22	0.36	0.15	0.18	0.22	0.21	0.24	0.18	0.23	0.18	0.22	0.25	0.18	0.18	0.12	0.37	0.21	0.27	0.32
L2		1	0.10	0.16	0.16	0.15	0.21	0.23	0.21	0.19	0.16	0.10	0.18	0.19	0.32	0.12	0.18	0.25	0.18	0.18	0.18	0.26	0.15	0.22	0.22	0.15	0.15	0.15	0.37	0.21	0.24	0.32
L3			1	0.13	0.09	0.10	0.16	0.21	0.16	0.20	0.14	0.11	0.18	0.17	0.32	0.07	0.16	0.19	0.16	0.13	0.10	0.18	0.16	0.16	0.19	0.13	0.10	0.13	0.27	0.14	0.18	0.27
L4				1	0.16	0.15	0.21	0.20	0.24	0.22	0.16	0.13	0.18	0.19	0.32	0.12	0.18	0.25	0.21	0.21	0.18	0.26	0.15	0.19	0.22	0.18	0.12	0.12	0.30	0.16	0.24	0.20
L5					1	0.07	0.10	0.15	0.16	0.17	0.08	0.14	0.13	0.14	0.32	0.10	0.21	0.19	0.14	0.13	0.10	0.18	0.13	0.16	0.19	0.13	0.16	0.16	0.27	0.14	0.18	0.29
L6						1	0.14	0.11	0.17	0.10	0.10	0.12	0.14	0.13	0.33	0.11	0.19	0.15	0.15	0.11	0.08	0.16	0.11	0.15	0.15	0.08	0.11	0.14	0.28	0.15	0.19	0.28
L7							1	0.14	0.17	0.21	0.12	0.18	0.14	0.16	0.23	0.14	0.25	0.21	0.15	0.17	0.14	0.19	0.17	0.21	0.21	0.17	0.19	0.19	0.25	0.17	0.19	0.33
L8								1	0.16	0.10	0.09	0.12	0.16	0.18	0.39	0.19	0.16	0.12	0.12	0.19	0.14	0.18	0.14	0.17	0.11	0.14	0.14	0.11	0.33	0.17	0.22	0.24
L9									1	0.16	0.12	0.12	0.19	0.13	0.33	0.17	0.11	0.12	0.09	0.11	0.11	0.13	0.14	0.15	0.15	0.11	0.14	0.14	0.28	0.17	0.19	0.28
L10										1	0.08	0.11	0.21	0.17	0.36	0.18	0.18	0.14	0.16	0.19	0.13	0.18	0.16	0.16	0.16	0.10	0.16	0.13	0.40	0.19	0.21	0.27
L11											1	0.08	0.12	0.17	0.31	0.12	0.15	0.14	0.08	0.16	0.12	0.14	0.13	0.16	0.13	0.12	0.15	0.10	0.32	0.13	0.15	0.23
L12												1	0.15	0.17	0.34	0.10	0.10	0.14	0.13	0.18	0.12	0.17	0.10	0.16	0.16	0.12	0.10	0.04	0.32	0.16	0.21	0.23
L13													1	0.13	0.23	0.11	0.19	0.15	0.09	0.14	0.17	0.19	0.11	0.21	0.24	0.17	0.19	0.19	0.21	0.15	0.22	0.25
L14														1	0.25	0.10	0.21	0.14	0.16	0.07	0.10	0.15	0.16	0.16	0.22	0.13	0.16	0.18	0.27	0.16	0.21	0.27
L15															1	0.27	0.40	0.34	0.33	0.31	0.33	0.33	0.31	0.32	0.43	0.33	0.40	0.40	0.22	0.30	0.33	0.37
L16																1	0.14	0.15	0.15	0.09	0.11	0.13	0.14	0.18	0.21	0.14	0.11	0.11	0.25	0.12	0.17	0.22
L17																	1	0.12	0.12	0.17	0.17	0.16	0.14	0.18	0.12	0.14	0.08	0.08	0.28	0.17	0.22	0.19
L18																		1	0.11	0.10	0.12	0.09	0.18	0.16	0.13	0.10	0.12	0.15	0.26	0.18	0.23	0.20
L19																			1	0.12	0.15	0.14	0.15	0.21	0.16	0.15	0.17	0.15	0.28	0.13	0.15	0.25
L20																				1	0.09	0.08	0.18	0.12	0.15	0.09	0.11	0.20	0.22	0.15	0.17	0.25
L21																					1	0.10	0.11	0.09	0.12	0.06	0.08	0.11	0.28	0.15	0.19	0.28
L22																						1	0.22	0.17	0.11	0.13	0.16	0.16	0.24	0.17	0.19	0.27
L23																							1	0.12	0.18	0.11	0.11	0.11	0.25	0.21	0.26	0.28
L24																								1	0.16	0.09	0.09	0.18	0.23	0.18	0.24	0.23
L25																									1	0.12	0.12	0.12	0.30	0.21	0.27	0.29
L26																										1	0.06	0.14	0.28	0.17	0.22	0.25
L27																											1	0.08	0.28	0.17	0.22	0.19
L28																												1	0.34	0.17	0.22	0.22
T1																													1	0.25	0.34	0.31
T2																														1	0.09	0.14
T3																															1	0.16
T4																																1

Correlation Between Genetic Distance and Hybrid Performance and Heterosis

Genetic distance (GD) based on SSR markers were computed (Table 4). Genetic distance (GD) ranged from 0.041 to 0.429, with an average of 0.183. The result implied that each cluster dendrogram substantially reflected its own genetic relationship among parents.

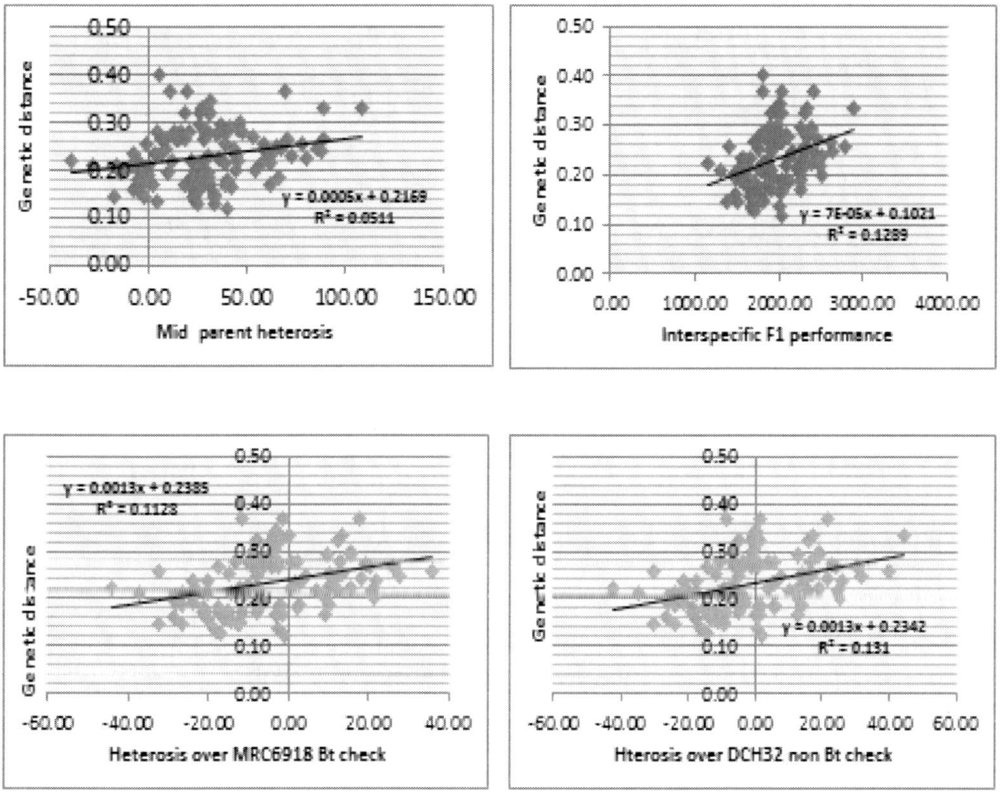

Figure 2. Relationship between genetic distance (GD) and interspecific F_1 performance, mid parent heterosis and heterosis over MRC 6918 and DCH 32 for seed cotton yield (kg/ha).

Overall, a low significant correlation of GD with hybrid performance and heterosis was detected (Table 5). The correlation between (number of bolls per plant, mean boll weight, seed index, fiber strength, fiber uniformity ratio and fiber elongation) and genetic distance (GD) were not significant. Highly significant positive correlation were found between genetic distance (GD) and ginning outturn for F_1 performance (0.277) and heterosis over MRC6918 and DCH32 (0.279), between genetic distance (GD) seed cotton yield (Figure 2) for F_1performance (0.359) and heterosis over MRC6918 (0.336) and over DCH32 (0.362) and between genetic distance (GD) and fiber micronaire value for mid parent heterosis (0.266). Significant positive correlation were found between genetic distance (GD) and ginning outturn for mid parent heterosis (0.237), between genetic distance (GD) and lint index for mid parent heterosis (0.227), F_1 performance (0.251), heterosis over MRC6918 and DCH32 (0.250), between genetic distance (GD) and seed cotton yield for mid parent heterosis (0.226), between genetic distance (GD) and fiber length for mid parent heterosis (0.210), between genetic distance (GD) and fiber micronaire value for F_1 performance and heterosis over MRC6918 and DCH32 (0.241).

Table 5. Correlation coefficients of genetic distance (GD) with F_1 performance and heterosis

Traits	Mid Parent Heterosis	F1 Performance	Heterosis over Mrc6918 check	Heterosis over DCH32 check
Number of bolls per plant	-0.347	-0.181	-0.177	-0.177
Mean boll weight (g)	-0.222	-0.297	-0.290	-0.290
Seed Index (g)	0.193	0.170	0.164	0.164
Ginning Outturn (%)	0.237*	0.277**	0.279**	0.279**
Lint Index (g)	0.227*	0.251*	0.250*	0.250*
Seed Cotton Yield (kg/ha)	0.226*	0.359**	0.336**	0.362**
Fibre Length (mm)	0.210*	0.120	0.120	0.120
Fibre Strength ((g/t)	-0.179	-0.130	-0.130	-0.130
FibreMicronair Value (µg/inch)	0.266**	0.241*	0.241*	0.241*
Fibre Uniformity Ratio %	-0.036	-0.056	-0.056	-0.056
Fibre Maturity Ratio	0.221*	0.141	0.148	0.148
Fibre Elongation %	-0.241	-0.119	-0.118	-0.118

* Significant at P = 0.05 ** Significant at P = 0.01.

G.hirsutum and *G. barbadense* are allotetraploid (2n=4x=52) cottons, which together represent the most extensively cultivated species worldwide. While *G.hirsutum*is the most widely- cultivated species –well- known for its higher yield ad wider environmental adaptation. It was recognized that the two species cross easily and produce vigorous F_1hybrids (Loden and Richmind 1915). Useful heterosis in interspecific Fhybrids which combined productivity and quality has been reported by many researcher (Davis and Palomo 1980, Roupakias et al. 1998, Galanopoulou- Sendouca and Roupakias 1999 and Zhang and Wang 2005).

DNA based molecular markers acted as a versatile tool to study variability and diversity in different plant species. The development of DNA based markers represent an alternative procedure of the identification of promising parental lines for superior performances of hybrids. The microsatellite (SSR's) markers have been widely used for the estimation of variation among closely related individuals due to its multiallelic nature and high polymorphism. Molecular markers based on polymorphism of DNA are especially useful for this purpose because they are not affected by environment (Tatineni et al., 1996 and Saghai-Maroof et al., 1984). Several examples of the application of molecular markers to estimate genetic distances have been reported in maize (Smith et al., 1990) and rice (Zhang et al., 1995). Thus, molecular markers like SSR's (microsatellite) could be used for germplasm classification and clustering to derive valuable information for heterosis prediction.

In this present study, the relationship of genetic distance based on SSR markers with hybrid performance and heterosis in interspecific hybrids of cotton was examined.

Correlations of GD with hybrid performance and heterosis in interspecific hybrids were highly significant positive correlation were found between genetic distance (GD) and ginning outturn for F_1 performance and heterosis over MRC6918 and DCH32, between genetic distance (GD) seed cotton yield for F_1performance and heterosis over MRC6918 and over DCH32 and between genetic distance (GD) and fiber micronaire value for mid parent heterosis. The correlation between (number of bolls per plant, mean boll weight, seed index, fiber strength, fiber uniformity ratio and fiber elongation) and genetic distance (GD) were not significant. Therefore, they were useful for heterosis prediction in seed cotton yield, lint index, ginning outturn and fiber micronaire. According to Bernardo (1992) inadequate genome coverage, random dispersion of molecular markers (unlinked to QTLs) and different levels of dominance could be the reason for low correlation between molecular distance and heterosis and/or F_1 performance. The existence of multiple allelism and epistasis could also cause the low correlation of GD and F_1 performance/ heterosis.

An assessment of the usefulness of molecular markers in breeding cotton for yield and fiber quality improvement may therefore need further consideration. More molecular markers covering all 26 chromosomes and at higher densities and molecular markers that are linked to QTL s for agronomic traits and fibre properties are needed for further studies.

REFERENCES

Beasley J.O., 1940. The origin of American tetraploid *Gossypium* species. Am. Nat. 74: 285–286.

Bernardo, R., 1992: Relationship between single – cross performance and molecular marker heterozygosity. *Theor. Appl. Genet.* 83, 628-634.

Davis, D. D. 1978. Hybrid cotton: Specific problems and potentials. *Adv. Agron.* 30:129-157. doi: 10.1016/ S0065-2113 (08) 60705-1.

Davis, D. D., and A. Palomo, 1980: Yield Stability of Interspecific Hybrids NX-1. *Proc. Belt. Cott. Prod. Res.* Conf. National Council of America, Memphis, TN.

Diers, B. W., P. B. E. McVetty, and T. C. Osborn, 1996: Relationship between heterosis and genetic distance based on RFLP markers in oilseed rape (*Brassica napus L.)Crop Sci.* 36, 79-83.

Galanopoulou- Sendouca, S., and D. Roupakias, 1999: Performance of cotton F1 hybrids and its relation to the mean yield of advanced bulk generation. *Eur. J. Agron.* 11, 53-62.

Gutierrez, O. A., S. Basu, S. Saha, J. N. Jenkins, D. B. Shoemaker, C.L. Cheatham, and J. C. McCarty, 2002: Genetic distance among selected cotton genotypes and its relationship with F2 performance *Crop Sci.* 42, 1841-01847.

Jones, D. F., 1945: Heterosis resulting from degenerative changes. *Genetics* 30, 527-542.

Loden, H. D., and T. R.. Richmond, 1915: Hybrid vigor in cotton – cytogenetic aspects and pratical applications. *Econ. Bot.* 5, 387-408

Meredith, W. R., Jr, and J. S. Brown, 1998: Heterosis and combining ability of cotton originating from different regions of the United States. *J. Cotton Sci.* 2, 77-84.

Mohammadi, S. A., Prassana, B. M., Sudan, C. and Singh, N. N., 2008, SSR heterogenic patterns of Maize parental lines and prediction of hybrid performance. *Journal of Biotechnology and Biotechnological Equipments,* 22(1): 541-547.

Nei, M., and W. Li, 1979: Mathematical model for studying genetic variation terms of restriction endonucleases. *Proc. Natl Acad. Sci.* USA 6, 5269-5273.

Riaz, A., G. Li, Z. Quresh, M. S. Swati, and C. F. Quiros, 2001: Genetic diversity of oilseed *Brassica napus*inbred lines based on sequence – related amplified polymorphism and its relation to hybrid performance. *Plant Breeding* 120, 411-415.

Rolf, F. J., 2001: *NTSYS-PC Numerical Taxonomy and Multivariate Analysis System.* Exeter Publ, Setauket.

Roupakias, D. G., E. Gouli- Vavdinoudi, M. Koutsika- Sotiriou, S. Galanopoulou- Sendouca, and A. S. Mavromatis, 1998: Heterosis in cotton. Biotechnology in agricultural and forestry. In: Y. P. S. Bajaj (ed), *Cotton*, Vol. 42, 140-172, Springer- Verlag, Berlin.

Saghaii-Maroof, M. A., Soliman, K. M., Jorgensen, R. A. and Allard, R. W., 1984, Ribosomal DNA spacer length polymorphism in barley: Mendelian inheritance, chromosomal location and population dynamics. *Proceedings of National Academy of Sciences,* USA, 81: 8014-8018.

Sheng, J. X., G. Y. Lu, T. D. Fu, and G.S. Yang, 2002: Relationships between genetic diversity and hybrid performance in Oilseed rape (*Brassica napus*). *ActaAgron. Sin.* 28, 622-627.

Smith, O. S., Smith, J.S.C., Bowen, S.L., Tenborg, R.A. and Wall, S.J., 1990, Similarities among a group of elite maize inbreeds as measured by pedigree, F1 grain yield, grain yield, heterosis and RFLPs. *Theor. Appl. Genet.,* 80: 833-840.

Tanksley, S. D., 1983: Molecular markers in plant breeding. *Plant Mol. Biol. Rep.* 1, 3-8.

Tatineni, V., Cantrell, R. G. and Davis, D. D., 1996, Genetic diversity in elite cotton germplasm determined by morphological characteristics and RAPDs. *Crop Sci.*, 36: 186-192.

Wu, Y. T., T. Z. Zhang, X. F. Zhu, and G. M. Wang, 2002: Relationship between F1, F2 hybrid yield, heterosis and genetic distance measured by molecular markers and parent performance in cotton. *Sci. Agric. Sin.* 1, 498-507.

Zhang, Q., Saghai, M. A., Yang, G. P., Liu, K. D., Zhou, K. A., Gravois, C. G., Xu and Gao, Y. G., 1995, Correlatoons between molecular marker polymorphism and hybrid performance in rice. *Plant Genome IV Conference*, San Diego, CA, p. 307.

Zhang, X. Q., and X. D. Wang, 2005: Preliminary study on heterosis of interspecific hybrid cotton (*Gossypiumhirsutum X G. barbadense*) based on cytoplasmic male sterility system. *ActaGossypii Sin.* 17, 79-83.

Zhang, X.Q., Wang, X. D., Jiang, P. D., Hua, S. J., Zhang, H. P. and Dutt, Y., 2007, Relationship between molecular marker heterozygosity and hybrid performance in intra- and interspecific hybrids of cotton. *Journal of Plant Breeding,* 126, 385-391.

In: Biotechnology and Conservation … (2 Volume Set) ISBN: 978-1-63117-869-6
Editors: S. A. Farooq, R. Abed and S. Baqir

Chapter 23

BIOLISTIC DELIVERY OF PLASMID DNA IN OMANI DATE PALM

S. A. Farooq, S. H. Al Rashdi and T. T. Farook*
Department of Biology, College of Science, Sultan Qaboos University,
Al Khoud, Oman

ABSTRACT

In the Arabian Peninsula Date palm tree suffers from many diseases which affect the productivity which emphasizes the demand of good date clones. Nowadays gene transformation methods are routinely used to overcome the problems to enhance the production, disease resistance and yield. This study focuses on tissue culture and production of transgenic date palm using gene gun device. It is possible to get thousands of similar clones of date palm using tissue culture/micropropagation technique using (MS) media with plant growth regulators which support callus formation and somatic embryogenesis (Farooq et al., 2012; Badawy et al., 2005). In this study attempts are made to standardize the preparation of cartridges to use in gene gun, by coating the nano particles of (tungsten/gold) with the plasmid DNA containing the desired gene along with a reporter gene and its delivery into the targeted date palm cells and to visualize the results using histochemical GUS gene assay in two varieties Aboedah (AB) and Alkhlas (ZK) of date palm. Different physical parameters were applied such as two different types of nano particles (tungsten/gold) with different particle sizes and different pressure regimes such as 900 and 1100 psi; with two different DNA concentrations (1 and 3 μg) at the shooting distance of 4.5 and 9 cm from the target tissue were used as variables. The data recorded on these parameters is used to evaluate the efficiency of microprojectile delivery into the target cells.

* Email: sfarook@squ.edu.om.

INTRODUCTION

Date palm (*Phoenix dactylifera* L.) is a major fruit plantation crop of the Arabian Peninsula. It is a primary fruit crop in Oman occupying 49% of cultivated area with a yield of 238,000 MT. There are about 180 varieties of date palm are grown in Sultanate. In recent years the production is declining due to Urbanization, Desertification, heavy insect infestation such as Dubas bug and Soil and water salinity. In vitro micro-propagation technology is successfully used to produce thousands of young plantlets. However the genetic transformation in date palm is still at the initial stages. In this study, we have been doing transient gene transformation in the callus cells of date palm using gene gun device. Gene gun uses particle bombardment for cell transformation in which high density sub-cellular sized gold/tungsten particles coated with DNA are accelerated to high velocity to deliver genes into living cells. The project aims to standardize the protocols for the biolistic gene transfer and the expression of the genes in the cells using luciferase (LUC) and glucuronidase (GUS) reporters.

Optimization of gene transfer using reporter marker genes is a pre-requisite for developing an efficient and reliable transformation system for any crop species. Furthermore, identification of suitable selection agents, such as antibiotic or herbicide, and its minimal concentration is important for isolating and regenerating transformed cells into whole transgenic plants. Successful transfer of foreign DNA into plant cells could be achieved after optimizing physical and biological parameters that effect DNA delivery and biological condition of the receiving cells. Among the parameters commonly optimized for DNA delivery into plants are helium pressure, distance from microcarrier carrying the DNA to the target tissue and the concentration of DNA used per bombardment. There are few reports on transient expression of transgene in date palm using biolistic. In this study, we describe the optimization of physical and biological parameters affecting DNA delivery into date palm and followed by regeneration of putative transgenic date palm plants.

Callus initiation and somatic embryogenesis: Meristem explants from the offshoots of date palm were obtained from three cultivars namely Aboedah and Alkhalas. Aseptically the callus was initiated in MS medium with 2, 4-D, 2-ip and NAA. The calli were sub-cultured on MS medium and the embryogenic callus was obtained with different concentrations of 2-ip. The reporter gene constructs used in this study contain *uidA* gene encoding β-glucuronidase under the control of the constitutive CaMv 35S promoter. This plasmid is optimized for transient expression using GUS histochemical assay (Jefferson et al., 1987).

Cartridge preparation: Gold/Tungsten particles were washed with ethanol rinsed with sterile double distilled water before coating with the plasmid DNA. To precipitate plasmid DNA onto the microcarrier, particle suspensions were mixed with spermidine and CaCl2 and injected in the tubing and dried with helium. The tube is cut into 2 Cm size cartridges.

The target callus cells were placed at a distance of 6, 9 and 12 cm from the stopping screen. Explants on each plate were bombarded at 1100 psi helium pressure. After bombardment, the cultures were maintained on the regeneration medium in 16 h photoperiod at 25°C for 24 h.

Histochemical GUS assay: Histochemical assay for glucuronidase (GUS) was conducted on explants 24 h after bombardment. The callus was incubated in X-Gluc staining at 37°C overnight. The GUS expression was recorded in terms of number of blue spot per explants.

Bombardment parameters: Biological and physical parameters mentioned below for biolistic transformation were optimized using β-glucuronidase (GUS) as the reporter gene. Biological parameters include the DNA concentrations (0.6, 1.2 and 1.8 μg per bombardment). For each parameter, the explants were bombarded at 9 cm target distance with 900 psi helium pressure. The physical parameters include rupture disk pressure (650, 900 and 1100 psi) and distance from stopping plate to target tissue (6, 9 and 12 cm), Gold or Tungsten microcarriers. Each of these treatments was conducted in three replicates and the experiment was repeated twice. All bombarded explants were histochemically tested for GUS expression for parameter optimization 24 h after bombardment.

The present studies showed that embryogenic callus is better suited for biolistic-mediated transformation. The data on the five physical and biological parameters used for biolistic-transformation, namely, helium pressure; microcarrier to target tissue distance, DNA concentrations is recorded and analyzed. The parameters for optimizing transient expression of GUS in callus of date palm are studied. Transient expression of GUS gene was demonstrated in several plants.

Effect of DNA concentration: The use of the appropriate amount of DNA is important in order to produce efficient DNA-gold particle binding. The precipitation of DNA on gold particles will also determine the potential amount of DNA for delivery. It was found that 1.2 μg DNA per bombardment gave the highest transient GUS expression in this study. The effect of DNA concentration per bombardment on GUS expression was examined in bombarded calli. The quantity of DNA coated with gold particles was varied from 0.6 to 1.8 μg per bombardment. The concentration of DNA at 1.20 μg per bombardment produced the highest number of GUS spots per explant as compared to 0.6 and 1.8 μg per bombardment. Aggregation of plasmid coated gold particles was observed under the microscope when the highest DNA concentration was used for bombardment. This observation could explain why transient expression did not increase when high DNA concentration was used.

Effect of helium pressure: The maximum GUS expression was observed when the DNA-coated gold particles were propelled at 1100 psi helium pressure. However, the differences to the other helium pressures were not significant at $p<0.05$.

Effect of target distance: The increase in the target distance caused reduction in the average number of GUS spots/callus and percentage of explants expressing GUS (Mousavi, 2009). The results showed that the target distance of 6 cm was significantly better than the other distances in terms of the number GUS expressing cells.

One of the most important parameters to optimize for biolistic transformation is the target distance between the microcarrier to the target tissue in the bombardment chamber. This directly affects the distance that the microcarriers travel to the target cells for microcarriers penetration and gene transfer into plant genome. In this study, 6 cm target distance is optimal in terms of GUS expression as compared to 9 and 12 cm. Increasing the distance causes reduction in the average number of explants expressing GUS.

Conclusion

In this study, the following biolistic parameters were successfully optimized: 1100 psi helium pressure, 6 cm target distance, concentration of DNA at 1.2 μg. The optimized condition was used to transform date palm which is followed by selection on specific antibiotic. Transformants were verified with GUS histochemical assay.

REFERENCES

Badawy, E. M., A. M.A Habib, A. El-Ban G. (2005) Propagation of date palm (Phoenix dactylifera) plants by using tissue culture technique. *Arabian Journal of Biotechnology* 8: 343–354.

Farooq S. A., Khan R.S and T.T. Farook. (2012) *Tissue culture studies in Date palm.* In Date: The genus Phoenix, Production, Processing, Food and Medicinal values. Chapter 2. Pages 13-28. CRC Press, Taylor & Francis USA

Jefferson RA, Kavanagh TA, Bevan MW (1987) GUS fusions: β-glucuronidase as a sensitive and versatile gene fusion marker in higher plants. *EMBO. J.* 6: 3901-3907.

Mousavi, M., Mousavi, A., Habashi, A., & Arzani, K. (2009) Optimization of physical and biological parameters for transient expression of uidA gene in embryogenic callus of date palm (phoenix dactylifera L.) via particle bombardment. *African J Biotech* 8: 3721-3730

In: Biotechnology and Conservation … (2 Volume Set) ISBN: 978-1-63117-869-6
Editors: S. A. Farooq, R. Abed and S. Baqir

Chapter 24

THE GENOMIC DNA STRUCTURE IN SOMATIC EMBRYOGENESIS OF DATE PALM (*PHOENIX DACTYLIFERA* L.) BY FLOW CYTOMETRIC, RAPD AND AFLP ANALYSIS

D. Yatta El Djouzi, F. Abed, B. Amara, M. S. Yakhou and F. Benhafsi

National institute of agronomic research of Algeria (INRAA).
Biotechnology and Plant Breeding Division CRP M. Boualem,

ABSTRACT

Somatic embryogenesis is an ideal technique for clonal propagation of date palm and has great potential for large-scale multiplication. In addition the plantlets issues from somatic embryos are viable and vigorous. Vegetative propagation in vitro method which allows the mass multiplication of elite genotypes of date palm, has been developed over several date palm varieties resistant bayoud and high quality. Protocols developed were validated by planting one hundred and sixty five plantlets of different varieties in 2008.

Several molecular approaches have been developed to assess the plantlets conformity from the date palm. It is the genomic DNA structure study, using flow cytometric and RAPD, and AFLP analysis. Flow cytometry analysis showed that the embryogenic callus at different stages and in vitro micropropagation plantlets have the same ploidy level. Molecular markers analysis was performed by comparing the électrophorétic profiles of the plantlet regenerated by somatic embryogenesis to those of the mother plant by RAPD and AFLP markers. Eight primers selected by the RAPD technique, reveal a high percent of common bands that varies 93% to 97%. It also shows the systematic presence of specific bands to the plant mother and to the vitroplants but their percent is low between of 3% to 7%. The 17 primers combination selected by the AFLP technique show a percent of conformity is between 91%to 98% and a variation between 2% to 9%.

The establishment of a reliable or control genetic test using these markers allow the micropropagation of date palm to move towards industry multiplication and have an early conformity test before planting. This Biotechnology is used on the date palm as a privileged way to establish a strategy to preserve and enhance this phoenicicole heritage.

Keywords: Date palm, cultivars, somatic embryogenesis, molecular markers, RAPD, AFLP, flow cytometry

INTRODUCTION

The date palm (*Phoenix dactylifera* L. Arecaceae family) is a monocotyledonous dioecious species grown particularly in the arid and semi arid areas of world. It is the source of income to the oases inhabitants, provides soil stability, moisture, shade, shelter from the wind, and creates favourable conditions for improving secondary crops as well as cereals, Alfalfa, vegetables and fruits. In Algeria the date palm plantations are spread in an area of about 160,000 ha. With more than 17 million date palms are planted and the total production is estimated at 5.5 million Qt, with1.8 million Qt of Deglet Noor.

Unfortunately, the date plants are prone to many diseases, the most dangerous is the vascular wilt caused by a pathogenic fungus *Fusarium oxysporum* fsp *albedinis* living in the soil. The disease has already killed more than 10 million trees in Marocco and 3 millions in Algeria. The problem is aggravated by the fact that the pathogen attacks the most vigorous and productive trees of the best commercial cultivars.

The replacement of dead or infested trees is problematic due to the lower fruit quality of the resistant cultivars and the limited potential of vegetative propagation through offshoots. The in vitro date palm propagation by somatic embryogenesis remains the most interesting to produce the necessary material and for the reconstruction and extension of the palm. It allows the production of embryos from somatic cell with enormous benefits including high multiplication rate and a period of rapid regeneration. Many works have been recorded in this way (Ammar et al., 1977; Reynolds and Murashige 1979; Bouguedera 1979 Mater 1986; Drira and Benbadis 1985; Daguin and Letouzé 1987; Zaid, 1989; Yatta 2007).

In vitro vegetative propagation technology, which allows the mass multiplication of elite genotypes, has been developed over several date palm cultivars resistant to bayoud and high quality. Genetic variation is higher plants regenerated from cell and tissue cultures has been reported (D'amato, 1965; Daguin et al., 1987). Nowadays, several molecular approach have been developed and can be used to assess the plantlets conformity of date palm, such as the genomic DNA structure study, using flow cytometric, the RAPD, and AFLP analysis.

This paper describes a current progress involved to regenerate plantlets through in vitro tissue culture and preliminary studies undertakes to verify the genetic stability of its plantlets by these markers. The establishment of a reliable genetic test using flow cytometric, the RAPD, and AFLP analysis allow the micropropagation of date palm to move towards industry multiplication and have an early conformity test before planting.

SOMATIC EMBRYOGENESIS

Plant material and medium composition: offshoots of 13 date palm CV. served as explants. Explants were cultured on modified MS (Murashige and Skoog 1962) medium, supplemented (in mg.l^{-1}) with 1thiamine, 100 myo-inositol, 170 NaH2PO4, KH2PO, 45000 sugar, 2000 Phytagel, 200 ammonium citrate, 100 glutamine, 200 to 3000 active charcoal, 1

IPA, and 25 and 100 2.4 D, 12.5 picloram. The cultures were placed under controlled condition at a temperature of 28°C for 16 hours and at 22°C for 8 hours. Cultures were incubated under complete dark condition until callus formation occurred. The embryogenic tissue had the highest capacity for somatic embryo maturation. Six to nine months after successive subcultures, portions of embryogenic tissue were transferred onto maturation medium devoid of growth regulators under 16 hours /day photoperiod.

DNA extraction: Total genomic deoxyribonucleic acid (DNA) was isolated from mother plants of three cultivars namely: Tazerzait, Tazoughart and Takermoust and also from leaf tissues of their several regenerated plants using DN easy method Plant Maxi Kit Qiagen.

Primers and RAPD assays: Nine universal primers (Operon, Alameda, USA) that gave a reproducible amplification which revealed an-inter varietal polymorphism with date palm (Sedra et al., 1998; Ben Abdallah et al., 2000; Trifi et al., 2000 2001 and yatta et saka, 2000) were tested. They are: OPA11, OPB08, OPB01, OPD18, OPD19, OPX4, OPD12, OPJ4. DNA amplification was carried in 25μl reaction that Contains 50ng of date palm DNA, 150 μM of d'NTP, 3mM of $MgCl_2$, 5u Taq DNA polymerase, and 30pM of primer. The relational mixture were overlaid with mineral oil. The amplification was carried on Perkin Elmer 480 thermal cycler programmed for 45 cycles of 1 min at 94°C, 1min at 36°C, 2min at 72°C. The last cycle was followed by 10min at 72°C. After the amplification was performed, the RAPD products were separated by electrophoresis using 1.8% agarose in 1x TAE buffer.

Primers and AFLP assays: 17 primer pairs tested through 24 offered better readability in this study these are identified as fellows: EAGT/MACA, EAGT/MAAG, EAGT/MACG, EAGT/MACT, EAGT/MAGT, EAGT/MAAT, EAGT/MAAC, EAAC/MAAT, EAAC/MAAC, EAAC/MAGT, EAAC/MGC, EAAC/MACT, EACG/ MACG, EACG /MACT, EACG /MAGT, EACG/ MAAT, EACG/ MAAT where E and M correspond to EcoRI and MseI restriction enzymes respectively. After amplification, the products are subjected to a migration 2h30mn to 50W In acrylamide gel(6%) and visualized after silver staining.

Flow Cytometric Analysis of Ploidy Level

Nuclei were extracted by chopping up leaves of vitroplants and leaves issued from plant mother from cvs. Taquerbuchet and Tegaza (approx. 300mg). The material was chopped with a razor blade in 1 ml extraction buffer of Dolezel et al. (1989) and stained with propidium iodide (P4170; Sigma). Following stirring in a vortex to separate nuclei clusters, the solution was left to incubate for 5min before the samples were analyzed by flow cytometry (CyFlow Partec). For the measurement of the absolute DNA value, Zea mays (CE 777) (C=5.43pg DNA) which leaf nuclei were used as internal standards.

Somatic Embryogenesis

The initiated callus was maintained on the initiation media (100mg.l^{-1} 2,4-D and picloram 12.5mg.l^{-1} associated respectively with 3 and 1 mg.l^{-1} IPA plus 3g Activated charcoal/L) to get considerable amount of callus. The callus obtained have a compact, granular and friable appearance and white to beige color. Their size varies from 0.5 to 4.5cm (Figure 2.a). However, with $25 mg.l^{-1}$ 2.4D, some callus become black and necrotic with time, but has the ability to regeneration. The induced calli were compact, friable and have a pale to brownish color. Their size is between 0.6 to 0.9 cm. Culture media M100 and P12.5 tested were favorable to the neoformation of granular friable callus, with a percentage 89% in the medium M100 (Deglet Nur) and 86% in the middle P12.5 (Tinaceur). with M25 medium, the maximum rate obtained is about 50% (Deglet Nur and hartane) (Figure 1). The produced callus was subsequently transferred on callus growth media ($25 mg.l^{-1}$ 2.4-D, $100 mg.l^{-1}$ 2,4-D and picloram $12.5 mg.l^{-1}$ associated respectively with 3 and 1 $mg.l^{-1}$ IPA plus 2g to 3g Activated charcoal/L) to produce embryogenic callus.

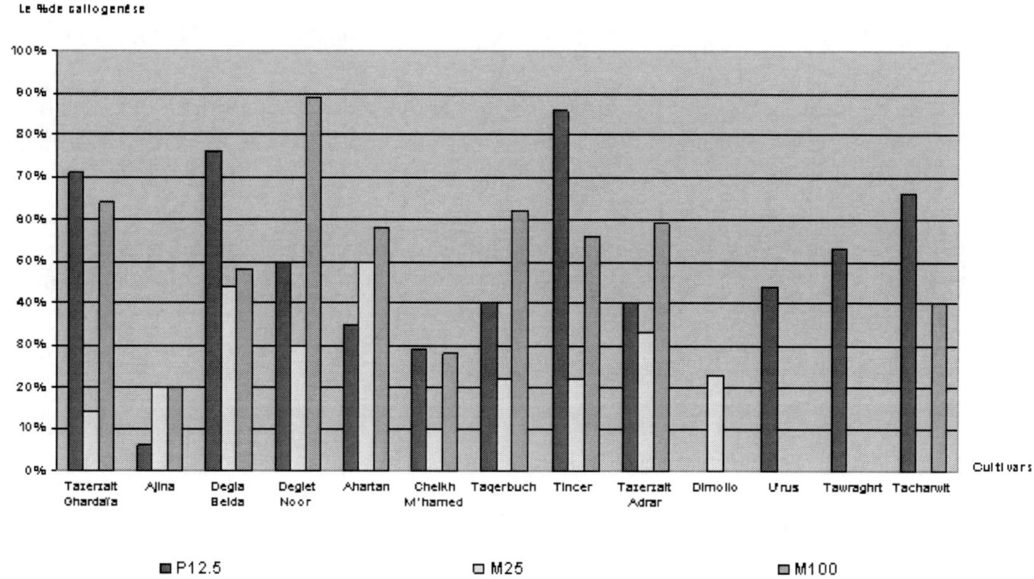

Figure 1. Callus Percentage obtained on culture media P12.5, M25 and M100 in different cultivars.

Proliferating callus are subdivided during subculture in two or more parts according to their size (0.5 to 1cm). We used the same induction media (M25, M100, P12,5) for callus multiplication. The stimulating effect of 2-4D and picloram on callus formation and their multiplication was reported in other date palm (Mater 1986, Daguin et Letouzé 1988, Bhaskaran et Roberta 1995, Hassan and roy 2005).

The embryogenic callus are transferred to hormone-free germination medium. callus began to redifferentiate and developed somatic embryos. The first somatic embryos appear after 6 months of transfer. They have asynchronous development and they are not all at the same stage. The percentage of germinated somatic embryos ranged from 5-70% and 33-100% for tuft of embryos.

The transfer of small plantlets 2 to 4 cm in length on GMP medium gave after two to four months plantlets average length of 8 to 9 cm. The highest number of plant is obtained with the cultivar Tazerzeit Adrar (770) with about 18 vitroplants per callus. The results reported here show that the initiation, development and germination of somatic embryos gave 1905 vitroplants acclimatization (figure 2).

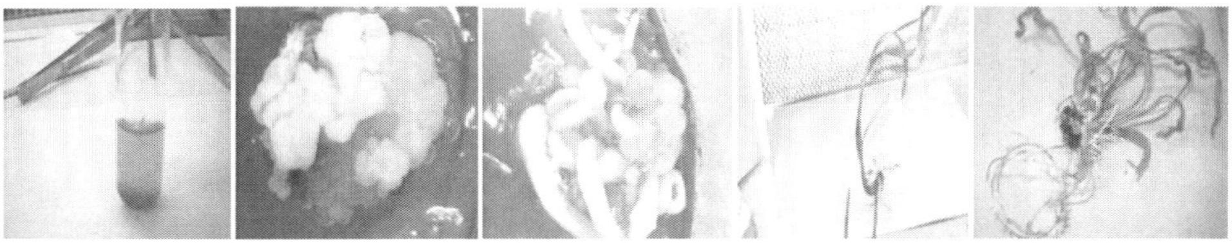

Figure 2. (a). Explants were cultured in the medium (P12.5), (b). granular embryogenic cals, (c). Phase nodulation and ripening, (d). Plant individual, (e). Tuft plantlets.

Table 1. Number of individual plants obtained per callus on GMP medium (D. Yatta 2008)

Cultivars	Callus number germinated	Plantlets number obtained	Average plants / cal
Tinaceur	29	31	01
Taqerbucht	48	562	11
Tazerzeit d'Adrar	42	770	18
Ahartan	30	80	02
Cheikh M'hamed	23	17	1
U'rus	18	120	06
Tazerzeit de Ghardaïa	48	359	07
Adjina	12	16	01
Degla Beida	27	00	00
Dimollo	10	15	01
Tawarght	9	05	00.55
Tacharwit	6	00	00
Deglet Nur	142	1457	10

GENETIC XONFORMITY WITH RAPD AND AFLP

Throughout this study, RAPD reactions were always duplicated and care was taken to ensure consistency in DNA banding profiles between three date palm vitroplants and their mother plants (Tazerzait, and Tazoughart Takermoust).

From the twenty primers tested, eight were selected after three repetitions of the bands obtained: A11, B8, B1, D18, D19, X4, D12, D4.

The RAPD profiles (Figure 3) were showed that primer D19 generated five fragments whose four fragments are monomorphic and one polymorphic fragment (472pb). In the cultivar Tazerzait, this fragment is observed just in the second vitroplant and his plant mother but it is absent in the first vitroplant.

All electrophoretic profiles of plantlets are similar to those of the mother plants with primers A11, x4, B1, et J4 (Figure 3), unlike D19 which the profiles are a little different in cultivar Tazerzait, primers D12 in cultivar Takermoust, and primers B8 and D18 in cultivar Tazoughart. These primers show in their total 97% of monomorphic band for Tazerzeit cultivar, 93% monomorphic band for Tazoughart cultivar and 88% monomorphic band for the cultivar Takermoust. Although the regenerating somatic embryogenesis through callus stage, different plants from this path are consistent with the mother plant.

Eight primers selected by the RAPD technique, reveal a high percent of common bands that varies 93% to 97%. It also shows the systematic presence of specific bands to the plant mother and to the vitroplants but their percent is low between of 3% to 7%.

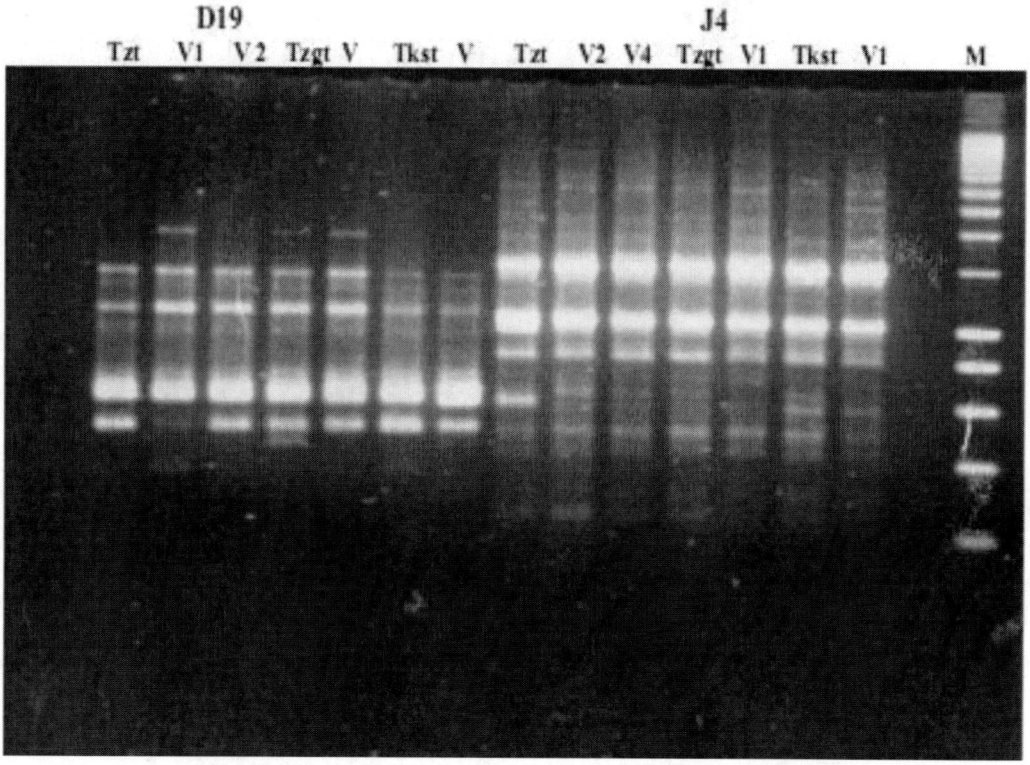

Figure 3. Electrophoretic profiles of RAPD fragments amplified with primers D19 and D 4 on the three cultivars and their vitroplants. M: 1 kb marker smart.

AFLP MARKERS: AMPLIFIED FRAGMENT LENGTH POLYMORPHISM

In this study, we limited also with three vitroplants and their mother choosing 17 primer pairs through 24 offered better readability. In fact, these results show the systematic presence of specific bands to the plant mother and specific bands to the vitroplants but the percentage is low and varies from 1% to 4%. The primer pairs selected show a high percentage of common bands ranging from 91% to 97%. Electrophoretic profiles plantlets are identical to cultivars from which Profiles vitroplants and mother plant were similar, with the exception of some primer combinations as EACG-ECAE which exhibited small variation with the cultivar Takermoust. Some primer combination showed no variation as EAGT-MACA, EAGT-MAAG in cultivar Tazerzait, EAGT - MACA, EAGT - MAAG, EAGT - MAGT, EAAC - MAAT, EAAC - MAGC in cultivar Tazoughart and EAAC-MAAG in cultivar Takermoust.

FLOW CYTOMETRIC ANALYSIS OF DNA CONTENT

Using flow cytometry, we estimated the nuclear genome size of P. dactylifera L. cv. and analysed the stability for this parameter in regenerated plants. We compared the ploidy level with vitroplants issued from somatic embryogenesis and their plant mother of two cvs. Taqerbuchet and Tegaza. Our experiment showed that all vitroplants of the cvs. Taqerbuchet and Tegaza, and their plant mother were diploid. No variation in genome size appeared. The DNA content of vitroplants issued from somatic embryogenesis of *P. dactylifera* L. cv. Tegaza was 1.66±0.05pg (Figure 5). cv. and plant mother Tegaza was 2C=1.63pb.

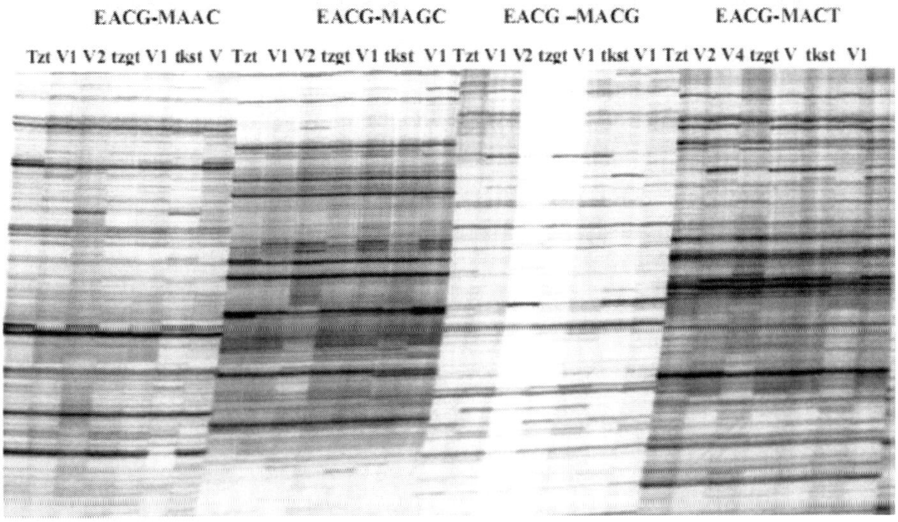

Figure 4. Portion of gel realized with primer pairs EACG-MAAC, EACG-MAGC, EACG –MACG, and EACG-MACT, showing the specificity of AFLP profiles in three cultivars of date palm and their vitroplants 1: Tazerzait; 2: Tazoughart, 3: takermoust v: vitroplants.

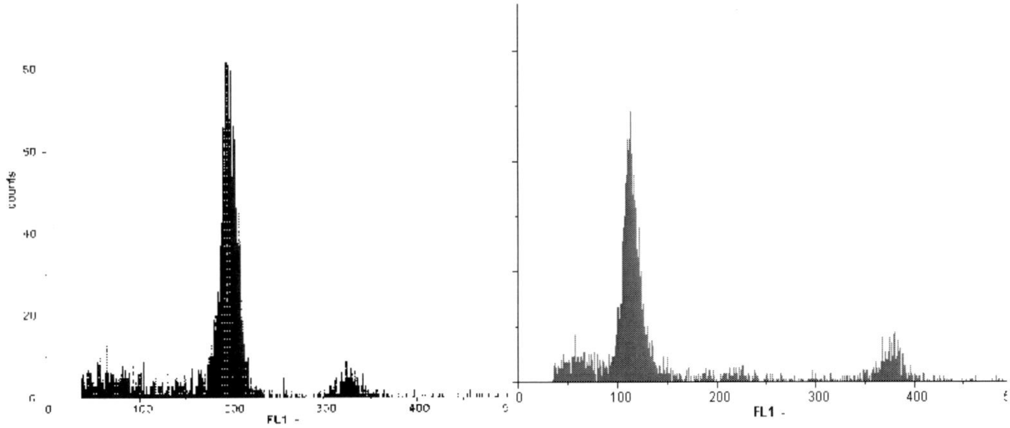

Figure 5. Histograms of fluorescence intensities of nuclei isolated from date palm leaves stained with propidium iodide.A vitroplants issued from somatic embryogenesis of cv. Tegaza (2C=1.66pb), B plant mother Tegaza (2C=1.63pb).

In vitro vegetative propagation method, which allows the mass multiplication of elite genotypes, has been developed over several date palm cultivars resistant to bayoud and high quality.

Our results showed that the most important callus was obtained in the presence of picloram and 2,4-D at concentrations of 12.5 to 100 $mg.l^{-1}$ associated with 1 and 3 $mg.l^{-1}$ IPA. These concentrations require the presence of 2 to 3 $g.l^{-1}$ of activated charcoal. The average callus formation was observed in the presence of 25 $mg.l^{-1}$of 2.4D associated 1 $mg.l^{-1}$IPA. Several authors (Reuveni1979, Reynolds and Murachige 1979, Huang and Muraching 1983, Zaid 1989, Saka and Abed 1989, Yatta 2007) reported the performance of these two growth hormones in the induction of callus. This is also in agreement with the results obtained by Tisserat 1979 on zygotic embryos.

The induction of somatic embryogenesis occured on medium without growth hormones and at the light. This result is already described by Masson and Tisserat 1980. Fergani 1998 showed that the induction is obtained in the dark and in a medium containing MG2 * 2$mg.l^{-1}$ 2,4-D and 3$mg.l^{-1}$IPA and in the presence of vitamins morel.

Maturation of somatic embryos and their germination is obtained on the medium without growth substances containing GMN200 1$mg.l^{-1}$ Thiamine, 100$mg.l^{-1}$ KH2PO4, 170$mg.l^{-1}$NH2PO4, 200 $mg.l^{-1}$ of ammonium nitrate. These results agreed with the previous reports Murashige and Reynolds, 1979; Tissérat and De Mason,1980; Daguin and Letouzé,1988; Al-khayri, 2011.

Our results showed that explants of Takerboucht, Tazerzeit cultivars of Adrar and Ghardaia and also the explants of Deglet Nur cultivar produced the highest number of somatic embryos. These embryos are either individually or in tuffs. These observations were described by Zaid 1989, and Veramendi Navarro1996.

The elongation and rooting plantlets were obtained on medium GMP containing 1001mg.l-1Thiamine hydrochloride,170mg.l-1 NH2PO4, 200mg.l-1of ammonium nitrate, 40 mg.l-1Adenine, 501mg.l-1nicotinic acid, and 50mg.l-1pyridoxine. The germinated vitroplants that developed a excellent root system were transferred into plastic pots containing soil and sand (1:1) in greenhouse for acclimatization. These data are in accordance with those obtained by (Wuidart W. and Konan K., 1989, Saka H. 1998).

Propagation of substantial number of palm species has been proposed by various authors using tissus culture technique. Genetic variation in plants regenerated from in vitro culture has been reported (Larkin and Scowcroft, 1981; Saker et al., 2000). Therefore, evaluation of tissue cultured palm for their clonal status, based on comparison of vegetative and productive characteristics, must wait for years. Our preliminary studies based on molecular tools undertaken to verify the clonal nature of palms produced through tissue culture. RAPD, AFLP, and flow cytometry analysis were conducted on plantlets derived from tissue culture to determine their clonal status. The results of polymorphic DNA (RAPD) clearly showed that the electrophoretic profiles of vitroplants regenerated through somatic embryogenesis are similar to those plant mother. Eight primers selected by the RAPD technique, reveal a high percent of common bands that varies 93% to 97%. It also shows the systematic presence of specific bands to the plant mother and to the vitroplants but their percent is low between of 3% to 7%. However, it will not be taken into account as the percentage difference is very small. It varies from 4% to 11%. This is showing that at least the 2,4-D and picloram didn't induce somaclonal variation in date palm. This result join those of Letouzé et al., 2000, subject to verification more number of plantlets.

We are also interested in the AFLP technique to continue this work. The seventeen primer combinations selected by the AFLP technique showed a percent of conformity is between 91%to 98% and a variation between 2% to 9%. The results show similarities perfect for RAPD and AFLP profiles of plantlets and the mother plant. Polymorphism revealed by these markers depends on cultivars and primers or primer combinations used. Similar results were observed (Letouze et al., 1998; Munshi and Osman, 2010; Othamani et al., 2010; Reda, E.A. M et al., 2011).

Our attention is also focusing on flow cytometry analysis estimate the nuclear DNA content and ploidy levels. our results has revealed that the vitroplants regenerated somatic embryogenesis and plants mother showed the same ploidy level. All cultivars were diploid, since the mean 2c nuclear DNA was between 1.63 and 1.66pg. Fki et al., 2003 obtained the same type of results when analyzing the ploidy level of plantlets regenerated from embryogenic suspension culture.

Molecular approaches are used to understand the mechanisms falling within the process of regeneration by tissue culture and genetic stability of date palm plantlets. Further study using a large number of plantlets is envisaged in order to ensure the reliability of the protocol developed on somatic embryogenesis technique used in our work. The establishment of a reliable or control genetic test using these markers allow an early conformity test before planting and to move the micropropagation of date palm towards industry multiplication. This way of multiplication is a top priority for the preservation and ex situ conservation of our phoenicicole heritage.

Acknowledgments

This research was supported by grants: 1) INRAA National Institute of Agriculture Research. 2) National Programme of research (Ministry of high Education and Research of Algeria).

References

Adawy SSH, Ebtissam HA, El khishin D, Saker MM and Eltriby HA. 2002. Genetic variability studies and molecular fingerprinting of some Egyptian date palm *(Phoenix dactylifera* L.) cultivars. RAPD and ISSR profiling. *Arab J. Biotech.*, 5(2): 225- 236.

Al-Khayri JM. 2011. Basal salt requirement differ according to culture stage and cultivar in date palm somatic embryogenesis. *Am. J. bioch. biotech*, 7(1): 32-42,2011.

Baaziz M and Saaidi M. 1988. Preliminary identification of date palm cultivars by esterase isoenzymes and peroxydase activities. *Can. J. Bot.* 66: 89-93.

Ben Abdallah A, Stiti A, Lepoivre Pand Patrick du jardin 2000. Identification de cultivars de palmier dattier *(Phoenix dactylifera* L.) par l'amplification aléatoire d'ADN (RAPD). *Cahier Agriculture* 200; 9:103-7.

Bouguedoura N.,1979. Contribution à la connaissance du palmier dattier (*Phoenix dactylifera* L.). Etude des productions axillaires. *Thèse de 3ème cycle, Université des Sciences et de la Technologie d'Alger,* Alger, 64 p.

Daguin F et Letouzé R. 1988. Régénération du palmier dattier (*Phoenix dactylifera* L.) par embryogenèse somatique: Amélioration de l'efficacité par passage en milieu liquide agité. *Fruits* - vol.43, n°3.

Daguin F. and Letouzé R. 1987. Vitreous plants in vitro: Relation –ship with ammonium content of the nutrient medium. *Acta Hort.*, 212, 259- 561.

D'Amato F. 1965. endoploidy as factor in plant development. P.

Drira N, Benbadis A. 1985. Multiplication végétative du palmier dattier (*Phoenix dactylifera* L.) par réversion, en culture in vitro d'ébauches florales de pieds femelles adultes. *J. Plant Physiol.* 119: 227-612.

Fki L, Masmoudi R, Drira N and Rival A. 2003. An optimised protocol for plant regeneration from embryogenic suspension cultures of date palm *Phoenix dactylifera* L. cv. Deglet Nour. *Plant Cell Reports* 21: 517-524.

Giogetti L, Ruffini MC, Alessandra T, NutiRonchi V and Chiara G. 2011. Cytogenetic and Flow cytometry approach for early detection of 'mantled' somaclonal variants of oil palm regenerated by somatic embryogenesis: first results on the characterization of regeneration system. *Caryologia*, Vol. 64, no.2: 223-234.

Huang LC et Muraching T.1983. Tissue culture investigations of Bamboo; 1-Callus culture of *Bambusa phyllostachys* and sasa. *Bot. Bull. Academia Sinica*, 24, 31-52.

Kriaâw BS, Hammami F, Masmoudi A, Masmoudi RB, Drira N. 2012. The date palm (*Phoenix dactylifera* L.) micropropagation using completely mature female flowers. *Plant biology and pathology.C.R.* Biologies, 335, pp.194-204.

Letouze R, Daguin F, Hamama L, Paquier K, Marionnet F, and Javouhey M. 2000. Mass micropropagation of date palm (*Phoenix dactylifera* L.) through somatic embryogenesis. Histological study of embryo formation and cultivar identification by RAPD markers. Proceedings of date palm international symposium. Held in Windhoek, Namibia 22- 25 Februry 2000, pp 55-64.

Letouze R, Daguin F, Satour P, Hamama L. and Marionnet F. 1998. Molecular characterization of Egyptian date palm: II. RAPD fingerprints. In: *Proceedings of the First International Conference on Date Palms, Al-Ain, United Arab Emirates*, pp. 158–167.

Mater AA. 1986. In vitro propagation of *Phoenix dactylifera* L. Date palm. J. 2, 57-77.

Munshi A and Osman G. 2010. Investigation on molecular phylogeny of date palm (*Phoenix dactylifera* L.) cultivars by protein, RAPD and ISSR markers in Saudi Arabia. *Australian Journal of Crop* Science 4(1): 23-28.

Murashige T et Skoog F. 1962. A revised medium for rapid growth and bio assays with tobacco tissue cultures. Physiol. Plant. Arum, 15, 473-497.

Reda EAM, Abel HA and Mohamed RA. 2011. Genetic stability among date palm plantlets regenerated from petiole explants. *Africa J. Biotech.*, vol 10 (65):134311- 14318.

Reuveni O. 1979. Embryogenesis and plantlets growth of date palm(*Phoenix dactylifera* L.) derived from callus tissue. *Plant Physiol. Suppl.* n°63,138p.

Reynolds J et Murashige T. 1979. Asexual embrogenesis in callus culture of palms. *In vitro*, 15, 5 383-387.

Saker M, Bekheet SA, Taha HS, Fahmy AS and Moursy HA. 2000. Detection of somaclonal variations in tissue culture-derived date palm plants using isozyme analysis and RAPD finger prints. *Biol. Plant.* 43:347-351.

Sedra MYH, Philippe L, Trouslot P, combes M and Hamon S. 1998. Identification and genetic analysis of date palm (*Phoenix dactylifera* L.) varieties from morocco using RAPD markers. *Euphytica*, 103: 75-82 1998.

Trifi M, Rhouma A, Marrakchi M, 2000. Phylogenetic relation in Tunisia date palm (*Phoenix dactylifera* L) Germplasm collection using DNA amplification fingerprinting Agronomie 20, (2000) 665-671.

Veramendi J., Navarro L.1996. Influence of physical conditions of nutrient medium and sucrose on somatic embryogenesis of date palm. *Plant Cell Tissue. Organ. Cult.* 45:159-164.

Wuidart W et Konan K. 1989. Le sevrage des vitroplants de palmier à huile. *Oléagineux.* Vol. 44. N° 12. pp 573-576.

Yatta D, Abed F, Bouguedoura N, et bennaceur M. 2006. Caractérisation de quelques variétés de palmier dattier (*Phoenix dactylifèra* L.) du Touat, du M'zab, et de Oued Rhir par marqueurs RAPD. Xèmes Journées Scientifiques du réseau des chercheurs « *Biotechnologies végétales: Amélioration des plantes et sécurité alimentaire*». *AUF.* p 105-106.

Yatta D. 2007. Etude de l'embryogenèse somatique chez le palmier dattier (*Phoenix dactylifera* L.): caractérisation moléculaire du matériel végétal initial en vue de l'étude de la conformité des vitroplants. *Thèse Magister.* USTHB, 117p.

Zaid A and Tisserat B.1983. Morphogenetic responses obtained from a variety of somatic explants tissues of date palm. *Bot. Mag.* Tokyo, 96:67-73.

BIOTECHNOLOGY AND CONSERVATION OF SPECIES FROM ARID REGIONS (2 VOLUME SET)

VOLUME 2

Biotechnology in Agriculture, Industry and Medicine

Additional books in this series can be found on Nova's website under the Series tab.

Additional e-books in this series can be found on Nova's website under the e-book tab.

Biotechnology in Agriculture, Industry and Medicine

Biotechnology and Conservation of Species from Arid Regions (2 Volume Set)

Volume 2

Sardar A. Farooq
Raeid Abed
and
Senan Baqir
Editors

nova publishers
New York

Library of Congress Cataloging-in-Publication Data

ISBN: 978-1-63117-869-6

Published by Nova Science Publishers, Inc. † New York

CONTENTS

BIOTECHNOLOGY AND CONSERVATION OF SPECIES FROM ARID REGIONS.

VOLUME 2

In: Biotechnology and Conservation … (2 Volume Set) ISBN: 978-1-63117-869-6
Editors: S. A. Farooq, R. Abed and S. Baqir

Chapter 25

AFLP Genotyping for Plant Genetic Resources Use in Crop Improvement: Models for Predicting Hybrid Performance among Accessions from the USDA Alfalfa Core Collection

A. H. Al Lawati [1] ***and I. M. Ray*** [2,*]
[1]The Research Council, Muscat, Sultanate of Oman
[2]Dep. of Plant and Environmental Sciences, New Mexico State University, Las Cruces, US

Abstract

Utilization of nuclear genetic distance (GD) information based on molecular markers such as AFLP has been proposed to help reduce the time required, and resources needed, for selecting proper parents that produce heterosis (Hybrid vigor) among their F1 progenies. In alfalfa (Medicago sativa. L), this process usually takes at least four years. In the first year, hybrids are generated between several parents. In the second, third, and fourth years, forage yield general combining ability (GCA) of the parents and specific combining ability (SCA) of the individual hybrids are estimated. The purpose of this study was to evaluate if regression models based on GD alone or in various combinations with the variables of GCA, mid-parent performance (VE), physiological parameters such as fall regrowth (FR) and winter-hardiness (WH), of nine parental populations (populations from the USDA National Plant Germplasm System alfalfa core collection) were able to predict the performance of their 36-diallel hybrids in Semi-arid region of Southern New Mexico State (U.S.A). These variables were also tested alone, or in various other combinations, to predict hybrid performance and heterosis. The results revealed that AFLP genetic distance was able to predict yield performance for 63% of 19 superior hybrids, superior than best check cultivars. The best single model that significantly predicts hybrid performance was GCA. It was able to identify 79% of superior hybrids followed by WH (74%), VE (68%), and FR (68%). Combining GD with

* Corresponding author email address: ali.allawati@trc.gov.om.

GCA or any other variable in regression models did not improve the prediction ability to identify superior hybrids. The implication of these results is that in abiotic stressed environment (average annual extreme minimum temperature of -12 to -9 C) evaluation of physiological parameters such as FR or WH, could offer a practical approach to predict hybrid performance and heterosis among the parents used in this study, because these parameters are critical indicators for recovering from winter freezing temperature. Thus, it is very important for evaluating plant genetic resources for crop improvement to evaluate physiological parameters that help to select parents/populations for crop improvement. In case of Oman, physiological parameters of water-use efficiency, salt tolerance, and tolerance to high temperature are very important parameters need to be evaluated for crop improvement beside population genotyping.

Keywords: Predicting models, alfalfa hybrid performance, AFLP, New Mexico

INTRODUCTION

Performance of alfalfa hybrids have been reported in several studies (Childers and Barnes, 1972; Brown and Bingham, 1984; Mckersie et al., 1989; Buchan et al., 1991). Until recently, however, the ability to produce large quantities of hybrid seed has limited the commercial development of hybrid alfalfa. Utilization of cytoplasmic male sterile and fertile restorer lines in producing commercial hybrids has been discussed (Viands et al., 1988). However, hybrid cultivar production using cytoplasmic male sterility was not considered economically practical because of difficulties in getting sufficient seed production of male-sterile lines. This has been attributed to discrimination of pollen-collecting insects against pollen-deficient male sterile lines.

Alternative approaches to commercially developing hybrid alfalfa populations have been proposed. Brummer (1999) suggested producing semi-hybrid cultivars that could capture half of the potential heterotic response. The semi-hybrid cultivar approach consists of mixing seed of two parent populations, planting this mixture, and then allowing the plants to pollinate each other. Approximately 50% of the seed produced in this manner would be from inter-population crosses with the other 50% of the seed generated from intra-population crosses. Thus, the inter-population hybrid progenies should exhibit additive and dominance genetic (i.e. heterosis) effects while the intra-population progenies would exhibit additive genetic effects.

In 2001, there was a breakthrough in commercial alfalfa hybrid production with the release of HybriForce-400 alfalfa by Dairyland Seed Company (West Bend, Wisconsin, www.dairylandseed.com). This hybrid was developed by using a male sterility system of hybridization called msSUNSTRA^{TM}. In operation, selected male sterile (i.e. female) alfalfa plants are grown together with other male fertile plants and allowed to cross-pollinate. The resultant seeds from this cross are predominantly an F_1, with some seed derived from self-pollination of pollen donor plants (Wiersma, 2001). In general, hybrid populations produced using the msSUNSTRA system have performed well in multiple yield trails conducted throughout the U.S.

Selection of proper parents that produce heterosis among their F_1 progenies is the key process for developing hybrid cultivars. This process is the most time-consuming and resource-intensive part of hybrid development because it requires the formation of multiple hybrids between numerous parents in order to estimate general combining ability (GCA) of the parents and the specific combining ability (SCA) of individual hybrids. It has been suggested that utilization of genetic distance (GD) based on molecular markers might help to reduce such expenditures by providing plant breeders with a mechanism to identify parents with different allele frequencies to maximize heterozygousity and linkage disequilibrium, and hence, hybrid performance (Melchinger and Gumbar, 1998; Lee et al., 1989; Melchinger et al., 1990). Thus, since SCA provides an estimate of nonadditive effects, and is dependent on parental allelic diversity, genetic diversity of parents may provide an independent approach for estimating SCA between individual parents. In oilseed rape (*Brasica napa*), Diers et al. (1996) found that when GCA was combined with molecular GD the ability to predict hybrid performance and heterosis increased.

Several studies have been conducted in different crop species to determine associations between genetic distance based on molecular markers and hybrid performance. The reports concerning these associations varied from study to study and even within a crop. In corn (*Zea mays* L.), Betrán et al. (2003) found a positive correlation between GD and F_1 performance. On the other hand, Lee et al. (2002) reported high correlations between GD based on Amplified Fragment Length Polymorphism (AFLP) and agronomic performance in the inter-population hybrids derived from 25 inbred lines of two groups of Lancaster Sure Crop and Reid Yellow Dent corn. The agronomic traits reported in their study were plant height, ear diameter, hundred seed weight, and grain yield. Other studies have also shown that correlations between Restriction Fragment Length Polymorphism (RFLP) genetic distance and hybrid performance were significant in maize (Smith et al., 1990; Godshalk et al., 1990). Ajmone Marsan et al. (1998) observed a positive correlation between parental GD and hybrid performance for grain yield of maize inbred lines, but the correlation was too small to be of predictive value. Melchinger et al. (1992) evaluated diallel crosses among six flint and six dent-inbred lines of European maize and noted that RFLP-based genetic distance measures seemed to be useful primarily for predicting forage yield of crosses between lines from the same germplasm/ heterotic group, and secondarily in some crosses between lines from different heterotic groups. Thus, genetic distance was not as useful as a predictor of hybrid forage-corn yield of crosses between lines from highly divergent groups. Boppenmaire et al. (1992), Godshalk et al. (1990), and Melchinger et al. (1990) have shown that molecular genetic distance was not a good predictor of maize hybrid performance.

In wheat (*Triticum aestivum* L.), El-Maghraby et al. (2005) found that genetic distance based on SSR markers was able to predict ($0.4 \leq r \leq 0.5$) grain yield among 28 hybrids developed from diallel crosses between eight cultivars. However, they reported that selection of crosses based solely on microsatellite data would miss many superior combinations.

In rice (*Oryza sativa* L.), Kwon et al. (2002) found that correlation values of F_1 performance with GD based on SSR and Random Amplification of Polymorphic DNA (RAPD) markers were significant for three of eight traits tested, including: grain yield, culm length, and spikelets per panicle. Their results also indicated that the level of correlation between rice hybrid performance and genetic divergence was dependent on the germplasm evaluated. They attributed their results to (1) narrow genetic diversity among the accessions tested, and (2) lack of linkage between QTL that controlled the measured traits and most of the markers which were randomly distributed throughout the genome. Finally, they recommended utilizing molecular markers that were linked to QTL to improve the correlation between genetic distance and hybrid performance.

Maroof et al. (1997) reported a significant association between genetic distance based on RFLPs and hybrid performance in 28 diallel cross hybrids generated from eight lines used in the U.S. southern long-grain rice breeding programs. However, this association was attributed mainly to high level of heterozygosity and high performance in crosses between an *indica* variety and the remaining seven varieties of *japonica*. They concluded that the level of correlations obtained between marker distance and hybrid performance was dependent on the germplasm used. A similar conclusion was drawn by Zhang et al. (1996). However, in an earlier study, Zhang et al. (1994) found a significant correlation between molecular genetic distance and hybrid performance in rice.

Reports on the association between molecular genetic distance and hybrid performance have also varied in other crops. In pepper (*Capsicum annuum* L.), Geleta et al. (2004) found no correlation between GD based on AFLP heterozygosity, and hybrid performance for fruit yield and days to flowering among diallel cross hybrids. However, other studies have shown significant association between genetic distance based on molecular markers and hybrid performance. In pepino (*Solanum muricatum*), the correlation between AFLP genetic distance and the yield of hybrids was highly significant ($r>0.90$) (Rodriguez-Burruezo et al., 2003). Riaz et al. (2001) reported a significant relationship between genetic distance of oilseed rape (*Brassica napus*) parent inbred lines and seed yield in their hybrids. Cheres et al. (2000) observed that performance of sunflower hybrids (*Helianthus annuus*) was significantly associated with AFLP genetic distance.

Segovia-Lerma (2000) found that GD based on AFLP molecular marker did not predict hybrid yield among 28 hybrids of *M. sativa* subspecies *sativa*. However, it did predict the performance among eight *sativa* × *falcata* subspecies hybrids. Kidwell (1994b, 1994c) showed that genetic distance and heterozygosity (as detected by RFLP) were not correlated with alfalfa forage yield in diploid crosses, but were correlated with yield of the isogenic tetraploid hybrids. It was then suggested that a better correlation at the tetraploid than diploid level could have resulted from complementary gene interactions, which are greater in autotetraploids than diploids (Bingham et al., 1994). Thus, deleterious recessive alleles were more likely to be expressed in diploids, but masked in tetraploids, because of differences in segregation relationships at the two-ploidy levels (Kidwell et al., 1994c). In general, the above-mentioned studies stated that the association of GD with hybrid performance was dependent on the genetic material used.

Several reasons have been proposed for low correlations between molecular genetic distance and hybrid performance. Based on computer simulation, Bernardo (1992) recommended that six conditions are required for genetic distance based on molecular markers to be useful. These are: (1) genetic effects for the trait of interest should exhibit a strong dominance effect (e.g., dominance or overdominance), (2) complementary alleles and allele frequencies at individual loci in the parents are negatively correlated, (3) heritability of the trait of interest must be high, (4) average parental alleles frequencies vary only within a narrow range, (5) at least 30-50% of the QTL influencing the trait of interest must be linked to molecular markers, and (6) not more that 20-30% of the molecular markers are randomly unlinked to the QTL.

Morpho-physiological traits may also be utilized in predicting hybrid performance. For example, Zaidi et al. (2003) reported that selection based on grain yield, plant height, 50% anthesis, and 50% silking of inbred parental lines *per se* might indirectly be used to predict the hybrid performance of corn. Simple cophenetic correlations between grain yield and plant height, 50% anthesis, and 50% silking were 0.72, 0.57, and 0.49, respectively. The combination of molecular genetic distance with phenotypic data may enhance prediction models compared with molecular genetic distance alone. For instance, Jordan et al. (2003) found that a model combining phenotypic trait data and parental molecular diversity explained 71% of the variation in hybrid grain yield of sorghum (*Sorghum bicolor*). Phenotypic traits such as days to maturity and plant height explained 0.40 to 0.47, respectively, of the variation in grain yield. Molecular diversity alone explained 42% of the grain yield variation. Therefore, the combination of phenotypic data with molecular diversity data had potential for use in the selection of parents for generating heterotic hybrids. In the case of alfalfa, fall regrowth, plant height and winter-hardiness may have potential usefulness in predicting alfalfa forage yield.

In a regression model to predict hybrid performance *Per se* performance of the parents (also called variety effects) may also be used. Based on computer simulation, however, Smith (1986) postulated that for traits controlled by large number of loci showing complete dominance, the correlations between lines *per se* and testcross performance were expected to be less than 0.5. This is due to the masking effects of favorable dominant alleles in the tester. However, if variation is mostly contributed from additive genetic variance, then *per se* performance could be utilized in selecting the best parental lines for further investigations (Hallauer and Miranda, 1988).

Few investigations have studied the relationship of molecular GD and/or morpho-physiological traits with hybrid performance for crosses generated from high performing accessions of alfalfa that possess widely different fall dormancy response. The nine alfalfa accessions used for the current study were previously characterized for their nuclear and chloroplast genetic diversity using DNA markers (Bhandari et al., 2011). In addition, they were among the best-performing populations of the alfalfa core collection in southern New Mexico within their respective fall dormancy categories. These nine populations also demonstrated that they imparted different genetic effects to their hybrids (i.e. dominance and additive effects, Al Lawati et al., 2010). The aim of the present investigation was to assess if regression models based on general combining ability (GCA), AFLP genetic diversity (GD), variety effects (VE), fall regrowth (FR), winter-hardiness (WH) alone, or in various combinations, were able to predict the performance of hybrids generated from a half-diallel mating among these nine alfalfa populations.

Plant Genetic Material and Field Evaluation

Plant genetic materials used in this study consisted of nine populations from the perennial *Medicago* core collection of the National Plant Germplasm (U.S. Department of Agriculture), and their 36 F_1 hybrids, as previously described in Al Lawati et al. (2010) and in Bhandari et al. (2011), see Table 1 for the PI numbers. These entries were planted during September 2002 at the Leyendecker Plant Science Research Center near Las Cruces, NM, using a randomized complete block design with 4 replications.

Table 1. AFLP-based genetic distance (GD) estimates using Jaccard's estimation of similarity/distinctness among nine alfalfa germplasms using bulked DNA templates†

Populations	P1†	P2	P3	P4	P5	P6	P7	P8	P9	Mean
P1	0.00									0.82
P2	0.80	0.00								0.86
P3	0.81	0.81	0.00							0.83
P4	0.85	0.85	0.84	0.00						0.81
P5	1.07	1.26	1.11	0.84	0.00					1.00
P6	0.82	0.83	0.80	0.81	0.93	0.00				0.77
P7	0.76	0.78	0.76	0.80	0.97	0.70	0.00			0.74
P8	0.74	0.84	0.76	0.75	0.87	0.68	0.57	0.00		0.72
P9	0.66	0.71	0.71	0.75	0.95	0.57	0.58	0.57	0.00	0.69

† Plant Introduction (PI) of NPGS-USDA, P1= PI 172188, P2= PI 183653, P3= PI 247789, P4= PI 340114, P5= PI 434600, P6= PI 445676, P7= PI 458916, P8= PI 467977, and P9= PI 502478.

Each entry was planted in a three-row plot of 1.5m length using 300 seeds per plot. Spacing between the rows within a plot was 30 cm, and plots were separated by 60 cm. The soil type of the field site was Glendale sandy clay loam (Typic Torrifluvent, pH 8.0). Prior to planting N-P-K fertilizer (11-52-0) was applied to the field at the rate of 224 kg ha^{-1} and cow manure at a rate of 45 Mg ha^{-1}. The field was irrigated every 14 to 18 days depending on precipitation received and evaporation rates during the growing season. Fertilization and irrigation management of the experimental field was similar to that practiced by farmers in southern New Mexico. Five harvests were taken in each of two years (2003 and 2004) at approximately 30 day intervals. The first harvest was collected in early May and the last harvest in mid October. Data of only three harvests from each year (i.e. first, second, and fourth) were used to estimate the effects of general combining ability (GCA), specific combining ability (SCA), parental *per se* performance, and hybrid performance as described in Al Lawati et al. (2010).

AFLP Analysis

AFLP profiles using genomic DNA from bulked leaf tissue of 30 individuals for each population were utilized. DNA preparation, data scoring, and analysis of genetic distance (GD) were performed as described in Bhendari et al. (2011). However, since M. falcata was not used as a parent in the diallel hybrid field study, the GD coefficients used in this study were recomputed without the Medicago sativa subspecies falcata fragment data. The new GD estimates of the nine parents were used to evaluate their ability to predict hybrid yield using linear regression models (Table 1).

Fall Regrowth and Winter-Hardiness

Fall regrowth (FR) was taken by measuring plant height (cm) in each of the field plots representing the nine alfalfa parents. The measurements were taken on 31 October 2003. Three measurements were taken from the middle row of each plot for each replication. Winter-hardiness was estimated by averaging the values of frost damage and winter injury of the parents as provided by the GRIN database (www.ars-grin.gov/npgs).

Statistical Analysis

Simple and multiple linear regression models were used to determine the influence of the AFLP marker diversity estimates, general combining ability (GCA), variety effects (VE), fall regrowth (FR), winter-hardiness (WH), and chloroplast genetic distance (Chlor. GD) on high-parent heterosis, mid-parent heterosis and hybrid performance for the forage-yield trait. A simple linear regression model describing the relationship between hybrid yield performance, as the dependent variable, and GCA, VE, FR, WH, or GD as the independent variable was constructed according to Neter et al. (1996):

$$Y_{jj'} = \beta_0 + \beta_1 X_{jj'} + \varepsilon_{jj'} \;,$$

where

$Y_{jj'}$: is the value of hybrid performance from the cross between parents j and j'.

β_0: is the value of the y-intercept parameter.

β_1: is the regression coefficient associated with the contribution of GCA, VE, FR, WH, or GD, hybrid performance.

$X_{jj'}$: is the constant-value of GD between parents j and j' or mean parental GCA effect or mean VE, FR, or WH effect.

$\varepsilon_{jj'}$: is the random error, which is assumed to be independently and identically and normally distributed with mean zero and constant variance σ^2_e.

A multiple linear regression model describing the relationship between hybrid performance, as the dependent variable, and GCA+GD, VE+GD, FR+VE, or FR+GD as the dependent variables, was a first-order model for two factors (Segovia-Lerma, 2000). As an example, the model including GCA and GD as independent variables is described:

$$Y_{jj'} = \beta_0 + \beta_1[1/2(gca_j + gca_{j'})] + \beta_2(gd_{jj'}) + \varepsilon_{jj'} \;,$$

where

$Y_{jj'}$: is the hybrid performance from the cross between parents j and j'.

β_0: is the value of the y-intercept parameter.

β_1: is the regression coefficient associated with the contribution of GCA hybrid performance.

gca_j: is the general combining ability of parent j.

$gca_{j'}$: is the general combining ability of parent j'.

β_2: is the coefficient of regression associated with the contribution of GD to hybrid performance.

$gd_{jj'}$: is the pairwise genetic diversity estimate between parent j and j'.

$\varepsilon_{jj'}$: is the random error, which assumed to be independently and identically normally distributed with mean zero and variance σ^2_e.

For the other two factor models, gca_j and $gca_{j'}$ were replaced by the variety effect of the parents ve_j and $ve_{j'}$, fall regrowth of the parents fr_j and $fr_{j'}$, or winter-hardiness of the parents wh_j and $wh_{j'}$.

A first-order model for three factors (FR+VE+GD) was also utilized for predicting hybrid performance. The regression model (Neter et al., 1996) is:

$$Y_{jj'} = \beta_0 + \beta_1(fr_{jj'}) + \beta_2(ve_{jj'}) + \beta_3(gd_{jj'}) + \varepsilon_{jj'} ,$$

where

$Y_{jj'}$: is the hybrid performance from cross $j\,j'$.

β_0: is the value of the y-intercept parameter.

β_1: is the coefficient of regression associated with the contribution of FR to hybrid performance.

β_2: is the coefficient of regression associated with the contribution of VE to hybrid performance.

β_3: is the coefficient of regression associated with the contribution of GD to hybrid performance.

$\varepsilon_{jj'}$: is the random error, which assumed to be independently and identically distributed with mean zero and variance σ^2_e.

A similar model for predicting hybrid performance, which included WH instead of GD (i.e. FR+VE+WH), was also evaluated.

Genetic diversity based on AFLP analysis (Bhandari et al., 2011) revealed that the nuclear genetic distances among most of these populations were relatively low. Combining nuclear genetic distance with chloroplast genetic distance enhanced the ability to distinguish between different/similar germplasms. Significant variety and general/specific combining ability effects were also detected among these nine parent populations based on diallel analysis (Al Lawati et al., 2010). Therefore, in this study we evaluated the usefulness of applying the above parameters, in addition to fall regrowth and winter-hardiness measurements of the parents, to predict forage yield among the 36-diallel hybrids.

Prediction of Forage Yield

The regression models for prediction of hybrid yield based on general combining ability effects (GCA) and AFLP genetic distance (GD) are presented in Table 2. All the models were significant. The combined model of GCA and GD effects was highly significant, and its coefficient of determination was 0.62. In this model, however, only the regression coefficient of GCA was significant and GD was not significant. When the prediction models were based on individual parameters, i.e. GCA alone or GD alone, both models were significant with coefficients of correlation of 0.79 and 0.49, respectively.

Table 2. Regression coefficients and their standard error (in parenthesis) of hybrid yield prediction models based on general combining ability effects (GCA) and AFLP marker based genetic distance estimates (GD)

Model	Intercept	Regression coefficient		Overall model significance	r[†]	r2[‡]
		GCA	GD			
(GCA+GD)	14.658	2.027*** (0.353)	-0.116ns (0.967)	< 0.0001	0.79	0.62
GCA	14.565	2.000*** (0.106)		< 0.0001	0.79	0.62
GD	11.812		3.431** (1.074)	0.0022	0.49	0.24

[†] Correlation coefficient between actual and predicted values (n=36).
[‡]Coefficient of determination.
ns, **, *** non-significant, regression coefficient significant at the 0.01 and 0.001 probability levels, respectively.

Regression models based on GCA provided the best prediction of hybrid performance. GCA reflects the presence of genetic factors in the parent populations that influence hybrid yield in an additive manner. Thus, additive gene action appears to provide the major contribution towards maximizing hybrid yield in this study. The importance of accumulating of alleles with additive effects in improving forage yield of alfalfa was reported in several studies for both population *per se* performance (Pfeiffer and Bingham, 1983), and for hybrid performance (Hill 1975; Dunbier and Bingham, 1975; Woodfield and Bingham, 1995; Segovia-Lerma et al., 2004). The importance of additive effects influencing other quantitative traits in alfalfa has also been detected (Campbell et al., 1993; Bolanos-Aguilar et al., 2001; Riday and Brummer, 2002; Ray et al., 2004). These results, with the current investigation findings, indicate that hybrid forage yield of alfalfa is mainly influenced by additive types of gene action.

The significance of GD in the single-factor model, but its non-significance in the combined model likely resulted from co-linearity between the GCA effect and GD. The correlation coefficient between the calculated mid-parent GCA value between the two parents of each hybrid, and the corresponding pairwise GD estimate was r=0.64, (p<0.01, n=36). This suggests that superior hybrids might have resulted from the combining of different sets of desirable alleles that influenced yield in an additive manner as discussed later in this study. The moderate association (r=0.49) between GD and hybrid yield was probably too small to be of predictive value, as mentioned by Bernardo (1992). Similar results were reported by Segovia-Lerma (2000) who evaluated the ability to predict alfalfa hybrid performance using multiple regression models containing GCA and GD.

Two conditions are required for predicting hybrid performance based on molecular marker heterozygousity. At least 30-50% of the QTL must be linked to molecular markers, and not more than 20-30% of the molecular markers should be randomly distributed or unlinked to QTL (Bernardo, 1992). Therefore, several reasons for the relatively low association between molecular markers and alfalfa hybrid performance could include: the random nature of the AFLP molecular markers, low heritability of the quantitative traits of alfalfa, absence of a high level of gametic disequilibrium between QTL and the molecular markers, and differences in linkage disequilibrium among markers and QTL between different populations. In other crops, the associations between genetic distance and hybrid yield have varied. Low associations ($r<0.35$) were reported for maize (Melchinger et al., 1990), rice (Zhang et al., 1994; Kwon et al., 2002), and wheat (Dreisigacker et al., 2005). Moderate associations ($0.35<r<0.69$) were also reported for maize (Melchinger et al., 1992), oilseed rape (Diers et al., 1996), and rice (Zhang et al., 1995). High associations ($r>0.69$) were found for sweet corn (Tracy et al., 2000), for field corn (Smith et al., 1990) and for rice (Saghai Maroof et al., 1997). These variations may have resulted from different plant genetic materials that were used in the above experiments.

As previously mentioned, associations between molecular marker diversity and quantitative traits may be due to the low to moderate heritability of most quantitative traits (Bernardo 1992). For example, Riday and Brummer (2005) found the heritability of biomass yield was 0.30. Brummer et al. (2000) found that the heritability of fall regrowth on a progeny mean basis was 0.29, and that for winter injury was 0.39. Neutral detergent-soluble fiber possessed a moderate heritability of 0.41 (Fonseca et al., 1999). Ray et al., (1999) found that carbon isotype discrimination, ash concentration, and forage yield were moderately heritable (h^2 =0.40-0.56). These data confirm that many quantitative traits in alfalfa possess low to moderate heritability, which could explain why we detected only a moderate association between molecular markers and hybrid yield in the current study.

Melchinger et al. (1990) suggested the importance of a high level of gametic disequilibrium between QTL and DNA markers for increasing the significance of association between them. In addition, Price et al. (1984) demonstrated that gametic (linkage) disequilibrium occurs more often in predominantly self-pollinating than in outcrossing populations. This is because the phenotype and allelic structure for inbred parents are essentially fixed. If such parents have contrasting phenotypes and allelic composition, this maximizes opportunities to detect both phenotypic and genetic differences and potential associations in their hybrid progeny. In the current study, the parental alfalfa populations represented multi-parent synthetics. Hence, the QTL for forage yield performance were not fixed as in inbred lines. Thus, the quantitative traits of the allogamous parents and their progeny are less precisely defined. Furthermore, genetic characterization of heterozygous allogamous parents is more difficult, particularly with dominant markers such as AFLPs, given the inability to distinguish between heterozygous/homozygous allelic states. Consequently, the ability to detect associations between genetic markers and quantitative traits is more difficult.

Charcosset and Essioux (1994) demonstrated that different heterotic groups might have different QTL that are responsible for hybrid performance. They recommended making crosses between related lines, or between no more than two heterotic groups, for improving the correlation between GD and hybrids performance. This conclusion was also reached by Betrán et al. (2003) for tropical maize. In our current study, the alfalfa populations used were from different dormancy classes, which appear to represent multiple heterotic groups. This may make it more difficult to detect associations between hybrid yield and marker diversity. Studies have shown that dormant and nondormant populations differ for many physiological traits. For instance, Cole and Dobrenz (1970) found that stomata density of nondormant cultivars was 20 to 55% less than that of a dormant-type cultivar. Ray et al. (1998) found that the dormant germplasms of *M. sativa* subspecies *varia* and Ladak had significantly lower carbon isotope discrimination values compared to the nondormant populations of African, Chilean, Indian, and Peruvian. Kalengamaliro et al. (2003) found that callus tissue cultured from nondormant alfalfa had more rapid sugar uptake and higher cell respiration rates when compared with dormant alfalfa cultivars. Thus, when the genetic mechanisms controlling these and other processes are integrated during hybrid development, markers associated with genes that contribute to hybrid performance are less likely to differ between hybrids derived from two nondormant parents, as compared to a hybrid derived from nondormant and dormant parents. Since forage yield represents the integrated response of many physiological and biochemical processes, it is likely that these different alfalfa heterotic groups have different QTL influencing forage yield performance, and consequently, have different molecular makers associated with them.

Prediction Models Based on *Per se* Performance of the Populations

Table 3 presents prediction models for hybrid yield, based on *per se* parental performance (i.e. variety effects, VE) and GD. The dependent variables of hybrid yield were significantly ($p<0.01$) predicted by the combined models of VE and GD, and the VE effect alone. The regression coefficients of GD were not significant ($p>0.05$) in the combined models.

Lonnquist and Lindsey (1964) suggested that parental performance *per se* could be utilized instead of GCA to predict hybrid performance in maize. They reasoned that *per se* performance has two main advantages over GCA. First, it saves resources. In the GCA approach, two years are needed to estimate GCA, where the first year is used for making crosses among parental populations, and the second year for field evaluation. However, in the *per se* performance approach only one year is needed for evaluating the parental population performance. In addition, the *per se* performance approach does not encounter the potential masking effect of testers that are necessary for estimating GCA (Lonnquist and Lindsey, 1964). Furthermore, VE can be used to predict the performance of the hybrids when the correlation between GCA and VE is high, as suggested by Segovia-Lerma (2000). In the current study, the correlation between GCA and VE was high ($r=0.88$; $p<0.01$; $n=9$).

Regarding the prediction of hybrid performance by variety effect, the power of the VE+GD combined model and the simple VE model to predict hybrid yield is considered as moderate to high. Our results were similar to those obtained in several studies including that in spring bread wheat (Dreisigacker et al., 2005), winter triticale (Oettler et al. 2005), tropical maize under low nitrogen fertility (Zaidi et al., 2003), and alfalfa (Segovia-Lerma, 2000). On the other hand, some studies in horticultural crops reported no significant association between hybrid yield performance and the mid-parent value for pepino (*Solanum muricatum*) (Rodriguez-Burruezo et al., 2003) and *Musa* germplasm (Tenkouano et al., 1998).

Hybrid Yield and Heterosis Prediction Models Based on Fall Regrowth (FR)

All the models utilizing fall regrowth to predict hybrid yield were significant (Table 4). In particular, prediction models for hybrid yield that contained both FR and VE as predictor variables had relatively high correlation coefficients (or medium range of determination coefficient) approaching the values obtained between GCA and hybrid yield. These results, however, were similar to those obtained for VE effects alone in predicting hybrid yield (Table 3.). Furthermore, the regression coefficients of FR in the combined models with VE were not significant. Initially, this might suggest that FR was not a useful predictor of hybrid yield.

Table 3. Regression coefficients and their standard error (in parenthesis) of hybrid yield prediction models based on variety effects (VE) and AFLP marker distance estimates (GD)

Model	Intercept	Regression coefficient		Overall model significance	$r^{\dagger}$	$r2^{\ddagger}$
		VE	GD			
Hybrid Yield						
(VE+GD)	13.421	0.464** (0.109)	1.426ns (0.968)	<0.0001	0.72	0.51
VE	14.565	0.542** (0.100)		<0.0001	0.69	0.48

† Correlation coefficient between actual and predicted values (n=36).

‡ Coefficient of determination.

ns, *, **, *** non-significant, regression coefficient significant at the 0.05, 0.01 and 0.001 probability levels, respectively.

Table 4. Regression coefficients and their standard error (in parenthesis) of hybrid yield prediction models based on fall regrowth (FR), variety effects (VE) and AFLP marker distance estimates (GD)

Model	Intercept	Regression coefficient			Overall model significance	$r^{\dagger}$	$r^{2\ddagger}$
		FR	GD	VE			
Hybrid yield							
(FR+GD+VE)	12.924	0.012^{ns} (0.029)	1.559^{ns} (1.031)	0.414^{**} (0.163)	<0.0001	0.72	0.51
(FR+GD)	10.025	0.067^{***} (0.023)	2.980^{***} (0.936)		0.0001	0.65	0.42
(FR+VE)	14.615	-0.001^{ns} (0.028)		0.547^{***} (0.139)	<0.0001	0.69	0.48
FR	12.079	0.077^{***} (0.024)			0.0026	0.49	0.24

† Correlation coefficient between actual and predicted values (n=36).

‡ Coefficient of determination.

ns, *, **, non-significant, regression coefficient significant at the 0.05, and 0.01 probability levels, respectively.

However, the simple model based on FR alone was able to predict hybrid performance ($r=0.49$, $p<0.01$). Subsequent analysis of FR and VE data indicated that they were collinear ($r=0.71$, $p<0.05$, $n=9$), which reflects that parental populations with less fall regrowth tend to yield less, (e.g., P7 and P8, see table 1).

We also observed in the prediction model based on FR and GD, that the coefficients of regression for both variables were significant in their contribution towards estimating hybrid yield ($r^2=0.42$, $p<0.01$).

The regression coefficients of these two variables were positive (Table 4), indicating that hybrid yield tended to increase when the fall regrowth of the parents and genetic distance between the parents increased.

Prediction Ability of the Models to Identify Superior Alfalfa Hybrids

Table 5 displays the ability of nine prediction models to identify the nineteen alfalfa hybrids that yielded statistically higher than, or similar to, the best check (CW68115). Most of the nine prediction models were able to identify at least two-thirds of the superior hybrids. GCA alone was able to identify 15 of the 19 (i.e. 79%) superior alfalfa hybrids. The lowest single predictor variable was AFLP genetic distance, which identified 63% of the superior hybrids. VE alone and FR alone were able to predict 68% of the superior hybrids. These results are similar in their performance to the findings of Segovia-Lerma (2000) where GCA and VE were able to predict 80% and 60%, respectively, of superior alfalfa hybrids.

Five prediction models based on multiple predictor variables are also presented in Table 5. Combining GD with GCA did not improve the ability to identify superior hybrids. This confirms the non-significant GD regression coefficient in the combined model (Table 2). An identical scenario was reported by Segovia-Lerma (2000) where combining GD with GCA did not improve the prediction power as compared to GCA alone. In general, similar observations were detected by adding GD with other variables, for example, combining of GD with VE and GD with FR. Winter-hardiness is a trait that provides information pertaining to freezing injury that causes disruption of the cell membranes, and the ability to survive under extended periods of cold (McKenzie et al., 1988). Winter-hardiness, as estimated by averaging the values of frost damage and winter injury of the parents (as provided by the GRIN database), was able to predict 74% of the best-performing hybrids. This variable was examined because the ability of parents to provide varying degrees of protection during winter months could influence hybrid yield over time. However, winter-hardiness essentially did not improve the prediction abilities when combined with GCA or VE. The differing ability of populations to re-grow in the fall is frequently referred to as fall dormancy. Fall dormancy is triggered by decreasing photoperiod and low temperatures in the fall. Populations that produce little top growth during the fall because of greater fall dormancy are generally more winter-hardy than nondormant populations (McKenzie et al., 1988). Populations with frost damage resistance and low winter injury are also more likely to survive in a growing region such as southern New Mexico, which routinely experiences low night-time temperatures (i.e. below 0 °C) over a period of several months. Interestingly, 89% of the best-performing hybrids could be detected by combining the prediction models of VE and FR (Table 6). From a breeding program perspective, these results may enable breeders to identify parents more efficiently for generating superior hybrids of alfalfa. Using a VE and FR approach could save one year that would normally be allocated for crossing potential parents for developing new populations that are required for estimating GCA. It also allows *[n (n-3)]/2* more parent populations to be evaluated as compared to the GCA approach. Low contribution of GD towards predicting hybrid performance among improved alfalfa parents indicates that resources used to measure GD may be directed elsewhere. This is important because the estimation of GD based on molecular markers involves a high investment at the beginning of the program towards purchasing electrophoresis and DNA sequencer equipments, as well as equipment maintenance costs. Hence, the most economical approach to identify parents for alfalfa hybrid development could involve measuring yield, fall regrowth and perhaps winter-hardiness of populations *per se*.

Table 5. Ability of nine prediction models to identify 19 alfalfa diallel hybrids that performed higher than, or similar to, the best check cultivar

Hybrid†	GCA‡	GD	VE	FR	GCA+GD	VE+GD	FR+GD	FR+VE	FR+VE+GD
25	×	×	×		×	×	×	×	×
45	×	×	×		×	×	×	×	×
67									
56	×	×	×	×	×	×	×	×	×
26	×	×	×	×	×	×	×	×	×
12	×	×	×	×	×	×	×	×	×
34	×	×		×	×	×	×		×
37				×					
35	×	×	×	×	×	×	×	×	×
24	×	×	×	×	×	×	×	×	×
39	×			×	×		×		
27	×		×		×	×		×	×
59	×	×	×		×	×	×	×	×
29	×		×	×	×	×		×	×
69	×		×	×	×			×	×
46	×	×	×	×	×	×	×		
13		×		×			×		
58		×							
49	×		×	×	×	×	×	×	×
Total	15	12	13	13	15	13	13	12	13
Identified (%)	78.9	63.2	68.4	68.4	78.9	68.4	68.4	63.2	68.4

† Hybrid populations designated as 12 indicate the hybrid between P1 (11) and P2 (22) ... etc.

‡ GCA, general combining ability; GD, genetic distance; VE, variety effect; FR, mid-parent value for fall regrowth.

×, indicates which hybrids were identified by the corresponding prediction models.

Table 6. Ability of nine additional prediction models to identify 19 alfalfa diallel hybrids that performed higher than or similar to, the best check cultivar

Hybrid†	WH	GD + WH	GCA+ WH	VE+WH	FR+WH	FR+VE+WH
25	×	×	×	×		×
45			×	×		×
67	×					
56	×	×	×	×	×	×
26	×	×	×	×	×	×
12	×	×	×	×	×	×
34	×	×	×		×	
37	×	×			×	
35	×	×	×	×	×	×
24		×	×	×	×	×
39	×	×	×		×	
27			×	×		×
59	×	×	×	×		×
29	×	×	×	×	×	×
69	×		×	×	×	×
46	×	×	×	×	×	×
13	×	×			×	
58						
49			×	×	×	×
Total	14	13	15	13	13	13
identified (%)	73.8	68.4	78.9	68.4	68.4	68.4

† Hybrid populations designated as 12 indicate the hybrid between P1 (11) and P2 (22) ... etc.

‡ GCA, general combining ability; GD, genetic distance; VE, variety effect; FR, mid-parent value for fall regrowth; WH, winter-hardiness.

×, indicates which hybrids were identified by the corresponding prediction models.

CONCLUSION

The main objective of this study was to determine the potential of using AFLP nuclear genetic distance (GD), general combining ability (GCA), variety effect (VE), fall regrowth (FR), and winter-hardiness (WH) variables to predict the performance of alfalfa hybrids. Each of these variables were used in simple and multiple linear regression models to predict hybrid performance among 36 half-diallel hybrids generated from nine alfalfa germplasms. The results achieved indicated that GCA alone was the best single model to predict hybrid performance, where it was able to identify 79% of superior hybrids. It was followed by WH of which identified 74% of superior hybrids. AFLP genetic distance was able to identify 63% of the nineteen superior hybrids. The implication of these results is that evaluation of VE and FR or WH could provide a viable approach to predict hybrid performance among the parents used in this study. Measuring variety effects and FR is much more efficient in resource utilization than measuring GCA, since they do not require the additional time and resources that are needed to cross parents to generate their F1 progeny. Furthermore, significantly more potential populations could be evaluated with the same resources that would otherwise be needed to evaluate GCA of relatively few populations.

REFERENCES

Ajmone Marsan, P., P. Castiglioni, F. Fusari, M. Kuiper, and M. Motto. 1998. Genetic diversity and its relationship to hybrid performance in maize as revealed by RFLP and AFLP markers. *Theoretical and Applied Genetics* 96:219-227.

Al Lawati, A.H., C.A. Pierce, L.W. Murray, and I.M Ray. (2010). Combining Ability and Heterosis for Forage Yield among Elite Alfalfa Core Collection Accessions with Different Fall Dormancy Response. *Crop Sci.* 50:150-158.

Bauchan, G.R., N.R. Campbell, N.R. O'Neil, and J.J.H. Elgin. 1991. Self-incompatibility in two alfalfa populations. *Crop Science* 30:1205-1210.

Bernardo, R. 1992. Relationship between single-cross performance and molecular marker heterozygosity. *Theoretical and Applied Genetics* 83:628-634.

Betran, F.J., J.M. Ribaut, D. Beck, and D. Gonzalez de Leon. 2003. Genetic diversity, specific combining ability, and heterosis in tropical maize under stress and nonstress environments. *Crop Science* 43:797-806.

Bhandari, H.S., A.H. Al Lawati, and I.M. Ray. 2011. Amplified Fragment Length Polymorphism Marker Diversity among High Yielding Perennial Alfalfa (*Medicago sativa* L.) Core Collection Accessions. *Crop Sciences* 51: 1110-1121.

Bingham, E.T., R.W. Groose, D.R. Woodfield, and K.K. Kidwell. 1994. Complementary gene interactions in alfalfa are greater in autotetraploids than diploids. *Crop Science* 34:823-829.

Bolanos-Aguilar, E.D., C. Huyghe, D. Djukic, B. Julier, and C. Ecalle. 2001. Genetic control of alfalfa seed yield and its components. *Plant breeding,* Zeitschrift fur Pflanzenzuchtung 120:67-72.

Boppenmaier, J, A.E. Melchinger, E. Brunklausjung, H.H. Geiger, and R.G. Herrmann. 1992. Genetic diversity for RFLPS in European maize inbreds. 1. Relation to performance of Flint X Dent crosses for forage traits. *Crop Science* 32:895-902.

Brown, D.E, and E.T. Bingham. 1984. Hybrid alfalfa seed production using a female-sterile pollenizer. *Crop Science* 24:1207-1208.

Brummer, E.C. 1999. Capturing heterosis in forage crop cultivar development. *Crop. Science* 39:943-954.

Brummer, E.C., M.M. Shah, and D. Luth. 2000. Reexamining the relationship between fall dormancy and winter hardiness in alfalfa. *Crop Science* 40:971-977.

Campbell, T.A., Z.L. Xia, P.R. Jackson, and V.C. Baligar. 1993. Diallel analysis of tolerance to aluminum in alfalfa. *Euphytica* 72 (3):157-162.

Cerna, F.J., S.R. Cianzio, A. Rafalski, S. Tingey, and D. Dyer. 1997. Relationship between seed yield heterosis and molecular marker heterozygosity in soybean. *Theoretical and Applied Genetics* 95:460-467.

Charcosset, A., and L. Essioux. 1994. The effect of population-structure on the relationship between heterosis and heterozygosity at marker loci. *Theoretical and Applied Genetics* 89:336-343.

Cheres, M.T., J.F. Miller, J.R. Crane, and S.J. Knapp. 2000. Genetic distance as a predictor of heterosis and hybrid performance within and between heterotic groups in sunflower. *Theoretical and Applied Genetics* 100:889-894.

Childers, W.R, and K.D. Barnes. 1972. Evolution of hybrid alfalfa. *Agricultural Sciences Review* 10:11-18.

Cole, D.F., and A.K. Dobrenz. 1970. Stomate density of alfalfa (*Medicago sativa* L.). *Crop Science* 10:61-63.

Diers, B.W., P.B.E. McVetty, and T.C. Osborn. 1996. Relationship between heterosis and genetic distance based on restriction fragment length polymorphism markers in oilseed rape (Brassica napus L). *Crop Science* 36:79-83.

Dreisigacker, S., A.E. Melchinger, P. Zhang, K. Ammar, C. Flachenecker, D. Hoisington, and M.L. Warburton. 2005. Hybrid performance and heterosis in spring bread wheat, and their relations to SSR-based genetic distances and coefficients of parentage. *Euphytica* 144:51-59.

Dunbier, M.W., and E.T. Bingham. 1975. Maximum heterozygosity in alfalfa: Results using haploid-derived autotetraploids. *Crop Science* 15:527-531.

El-Maghraby, M.A., M.E. Moussa, N.S. Hana, and H.A. Agrama. 2005. Combining ability under drought stress relative to SSR diversity in common wheat. *Euphytica* 141:301-308.

Fonseca, C.E.L., D.R. Viands, J.L. Hansen, and A.N. Pell. 1999. Associations among forage quality traits, vigor, and disease resistance in alfalfa. *Crop Science* 39:1271-1276.

Geleta, L.F., M.T. Labuschagne, and C.D. Viljoen. 2004. Relationship between heterosis and genetic distance based on morphological traits and AFLP markers in pepper. *Plant Breeding* 123:467-473.

Godshalk, E.B., M. Lee, and K.R. Lamkey. 1990. Relationship of restriction-fragment-length-fragment-polymorphisms to single-cross hybrid performance of maize. *Theoretical and Applied Genetics* 80:273-280.

Hallauer, A.R., and J.B. Miranda. 1988. Quantitative genetics in maize breeding Iowa State University Press, Ames, Iowa.

Hill, R.R.J. 1975. Parental Inbreeding and Performance of Alfalfa Single-Crosses. *Crop Science* 15:373-375.

Jordan, D.R., Y. Tao, I.D. Godwin, R.G. Henzell, M. Cooper, and C.L. McIntyre. 2003. Prediction of hybrid performance in grain sorghum using RFLP markers. *Theoretical and Applied Genetics* 106:559-567.

Kalengamaliro, N.E., S.M. Cunningham, and J.J. Volenec. 2003. Growth, sugar accumulation, and dark respiration of suspension cell cultures derived from contrasting alfalfa cultivars. *Plant Cell Tissue and Organ Culture* 72:163-171.

Kidwell, K.K., D.F. Austin, and T.C. Osborn. 1994a. RFLP evaluation of 9 *Medicago* accessions representing the original germplasm sources for North-American alfalfa cultivars. *Crop Science* 34:230-236.

Kidwell, K.K., D.R. Woodfield, E.T. Bingham, and T.C. Osborn. 1994b. Molecular marker diversity and yield of isogenic 2x and 4x single crosses of alfalfa. *Crop Science* 34:784-788.

Kidwell, K.K.u.U.o.W.M.W.I., E.T. Bingham, D.R. Woodfield, and T.C. Osborn. 1994c. Relationships among genetic distance, forage yield and heterozygosity in isogenic diploid and tetraploid alfalfa populations. *Theoretical and Applied Genetics* 89:323-328.

Kwon, S.J., W.G. Ha, H.G. Hwang, S.J. Yang, H.C. Choi, H.P. Moon, and S.N. Ahn. 2002. Relationship between heterosis and genetic divergence in 'Tongil'-type rice. *Plant Breeding* 121:487-492.

Lee, J.K., H.K. Min, J.Y. Park, N.K. Huh, K.J. Park, J.K. Choi, K.C. Park, and N.S. Kim. 2002. Molecular genetic distances and hybrid performances among maize inbreds and their hybrids. *Korean Journal of Genetics* 24:281-290.

Lee, M., E.B. Godshalk, K.R. Lamkey, and W.W. Woodman. 1989. Association of restriction fragment length polymorphisms among maize inbreds with agronomic performance of their crosses. *Crop Science* 29:1067-1071.

Lonnquist, J.H., and M.F. Lindsey. 1973. Topcross versus S_1 line performance in corn (*Zea mays* L.). Crop Science 13:580-584.

Maroof, M.A.S., G.P. Yang, Q.F. Zhang, and K.A. Gravois. 1997. Correlation between molecular marker distance and hybrid performance in US southern long grain rice. *Crop Science* 37:145-150.

McKenzie, J.S., R. Paquin, and S.H. Duke. 1988. Cold and heat tolerance, p. 1084, *In* A. A. Hanson, et al., eds. Alfalfa and alfalfa improvement, Vol. 29. *Crop Science Society of America*, Inc., Madison, Wisconsin.

Mckersie, B.D., T. Senaratna, S.R. Bowley, D.C.W. Brown, J.E. Krochko, J.D. Bewley. 1989. Application of artificial seed technology in the production of hybrid. In-vitro cellular & developmental biology 25:1183-1188.

Melchinger, A.E., and R.K. Gumber. 1998. Overview of heterosis and heterotic groups in agronomic crops, p. 29-44, *In* K. R. Lamkey and J. E. Staub, eds. Concepts and breeding of heterosis in crop plants, Vol. Special publication No. 25. American Society of Agronomy, Inc.; *Crop Science Society of America*, Madison, Wisconsin.

Melchinger, A.E., M. Lee, K.R. Lamkey, A.R. Hallauer, and W.L. Woodman. 1990. Genetic diversity for restriction-fragment-length-polymorphisms and heterosis for 2 diallel sets of maize inbred. *Theoretical and Applied Genetics* 80:488-496.

Melchinger, A.E., J. Boppenmaier, B.S. Dhillon, W.G. Pollmer, and R.G. Herrmann. 1992. Genetic diversity for RFLPS in European maize inbreds. 2. Relation to performance of hybrids within versus between heterotic groups for forage traits. *Theoretical and Applied Genetics* 84:672-681.

Neter, J., M.H. Kutner, C.J. Nachtsheim, and W. Wasserman. 1996. Applied linear regression models. 3RD ed. IRWIN, Chicago.

Oettler, G., S.H. Tams, H.F. Utz, E. Bauer, and A.E. Melchinger. 2005. Prospects for hybrid breeding in winter triticale: I. Heterosis and combining ability for agronomic traits in European elite germplasm. *Crop Science* 45:1476-1482.

Pfeiffer, T.W., and E.T. Bingham. 1983. Improvement of fertility and herbage yield by selection within two-allele populations of tetraploid alfalfa Medicgo sativa, chromosome doubling, heterozygosity. *Crop Science* 23:633-636.

Price, S., K.M. Shumaker, A.L. Kahler, R.W. Allard, and J.E. Hill. 1984. Estimate of population differentiation obtained from enzyme polymorphisms and quantitative characters. *Journal of Heredity* 75:141-142.

Ray, I.M., A. Segovia-Lerma, and L.W. Murray. 2004. Diallel analysis of carbon isotope discrimination and its association with forage yield among nine historically recognized alfalfa germplasms. *Crop Science* 44:1970-1975.

Ray, I.M., M.S. Townsend, C.H. Muncy, and J.A. Henning. 1999. Heritabilities of water-use efficiency traits and correlations with agronomic traits in water-stressed alfalfa. *Crop Science* 39:494-498.

Ray, I.M. M.S. Townsend, and J.A. Henning. 1998. Variation for yield, water-use efficiency, and canopy morphology among nine alfalfa germplasms. Crop Science 38:1386-1390.

Reif, J.C., A.E. Melchinger, X.C. Xia, M.L. Warburton, D.A. Hoisington, S.K. Vasal, G. Srinivasan, M. Bohn, and M. Frisch. 2003. Genetic distance based on simple sequence repeats and heterosis in tropical maize populations. *Crop Science* 43:1275-1282.

Riaz, A., G. Li, Z. Quresh, M.S. Swati, and C.F. Quiros. 2001. Genetic diversity of oilseed Brassica napus inbred lines based on sequence-related amplified polymorphism and its relation to hybrid performance. Plant breeding = Zeitschrift fur Pflanzenzuchtung 120:411-415.

Riday, H., and E.C. Brummer. 2002. Forage yield heterosis in alfalfa. *Crop science* 42:716-723.

Riday, H., and E.C. Brummer. 2005. Heterosis in a broad range of alfalfa germplasm. *Crop science* 45:8-17.

Riday, H., E.C. Brummer, T.A. Campbell, D. Luth, and P.M. Cazcarro. 2003. Comparisons of genetic and morphological distance with heterosis between Medicago sativa subsp sativa and subsp falcata. *Euphytica* 131:37-45.

Rodriguez-Burruezo, A., J. Prohens, and F. Nuez. 2003. Performance of hybrid segregating populations of pepino (Solanum muricatum) and its relation to genetic distance among parents. *Journal of Horticultural Science & Biotechnology* 78:911-918.

Segovia-Lerma, A. 2000. Heterosis, AFLP marker diversity, and their association among nine alfalfa germplasms, Ph.D. dissertation, New Mexico State University, Las Cruces.

Segovia-Lerma, A., L.W. Murray, M.S. Townsend, and I.M. Ray. 2004. Population-based diallel analyses among nine historically recognized alfalfa germplasms. *Theoretical and Applied Genetics* 109:1568-1575.

Skinner, D.Z. 2000. Non random chloroplast DNA hypervariability in *Medicago sativa*. Theoretical and Applied Genetics 101:1242-1249.

Smith, O.S. 1986. Covariance between line per se and testcross performance. *Crop Science* 26:540-543.

Smith, O.S., J.S.C. Smith, S.L. Bowen, R.A. Tenborg, and S.J. Wall. 1990. Similarities among a group of elite maize inbreds as measured by pedigree, F1 grain yield, grain yield, heterosis, and RFLPs. *Theoretical and applied genetics* 80:833-840.

Tenkouano, A., J.H. Crouch, H.K. Crouch, and R. Ortiz. 1998. Genetic diversity, hybrid performance, and combining ability for yield in Musa germplasm. *Euphytica* 102:281-288.

Tracy, W.F., L.E. Talbert, and J.T. Gerdes. 2000. Molecular variation and F1 performance among strains of the sweet corn inbred P39. *Crop science* 40:1763-1768.

Viands, D.R., P. Sun, and D.K. Barnes. 1988. Pollination control: Mechanical and sterility, p. 1084, *In* A. A. Hanson, et al., eds. Alfalfa and Alfalfa Improvement, Vol. 29. Crop Science Society of America, Inc., Madison, Wisconsin.

Wiersma, D.W. 2001. Are hybrids the new yield force in alfalfa? A summery of alfalfa hybrid performance in university variety trials. *Focus on forage* 3:1-4.

Woodfield, D.R., and E.T. Bingham. 1995. Improvement in 2-allele autotetraploid populations of alfalfa explained by accumulation of favorable alleles. *Crop Science* 35:988-994.

Zaidi, P.H., G. Srinivasan, and C. Sanchez. 2003. Morpho-physiological traits associated with variable field performance of different types maize germplasm across multiple environments. *Maydica* 48:207-220.

Zhang, Q.F., Y.J. Gao, M.A.S. Maroof, S.H. Yang, and J.X. Li. 1995. Molecular divergence and hybrid performance in rice. *Molecular Breeding* 1:133-142.

Zhang, Q.F., Y.J. Gao, S.H. Yang, R.A. Ragab, M.A.S. Maroof, and Z.B. Li. 1994. A diallel analysis of heterosis in elite hybrid rice-based on RFLPs and microsatellites. *Theoretical and Applied Genetics* 89:185-192.

Zhang, Q.F., Z.Q. Zhou, G.P. Yang, C.G. Xu, K.D. Liu, and M.A.S. Maroof. 1996. Molecular marker heterozygosity and hybrid performance in indica and japonica rice. *Theoretical and Applied Genetics* 93:1218-1224.

In: Biotechnology and Conservation … (2 Volume Set)
Editors: S. A. Farooq, R. Abed and S. Baqir
ISBN: 978-1-63117-869-6

Chapter 26

SOMATIC EMBRYOGENESIS FROM OFFSHOOT OF DIFFERENT CULTIVARS OF DATE PALM (*PHOENIX DACTYLIFERA* L.) OF THREE REGIONS OF SOUTHERN ALGERIA

F. Abed, D. Yatta, B. Amara, M. S. Yakhou and F. Benhafsi
Division Biotechnologies et Amélioration des plantes,
INRAA, CRP Mehdi Boualem, Baraki Alger, Algérie

ABSTRACT

Offshoot explants of different cultivars of three southern regions are introduced in vitro. Induction of callogenesis is studied on modified MS (1962) supplemented with 3 growth regulators combinations, 2,4D (25 and 100mg/l), Picloram at (12,5mg/l) and 2IP (1 and 3mg/l). Various calli types are obtained but only the nodular and granular friable calli are embryogenous. The callogenesis is genotype dependant and required at least 3 months. A phase in stirred liquid medium is used for individualizing the embryos. The filtration at 400 μm allows mature embryos can germinate and growth in full on plantlets. The acclimatization of vitroplants is obtained when all the conditions are met; Humidity at 90%, Température at (26-27°C), and the plantlets were vigorous.

Keywords: Date palm, cultivars, somatic embryogenesis, callogenesis, synchronization, acclimatizaton, genetic

INTRODUCTION

Date palm, Phoenix dactylifera L. is a monocotyledon species, dioecious, from Arecacae family and the oldest angiosperm in the world. It is one of the most ecological, economically important for many centuries and is a rich source of nutrition, contributing to food security. It is well-known for its ability to withstand extremes of temperature and can tolerate very high summer temperatures as well very cold winters. About 90,000,000 date palm are grown over a million hectares worldwide and over 10,000,000 of population are concerned. In Algeria the 17,000,000 of date palm are grown over 160,000 ha, and the genetic diversity is exceptionally because the higher diversity of regions. The species can be propagated through seeds, takes 4–7 years after planting to bear fruit (Zaid 2003). However, the seedlings are heterogeneous and take 7 years to fruit. Furthermore, 50% of the seedlings may turn out to be males. Also, date fruits from seedling plants are often smaller and with poorer quality. The common means of propagation is by transplanting offshoots when they are 3–5 years old. Plants grown from offshoots will fruit 2–3 years earlier than seedling plants. However, offshoots develop slowly and the numbers produced are limited no more than 15 or 20 offshoots per tree. The main danger for the species is the Bayoud disease caused by *Fusarium oxysporum* f.sp. albedinis fungus. It has destroyed more than ten million trees of the Moroccan palm plantations and more than three millions trees in Algeria (Djerbi, 1988). The second disease is the brittle leaf disease, has infected more than 36,000 trees in Tunisia (Triki et al., 2003) because the causal agent is not identified, and the red palm weevil, *Rhynchophorus ferrugineus* recently identified in Morocco are a serious threats for Algerian date palm cultivation. In view of this critical situation, tissue culture multiplication technique has an important interest. This technique allows a secure exchange of healthy material.

Protocols for *in vitro* somatic embryogenesis in date palm have been developed by several researchers (Tisserat, 1979; 1984; Sharma et al., 1986, Daguin et letouze, 1988, Bouguedoura, 1991; Scarnec 1993, El Hadrami et al. 1995, saka et al., 1997). The explants are incubated in complete darkness for 3–10 months, in culture rooms for the production of embryogenic callus. The phenolics interference is overcome by the use of activated charcoal, polyvinylpyrrolidone, ascorbic acid and citric acid (Zaid and Tisserat 1983). Frequent subculturing is also adopted to overcome a browning problem. The basal medium is supplemented with amino acids (adenine and glutamine), vitamins (inositol, pyridoxine, nicotinic acid and thiamine), ammonium citrate, potassium phosphate, sucrose, etc. Various auxins (2,4-dichlorophenoxy acetic acid, indole acetic acid, picloram,) and cytokinins (2-iP) are also being used. The 2,4D occurs a higher rate of callogenesis but is known as an auxin that shown undesirable traits and is hampered by the loss of cell totipotentcy in relatively short time (Eapen et George, 1990, Fitch and Moore 1990). Picloram is used with success to maintain callus proliferation and totipotency in sugar can and date palm (Fitch et Moore, 1990, Omar et Novak, 1990). Murashige and Skoog (1962) medium is the most widely used for date palm tissue culture.

Suspension culture of date palm friable callus for rapid somatic embryogenesis has been established (Sharma et al., 1986, Saka 2001, Fki et al., 2003). Hundreds of somatic embryos can be developed from suspension culture within a short time. Regeneration of somatic embryos occurs when the callus is subcultured using suitable media, often hormone-free (Daguin et letouze, 1988).

Twenty five genotypes were used in this experiment: Tinaceur, Tegaza, Takerboucht el kahla, Takerboucht es safra, Tazarzait, tilemsou, Ahartan, Cheikh M'hamed, Aghamou, Tadelt, outkbala, deglet nour (middle), Tazrzait, Tazoughart, Tantboucht, Tanaslit, Amari, Tinissin, Adjina, Halwa, Ghers, Degla beida, Dimollo, tachrwit, Deglet Nour (East). The offshoot after dissection are rinsed with water and were surface sterilized by immersion of (Benlate, 6g in 100ml) followed by three rinses in water. The offshoot were dipped on 12% of sodium hypochorit and 100mg/l solution for 20 minutes on orbital shaker, rinsed by three rinses in sterile distilled water under laminar flow.

Callus Initiation and Maintenance

Eight to twelve (8 to 12) explants from offshoot was placed on modified MS medium (Reynolds et Murashige, 1978) to induce embryogenic calli. The MS (Murashige et Skoog 1962), modified is supplemented (in mg/L): 1 thiamine, 100 myo-inositol, 170 NaH2PO4, 100 KH2PO4, 45000 sucrose, 2000 phytagel, 200 citrate d'ammonium, 100 glutamine, 200 à 3000 activated charcoal. The media is supplemented with plant regulators at various concentrations; 2,4D at (25 and 100mg/l), and Picloram at (12,5mg/l), 2ip at (1 to 3mg/l). The pH is adjusted to 5,8 before sterilization of medium. 5 replications are done for each treatment. The cultures initiating embryogenic callus were cultured in tubes containing 20ml of medium at 27 + – 2°C in dark and subcultured every 5 weeks.

After 12 to 36 weeks of culture, explants producing compact or friable callus were transferred in a 250 ml flask in the same basal medium with lower growth regulators or free hormones to sustain further growth.

Initiation of Cells Suspension Culture

Suspension cultures were derived by transfer of small portions of calli, 5 to 10g were chopped and inoculated on liquid medium in 250ml Flask. The cultures were grown on the proliferation or germination media, in photoperiod of 16 h at 100rpm to favour suspension culture growth. Cell suspension and cell aggregate were subcultured every 2weeks and were filtred with 400 mm mesh. At each transfer, biomass is weighed and cell aggregate and cell suspension removed until a suspension is obtained. The proliferation medium is containing 5mg/l of picloram and used for cell suspension<400 mm mesh, and the germination medium is devoid of growth regulator.

The histological techniques for friable callus used is the on of Bagniol et al. (1992).

Regeneration

To control the totipotency of cells, nodules transferred on liquid media for 6 cycles are plated on Petri dishes containing the same germinated media solidified with 2g/l of gelrite, under a 16h of photoperiod. The germinated embryos were noted.

Acclimatization

For a better control of acclimatization, we have studied several factors, including the physiology of vitroplants and the physicochemical conditions of acclimatization.

The primary cellular proliferations are obtained 8 weeks for all the genotype. These cellular proliferations generally appear on either upper or lower surfaces of primordial leaves; of apical buds.

Table 1. Effect of growth regulators on embryonic calli of genotype of East south region (Touggourt)

Media	Tanaslit		Amari		Tinissin		Adjina		Halwa		Ghers		Degla Beida		Dimollo		Tacharwit		Deglet Nour	
	T	%	T	%	T	%	T	%	T	%	T	%	T	%	T	%	T	%	T	%
P12.5	3	10	6	6	7	37	7	13	7	55	6	37	9	18	6	2	6	30	4	62
M25	6	4	7	6	9	20	9	9	7	35	8	9	10	19	8	23	6	20	7	30
M100	4	58	6	41	6	24	6	30	6	55	6	2	9	6	6	48	6	10	6	37

T: time of response to induce calli, %: rate of calli obtained; P12.5: medium containing 12,5 mg/l of Picloram, M25: medium containing 25 mg/l of 2.4D; M100: medium containing 100 mg/l of 2.4D.

Table 2. Effect of growth regulators on embryonic calli of genotypes of t southern west region (Adrar)

Media	Tinacer		TagazaIT		Takerboucht kahla		Takerbouch t safra		Tazerzait		Tilemsou		Ahartan		Cheikh M'hamed		Aghamou	
	T	%	T	%	T	%	T	%	T	%	T	%	T	%	T	%	T	%
P12.5	3	58	4	50	2	85	7	9	7	5	6	40	9	44	6	37	6	48
M25	6	31	7	12	6	8	11	6	7	5	10	14	10	12	8	25	6	39
M100	4	77	5	69	4	58	6	45	6	41	8	49	9	32	6	45	6	67

T: time of response to induce calli; %: rate of calli obtained, P12.5: medium containing 12,5 mg/l of Picloram, M25: medium containing 25 mg/l of 2.4D; M100: medium containing 100 mg/l of 2.4D.

Table 3. Effect of growth regulator on embryonic calli of genotype of middle south region (Ghardaia)

Media	Tadelt		Outakbala		Deglet Nour		Tazerzait		Tazoughart		Tantboucht	
	T	%	T	%	T	%	T	%	T	%	T	%
P12.5	3	17	6	49	5	45	7	24	7	3	8	12
M25	6	9	10	8	6	33	9	1	9	3	10	6
M100	4	29	7	54	4	57	6	42	6	42	8	12

T: time of response to induce calli, %: rate of calli obtained, P12.5: medium containing 12,5 mg/l of Picloram, M25: medium containing 25 mg/l of 2.4D; M100: medium containing 100 mg/l of 2.4D.

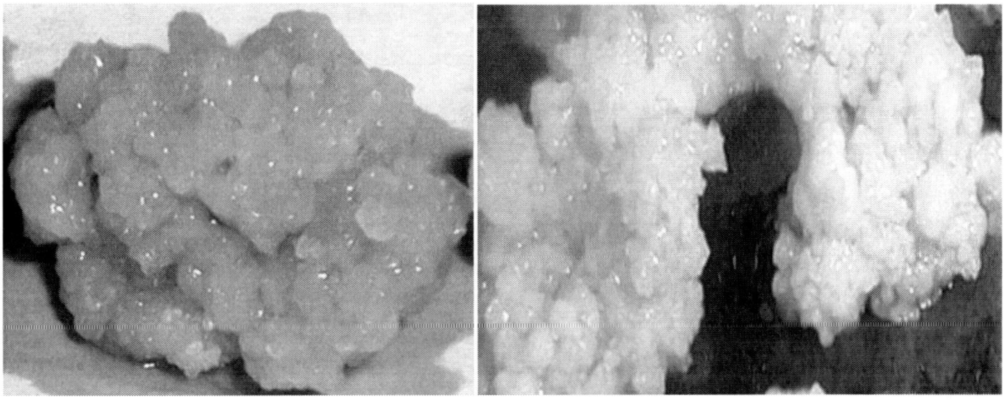

a b

Figure 1. Date palm embryogenic callus: (a) callus observed on 2,4D medium, (b) callus observed on P12,5 of Picloram.

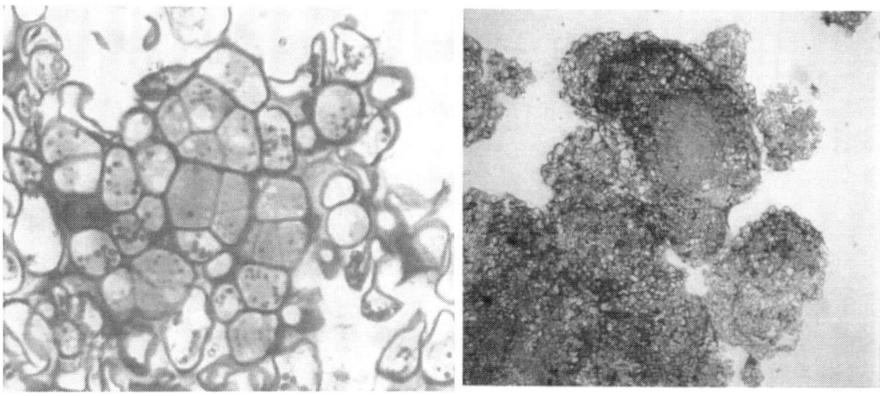

a (25 μm) b (100 μm)

Figure 2. Histological section of embryogenic callus on media containing Picloram (a) nodular friable callus: Cells in division, uniformly colored with large vacuole surrounded of pre- vascular cells (x40); (b) nodular callus on maturation step: organisation of cluster of smooth spherical cells indicated formation of proembryons (x40).

Independent of the genotype, the time response of explants producing calli is different. Our experiment showed that Takerboucht el Kahda produces the calli only on two months.

The slow response has been obtained at 10 to 11 months, for cvs Tilemsou, Ahartan, Outkbala, Degla beida, and Takerboucht es safra, on M25 medium. In all the genotypes and media used, the formation of callus requires 4 to 8 months for the three media.

The data shown that is the very slow culturing step has obviously a negative impact on the rate of embryonic calli. The genotype has a strong influence in the callusing frequency were observed between genotypes. Takerboucht el kahla (85%) on the medium containing 12,5 of picloram, and (77%) for Tinacer on the medium containing 100mg/l of 2,4D.

Indeed, using the same concentration of 2,4-D (100 mg/l), the frequencies of callus induction obtained from offshoot for all cultivars of the different regions ranged from 2, to 77%. The frequencies of callus obtained from offshoot of the cultivars on medium containing 25mg/l of 2,4D showed that the lower rate is at 1% and the higher rate at 39% respectively for Tazerzait and Aghamou. For media containing 12,5 of Picloram, the cultivars testing ranged from 2 to 85% respectively for Dimollo and Takerboucht el Kahla. Those of cultivars reacted invariably to all the growth regulators.

Our study has shown that Piclorm is a very higher stimulus for embryogenic callus (Figure 1b, Figure 2 (a et b)). This auxin is able to produce embryogenic callus as well as 2.4D for both concentrations used.

The histological studies showed meristematic zones on differentiation (Figure 2b). The cells present an important accumulation of protein, polysaccharides and dense nucleus. However, the concentrations of the growth regulators tested showed that the optimal embryogenic callus induction medium with regard to auxin type and concentration were defined for 25 cultivars.

Cell Suspension

Attempt is made to adapt the regeneration capability for suspension culture of one Tegaza cultivar. Friable calli is chopped and used to initiate the cell suspension. From 1 week in a stirred medium, embryogenic masses started to detach from mother callus tissue. The diameter of the particles varied from 1 to 5 mm. For each initial callus ranging from 5 to 10 g/l, a final weight is obtained. Our study, showed that the both density of suspension tested allowed an increase in cell weight fresh (Figure 3). We observed that inoculating with 10g/l increased 3 times after 5 weeks and decreased after. The best results are obtained with 5g inoculation density. The weight increased 2 times after 2 weeks and the weight increased at the final at 80g/l on medium devoid growth regulators. The cluster cells < 400mm mesh filtration was inoculate on fresh medium containing 5 mg/l for maintaining and proliferation of the suspension. At 8 to 10 weeks various stages of embryo development, namely: spherical, elongated (after the cotyledon appears) and cotyledonary leaf formation were observed in a given flask (Figure 4).

Regeneration

At the end of 6 to 11 month culture on solid media, only the calli containing smooth globular structures were recorded as embryonic callus. The calli were transferred on germination media devoid of growth regulators.

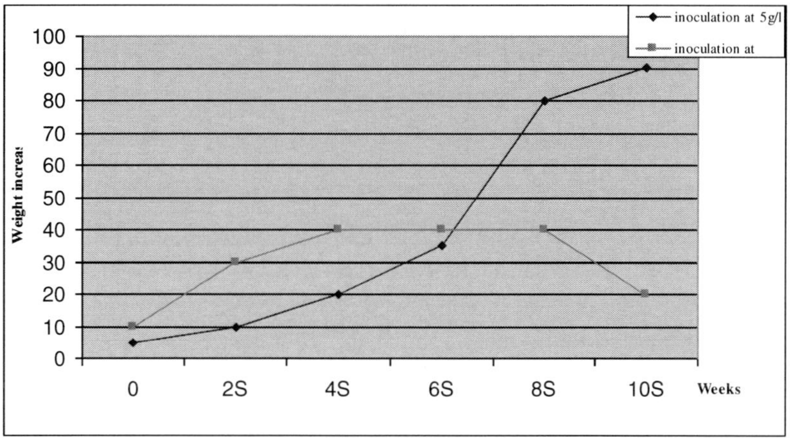

Figure 3. Effect of callus density in weight increase after 6 Weeks of cv Tegaza.

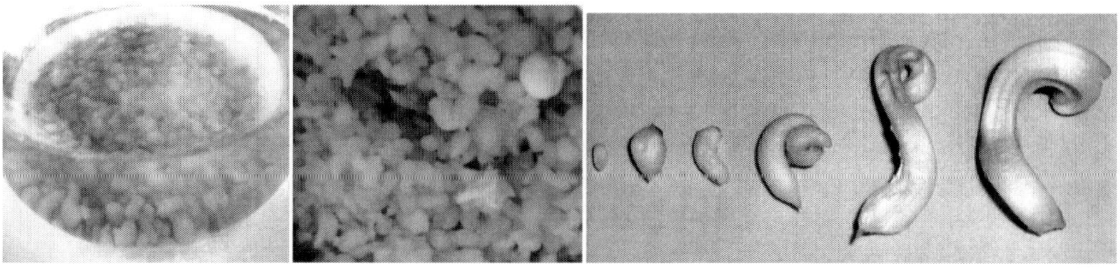

(a) (b) (c)

Figure 4. Mass production of date palm proembryos through liquid suspension culture (a and b) and (c) different stage of somatic embryos.

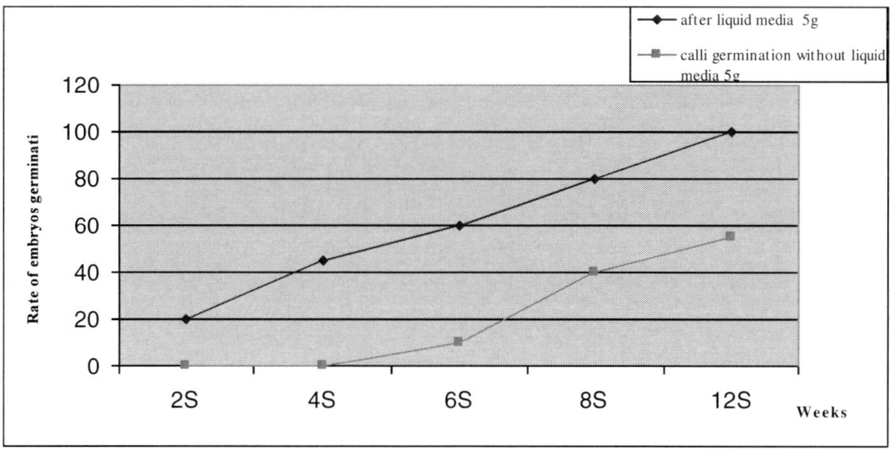

Figure 5. Effect of stirred medium on germination of date palm embryos.

The germination occurred after 6 weeks with 10% only. The embryos transferred on solidified germination medium from the stirred medium showed a higher germination only after 4 weeks with 40%. The germination rate in the final time 12 weeks is 100%. When the calli were transferred directly on solidified germination medium the rate obtained at 12 weeks is 58%. This long and low germination could be negative for mass propagation processes.

Acclimatization

The acclimatization phase is the most important stage in the protocol of date palm micropropagation, when not properly controlled; the whole process will be inefficient. Plants transplanted into the greenhouse should progressively resist to higher luminosity; lower relative humidity; fluctuation of temperature and biotic stresses. For a better control of acclimatization, we have studied several factors, including the physiology of vitroplants and the physicochemical conditions of acclimatization.

Our study showed that rate of survival of plantlets is obtained when the humidity is (95 to 100) and temperature about 26 and 27°C. However, successful plant acclimatization of date palm should be started at the physiological and morphological of plantlets. Those to be transferred to the acclimatization stage should have certain important characteristics that enable them to succeed in the greenhouse. In fact, plantlets must be at least 12–15 cm in length with a well-formed and closed crown, two or three fully opened leaves and more than three roots in lengh 4cm. The substrate is also important, the best results obtained is the equal part of peat and sand (1:1). The survival percentage was 70% after 3months.

Optimal results were obtained with Picloram at 12.5 mg/l and 100mg/l of 2.4D. All the concentrations of various growth regulators tested allowed callogenesis at various rate and genotype effect have been demonstrated. The histological examination has confirmed that the friable calli obtained is embryogenic. We showed also that embryonic cell suspension increase the rate of regenerating plantlets. However, Callogenesis is a very slow process which may required for some genotype 11 months. Several papers were reported that in date palm 4–8 months were required to obtain calli (Saka et al., 1997, Fki, 2009). This seems to be a generic characteristic of *in vitro* cultivated Arecaceae, as it was also described for the coconut palm (Verdeil and Buffard-Morel 1995) and the African oil palm *Elaeis guineensis acq.* (Duval et al., 1995). Genotypic effects have been previously reported in various species (Lazzeri et al., 1987, Saka et al., 1997, Fki, 2005). The both growth regulators used picloram at 12.5 mg/l and 100 mg/l of 2,4-D auxin produced callogenesis but the rate is different for all the genotype used. Our study makes in evidence that Picloram produce embryogenic callus with 12.5 mg/l for several cultivars. This finding is with disagreement with Fki, (2005) who reported that Picloram (0.2–0.5 mg/L) induced callogenesis although it generated non-embryogenic calli or abnormal somatic embryos. The 2,4D auxin is reported by (Tisserat 1979; El Hadrami and Baaziz 1995; Fki et al., 2003) is the most suitable plant growth regulator for the initiation of callogenesis in date palm. Picloram is the auxin-like used with success to maintain callus proliferation and totipotency in sugar cane and date palm.

The solid media for germination of somatic embryos could not be used to ensure large-scale propagation. For this reason, extensive efforts have been deployed to establish embryogenic suspension cultures with high morphogenetic potentialities (Daguin and Letouzé 1988; Sharma et al., 1986; Fki et al., 2003).

The resulting suspensions were highly heterogeneous, containing cells at various stages of differentiation after 400mm mesh filtration. The best results are obtained with 5g/l inoculation density. The weight increased 2 times after 2 weeks on medium devoid growth regulators and the exponnentiel curve is obtained after 10 weeks comparing to 10g/L of inoculating. This is due to a higher concentration of calli witch have producing a higher proportion of a non regenerating cells rich in amidon (Georget et al., 2000). Our study confirmed the optimized protocol for plant regeneration from embryogenic suspension cultures of date palm cv. Deglet Nour described by (Fki et al., 2003). Chopping the callus into pieces favoured the formation of PMEs. These results confirmed those obtained by Kreuger et al. (1995) with cultures of *Cyclamen persicum*. Sané et al. (2006) and Othmani et al. (2009) recently confirmed the positive effect of callus chopping on the differentiation of date palm somatic embryos.

The acclimatization procedure can be successfully used when transplanting date palm from *in vitro* to *in vivo* environment. Our study showed that the survival percentage was 70% after 3months when all the conditions are required. This result is agreed with those obtained by (Abahmane, 2009).

Conclusion

Somatic embryogenesis has a remarkable potential for date palm development. This technology can be used for large-scale propagation. In fact, higher efficiency somatic embryogenesis has been obtained from offshoot explants for several cultivars of three regions of the Sahara. Several papers describe a suspension culture as an efficient technique in embryogenesis processes. Indeed, this technique is suitable to increase the rate of regeneration plantlets. However, the factors controlling callus induction still need to be improved and the quality of embryogenic calli requires further improvement. It is important to master the cell culture of date palm because the use of somatic cell genetic techniques, such as protoplast regeneration, cell hybridization and cell selection exhibiting resistance to applied stresses has been widely recognized as a tool to improve species (Karp et al., 1990). The acclimatization of micropropagation date palm from *in vitro* to *in vivo* environment concerned with physiological and morphological plantlets, humidity control and temperature control. Plant regeneration recovery was reliable and efficient and all plants recovered are normal in appearance.

Acknowledgments

This work was supported by the National institute of Agricultural research and the Ministry of Education, Scientific Research and Technology of Algeria, the International Atomic Energy Agency (TC Projects RAF/5/049 ALG/5/023).

References

Abahmane L. 2009 Date Palm Micropropagation via Organogenesis in S. M. Jain et al. (eds.), Date Palm Biotechnology, DOI 10.1007/978-94-007-1318-5_5, © Springer Science+Business Media B.V. 2011.

Bagniol S., Engelmann F., Michaux Ferriere M., 1992. Histocytological study of a peaces from *in vitro* plantlets of date palm (*Phoenix dactylifera* L.) during a cryoconservation process. *Cryo. Letters* 13, 405-412.

Bouguedoura N., 1991. Connaissance de la morphogenèse du palmier dattier (*Phoenix dactylifera* L.). Etude *in situ* et *in vitro* du développement morphogénétique des appareils végétatifs et reproducteurs. Thèse de doctorat d'état, USTHB, Alger. 201p.

Daguin F, et Letouze R 1988 Régénération du palmier dattier (Phoenix dactylifera L.) par embryogenèse somatique: amélioration de l'efficacité par passage en milieu liquide agité. Fruits 43:191–194.

Daikh H, Demarly.

Djerbi. M. 1988. Les maladies du Palmier dattier. Projet Régional de lutte contre le Bayoud RAB/84/018. 127 pp., 30 planches.

Duval Y, Engelmann F, Durand-Gasselin T 1995 Somatic embryogenesis in oil palm (Elaeis Guineesis Jacq). In: Bajaj YPS (ed.) *Somatic embryogenesis and synthetic seed,* vol 1. Springer Verlag, Berlin/Netherlands, pp 235–352.

Eapen S, Geoge L., 1990: Influence of phytohormones, carbohydrates, aminoacids, growth supplements and antibiotics on somatic embryogenesis and plant differenciation in finger millet. *Plant cell, tissue and organe culture*, 22:87-93.

El Hadrami I, Baaziz M 1995 Somatic embryogenesis and analysis of peroxydases in Phoenix dactilifera L. *Biol. Plant* 37:179–203.

Fish and MMM, Moore PH, 1990 Comparison of 2,4D and Picloram for selection of long term totipotent green callus of sugar cane. *Plant cell and tissue organe culture* 20 (3):157-164.

Fki L, Masmoudi R, Drira N, Rival A 2003 An optimised protocol for plant regeneration from embryogenic suspension cultures of date palm (Phoenix dactylifera L.) cv. Deglet Nour. *Plant Cell Rep.* 21:517–524.

Fki L. 2005. Application des suspension cellulaires embryogènes au clonage et à l'amélioration *in vitro* du palmier dattier *(Phoeuix dactylifern* L.). Thèse de doctorat en sciences biologiques. Faculté de sciences de Sfax.

Kreuger M, Postma E, Brouwer Y, Holst GJ 1995 Somatic embryogenesis of Cyclamen persicum in liquid medium. *Phys. Plant* 94:605–612.

Lazzeri PA, Hilderbrand DB., Collins GB., 1987. Soybean somatic embryogenesis: Effect of hormones and culture manipulation. *Plant cell, tissue and organe culture*, 10: 197- 208.

Murachige, T. et Skoog, F. 1962. A revised medium for rapid grouth and biossays with tabacco tissue culture. *Physiol.* 15.pp:473-497.

Omar MS, Novak FJ., 1990. In vitro regeneration and ethylmethane sulfonate (EMS) uptake in somatic embryogenesis of date palm, *Plant cell, Tissue org Cult*,20/185-190.

Othmani A, Bayoudh C, Drira N et al., 2009. Somatic embryogenesis and plant regeneration in date palm Phoenix dactylifera L., cv. "Boufeggous" is significantly improved by fine chopping and partial desiccation of embryogenic callus. *Plant Cell Tissue Org. Cult.* 97:71–79.

Saka H., Abed F., Amara B., kermiche A. 1997 Embryogenèse somatique du palmier dattier (*Phoenix dactylifera* L.) –I- Induction de la callogenèse à partir d'organes de rejets de quelques cultivars. *Rev. Rech. Agr.* 0: 1-8p.

Sané D., Aberlenc-Bertossi F., gassama-Dia Y. K., Sagna M., Duval Y., Borgel A. 2006. Obtention de clônes acclimatées par embryogenèse somatique à partir de suspensions cellulaires de palmier dattier (*Phoenix dactylifera* L.).

Recueil Conf. Int. Sur «Mutagénèse induite et biotechnologies d'appui pour la protection du palmier dattier contre le bayoud. 74p.

Sharma D. R., Deepak S. et Chowdury, J. B. 1986. Regeneration of plantlets from somatic tissues of the date palm *Phoenix dactylifera* L. *Indian J. Exp. Bot.* 24:763-766.

Scoarnec A. 1991. Régénération du palmier dattier (*Phoenix dactylifera* L.) par organogenèse et embryogenèse somatique indirectes. Caractéristiques du matériel végétal en culture de tissus et en acclimatation. Thèse Doct. Univ. Angers, France, 160p.

Tisserat B, 1979 Propagation of date palm (Phoenix dactylifera L.) in vitro. *J. Exp. Bot.* 30:1275–1283.

Triki M. A., Zouba A., Khouldia O., Ben Mahmoud O., Takrouni M. I., Garnier M., Bové J. M., Montarone M., Poupet A., Flores R., Daros J. A., Fadda Z. G. N., Moreno P., Duran-Villa N. 2003. «Maladie des feuilles cassantes» or a Brittle leaf disease of date palms in tunisia/ Biotic or abiotic disease? *Journal of plant Pathology* 85 (2): 71-79.

Zaid. M., 1989. Embryogenèse somatique chez le Palmier dattier *(Phoenix dactylifera* L.) Thèse de Docteur ès Sciences, Université de Paris Sud, univ. d'Orsay, 92 pp.

Zaid ZE 2003 *Comparative studies on the production of date palm cultivars via tissue culture technique*. Ph.D. thesis, Faculty of Agri., Cairo Univ., Egypt.

In: Biotechnology and Conservation ... (2 Volume Set) ISBN: 978-1-63117-869-6
Editors: S. A. Farooq, R. Abed and S. Baqir

Chapter 27

Managing High Altitude Forests for Climate Change Mitigation and Biodiversity Conservation: Lessons for the Management of Forests in Al Jebel Al Akhdar Mountain Range in Oman

Mekuria Argaw**, Adugna Feyissa and Tulu Tola
Addis Ababa University, College of Natural Science,
Center for Enviornmental Science, Ethiopia

Abstract

High altitude forests play a key role in mitigating climate change through sequestration of atmospheric carbon and the conservation of biodiversity. Forests sequester and store the largest amount of atmospheric carbon in terrestrial ecosystems. Studies of estimating carbon stocks in different carbon pools were conducted in selected high altitude sacred site forests in the highlands of Ethiopia. A systematic transect and quadrate plot sampling technique was employed to collect vegetation data, soil and litter samples. The influence of environmental variables on carbon stock density was tested using one-way ANOVA. The results show that the sacred site forests have as large number of species as in other similar contiguous dry afro-montane forests. Small and isolated sacred site church forests (those found around cities) have poor regeneration of indigenous species and very low seedling and sapling stocks. However, such forests were found to sequester a total carbon stock ranging between 297±44.3 and 616 ± 42.5 t ha^{-1}, of which about 54 % is stored in the above and below ground biomass while 45 % is stored in soil organic matter. The carbon stock in the litter biomass ranges barely between 3.5 ± 0.2 and 4.9 ± 2 t ha^{-1}. Carbon sequestration in the above and below ground biomass, litter and soil were positively influenced by altitude and slope aspect. However, with increasing slope gradient, the above and below ground carbon, soil carbon stock showed a decreasing trend. The results suggest that well managed sacred site forests can significantly contribute to carbon emission reduction and enhancing in-situ

* Corresponding author email address: mekuria.argaw@aau.edu.et.

conservation of forest biodiversity. Generally, from this study, it was learned that the more the diversity of the vegetation is the more opportunity for increased carbon sequestration. The results of the studies also illustrate that well managed high altitude forests contribute to reducing the impacts of climate change by absorbing and storing significant amount of atmospheric carbon in the standing biomass as well as the soil pool. Such experiences and research outcomes can serve as cursory indicators for the management of the forest resources of the high altitude degraded forest ecosystems of the Al Jebel Al Akhdar Mountain Range in Northern Oman both for climate change mitigation and biodiversity conservation.

Keywords: High altitude forests, climate change, Ethiopia, environmental variables, soil carbon

INTRODUCTION

Plants and soils are the largest biological sinks of carbon on terrestrial ecosystems (Han et al., 2007). Carbon in such ecosystems (unlike in geologic and ocean systems) exists in organic compounds produced by living things, including leaves, wood, roots, dead plant material and the brown organic matter in soils (USNETL, 2000; Herzog, 2001; Han et al., 2007). Forest ecosystems, with the largest accumulated biomass, constitute the largest above ground carbon pool. Global estimates show that forest ecosystems sequester and store more than 80% of all terrestrial above ground carbon and more than 70% of all soil organic carbon (Jandl et al., 2006; Perschel et al., 2007; Sundquist et al., 2008). According to the IPCC (2007) report, forests have an average biophysical mitigation potential of 5,380 Mt CO_2/yr until 2050.

Forest degradation and deforestation in the tropics contributes to over 20 % of the global carbon release to the atmosphere (IPCC, 2007) and hence causes global warming. According to Broadmeadow and Robert (2003), an estimated 1.6 GtC per year has been released to the atmosphere during the 1990s because of deforestation while the sequestration of carbon by terrestrial vegetation was between 2 and 3 GtC per year in the same period. This indicates disturbances in the forest due to natural and human influences lead to substaitnial carbon release into the atmosphere compared to the amount used by vegetation during photosynthesis (Brown, 2002). Thus, sustainable management strategies are, therefore, necessary to keep forests as important carbon sink rather than source. Forest management practices targeting biodiversity improvement and related ecosystem services enhance the carbon sequestration potential of both the soil and the vegetation. Forest and soil conservation interventions through reversing land use patterns, restoring and establishing new forests, wetlands, and grasslands or reducing CO2 emissions through reducing deforestation and further degradation can enhance carbon storage in such ecosystems (Sundquist et al., 2008).

The Ethiopian highlands forests suffered from years of unabated deforestation that led to the reduction of the once 40 % forest cover to the current 2.4 % mainly due to permanent conversion to agriculture and other forms of land use. Currently only small patches of remnant forests are found in the mountain regions of the highlands (eastern and south western highlands) and moist lowlands. In some parts of the highlands, these remnant forests are restricted to high altitude mountains, sacred sites and sanctuaries. However, the biodiversity (mainly the flora) is uniquely rich and most of the remnant high altitude forest ecosystems are part of the Horn of Africa biodiversity hotspots. The rich flora of these forest vegetation contain about 7000 different flowering plants out of which about 12% are endemic. As reported by Yitebitu Moges et al. (2010), the estimated forest carbon stock in Ethiopia is about 272 million metric tons, which is almost 83% of the country's global annual carbon emission (333 Mega tone of carbon per year). Forest management activities are increasingly taking into consideration the role of forests as carbon sinks and there is a pressing need for data and scientific information on factors that determine carbon storage potential in forests (McEwan et al., 2011). It is obvious that forest carbon sequestration and storages are influenced by both biotic (e.g., species diversity) and abiotic (altitude, slope, aspect, soil type) factors. Besides, individual tree species have unique functional traits, such as specific leaf area and wood density that convey a particular capacity for carbon capture and sequestration. Thus, species rich forests may have an increased capacity for carbon storage (Brown et al., 1999; McEwan et al., 2011) and this study tries to explore how forests managed for biodiversity conservation contribute to climate change mitigation. The Juniper-Olive forest of the Al Jebel Al Akhdar in Northern Oman is one of the unique Arabian mountain vegetation ecosystems that have gone through years of uncontrolled exploitation resulting in sever degradation. However, the mountain ecosystem is unique in its biodiversity and has great resilience potential for restoration. There is an opportunity to manage the forest to enhance the carbon sequestration potential along with the restoration efforts. The results from the Ethiopian highland forest studies can draw a lesson to meet this objective in the mountain forests in Northern Oman.

The current study was conducted in selected high altitude remnant forests in the central highlands of Ethiopia, in sacred sites where patches of natural forests have survived as a result of the traditional conservation system and protective patronage of religious institutions. The contribution of sacred site forests to biodiversity conservation and climate change mitigation offer a model for conservation and environmental protection (Camara, 1994). There is an increasing interest by international organizations such as UNESCO and WWF to explore the role of natural sacred site forests in addressing issue such as climate change aprt from biodiversity conservation (Alemayehu, 2002). Such forests have significant relevance for the implementation of the CBD which emphasizes the use of traditional wisdom and practices for conservation and sustainable use of biological diversity (Chandrashekara and Sankar, 1998). The major aim of this study was to examine the composition and structure of the sacred site forests, to estimate the carbon stock in the different carbon pools, to assess the influence of environmental factors on carbon storage and to discuss the role of such forests for biodiversity conservation and climate change mitigation in the perspective of the degraded highland forests in Ethiopia and in Northern Oman.

The study was conducted in two locations. The first location is a mountain remnant forest surrounding a sacred religious site called Menagesha Mariam in Welmera district at a distance of 30 km west of Addis Ababa. The altitude ranges between 2580 to 2910 m.a.s.l. The forest vegetation is dominated by indigenous species of trees and shrubs that include species such as *Juniperus procera* and *Olea europaea* in the upper canopy structure. The area receives a mean annual rainfall of 1028 mm and has a mean annual temperature of 200C. The second location is within Addis Ababa, in selected religious sites (church forests) between the geographical locations of 90 N to 90 5'N and 38042'E to 380 45'E (Figure 1). The altitude in these sites ranges between 2381 to 2984 m.a.s.l. The mean annual rainfall and temperature are 1128 mm and 210C, respectively. The selected church forests were those established between the period 1370 and 1924 G.C, each occupying a small patch of 0.08 to 1.8 ha. The vegetation is a mix of naturally grown tree species such as *Juniperus procera* and *Olea europea* in and some buffer rows of planted exotic species such as *Cupressus lustanica* and *Eucalyptus globules*.

Data were collected for the four major carbon pools such as above ground carbon, below ground carbon, litter biomass carbon and soil carbon. The dead wood was excluded because the forest management systems in the studied forests permit removal of dead wood for fuel and thus data was not available to examine the carbon stock in dead wood. Samples were collected for vegetation, litter and soil.

In both study sites, a systematic transect and quadrat plot sampling method was applied to collect vegetation and litter samples. Quadrat plots with a size of 10m x 20m were distributed along transects at fixed distance intervals. In each plot, trees with a DBH of > 5cm were measured for DBH at 1.3 m above the ground and height using diameter tape and Haga-hypsometer, respectively.

Litter samples were collected from a 1m x 1m sub-quadrat plots laid at the four corners and in the middle of the quadrats used for vegetation sampling. A composite sample of 100 g was taken from each quadrate for laboratory analysis. All the composite samples were sealed in ziplock plastic bags and weighted for wet weight. Each 100 g sub-sample was oven dried at 105^0C and weighted for oven dry weight and the weighted dry samples were analysed for carbon using the Walkley-Black method.

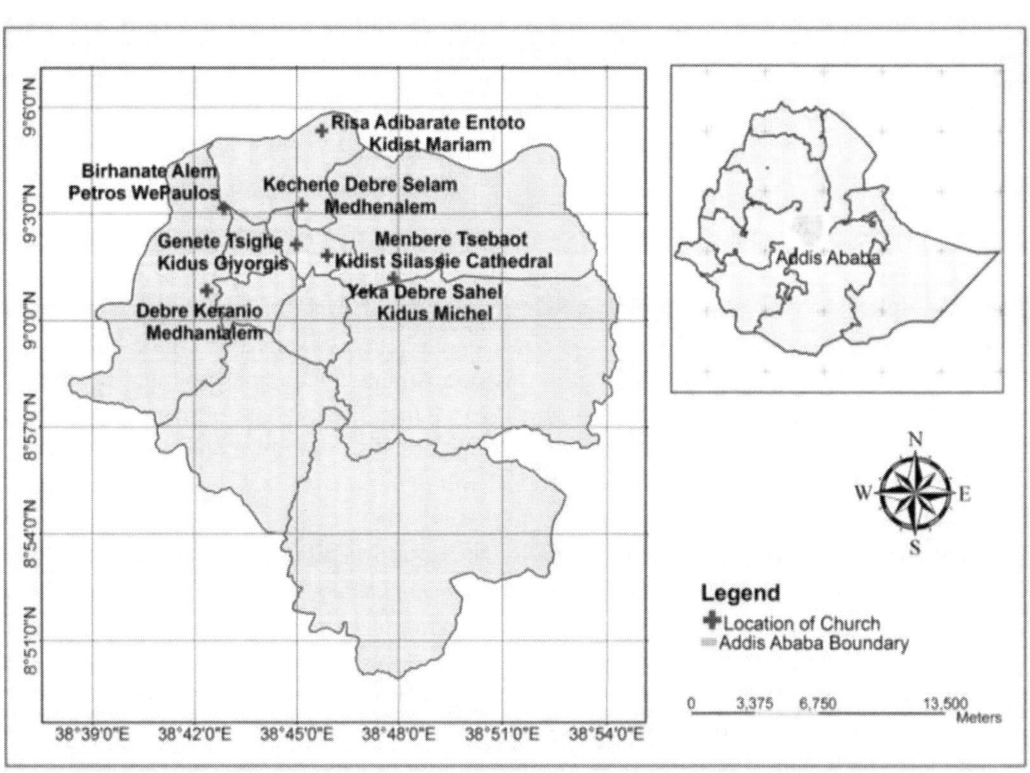

Figure 1. Location of the selected church forest sites in North Western parts of Addis Ababa.

Soil samples were collected from the same sub-quadrats (at the four corners and in the middle) those used for litter sampling, using a soil core sampler with a depth and diameter of 30 cm and 5cm, respectively. A composite sample of 100 g was taken from each sub-quadrat, air-dried and submitted for soil carbon analysis in the laboratory. Organic carbon was determined using the Walkley-Black Method.

Allometric equations are commonly used to estimate above ground biomass in the absence of species-specific estimation models. In this study, the equation shown below (*equ. 1*) has been used to calculate the above ground biomass because of its suitability to the climatic conditions in the study sites (Pearson et al., 2005).

$$Y\ (AGB) = 34.4703 - 8.0671(DBH) + 0.6589(DBH2) \qquad \text{(equ. 1)}$$

where, Y is above ground biomass, DBH is diameter at breast height, while the below ground biomass (BGB) was estimated by using the equation below:

$$BGB = AGB \times 0.27 \qquad \text{(equ. 2)}$$

The carbon content in the biomass was determined by multiplying biomass values by the default conversion value of 0.47. The CO_2 equivalent was calculated by using the molecular weight fraction of 3.67 (Brown et al., 1989).

Estimation of Carbon in Litter Biomass

$$LB = \frac{W_{field}}{A} * \frac{Wsub\_sample\ (dry)}{Wsub\_sample\ (wet)} + \frac{1}{1000} \qquad \text{(equ. 3)}$$

where *LB* is litter biomass in t ha^{-1}, W_{field} is weight of wet field sample of litter in g; A is size of the area in which litter was collected in ha; *W sub-sample dry* is the weight of the oven-dry sub-sample of litter in g, and *W sub-sample, fresh* is weight of the fresh sub-sample in g (Pearson et al., 2005).

$$C_L = LB \times \%\ C \qquad \text{(equ. 4)}$$

where, CL is total carbon stocks in the dead litter in t ha^{-1}, % C is carbon fraction determined in the laboratory (Pearson et al., 2005).

Soil Carbon Estimation

The soil carbon stock was calculated according to Pearson et al. (2005). The bulk density, sampling depth and the percent carbon content in the sample soil were used to calculate the carbon stock.

$$SOC = BD * d * \%\ C \qquad \text{(equ. 5)}$$

where, SOC is soil organic carbon stock per unit area (t ha^{-1}), BD is soil bulk density (g cm^{-3}), D is the total depth at which the sample was taken (30 cm), and %C is Carbon concentration (%)

Table 1. Types of species and stem density (stem ha^{-1}) recorded in the Menagesha Mariam Forest

Species name	No of individuals	Frequency (%)	Density
Juniperus procera	1510	92.75	1094.20
Olea europaea	1182	76.81	856.52
Olinia rochetiana	633	68.12	458.70
Acacia mearnsii	545	26.09	394.93
Maytenus arbutifolia	535	49.28	387.68
Rhamnus staddo	444	30.43	321.74
Erica arborea	331	37.68	239.86
Pittosporum viridiflorum	184	37.68	133.33
Myrica salicifolia	170	23.19	123.19
Acacia abyssinica	169	24.64	122.46
Maytenus obscura	163	42.03	118.12
Rhus vulgaris	161	43.48	116.67
Osyris quadripartita	147	36.23	106.52
Nuxia congesta	115	20.29	83.33
Rutty fruticosa	106	18.84	76.81
Harrisonia abyssinica	38	10.14	27.54
Carissa spinarum	35	14.49	25.36
Podocarpus falcatus	31	11.59	22.46
Dovyalis abyssinica	22	8.70	15.94
Dombeya torridaea	14	2.90	10.14
Maesa lanceolata	13	5.80	9.42
Myrsine africana	13	7.25	9.42
Prunus africana	10	5.80	7.25
Bersama abyssinica	10	8.70	7.25
Allophylus abyssinicus	7	11.59	5.07
Rosa abyssinica	5	5.80	3.62
Helichrysum foetidum	4	1.45	2.90
Ekebergia capensis	2	1.45	1.45
Hypericum revolutum	2	2.90	1.45
Exotic Species			
Cupressus lusitanica	318	21.74	230.43
Eucalyptus globulus	227	8.70	164.49
Eucalyptus camaldulensis	38	1.45	27.54
Casuarina cunninghamiana	13	1.45	9.42
Pinus patula	2	1.45	1.45
Total	7199		

Vegetation data were analyzed using descriptive statistics. The carbon stock variation in the different carbon pools as influenced by environmental variables was tested using one-way ANOVA. Data analyses were carried out using SPSS ver. 15.0.

COMPOSITION AND STRUCTURE

The high altitude forest of Menagesha Mariam is composed of 29 indigenous and 5 exotic species in which *Juniperus procera* and *Olea europaea* were found to be dominant and co-dominant species having a total stem density of 1094 stems/ha and 856 stems/ha, respectively (Table 1). These two species were also the most frequently encountered in the study plots. *Eucalyptus camaldulensis, E. globulus, Cupressus lusitanica, C. cunninghamiana and Pinus patula* were recorded exotic species, mainly in the lower altitude buffer belt of the mountain forest.

Table 2. Types species and stem density (stem ha^{-1}) in selected church forests in and around Addis Ababa

Species name	No of individuals	Frequency (%)	Density
Juniperus procera	763	91	884
Olea europaea	281	76	291
Cordia africana	8	7	8
Millettia ferruginea	3		3
Dracaena steudneri	38	45	42
Ficus sur	22	51	22
Allophyllus abyssinica	12	34	11
Phoenix reclinata	12	28	12
Croton mcarostachyus	27	48	27
Euphorbia abyssinic	6	12	6
Erythrina brucei	3	6	3
Acacia abyssinica	4	8	4
Exotic			
Cupressus lusitanica	8	7	25
Eucalyptus saligna	82	71	88
Pinus radiate	2	1.5	2
Acacia melanoxylon	18	39	18
Casuarina cunninghamiana	25	52	26
Dalbergia melanoxylon	7	23	8
Spathodea nilotica	2	2.5	2
Araucaria jussieu	2	1.8	2
Callistemon citrinus	9	11	9
Gravillea robusta	26	43	26
Total	1360		

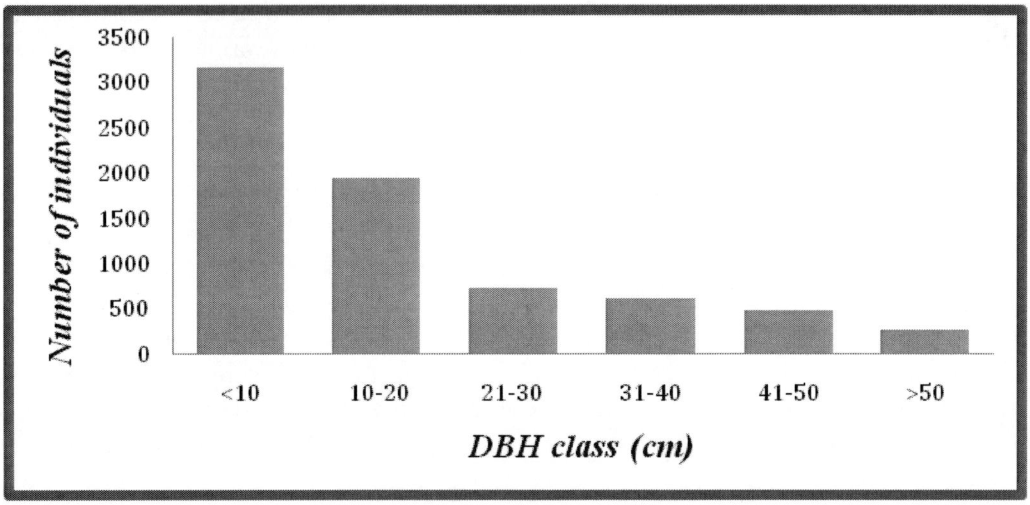

Figure 2. Size class distribution (DBH) of trees recorded in the sampled areas of the Menagesha Mariam sacred site forest.

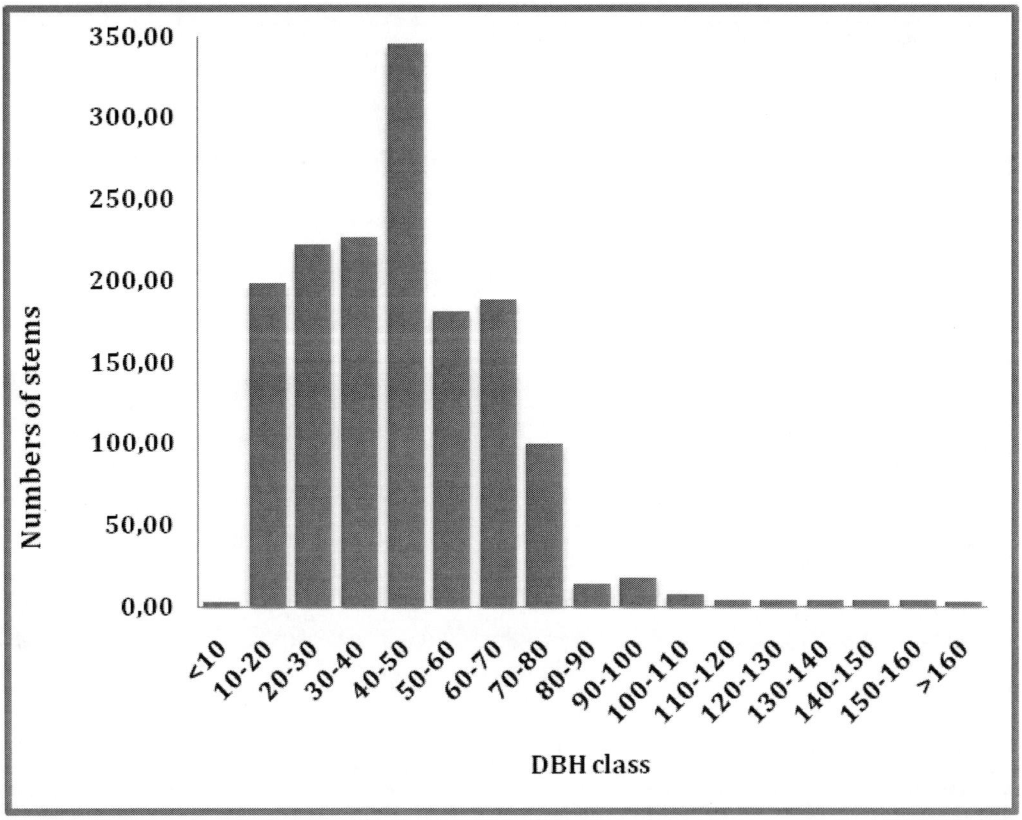

Figure 3. Size class distribution (DBH) of trees recorded in the sampled areas of the studied sacred sites forests in Addis Ababa.

The species composition in the seven sacred sites of church forests in the second study area is largely dominated by upper storey indigenous and exotic species. A total of 22 species (12 indigenous and 10 exotic) were recorded from the sites (Table 2). The most commonly encountered and dominant species in all the seven sites were *Juniperus procera and Olea europaea* with a density of 884 and 291 stems/ha, respectively. However, the number and composition of species in the seven sites differ with the time of establishment of the churches. The older church sites have more indigenous species than the recent ones. There is hardly distinctive storey difference in the forest structure other than some recent plantings of exotic species, saplings and seedlings of the already existing indigenous species in the middle storey.

The diameter distribution of trees recorded in both study sites show that there are more number of individuals in the lower diameter class and the pattern shows a gradual decline towards the largest diameter class (Figure 2, Figure 3). This is what is called an inverted "J" distribution, which is an indicator of a healthy population distribution with good stock of individuals in the sapling and seedling stocks.

Carbon Stock in the Different Carbon Pools

As shown in Table 3 below, the carbon stock in the above and below ground biomass and in the soil of the Menagesha Mariam forest site was found to be higher than those of the sacred church forest sites in Addis Ababa, except for the litter biomass carbon. Much of the carbon in the above ground biomass is held up in the large sized tree species of the forest.

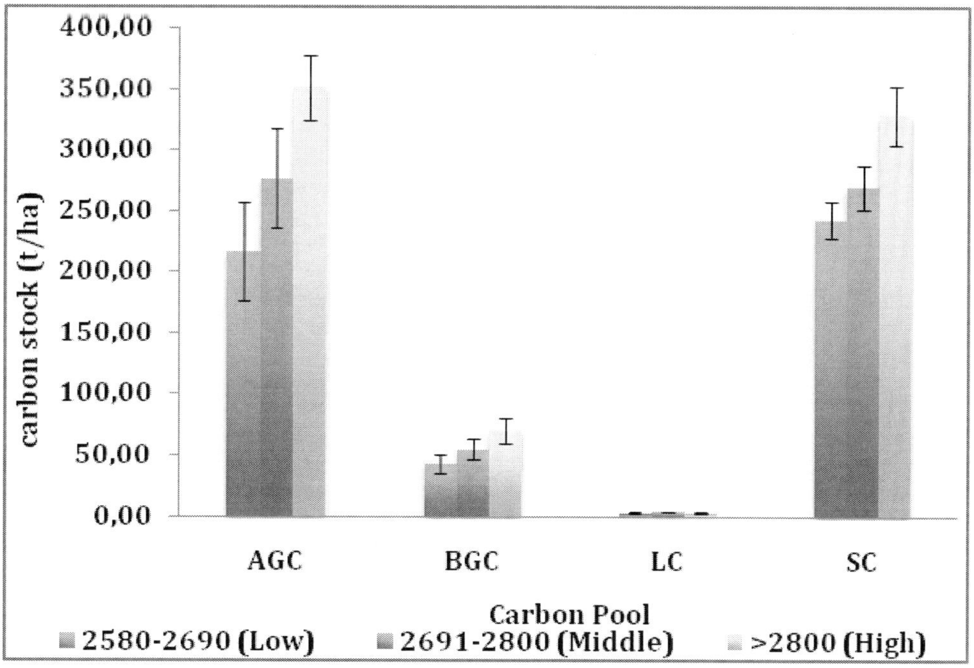

Figure 4. Carbon stock in different carbon pools along altitudinal gradient.

Carbon Stock Along Environmental Gradients

This analysis was done for the first study site since there was a natural difference in the topography of the forest mountain stretching from the foothill to the top plateau whereas the forest patches in the second study sites were found in similar altitudinal ranges and topography. Accordingly, as shown in Figure 4 below, except the carbon stock in the litter pool, estimated carbon stock in the above ground and below ground biomass as well as in the soil has shown a steady increase along the altitudinal gradient. However, though the pattern shows a gradual increase, the difference is not statistically significant (Table 4).

The analysis of carbon stock along slope gradient and aspect shows that slope aspect makes a marked difference in carbon sequestration in different carbon pools while gradient does not affect carbon stock in any of the pools. As shown in Table 4 below, east and southeast facing areas accumulate less carbon in the above and below ground carbon pools than any of the other directions. On the other hand, north and northwest facing slope areas were found to have the highest carbon stock in the above and below ground biomass as well as the soil carbon pools. Aspect, thus, showed a highly significant difference in carbon stock ($F=7.64$, $P=0.000$) (Table 5).

Table 3. Estimated carbon stock in the different carbon pools at Menagesha Mariam and in sacred church forest sites in Addis Ababa

Study site	Carbon pool	Mean ± SE (t ha^{-1})	Minimum	Maximum
Menagesha	Above ground	278.08 ± 25.72	19	782.28
Mariam sacred	Below ground	55.61 ± 5.14	3.79	156.46
forest	Litter	3.47 ± 0.2	0.33	7.53
	Soil	277.56 ± 11.56	148.74	551.3
Total		616 ±42.3		
Sacred church	Above ground	129.86 ± 24.1	20.03	444.15
forests in Addis	Below ground	25.97 ± 6.82	12.7	57.3
Ababa	Litter	4.95 ± 2.1	1.3	6.4
	Soil	135.94 ± 21.26	87.9	179.5
Total		297.44.3		

Table 4. Carbon stock in different carbon pools as influenced by topography (slope aspect and gradient)

Slope Aspect	AGC	BGC	LC	SC
S	108.11 ± 26.91	21.67 ± 5.44	5.78 ± 0.68	298.33 ± 34.74
SE	70.25 ± 14.59	14.25 ± 2.88	3.63 ± 0.63	241.00 ± 29.58
E	112.43 ± 27.35	22.57 ± 5.52	3.57 ± 0.75	247.57 ± 30.24
NE	285.13 ± 84.76	56.88 ± 16.98	3.00 ± 0.8	339.88 ± 42.87
N	492 ± 41.30	98.3 ± 8.22	2.88 ± 0.29	346.38 ± 43.93
NW	452.75 ± 84.11	90.38 ± 16.79	2.5 ± 0.38	263.88 ± 37.41
W	293.02 ± 83.22	58.63 ± 16.64	2.4 ± 0.22	241.2 ± 17.23
SW	336.21 ± 34.45	67.31 ± 6.90	3.27 ± 0.27	253.64 ± 15.34
Slope Gradient				
Low (10-30 %)	304.27 ± 67.33	60.85 ± 13.47	3.00 ± 0.46	286.33 ± 29.86
Medium (30-60%)	278.41 ± 34.61	55.68 ± 6.92	3.47 ± 0.28	279.72 ± 16.08
High (>60 %)	252.75 ± 43.26	50.55 ± 8.65	3.91 ± 0.37	264.19 ± 17.40

Table 5. Result of ANOVA in carbon stock in the different pools along slope gradient and aspect

Environmental variables	Carbon pools	F-value	p-value
	AGC	1.344	0.278
	BGC	1.344	0.278
	LC	1.329	0.287
Altitude	SOC	1.288	0.311
	AGC	7.640	0.000
	BGC	7.589	0.000
	LC	3.891	0.001
Aspects	SOC	1.980	0.065
	AGC	0.710	0.825
	BGC	0.710	0.825
	LC	0.836	0.684
Slope gradient	SOC	0.480	0.976

The species composition of the sacred site forests of Menagesha Mariam mountain forest and the church forests around Addis Ababa was found to be characteristically similar to a typical highland dry afro-montane forest with a relatively high density of the two common species *Juniperus procera* and *Olea europaea*. At Menagesha Mariam forest, some of the upper and middle storey species such as *Prunus africana*, *Bersama abyssinica, Ekebergia capensis and Hypericum revoltum* were found to be the least dominant species. This might be due to the fact that these species are very poor in regeneration partly due to the recalcitrant nature of the seeds. However, with a high density of *Erica arborea* species in the higher altitudes and relatively dense lower storey species in the middle and lower altitudes (*Olinia rochetiana, Maytenus arbutifolia, Rhamnus staddo, Myrica salicifolia, Rhus vulgaris, etc...)* the forest has maintained comparably higher number of species in an isolated sacred site. Often, similar isolated patches of forests in mountains and inaccessible ridges suffer from relentless deforestation and subsequent decline in number of species. Despite the small size of land that these sacred sites occupy, the species density per unit area is far larger than those of similar dry afro-montane contiguous forests such as Munesa Shashemene forest, where the density of trees was found to be only 306 ha^{-1} (Getachew Tesfaye, 2007). Those small patches of church forests occupying land size ranging from 0.08 ha to 1.8 ha were found to have a significantly high number of indigenous species, as large as 884 stems ha^{-1}, which simply shows the primary contribution of sacred site forests for biodiversity conservation in the study sites.

The size class distribution of stems is an indicator of the forest dynamics and the regeneration potential of individual species as well as structure of the population in the vegetation at large. The pattern of the DBH distribution in the sacred site of Menagesha Mariam forest showed a typically inverted "J" distribution having large numbers of recruitments (saplings and seedlings) in the lower size class tapering towards the larger size classes. The presence of large numbers of individuals in the seedling and sapling stocks is a good indicator of a healthy regeneration status of the forest at large and those of the key species that make up the forest ecosystem. The lower size classes (seedling and saplings) were dominated by *Juniperus procera* and *Olea europaea* suggesting good conservation status of the key species of the forest vegetation. Whereas, in the sacred sites of the church forests in Addis Ababa, the regeneration stock is virtually absent and many of the key species are not represented in the lower size class. This might be due to the fact that such sites are often exposed for trampling and litter removal because of the proximity to the city. This hinders regeneration and disfavours germination of seeds. However, the planting of exotic species has become more common in the recently established sacred site church forests may be due to the poor regenerating of the indigenous species.

The conservation and management of sacred site forests in high altitude areas is beneficial not only for biodiversity conservation but also in regulating the local climate and sequestering atmospheric carbon. The estimated carbon stock in the different carbon pools in Menagesha Mariam forest site and church forests around Addis Ababa showed that much of the forest carbon is held up in the standing biomass and the soil organic matter. As discussed in Houghton (2001), forests store carbon in the standing large sized and long lived tree species with dense wood; and in the soil in the form of organic carbon accumulated over years of biomass return. Accordingly, much of the above ground biomass carbon in both study sites is attributed to the dominant, densely populated and large sized tree species such as *Juniperus procera* and *Olea europaea*. The amount of carbon in the standing biomass and in the below ground biomass in both study sites is comparable to other similar studies in the highlands. For instance, Mesfin (2011) reported that the carbon stored in the standing and below ground biomass in Menagesha Suba forest was 133 t ha^{-1} and 27 t ha^{-1}, respectively. The current result is in conformity with further to the global dry forest carbon estimate, which is about 123 t ha^{-1} (Murphy and Lugo, 1986). The higher carbon stock in such isolated patches of forests (as compared to large contiguous forests) is because of the higher tree density resulting from the protection of the sites from human and livestock interference preventing deforestation, enhancing sequestration and reducing emission (Sharma et al., 2011). The carbon in the litter pool was very low compared to the soil and biomass. This is because the carbon stock in the litter and soil pool is largely influenced by a number of biological and physical factors that determine litter fall and rate of decomposition (Fisher and Binkly, 2000). Litter fall depends on species, age and density of the vegetation. In addition, the decomposition and subsequent addition to the soil is determined by the climate (temperature and moisture). Thus, rate of decomposition of biomass is high in the tropic and carbon is stored more in the soil organic matter than the surface litter (Fisher and Binkly, 2000; Sheikh et al., 2009).

Carbon stock in the different pools varies with environmental and biological variables. Altitudinal gradient largely determines species diversity, composition and tree density and thus the carbon stock in forest ecosystem (Luo et al., 2005; Mosser et al., 2007; Alves et al., 2010). The current study showed a gradual increase (though not statistically significant) in carbon stock along the altitudinal ascent. This result is in agreement with studies in the tropical moist forest in Brazil (Castilho et al., 2006; Alves et al., 2010), in the moist temperate valley slopes of the Garhwal Himalaya of India (Gairola et al., 2011) and in the Mt Changbai of china (Zhu et al., 2011) where there were evidences of carbon stock increase along with an altitudinal gradient. Other studies, however, showed a gradual decline in carbon stock as altitude increases (Luo et al., 2005, Leuschner et al., 2007; Mosser et al., 2007; Zhu et al., 2011). Naturally, number of tree species, stem density and cover decreases with altitude in High Mountain or high altitude forest areas. In the current study, the highest point (2910) was still within the treeline and the presence of large sized indigenous trees contributed to the increase in carbon stock in higher altitude areas.

Slope aspect is another important environmental determinant affecting carbon stock in a forest ecosystem (Bayat, 2011) and can be used as a useful predictor of forest carbon stock. The present study showed larger biomass carbon in north and north-western directions while very low values were recorded on the east and south-east directions. Similar results were reported from a carbon stock study of Beech Forest (Bayat, 2011). The same study reported a 20% variation in above ground carbon between the north facing aspect and the rest. This might be due to the difference in net primary production because of the extended sunshine hours for the north facing slopes resulting in increased carbon sequestration.

Zhu et al. (2011) state that studying carbon stock in forest carbon pools along slope gradients is important for better understanding of possible changes in carbon stock and thus carbon the sequestration potential in response to the future climate change in mountain regions. Though variations were not statistically significant, carbon stock in the different carbon pools (above ground, below ground and soil pool) showed decreasing trend with increasing slope gradient. This might be true to the fact that highly steep areas (> 60 %) usually have less vegetation cover and large sized trees do not inhabit such steep slope areas and thus a decreased in the carbon stock (Maggi et al., 2005). The other reason could be that the availability of water and nutrients, which allows more plant growth and litter accumulation, in lower slope areas than in higher slope areas (Casado et al., 1985 cited in Maggi et al., 2005).

Similar to the Ethiopian highland forests, the Al Jebel Al Akhdar (the green mountain) forest ecosystem in Northern Oman is a centre of plant endemism and high species diversity. Its particular resemblance is that the vegetation at the higher altitude areas (above 2000 m.a.s.l) is dominated by *Juniperus excelsa (procera)* and *Olea europaea* (Brinkmann et al., 2007). Though dominant, most of the *Juniperus* trees are found in poor conditions (degrading) and regeneration is minimal due to human disturbance, grazing pressure from goats and feral donkeys and climate change (particularly increased temperature and reduced rainfall) (Robinson and Leila, unpublished). However, the ecosystem has a high potential for biodiversity restoration and enhancing carbon sequestration through planned intervention.

CONCLUSION

High altitude forests those managed as sacred sites in Addis Ababa and the surrounding high mountain areas such as Menagesha Mariam mountain have a typical characteristic feature of the dry afro-montane forest vegetation type with proportionally high number of species and dense coverage of tree stands per unit area comparable to similar forests in other parts of the country. Over 27 indigenous species were recorded in Menagesha Mariam mountain forest in about 450 ha area and over 12 indigenous species were recorded in seven church forests (in 3.7 ha area) around Addis Ababa. The density of the two typically dominant species, *Juniperus procera* and *Olea europaea*, was 1059 and 884 stems ha^{-1}, respectively. These figures are very high compared to large contiguous forests in other similar highland areas of the country, which in some cases stands at about 309 stems ha^{-1}. This indicates that, though such sacred site forests are found in isolation and in small patches, the cessation of human/livestock pressure and the management through nature has contributed to the conservation of biodiversity and several of the indigenous species in the highland remnant forests. Unlike the Menagesha Mariam forest, the absence of recruitment and lower size class individuals of the key species in the church forests are indicators of poor regeneration potential and the need for management intervention. These forests are not only contributing to biodiversity conservation but also they are serving as important carbon sinks for climate change mitigation. As shown from the results of the carbon stock estimation in the different pools, substantial amount of carbon is stored in the standing biomass of trees, in the root biomass and in the soil organic matter. The figures are comparably higher than the carbon stocks of those of similar forest areas in the highlands and elsewhere. The results in this study suggest that among the key environmental variables, slope aspect is an important source of variation in carbon stock. Slopes facing longer sunshine hours were found to have higher carbon stock than the others. The study generally indicates that high altitude forests (small or large) managed for biodiversity conservation can also serve as important carbon sinks to mitigate climate change and can be considered potential sources of carbon benefits to local communities. From this study, the lesson that can be drawn for the Al Jebel Al Akhdar forest in Oman is that the rehabilitation and restoration of the forest ecosystem can be combined with managing the forest for enhancing the carbon sequestration potential so as to contribute to climate change mitigation and at the same time to link it to the global carbon market.

ACKNOWLEDGMENTS

The authors would like to thank the Centre for Environmental Science of Addis Ababa University for providing financial assistance for the fieldwork of two of the authors and the Holeta Research Centre for allowing the authors to conduct the soil organic matter analysis in their research lab.

REFERENCES

Alemayehu Wassie. 2002. Opportunities, Constraints and Prospects of the Ethiopian Orthodox Tewahido Churches in Conserving Forest Resources: The Case of Churches in South Gonder, Northern Ethiopia.

Alves, L. F., Vieira, S. A., Scaranello, M. A., Camargo, P. B., Santos, F. A. M., Joly, C. A. and Martinelli, L. A. 2010. Forest structure and live aboveground biomass variation along an elevational gradient of tropical Atlantic moist forest (Brazil). *Forest Ecology Management* 260: 679–691.

Bayat, T. A. 2007. Carbon Stock in an Apennine Beech Forest. M.Sc. Thesis, University of Twente, Enschede, the Netherlands.

Brinkmann K., Gebauer J., Patzelt A., Schlecht E., and Buerkert A. 2007. The Flora of the Sultanate of Oman, Effects of Altitude and Grazing on Al Jebel Al Akhdar, Muscat, Oman.

Broadmeadow, M. and Robert, M. 2003. *Forests, Carbon and Climate Change: The UK Contribution.* Forestry Commission Bulletin 125. Forestry Commission, Edinburgh.

Brown, S. 2002. Measuring carbon in forests: current status and future challenges. *Environmental Pollution*, 116(3): 363-372.

Brown, S. A. J., Gillespie, J. R. and Lugo, A. E. 1989. Biomass estimation methods for tropical forests with application to forest inventory data. *Forest Science,* 35(4): 881–902.

Brown, S. L., Schroeder, P. E. and Kern, J. S. 1999. Spatial distribution of biomass in forests of the eastern USA. *Forest Ecology Management* 123: 81–90.

Camara, T. 1994. *Biodiversite et forests sacrees en Casamance region de Ziquichor: Afrinet Report 10,* UNESCO- Rosta, Dakar.

Chandrashekara, U.M. and Sankar S, 1998. Ecology and management of sacred groves in Kerala, India: *Forest Ecology and Management*. Vol 112, Issues 1-2: 165-177.

Castilho, C. V., Magnusson, W. E., de Araújo, R. N. O., Luizão, R. C. C., Luizão, F. J., Albertina, P., Higuchi, N. 2006. Variation in aboveground tree live biomass in a central Amazonian Forest: effects of soil and topography. *Forest Ecology Management* 234: 85–96.

Fisher, R. F. and Binkley, D. 2000. *Ecology and Management of Forest Soils.* John Willey & Sons, Inc. New York, USA.

Gairola, S., Sharma, C. M., Ghildiyal, S. K. and Sarvesh Suyal. 2011. *Live tree biomass and carbon variation along an altitudinal gradient in moist temperate valley slopes of the Garhwal Himalaya (India).* Department of Botany, HNB Garhwal University, Srinagar Garhwal 246 174, India.

Getachew Tesfaye. 2007. Structure, Biomass and Net Primary Production in a Dry Tropical Afromontane Forest in Ethiopia. *Final Project Report,* Department of Biology, Addis Ababa University, Ethiopia. Funded by US National Science Foundation through International START, Pp. 4-20.

Han, X. F., Plodinec, J. M, Yi Su, David, L., Monts, L. D. and Li, Z. 2007. Terrestrial carbon pools in southeast and south-central United States. Institute for Clean Energy Technology (ICET), Mississippi State University, USA. Climatic Change 84:191–202.

Houghton, R. A., Lawrence, K.T., Hackler, J. L. and Brown, S. 2001. The spatial distribution of forest biomass in the Brazilian Amazon: a comparison of estimates *Glob. Change Biol.* 7, 731–46.

Herzog, H. J. 2001. What Future for Carbon Capture and Sequestration? *American Chemical Society journal.* U.S. Government Printing Offices, Washington, DC. Feature Issue 7 35:148 A – 153 A.

IPCC. 2007. Highlights from Climate Change 2007. *The Physical Science Basis. Summary for Policy Makers. Contribution of Working Group I to the Fourth Assessment Report of the Intergovernmental Panel on Climate Change* (Solomon, S., Qin, D., Manning, M., Chen, Z., Marquis, M., Averyt, K. B., Tignor, M., Miller, H. L., eds). Cambridge University Press, Institute of Terrestrial Ecology, Edinburgh, Pp. 545-552.

Jandl, R., Rasmussen, K., Tomé, M. and Johnson, D.W. 2006. The role of Forests in carbon Cycles, Sequestration and Storage. Issue 4. *Forest management and carbon sequestration. Federal Research and Training Centre for Forests, Natural Hazard and Landscape (BFW),* Vienna, Austria.

Leuschner, C., Moser, G., Bertsch, C., Roderstein, M. and Hertel, D. 2007. Large altitudinal increase in tree root/shoot ratio in tropical mountain forest of Ecuador. *Basic Applied Ecology* 8: 219– 230.

Luo, T. X., Brown, S., Pan, Y. D., Shi, P. L., Ouyang, H., Yu, Z. L., Zhu, H. Z. 2005. Root biomass along subtropical to alpine gradients: global implication from Tibetan transects studies. *Forest Ecology Management* 206:349–363.

Maggi, O., Persiani, M. A., Casado, M. A. and Pineda, F. D. 2005. Effects of elevation, slope position and livestock exclusion on microfungi isolated from soils of Mediterranean grasslands. The Mycological Society of America, Lawrence. *Mycologia* 97(5): 984–995.

McEwan, W. R., Lin, Y., Sun, I. F., Hsieh, C., Su, S., Chang, L., Song, G. M., Wange, H., Hwong, J., Lin, K., Yang, K., Chiang, J. 2011. Topographic and biotic regulation of aboveground carbon storage in subtropical broad-leaved forests of Taiwan. *Forest Ecology and Management* 262:1817–1825.

Mesfin Sahile. 2011. Estimating and Mapping of Carbon Stocks based on Remote Sensing, GIS and Ground Survey in the Menagesha Suba State Forest. M.Sc. Thesis, Addis Ababa University, Addis Ababa.

Moser, G., Hertel, D. and Leuschner, C. 2007. Altitudinal change in LAI and stand leaf biomass in tropical montane forests: a transect study in Ecuador and a pan-tropical meta-analysis. *Ecosystems* 10: 924–935.

Pearson, T., Walker, S. and Brown, S. 2005. *Sourcebook for land-use, land-use change and forestry projects.* Winrock International and the Bio-carbon fund of the World Bank. Arlington, USA, Pp. 19-35.

Perschel, R. T., Alexander, M. E. and Marcia J. S. 2007. *Climate Change, Carbon, and the Forests of the Northeast.* Forest Guild. Funded by Orchard, Davis Conservation, French, Merck and Surdna Foundation, Santa Fe, Pp. 1-10.

Sheikh, M. A., Kumar, M. and Rainer, W. and Bussmann, R. W. 2009. Altitudinal variation in soil organic carbon stock in coniferous subtropical and broadleaf temperate forests in Garhwal Himalaya. Department of Forestry, HNB Garhwal University, Srinagar Garhwal, Uttarakhand, India. *Carbon Balance management* 4: 1-6.

Sundquist, E., Robert, B., Stephen, F., Robert, G., Jennifer, H., Yousif, K., Larry, T. and Mark W. 2008. *Carbon Sequestration to Mitigate Climate Change.* U.S. Geological Survey, science for a changing world, New York, USA, Pp. 1-4.

USNETL (U.S National Energy Technology Laboratory). 2000. Terrestrial Sequestration Program. Carbon and Storage of Carbon in Terrestrial Ecosystems. Sequestration Program Facts. U.S department of Energy publishing, Washington, DC, Pp. 12-18.

Yetebitu Moges, Zewdu Eshetu and Sisay Nune. 2010. *Manual for assessment and monitoring of carbon in forest and other land uses in Ethiopia (Draft).* Addis Ababa, Ethiopia.

Zhu B, Xiangping Wang, Jingyun Fang, Shilong Piao, Haihua Shen, Shuqing Zhao, Changhui Peng. 2011. Altitudinal changes in carbon storage of temperate forests on Mt Changbai, Northeast China. Carbon cycle process in East Asia, The Botanical Society of Japan and Springer 2010. *Journal of Plant Resource* 10:1-14.

In: Biotechnology and Conservation … (2 Volume Set) ISBN: 978-1-63117-869-6
Editors: S. A. Farooq, R. Abed and S. Baqir

Chapter 28

ADAPTIVE STRATEGY OF *PISTACIA ATLANTICA* DESF. SUBSP. *ATLANTICA* POPULATIONS: FOLIAR EPIDERMAL MICROMORPHOLOGY

***M. Tirse*[1,*], *H. Benhassaini*[1], *K. Sail*[2] *and G. Bassou*[2]**
[1]Laboratory of Plant Biodiversity: Conservation and Valorisation. University Djillali Liabes of Sidi Bel Abbès, Faculty of Science, Sidi Bel Abbès, Algeria
[2]Laboratory L2MSM, University Djillali Liabes of Sidi Bel Abbès, Faculty of Science, Sidi Bel Abbès, Algeria

ABSTRACT

Atlas pistachio: *Pistacia atlantica* Desf. subsp. *atlantica* is a tree of the arid and semi-arid areas. It is subjected to severe stress conditions that directly affect the anatomy of the leaf. This work focuses on the study of foliar epidermal micromorphology of *P. atlantica* collected from two sites (Ouled Mimoun and Ben Badis) in arid and semi-aride bioclimates respectively through light microscopy and scanning electron microscopy (SEM). Six stomatal types have been observed where staurorocytic and cyclocytic were never cited in *Pistacia atlantica*. Epidermis structure varied among the two studied populations on the two faces. Stomata are slightly sunken and covered with wax resin. Samples of Ben Badis' site reveal a high diversity of trichomes.

Keywords: Arid, semi-arid, *Pistacia atlantica* Desf. ssp. *atlantica*, stomatal types, trichomes

INTRODUCTION

Pistacia (*Anacardiaceae* family) is a xerophytic genus, characterised by many adaptations to aridity, such as advanced development of a palisade tissue and a very strong root system (Al-Saghir and Porter 2005).

* Corresponding author: M_tirse@yahoo.fr.

P. atlantica Desf. subsp. *atlantica* is an endemic species in the north of Africa. It grows in very harsh and dry areas with a low rainfall. As the majority of xerophyts, the species of the arid and semi-arid zones show in many cases, stomata sunken at various depths overarched by papillae, surrounded by rings of waxes or hidden by epidermal convexities (Rotondi et al., 2003). The presence of trichomes, covering the abaxial surfaces of some leaves species, is a xeromorphic feature (Rotondi et al., 2003). Botanists have also shown a great interest towards the diversity of plants' trichomes because of their adaptive and taxonomic value (Theobald et al., 1979; Baran et al., 2010) especially at the specific level (Theobald et al., 1979). Trichomes can have a variety of morphologies and they may provide some shade to protect plants from UV irradiation or to reduce transpiration (Ishida et al., 2008).

Previous studies have demonstrated that *P. atlantica* adopts different strategies involving micro-morphological characters to adapt to the environmental conditions (Kadi-bennane et al. 2005; Ait-Said et al. 2011; Belhadj et al., 2011).

This paper is the first contribution to the description of the foliar micro-morphological features of *P. atlantica* subsp. *atlantica* along a latitudinal transect (west-east) in the north western region of Algeria, at different altitudes and bioclimatic sites.

The leaves were harvested in April 2010 from two sites chosen along a latitudinal transect (west- east transect) (Figure 1): Ouled Mimoun (OM) and Ben Badis (BB). Specimens were deposited at the herbarium of the University of Sidi Bel-Abbès. For each site, climate and geographical factors are described (Table 1).

Table 1. Main climatic data of the sampling sites

Site	Latitude	Longitude	Altitude	Ann. Rain (mm)	Q3	Climate typ
OuledMimoun (OM)	34° 53̀ 51 N	00° 03̀ 18 O	720	127,48	39,78	Arid
Ben Badis (BB)	34° 35̀ 22 N	00° 56̀ 09 O	756	141,88	44,72	Semi-arid

Figure 1. Geographical location of the investigated *P. atlantica* populations.

The investigations focused on the epidermal micromorphology characters of the leaflets. That is why; a thick layer of clear nail polish was deposited on both sides of the leaf. The epidermal peels were carefully spread on slides which were examined and photographed with Zeiss light microscope at (x10) and (x40) magnification.

The main anatomical leaf features are reported together with the light and electron microscope images. The leaflets were cleaned with alcohol, dried and preserved from air pollution. The excitea were glued on a strip of double sided tape then, they were coated with a layer of carbon. Micrographs of the samples were performed by scanning electron micrographs SEM (type Zeiss DMS 982 Gemini).

The leaflets of the tow studied sites are hypo-amphistomatics. Although the stomata are located on both of the leaflets sides. Numerous stomata are dispersed over the abaxial face (Figure 2.a). On the adaxial face, few stomata are aligned along the midrib. The closest are arranged parallel to the midrib. The stomata are inclined and inserted perpendicularly (Figure 2. b). For the two sites, microphotographs show a rough and striated structure of the epidermis on abaxial face. At (10000) magnification, the stomata surface in (OM) leaflets shows a rough structure and festooned outer stomatal ledge covered with resin particles. The ridge of the outer stomatal ledge aperture has a highly pleated structure (Figure 3.a). For (BB) leaflets, the parallel ridges projected from the stomata form a complicated and concentric network reticulated and well striated (Figure 3.b). On the adaxial face, at (3000) magnification, the surface leaflets of (OM) site show a strongly rough surface of epidermis covered with particles of resin (Figure 4. a). However, it's smooth coated by resin particles near the midrib on those of (BB) samples (Figure 4. b).

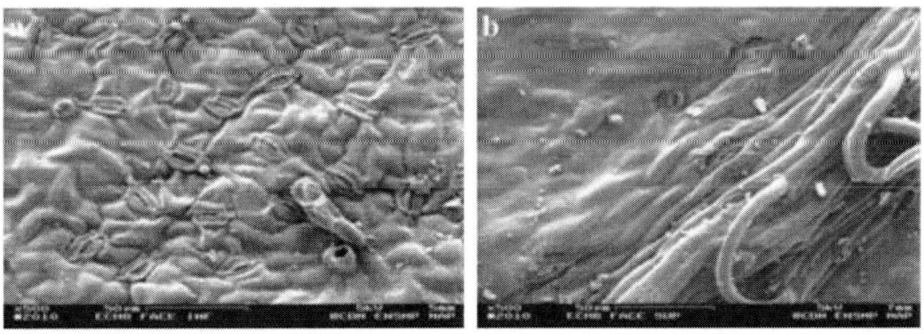

Figure 2. SEM photographs showing dispersion of stomata on the surface of the leaflets (x500) (a) abaxial face (b) adaxial face.

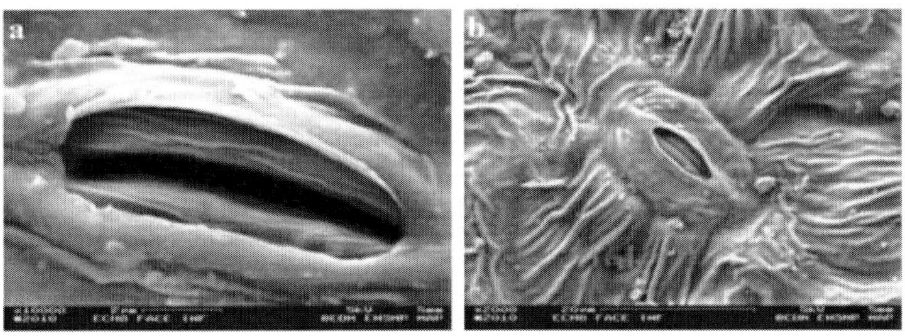

Figure 3. SEM photographs showing the structure of stomata on the abaxial leaflet side (a) Ouled Mimoun (x 10000) showing the outer stomatal ledge (b) Ben Badis (x 2000).

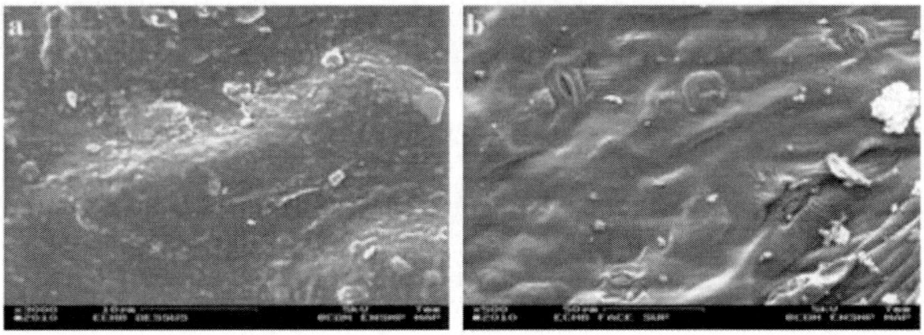

Figure 4. SEM photographs of epidermis structure on adaxial side (a) Ouled Mimoun (x 3000) (b) Ben Badis (x500).

Table 2. Stomatal and trichomes types

	Ouled Mimoun (OM)		Ben Badis (BB)
Stomatal types		Diacytic Actinocytic	
	Staurocytic * Cyclocytic *		Paracytic Laterocytic
Trichomes		Long and narrow unicellular Bifurcate Unicellular hooked Capitates trichomes	
	Unicellar courbed. Tricellular		Biseriated Multicellular Amoeboid shaped Stellate

* New stomatal types recorded.

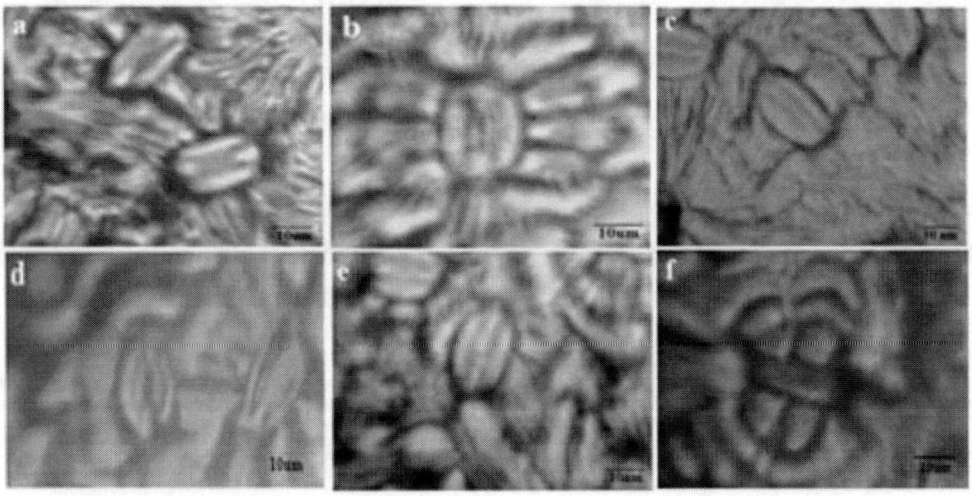

Figure 5. Stomatal types accorded in leaflets of *P. atlantica* subsp. *atlantica* (a) actinocytic (b) paracytic (c) laterocytic (d) diacytic (e) staurocytic (f) cyclocytic (Light microscopy).

A diversity of stomatal types was recorded (Table 2). Diacytic and actinocytic stomata are observed on both of two studied sites (Figure 5. a.b). Paracytic and laterocytic stomata are found in Ben Badis (BB) samples (Figure 5.c.d). However, new stomatal types (staurocytic and cyclocytic) were identified on the leaflets of (OM) site (Figure 5. e.f). Two contiguous stomata were observed on the adaxial surface of (BB) samples.

In the present study, two types of trichomes are observed: the glandular and non-glandular trichomes (Table 2). A high diversity of non-glandular trichomes (Long and narrow unicellular, bifurcate, unicellular hooked) is recorded (Figure 6.a.b.c). Other forms of non-glandular trichomes are observed such as unicellular courbed, tricellular in (OM) samples and biseriated, multicellular, amoeboid shaped in (BB) samples (Figure 6.d.e.f.g.h).

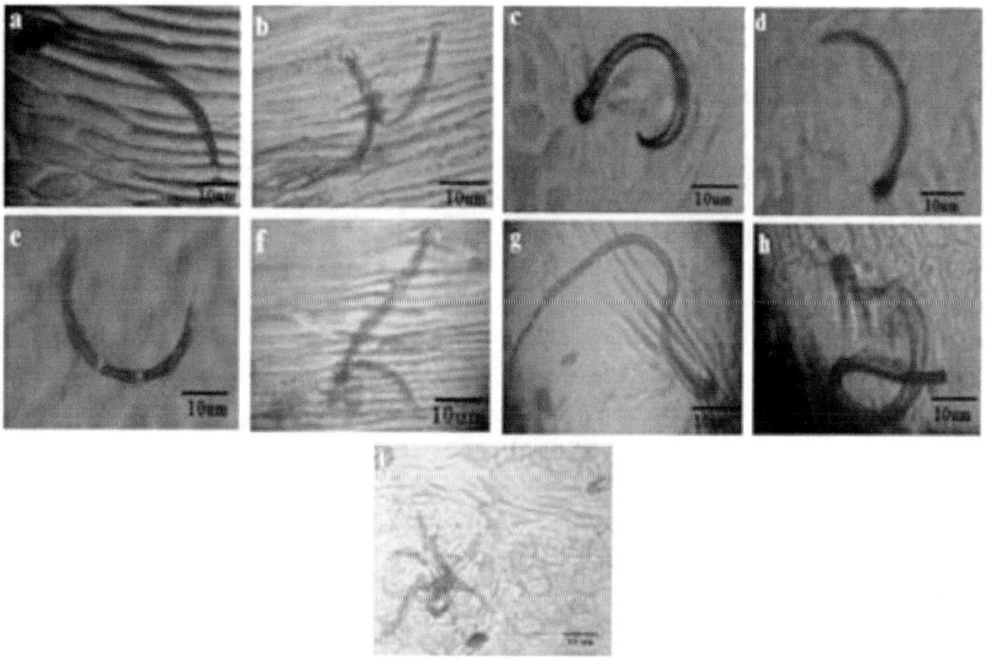

Figure 6. Types of non-glandular trichomes (a) Long and narrow unicellular (b) bifurcate (c) unicellular hooked (d) unicellular rolled (e) tricellular (f) biseriated (g) multicellular (h) amoeboid shaped (i) stellate trichome (Light microscopy).

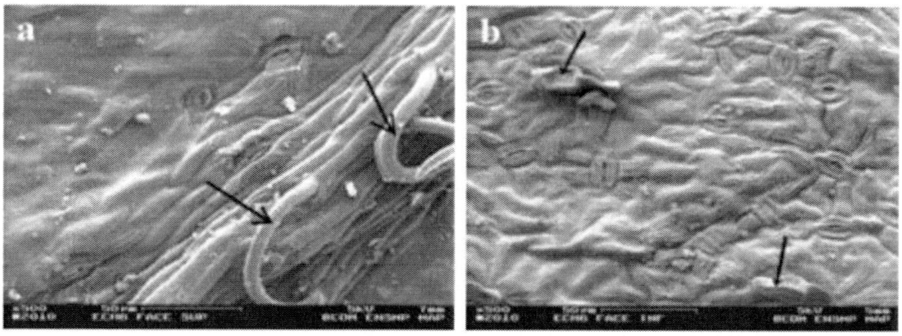

Figure 7. SEM photographs showing insertion of trichomes on epidermis surface (a) ciliated trichomes on and along the midrib on the adaxial face (b) capitates trichomes on the abaxial face.

A non-glandular stellate trichome rarely cited in *P. atlantica* is noted in (BB) site (Figure 6.i). For the two sites, a numerous of non-glandular ciliated trichomes are found on the adaxial face, inserted irregularly along the midrib, and occasionally projected on the secondary veins, very rarely over the remaining leaf area (Figure 7.a). The other face is covered by glandular capitates trichomes (Figure 7. b).

According to the terminology of the stomatal complex of Metcalfe and Chalk (1950), the leaflets of the two studied sites are hypo-amphistomatics. However, in their comparative study on stomatal traits in populations of *P. atlantica* of south Algeria, Belhadj et al. (2011) reported the amphistomatic character. The amphistomaty is often mentioned in *P. atlantica* (Al-Saghir and Porter 2005; Özeker and Misirli 2005; Belhadj et al., 2007; Ait-Said et al., 2011; Belhadj et al., 2011). However, stomata are absent on adaxial surface, one possible reason of the taxon adapts to the lack of water (Smail Saadoun 2005). In effect, the amphistomatal distribution may be a feature of the response to increasing aridity during the Tertiary. It is considered as a strongly xeromorphic trait (Rotondi et al., 2003). The presence of stomata on both sides of the leaflets is a plesiomorphic character and their presence on certain face of *Pistacia* species is an evolved character. Thus, the loss of stomata can be reflected as an adaptation to climate change (Al-Saghir and Porter 2005). Moreover, the stomata are perhaps the most significant from the point of view of systematics and phylogeny (Shiva Kameshwari 2011).

Our results concerning the structure of the epidermis covered with particles of resin are in accordance with those given recently by Belhadj et al. (2011). This may suggest a protective role against excessive water loss in arid regions with low and / or irregular precipitation.

In agreement with several authors, actinocytic stomata are frequently observed in *Pistacia atlantica* (Belhadj et al., 2007; Ait-Said et al., 2011; Belhadj et al., 2011). The presence of five ontogenetic stomatal groups qualifies *P. atlantica* as a polytypic species (Kadi-Bennane et al., 2005). However, the increased aridity involves a high frequency of paracytic mésopérigenous type (Smail Saadoun 2005). The occurrence of contiguous stomata is a character associated with polyploidy (Agbagwa and Okoli 2006). They can be the result of two meristemoids developing in single stomata (Reddy and Shah 1979).

The data concerning the presence/absence of trichomes in *Pistacia* leaflets are conflicting. Al-Saghir and Porter (2005) indicated that all *Pistacia* species are hairless and have no trichomes. Monjauze (1989) noted the presence of the microscopic hairs on the leaflets of pistachio (*P. vera*), but also the existence of pistachio trees without hairs in Algeria. The differences in trichomes abundance on adaxial and abaxial epidermis may be due to environmental and / or genetic diversity between the samples collected from different localities. However, the number of glandular trichomes is generally, highly variable, since it is strongly influenced by external factors such as temperature and light intensity (Krstic et al., 2006). The presence of different types of trichomes is an adaptive strategy (Chmielewska and Chermetsky 2005), since the abundance of ciliated trichomes is an important xeromorphic character (Baran et al., 2010). This may accord with our results indicating the diversity of types and forms of trichomes in (OM) and (BB) sites. Specifically, the ciliated trichomes ensure a protective role against excessive transpiration to the stellate non-glandular trichomes that create a downy surface. The presence of stellate trichomes only in (BB) samples may confirms the protective role. Our results are in agreement with the samples collected from site with low aridity (Ait Said et al., 2011). However, no stellate trichome was detected in the populations collected along an increasing aridity (Belhadj et al., 2007). Shaheen et al. (2009)

suggested that the contradiction of the presence and/or absence of conical non-glandular trichomes, peltates and capitates glandular trichomes can be attributed to the different geographic localities. The presence of hairs and their types can be of taxonomic use in delimiting some genera (Taïa 2004). However, length, size and density of trichomes are variable and may be influenced by environmental conditions (Metcalfe and Chalk 1950).

ACKNOWLEDGMENTS

The authors wish to thank the Microscopy Laboratory Team. Centre des Matériaux Pierre Marie Fourt d'Evry. Paris. France.

REFERENCES

Agbagwa IO and Okoli BE. 2006. Leaf Epidermal Micromorphology in the systematic of *Abrus* (*Papilionaceae*) in parts of Tropical West Africa. *Asian Journal of Plant Sciences*. 51: 41-49.

Ait-Said S, Fernandez C, Greff S, Derridj A, Gauquelin T and Mevy JP. 2011. Inter-population variability of leaf morpho-anatomical and terpenoid patterns of *Pistacia atlantica* Desf. ssp. *atlantica* growing along an aridity gradient in Algeria. *Flora*. 3: 1-9.

Al-Saghir GM and Porter DM. 2005. Stomatal distribution in *Pistacia* sp. (*Anacardiaceae*). *International J. of Botany*. 2:183-187.

Al-Saghir GM and Porter DM. 2012. Taxonomic Revision of the Genus *Pistacia* L. (*Anacardiaceae*). *American J. of Plant Science*. 3:12-32.

Baran P, Özdemir C and Aktas K. 2010. Structural investigation of the glandular trichomes of *Salvia argentea*. *Biologia*. 65: 33-38.

Belhadj S, Derridji A, Aigouy T, Gers C, Gauquelin T and Mevy JP. 2007. Comparative morphology of leaf epidermis in Eight populations of Atlas pistachio (*Pistacia atlantica* Desf., *Anacardiaceae*). *Microscopy Research and Technique*. 70: 837- 846.

Belhadj S, Derridj A, Moriana A, Gijon MDC, Mevy JP and Gauquelin T. 2011. Comparative analysis of stomatal characters in eight wild Atlas pistachio populations (*Pistacia atlantica* Desf. *Anacardiaceae*). *International Research J. of Plant Science*. 2: 060-069.

Chmielewska EW and Chermetsky M. 2005. Structure of trichomes from the surface of leaves of some species of *Kalanchoë adans* (*Crassulaceae*). *Acta Biologica Cracoviensia Series Botanica*. 47: 15-22.

Ishida T, Kurata T, Okada K and Wada T. 2008.A genetic regulatory network in the development of trichomes and root hairs. *Annual Review of Plant Biology*. 59: 365-86.

Kadi-Bennane S, Ait-Said S, Smail Saadoun N. 2005. Etude adaptative de trois populations de *Pistacia atlantica* Desf. ssp. *atlantica* (Ain Oussera- Messaad- Taissa) par le biais du complexe stomatique. *Options Méditerranéennes*. 63: 365-368.

Kafkas S and Perl-Treves R. 2001. Morphological and molecular phylogeny of *Pistacia* species in Turkey. *Theoretical and Applied Genetics*. 102: 908-915.

Krstic L, Malencic D and Anackov G. 2006. Structural investigations of trichomes and essential oil composition of *Salvia verticillata*. *Botanica Helvetica*. 116: 159-168.

Metcalfe CR and Chalk L. 1950. Anatomy of the Dicotyledons. *Systematic anatomy of leaf and stem, with a brief history of the subject.* Oxford U. K.

Monjauze A. 1980. Connaissance du Betoum: *Pistacia atlantica* Desf. *Revue Forestière Française*. 32: 356-363.

Özeker E and Misirli A. 2005. Research on leaf properties and stomata distribution of some *Pistacia* ssp. *Cahiers Options Méditerranéennes*. 56: 237- 241.

Reddy PKR and Shah GL. 1979. Observations on the structure and Ontogeny of stomata and trichomes on developing and mature pericarp of *Cassia occidentalis*. *Biologia Plantarum*. 21: 321-327.

Rotondi A, Rossi F, Asunis C and Cesaraccio C. 2003. Leaf xeromorphic adaptations of some plants of a Coastal Mediterranean *machia* ecosystem. *J. of Mediterranean Ecology*. 4: 25-35.

Shiva Kameshwari MN. 2011. Epidermal Micromorphology in Populations of *Urginea indica* Kunth. (*Liliaceae*). *International J. of Engineering Science and Technology* (IJEST). 3: 3816-3824.

Shaheen N, Yasmin G, Quazim Hayat M. 2009. Diversity of foliar trichomes and their systematic relevance in the Genus *Hibiscus* (*Malvaceae*). *International. J. agriculture & biology*.09: 279-284.

Smail Saadoun N. 2005. Types stomatiques du genre *Pistacia*: *Pistacia atlantica* Desf. ssp. *atlantica* et *Pistacia lentiscus* L. *Options Méditerranéennes*. 63: 369-371.

Taia WK. 2004. Leaf characters within tribe *Trifolieae* (Family *Leguminoseae*). *Pakistan J. of Biological Sciences*.8: 1463-1472.

Theobald WL, Krahulik JL and Rollins RC. 1979. Trichome description and classification. In *Metcalfe CR and Chalk L (Anatomy of the Dicotyledons)*. 2nd edition. Clarendon Press, Oxford.1: 40-53.

In: Biotechnology and Conservation … (2 Volume Set) ISBN: 978-1-63117-869-6
Editors: S. A. Farooq, R. Abed and S. Baqir

Chapter 29

ANTIMICROBIAL COMPOUNDS FROM ETHANO-MEDICINAL PLANTS OF OMAN

Khamis S. Al-Dhafri, Sardar A. Farooq*, Elsadig Eltayeb and Saif N. Bahry
Department of Biology, College of Science,
Sultan Qaboos University, Al Khoud, Oman

ABSTRACT

Eight species of Omani ethno-medicinal plants belonging to Zygophyllaceae viz. Fagonia bruguieri, Fagonia indica, Fagonia socotrana, Seetzenia lanata, Tribulus pentanadrus, Tribulus tertrestris, Tetraena qatarense, Tetraena simplexare traditionally used in the treatment of various illnesses and body disorders. Ethanolic extracts were screened for antimicrobial activity against different strains of human pathogenic bacteria and fungi such as Staphylococcus aureus, Bacillus subtilis, Escherichia coli, Pseudomonas aeruginosa, Enterobacter faecalis and Cladosporum herbarum. The tests were carried out using agar diffusion method and bioautography. The percentage yield and inhibition zone diameter of the ethanol extract and its fractions were determined and the data were statistically analyzed. The efficiency of different fractions was compared by measuring the clearing zones. Of all the plants tested, Fagonia indica, Tribulus pentandrus, and Tetraena simplex were found to be more effective against bacterial and fungal strains. Brine Shrimp lethality test was used to determine LC50 of plant extracts. For ethanol extract LC50 indicated that Fagonia species and Seetzeinia lanata are more effective than Tribulus and Tetraena species. All selected species were found to have antimicrobial activity on at least two microbial strains. The antimicrobial activity profile showed that E. coli is the most resistant among the tested strains. The results indicate the potential of these medicinal plants in treating microbial infections of the body.

Keywords: Zygophyllaceae, antimicrobial compounds, inhibition zone

* Corresponding author.

INTRODUCTION

Sultanate of Oman is situated at the southern part of Arabian Peninsula and has a tropical climate. Due to different climate during the year, there is considerable diversity in the vegetation between the southern region which has flora similar to that of North and East Africa, and Northern Hajar mountain, which has plants closer to the South-West Asian flora. However, many of these plants have not yet been investigated phytochemically and pharmacologically; therefore, the study on the chemical constituents of Omani medicinal plants for bioactive compounds assumes great importance and is worth undertaking.

Medicinal plants are of great importance to the health of individuals and communities. However, they play a significant role in the defense of plants against pathogens and act as chemical signaling agents to attract insects for pollination (Taylor, 2000). The appearance of multi-resistance bacteria become a major cause of failure of the treatment of infectious diseases. This required continuous updating of knowledge concerning treatment. Therefore, there is a need to look for alternative strategies for the disease resistant bacteria. To do that plants are screened for antibiotic compound and tested through bioassay methods against resistant pathogens utilizing traditional medicine for the primary of plants in health care needs (Mohana et al., 2008).

Sultanate of Oman has a 1,700 km of coastal line and approximately an area of 309,500 km. According to Ghazanfar (2003), in Oman there are 1,208 species of vascular plants, out of which 1,182 are angiosperms, 4 species are gymnosperms, and 22 species are vascular cryptogams. However, many of these plants have not yet been investigated phytochemically and pharmacologically. Out of the 1208 species, 73 are endemic and 400 have been used in folkore medicine (Amin, and Mousa, 2007).

Medicinal plants are of great importance to the health of individuals and communities. The medicinal value of the plants lies in the chemical substances/metabolites produced by them which can act on the physiology of human body. Plant metabolites can be classified generally into two major groups. Primary metabolites are essential for the growth and reproduction of plants e.g., sugars, amino acids, and nucleotides etc. and these are transformed into secondary metabolites or natural products by complex biochemical pathways and are not known to have significant role in the metabolism of the plants. However, they play a significant role in the defense of plants against pathogens and act as chemical signaling agents to attract insects for pollination (Taylor, 2000). The Natural products with bioactive constituents are used as medicines for many centuries in different countries and are classified as alkaloids, tannins, terpenes, flavanoids, steroids, lignins, phenolic compounds, and organic cyanides (Fatope et al., 2000). The appearance of multi-resistance bacteria became a major cause of failure of the treatment of infectious diseases. This required continuous updating of knowledge concerning treatment. Therefore, there is a need to look for alternative strategies for the disease resistant bacteria.

THE HISTORY OF HERBAL MEDICINE

Man has used some plants for food and others for medicine. The search for medicinal plants has been an integral part of human society since the earliest recorded history. The use of different plants (phytochemicals) in the treatment and prevention of human illness has started since the beginning of human civilizations. Before the advent of herbal medicine, the superstition dominated the human minds and the diseases were thought to be caused by devils or mythical gods. To treat such diseases, the common practice was to please the gods by sacrifice in those days. However, with the advancement of human thinking, logic and scientific reasoning the foundation was laid to the analysis of the causes of different diseases. Man has tried nearly everything in the vicinity that he laid his hand to cure different diseases; as a result a large number of medicinal plants were discovered by accidental trials in different societies. Ayurveda is the traditional Indian medicine which has a history of over 3000 years; Unani (Greek) medicine was popular in 18th century. Similarly Chinese folk medicine has a long history. Chinese monks used green tea as a stimulant, others used coffee. Some plants were considered sacred and worshiped e.g., *Lotus* and *Nigella*, some are used as poisons. Medicinal plants discovered by traditional societies are proving to be an important source of potential therapeutic drugs. Exploring the medicinal plants of indigenous people from remote areas and with the use of sophisticated assays, it is possible to discover bioactive molecules produced by these plants and use them in treatment of diseases. Flowering plants and ferns have contributed to more than 120 commercially sold drugs and account for over 25% of all prescriptions issued in China (Dagmar et al., 2003) and North America (Cox and Balick, 1994). Traditional healers claim that their medicine is cheaper and more effective than modern medicine. In developing countries, low-income people such as farmers, people of small isolated villages and native communities use folk medicine for the treatment of common infections (Rojas et al., 2006).

The history of Chemotherapy can be divided into three historical periods. It dates back to 16th century, where plants were used as main source of therapeutic drugs e.g., Peruvian Indians, used *Cinchona* bark, containing quinine alkaloid to treat malaria. This medicine was later introduced in Europe, by Spaniards. The root bark of *Ipecacuanha* was introduced to Europe, for its medicinal qualities for treating diarrhea and amoebic dysentery (Horton, 2003).

In Arabian region, the mercuric chloride was used for preventing infection in open wounds as antiseptic. Antiseptics came in nineteenth century as general use medicine in the form of Chlorinated Soda (hypochlorite) and Iodine tincture in 1839. Louis Pasteur started the second period of chemotherapy with Robert Koch by establishing the role of microbe's infectious diseases and killed microorganisms by various antiseptics and disinfectants (Collette, 1999).

Herbal remedies used in the traditional folk medicine provide an interesting and still largely unexplored source for the production and development of potentially new drugs for chemotherapy which might help to overcome the growing problem of resistance pathogens, and also the toxicity of the currently available commercial antibiotics. The use of medicinal plants as traditional medicines is well known in rural areas of many developing countries (Gupta et al., 2005). The traditional medicinal methods, especially the use of medicinal plants, still play a vital role to cover the basic health needs in the developing countries in general and in Arabian Peninsula in particular (Ghazanfar, 1993).

ARAB MEDICINE

Herbal medicine is one of the complementary and alternative medicine, which is admired by the public worldwide. There is increasing interest in new therapies and the traditional use of Arab medicine, as well. Historically, Eastern region of the Mediterranean has been distinguished throughout the generations with a rich inventory of natural herbal medicine. It is very well known and documented that native Arab medicine has contributed to a great extent to the development of modern medicine in Europe, as well worldwide, and remains one of the closest forms of original world medicine. There is also a popular belief that the herbal products do not have side effects but slow to act with great effectiveness (Saad et al., 2005).

In the period of golden age of the Muslims (AD 632-1150), the Arabs having conquered Egypt, Syria, Iraq and Persia extended their empire from the borders of India to Spain, and during this period they had 4000 remedies using medicinal plants. The Arab rulers of Middle East included Jafar Al-Mansur 754-775; Harun al Rashid (780-809) and his son Al-Mamun (813-833) have contributed significantly in the development of herbal medicine. In the reign of al-Ma'mun an institution, 'The House of Wisdom' was built and learned from Greek Science through the translators, and the most famous of all the translators was Hunayn Ibn-Is'haq, a Nestorian Christian who became court physician to the Khalif al-Mutawakkil. Hunayn and his team translated a large number of Medical works of Hippocrates and Galen, as well as philosophical works by Plato and Aristotle and mathematical works of Euclid and Archimede (Aarmoosh and AlOmari, 2005).

The writers during this era Ibn Hayyan or Jabir (776), Al Razi (865-925) have given comprehensive literature in medicine. Abu Mansur (970), Ibn Sina (980-1036) wrote books in medicine. Ibn Al-baitar (1197-1248) and Abu Zakaria Yahya ibn al-Awwam (1864-1867) were proclaimed pharmacists. Al Idrissi (1100-1166) wrote a comprehensive botanical-pharmacological book. Abu Hamid Muhamed bin Muhamed ibn Taus Ahmed AlTusi Al Shafii, who was born in 450/1058, and generally known al-Gazali often Al-Gazel in English has contributed a lot in herbal medicine. Al-Ghazali classified medicine as one of praiseworthy (*Mahmud*) secular science, he wrote number of pages of detailed anatomical exposition of the part of human body in his book meditation (tafakkur) (Sharif, 1963).

TAXONOMY OF ZYGOPHYLLACEAE

Generally speaking family of *Zygophyllaceae* comprises of trees, shrubs, sub-shrubs or annual or perennial herbs. But Omani members are usually herbaceous. The plants are regularly found with jointed branches that are swollen at the nodes; axillary or stipular thorns are present. The leaf is petiolate, stipulate, opposite or less often alternate, and it is bi- or trifoliolate or pinnately multifoliolate, but rarely simple; usually with glandular dots, leaf (let) lamina entire, often asymmetric, flattened, fleshy. The flowers are solitary, and set as paired or in few-flowered cymes, axillary or terminal, bisexual, actinomorphic or rarely slightly zygomorphic; sepals 4–6, free, and they are rarely connate at base. Although they are usually imbricate, valvate in *Seetzenia*; petals free, often clawed, mostly as many as sepals, disc often present; stamens 8–12, as many as or twice the number of petals; filaments with basal scales or appendages; anthers introrse, dorsifixed, 4-sporangiate, with longitudinal dehiscence;

ovary syncarpous, superior, sessile, angular, ribbed or winged, 4–5 locular; style filiform; stigma capitate, clavate or slightly lobed; ovules 1 to many per locule, pendulous, usually with axile placentation. The fruit is in a loculicidal or septicidal capsule or split into mericarps which may be winged, lobed or angled, spiny. The seeds are with or without endosperm; embryo straight or slightly curved. The flowers are insect pollinated. *Zygophyllaceae* comprises 30 genera and 230–240 species that grow in hot dry regions of Asia, Africa, Europe, Australia and the Americas.

Zygophyllaceae is one of the 18 families in which the C4 photosynthetic pathway is found. All *Tribulus* sp. so far examined have shown the C4 pathway. *Zygophyllum simplex* has so far been categorized as C4 (Welkie and Caldwell 1970; Sheahan and Cutler 1993). All C4 species exhibit typical Kranz anatomy, except *Z. simplex* which has centric leaves, and the Kranz cells form an incomplete sheath around only the outer part of the vein. (Sheahan, 2007).

Domain: Eukaryota; Kingdom: Plantae; Subkingdom: Viridaeplantae; Phylum: Tracheophyta; Subphylum: Euphyllophytina; Infraphylum: Radiatopses; Class: Magnoliopsida; Subclass: Rosidae; Superorder: Geranianae; Order: Zygophyllales; Suborder: Zygophyllineae; Family: Zygophyllaceae

The following are the Subfamilies

- Larreoideae
- Morkillioideae
- Seetzenioideae
- Tribuloideae
- Zygophylloideae

Zygophyllaceae has about 230-240 species, distributed within, about 30 Genera. In Oman there are only six genera and 24 species (Gazanfar, 1993). The selected species of *Zygophyllaceeae* are as follows:

1) Fagonia bruguieri (Figure 1 A and B)
2) Fagonia indica (Figure 1 C)
3) Fagonia socotrana (Figure 1 D)
4) Seetzenia lanata (Figure 1 G)
5) Tribulus pentanadrus (Figure 1 F)
6) Tribulus tertrestris (Figure 1 E)
7) Tetraena (Zygophyllum) qatarense (Figure 1 H)
8) Tetraena (Zygophyllum) simplex (Figure 1 I)

The taxonomy and botanical description of the selected plants is as stated below:

Fagonia sp.

Diffusely branched small shrubs, sub shrubs or herbs reaching up to 80 cm high, glabrous, pubescent or glandular, often with spinose stipules. Leaves are opposite, mostly trifoliolate, occasionally unifoliolate by abortion of lateral leaflets. Flowers pink to purple, rarely whitish, solitary, axillary; sepals and petals 5; disc inconspicuous; stamens 10, without appendages; stigma simple to minutely lobed; ovary sessile, 5-angled, 5-locular with two or more ovules per locule; style subulate, 5-sided. Fruit a deeply 5-angled or -lobed capsule with persistent style; each carpel ventrally dehiscent. Seeds ellipsoid, pendulous, with bony endosperm, mucilaginous. 2n = 18, 20, 22. About 30–40 species in dry regions of Africa, Asia, the Mediterranean basin and North America, as well the same can be seen in Oman (Miller and Morris, 1988).

Fagonia Bruguieri DC. (Arabic Name: Durayma, Shukka'a, Dhreima, Aaqul, Shayeet)

F. bruguieri (Figure 1 A and B) is most common after rains in nearly all regions of Oman especially Northern region wadi and gravel area (Ghazanfer, 1993). *F. bruguieri* is a perennial, has a woody base, and has horizontal stems branching from base, reaching up to 50 cm, spiny, dark green to purplish, the leaves near the base are trifoliate and upper ones are single, opposite, ovate to elliptic. Flowers start as white change to pink with fragrance and solitary. It has onion shaped fruit capsule (Jongbloed, 2003). *F. bruguieri* is widely spread in sand, hillside, gravel plain and low elevation. This plant can be used as camel feed, used to treat obstruction in the bowels, and also as a remedy for venereal diseases (Yaser et al., 2000). While in Pakistan, the aerial parts of *F. Bruguieri* are dried, powdered and orally used with water to treat ear infections and other ear illness (Mushtaq, et al., 2007). According to Hamiche and Maiza (2006), *F. Bruguieri* is used in central Sahara to treat Jaundice, liver insufficiency, Palpitations, and anxiety.

Figure 1. A, B) Fagonia bruguieri C) Fagonia indica D) Fagonia socotrana E) Tribulus terrestris F) Tribulus pentanadrus G) Seetzenia lanata H) Tetraena qatarensis and I) Tetraena simplex.

Fagonia indica Burm.f. (= F. parviflora Boiss.) (Arabic Name: Shekka'a, Dhreima, Hulaymah)

F. indica (Figure 1 C) is one of common shrub found throughout the year by the foothills, sandy wadis, between mountains, at low elevations, and gravel plains. However, it is a perennial and has a woody base, and the stems are much branched, pale green, grooved, reach to more than 80 cm with unequal spines. Leaves are opposite, elliptic to linear with rounded tips, forming whorl spine at nodes. Flowers are pink, solitary with fragrance, made of 5 backward folded petals. It has an onion shaped fruit with 5 segmented capsule covered with fine short hairs (Jongbloed, 2003). *F. indica* is found to flower after the rainy season, and during winter.

F. indica has many traditional medicinal uses. In Pakistan, aerial parts are dried under shade, grounded to powder, mixed with equal quantities of fruit powder of *Trachyspermum copticum* (Ajwain), and *Terminalla chebula*, the mixture used to treat ear diseases (Mushtaq, et al., 2007). In Oman *F. indica* is used to treat colic, fever combined with *Blepharis ciliaris*, and mixed with thyme to treat kidney stone and eye illness (Jongbloed, 2003). The indigenous knowledge of plant-based remedies from local people and herbal doctors (Hakims) indicated that *F. indica* has shown a good cure for liver and spleen inflammation, by taking its water extract orally and the paste after grinding is used on top of abdominal area of inflamed organ.

Fagonia Socotrana (Balf. f.) Schweinf

F. socotrana (Figure 1 D) is an annual shrub, with opposite leaves, trifoliolate, and becomes unifoliolate by reduction. It has no spines like other *Fagonia* sp., its fruit is an obconical capsule with five lobes. It inhabits the coastal plain, and in Oman it is found only in Southern Dhofar region.

Seetzenia R. Br. can be described as low, creeping annual or perennial herb branching from a woody base, with lax, brittle, jointed branches and pubescent nodes. Leaves are small, opposite, succulent, trifoliolate, petiolate with apiculate leaflets. Flowers solitary, axillary, pedicellate; sepals 5, somewhat fleshy, valvate; petals absent; disc inconspicuous, 5-lobed; stamens normally 5, without appendages; ovary sessile, oblong-clavate, 5-locular with 1 ovule per locule; styles normally 5, stigmas capitate. Fruit a 5-lobed septicidal capsule separating into 5 one-seeded mericarps from a 5- angled central axis. Seeds compressed, with scanty endosperm. One species, *S. lanata* R. Br., is found in sandy and saline deserts, disjunct in North and South Africa, also in the Middle East and Asia.

Seetzenia lanata (Willd.) Bullock (Syn: S. orientalis Decne.) (Arabic Name: Abu Showka, Habein)

Seetzenia lanata (Figure 1 G) is a perennial prostrated herb with woody base and branched stem. It has trifoliate fleshy leaves; on stalk spoon shaped leaflets tapering to base with pointed tips are present. Flowers in *Seetzenia* are solitary, axillary, pedicellate; sepals 5, somewhat fleshy, valvate; petals absent; disc inconspicuous, 5-lobed; stamens normally 5, without appendages; ovary sessile, oblong-clavate, 5-locularwith 1 ovule per locule; styles normally 5, stigmas capitate. However in *S. lanata* fruits are angled distinctive 'lampion', with yellow capsule (Jongbloed, 2003). It is found in graveled wadi beds, and at low elevation hills. It is very rare, and the only place *S. Lanata* was found off the road to the south of SQU campus, by the road going to Wahat Muscat, which might indicate the specificity of this plant in growth habitat. According to Denham and Clapperton, *S. lanata* is found in sandy and saline deserts, disjunct in North and South Africa, also in the Middle East and Asia.

Tribulus Pentanadrus Forsskal (Arabic Name: Halwat albidar, Shershir)

T. pentanadrus (Figure 1 F) is found growing wherever there is a source of water. Its profound area is in date palm plantations and alfalfa farms especially in eastern region of Oman (Welayat Mudaibi), or after the rain in sandy plains and wadis. It is a perennial, branched from the base, prostrate or spreading, covered with short white hairs. The leaves are paired with 12 leaflets. Flower solitary, yellow and fruits are found along the underside of stems consist of 5 segments.

Tribulus Terrestris L. (Arabic Name: Shershir, Um qreen, Qatb, Hisek, Shiqshiq)

T. terrestris (Figure 1 E) is a prostrating plant with stem branching from the base and reach up to a meter or more, depending on the place where it is found. It is an annual or biennial, and leaves are paired with one shorter than the other, 8-12 oval to oblong leaflets, with rounded tip. Flower yellow, solitary with long stalk, arise from leaf nodes with pointed sepals, petals are double the size of sepals. Hairy, thorny fruits, consisting of five segments, each with distinctive lateral five spines. *T. terrestris* can be found everywhere in urban areas, roadsides, wadis, sand and gravel plains. *T. terrestris* is used in folk medicine as tonic, aphrodisiac, analgesic, astringent, stomachic, anti-hypertensive, diuretic, lithon-triptic and urinary anti-infective (Majeed and Mahmood, 1988; Saad Aldein, 1986). *T. Terrestris* been reported to have antibacterial activity and is used in different parts of Turkey and Iran (Abbasoglu and Tosun, 1994; Kianbakht and Jahaniani, 2003). In North Africa its fruits are used to treat oral inflammations, as a tonic, and as a cure for bladder disorders (Jongbloed, 2003). In addition, Central Sahara residents use *T. terrestris* for Kidney diseases: stones, urinary decrease, intercostal pains, rheumatism, aphthae, mycosis, palpitations, helminthiases, dysmenorrhea, aphrodisiac, and spleen illness (Hammiche and Maiza, 2006).

Tetraena Qatarensis (Hadidi) Beier and Thulin (Syn. Tetraena Qatarense Hadidi) (Arabic Name: Harm, Rotreet, Balbal)

Z. qatarense (Figure 1 H) is a shrub with cylindrical succulent green glabrous compound leaves, cylindrical fleshy leaflets, with at least 10 mm long, terminating a slightly longer fleshy petiole. Capsules are club shaped, with obtuse apex. The stem is much branched, and erect. It is a desert shrub, up to 75 cm high, green and glabrous. The plant is grazed by camels, and the seeds are employed as a substitute of pepper. In traditional medicine it is used as anthelmintic and diuretic. Fruits are used in the treatment of asthma, diabetes. An aqueous extract of the plant exhibited antihistaminic, antipyretic, diuretic, hypotensive and local anesthetic activities, and produced stimulation and depression of isolated amphibian heart, relaxation of isolated intestine, contraction of the uterus, and vasodilation. The extract antagonized acetylcholine action on skeletal muscle and acted additively with the muscle relaxant effect of d-tubocurarine. The LD50 when injected subcutaneously into toads was found to be 25 ml/100g (Saad et al., 1967).

Tetraena simplex (L.) Beier and Thulin (Syn. Tetraena simplex L. and Z. Coccineum L. (Arabic Name: Harm, Rotreet, Abu rukaiba, (Qirm, Qarmal) (Figure 1 I)

The plant description is similar to ***Tetraena qatarensis***, except for its traditional medicinal use. *Tetraena simple* (L.) is used for treating rheumatism, asthma, diuretic, cough, and conjunctivitis (Saad et al., 1967).

Phytochemistry

Phenolic compounds including methylated flavonoids and lignans are frequent in the family (Table 1).

Table 1. Identified Compounds in *Zygophyllaceae*

Species	Compounds	Reference
Fagonia bruguieri DC	15,16-dihydroxy-7-oxo-cis-ent-erythrox-3-ene (fagonone) and 16-O-acetylfagonone	Abdel-Kader, et al., 1993
Fagonia indica Burm.f.	Triterpenoid	Shaker et al., 1999
Fagonia socotrana (Balf. f.) Schweinf.		
Seetzenia lanata (Willd.) Bullock		
Tribulus pentandrus Forssk.	Kaemperol 1a-f ,Quercetin 2a-h, Isorhamnetin 3a-d and Tricin	AlBar
Tribulus terrestris L.	Kaempterol 1a, 1g, 1c, 1d ,Quercetin 2a,2c d,2g,2h, Steroid sapogenin, Spirostanol Saponin 47-38, and Isorhamnetin 3a,3f Steroidal	AlBar, Xu et al., 2000, Cai et al., 2001, Kostova et al., 2002, de Combarieu et al., 2003
Tetraena qatarense Hadidi	Zygophyloside F [a] + 23 (MeOH, c 0.25), Zygophyloside G mp 240-244, Zygophyloside H mp 215-219, 3-O-[beta-D-glucopyranosyl]-quinovic acid 28-O-[beta-D-glucopyranosyl] ester, 27-Methyl Zygophyloside F, 3-O-[alpha-L-arabinopyranosyl-(1 -- 2)-beta-D-quinovopyranosyl]-quinovic acid, 3-O-[beta-D-2-O-sulphonylquinovopyranosyl]-quinovic acid, 3-O-[beta-D-glucopyranosyl[-quinovic acid, 3-O-[beta-D-quinovopyranosyl[-quinovic acid 28-O-[beta-glucopyranosyl] ester, 3-O-[beta-D-quinovopyranosyl]-quinovic acid, Quinovic acid, quinovic acid 3-alpha-L-rhamnoside, Kaempferol 3-O-glucoside	Elgamal, M. et al. (1995) Pollmann, K. et al. (1997), H. A. Hassanean and E. K. Desoky, (1992), A. M. Attia, (1986), H. A. Hassanean, H., et al. (1989)
Tetraena simplex L.	6''-(2-E-butenoyl)isohamnetin-3-O-glucoside, Isorhamnetin, Isorhamnetin-3-O-glucoside, Kaempferol 3-O-glucoside, Sitosterol glucoside, and quinovic acid 3-alpha-L-rhamnoside	H. A. Hassanean and E. K. Desoky, (1992), A. M. Attia, (1986), H. A. Hassanean, H., et al., (1989)

Lignans and neolignans abound both in terms of different compounds and quantity. The wood of *Guaiacum* contains 15–20% resin, mainly constituted of lignans (Hegnauer, 1990). The neolignan nordihydroguaiaretic acid (NDGA) is known from *Bulnesia, Guaiacum, Porlieria* and *Larrea*. In *Larrea*, the leaves are covered by wax and appear "varnished". The wax is a complex mixture containing much NDGA (up to 10% of the dry weight of the leaves of *L. tridentata*) and various methylated flavonoid glycones. The high reactivity of NDGA to oxygen and especially the reactivity of the oxidized NDGA to hydroxyl and amino groups probably account for its effectiveness as a defense substance against herbivores (Mabry et al., 1977 and Sheahan, 2007). *Zygophyllacea*e are among the relatively few families which produce steroid and triterpenoid saponins. According to Hegnauer (1990), these may also be responsible for the observed resistance to herbivore activity. The family also produces the quinazoline alkaloids harman (e.g., in *Fagonia cretica* and *Tribulus terrestris*), harmin and harmol (in *Tetraena fabago*). Mucilage has been found in the leaf epidermis of *Augea* and is reported in the epidermis of *Plectrocarpa* (Rojas et al., 2001); they may also be abundant in seeds. Calcium oxalate crystals are frequent, sometimes very abundant; these are mostly prismatic and acicular. Recent molecular work has led to a review of the taxonomy of the family (Sheahan and Chase 1996, 2000). According to these studies, the inclusion of *Augea* in *Zygophylloideae* has been confirmed, in spite of many morphological and anatomical autapomorphies, and this relationship is supported by chemical similarities (Maksoud and El-Hadidi 1988; Narayana et al., 1990). It has also been proposed that Engler's *Zygophylloideae* should be divided into four well-supported monophyletic subfamilies: Seetzenioideae, Larreoideae, Tribuloideae and Zygophylloideae sensu stricto (Sheahan and Chase 1996, 2000). *Seetzenia* was previously thought to be related to *Fagonia*, chiefly on account of its trifoliolate leaves, but palynological and chemical evidence (Lahham and Al-Eisawi 1986) as well as molecular data supports its isolated position in a separate subfamily. A substantial revision of this subfamily has recently been made, using both molecular and morphological characters, by Beier et al. (2003). They found six monophyletic groups within the subfamily: *Fagonia* and *Augea* retain their generic status.

Antimicrobial Compounds

An anti-microbial is a substance that kills or inhibits the growth of microorganisms such as bacteria, fungi, or protozoans. Antimicrobial drugs either kill microbes (microbiocidal) or prevent the growth of microbes (microbiostatic). The history of antimicrobials begins with the observations of Pasteur and Joubert, who discovered that one type of bacteria could prevent the growth of another. They did not know at that time that the reason one bacterium failed to grow was that the other bacterium was producing an antibiotic. Technically, antibiotics are only those substances that are produced by one microorganism that kill, or prevent the growth, of another microorganism. Of course, in today's common usage, the term antibiotic is used to refer to almost any drug that attempts to rid your body of a bacterial infection. Antimicrobials include not just antibiotics, but natural compounds and synthetically formed compounds as well. However, with the development of antimicrobials, microorganisms have adapted and become resistant to previous antimicrobial agents. The old antimicrobial technology was based on poisons, which may not have killed the microbe completely, allowing the microbe survive, change (mutate), and become resistant to the poisons. Even though pharmacological industries have produced a number of new antibiotics in the last three decades, resistance to these drugs by microorganisms has increased. In general, bacteria have the genetic ability to transmit and acquire resistance to drugs, which are utilized as therapeutic agents (Kumar, 2004). Such a fact is cause for concern, because of new multi-resistant bacterial strains. Consequently, new infections can occur resulting in high mortality. From 1980 to 2010, a high incidence of resistant microorganisms is documented in scientific literature.

Traditional healers long have used plants to prevent or cure infectious disease. Many of these plants have been investigated scientifically for antimicrobial activity, and a large number of plant products have been shown to inhibit the growth of pathogenic microorganisms. A number of these agents appear to have structures and modes of action that are distinct from those of the antibiotics in current use, suggesting that cross-resistance with agents already in use may be minimal. So, it is worthwhile to study plants and plant products for activity against resistant bacteria. The use of plant extracts and phytochemicals, with known antimicrobial properties, can be of great significance in therapeutic treatments (Thulin, 1993).

The antimicrobial properties of plants have been investigated by a number of researchers' worldwide. Among the compounds extracted from the plants, several inhibited the growth of *Staphylococus aureus, Escherichia coli,* and *Aspergillus niger*, besides other bacterial and fungal species (Aarmoosh, H., and M., AlOmari. 2005, Al-Yahya et al., 1985). In this study, Zygophyllaceae plants are chosen, that are used in folk medicine to determine their antimicrobial activity. In general, these plants are used in traditional medicine in the treatment of skin diseases, pain relievers, venereal diseases, respiratory problems and nervous disorders. The active ingredients in most medicinal plants from the Arabian Gulf area are mainly flavonoids and tannins, which are present in more than ninety percent of the recorded species (Al-Yahya et al., 1985; Rizk, 1986; Rizk and Al-Nowaihi, 1989).

The Mechanism of Antimicrobial Activity

The mechanism of antimicrobial activity of compounds were found to correlate with the lysis of liposomes, where binding followed by slow disruption of the lipid membrane was observed Steiner et al., (1988) Rojas (2001). Lysis of bacteria occurs in the micromolar range, which is just sufficient peptide to form a monolayer on all bacterial cells. According to a study done by Christensen et al., (1988) on the electrical conductance of the peptides in planar lipid bilayers, they suggested that the mechanism of antibacterial action involves the following: 1) positively charged peptide makes an electrostatic contact with the negative bilayer of phosphatidylserine and diphytanoyl-phosphatidylcholine) in this a polar lipid environment, the peptide assumes an amphipathic helical conformation that inserts itself into the lipid; 3) under a voltage gradient, several of these amphipathic helices would then aggregate to form a pore, with the outside hydrophobic and the peptide in van der Waals' contact with the lipid, while the inside would be water filled and more hydrophilic and positively charged (Rahman and Ansari, 1984).

One of the major problems occurring these days is the antibiotic resistance phenomena. The prior discovered antibiotics seem to be degrading their medicinal and clinical values, due to the antibiotic resistance occurred among users and the surrounding environment (Gislene et al., 2000). Through this research work attempt to come up with alternative antimicrobial compounds from medicinal plants of Oman. This would increase the potential of finding novel antibiotics from different kinds of medicinal plants. Another problem is the cultivation of the diverse bacteria in the *in-vitro* condition. This is because it is not easy to replicate the same condition that the bacteria are living in, on agar or liquid culture. There is still a lot of optimization to be performed and parameters to be considered in order to fit the demand of various types of bacteria, in terms of finding out their antibacterial properties and the active site of the biocatalysts or bioactive compounds.

The screening of biologically active natural compound needs precise bioassays. However, to detect active compounds in complex plant extract depends on the sensitivity of the test systems. Once compounds are extracted, and before being used in new therapeutic treatments, they should have their toxicity tested *in vivo*. Bioassays have to examine the toxicity of extracts from different plants. Therefore, the methods used in bioassay must be simple, inexpensive and rapid as the number of tested samples is large. Besides, they ought to be quite sensitive in order to detect active principle which normally is found in small quantity in crude extracts. There are many kinds of bioassays used to isolate different types of bioactive compounds. A general bioassay that appears capable of detecting a board spectrum of bioactivity present in crude extracts is the brine shrimp lethality bioassay. The technique is easily mastered, costs little, and utilizes small amount of test material. The aim of this method is to provide a front-line screen that can be backed up by more specific and more expensive bioassays once the active compounds have been isolated. It appears that BST is predictive of cytotoxicity and pesticidal activity (Ghisalberti, 1993). The brine shrimp assay was proposed by Michael et al., (1956) and later developed by Vanhaecke et al. (1981); Meyer et al., (1982); Sleet and Brendel (1983). The assay is considered a useful tool for preliminary assessment of toxicity and it has been used for the detection of plant extract toxicity (McLaughlin et al., 1991), cytotoxicity testing and for fungal toxins. In this study brine shrimp lethality test (BST) and antimicrobial bioassay are used.

Antimicrobial activities can be assessed by different types of screening assays. One of these methods is bioautography, which is very convenient and easy procedure for testing plant extracts as well pure substances for their effect on human pathogenic and plant pathogenic microbes. In isolation of these mixture compounds it is a usual practice that a number of different separation techniques (TLC, column chromatography, flash chromatography, Sephadex chromatography and HPLC) need to be used in order to obtain pure compounds for the structure and biological activity determination (Oleszek, and Bialy, 2006).

However, two of these techniques are going to be used as bioassay of antimicrobial effects in our present study: i) direct by TLC bioautographic detection, and ii) solid agar (diffusion method). In TLC bioautography, *Cladosporium herbarum* was grown directly on the TLC plates with the plant samples. This procedure was recommended for spore producing fungi like the one we used and bacteria. The technique is a sensitive test for locating bioactive compounds on the chromatogram, and its antifungal activity by growing the fungus directly on the TLC plate (Rahalison et al., 1991, Nostro et al., 2000). Marston et al. (1993) tested several hundred plant extracts using bioautography on TLC plates with the fungi *Cladosporium cucumerinum* and *Candida albicans* to detect molluscicidal activity. Bioautography is a postchromatographic detection method that is widely used for the bioassay of antimicrobial effects. This technique is based on paper or thin-layer chromatography (TLC) separations. Bioautography detects the growth-inhibiting or growth-promoting biological effects of the applied substances. Various bioautographic assays may be used to detect antibacterial, antifungal, antiprotozoal, and cytotoxic substances. The bioactivity-based bioautographic analysis is the most effective method to detect antimicrobial compounds because it detects activity even in complex sample material. Hostettmann and Marston, (2002) and Kanatiwela and Adikaram, (2009) have used this technique effectively to detect antifungal compounds and fungicide residues.

In the diffusion method, the filter paper discs are soaked in a specific concentration of extract and dried and then applied on top of inoculated solid agar plates with known tested microbes. After incubation the diameter of the clear inhibition zones around the discs can be measured.

Phytochemistry investigation revealed the presence of more than seventy compounds in the selected species which include phenolics, methylated flavonoids. Sheahan and Cutler (1993), and Hegnauer (1973, 1990) have also reported a large number of compounds. The molecular weight was used to check the presence of the compound in each spectrum. Among the identified compounds the following are noteworthy: isorhamnetin-3-glucoside, isorhamnetin-3-O-rutinoside, herbacetin-8-methyl-ether-3-Orutinoside, herbacetin-8-methylether-3, 7-O-diglucoside and herbacetin-8-methylether-3-rutinoside-7-glucoside. Some of these were identified earlier in Fagonia arabica as well as other species (El-Negoumy et al., 1986; Saleh et al., 1990). Several glycosides belonging to kaempferol, quercetin and isorhamnetin were isolated and identified from Tribulus terrestris (Saleh and El-Hadidi 1977; Saleh et al., 1982). Quercetin- 3-O-rutinoside, isorhamnetin-3-0-glucoside and isorhamnetin-3-O-rutinoside were isolated and identified from Z. album (Saleh and El-Hadidi 1977, Sparg et al., 2004).

The ethanol fraction of Fagonia indica showed significant effects onto E. coli with clear wide zone of 17 mm. The compounds identified in this fraction are: Cis-O-Coumarinic acid lactone, 3-O-Methylbatatasin III, Lactaroviolin, 12-aminododecanamide. The petroleum ether, ethyl acetate, n-butanol, and aqueous fractions exhibited activities on all other three bacteria used in the test viz. St. aureus, Ps. aeroginosa, and B. subtilis, and in each fraction there is a close similarity in the inhibition zone diameter between species inhibited, except in case of aqueous fraction which has the highest inhibition zone against St. aureus compare to others. In addition, aqueous fraction had a large number of identified active compounds (kaempferol 3-O-rutinoside, gossypetin 3, 7, 3'-trimethyl ether, 6-Isohexenyl-α-naphthoquinone), (4-Dimethylamino-2(5H)-Furanone (Isoguvacine), 2,3-Benzofuran, 3-Nor-4-patchoulene). These were isolated from Z. album by Saleh and El-Hadidi (1977).

The ethanol and aqueous fraction of all tested species of plants are very effective and exhibited the inhibition zones. Similarly bioautography studies and brine shrimp lethality tests support the same view indicating the possible use of the plants from *Zygophyllaceae*, can be used in pharmaceutical industry. The aqueous fraction proved by Shirfule et al. (2009) as well this study adds further proof for the effectiveness of these ethano-botanical plants in the traditional treatment of various diseases and illness. Definitely it is quite possible to obtain pure isolated compounds from the species studied provided the time and right equipment are provided.

Conclusion

Through this research work an attempt was made to identify the alternative traditional methods for antimicrobial compounds from medicinal plants. After the screening process and detection of antimicrobial compounds, the best fractionation noted was the aqueous fraction in comparison with the rest of the fractions. More purification and understanding of antibacterial compounds properties are needed. The ability of its preparative-scale production through "optimization" in various media for the isolation of antimicrobial compounds would be added advantage. Further study is needed in order to get compounds available for pharmaceutical industry and public use.

Acknowledgments

A great thanks to the university administration, the research team, and my colleagues who were of good help in completing this work on some of Omani medicinal plant Zygophyllaceae sps.

References

Aarmoosh, H., and M., AlOmari. 2005. *Herbs in a Book*, Dar Al-Nafaes publishing. Syria-Damascus. (in Arabic).

Abbasoglu, U., Tosun, F. 1994. Antimicrobial activity of *Tribulus terrestris* L. growing in Turkey. *Hacettepe Universitesi Eczacilik Fakultesi Dergisi*, 14:81-85.

Al-Yahya MA, Al-Meshal IA, Mossa JS, Al-Badr A, Tariq M 1985. Saudi Plants: A phytochemical and Biological Approach. King Saud University- Riyadh Saudi Arabia.

Beier, B.-A., Chase, M. W., Thulin, M. 2003. Phylogenetic relationships and taxonomy of subfamily Zygophylloideae (Zygophyllaceae) based on molecular andmorphological data. *Pl. Syst. Evol.* 240:11–39.

Collette, Sheila. 1999: *Wildflowers of Saudi Arabia- Riyadh.* National Commission for Wildlife Conservation and Development (NCWCD) Publishing. K. S. A. Riyadh.

Cox, P.A., Balick, M. J. 1994. The ethnobotanical approach to drug discovery. *Scientific American.* 270, 82-87.

Dagmar J., K. anovska, Katerina Kubikova and Ladislav Kokoska. 2003. Screening for antimicrobial activity of some medicinal plants species of Taditional Chinese medicine. Csech. *J. food. Sci.* 2003. 21:107-110.

Fatope, M. O.; Al-Kindy S. M. Z. and Abdulnour A. O., 2000, Natural Products as pest, Microbial Disease and tumor Control Agents, *Science and Technology, Research Trends (Special Review),* 55-71.

Ghazanfar, S. A., 2003. Flora of the Sultanate of Oman. Vol.1: Piperaceae – Primulaceae (text + photo CD). National Botanic Garden of Belgium, Meise, Belgium.

Gislene G. F. N., J. Locatelli1, P. C. Freitas, Giuliana L. S. 2000. Antibacterial activity of plant extracts and phytochemicals on antibiotic resistant bacteria. *Brazilian Journal of Microbiology*, 31:247-256.

Gupta MP, Solis PN, Calderon AI, Guionneau-Sinclair F, Correa M, Galdames C, Guerra C, Espinosa A, Alvenda GI, Robles G, Ocampo R. 2005. Medical ethnobotany of the Teribes of Bocas del Toro, Panama. *J. Ethnopharmacol.* 96:389-401.

Hammiche V. and Maiza, K. 2006. Traditional medicine in Central Sahara: Pharmacopoeia of Tassili N'ajjer. *Journal of Ethnopharmacology*. 105, 358-367.

Hegnauer, R., 1990. Chemotaxonomie der Pflanzen.pflanzenEine U¨ bersicht u¨ber die Verbreitung und die systematische Bedeutung der Pflanzenstoffe, vol. 9. Birkha¨ user Verlag, Basel/Boston/Stuttgart.

Horton J. 2003. Global anthelmintic chemotherapy programs: learning from history. *Trends Parasitol*; 19: 405–99.

Hostettmann K. and A. Marston. 2002. Twenty years of research into medicinal plants: Results and perspectives. *Phytochemistry Reviews* 1: 275–285,

Jongbloed, M. V. The comprehensive guide to the Wild Flowers of the United Arab Emirates. Environmental Research and Wildlife Development Agency (ERWDA), Abu Dhabi, 2003.

Kanatiwela H. M. C. K. and N. K. B. Adikaram. 2009. A TLC-bioassay based method for detection of fungicide residues on harvested fresh produce. *J. Natn. Sci. Foundation Sri Lanka* 2009 37 (4):257-262.

Kianbakht, S., Jahaniani, F., 2003. Evaluation of antibacterial activity of *Tribulus terrestris* L. growing in Iran. *Iranian J. Pharmacol. Ther.*, 2: 22-24.

Kumar, P., 2004. Valuation of medicinal plants for pharmaceutical uses. *Current Sci.*, 86: 930-937.

Lahham, J. N., Al-Eisawi, D. 1986. Pollen morphology of Jordanian Zygophyllaceae. *Candollea* 41:325–328.

McLaughlin, J. L., 1991, *Crown gall tumours on potato discs and brine shrimp lethality*. Two simple bioassays for higher plant screening and fraction.

Meyer B. N., Ferringni N. R., Putnam L. B., Nichols D. E., McLaughlin J. L., 1982, Brine shrimp: A convenient general bioassay for active plant constituents. *Planta Med.*, 45, 31-34.

Mohana, D. C., S. Satish and K. A. Raveesha, 2008: Antibacterial Evaluation of Some Plant Extracts Against Some Human Pathogenic Bacteria, *Advances in Biological Research* 2 (3-4): 49-55.

Mushtaq A., M. A. Khan, M. Zafar and S. Sultana, 2007: Treatment of common ailments by plant-based remedies among the people of district attock (Punjab) of northern Pakistan, *Afr. J. Trad. CAM* (2007) 4: 112 – 120.

Nostro, A., Germano, M. P., D'Angelo, V., Marino, A., Cannatelli, M. A., 2000: Extraction methods

Oleszek, W., Z. Bialy. 2006. Chromatographic determination of plant saponins—An update (2002–2005). *Journal of Chromatography A*, 1112. 78–91.

Rahalison, M. Hamburger, K. Hostettmann, M. Monod and bioautography for evaluation of medicinal plant antimicrobial activity. Letters in Applied Microbiology 30, 379–384E. Frenk, 1991 A bioautographic agar overlay method for the detection of antifungal compounds from higher plants, *Phytochemical Analysis* 2:199–203.

Rahman, A. U., and Ansari, A. A. 1984. *Journal of Natural Products*, 47, 186.

Rojas, G., Levaro, J., Tortoriello, J., Navarro, V., 2001. Antimicrobial evaluation of certain plants used in Mexican traditional medicine for the treatment of respiratory diseases. *Journal of Ethnopharmacology* 74, 97–101.

Rizk AM 1986 The Phytochemistry of the flora of Qatar. Doha, State of Qatar Scientific and Applied Research Centre University of Qatar.

Rizk AM, Al-Nowaihi A (1989) the phytochemistry of the horticultural plants of Qatar. Doha, Qatar Scientific and Applied Research Centre, University of Qatar.

Rojas J. J, Ochoa VJ, Ocampo SA, Muñoz JF. 2006. Screening for antimicrobial activity of ten medicinal plants used in Colombian folkloric medicine: A possible alternative in the treatment of non-nosocomial infections. *BMC Compelmentary and Alternative Medicine* 6:2

Saad Aldein, S. M., 1986. Medicinal Herbs, 1st Ed. Dar Al-Shoun Al-Thaqafia Al-Aama for Publishing, Baghdad, p.70 (in Arabic).

Sharif, M. M. 1963. A History of Muslim Philosophy. Volume I, Pakistan Philosophical Congress.

Sheahan, M. C., Cutler, D. C. 1993. Contribution of vegetative anatomy to the systematics of the Zygophyllaceae *R. Br. Bot. J. Linn. Soc.* 113:227–262. Sharaf M, El-Ansari MA, Matlin SA, Saleh NAM (1997) Four flavonoid glycosides from Peganum harmala. *Phytochemistry* 44: 533–536.

Sheahan M. 2007. Zygophyllaceae. In: Flowering Plants: Eudicots. The Families and Genera of Vascular Plants. 9, 488-500.

Sheahan, M. C., Chase, M. W. 1996. A phylogenetic analysis of Zygophyllaceae R. Br. based on morphological, anatomical and rbcL DNA sequence data. *Bot. J. Linn. Soc.* 122:279–300.

Sheahan, M. C. and M. W. Chase, M. W. 2000. Phylogenetic Relationshipsrelationships within Zygophyllaceae Based based on DNA Sequences of Three Plastid Regionssequences of three plastid regions, with Special Emphasisspecial emphasis on Zygophylloideae. *Systematic Botany*, Vol. *Syst. Bot.* 25, No. 2, pp. 371-384.

Sleet, R. B. and Brendel, K. 1983. Improved methods for harvesting and counting synchronous populations of Artemia nauplii for use in developmental toxicology. *Ecotoxicologyl and Environmental Safety*, 7: 435-446.

Steiner, H., Andreu, D. and Merri¢eld, R. B. 1988 Binding and action of cecropin and cecropin analogues: antibacterial peptides from insects. *Biochim. Biophys. Acta* 939, 260-266.

Taylor, L., 2000. *Plant based drugs and medicine,* 2000, www.rain-tree.com

Thulin, M. 1993. Zygophyllaceae. Flora of Somalia, vol. I. Royal Botanic Gardens, Kew.

Vanhaecke, P., Persoone, G., Claus, C. and Sorgeloos, P. 1981. Proposal for a short-term toxicity test with Artemia nauplii. *Ecotoxicologyl and Environmental Safety*, 5: 382-387.

Welkie, G. W., Caldwell, M. 1970. Leaf anatomy of species in some dicotyledon families as related to the C3 and C4 pathways of carbon fixation. *Canad. J. Bot.* 48:2135–2146.

Zhao, G., Hui, Y., Rupprecht, J. K., McLaughlin, J. L., and Wood, K. V. (1992). Additional bioactive compounds and trilobacin, a novel highly cytotoxic acetogenin, from the bark of Asimina triloba. *Journal of Natural Products*, 55, 347-356.

In: Biotechnology and Conservation … (2 Volume Set) ISBN: 978-1-63117-869-6
Editors: S. A. Farooq, R. Abed and S. Baqir

Chapter 30

ANTIOXIDANT CONTENT OF HOT PEPPER FROM THE USDA NATIONAL COLLECTION

***T. S. Kochhar*[1*] *and G. F. Antonious*[2]**

[1]Natural Sciences Division, Blue Grass Community and Technical College, Lexington, KY, US

[2]College of Agriculture, Food Science, and Sustainable Systems, Kentucky State University, Frankfort, KY, US

ABSTRACT

The search for beneficial phytochemicals is growing worldwide. Phenols, ascorbic acid, capsaicin, and β-carotene are some of the several classes of naturally occurring compounds having antioxidants activity. Seeds of several accessions (genotypes) of hot pepper (*Capsicum chinense*) were collected from Belize, Brazil, Colombia, Ecuador, Mexico, Peru, Puerto Rico, and United States. Seeds were field grown in a silty-loam soil. Mature fruits of hot pepper were analyzed for capsaicin, ascorbic acid, β-carotene and phenol contents, which are important antioxidants of a number of benefits for human health. The main objective of this investigation was to select candidate accessions of hot pepper having high concentrations of these phytochemicals for use as parents in breeding for these antioxidant compounds. Fruits of C. *chinense* accessions PI-640900 (USA) contained the greatest concentration of capsaicin (1.52 mg g^{-1} fresh fruit) and dihydrocapsaicin (1.16 mg g^{-1} fresh fruit), while total major capsaicinoids (capsaicin and dihydrocapsaicin) in the fruits of PI-438648 (Mexico) averaged 2 mg g^{-1}fresh fruit. PI-152452 (Brazil) and PI-360726 (Ecuador) contained the greatest concentrations of ascorbic acid (1.2 and 1.1 mg g^{-1} fresh fruit, respectively), while PI-438648 (Mexico) contained the greatest concentration of total phenols content (349 μg g^{-1} fresh fruit). PI-355817 from Ecuador contained the greatest concentration of β-carotene among the other 63 accessions tested. These selected accessions were identified as potential candidates for use in hot pepper breeding programs and for mass production of fruits with value-added traits.

Keywords: Ascorbic acid, β-carotene, Total phenols, Capsaicin, Hot pepper genotypes

* Corresponding author: tkoch403@gmail.com.

INTRODUCTION

Currently there is considerable interest in the enhancement of compounds in food that possess health-promoting attributes such as antioxidant properties, which were previously regarded as non-nutritive (Van der Sluis et al., 2002). Such bioactive compounds are referred to as "phytochemicals". Compounds acting as antioxidants are likely to be beneficial to life by promoting cellular defenses and by acting as free radical scavengers and reducing agents (Oboh, 2006). The fruits of pepper (*Capsicum* spp.) contain numerous compounds with antioxidant activity such as capsacinoids, β-carotene, phenols, and ascorbic acid (Shahidi, 2000). The hydrocarbon carotenoid β-carotene is found widely in chloroplasts of higher plants. This compound has been known to possess provitamin-A activity and also is a powerful antioxidant.

The cancer-preventive activities of carotenoids have been associated with their antioxidant properties (Weissenberg et al., 1997). Carotenoids are synthesized in plants, fungi, bacteria, and algae, but not in humans (Tapiero et al., 2004). Perera and Yen (2007) reported that consumption of careteniod-rich foods reduced the incidence of several diseases such as cancers, cardiovascular diseases, age-related macular degeneration, cataracts, disorders related to low immune functions, and other degenerative diseases.

Capsaicinoids are alkaloids important in pharmaceutical industry for their neurological effects. In the fruit of *Capsicum*, they account for pungency or 'heat'. When used at low levels in the diet, capsaicinoids significantly decrease serum, myocardial and aortic total cholesterol levels. Two major capsaicinoids, capsaicin and dihydrocapsaicin are responsible for about 90% of the pungency in hot pepper (Govindarajan and Sathyanarayana, 1991).

Phenols may interfere with various stages of the cancer process, resulting in reduction of overall cancer risk. They prevent oxidative damage to biological molecules such as DNA, lipids and proteins which play a role in chronic diseases such as cancer and cardiovascular disease. Plant phenols include simple phenols, flavonoids, anthocyanins, lignans and lignins, stilbenes and tannins (Hollman, 2001). The role of phenols as antioxidants with properties similar to vitamins C, E, and β-carotene have prompted a number of studies of these compounds. A wide variety of phenolic compounds derived from spices, like capsaicin possess potent antimutagenic and anticarcinogenic properties. Some mammalian metabolites of polyphenols and tannins (PPT) may protect the vascular endothelium (Surh and Seoul, 2002). Diets rich in PPT may have the ability to protect against type-2 diabetes through their effects on glucose absorption and associated hormones (Clifford, 2004).

When derivatives of ascorbic acid were tested on cancer cells, ascorbic acid esters revealed promising anticancer activity. Additionally, ascorbic acid found in most fruits and vegetables protects against heart disease, high cholesterol, high blood pressure, and cancer (Naidu, 2003). Some pepper varieties have significant inhibitory effects on carbohydrate-degrading enzymes such as intestinal α-glucosidases that are related to glucose absorption. Hydrolysis of dietary carbohydrates, such as starch, is the major source of glucose in the blood (University of Maryland Medical Center, 2008). It has been previously suggested that screening of pepper accessions for their ability to control/regulate intestinal glucose absorption may result in an alternative means for the dietary management of type-2 diabetes (Kwon et al., 2007).

Capsicum chinense has been referred to as the most cultivated pepper in South America (DeWitt and Bosland, 1996). Pepper varieties are grown for their food value, health-promoting properties (Padilla and Yahia, 1998), and as source of capsaicinoids for medicinal uses (Sicuteri et al., 1990). However, at present, limited information is available on variability in the composition of the fruit of this species for health-promoting compounds. Variability in the presence and concentration(s) of phytochemicals in pepper fruit could be a factor affecting their selection for the use of as parents in an improvement program.

The objectives of this investigation were: 1) to determine the concentration of β-carotene, total phenols, ascorbic acid, and capsaicinoids in 63 hot pepper accessions of *Capsicum chinense* collected from eight countries; and 2) to identify accessions within those with greatest concentrations of β-carotene, ascorbic acid, and total phenols for potential use in crop improvement.

MATERIALS AND METHODS

Seeds of 63 accessions of Capsicum chinense were obtained from USDA/ARS Capsicum germplasm collection in Griffin, GA. These accessions represented cultivars and landraces originally acquired from a variety of locations including: Belize (n=9), Brazil (n=7), Colombia (n=8), Ecuador (n=6), Mexico (n=10), Peru (n=10), Puerto Rico (n=6), and United States (n=7). Seeds were sown in March and the seedlings were transplanted in the field in May into rows about 1.5 m apart and 0.25 m between plants within rows. Plants were fertilized and weeded as needed. Randomly selected fruits of each accession were harvested at full maturity.

At harvest, pepper fruits were cut into small pieces and 30 g representative subsamples were blended in a household blender at high speed with 100 mL of acetone for 2 min in dim light to extract β-carotene (Antonious and Kasperbauer, 2002). The homogentate was filtered with suction through a Buchner funnel containing Whatman filter paper No.1 (Fisher Scientific, Pittsburg, PA).

The resulting thick paste was extracted twice with acetone until the extract was colorless. The filtrates were combined, transferred to separatory funnel containing 50 mL of 4% aqueous NaCl and 100 mL of petroleum ether (BP 40-60O C). Absorption of the petroleum ether layer was measured at 450 nm in dim light. A calibration curve was prepared for each group of samples using 99% pure β-carotene in the range of 10-100 μg mL-1.

Representative fruit samples (20 g) were blended with 150 mL of ethanol to extract phenols. Homogenates were filtered through Whatman No. 1 filter paper and one mL aliquots of filtrate were used for determination of total phenols (McGrath et al., 1982) using a standard calibration curve (1 to16 μg mL-1) of chlorogenic acid. Ascorbic acid was extracted by blending 20 g of fruit with 100 mL of 0.4 % (w/v) oxalic acid solution and determined by the dichlorophenolindophenol method (Association of Official Chemists, 1970).

Purified standards of β-carotene, ascorbic acid, and chlorogenic acid were obtained from Sigma-Aldrich Inc. (Saint Louis, MO 63103, USA) and used to obtain calibration curves. Concentrations of each compound, expressed on a fresh weight basis, were statistically analyzed using ANOVA procedure. Means were compared using Duncan's multiple range test (SAS Institute, 2003).

RESULTS AND DISCUSSION

The concentration of the phytochemicals analyzed varied significantly among accessions from the same country of origin, and between countries of origin. Concentrations of ascorbic acid in two accessions, PI 152452 (Brazil) and PI 360726 (Ecuador), were significantly higher (1,224 and 1,139 μg g-1 fresh fruit, respectively) (Figure 1, upper graph) than other accessions analyzed. These accessions could be useful as parents in hybridizations to produce high ascorbic acid containing varieties. Fruits of accessions that originated in Ecuador contained the greatest concentration of ascorbic acid (Figure 1, lower graph), while PI 281424 from Peru contained the lowest (266 μg g-1 fresh fruit) (Figure 2).

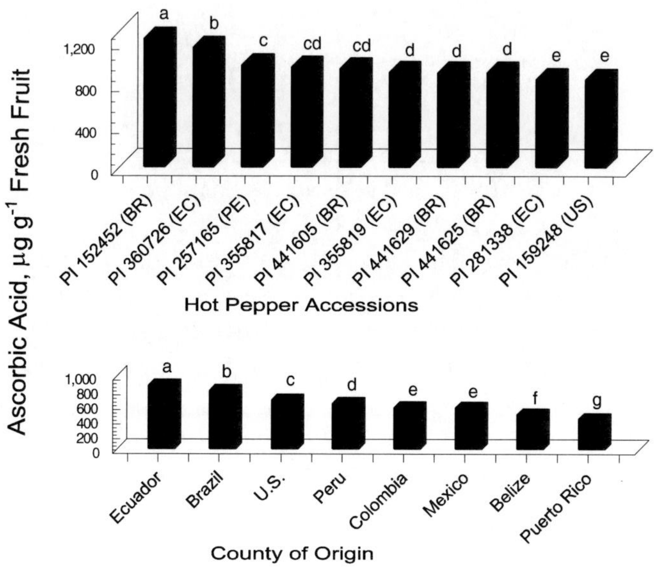

Figure 1. Concentrations of ascorbic acid in 10 top accessions of *Capsicum chinense* (upper graph) grown from seeds of different countries of origin; Brazil (BR), Ecuador (EC), Peru (PE), and United States (US) having greatest concentrations of ascorbic acid among 63 accessions tested and concentrations of ascorbic acid among hot pepper countries of origin (lower graph). Bars accompanied by different letter(s) indicate significant differences ($P < 0.05$) using Duncan's multiple range test (SAS Institute, 2003).

Among the 63 accessions analyzed, concentrations of total phenols were significantly higher in PI 438648, PI 159248, and PI 360900 (Figure 3, upper graph). Seeds of these accessions originated in Mexico and the US, respectively (Figure 3, upper graph). Total phenols concentrations were generally low in fruits of accessions from Belize, especially in PI 22445 (Figure 4). Figure 5 illustrates the variability among 10 accessions with the greatest concentrations of β-carotene (Figure 5, upper graph) and among the eight countries of origin included in this investigation. Statistical analyses revealed that greatest concentrations of β-carotene were found in fruits of accessions from Ecuador. Figure 6 illustrates the relative concentration of beta carotene in all of the 63 accessions analyzed. Previous work conducted in our laboratory on capsaicinoids in *C. chinense* (Antonious et al., 2009) revealed pronounced variability in total capsaicinoids (capsaicin plus dihydrocapsaicin) concentrations among the 63 *C. chinense* accessions. Fruits of accession PI 640900 (USA) contained the greatest concentration of capsaicin (1.5 mg g-1 fresh fruit) and dihydrocapsaicin (1.2 mg g-1 fresh fruit), while total major capsaicinoids in the fruits of PI 438648 (Mexico) averaged 2 mg g-1 fruit (see Figure 7).

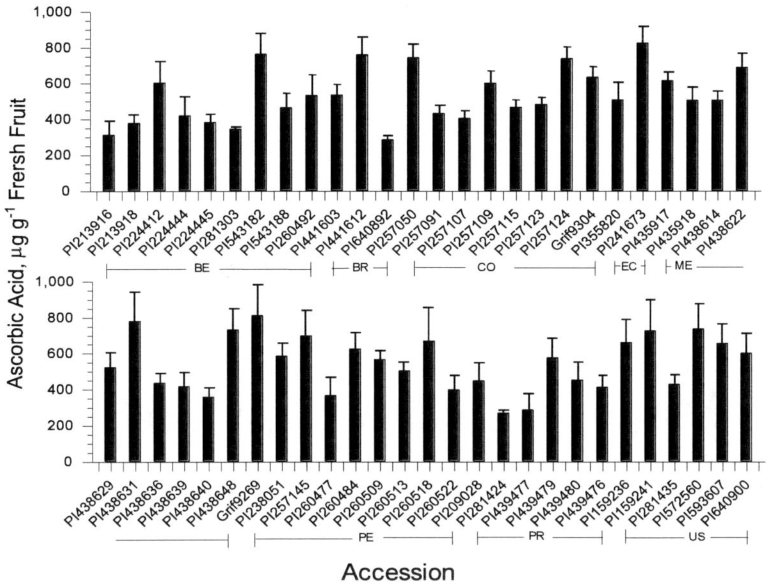

Figure 2. Concentrations of ascorbic acid in 53 accessions of *Capsicum chinense* grown from seeds of different countries of origin; Belize (BE), Brazil (BR), Colombia (CO), Ecuador (EC), Mexico (ME), Peru (PE), Puerto Rico (PR), and the United States (US). Vertical bars indicate ± standard error.

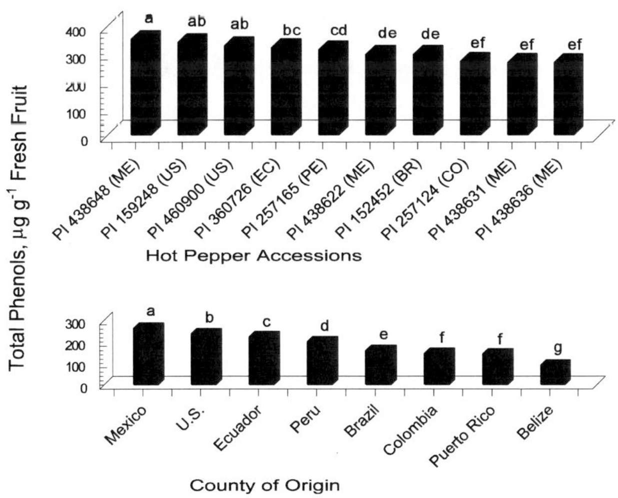

Figure 3. Concentrations of phenols in 10 top accessions of *Capsicum chinense* (upper graph) grown from seeds of different countries of origin; Mexico (ME), United States (US), Ecuador (EC), Peru (PE), Brazil (BR), and Colombia (CO) having greatest concentrations of total phenols among 63 accessions tested and concentrations of phenols among hot pepper countries of origin (lower graph). Bars accompanied by different letter(s) indicate significant differences ($P< 0.05$) using Duncan's multiple range test (SAS Institute, 2003).

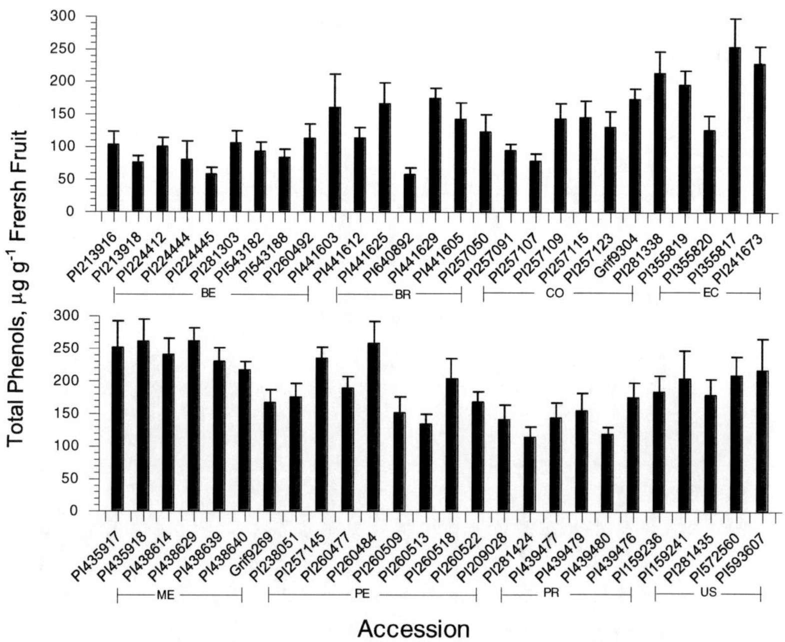

Figure 4. Concentrations of total phenols in 53 accessions of *Capsicum chinense* grown from seeds of different countries of origin; Belize (BE), Brazil (BR), Colombia (CO), Ecuador (EC), Mexico (ME), Peru (PE), Puerto Rico (PR), and the United States (US). Vertical bars indicate ± standard error.

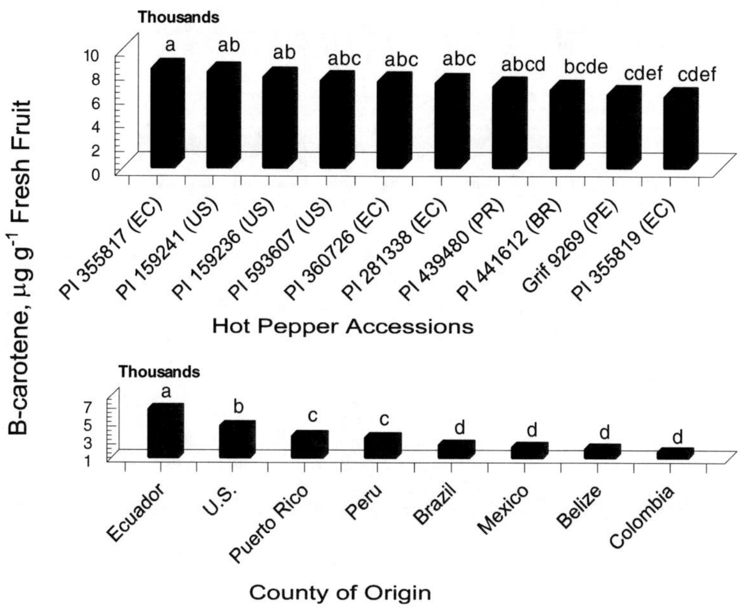

Figure 5. Concentrations of β-carotene in 10 top accessions of *Capsicum chinense* (upper graph) grown from seeds of different countries of origin; Ecuador (EC), United States (US), Puerto Rico (PR), Brazil (BR), and Peru (PE) having greatest concentrations of β-carotene among 63 accessions tested and concentrations of β-carotene among hot pepper countries of origin (lower graph). Bars accompanied by different letter(s) indicate significant differences ($P< 0.05$) using Duncan's multiple range test (SAS Institute, 2003).

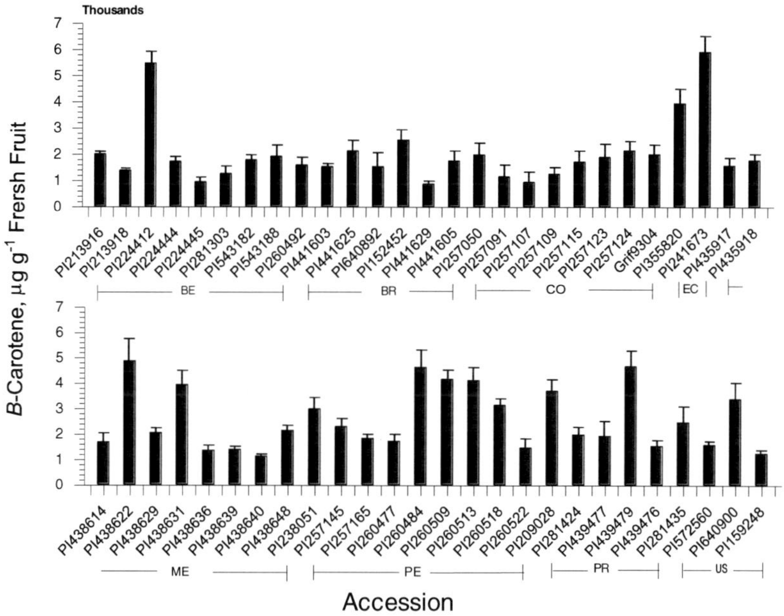

Figure 6. Concentrations of β-carotene in 53 accessions of *Capsicum chinense* grown from seeds of different countries of origin; Belize (BE), Brazil (BR), Colombia (CO), Ecuador (EC), Mexico (ME), Peru (PE), Puerto Rico (PR), and the United States (US). Vertical bars indicate ± standard error.

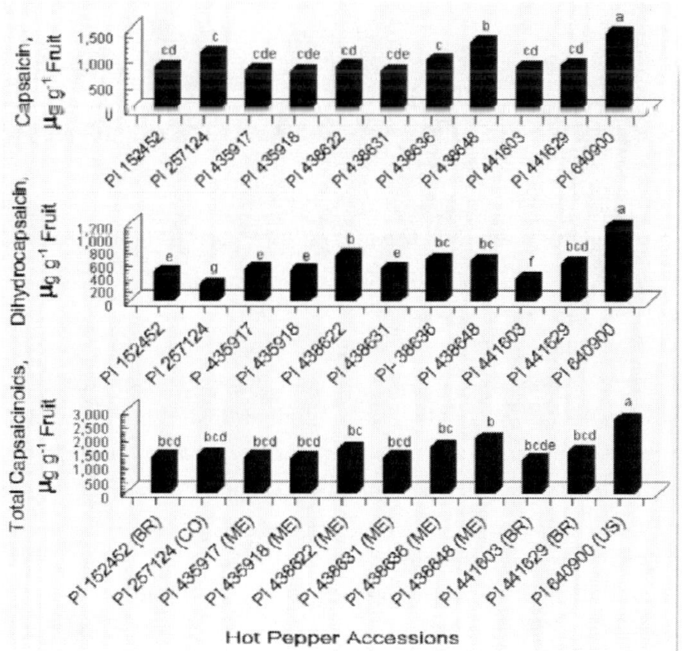

Figure 7. Levels of capsaicin, dihydrocapsaicin, and total capsaicinoids in 11 accessions of *Capsicum chinense* grown from seeds originated from different countries of origin (BR=Brazil, CO=Columbia, Me=Mexico, US=United States) having the greatest levels of capsaicinoids among 63 accessions analysed. Bars accompanied by different letter(s) indicate significant differences using Duncan's multiple range test.

The present investigation is a continuation of our work on *C. chinense* of the world which suggests that a great variability exists within *Capsicum chinense* for total phenols, ascorbic acid, β-carotene, and capsaicinoids. These compounds have antioxidant properties and are thus important quality attributes. Variability for these traits might be utilized via plant breeding approaches to produce fruit desirable to the consumer for their value-added health-promoting characteristics.

ACKNOWLEDGMENTS

This investigation was supported by grant No.KYX-10-08-43P from USDA/CSREES and grant from KY-WVA LSAMP Program to Kentucky State University.

REFERENCES

Antonious GF and Kasperbauer MJ. 2002. Color of light reflected to leaves modifies nutrient content of carrot roots. *Crop Sci.* 42: 1211-1216.

Antonious GF, Berke T, Jarret RL. 2009. Pungency in Capsicum chinense: Variation among countries of origin. *J. Environ. Sci. Health.* 44: 1-6.

Association of Official Analytical Chemists (AOAC). 1970. *Official methods for analysis of the Association of Analytical Chemists*, 11th ed., Washington, DC.

Clifford MN 2004. Diet-derived phenols in plasma and tissues and their implications for health. *Planta Medica* 70: 1103-1114.

DeWitt D and Bosland PW. 1996. *Peppers of the World: An identification Guide.* Ten Speed Press: Berkeley, CA

Govindarajan VS and Sathyanarayana MN. 1991. Capsicum production, technology, chemistry, and quality, part V. Impact on physiology, pharmacology, nutrition and metabolism; structure, pungency, pain and desensitization sequences. *Crit. Rev. Food Sci. Nutri.* 29: 435-471.

Hollman PCH. 2001. Evidence for health benefits of plant phenols: Local or systemic effects? *J. Sci. Food Agric.* 81: 842-852.

Kwon YI, Apostolidis E and Shetty K. 2007. Evaluation of pepper (Capsicum annuum) for management of diabetes and hypertension. *J. Food Biochem.* 31: 370-385.

Naidu KA. 2003. Vitamin C in human health and disease is still a mystery? *An overview. Nutrition Journal.* 2: 2-7.

NIH (National Institutes of Health), National Diabetes Education Program. 2006. http://www.ndep.nih.gov/diabetes/Mealplanner/pyramid.htm

Oboh G. 2006. Nutritive value, antioxidant and antimicrobial properties of Struchium sparganophora leaves. *J. Med. Food.* 9: 276-280.

Perera CO and Yen GM. 2007. Functional properties of carotenoids in human health. *International J. Food Properties.* 10: 201-230.

SAS Institute Inc., SAS/STAT Guide, version 9.1.3. 2003. SAS Inc., Campus Drive, Cary, NC 27513, USA.

F Shahidi. 2000. Antioxidant factors in plant foods and selected oilseeds. *BioFactors.* 13:179-185.

Sicuteri F, Fanciullacci M, Nicolodi M, Geppetti P, Fusco BM, Marabini M, Alessandri M, Campagnolo V. 1990. Substance P theory: A unique focus on the painful and painless phenomena of cluster headache. *Headache* 30: 69-70.

Surh YJ and Seoul SK 2002. Anti-tumor promoting potential of selected spice ingredients with oxidative and anti-inflammatory activities. *Food Chem. Toxicol.* 40: 1091-1097.

USDA (United States Department of Agriculture). 2006 http://www.mypyramid.gov/pyramid/vegetables.html.

Tapiero H, Townsend DM, Tew KD. 2004. The role of carotenoids in the prevention of human pathologies. *Biomedicine and Pharmacotherapy* 58: 100-110.

University of Maryland Medical Centre. 2008. Vitamin C (Ascorbic acid), www.umm.edu/altmed./articles/vitamin-c-000339.htm

Van der Sluis AA, Dekker M, Skrede G, MN Jongen MN. 2002. Activity and concentration of polyphenolic antioxidants in apple juice. *J. Agric. Food Chem.* 50: 7211-7219.

Weissenberg A, Levy A, Schaeffler I, Menagem E, Barzilai M. 1997. Rapid isocratic HPLC analysis of beta-carotene in red peppers (Capsicum annuum L.) and food preparations. *Chromatographia* 46: 339-403.

In: Biotechnology and Conservation … (2 Volume Set)
Editors: S. A. Farooq, R. Abed and S. Baqir
ISBN: 978-1-63117-869-6

Chapter 31

The Bio-Toxic Effect of Essential Oils of *Peganum harmala* L. (Zygophyllaceae) on the Schistocerca gregaria (Forskål, 1775) (Orthoptera-Acrididae)

Kemassi Abdellah**, *Ould El Hadj-Khelil Aminata, Boual Zakaria, Hamid Oudjani Aicha and Ould El Hadj Mohamed Didi

Ecosystem Protection in Arid and Semi Arid Laboratory
Kasdi Merbah University, Ouargla, Algeria

Abstract

The biological activity of crude leaf essential oil of Peganum harmala L. collected from Oued M'Zab in Ghardaïa region (Algerian septentrional Sahara), on the larvae L5 and adult individuals of desert locust, showed a toxic effect in the desert locust. After treatment, the fifth stage larvae and imagos of S. gregaria, by crude extracts of P. harmala leaf essential oils, problems of imbalances and convulsive movements are observed. These are the same symptoms noted, in insects treated with insecticides. The lethal time 50 (LT50) measured immediately after treatment, are of the order of 06 mn 12 ' of L5 larvae and for 19 mn 21' for imagos of this insect. The fifth stage larvae of the desert locust seem more sensitive to the action of essential oils as imagos.

Keywords: S. gregaria, toxicity, P. harmala, Sahara, essential oils

1. Introduction

Looking for new techniques to protect crops against pests in order to increase agricultural production for a growing world population while preserving the environment, organizations and research institutions are moving towards biological control (Appert and Deus 1982,

* akemassi@yahoo.fr.

Anthelme et al., 2006). The possibility of using plant secondary metabolite against insect pests in general and against the desert locust in particular has generated a lot of work. The most recent are those of Abbassi et al. (2003a, 2003b, 2004, 2005), Ould El Hadj et al. (2006), Zouiten et al. (2006), Idriss and Hermes (2008), and Doumandji-Mitiche Doumandji (2008), and Ammar N'cir (2008) and Kemassi et al. (2010).

The Sahara has a unique plant biodiversity, is about 480 species (Mair, 1933), of which there are 162 endemic species in the Northern Sahara and one which is added a centuries-old tradition of traditional medicines. Several species are known for their remarkable healing properties (Quézel, 1963). The wild plants of arid region are considered one of the plant genetic resources that are of agronomic, economic and ecological as well as strategic (UNESCO, 1960). Given this fact, and to better characterize the potential of the Saharan flora, this study research from Peganum harmala L. (Zygophyllaceae), a wild plant of septentrional Sahara of Algeria, untouched by the desert locust, its acridicides characteristics.

2. Methodology

2.1. Biological Material

The biological material is made up of fifth stage larvae (L5) and imagos of locusts and leaves of Peganum harmala L., harvested from Oued M'Zab (region of Ghardaïa, Algerian septentrional Sahara).

2.1.1. Breeding of Schistocerca Gregaria

The fifth stage larvae and imagos of locusts experienced result from a mass breeding maintained in protection ecosystems in arid and semi-arid areas laboratory, University Kasdi Merbah-Ouargla.

2.1.2. Plant Material

Peganum harmala L. is a perennial herb of the family Zygophyllaceae, usually little branched stems, 30 to 90 cm high, with rather short internodes. It has elongated leaves and irregularly divided into multiple strips very thin, dirty white flowers with large sepals unequal persistent beyond the corolla and petals cream washed pink-orange-yellow ribbed, oblong subsymétriques (photo 1). This plant grows in Southern Europe and Austro-eastern Asia Minor, Tibet, Iran, Turkestan, Syria, Arabia, Egypt and North Africa. In Algeria, P. harmala is common to the highlands, northern and southern Sahara, and the mountains of the central Sahara (Maire, 1933; Ozenda, 1991; UICN 2001). It is used by local people as a fumigant for treating seizures in children in decoction and ointment for the treatment of fevers in frictions for rheumatism. P. harmala has properties anthelmintic, antimalarial, antispasmodic, sudorific and intoxicating. This plant is not grazed by animals (Ozenda, 1991; UICN, 2001).

2.2. Extraction of Essential Oils

The leaves of P. harmala subjected to extraction are taken from seedling stage plants, harvested from their natural habitat of existence far places by man. With the help of a simple hydrodistillation assembly, the fresh leaves of P. harmala are brought to a boil for 6 hours, decanting is then performed. The resulting product was dried using anhydrous sodium sulphate to remove the little water that may have been retained in the organic phase. The resulting product is a pure essential oil, used for the treatment of insects.

2.3. Study of the Toxicity

With the help of a micro-spray (Ultra Low Volume), pure essential oils are sprayed directly on the fifth stage juveniles and adults of S. gregaria to their action studied by contact. It was noted after treatment motor activity and the rate and time of death. The experiment is followed until the death of all individuals in the treated groups. To this end, four lots of insects for 60 individuals including 30 males and 30 females per lot are made, making a total of 240 individuals. Two lots are fifth stage larvae including one for the control and the other for the treatment and the other two consist of imagos which one for the control and the other for treatment.

2.4. Determination of Lethal Time 50 (LT50)

The median lethal time 50 (LT50), is the time required for 50% of individuals of a population die from treatment with any substance. It is calculated from the probit regression line corresponding to the percentage of mortality adjusted for log processing time. It is used Schneider formula and the probits table.

Schneider Formula

MC = [M2-M1/100-M1] x 100

MC: corrected mortality (%); M2: mortality in the treated population (%); M1: mortality in the control population (%).

3. Results

3.1. Effect of Essential Oil of Peganum Harmala on the Mortality of Larvae L5 and Adults of Schistocerca Gregaria

Showed the results of Figures 1 and 2, it appears that the essential oils of P. harmala have a lethal effect on larvae L5 as well as the adult of desert locust. The fifth stage larvae seem more sensitive than adults; they die first, L5 larvae, mortality was recorded from the third minute, after treatment and 100% mortality was reached after 08 mn 30', while it is in 30 mn 18' in adults. But the early deaths are visible from the twenty-second minute at the beginning of treatment in adults. However, no deaths were recorded in individuals of control groups and that all larvae L5 have completed fledge after 7±1 days.

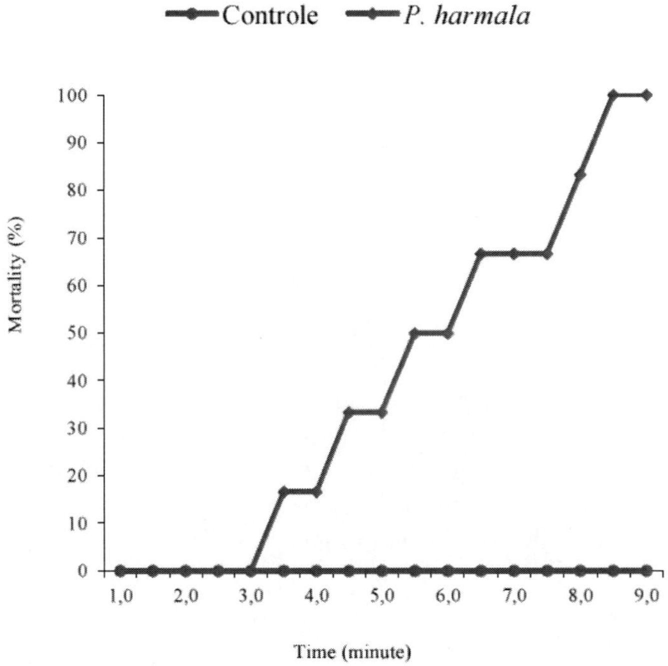

Figure 1. Kinetics of the cumulative mortality of S. gregaria larvae L5 control and treated with essential oils of Peganum harmala.

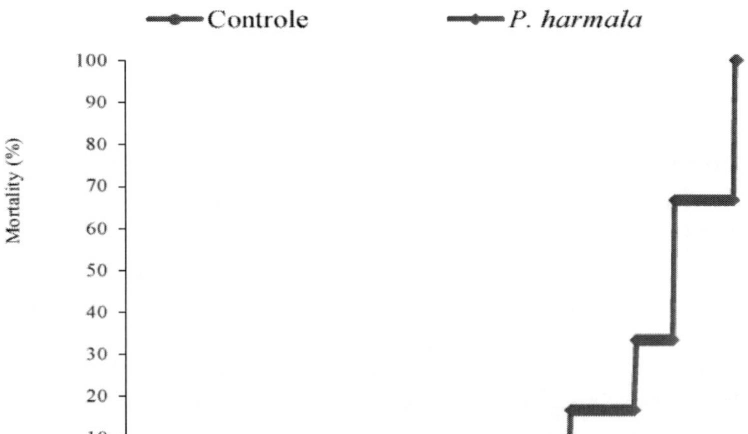

Figure 2. Kinetics of the cumulative mortality of adult S. gregaria witnesses and processed by the essential oils of Peganum harmala.

Furthermore, it is important to note that males either their stages of development, larvae L5 or adults seem more sensitive, they also die sooner than females. In addition, balance disorders, convulsive movements, intense defecation loss of the ability to perch with a support following the inability to join tarsique, trembling of appendages and increased respiratory rate are observed. These events demonstrate the neurotoxic action and organohalogen effects of P. harmala essential oils on locust is probably the consequence of the effect of its extracts on the nervous system of desert locust. The neurotoxic action of crude P. harmala essential oils on the fifth stage larvae and adults of the desert locust stems probably from the effect of different chemical compounds that are particularly alkaloids on the nervous system of the desert locust (Abbassi et al., 2005). Similar events are observed in locusts treated with insecticides organohalogens used, for fight against locusts swarm (Chauvin, 1956; Moreteau, 1991). An examination of dead locusts treated under a dissecting microscope, revealed the absence of lesions in the cuticle, for that, the essential oils of the plant exert little effect on the cuticle and inhalation action may be considered because of the volatile character of its plant species. ISMANS (2000) and Chiasson and Beloin (2007), studying the biological activity of essential oils of many plants, including oregano, basil, marjoram, thyme, sage, bay leaf, rosemary, lavender and others on many insects that thrips, aphids the Coleoptera and Hymenoptera whiteflies, note that essential oils act directly on the cuticle of insects and mites, soft-bodied as thrips, aphids, whiteflies and some mites. By cons, they have been less effective on insect cuticle lasts such as Coleoptera and Hymenoptera adults and some predatory mites. However, it should also be noted that individuals either male developmental stage larvae or adults die before females, this is probably related to the difference in weight between males and females of the locust. Usually in locusts sexual dimorphism is apparent, the males weigh less compared to females. Meanwhile, it is recognized that resistance to toxic differs from one species to another, and for the same species, it is relative to several factors whose weight is critical.

3-2. Lethal time 50 (LT50) of Peganum Harmala Essential Oils of on Larvae L5 and Adults Schistocerca Gregaria

Table 1 and Figure 2 (A, B) group the regression equations and, regression coefficients and values of TL50 and TL90 evaluated for essential oils of this Saharan plant on the L5 larvae and imagos of desert locusts.

Table 1. Regression equation, regression coefficient and TL50 and TL90 calculated for the essential oil of Peganum harmala

Stage	*Regression equation*	*Régression coefficient (R^2)*	*Lethal time 50 (TL_{50}) (minute)*	*Lethal time 90 (TL_{90}) (minute)*
larvae L_5	y = 9,0624x – 1,8647	0,9042	06 min 12’	07 min. 56’
Adults	y = 3,8067x – 2,8485	0,3542	19 min 21’	41min. 43’

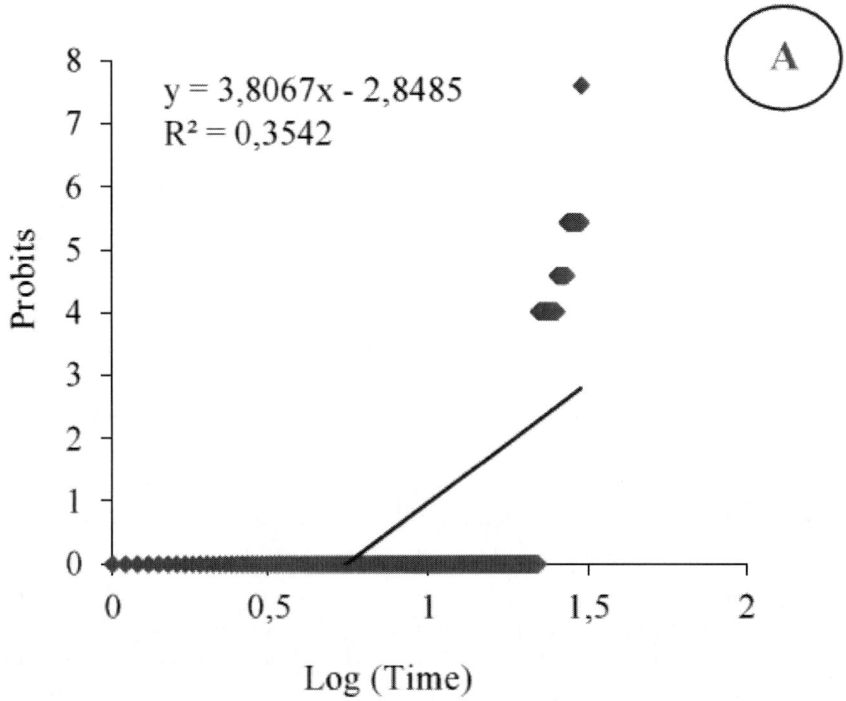

Essential oils of P. harmala in time for adults of S. gregaria

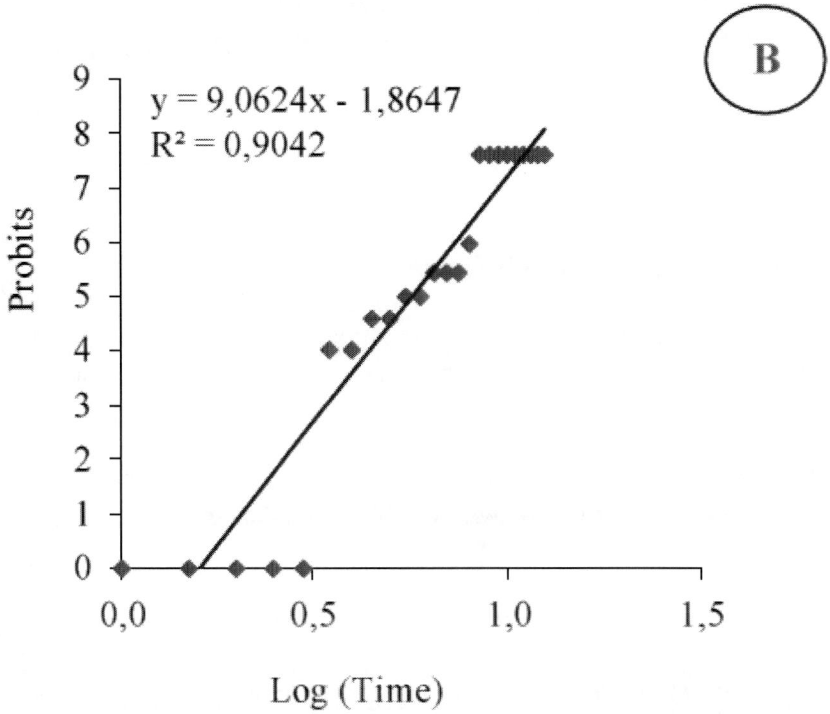

Essential oils of P. harmala time of L5 larvae of S. gregaria

Figure 2. (A,B)- Action of Peganum harmala essential oils on the fifth stage larvae and adult of Schistocerca gregaria in time.

Evaluation of lethal time 50 (LT50) and 90 (TL90) for P. harmala essential oils on L5 larvae and adults of S. gregaria, has to confirm the speed of action of these extracts on L5 larvae compared to adults. The LT50 reported for the fifth stage larvae being shorter, it is of the order of 06 min 12', as for the assessment for adults is the 19 min 21'. As for the lethal time 90 (TL90) evaluated, they are 07 min. 56' and 41min 43'for the fifth stage larvae and adults respectively. This confirms the sensitivity of larvae per intake for adults. Moreover, it is commonly accepted that insect resistance to toxins increases depending on the stage of development, and that adults are generally more resistant than larvae (Chauvin, 1956).

CONCLUSION

The study of the toxicity of essential oils of Peganum harmala on the fifth stage larvae and adults of Schistocerca gregaria demonstrate their power to put insecticide on the desert locust. The fifth stage larvae are more susceptible to the biocidal action of essential oils compared to adults, lethal time 50 (LT50) estimated for L5 larvae are shorter to those reported for adults of S. gregaria. The neurotoxicity symptoms were reported, then disorders and convulsive movements, inability to perch in the support is also noted in the larvae and adults exposed to essential oils of P. harmala, this reflect the neurotoxic effect of its crops extrqcts on desert locus. From this perspective, the use of P. harmala essential oils against the locusts could be considered. These natural compounds could be a building block for the synthesis of new molecules with particular effectiveness of locusts and without risk of environmental poisoning. However, in advance, it is appropriate to refine knowledge of the chemical composition of the essence plant of the functional properties and to determine the terms and possible applications without harming the ecosystem and in compliance with health human.

REFERENCES

Abbassi K., Atay-Kadiri Z. and GHAOUT S., 2003a.- Biological effects of alkaloïds extracted from three plants of Moroccan arid areas on the desert locust. The Royal Entomological Society, *Physiological Entomology*, (28): 232-236.

Abbassi K., Mergaoui L., Atay Kadiri Z., Stambouli A. Et Ghaout S., 2003b.- Effet des extraits de Peganum harmala L. (Zygophyllaceae) sur le Criquet pèlerin Schistocerca gregaria (Forskål, 1775). *Zool. Baetica*, vol. 13 et 14: 203-217.

Abbassi K., Atay-Kadiri Z. Et Ghaout S., 2004.- Activité biologique des feuilles de Calotropis procera (AIT. R. BR.) sur le criquet pèlerin Schistocerca gregaria (Forskål, 1775*). Zool. Baetica*, vol. 15: 153-166.

Abbassi K., Mergaoui L., Atay-Kadiri Z., Ghaout S. Et Stambouli A., 2005.- Activités biologiques des feuilles de Peganum harmala (Zygophyllaceae) en floraison sur la mortalité et l'activité génésique chez le criquet pèlerin. *Zool. Bœtica,* vol.16: 31-46.

Ammar M. and N'CIR S., 2008.- Incorporation of Cestrum parquii (Solonaceae) leaves in an artificial diet affected larval longevity and gut structure of the desert locust Schistocerca gregaria (Forskål, 1775). *Tunisian journal of plant protection*, vol. 3 (1): 27-34.

Anthelme F., Waziri Mato M., De Boissieu D. Et Giazzi F., 2006.- Dégradation des ressources végétales au contact des activités humaines et perspectives de conservation dans le massif de l'Aïr (Sahara, Niger). La revue en sciences de l'environnement, vol. 7 (2): 1-12.

Appert J. et DEUS J. 1982.- Les ravageurs des cultures vivrières et maraîchères sous les tropiques. Ed. Maison neuve et Larose, Paris, 419 p.

Chauvin R., 1956.- Physiologie des insectes. Le comportement, les grandes fonctions, écophysiologie. Ed. INRA., Paris, 917 p.

Chiasson H. et Beloin N., 2007.- Les huiles essentielles, des biopesticides «Nouveau genre». Revue de littérature "Antennae", vol. 14 (1): 3-6.

Doumandji-Mitiche B. Et Doumandji S., 2008.- Quelques agents biologiques susceptibles d'être utilisés en lutte anti-acridienne. Revue des régions arides, vol. 3(21):1154-1158.

Idrissi Hassani L. M. et Hermas J., 2008.- Effet de l'alimentation en Peganum harmala L. (Zygophyllaceae) sur le tube digestif du criquet pèlerin Schistocerca gregaria Forsk. (Orthoptera, Acrididae). *Zool. Baetica,* vol.19: 71-84.

Isman, M.B., 2000.- Plant essential oils for pest and disease management. *Crop. Prot.,* vol. 9: 603-608.

Kemassi A., Boual Z., Ould El Hadj-Khelil A., Dadi Bouhoun M. et Ould El Hadj Md., 2010.- Activité biologique de l'extrait d'Euphorbia guyoniana (Boiss. & Reut.) (Euphorbiaceae) chez le Criquet pèlerin Schistocerca gregaria (Forskål, 1775) (Orthoptera-Acrididae). Annales de Sciences et Technologie, Université Kasdi Merbah-Ouargla, 2 (1): 60-71.

Maire R., 1933.- Études sur la flore et la végétation du Sahara central. Mémoire de la société d'histoire naturelle de l'Afrique du nord n°3, Mission du Hoggar II, Alger, 361 p.

Moreteau B., 1991.- Etude de certains aspects de la physio-toxicologie d'insecticides de synthèse chez le Criquet migrateur Locusta migratoria (R. & F.). La lutte antiacridienne. Ed. AUPEL-UREF, Paris: 167-178.

Ould El Hadj M. D., Tankari Dan-Badjo A., Halouane F. Et Doumandji S., 2006- Toxicité comparée des extraits de trois plantes acridifuges sur les larves du cinquième stade et sur les adultes de Schistocerca gregaria (Forskål, 1775) (Orthoptera-Cyrtacanthacridinae). Sécheresse, vol.. 17(3): 407-414.

Ould El Hadj M. D., Dan-Badjo A.T., Halouane F. Et Doumandji S., 2006.-Toxicité comparée des extraits de trois plantes acridifuges sur les larves de cinquième stade et sur les adultes de Schistocerca gregaria (Forskål, 1775). (Orthoptère, Acrididae). Sécheresse, vol. 17(3): 407-414.

Ozenda P., 1991.- Flore et végétation du Sahara. (3éme édition, augmentée). Ed. CNRS, Paris: 662 p.

Quezel P., 1963 - La végétation au Sahara. Ed. Masson et Cie, Paris, 33 p.

U.I.C.N., 2001.- Connaissance, Valorisation et Contrôle de l'Utilisation de la Flore Sauvage en Médecine Traditionnelle (Plantes Médicinales). Programme Union Internationale pour la Conservation de la Nature pour l'Afrique du Nord. Ministère de l'Agriculture Algérienne, 153 p.

UNESCO, 1960.- Les plantes médicinales des régions arides. Recherche sur les zones arides XIII. Ed. UNESCO, Rome, 97 p.

Zouiten H., Abbassi K., Atay-Kadiri Z., Mzari M., El Mahi M. and Essassi E. M., 2006.- Insecticidal activity of Solanum sodomaeum (solonaceae) extracts on Schistocerca gregaria (Forskål) larvae. *Journal of orthoptera research*, 15 (2):171-173.

In: Biotechnology and Conservation … (2 Volume Set) ISBN: 978-1-63117-869-6
Editors: S. A. Farooq, R. Abed and S. Baqir

Chapter 32

EFFECT OF TANNINS ON *IN VITRO* GAS PRODUCTION AND DIGESTIBILITY OF FEEDSTUFFS FROM NORTH AFRICAN ARID ZONE

***R. Arhab*[1,], *D. Macheboeuf*[2], *M. Aggoun*[2], *D. Viala*[2] *and J. M. Besle*[2]**

[1]Natural Sciences and Life Departement, Exact Sciences and Natural Sciences and Life Faculty, Larbi Ben Mhidi University, Oum El Bouaghi, Algeria
[2]URH, INRA, Theix-Champanelle, Clermond Ferrand, France

ABSTRACT

The influence of tannins present in arid zone forages from North Africa: *Aristida plumosa*, *Danthonia forskahlii*, *Astragalus gombiformis*, *Genista saharae*, two date palm fractions (leaves and racemes), and vetch-oat hay taken as control on *in vitro* gas production and *in vitro* organic matter digestibility (IVOMD) was evaluated. Chemical analysis revealed the low nutritional quality of these forages. They were high in NDF, ADF and lignin (679.5, 455.7 and 86 g/kg DM, respectively) and low in nitrogen (< 12 g/kg DM) except for *Astragalus* that had 20 g N/kg DM. Phenolic compounds (total phenols, total tannins and total condensed tannins) were 61.8, 49.1 and 36.2 g/kg DM for palm leaves followed by *Astragalus*, racemes, *Genista*, *Aristida* and *Danthonia*. Gas production ranged between 55.2 and 152.6 mL/g DM whereas IVOMD ranged between 21 and 56.5%. Addition of PEG resulted in an overall increase in both gas production (20.2%) and IVOMD (30.7%), with the exception of *Danthonia* and *Aristida*. The largest increment for gas production was recorded for *Aristida* (low tannins content). However, the higher increase in IVOMD was noted for racemes, *Astragalus* and palm leaves (high tannins content). The variable responses among forages studied suggest that factors other than phenolic compounds also affect *in vitro* fermentation.

Keywords: *In vitro* gas production; North African forages; polyethylene glycol; *in vitro* digestibility

Introduction

Animal production in Algeria, particularly in arid regions, is almost exclusively based on pasture of native plants. These plants can be classified into two main groups (Longuo et al., 1989): ephemeral plants, which germinate and remain green for only a few weeks after rain, and perennial plants, characterized by a slow vegetative cycle with a growing period from March to June (Haddi et al., 2003). Moreover, the arid regions are represented in part by oasis where the cultivation of date palm trees is preponderant. Local farmers use date palm fractions, principally discarded dates, leaves and racemes for ruminant feeding supplementation (Genin et al., 2004). Most of these forages contain antinutritive factors such as tannins in response to their harsh environmental conditions. However, no studies have been done on the effect of tannins present in this type of fodders on *in vitro* fermentation.

The *in vitro* gas production method is a relatively simple and inexpensive tool to study potential effects, mechanisms of action and fate of phytochemicals in the rumen (Makkar, 2005). This method, coupled with polyethylene (PEG) used as a specific binding agent, provides useful information on the biological activity of tannins (Ammar et al., 2004a). However, the *in vitro* gas production must be completed by other measures of end-products fermentation such as the amount of degraded matter for obtaining a more complete information. This study was conducted for assessing the effect of tannins, using PEG (MW 4000), on *in vitro* gas production and degradability of four plants collected from the South-East of Algeria and two date palm fractions, comparatively to vetch-oat hay taken as control.

Material and Methods

Study Area

Forages were collected from the administrative districts of El-Oued, located in the South-East of Algeria. El-Oued region is situated at 6°53’E and 32°20’N. Its average altitude above sea level is 67 m. The climate of the region is arid with annual mean rainfall of 75 mm, and mean temperature comprised between 1°C in January and 43°C in July.

Forage Samples Collection and Processing

Experimental feedstuffs consisted of four autochthonous North African species and two date palm fractions, widely utilized by local farmers. The plants were selected based on herdsmen knowledge that are consumed by dromedary, goats and sheep.

The gramineous (*Aristida plumosa*: locally named *Ksiba* and *Danthonia forskahlii:* locally named *Sfar*) and leguminosea (*Genista saharae*: locally named *Merck* and *Astragalus gombiformis*: locally named *Foulet El Ibel*) plants were harvested at maturity and flowering stages, respectively. The edible plant samples (leaves, stems and flowers) were hand plucked and chopped to 2cm length. The date palm fractions were: racemes (stems floral without dates) and leaves (leaflets and rachis). The racemes were sampled from the dates conditioning factory. For leaves, the samples consisted on leaves removed at senescence stage from date palm trees. Vetch-oat hay, taken as control, was provided by ITELV (Technical Institute of Breeding, Ain Mlila, Algeria). The samples were dried at 60°C in forced air oven for 48h, except samples for tannins determination that were sun dried. The forages were then ground to pass a 1mm sieve and used for chemical analysis, *in vitro* gas production and *in vitro* digestibility.

Chemical Analysis

Samples were analysed for dry matter (DM) and organic matter by the methods of the AOAC (1990; method ID 942.05). Neutral detergent fibre (NDF), acid detergent fibre (ADF) and lignin (ADL) were determined using the reagents described by Van Soest et al. (1991). The crude protein content of plants and crude protein bound to NDF fraction (NDICP) were determined by the Kjeldhal procedure.

Phytochemical Analysis of Antinutritive Factors

The phytochemical analysis, based on colorimetric reactions, was carried out according to the procedures described by Larrahondo (1985) and Rosales et al. (1989). The dried samples (15 g) were extracted with methanol solution (90%, 30 mL) and ether solution (30 mL) for 40 min. The extraction was repeated three times, and the mixture was filtered. At the end of the extraction procedure and decantation, two phases were obtained; the lower layer is the methanol-water (polar phase) and the top non-polar fraction being formed with ether.

For determination of saponins, 9 mL of water were added to 1 mL of the methanol fraction and then filtered. 1 mL of this solution was vigorously shaken in a small test tube for 30 seconds. After 15 minutes, the height of foam was measured, giving an indication of the levels of saponins in the forage using the following criteria: < 5 mm absence of saponins, 5-9 mm low content of saponins, 10-14 mm medium content of saponins and > 15 mm high content of saponins.

The phenolic compounds were determined by addition of two or three drops of ferric chlorure??? ($FeCl_3$) to 1 mL of methanol extract which was diluted to 50% by distilled water. The change in colour indicated the presence of phenolic compounds as follows: none change absence of phenolic compounds, dark blue presence of phenols or hydrolysable tannins, and dark green presence of condensed tannins.

For steroids, 10 mL of aqueous extract treated with chlorhydrique acid to drop pH to 2.0 units, were added to 10 mL of ether extract. After evaporation, 0.5 mL of chloroform and 0.5 mL anhydrous acetic acid were added. The addition of 1 to 2 mL of sulfuric acid induced the apparition of greenish blue colour, which indicates the presence of steroids.

Quantitative Analysis of Phenolic Compounds

Analysis of phenolic compounds was carried out in three replicates. Dried plant material (200 mg) was extracted with 10 mL of acetone (70% v/v) and subjected to ultrasonic treatment for 20 min at room temperature. The content was centrifuged (4°C, 10 min, 3000 g) and the supernatant was kept on ice until analysis. The pellet was treated a second time as described above. Total phenols (TP) were estimated by the Folin-Ciocalteau reaction (Makkar et al., 1993). A calibration curve was prepared using tannic acid. Total phenols were calculated as tannic acid equivalent and expressed as tannic acid eq-g/kg DM. For Total condensed tannins (TCT); the extract (0.5 mL) was treated with n butanol-HCl (3 mL, 95%) in the presence of ferric ammonium sulfate (0.1 mL). Reagents were heated in a boiling water bath for 60 min. Absorbance was read at 550 nm. TCT were expressed as leucocyanidins equivalents following the equation:

$$TCT = \frac{A_{550nm} \ x\ 78.26\ x\ dilution\ factor}{weight\ of\ sample\ on\ DM}$$

where $A_{550\ nm}$ is absorbance at 550 nm assuming that the effective $E^{1\%,\ 1\ cm,\ 550\ nm}$ of leucocyanidin is 460 (Porter et al., 1986). Total tannins (TT) were determined as the difference in total phenolics (measured by Folin-Ciocalteau reagent) before and after treatment with insoluble polyvinylpyrrolidone (Makkar et al., 1993). They were also determined according to the radial diffusion assay (Hagerman, 1987). Petri dishes were prepared with agarose and acetate buffer (pH= 5, 0.05 M). The acetone extract (40 µL) was added to wells (6 mm diameter) bored in the solidified agarose and the Petri dishes were placed in an incubator at 30°C, for 72 h. The diameters of the resulting rings were measured. The amount of tannins is proportional to the square of the diameter of the ring. A calibration curve was prepared using tannic acid. Values were expressed as tannic acid eq-g/kg DM.

In Vitro Study

Animals and Management

Rumen fluid, used as inoculum for *in vitro* fermentation, was obtained from three sheep fitted with permanent cannulae, that were fed a daily ration of 700 g vetch-oat hay (chemical composition illustrated in table 1) and 300 g of concentrate (barley 58%, wheat bran 38%, mineral and vitaminic premix 1%, NaCl 1% and limestone 2%) divided into two equal meals at 8:00 and 16:00 h. The sheep had free access to water throughout the experiment.

Polyethylene Glycol (PEG) Tannin Bioassay

Effect of PEG on the *in vitro* fermentation of feedstuffs was determined using the procedure described by Menke et al. (1979). The gas production parameters and metabolisable energy (ME) were determined in incubation with 200 mg of sample, but *in vitro* organic matter digestibility (IVOMD) and partitioning factor (PF: organic matter truly degraded (mg) and gas volume produced (mL) ratio) were measured in incubation with 350 mg of forage. The two incubations were made at the same time, in absence or presence of 1 g of PEG (MW, 4000) (Makkar et al., 1995). Rumen fluid was obtained from the three sheep in the morning 1 hour before feeding, flushed with CO_2, filtered through four layers of gauze and mixed with an anaerobic mineral solution as described by Makkar et al. (1995). The buffer solution was prepared without addition of nitrogen ($NaHCO_3$, 39 g/L). A portion (40 mL) of the buffered rumen fluid was transferred into the syringes and incubated at 39°C in an isothermal incubator equipped with a rotor, which run continuously at 9 rpm. All incubations were in triplicate. Gas production readings were recorded after 3, 6, 9, 24, 48, 72 and 96 h for the first incubation. Whereas for the second incubation, the fermentation was stopped at 24 h. Syringe contents were quantitatively transferred into a beaker by rinsing syringes with a total of 50 mL of neutral detergent solution (double strength; Blümmel and Becker, 1997a) and refluxed 1 h. The residue was filtered through a sintered glass crucible No.2 (pore size 40-90 µm). Residual organic matter was determined as outlined earlier.

Calculations and Statistical Analysis

For the first incubation, values of gas production, recorded at 96h and corrected for blank incubation that contained only rumen fluid plus buffer solution, were treated by the exponential model of Orskov and Mc Donald (1979): $y = a + b\ (1 - \exp^{-ct})$, where y (mL/g DM) is the gas production at time t, a (mL/g DM) is the gas production from the immediately degradable fraction, b (mL/g DM) is the gas production from the insoluble fraction, a+b (mL/g DM) is the potential gas production, c (h^{-1}) is the rate of gas production, and t (h) is the incubation time. The metabolisable energy (ME) was calculated using the equation as follows:

ME (MJ/kg DM) = 2.20 + 0.136 GP + 0.057 CP (Makkar and Becker, 1996)

where CP is crude protein expressed as g/kg DM and GP is mean gas production.

Data on chemical composition and tannins were subjected to analysis of variance (ANOVA) using the General Linear Model (GLM) procedure of SAS (1990), and were analysed based on the statistical model $Y = \mu + F + e$, where y is the general observation on nutrient compounds and tannins, µ is the general mean common for each parameter under investigation, F is the effect of substrate on the observed parameter, and e is the standard error term. Data on in vitro gas production, metabolisable energy, in vitro true organic matter digestibility, and partitioning factor were analysed based on the statistical model $Y = \mu + F + P + (F\times P) + e$, where, Y is the general observation, F is the effect of substrate on the observed parameters, P is the effect of PEG, $(S\times P)$ represents the interaction effects of substrate and PEG, and e is the standard error term common for all observations. Differences between treatments were compared using student's Newmann-Keuls test, and were considered statistically significant at $P < 0.05$. Standard errors of means were calculated from the residual means square in the analysis of variance.

RESULTS

Chemical Analysis

Chemical composition of the six substrates and the control is shown in table 1. There was a wide variation in all nutrient compounds of the forages. Except *G. saharae*, racemes and vetch-oat hay, all forages had high ash content (> 100 g/ kg DM). The highest value was noted for *A. gombiformis* (231.7 g/kg DM) and the lowest for *G. saharae* (36.4 g/kg DM).

The crude protein content was also highly variable, ranging from 25 g/kg DM in racemes to 125 g/kg DM in *A. gombiformis*, compared to vetch-oat hay (67.5 g/kg DM). Substantial quantities of CP were associated with NDF in *D. forskahlii* and date palm fractions, decreasing available crude protein significantly. The highest contents of NDF and ADF were found in *D. forskahlii* and the lowest ones were recorded for date palm leaves and vetch-oat hay (586.1 and 327.7 g/kg DM, respectively). All the samples had high lignin content, except vetch-oat hay. The highest value corresponding to *G. saharae* (142.4 g/kg DM) and the lowest value to vetch-oat hay (43.6 g/kg DM).

Table 1. Nutrient composition (g/kg DM) of selected feedstuffs from arid zone of North Africa

Substrates	DM	OM	CP	NDF	ADF	Lignin (sa)	NDICP
A. plumosa	846.6^{c}	885.4^{d}	74.4^{b}	747.9^{b}	413.8^{e}	64.5^{d}	23.3^{c}
D. forskahlii	899.8^{b}	888.3^{d}	60.6^{d}	824.4^{a}	562.3^{a}	79.2^{c}	39.5^{a}
A. gombiformis	551.7^{e}	768.1^{e}	125.0^{a}	614.9^{e}	445.2^{d}	78.1^{c}	40.2^{a}
G. saharae	894.0^{b}	963.6^{a}	74.4^{b}	621.9^{d}	499.1^{c}	142.4^{a}	24.5^{c}
Palm leaves	896.3^{b}	890.5^{d}	59.4^{e}	586.1^{f}	422.1^{e}	97.1^{b}	33.13^{b}
Racemes	923.4^{a}	936.1^{c}	25.0^{f}	745.1^{c}	519.9^{b}	97.3^{b}	13.3^{d}
Vetch-oat hay	891.1^{b}	942.2^{b}	67.5^{c}	616.2^{e}	327.7^{f}	43.6^{e}	31.2^{b}
S.E.M.	0.83	0.25	0.4	2.3	8.4	4.3	1.87

DM, dry matter ; OM, organic matter ; CP, crude protein ; NDF, neutral detergent fibre assayed without a heat stable amylase and expressed inclusive of residual ash ; ADF, acid detergent fibre expressed inclusive residual ash ; NDICP, crude protein bound to NDF fraction ; lignin (sa), lignin determined by solubilisation of cellulose with sulphuric acid ; S.E.M., standard errors of means ; $^{a, b, c, d, e, f}$ means with different superscripts within a same column are significantly different ($P < 0.05$).

Phytochemical analysis shows that feedstuffs contained antinutritive factors such as steroids, saponins, and phenols (table 2). The quantitative analysis of phenolic compounds is also illustrated in Table 2. The lowest values were recorded for the *Poaceae* family, whereas the highest levels were observed in the *Fabaceae* family and date palm fractions. The TP ranged from 4.5 g/kg DM in vetch-oat hay to 61.8 g/kg DM in date palm leaves, which had also the higher TT and TCT fraction, 49.1 and 36.2 g/kg DM, respectively. *A. plumosa* and *D. forskahlii* did not contain TCT. The highest PPC value was also recorded for date palm leaves followed by racemes, whereas, the other forages had not any PPC action. The ranking of forages according to tannins content was as follows: palm leaves > *A. gombiformis* > *G. saharae* = racemes > *A. plumosa* > *D. forskahlii* > vetch-oat hay. There was a negative correlation between NDF and TEP ($r = 0.661$), and TET ($r = 0.632$), but there was no significant correlation between phenolic compounds and ADF, lignin and crude protein. The protein precipitation capacity (PPC) and TT were positively correlated with TCT ($r = 0.759$, and 0.856, respectively, $P < 0.05$).

Table 2. Phytochemical analysis of antimicrobial factors and quantitative analysis of phenolic compounds (g/kg DM, standard equivalent) of selected feedstuffs

Substrates	Phytochemical analysis			Quantitative analysis			
	Steroids	Saponins	Phenols	TP	TT	TCT	PPC
A. plumosa	-	-	+	6.8[d]	4.4[d]	0.00[e]	0[c]
D. forskahlii	-	+	+	5.4[e]	3.2[e]	0.00[e]	0[c]
A. gombiformis	+	+	+	34.0[b]	21.3[b]	4.0[c]	0[c]
G. saharae	-	+	+	24.5[c]	18.2[c]	0.70[e]	0[c]
Palm leaves	-	+	+	61.8[a]	49.1[a]	36.2[a]	55.45[a]
Racemes	-	-	+	24.5[c]	18.2[c]	21.3[b]	32.25[b]
Vetch-oat hay	-	-	+	4.5[f]	2.2[f]	1.80[d]	0[c]
			S.E.M.	0.5	0.5	0.5	

TP, total phenols ; TT, total tannins ; TCT, total condensed tannins ; PPC, protein precipiting capacity ; S.E.M., standard errors of means ; [a, b, c, d, e, f] means in the same column affected with different letters are significantly different ($P < 0.05$) ; -, absence of colour or foam in the test tube which reveals that the feedstuff don't contain an antimicrobial factor ; +, colour development or foam apparition in the test tube that indicate the presence of antimicrobial factor.

Effect of Polyethylene Glycol on In Vitro Gas Production and Metabolisable Energy

The gas production over 24-h and ME, without and with PEG, are presented in table 3. In absence of PEG, the lowest gas volume produced was recorded for racemes (55.28 mL/g DM) and the highest value was noted for *A. gombiformis* (152.6 mL/g DM). The ranking of feedstuffs on the basis of gas production was as follows: *A. gombiformis* > vetch-oat hay > *G .saharae* > *D. forskahlii* > *A. plumosa* > palm leaves > racemes. As expected, addition of PEG increased gas volume after 24-h in all forages except for *D. forskahlii*. The highest increment was recorded for *A. plumosa* (54.3 units) and the lowest for *A. gombiformis* (9.10 units). In presence of PEG, the ranking of feedstuffs on the basis of gas production was as follows: vetch-oat hay > *A. gombiformis* > *G. saharae* > *A. plumosa* > *D. forskahlii* > palm leaves = racemes. The result indicate that *A. plumosa* tannins, whose concentration were low, reduced gas production significantly, but that of *A. gombiformis* appeared to be less sensitive to PEG. In presence of PEG, the gas production from *A. plumosa*, *A. gombiformis* and *G. saharae* come closer to vetch-oat hay. Furthermore, in presence of PEG the differences in gas production between leaves and racemes became insignificant. Besides, there were several unexpected results. In palm leaves, PEG addition had small effect compared to *A. plumosa*, despite their high tannins content which was detected by all chemical and biological assays. Similarly, PEG increased gas from *A. plumosa*, but not from *D. forskahlii,* although the two forages had approximately similar tannins content. Addition of PEG resulted also in an increase of ME content. Estimated ME was influenced ($P<0.05$) by substrate and PEG addition ($P<0.05$). However, the interaction of substrate and PEG effects was not significant ($P>0.05$). Racemes had the lowest ME content (5.14 MJ/kg DM) in the absence of PEG compared to *A. gombiformis* that had the highest ME of 13.42 MJ/kg DM. The presence of PEG caused a little increase in ME in all forages. The highest increment was recorded for *A. plumosa* (1.48 units), comparatively to *A. gombiformis* (0.3 units) for which the PEG addition induce a low increase in ME that was already higher.

Table 3. Corrected gas volume (mL/g DM) after 24 h and metabolisable energy (MJ/kg DM) in the absence and presence of polyethylene glycol of the selected feedstuffs

Substrates	Effect of PEG on GP response		Effect of PEG on ME response	
	- PEG	+ PEG	- PEG	+PEG
A. plumosa	84.2^{ef}	138.5^{c}	8.73^{c}	10.21
D. forskahlii	101.1^{e}	91.5^{e}	8.40^{d}	8.14
A. gombiformis	152.6^{abc}	161.7^{ab}	13.42^{a}	13.72
G. saharae	118.4^{d}	142.2^{c}	9.51^{bc}	10.32
Palm leaves	71.1^{f}	87.7^{ef}	7.52^{e}	7.97
Racemes	55.28^{g}	85.86^{ef}	5.14^{f}	5.49
Vetch-oat hay	146.6^{bc}	168.4^{a}	10.04^{b}	10.65
Species	*		*	
PEG treatment	*		*	
Species x PEG	*		NS	

ME, metabolisable energy ; PEG, polyethylene glycol ('+': with, '-': without); $^{a, b, c, d, e, f, g}$ means with different letters In the same row are significantly different ($p < 0.05$); *, denotes significant effect of species, PEG treatment and their interaction on in vitro gas production and ME of feedstuffs ($P < 0.05$) ; NS, non significant ($P > 0.05$).

Table 4. Effect of polyethylene glycol on in vitro gas production parameters over 96 h of the selected feedstuffs

Substrates	a (mL/g DM)		a + b (mL/g DM)		c ($\%h^{-1}$)	
	- PEG	+ PEG	- PEG	+ PEG	- PEG	+ PEG
A. plumosa	6.81	11.7	122.4^{g}	190.2^{bc}	5.09^{ef}	5.14^{ef}
D. forskahlii	-19.2	-19.46	151.2^{ef}	160.4^{de}	4.87^{ef}	4.51^{f}
A. gombiformis	9.73	3.53	176.9^{cd}	191.01^{bc}	8.22^{bc}	7.66^{bcd}
G. saharae	29.1	20.95	112.4^{f}	170.75^{d}	6.64^{bcde}	7.70^{bcd}
Leaves	-5.9	32.21	80.7^{i}	113.4^{g}	7.67^{bcd}	6.16^{cde}
Racemes	12.4	7.56	65.2^{j}	94.73^{h}	8.41^{b}	12.64^{a}
Vetch-oat hay	6.81	19.0	194.65^{b}	213.7^{a}	5.72^{def}	6.43^{bcde}
Significance						
Species			*		*	
PEG treatment			*		NS	
Species x PEG			*		*	

a, the gas production corresponding to the soluble fraction; a + b, potential gas production; c, rate of gas production ; $^{a, b, c, d, e, f, g, h, I, j}$ means in the same row without a common letter differ ($P < 0.05$); *, denote significant effect of species, PEG treatment and their interaction at $P < 0.05$; NS, non significant ($P > 0.05$).

Effect of Polyethylene Glycol on Gas Production Parameters

There were significant effects of substrate, PEG addition, and interaction of substrate and PEG effects ($P<0.05$) on potential gas production (a+b), but addition of PEG did not influenced the rate of gas production (Table 4). Vetch-oat hay had the higher potential gas production, both in the presence and the absence of PEG than the other forages. Whereas, racemes had the higher rate of gas production, both with and without PEG. For some forages (leaves, *D. forskahlii* and *A. gombiformis*), rate of gas production was slightly decreased in the presence of PEG.

Effect of Polyethylene Glycol on In Vitro Truly Digestibility and Partitioning Factor

There was a significant effect of substrate, PEG addition, and interaction of substrate and PEG addition on IVOMD (P<0.05) (Table 5). The IVOMD values ranged from 21.0% in racemes to 56.4% in vetch-oat hay in incubation without PEG. The ranking of forages on the basis of IVOMD was: vetch-oat hay > *A. plumosa* = *G. saharae* > *D. forskahlii*= palm leaves > *A. gombiformis* > racemes. Addition of PEG increased highly IVOMD in feedstuffs with high tannin contents (racemes, *A. gombiformis* and palm leaves, 25.7, 24.9 and 20.0 units, respectively) but moderately in forages with low tannin contents (vetch-oat hay, *D. forskahlii* and *A. plumosa*, 3.57, 3.25, 1.20 units, respectively). In presence of PEG, the ranking of forages became as follows: vetch-oat hay = palm leaves > *A. gombiformis* = *G. saharae* > racemes = *A. plumosa* = *D. forskahlii*. Partitioning factor (PF) was influenced (P<0.05) by substrate, and interaction of substrate and PEG inclusion (Table 5). In the absence of PEG, the PF values ranged from 2.56 mg/mL in *A. gombiformis* to 7.20 mg/mL in palm leaves. Addition of PEG mainly decreased PF in two forages (*A. plumosa* and palm leaves) indicating that PEG addition in these forages promote gas production but not organic matter digestibility. Whereas, addition of PEG did not alter PF in *G. saharae, D. forskahlii* and vetch-oat hay, but induced an increase in PF values for *A. gombiformis* and racemes.

Table 5. Effect of polyethylene glycol on in vitro truly organic matter digestibility and partitioning factor of forages sampled from the North African arid zone

Substrates	IVTOMD			PF	
	- PEG	+ PEG		- PEG	+ PEG
A. plumosa	45.58^{c}	46.78^{c}		5.93^{b}	3.92^{c}
D. forskahlii	41.13^{d}	44.38^{c}		5.82^{b}	5.12^{b}
A. gombiformis	26.83^{e}	51.73^{b}		2.56^{d}	3.47^{c}
G. saharae	44.96^{c}	50.94^{b}		5.31^{b}	5.29^{b}
Palm leaves	39.65^{d}	59.72^{a}		7.20^{a}	6.15^{b}
Racemes	21.08^{f}	46.80^{c}		3.93^{c}	5.40^{b}
Vetch-oat hay	56.44^{a}	60.01^{a}		5.48^{b}	5.28^{b}
Significance					
Species	*			*	
PEG treatment	*			NS	
Species x PEG	*			*	

IVOMD (g/100g DM), in vitro true organic matter digestibility ; PF (mg/mL), partitioning factor ; a, b, c, d, e, f means in the same row with different letters are significantly different (P < 0.05) ; * denotes significant effect of species, PEG treatment and their interaction at P < 0.05 ; NS, non significant (P > 0.05).

DISCUSSION

Chemical composition of plants from arid zone of North Africa that are resistant to drought and salinity is poorly documented. High ash content is a characteristic of desertic plants (Bokhari et al.,1990; Stringi et al.,1991), ash content was also relatively high (>100 g/kg DM) in most forages tested in this work except for *G. saharae*, racemes and vetch-oat hay. This was probably due in all of them to the contamination of aerial part of plants with sand. In the present study, CP content was particularly low (< 75g/kg DM) in almost all forages in agreement with reported data (Bahman et al., 1997; Pascual et al., 2000; Genin et al., 2004; Ramirez et al., 2004 and Ammar et al., 2004b). This may be due to the influence of soil type, environmental conditions and genotypic characteristics, factors that affect the nutritional proprieties of forages. Whereas, the relatively high CP in *A. gombiformis* (> 100g/kg DM) indicate its possible use as protein supplement. Crude protein associated with NDF fraction appeared to be higher in palm fractions and *D. forskahlii* which reduces the available nitrogen for ruminal microbiota, and implies that these forages would require a protein supplement for their use in ruminant feeding. All forages contained high NDF, ADF and lignin. These results were similar to that observed by Pascual et al. (2000) and Genin et al. (2004) for date palm fractions, by Ramirez et al. (2004) for *Aristida* genus, and by Ammar et al. (2004b) for *Genista* genus. The high cell wall content recorded could be due to the arid zone climate. In general, high temperature and low rainfall tend to increase cell wall polysaccharides and to decrease the soluble carbohydrates (Pascual et al., 2000).

Concentration of phenolic compounds varied widely among plant species. The lowest values were recorded for the *Poaceae* family, whereas the highest levels were observed in the *Fabaceae* family and date palm fractions, consistently with the results pointed out in the literature (Tisserand, 1990). The PPC values in all forages except date palm fractions were very weak. These results were not related to the quantity of tannins in the samples. This could be due to the fact that radial diffusion method, based on the measure of the potential biological activity of tannins in feeds, depend upon binding strength of tannins and their mode of binding to protein (Frazier et al., 2003), whereas chemical methods, based on chemical properties of tannins, indicate only the chemical nature of tannins (Silanikove et al., 1996)

The inclusion of PEG in fermentation media of the feedstuffs resulted in a marked increase in *in vitro* gas production with all forages except *D. forskahlii* for which this parameter was reduced in presence of PEG. The increases in gas production when samples were incubated with PEG were also reported for different forages by other authors (Baba et al., 2002; Rubanza et al., 2005 and Singh et al., 2005). As in this study, Singh et al. (2005) have also noted that addition of PEG reduced the gas production for two forages: *Leucaenea leucocephala* (-18.59 and –18.56%) and paddy straw (-7.39 and –6.52%) at 24 and 48h, respectively. The highest responses on *in vitro* gas production due to the inclusion of PEG in *A. plumosa* could be due to the inhibition of tannin effects. All tannin assays showed that *A. plumosa* and *D. forskahlii* had low content in phenolic compounds. However, *A. plumosa* tannins appeared to be more active than that of *D. forskahlii* because they produced large responses to PEG in the gas test. The discrepancies between the two forages may be likely attributed either to the limited ability of PEG to completely inhibit the negative effects of tannins (Baba et al., 2002; Frutos et al., 2004), which depends mainly both on stereochemistry and chemical structure of tannins, or to other factors that may be more important than tannins in limiting fermentation (Ndlovu and Nherera, 1997), which in the case of *D. forskahlii* could be the limited available nitrogen for ruminal microbiota, the higher NDF, ADF and lignin contents, and the saponins detected in this species.

Min et al. (2003) reported that condensed tannin levels of 20-40 g/kg DM produce beneficial effects. Getachew et al. (2002) concluded in an other study that samples containing total phenols and total tannins (tannic acid equivalent/kg DM) up to 40 and 20 g/kg DM, respectively, are not expected to induce an increase in gas production on addition of PEG. In contrast, almost all forages in this study had TT and TCT within the two ranges but gives a positive responses to PEG supplementation. These results showed that effects of tannins seem to be depended on several factors such as forage species, chemical nature and structure of tannins, biochemical interaction among tannins and proteins, than the tannins level itself. On the other hand in palm leaves, PEG addition had small effect despite their relatively high tannin content.

Gimárãez-Beelen et al. (2006) have reported that PEG treatment of three Brazilian legumes reduced the astringency by approximately 70%. Thus, the increase in gas production, in racemes in presence of PEG could possibly due to an increase in the available nutrient to ruminal microbiota especially nitrogen. For the leguminous plants and especially *G. saharae,* the effect of PEG on *in vitro* fermentation could be probably masked by the high lignin content in this forage.

The effect of PEG addition is more pronounced on potential gas production, measured at 96 h of incubation. The effects of tannins on nutrient degradability depends essentially on the formation of complexes between tannins and the components of diets, primarily proteins and to a lesser extent with amino acids, polysaccharides and minerals, as well as on their effects on the microbial population and on its enzymatic activity (Mc Sweeney et al., 2001). For racemes, addition of PEG resulted in an important increase of both potential gas production and rate of gas production. This result could suggest that tannins in this case are binding to fibres and the presence of PEG increased microbial plant adhesion and/or the fibrolytic microbial activity. However, the PEG supplementation induces a decrease in rate gas production in some forage, especially date palm leaves. This result has also been reported by Frutos et al. (2004) and Gimárãez-Beelen et al. et al. (2006). The latter authors have noted that for species, which the rate of gas production is reduced, the bacteria colonisation is restricted. This could suggest that complexes forming between tannins and PEG generate steric obstruction which do not permit and/or limit the fixation of adherent bacteria to the feeds.

In absence of PEG, *A. gombiformis* had the highest gas production, and the lowest IVOMD and PF values. Whereas, *A. plumosa* had lower gas production, and higher IVOMD and PF values. For *A. gombiformis*, the lowest IVOMD could be explained by the fact that tannins binding to proteins form complexes which are largely insoluble in ND solution, thus forming precipitates which will overestimate the undegradable fraction (Makkar et al., 1995). This situation could also suggest that fermentation process and gas production are probably affected by tannins in a complex fashion. In contrary to results obtained by Makkar et al. (1998), PEG addition in this study resulted mainly in an overall increase in IVOMD of 30.7% compared with an overall increase in gas production of 20.2%. This result indicates that effect of PEG addition improved IVOMD at the cost of gas production, which leads certainly in an improvement in microbial protein synthesis. As has been reported by Baba et al. (2002), the PEG addition resulted in decrease of PF values. However for some forages (racemes and *A. gombiformis*), addition of PEG caused an increase in PF values. The same observations have been also reported by Singh et al. (2005) for two forages (*F. roxburghii* and *R. pseudoacacia*). This was due to increased substrate degradability with lower gas production. Blümmel et al. (2003) suggested selection of forages for high degradability but proportionally low gas production.

The theoretical range for PF values for tannins free plants was suggested by Blümmel et al. (1997b) to be between 2.75 and 4.41. PF values of five forages used in this study were higher to the theoretical maximum value. According to Blümmel et al. (1997b), plants with high PF are in general highly digestible and the values correlate well with dry matter intake in ruminants. Thus, these results could suggest that these forages had a potential nutritive value which tends to enhance microbial synthesis rather than gas production.

CONCLUSION

On the basis of *in vitro* fermentation without PEG addition results, the two leguminosea forages *A. gombiformis* and *G. saharae* were judged to be nutritionally good, followed by *A. plumosa* and *D. forskahlii*. However, the weak fermentability of date palm fractions limits their utilisation by ruminants. Addition of PEG inactivated effects of tannins at different levels in forages studied. In vitro fermentation of *D. forskahlii* was not influenced by PEG inclusion despite their tannin level comparatively similar to *A. plumosa,* in which the effect of PEG is more pronounced. These discrepancies between the two forages could be due to the limited available nitrogen for ruminal microbiota, the higher NDF, ADF and lignin content detected in *D. forskahlii*. At same, the effect of PEG in *G. saharae* was probably concealed by its high lignin content. However for date palm fractions, the inclusion of PEG is clearly detectable. These results demonstrated that the *in vitro* fermentation of these substrates was not only associated to the quality and proportion of tannin, but other plant factors also affect *in vitro* gas production and mask the effect of phenolic compounds in the bioassay test.

REFERENCES

Ammar H, López S, González JS and Ranilla MJ 2004a. Comparison between analytical methods and biological assays for the assessment of tannin-related antinutritive effects in some Spanish browse species. *J. Sci. Food Agric.* 84 (11): 1349-1356.

Ammar H, López S, González JS and Ranilla MJ 2004b. Seasonal variations in the chemical composition and in vitro digestibility of some Spanish leguminous shrub species. *Anim. Feed Sci. Technol.* 115: 327-340.

AOAC 1990. Official Methods of analysis, 15th ed. Association of Official Analytical Chemists, AOAC, Arlington, VA.

Baba ASH, Castro FB and Orskov ER 2002. Partitioning of energy and degradability of browse plants in vitro and the implication of blocking the effects by the addition of polyethylene glycol. *Anim. Feed Sci. Technol.* 95: 93-104.

Bahman AM, Topps JH and Rooke JA 1997. Use of date palm leaves in high concentrate diets for lactating Friesian and Holstein cows. *J. Arid Environment.* 35: 141-146.

Blümmel M and Becker K 1997a. The degradability of fifty-four roughages and neutral detergent fibre as described by gas production and their relationship to voluntary feed intake. *British Journal of Nutrition.* 77: 757-768.

Blümmel M, Karsli A and Russel JR 2003. Influence of diet on growth yields of rumen microorganisms in vitro and in vivo: Influence of variable carbon fluxes to fermentation products. *British Journal of Nutrition*. 90: 625-635.

Blümmel M, Steingass H and Becker K 1997b. The relationship between in vitro gas production, in vitro microbial biomass and N-15 incorporation and its implication for the prediction of voluntary feed intake of roughages. *British Journal of Nutrition*. 77: 24-34.

Bokhari UG, Alyaeesh F and Al-Noori M 1990. Nutritional characteristics of important desert grasses in Saudi Arabia. *Journal Range Management*. 43: (3), 203-204.

Frazier RA, Papadopoulou A, Mueller-Harvey I, Kissoon D and Green RJ 2003. Probing protein-tannin interactions by isothermal titration microcalorimetry. *Journal of Agriculture and Food chemistry*. 51: 5189-5195.

Frutos P, Hervás G, Giráldez FJ and Mantecōn AR 2004. An in vitro study on the ability of polyethylene glycol to inhibit the effect of quebracho tannins and tannic acid on rumen fermentation in sheep, goats, cows, and deers. *Australian Journal of Agricultural Research*. 55 (11): 1125-1132.

Genin D, Kadria A, Khorchani T, Sakkal K, Belgacem F and Hamadi M 2004. Valorization of date palm by-products (DPBP) for livestock feeding in southern Tunisia. Potentialities and traditional utilization. In: Ben Salem H, Nefzaoui A and Morand-Fehr P (Eds), Nutrition and feeding strategies of sheep and goats under harsh climates. *CIHEAM, Options Méditerranéennes*. 59: 221 226.

Getachew G, Crovetto GM, Fondevila M, Krishnamoorthy U, Singh B, Spanghero M, Steingass H, Robinson PH and Kailas MM 2002. Laboratory variation of 24 h in vitro gas production and estimated metabolisable energy values of ruminant feeds. *Animal Feed Science and Technology*. 92: 51 57.

Guimarães-Beelen PM, Berchielli TT, Beelen R and Medeiros AN 2006. Influence of condensed tannins from Brazilian semi-arid legumes on ruminal degradability, microbial colonization and ruminal enzymatic activity in Saanen goats. *Small Ruminant Research*. 61, 35-44.

Haddi ML, Filacorda S, Meniai K, Rollin F and Susmel P 2003. In vitro fermentation kinetics of some halophytes shrubs sampled at three stages of maturity. *Animal Feed Science and Technology*. 104: 215-225.

Hagerman AE 1987. Radial diffusion method for determining tannin in plants extracts. *Journal of Chemistry and Ecology*. 13: 437-449.

Larrahondo JE 1985. Productos naturales: pruebas químicas iniciales en una planta. Guía de estudio del Departamento de Química, Universidad del Valle, pp. 10.

Longuo HF, Chelma A and Ouled Belkher A 1989. Quelques Aspects botaniques et nutritionnels des pâturages du dromadaire en Algérie. In: Séminaire sur la digestion, la nutrition et l'alimentation du dromadaire. *CIHEAM, Options Méditerranéennes*. 2: 47-53.

Makkar HPS 2005. In vitro gas methods for evaluation of feeds containing phytochemicals. *Animal Feed Science and Technology*. 123-124: 291-302.

Makkar HPS and Becker K 1996. A bioassay for polyphenols (tannins), in: Vercauteren J, Cheze C, Dumon MC and Weber J.F. (Eds.), Proceedings of polyphenols communications 96, 18th International conference on polyphenols, vol. 1, Bordeaux, France, pp. 197-198.

Makkar HPS, Blümmel M and Becker K 1995. Formation of complexes between polyvinylpyrrolidones or polyethylene glycol and their implication in gas production and true digestibility in vitro techniques. *British Journal of Nutrition*. 73: 897-913.

Makkar HPS, Blümmel M and Becker K 1998. Application of an in vitro gas method to understand the effects of natural plant products on availability and partitioning of nutrients. In: In vitro Techniques for Measuring Nutrient Supply to Ruminants. British Society of Animal Science, Occasional publication No. 22, pp.147-150.

Makkar HPS, Blümmel M, Borowy NK and Becker K 1993. Gravimetric determination of tannins and their correlations with chemical and protein precipitation methods. *Journal of Science and Food Agriculture*. 61: 161-165.

McSweeney CS, Palmer B, McNeill DM and Krause DO 2001. Microbial interaction with tannins: nutritional consequences for ruminants. *Animal Feed Science and Technology*. 91: 83-93.

Menke KH, Raab L, Salewski A, Steingass H, Fritz D and Shneider W 1979. The estimation of the digestibility and metabolizable energy content of ruminant feedingstuffs from the gas production when they are incubated with rumen liquor. *Journal of Agricultural Science Cambridge*. 97: 217-222.

Min BR, Barry TN, Attwood GT and McNaab WC 2003. The effect of condensed tannins on the nutrition and health of ruminants feed fresh temperate forages: a review. *Animal Feed Science and Technology*. 106: 3-19.

Ndlovu lR and Nherera FV 1997. Chemical composition and relationship to in vitro gas production of Zimbabwean browsable indigenous tree species. *Animal Feed Science and Technology*. 69: 121-129.

Orskov ER and Mc Donald I 1979. The estimation of protein degradability in the rumen from incubation measurements weighted according to rate of passage. *Journal of Agricultural Science Cambridge:* 92: 499-503.

Pascual JJ, Fernandez C, Diaz JR, Garces C and Rubert-Aleman J 2000. Voluntary intake and in vivo digestibility of different date-palm fractions by Murciano-*Granadina (Capra Hircus)*. *Journal of Arid Environments*. 45: 183-189.

Porter LJ, Hrstich LN and Chan BJM 1986. The conversion of procyanidins and prodelphinidins to cyanidin and delphinidin. *Phytochemistry*. 25: 223-230.

Ramirez RG, Haenlein GFW, García Castillo CG and Núñez-Gonzalez M A 2004. Protein, lignin and mineral contents and in situ dry matter digestibility of native Mexican grasses consumed by range goats. *Small Ruminant Research*. 52: 261-269.

Rosales M, Galindo W, Murgueitio E, Larrahondo J 1989. Sustancias antinutricionales en las hojas de árboles forrajeros. Livestock Research for Rural Development. 1 (1): 79–91.

Rubanza CDK, Shem MN, Otsyina R, Bakengesa SS, Ichinohe T and Fujihara T 2005. Polyphenolics and tannins effect on in vitro digestibility of selected Acacia species leaves. *Animal Feed Science and Technology*. 119: 129-142.

Silanikove N, Shinder D, Gilboa N, Eyal M and Nitsan Z 1996. Binding of polyethylene glycol to samples of forages plants as an assay of tannins and their negative effects on ruminal degradation. *Journal of Agriculture and Food chemistry*. 44: 3230-3234.

Singh B, Saho A, Sharma R and Bhat TK 2005. Effect of polyethylene glycol on gas production parameters and nitrogen disappearance of some tree forages. *Animal Feed Science and Technology*. 123-124: 351-364.

Statistical Analysis System Institute Inc., 1990. SAS/STAT® user's guide Int Vol 1, version 6, Fourth Edition, Cary, NC, USA.

Stringi L, Amato G and Gristina L 1991. The role of some forage shrubs in optimizing forage production in a semi-arid Mediterranean area. *Agriculture Mediterranean*. 121: 16-23.

Tisserand JL 1990. Les ressources alimentaires pour le bétail. Cahiers Options Méditerranéennes. 11: 237-248.

Van Soest PJ, Robertson JB and Lewis BA 1991. Methods for dietary fiber, neutral detergent fiber and non starch polysaccharides in relation to animal function. *Journal of Dairy Science*. 74: 3583-3597.

In: Biotechnology and Conservation ... (2 Volume Set) ISBN: 978-1-63117-869-6
Editors: S. A. Farooq, R. Abed and S. Baqir © 2014 Nova Science Publishers, Inc.

Chapter 33

POLYPHENOLS CONTENT AND ANTIOXIDANT PROPERTIES OF EXTRACTS FROM SELECTED HYPOTENSIVE PLANTS IN ALGERIA

Amel Bouaziz, Seddik Khennouf*, Assia Bentahar, Saliha Djidel, Saliha Dahamna and Smain Amira

Laboratory of Phytotherapy Applied to Chronic Diseases, Department of Biology and Animal Physiology, Faculty of Nature and Life Sciences, University Ferhat Abbas, Setif, Algeria

ABSTRACT

The oxidative stress is involved in various pathologies such as hypertension and inflammatory diseases. Various medicinal plants are consumed in the Mediterranean areas for hypertension. These plants are a natural source of phytochemicals resulting from the secondary metabolism and having antioxidant activity which counteract oxidative stress. The present study evaluates the antioxidant activity and polyphenolic content in the aqueous extract of four medicinal plants: *Laurus nobilis*, *Eucalyptus globulus*, *Rosmarinus officinalis* and *Petroselinum crispum*. These plants were selected based on an ethnopharmacological survey of plants largely used in folk medicine in Setif region (eastern Algeria) for the treatment of hypertension.

Total phenolic compounds contents vary from 7.15 ± 0.30 mg GA E/g dry weight to 39.61 ± 0.23 mg GA E/g of dry weight for *P. crispum* and *R. officinalis* respectively. The highest flavonoids content was recorded in *L. nobilis* extract (5.009 ± 0.110 mg QE/g of dry weight) and the weakest concentration was recorded for *P. crispum* (1.66 ± 0.06 mg QE/g of dry extract). Tannins content vary from 20.74 ± 1.882 mg TA E/g of dry extract in *R. officinalis* to 25.64 ± 1.721 mg TA E/g of dry extract in *P. crispum*. In β-carotene/ linoleic acid bleaching assay, the antioxidant activity of extracts lies between 22,75 ± 0,30% (for *P. crispum*) and 69,82 ± 3,14 % (for *E. globulus*).

* Corresponding author: Seddik Khennouf. Laboratory of Phytotherapy Applied to Chronic Diseases, Department of Biology and Animal Physiology, Faculty of Nature and Life Sciences, University Ferhat Abbas, Setif 1, 19000, Algeria. E-mail: Khennouf_sed@yahoo.fr.

The results of this present study indicate that these selected hypertensive plants are rich in antioxidants such as phenolic compounds which have the property to trap free radicals and to reduce oxidants which are involved at least in part in hypertension.

Keywords: Medicinal plants, Polyphenols, Flavonoids, tannins, Hypertension, Antioxidant activity, β - carotene

INTRODUCTION

Increased vascular oxidative stress could be involved in the pathogenesis of hypertension, a major risk factor for cardiovascular disease mortality (Rodrigo et al., 2007; Miyajima et al., 2007).

Oxidative stress occurs when there is an imbalance between the generation of reactive oxygen species (ROS) and the antioxidant defense systems so that the latter become overwhelmed (Becker et al., 2004; Clarice Yang et al., 2010). In human essential hypertension, ROS may increase due to a diminution of the activity of antioxidant enzymes (Pedro-Botet et al., 2000).

It is known that superoxide rapidly inactivates endothelium-derived nitric oxide (NO), the most important endogenous vasodilator, thereby promoting vasoconstriction (Dusting et al., 1999; Zicha et al., 2001) and leading to endothelial dysfunction (Johnstone et al., 1993; Panza et al., 1995).

Endothelial dysfunction is commonly described as the impairment of endothelium-dependent vasorelaxation caused by a loss of NO bioavailability in the vasculature (Cai and Harrison, 2000; Schulz et al., 2004). Attempts to counteract the hypertensive effect of ROS have led to the use of exogenous administration of antioxidants thought to improve the vascular function and reduce the blood pressure.

In fact, in recent years, research has focused on medicinal plants to extract natural and low-cost antioxidants including polyphenols, flavonoids that can replace synthetic additives such as butylated hydroxyanisole (BHA) and butylated hydroxytoluene (BHT) that might be carcinogenic (Whysner et al., 1994) and even toxic (Moure et al., 2001).

Furthermore, a large number of flavonoids and phenolic compounds have been isolated from crude plant extracts with proven cardiovascular potential (Dhiman et al., 2012) and a strong antioxidant activity (Baratto et al., 2003; Katalynic et al., 2006) and a powerful scavenger activity against free radicals (Kumaran and Karunakaran, 2007).

In Algeria, as in many countries, medicinal plants are widely used in the folk medicinal system to treat diseases including arterial hypertension. Ethnopharmacological survey carried out in Setif region (eastern Algeria) indicates that *Laurus nobilis*, *Eucalyptus globulus*, *Rosmarinus officinalis* and *Petroselinum crispum* are extensively used in folk medicine for the treatment of hypertension (data not published).

However, so far, there are no studies on the antioxidant potential of these plants decoctions.

Thus, the present work was undertaken with the aim to examine the antioxidant capacity of the aqueous extract of the four plants and to evaluate their polyphenols, flavonoids and tannins contents.

Plant Material

Plants were collected from different herbal vendors in Setif region and identified by Pr. Laouar H (Department of Ecology and Vegetal Biology, University of Setif). Bovine blood was obtained from a local abbattoir. All other reagents were purchased from Sigma Chemicals (Germany), Fluka and Prolabo.

Extraction Procedure

Dried plant materials (6 g) were boiled in 200 ml distilled water for 10 min. The decoctions were cooled at room temperature and filtered according to the traditional method preparation (Djidel et al., 2009). The solutions (3%) were stored refrigerated until shortly before use.

Determination of Total Polyphenols Content

The total polyphenols content was determined by the Folin–Ciocalteu method as described by Cliffe et al. (1994). In brief, 0.1 ml of each extracts was well mixed with 2.5 ml of distilled water and 0.5 ml of the Folin–Ciocalteu stock reagent, after 4 min 1.0 ml of Na_2 CO_3 reagent (20%) was added to the mixture. They were then incubated at room temperature for 1.5 h. The mixture absorbance was spectrophotometrically measured at wavelength 760 nm. The amount of total polyphenols in different extracts was determined from a standard curve of Gallic acid then the results were expressed in milligrams of gallic acid equivalents per gram of dry weight.

Determination of Total Flavonoids Contents

Flavonoids were quantified using aluminium chloride reagent ($AlCl_3$) (Bahorun et al., 1996). Each sample (1 ml) was mixed with 1 ml of aluminium chloride ($AlCl_3$) solution (2%) and allowed to stand for 10 min. Absorbance of the mixture was then determined at 430 nm versus prepared methanol blank. Results were expressed as mg equivalent Quercetin per gram dry weight.

Determination of Tannins Contents

The capacity to precipitate haemoglobin was determined by using bovine fresh blood according to the method described by Batesmith (1973). Briefly, one milliliter of extract was mixed with 1 ml of heamolysed bovine blood (absorbance equal to 1.6). After 20 min, the mixture was centrifuged at 4000 rpm for 10 min, and the absorbance of the supernatant was measured at 756 nm. Results were expressed as mg equivalent Tannic acid per gram dried weight.

β-CAROTENE BLEACHING ASSAY

In this assay, antioxidant capacity is determined by measuring the inhibition of the volatile organic compounds and the conjugated diene hydroperoxides arising from linoleic acid oxidation (Dapkevicius et al., 1998). A stock solution of *β*-carotene-linoleic acid mixture was prepared as follows: 0.5 mg *β*-carotene was dissolved in 1 ml of chloroform and 25 μl linoleic acid and 200 mg Tween 40 were added. Chloroform was completely evaporated. Then, 100 ml distilled water saturated with oxygen (30 min, 100 ml/min) were added with vigorous shaking. 2500 μL of this reaction mixture were dispensed into test tubes and 350 μl of the plants extracts, prepared at 2 mg /ml concentrations, were added and the emulsion system was incubated for 48 h at room temperature. The same procedure was repeated with synthetic antioxidant, BHT, as positive control, and a blank. After this incubation period, absorbances of the mixtures were measured at 490 nm. Antioxidative capacities of the extracts were compared with those of BHT and blank.

STATISTICAL ANALYSIS

Values are expressed as mean ± SD. One-way analysis of variance (ANOVA) followed by Dunnett's *t*-test for multiple comparisons was used and values of *P* less than 0.05 were considered significant.

TOTAL POLYPHENOLS, FLAVONOIDS AND TANNINS IN PLANT EXTRACTS

Phenolic compounds are commonly found in both edible and non-edible plants, and they have been reported to have multiple biological effects, including antioxidant activity (Kahkonen et al., 1999). From the results summarized in Table 1, we can easily conclude that the amount of total phenolics varied in different plants and ranged from 7.15 to 39.61 mg Gallic acid equivalent /g of dry material. The highest total phenolic levels have been detected in *R. officinalis* and the lowest in *P. crispum* (Table 1).

Furthermore, we can deduce that all these plants were rich in flavonoids and tannins. We outline that the amount of flavonoids in the aerials parts of the plants varies from 1.66 to 5.009 mg/g quercetin equivalent of the dry extract, whereas tannins content vary from 20.74 mg tannic acid equivalent/g of dry extract in *R. officinalis* to 25.64 mg tannic acid equivalent /g of dry extract in *P. crispum.*

β-CAROTENE/LINOLEIC ACID ASSAY

In the *β* -carotene bleaching assay, linoleic acid produces hydroperoxides as free radicals during incubation. The presence of antioxidants in the extracts will minimize the oxidation of *β* -carotene by hydroperoxides.

Table 1. Total polyphenols, flavonoids and tannins contents for the studied plants mg/g dry weight

Plant extract	Total polyphenols	Total flavonoids	Total tannins
Petroselinum crispum	7.15 ± 0.30	1.66 ± 0.06	25.64 ± 1.72
Laurus nobilis	21.05 ± 1.08	5.00 ± 0.11	22.87 ± 3.79
Eucalyptus globules	39.28 ± 0.69	3.65 ± 0.21	22.12 ± 2.12
Rosmarinus officinalis	39.61 ± 0.23	4.99 ± 0.21	20.70 ± 1.88

Each value represents the mean ± SD (n = 3). Total polyphenols content was expressed as mg gallic acid equivalent/g dried extract. Total flavonoids content was expressed as mg quercetin equivalent/g dried extract. Total tannins was expressed as mg tannic acid equivalent/g dried weight.

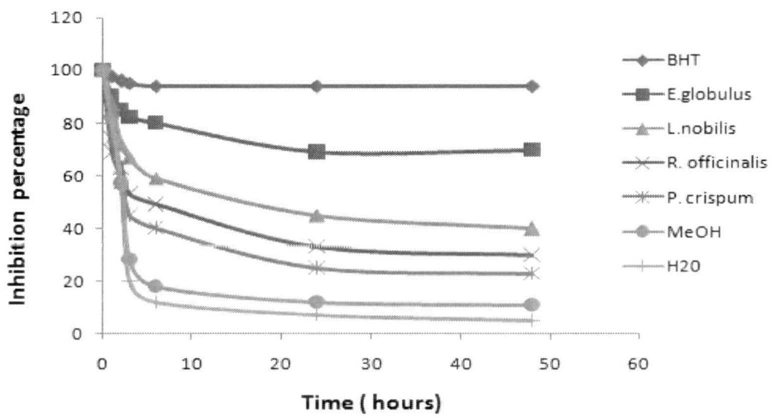

Figure 1. Changes in the percentage of the linoleic acid oxidation inhibition ratios under the influence of plants extracts (2 mg/ml), compared to BHT as a positive control during 48 h. Results are means of three different experiments.

Hydroperoxides formed in this system will be neutralized by the antioxidants from the extracts.

The results for the antioxidant activity through β-carotene-linoleate model system of the various plants extracts (Figure 1, Figure 2) indicate that all the species studied exhibit free radical scavenging activity in a wide range. *E. globulus* extract is the best inhibitor of the oxidation of β-carotene in 48h (69,82 ± 3,14 %), followed by *L. nobilis* (39,82 ± 1,14%) and finally the lowest effect was seen with *R. officinalis* and *P.crispum* (29,92 ± 0,89%, 22,75 ± 0,30% respectively).

L. nobilis, *E. globulus*, *R. officinalis* and *P. crispum* are largely used in the Algerian traditional medicine to treat hypertension. In this study, we report, for the first time, the evaluation of polyphenols, flavonoids and tannins contents of these plants decoctions, and their potential antioxidant capacity.

Our results revealed that the studied medicinal herbs exhibited clearly a high antioxidant activity and contain high amounts of phenolics. This result suggests that the antioxidant capacity of these hypotensive plants is due to the contribution of phenolic and flavonoid compounds.

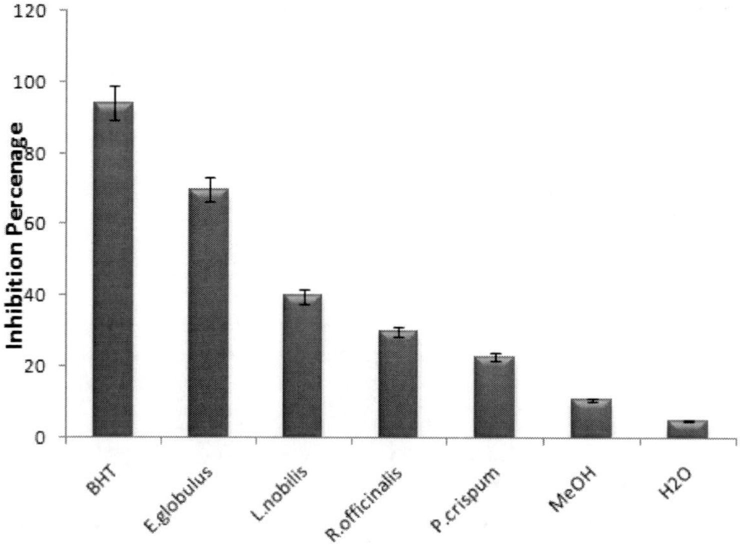

Figure 2. Percentage inhibition of the linoleic acid oxidation by plants extracts compared to BHT as a positive control.

Furthermore, the obtained results are in good agreement with those obtained by Tom (2011), who reported that the antihypertensive effect of the extract of *Terminalia superba* is associated with its antioxidant properties in hypertensive rats.

Also, it can be concluded that the antioxidant activity of plants extracts; is not the result of these compounds only, but may be also related to the presence of some individual active phenolic compounds (Djeridane et al., 2006).

Several studies have evaluated the relationships between the antioxidant activities of plant products and their phenolic contents.

However, these relationships were difficult to explain on the basis of quantitative analyses alone (Amira et al., 2012). The unclear relationship between the antioxidant activity and total phenolics may be explained in various ways, in fact, total phenolics content does not incorporate all the antioxidants.

In addition, the synergism between the antioxidants in the mixture makes the antioxidant activity not only dependant on the concentration, but also on the structure and the interaction between the antioxidants (Shahidi and Marian, 2003; Djeridane et al., 2006). This is the reason why samples such as *E. globulus* and *R. officinalis*, with similar concentrations of total phenolics, may vary in their antioxidant activity (β-Carotene Bleaching Assay).

In conclusion, the studied hypotensive plants contain high amounts of phenolics which exhibit antioxidant activity. Upon this study, we can state that further *in vivo* studies and other kinds of antioxidant assays are needed to confirm the advantageous quality of these plant extracts to treat hypertension.

Therefore, the identification of specific phenolic compounds responsible for the high antioxidant activities which can be very beneficial for use as food additives represents one of our future aims.

REFERENCES

Amira, S., Dade, M., Schinella, G., and Ríos, J. L. 2012. Anti-inflammatory, anti-oxidant, and apoptotic activities of four plant species used in folk medicine in the Mediterranean basin. *Pak. J. Pharm. Sci.* 25(1): 65-72.

Atmani, D., Chaher, N., Berboucha, M., Ayouni, K., Lounis, H., Boudaoud, H., Debbache, N., and Atmani, D. 2009. Antioxidant capacity and phenol content of selected Algerian medicinal plants. *Food Chemistry*. 112: 303–309.

Bahorun, T., Gressier, B., Trotin, F., Brunete, C., Dine, T., Vasseur, J., Gazin, J. C., Pinkas, M., Luycky, M., and Gazin, M. 1996. Oxygen species scavenging activity of phenolic extract from Hawthorn fresh plant organs and pharmaceutical preparation. *Arzneim Forsch/Drug Res*. 1-6.

Baratto, M. C., Tattini, M., Galardi, C., Pinelli, P., Romani, A., and Visiolid, F. 2003. Antioxidant activity of Galloyl quinic derivatives isolated from *Pistacia lentiscus* leaves. *Free Radical Research*. 37(4): 405–412.

Batesmith, E. C. 1973.A haemaanlysis of tannin: The concept of relative orstingency. *Phytochemistry*. 12: 907-912.

Becker, L. B. 2004. New concepts in reactive oxygen species and cardiovascular reperfusion physiology. *Cardiovasc. Res*. 61: 461–470.

Cai, H. and Harrison, D. G. 2000. Endothelial dysfunction in cardiovascular diseases: the role of oxidant stress. *Circ. Res*. 87: 840–844.

Clarice Yang, H. H., Breemen, C. V. and Chung, A. Y. 2010. Vasomotor dysfunction in the thoracic aorta of Marfan syndrome is associated with accumulation of oxidative stress. *Vascular Pharmacology*, 52: 37–45.

Cliffe, S., Fawer, M. S., Maier, G., Takata, K., Ritter, G. 1994. Enzyme assays for the phenolic content of natural juices. *Journal of Agricultural and Food Chemistry*. 42, 1824-1828.

Dapkevicius, A., Venskutonis, R., Van Beek, T. A., and Linssen, P. H. 1998. Antioxidant Activity of Extracts Obtained by Different Isolation Procedures from Some Aromatic Herbs Grown in Lithuania. *Journal of the Science of Food and Agriculture*, 77(1): 140-146.

Dhiman, A., Nanda, A. and Ahmad, S. 2012. A quest for staunch effects of flavonoids: Utopian protection against hepatic ailments. *Arabian Journal of Chemistry*. http://dx.doi.org/10.1016/j.arabjc.2012.05.001.

Djeridane, A., Yousfi, M., Nadjemi, B., Boutassouna, D., Stocker, P., and Vidal, N. 2006. Antioxidant activity of some algerian medicinal plants extracts containing phenolic compounds. *Food Chemistry*, 97: 654–660.

Djidel, S., Khennouf, S., Baghiani, A., Harzallah, D., and Arrar, L. 2010. Medicinal plants used traditionally in the algerian folk medicine for gastrointestinal disorders and hypertension: total polyphenols, flavonoides and antioxidant activity. *Acta Hort. (ISHS)*. 854: 59-65.

Dusting, G. J., Akita, K., Hickey, H., Smith, M., and Gurevich, V. 1999. Cyclosporin A and tacrolimus (FK506) suppress expression of inducible nitric oxide synthase *in vitro* by different mechanisms. *Br. J. Pharmacol*. 128: 337–344.

Johnstone, M. T., Creager, S. J., Scales, K. M., Cusco, J. A., Lee, B. K., and Creager, M. A., 1993. Impaired endothelium-dependent vasodilation in patients with insulin-dependent diabetes mellitus. *Circulation*. 88: 2510–2516.

Kahkonen, M. P., Hopia, A. I., Vuorela, H. J., Rauha, J. P., Pihlaja, K., Kujala, T. S., Heinonen, M. 1999. Antioxidant activity of plant extracts containing phenolic compounds. *Journal of Agriculture and Food Chemistry*. 47: 3954–3962.

Katalynic, V., Milos, M., Kulisic, T., and Jukic, M. 2006. Screening of 70 medicinal plant extracts for antioxidant capacity and total phenols. *Food Chemistry*. 94: 550–557.

Kumaran, A. and Karunakaran, R. J. 2007. Activity-guided isolation and identification of free radical-scavenging components from an aqueous extract of *Coleus aromaticus*. *Food Chemistry*. 100(1):356–361.

Miyajima, K., Minatoguchi, S. and Ito, Y. 2007. Reduction of QTc dispersion by the angiotensin II receptor blocker valsartan may be related to its anti-oxidative stress effect in patients with essential hypertension. *Hypertens. Res.* 30: 307–313.

Moure, A., Cruz, J. M., Franco, D., Dominguez, J. M., Sineiro, J., and Dominguez, H. 2001. Natural antioxidants from residual sources. *Food Chemistry*. 72(2): 145–171.

Panza, J. A., Garcia, C. E., Kilcoyne, C. M., Quyyumi, A. A., and Cannon, R. O., 1995. Impaired endothelium-dependent vasodilation in patients with essential hypertension. evidence that nitric oxide abnormality is not localized to a single signal transduction pathway. *Circulation*. 91: 1732–1738.

Pedro-Botet, J., Covas, M. I., Martin, S., and Rubies-Prat, J. 2000. Decreased endogenous antioxidant enzymatic status in essential hypertension. *J. Hum. Hypertens.* 14: 343–345.

Rodrigo, R., Brat, H., Passalacqua, W., Araya, J., Guichard, C., and Bachler, J. P. 2007. Relationship between oxidative stress and essential hypertension. *Hypertens. Res.* 30: 1159–1167.

Schulz, E., Anter, E. and Keaney, J. F. 2004. Oxidative stress, antioxidants, and endothelial function. *Curr. Med. Chem.* 11: 1093–1104.

Tom, E. N. L., Demougeot, C., Mtopia, O. P., Dimoa, T., Djomenia, P. D. D., Bilanda, D. C., Girard, C., and Berthelot, A. 2011. The aqueous extract of *Terminalia superba* (Combretaceae) prevents glucose-induced hypertension in rats. *J. of Ethnopharmacol.* 133: 828–833.

Shahidi, F. and Marian, N. 2003. *Phenolics in Food and Nutraceuticals*. CRC Press, Boca Raton. 1:144-150.

Whysner, J. Wang, C. X., Zang, E., Iatropoulos, M. J., and Williams, G. M. 1994. Dose response of promotion of butylated hydroxyanisole in chemically initiated tumors of the rat stomach. *Food and Chemical Toxicology*. 32: 215–222.

Zicha, J., Dobesova, Z. and Kunes, J. 2001. Relative deficiency of nitric Oxide dependent vasodilation in salt-hypertensive Dahl rats: the possible role of superoxide anions. *J. Hypertens*. 19: 247–254.

In: Biotechnology and Conservation … (2 Volume Set)
Editors: S. A. Farooq, R. Abed and S. Baqir
ISBN: 978-1-63117-869-6

Chapter 34

ANTIMICROBIAL BIOACTIVE COMPOUND ISOLATED FROM *PODAXIS PISTILLARIS* IN SOUTHERN IRAQ

Tawfik M. Muhsin[1,*], Abdul-Hafiz A. Al-Duboon[2] and Kawther T. Khalaf[3]

[1]Department of Biology, College of Education,
University of Basrah, Basrah, Iraq
[2]Marine Science Center, Basrah University, Basrah, Iraq
[3]College of Pharmacy, Basrah University, Basrah, Iraq

ABSTRACT

Podaxis pistillaris is a xerophilic mushroom widely distributed in the southern Iraqi desert. The aim of this study was to examine the antimicrobial bioactivity of metabolic compound isolated from the fungal extract using liquid culture media under optimized growth conditions *in vitro*. The results indicated that a single bioactive compound P1 was extracted from the fungal culture filtrate. The identification of this compound was confirmed by using GC-Mass and H^1NMR. Solubility, toxicity and the molecular weight of the isolated bioactive compound were determined. The isolated compound P1 belongs to aliphatic group with a molecular weight of 340 kd. The antimicrobial bioactivity of the extracted compound P1 against the bacterial strains *E. coli* and *S. aureus* and the dermatophytic fungus *Microsporum gypseum* was tested using a disc diffusion agar method. The minimal inhibitory concentration (MIC) was also performed. This compound exhibited a good bioactivity against the tested bacteria and revealed inhibition zones of 25 mm and 21 mm diam against *E. coli* and *S. aureus*, respectively. However, low bioactivity of this compound against the tested fungal dermatophyte was detected.

Keywords: Antibacterial, antifungal, bioactive compound, Podaxis

[*] Corresponding author: Prof. Dr. Tawfik M. Muhsin. E-mail: tmuhsin2001@yahoo.com.

1. Introduction

Fungi are considered to be good natural sources for secondary metabolites which include a diverse antimicrobial bioactive agents. The significances of bioactive compounds isolated from mushrooms in medical therapy and pharmaceutical applications have been reviewed by Lindequist et al. (2005). Nevertheless, in the last decade more research interests have been focused on the pharmaceutical applications of mushrooms as anticancer, antitumor, antioxidants and antimicrobial agents (Jonathan and Fasidi 2003; Anke et al., 2004; Kettering et al., 2005; Al-Fatimi et al. 2006, Jonathan and Awotona, 2010; Muhsin et al., / Biotechnology and Conservation Muhsin et al., 2011). However, a number of fungal species belong to the class Gasteromycetes (Basidiomycota) produce a wide range of bioactive chemical compounds. Among these fungal mushrooms, *Podaxis pistillaris* is quite common in desert areas and medicinally used in folk traditional treatments in different parts of the world. The edibility of this mushroom in many places over the world has also been reported as food in countries such as India and Afghanistan (Gupta and Singh 1991; Jiskani 2002), in Saudi Arabia (Hashem and AL-Rahmah 1993), in Yemen (Al-Fattimi et al., 2006) and in Iraq (Muhsin et al., 2012). Recently it has been demonstrated that *P. pistillaris* exhibits a bioactive compounds against bacteria (Al-Fattimi et al., 2006; Khalaf 2008). The aim of this study is to examine the bioactive chemical compounds produced by this desert mushroom against Gram positive and Gram negative strains of bacteria and fungal dermatophytes.

2. Materials and Methods

2.1. Fungal Mushroom Culture

Freshly fruit bodies of the desert mushroom *Podaxis pistillaris* were collected from the sandy soils in southern Iraq during spring 2010. In the laboratory, small pieces (0.5 cm long) were cut from the fruit body, surface sterilized with 3 % Sodium Hypochlorate for 3 min, washed with sterile distilled water and placed on Malt Extract Agar (MEA) in Petri dishes. Plates were incubated at 25 °C for two weeks. After cultivation, the mycelium was removed from the agar medium surface and amended into a fermentation liquid culture medium consisted of (50 g glucose and 50 g malt extract in 1L DW adjusted at pH 5.4) and incubated at 25 °C on a rotary shaker for 3 weeks (Al-Fatimi et al., 2005).

2.2. Extraction and Isolation of Bioactive Compounds

The fungal culture was filtered on Whatman No. 1 filter paper, the pH was adjusted at 3 using 2N HCl. The filtrate was extracted three times with Ethyl acetate: Toluene: Acetic acid (70:30:1 vol) using a separating funnel.

The organic layer was collected and dehydrated with Na2SO4 then placed in Petri dishes and dried at room temperature. Thin Layer chromatography (TLC) was applied for the isolation of the extracted metabolites using Silica gel of 2x 10 cm (Silica gel GF243, Merck, Germany) and Rf value was measured.

Further extraction was made on Silica Gel Column Chromatography (Silica gel G-60, Merck, Germany) (Figure 1). The identification of the purified compounds was made by using GC-mass and HNMR techniques.

2.3. Antibacterial and Antifungal Bioactivity Assay

Discs diffusion agar method as described by Casals (1979) was used to examine the antimicrobial activity of the isolated compound P1. Two strains of bacteria; *Escherichia coli* (ATCC 25922) and *Staphylococcus aureus* (ATCC 25923) were used for this purpose. 2.5 mg of the dried fungal extract was dissolved in 1 ml of DMSO solvent used as stock solution for this test. Discs of 0.6 mm diam. Whatman No1 filter paper were sterilized and soaked in the fungal extract and placed on plates containing Muller Hinton Agar (MHA) medium inoculated with 0.1 ml suspension of bacterial strains by streaking method. Similarly, the antifungal bioactivity of this compound was also tested against an isolate of dermatophytic fungus *Microsporum gypseum* using Sabouraud Dextrose Agar (SDA) medium. Bacterial cultures were incubated at 37 °C and the dermatophytic culture was incubated at 27 °C. Fungal culture was obtained from the Basrah General Hospital, Dermatology section.

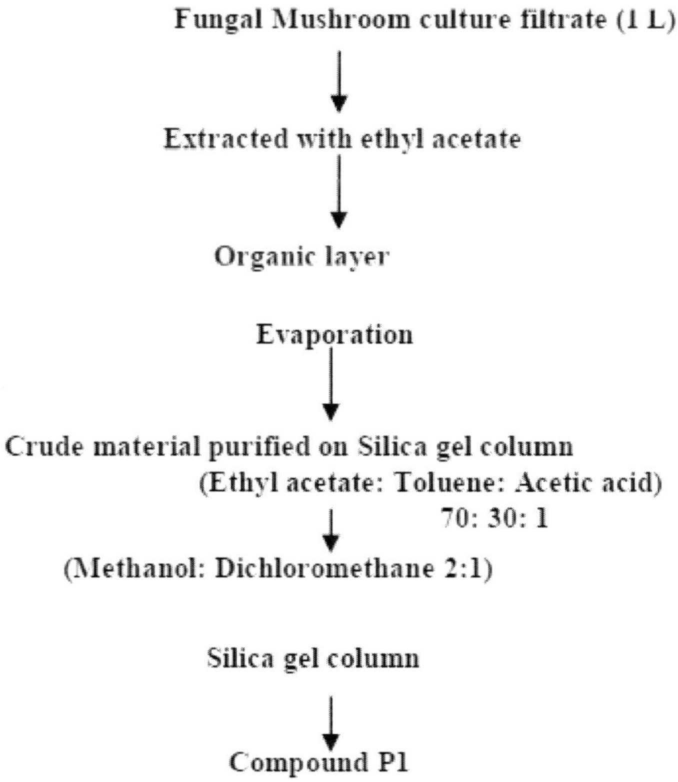

Figure 1. Steps of extraction of bioactive compound P1 from *P. pistillaris*.

2.4. The Minimum Inhibitory Concentration (MIC) Test

The MIC values were determined by the standard serial dilution assay (McGinnis, 1980) using serial dilutions of the fungal extract (100, 50, 25, 12, 6.5, 3.13, 1.56, 0.78, 0.39, 0.2, 0.1, 0.05 and 0.025 ug/L).

The MIC values in this assay were indicated by the absence of bacterial or fungal growth at the minimal concentration of the compound. Emmons Sabouraud's dextrose broth (ESDB) medium was used for this test.

2.5. Cytotoxic Test

Cytotoxicity of the isolated compound P1 was examined by using human RBC as previously described method (Xian-guo and Ursula, 1994).

2.6. Solubility Test

The solubility of the isolated bioactive compound P1 in various solvents (ethyl acetate, ethanol, methanol, chloroform, hexane, dimethyl sulfuoxide (DMSO) and water) was carried out.

2.7. Identification of Bioactive Compounds

Ultra violet (UV) spectrum (LKB-Sweden UV), Infra-red spectrum (IR) (Pye-Unicam sp 3-3005 UK), Gas chromatography Mass (GC) and H^1NMR methods were applied for the extraction and isolation and determination of the molecular weights and chemical group of the bioactive compound.

3. RESULTS

The fungal extract of *P. pistillaris* showed a single spot by using TLC and referred as compound P1 with Rf value of 0.72. Solubility test of the compound indicated that it is insoluble in water but soluble or partially soluble in the other examined solvents. Ultra violet (UV) spectra showed that the absorbency value of P1 compound was 2.5 A at wave length 250 nm (Figure 2).

Infra Red (IR) spectrum revealed that the compound P1 composed of various functional molecules structures (Table 1). The isolated compound is belong to the aliphatic group which contains CH functional molecule between 2863-2955 cm^{-1}.

IR spectra showed that compound P1 exhibited different spectra bands which represent various chemical functional groups (Figure 3).

The results of GC-mass (Figure 4) and HNMR (Figure 5) methods indicated that the molecular weight of compound P1 is 340 kd and belongs to the aliphatic group.

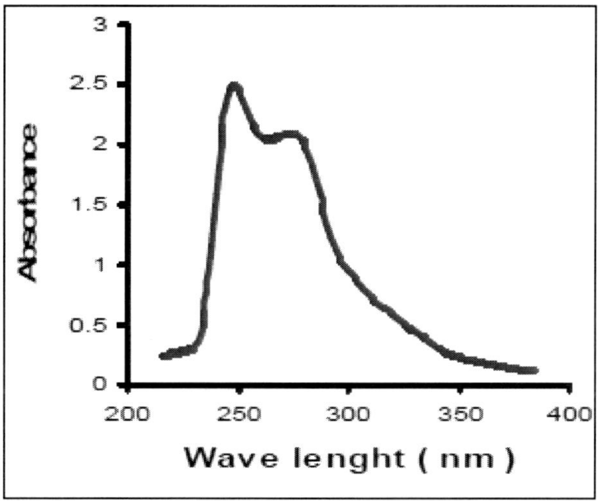

Figure 2. Absorbency of the purified compound P1 from the fungus *P. pistillaris.*

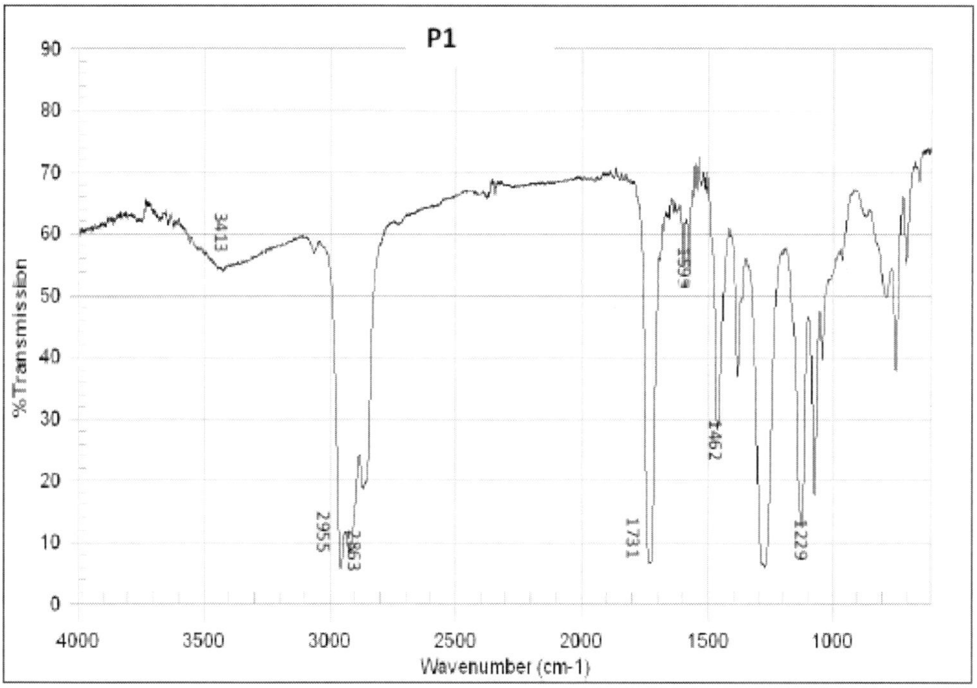

Figure 3. Infra red spectra of the purified compound P1 from *P. pistillaris.*

The MIC values of the bioactive compound P1 were 6.25 u/ml and 12.5 u/ml against *E. coli* and *S. aureus*, respectively. The inhibition zone diameters exhibited by the isolated compound were 25 mm and 21 mm against *E. coli* and *S. aureus*, respectively (Figure 6). However, low bioactivity was rendered by the purified compound against the tested fungal dermatophyte.

Nonetheless, the isolated bioactive compound did not show any toxicity against human RBC.

Table 1. Functional molecule groups of compound (P1) isolated from the mushroom *P. pistillaris*

Functional groups	
N-H, O-H	3413 SBr
CH,CH2, CH	-2955 2863 MBr
C=O	1731 V S
C=C	-1599 1462 WBr

SBr.=Strong band, VS=Very strong band, MBr =Medium Band, WBr = weak band.

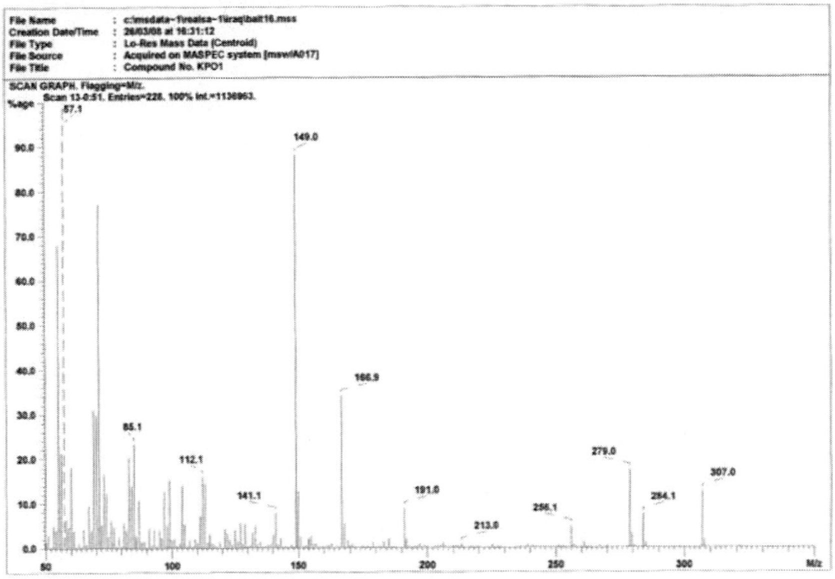

Figure 4. GC-Mass spectrum of compound P1 isolated from *P. pistillaris.*

4. Discussion

Fungi have been shown to be one of the good natural sources for a production of a diverse useful products (Janes et al., 2007; Davis et al., 2008) with a different chemical compounds (Schulz et al., 2002; Strobel and Daisy, 2003). Nevertheless, the production of metabolic substances by fungi is often affected by various growth conditional factors mainly the fermentation medium composition (Jonathan and Fasidi 2003; Vahidi et al., 2004).

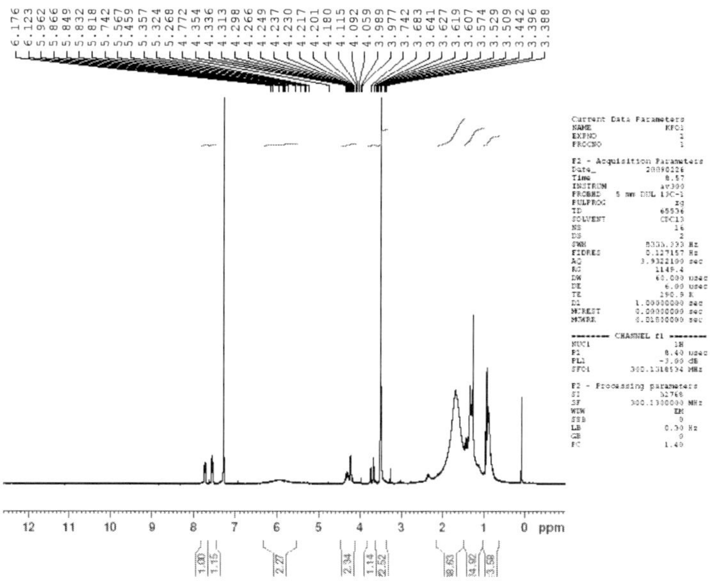

Figure 5. H[1]NMR spectrum of compound P1 isolated from *P. Pistillaris.*

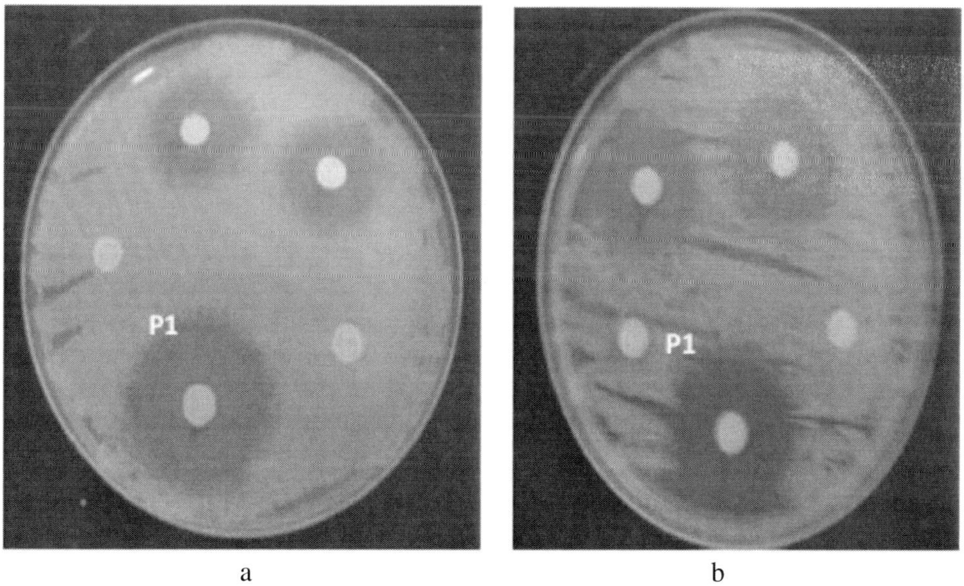

Figure 6. Inhibition zones exhibited by the compound P1 from *P. pistillaris* against *E. coli* (A) and *S. aureus* (B).

In the present study the fermentation liquid medium used was efficient for a production of bioactive metabolites by the selected mushroom. This finding is in concomitant with other studies which examined some basidiomycetous fungi (Stamets 2002; Roberts, 2004; Janes et al., 2007).

The results indicated that the filtrate extract of *P. pistillaris* exhibited a high inhibitory action against both bacterial strains *E. coli* and *S. aureus.*

However, the isolated compound showed a lower bioactivity against the tested dermatophyte *M. gypseum*. In a previous investigation (Muhsin et al., 2011) found that a purified extract of *Ganoderma applanatun* (Aphyllophorales) also exhibited a potential inhibitory action against both bacterial strains. The antibacterial inhibitory effects of fungal metabolites can be related to the action of chemical compounds that binding with the protein of the bacterial cell walls or to the inhibition of DNA synthesis by a specific compound (Shihabudeen et al., 2010). A verification of non toxicity of the fungal extracts of *P. pistillaris* against human blood revealed a negative test. Chemically the isolated compound P1 is related to the aliphatic group as indicated by the IR spectra with a low molecular weight. Other study (Al-Fattimy et al., 2006) reported that *P. pistillaris* extract contains three types of Epicorazine compound which exhibit an antibacterial activity. It can be speculated that the compound P1 might be identical to Epicorazine group, however, a further identification of the purified p1 compound is required.

A conclusion can be derived from this study that *P. pistillaris* possessing a potential compounds that would be a promising as antimicrobial agents.

ACKNOWLEDGMENTS

We are thankful to the authorities of the College of Education, Basrah University for supporing this research work.

REFERENCES

Al-Fatimi, M. A., W. D. Julich, R. Jansen, U. Lindequist 2006. Bioactive components of the traditionary used mushroom *Podaxis pistillaris*. *eCAM*. 3: 87-92.

Anke, T., A. Werle, R. Kappe, O. Sterner 2004. Laschiatrion, a new antifungal agent from a *Favolaschia* species (Basidiomycetes) active against human pathogen. *J. Antibiot*. 57: 496-501.

Davis, R. A., J. Longden, V. A. Avery, P. C. Healy 2008. The isolation, Structure determination and cytotoxicity of the new fungal metabolite, Trichodermamide C. *Bioorgan. Medicin. Letters* 18: 2836-2839.

Gupta, S. and S. P. Singh 1991. Nutritive value of mushroom *Podaxis pistillaris*. *Indian J. Mycol. Plant Pathol*. 21:273-276.

Hashem, A. R. and A. N. Al-Rahmah 1993. Growth of *Podaxis pistillaris* collected from Saudi Arabia at different concentrations of Cadmium and lead. *J. King Saud. Univ*. 5: 127-135.

Janes, D., S. Kreft, M. Jurc, K. Seme, B. Strukelj 2007. Antibacterial activity in higher fungi: Mushrooms and endophytic fungi from Slovenia. *Pharma. Biol*. 45: 700-7006.

Jiskani, M. M. 2002. Growing mushrooms: Step ahead to boost up the economy of Pakistan. *Pakis. J. Econom. Manag*. p. 15-17.

Jonathan, S. G. and I. O. Fasidi 2003. Effect of carbon, nitrogen and mineral sources on growth of *Psathyerella atroumbonata*. *Afric. J. Biomed. Res*. 6: 85-90.

Jonathan, S. G. and F. E. Awotona 2010. Studies on Antimicrobial Potentials of three *Ganoderma* species. *Afric. J. Biomed. Res*. 13:119-125.

Kettering, M., C. Valdivia, O. Sterner, T. Anke, E. Thines 2005. Heptemerones A-G, seven novel diterpenoids from Coprinus heptemerus; producing organism, fermentation, isolation and biological activities. *J. Antibiot*. 58: 390-396.

Khalaf, K. T. 2008. Extraction, purification and identification of bioactive substances from basidiomycetes. *PhD Thesis*, University of Basrah, Iraq. pp. 126.

Lindequist, U., H. J. Tim, W. G. Julich 2005. The pharmacological potentials of mushrooms. *eCAM* 2: 285-299.

McGinnis, M. R. 1980. *Laboratory Handbook of Medical Mycology*, Academic Press, New York.

Muhsin, T. M., Al-Duboon, A. A., K. T. Khalaf 2011. Bioactive compound from a polypore fungus *Ganoderma applanatum* (Pers. Ex Wallr.) *Pat. Jordan J. Biol. Sci*. 4: 205-213.

Roberts, L. M. 2004. Australian *Ganoderma* identification, growth and antibacterial properties. *Ph.D. Thesis,* Swinburne University of Technology. Australia.

Schulz, B., C. Boyle, S. Draeger, A. K. Römmert, A. Krohn 2002. Endophytic Fungi; a source of novel Biologically Active secondary metabolites, *Mycol. Res*. 106: 996-1004.

Shihabudeen, M. S., H. B. Priscilla, B. K. Thirumurugan 2010. Antimicrobial activity and phytochemical analysis of selected Indian folk medicinal plants. *Internat. J. Pharma. Sci. Res*. 10: 430-434.

Stamets, P. 2002. Novel antimicrobials from mushroom. In: *Herbal Gram. Issue*. Am. Bot. Council. 54:155-165.

Strobel, G. and B. Daisy 2003. Bioprospecting for microbial endophytes and their natural products. *Microbiol. Molecul. Biol. Rev*. 67: 491-502.

Vahidi, H., F. Kobarfard, F. Namjoyan 2004. Effect of cultivation conditions on growth and antifungal activity of *Mycena leptocephala*. *Afric. J. Biotech*. 3:608-9.

Xian-guo, H. and M. Ursula 1994. Antifungal compound from *Solanum nigrescens*. *J. Enthopharm*. 43: 173-177.

In: Biotechnology and Conservation … (2 Volume Set)
Editors: S. A. Farooq, R. Abed and S. Baqir
ISBN: 978-1-63117-869-6

Chapter 35

BIOCIDE EFFECTS OF ALKALOIDS AND SAPONINS EXTRACTED FROM *NARCISSUS BROUSSONETII* (LARG.) (AMARYLLIDACEAE) ON *BEMISIA TABACI* (GENNADIUS) (HOMOPTERA:ALEYRODIDAE)

Mina Moussaid[1]*, Chadi Berhal[2] and Hassane Moussaid[3]

[1]Laboratory of biochemistry, Cellular and molecular biology,
Faculty of science, University Hassan ii, Casablanca, Morocco
[2]Laboratory of phytosanitary, Faculty of agriculture,
Lebanese university, Beirut, Lebanon
[3]Laboratory of pharmaceutical chemistry,
Institute of pharmacy, Brussels, Belgium

ABSTRACT

The plants of arid and semi arid area, develop mechanisms to defend through the secondary metabolisms which emphasize the importance of these plants, we are interested in the insecticidal activity of a plant endemic to Morocco: *Narcissus broussonetii (Larg.)*

The fight against insect pests and diseases of crops with pesticides "Bio" is not impossible, because it mobilizes traditional techniques that are a part of organic agriculture. Today, chemical pesticides are accused of many evils; however there are some alternatives already used or tested. Our research was part of the prospect of developing a strategy for integrated management of pests associated with vegetable crops in Morocco; it is proposed to evaluate the insecticidal potential of a *Narcissus* of wetlands from Morocco: *Narcissus broussonetii (Larg.)*, for the protection of tomato culture *Lycopersicon esculentum Mill. Var. Daniella,* contrary to *Bemisia tabaci*, the extract of total alkaloids and saponins dissolved in ethanol at 1%, were applied at 0, 5, 10 and 20 g/l. Compared with controls, the compounds affected the survival of stages significantly, depending on the concentration and duration of exposure.

* Email: noune_moussaid@yahoo.fr.

The ovicide effect is spread from 37% to 69% and 17% to 41% respectively for alkaloids and saponins, and thus reduction by half the rate of adults emerging plants. These extracts also exert a large insecticidal activity on the adult form of the insect; the corrected mortality among these adults is estimated according to the concentrations of 30% to 86% for alkaloids, and 21% to 53% for saponins. Ovicide and insecticidal effects replicate linearly in the same direction, but the eggs require higher concentrations than adults, but the estimated LC_{50} does not avoid the inoculation of TYLCV by adult carriers of the virus.

Keywords: *Narcissus broussonetii (Larg)*, alkaloids, saponins, plant-derived insecticidal, *Bemisia. tabaci*, Tomato under greenhouse

INTRODUCTION

Whiteflies are tiny insects, they are real evils to be feared in the protected crops, a sort of tobacco whitefly recently appeared: *Bemisia tabaci (Homoptera, Aleyrodidae).* Virtually, it colonizes all plants, but it has a special fondness for the celandine, fuchsias, cucurbits and solanaceous (tomatoes and aubergines). The insect and larvae live under the leaves where they form colonies endless. These whiteflies sting plants to suck sap which can cause the complete withering away, as a result, they produce honeydew that disrupts chlorophyll assimilation and which develop destructive bacteria and fungi. (Brown, 1994; Schulten, 1997).

In addition to the direct trophic damage, the insect vehicle numerous viruses, such as virus TYLCV (Tomato Yellow Leaf Curl Virus) that causes the disease yellow leaf curl of tomato, which affects mainly the Middle East and Asia South East. (Messiaen et al., 1991)

In Morocco, the proliferation of *B. tabaci* with the TYLCV causes economic damage to real tomato crops in greenhouses and open field (Jabbour et al., 2001). The fight against the insect is mainly based on foliar sprays of synthetic insecticides, but the intensive the use compounds often leads to harmful effects on the environment and human health. (Van der Werf, 1997).

Today, chemical pesticides are accused of many evils, yet there are alternative products that have been used or tested. From this perspective the use of extracts of plants with insecticidal activity has some potential (Gomez et al., 1997).

Many studies have demonstrated the negative effects of plant extracts on phytophagous pests, so for example, extracts of *Melia azaderac H L., Azadirachta indica A. Juss.* affect fertility and mortality of *B. tabaci* (Nardo et al., 1997; DeSouza and Vandramin, 2000), soybean oil and cotton seeds have also proved toxic to the whitefly (Butler et al., 1988).

In the context to explore other alternatives biodegradable and not deleterious to the environment, compatible with the integrated pest management including organic farming, we conducted a study under controlled conditions, to demonstrate a negative effect (ovicidal and insecticide) of the saponins and alkaloids extract of a wetlands Moroccan *Narcissus: Narcissus broussonetii (Larg.)* about the emergence and survival of the adults of *B. tabaci.*

Narcissus broussonetii (Larg.) (Figure 1) is collected during the months of December and January in Tamaris (27 km south of Casablanca, Morocco). For each targeted compound (alkaloid or saponins), 200 g (whole plant) of *Narcissus broussonetii (Larg.) bulbs,* were dried in an incubator adjusted to 40 ° C until constant weight, they were then ground in a mortar. Selective extractions of ground materials were conducted specifically to each family of compounds studied (see below).

a b

Figure 1. *Narcissus broussonetii (Larg.)*: (a) flowers; (b) bulbs.

For measuring yield of alkaloids and saponins in the plant, three replicates were performed for each group of compounds. The final solids were weighed and the levels of desired compounds in g of compound per 100 g dry weight of plants used. To meet the needs of biological tests, several extractions were performed. For each compound, the concentrations tested were 0, 5, 10 and 20 g of alkaloids or saponins per liter of ethanol 1%.

Plans of tomatoes *L. esculentum Mill.* used as a carrier for biological assays are free from any infestations by an extended stay in isolated frames, from the external environment under controlled conditions at 26 ± 1 ° C, 65 ± 5% relative humidity and a photoperiod of L / D: 14/10 hours, each plant with eight leaves, aged 4-5 weeks, was transplanted into an earthenware pot (vol = 2.5 l) filled with local soil.

The stump of *B. tabaci* used for bioassays was made from insects that emerged from tomato plants *L. esculentum Mill.* grown under cover. Upon emergence, adults were released on plants of the same plant, *Daniella* variety, grown in breeding rooms set in the same conditions above.

Extraction

Total alkaloids: They were obtained by triple liquid - liquid extraction method of Ross and Rain (1977) in Harborne (1998), Two hundred grams of ground plant were extracted in Soxhlet, for six hours at a temperature of 65 ° C with the absolute ethanol for analysis (Merck type).

The crude extract obtained was then evaporated to dryness under vacuum at 40° C in a rotary evaporator (Bruchi type) to speed 4. The dry residue was taken up in 150 ml of chloroform for analysis (Riedel-deHaën type) and acidified with HCl (SOCHID type) to 5% at pH 3; it was allowed to stand for 30 minutes at room temperature. The acidic aqueous phase was extracted with 150 ml of chloroform, basified by the NH_3OH (SOCHID type) to 10% at pH 8-9 and allowed to stand for 15 minutes at room temperature. The chloroform phase was evaporated to dryness under vacuum. The dry residue, consisting of total alkaloids, was weighed and then taken up by 1% ethanol for biological tests.

Saponins: Saponins were extracted using the method developed by Applebaum et al., (1969), modified slightly. The ground material was kept for two hours by pure n-hexane for analysis (Riedel-deHaën type).

After removal of the organic phase, the resulting precipitate was macerated in absolute ethanol under magnetic stirring at room temperature for 24 hours. The ethanol phase was evaporated to dryness under vacuum at 40 ° C by rotary evaporator. The dry residue was extracted three times with 100 ml of the mixture distilled water / petroleum ether (v: v) heated at 50 ° C in a water bath for 30 minutes. The aqueous phases were mixed and then taken up by n-butanol for analysis (Riedel-deHaën type) and then allowed to stand for 30 minutes. The organic phase, rich in saponins, it evaporated to dryness at 40 ° C by rotary evaporator, and weighed, then taken up by 1% ethanol for biological tests.

Biological Tests

Extracts tests: They are prepared by dissolving the dry extracts (extract alkaloids and saponins) in ethanol at 1%; the indicator is composed of only 1% ethanol.

Mechanism study: For the detection of insecticidal activity, Twenty females of *B. tabaci*, newly emerged (<24 h), were released on each tomato plant with eight leaves, aged 4-5 weeks, transplanted in a clay pot of about 2.5 l filled with local soil and covered with a canvas for 48 hours. They were then taken out of plants. Leaves them with the eggs of the insect were dipped in different extracts of alkaloids and saponins at concentrations of 0, 5, 10, and 20 g of extract per liter of ethanol to 1% for 20 seconds then dried in the open air for two hours under a stream of air to evaporate the solvent dilution. The pots were returned to the breeding room under the same growing conditions of *B. tabaci*. The tomato plants were watered regularly. After eight to ten days, the total number of eggs hatched and unhatched was counted for each treatment, an average of 110 eggs were used for each concentration of test compound.

Mortality of adults: Twenty adults, newly emerged (<24 h), were released on a tomato plant in the same stage of development and transplanted in the same manner as before. The plants with adults were returned to the breeding room. The individuals killed and living after six, 24, 48 and 96 hours were counted, an insect is considered dead when lightly shaken by a brush, it does not move. For each concentration (0, 5, 10 and 20 g. l-1), the tests were replicated three times.

Mortality of eggs larvae and impact on the rate of emergence: For the detection of ovicidal and larvicidal activities, and the determination of LD_{50} and LD_{90} (Table 1), 1 ml of each concentration is used for a tomato plant. Each test is repeated three times. On each plant we spawning, freshly 30 to 40 lay eggs on the leaves impregnated with the extract to be tested.

Table 1. Dose-effect response of extracts from *N. serotinus L.* on the eggs and dead adults of *B.tabaci*

Alcaloïdes			Saponines		
Dose[1]	% Ovicide	% Insecticide	Dose[1]	% Ovicide	% Insecticide
5	10 ± 3	54±1	5	n.s	35±2
10	48 ±3	72±4	10	31 ±2	53±2
20	79 ± 2	93±0	20	53 ±4	75±1
Y ovicide = $\mathbf{0.5379 + 1.0320\ Log[Dose].\ R^2 = 0.78}$ **Y insecide =** $\mathbf{1.9892 + 1.0742\ Log[Dose].\ R^2 = 0.85}$			Y ovicide = $0.7656 + 1.0972\ Log[Dose].\ R^2 = 0.51$ Y insecticide = $1.5782 + 1.0342\ Log[Dose].\ R^2 = 0.65$		
DL_{50}	97.5	128.5	DL_{50}	42.6	69.5
DL_{90}	147.3	204.9	DL_{90}	76.7	112.5

[(1)] Dose of treatment expressed as weight of extract (mg).
n.s = non significant: non significant difference compared to the control;
Y = equation of the Log-probit trendline; R^2 = correlation coefficient.

A week after oviposition, It is expected with a magnifying glass the eggs hatched and unhatched. From the first emergence (third or fourth week), adults are removed as and when they appear to total cessation of emergence (ie a. D: absence of emergence over a period of two weeks) (Mbata and Eppendu, 1992). Mortality rates of eggs and the rate of adult deaths are calculated as follows:

$$\text{Mortality rates of eggs} = \frac{\text{number of eggs laid} - \text{number of eggs hatched}}{\text{number of eggs laid}} X\ 100$$

$$\text{Rates of adult deaths} = \frac{\text{number of adult deaths}}{\text{number of adults used}} X\ 100$$

The different types of insecticidal activity are expressed in insecticidal efficacy according to Abbott's formula (Moretti et al., 1998) that gives the corrected values of mortality based on the percentage mortality of the treated samples and the control. This correction to exclude bias due to natural mortality observed in our experimental conditions.

$$\%\ \text{efficiency} = \frac{\text{test sample mortality} - \text{mortality control}}{100 - \text{mortality control}} X\ 100$$

Statistical analysis is performed with the software Graph Pad Prism version 2.01. The data are subject to an analysis of variance (ANOVA), means (± SD) are compared with multiple comparison test of Newman-Keuls (Sanon et al., 1998). P values below 0.05 were considered significant.

The LD_{50} and LD_{90} were calculated from the regression log-probit.

Y = a + Log (X) (Y: probit value corresponding to the insecticidal effect

and X: Dose extract test).

The cycle of *B. tabaci* depends on temperature, relative humidity, photoperiod and host plant (Gerling and Howoritz, 1986). Data on the duration of the development cycle of tomato are very heterogeneous. And Lopez-Avila (1986) located at 26.7 ° C for a period of 27 days, while Tsai and Wang (1996) found at 25 ° C for a total of 18 days. We have considered only the first four days (zero hours to 96 hours) interpreting the data.

Compound Contents

Extractions specific from N. broussonetii bulbs, have clarified the contents desired compounds (Table 2). Dry weights are alkaloids that are most abundant, with a content equal to 2.36% of the dry weight of bulbs. Saponins represent average 1.22% of the dry weight of the same bulbs.

Table 2. Content of alkaloids and saponins extracted from N. broussonetii bulbs

Compound	Mean ± standard deviation ($g.100g^{-1}$ of dry weight) (N = 3)
Alkaloids	2.36±0.03
Saponins	1.22±0.07

Effects of Alkaloids or Saponins on Eggs

The effect of compounds extracted from *N. broussonetii* bulbs on embryo viability of *B. tabaci* varies between compounds tested and for the same compound; the percentage of unhatched eggs varies with the concentration tested. The both compounds tested reduced significantly egg hatch compared to the control (= 1% ethanol) (Figure 2).

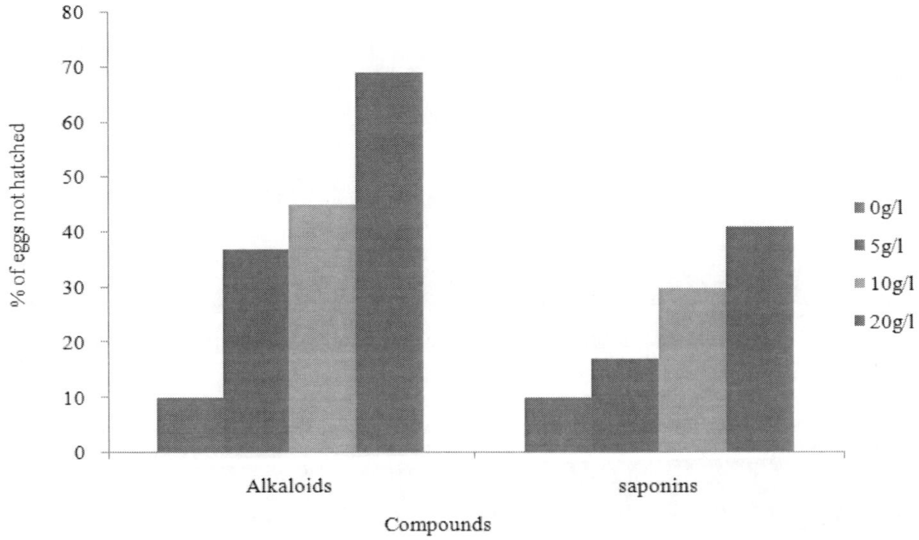

Figure 2. Effect of alkaloids or saponins extracted from N. broussonetii bulbs on Bemisia tabaci eggs (N>105 per compound and per concentration).

The number of unhatched eggs is higher when they are exposed to alkaloids, on average, these compounds inhibit alone, and hatching eggs treated 37% with 5 gl^{-1} at 69% with 20 gl^{-1}. Depending on the concentration in question, the saponins cause the average corrected mortality from 17% to 41% of embryos treated.

Effects of Alkaloids and Saponins Adults

The response of adult B. tabaci varies between compounds, and with the same compound depending on the concentration tested. As in the case of eggs, are alkaloids that cause the highest mortality, but the saponins affect relatively low adult survival of *B. tabaci*.

Effect of alkaloids: When adults of *B. tabaci* are treated with alkaloids, their replies vary depending on the concentration tested, the duration of exposure, with an interaction between these two parameters. As shown in Table 3, alkaloids cause adult mortality of *B. tabaci* significantly higher than that obtained with adults released from plants treated with 1% ethanol.

Table 3. Cumulative means (± SD) of adult mortality of *B. tabaci* due to Alkaloids extracted from *N. broussonetii (Larg.)* bulbs

hours after treatment	Concentration ($g.l^{-1}$)			
	0	5	10	20
6	10.00a±4.00	30.67a±2.89	36.33a±2.89	42.67a±7.64
24	15.00a±4.00	38.33a±2.89	44.33b±5.77	63.00b±5.00
48	23.33b±2.00	51.33b±7.64	59.67c±2.89	81.67c±2.89
96	26.67b±2.00	59.67b±2.89	71.67c±2.89	86.33c±2.89

abc: Mean values from the same column with different letters in superscript are significantly different according to Newman keuls test ($P < 0.05$).

The number of deaths then increases with the concentration applied, it goes about 30 to 59%, 36 to 7% and 42 to 86% of adults treated respectively with 5, 10 and 20 gl-1. It should also be noted that there is a strong individual variability on the response to applied concentrations and duration of exposure to these.

In addition, for each concentration tested the adult mortality of the insect increases linearly with the duration of the exposure, (Figure 3). The TL_{50} is estimated at 48, 36 and 10 hours respectively for 5, 10 and 20 gl-1, and decreases gradually as the concentration increases tested.

Effect of saponins: As in the case of alkaloids, adult mortality of *B. tabaci* due to saponins varies with the applied concentration and duration of exposure, but no interaction between these two parameters. Saponiques compounds affect more adults of whitefly than the control (Table 4). Here too, the proportion of adult deaths is increasing with concentrations studied, it ranges from about 21 to 46% with 5 $g.l^{-1}$, 23 to 50% with 10 $g.l^{-1}$and 38 to 53% with 20 $g.l^{-1}$.

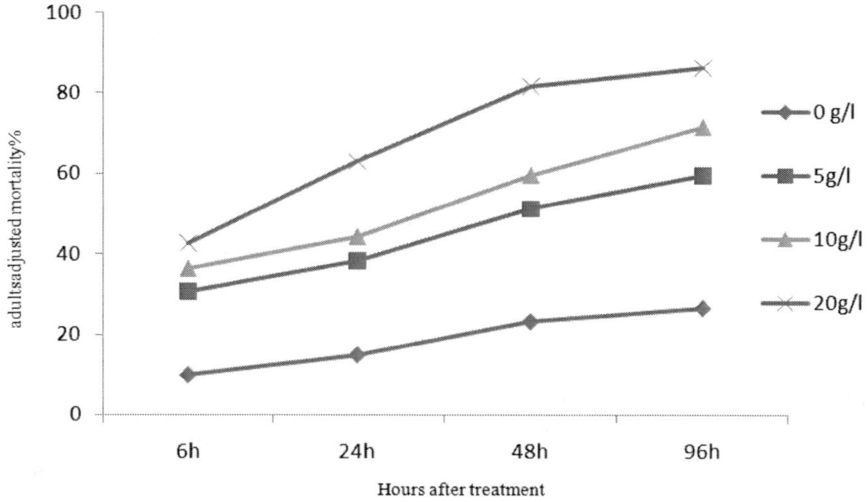

Figure 3. Chronological evolution of the adults adjusted mortality of B. tabaci caused by alkaloidss extracted from N. broussonetii (Lang.) bulbs.

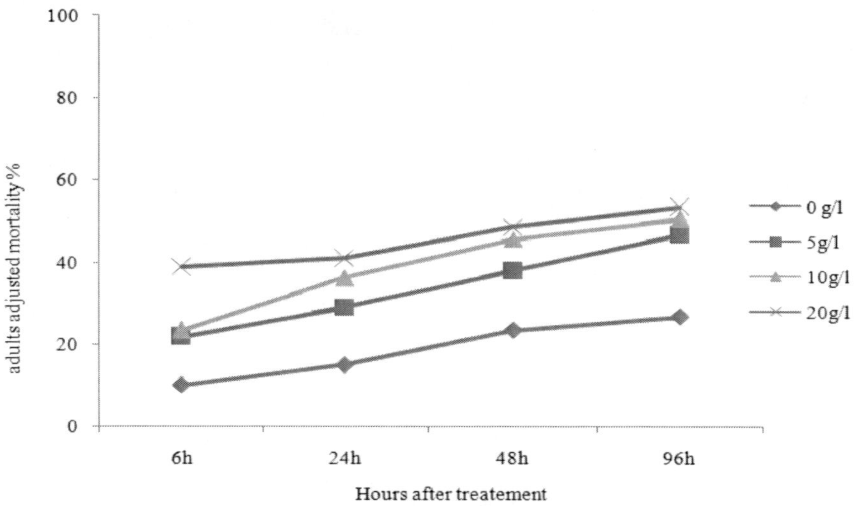

Figure 4. Chronological evolution of the adults adjusted mortality of B. tabaci due to saponins extracted from N. broussonetii (Lang.) bulbs.

Table 4. Cumulative means (± SD) of adult mortality of *B. tabaci* due to saponins extracted from *N. broussonetii (Larg.)* bulbs

hours after treatment	Concentration ($g.l^{-1}$)			
	0	5	10	20
6	$10.00^{a} \pm 4.00$	$21.67^{a} \pm 2.89$	$23.33^{a} \pm 2.89$	$38.75^{a} \pm 5.00$
24	$15.00^{a} \pm 4.00$	$29.00^{a} \pm 5.00$	$36.33^{b} \pm 2.89$	$40.96^{b} \pm 5.77$
48	$23.33^{b} \pm 2.00$	$38.00^{b} \pm 5.00$	$45.67^{b} \pm 5.00$	$48.67^{c} \pm 2.89$
96	$26.67^{b} \pm 2.00$	$46.67^{b} \pm 2.89$	$50.67^{b} \pm 2.89$	$53.33^{c} \pm 2.89$

abc: Mean values from the same column with different letters in superscript are significantly different according to Newman keuls test ($P < 0.05$).

With saponins, adults of *B. tabaci* respond in a very heterogeneous depending on concentrations and residence time with these compounds. We find that adult mortality is chronologically less dependent (Figure 4), LT_{50} was being estimated at 34 and 12 hours respectively for 10 and 20 $g.l^{-1}$.

Comparative analysis of the effect of alkaloids and saponins on eggs and adults of *B. Tabaci*: At the end of the experiment, 96 hours after application of the test compounds at different concentrations on adults it are therefore alkaloids that have proven the most toxic and, whatever the concentration considered, saponins are relatively less harmful to adults of *B. tabaci* (Table 5).

Table 5. Cumulative means (in %± SD) of adult adjusted mortality (N=60 per compound and per concentration) of B. tabaci observed 96 hours after treatment by Alkaloids or saponins extracted from Narcissus *broussonetii (Larg.)*

Concentration g.l-1	Alcaloïdes	Saponines
5	34 .54a±4.65	26.10a±7.04
10	67.80a±2.54	35.10b±3.23
20	89.32a±3.66	67.20b±7.98

a,b Mean values from the same row with different letters in superscript are significantly different according to Newman-keuls test(P<0.005).

Moreover, by comparing the response of eggs and adults compounds tested in this study, two developmental stages of *B. tabaci* respond in the same direction, the percentages of eggs and dead adults are positively correlated, but the eggs are relatively less vulnerable than adults.

Discussion

Compound content: The contents of the tested compounds vary by product considered; alkaloids appear more abundant in bulbs of *Narcissus broussonetii (Larg.)* as saponins. Levels found in our work can not be compared since the extraction of these compounds is made for the first time for *Narcissus broussonetii (Larg.)*.

Effect of alkaloids and saponins on eggs and adults of B. Tabaci: Under the conditions of study, alkaloids and saponins extracted from bulbs of *Narcissus broussonetii (Larg.)*, significantly affect the viability of eggs and adult survival of *B. tabaci* adults were more sensitive than eggs. Alkaloids were more effective than saponins.

Alkaloids can be toxic in contact embryos, contact and / or ingestion in adult's *B. tabaci* for adults, it is not impossible that insects are killed by starvation consecutive repellent effects or anti-palatable as in other invertebrates (El-Lakwah et al., 1997). The weak response of *B. tabaci* to meet saponins, particularly eggs, may be related to the high molecular weight to such compounds through the chorion of the egg, as determined by Magalhaes et al. (2003) in the case of *Biomphalaria glabrata* treated by saponins extracted from seeds *Swartizia langsdorffii.* In addition, the toxicity of saponins compounds is shown by Applebaum et al. (1969) during their incorporation in the midst of raising Callosobruchus chinensis where they observed an inhibition of insect development, and their application against Ostrinia nubilalis, significantly affects the growth and development of larvae of the insect (Nozzolillo et al., 1997). For non viruliferous adults, alkaloids to 20 gl^{-1} may reduce the degree of infection of plants by *B. tabaci* to an acceptable level in glasshouse, although the estimated LT_{50} is almost similar for both compounds at higher concentrations (20 gl^{-1}), it does not seem short enough to kill adult viruliferous before transmitting the TYLC to healthy plants, as the minimum time required for inoculation of the virus is approximately 40 min (Rubinstein et al., 1999).

Conclusion

Narcissus broussonetii (Larg.) may advantageously be used as a botanical insecticide for the protection of glasshouse by ovicidal and insecticidal activity of the chloroform fraction rich in alkaloids and butanol fraction rich in saponins.

The results show that the alkaloids are more active than the saponins of *B. tabaci*. In the short and medium terms, alkaloids at 20 gl^{-1}, can be integrated into a management program *B. tabaci* in glasshouse, especially at risk of resistance to other pesticides. Remains to evaluate their potential effects on the natural allies (enemies and pollinators) associated with protected cultivation and it in spite of the safety of these compounds against other parasitoids.

It should be noted that the larvicidal activity was not observed for both extracts. It will be for us to assess the toxicity and germination of tomato plants treated and structural elucidation and identification of compounds responsible for the biocidal activity, indeed a previous study, in the crude extract of bulbs, we able to determine the presence of two major alkaloids: Tazettine and Lycorine, we could determine the presence of two major alkaloids, and the toxicity is probably caused by the presence of the latter alkaloid Lycorine, which is a highly toxic alkaloid with a chemical structure it says an oxidative process particularly vulnerable (Moussaid et al., 2010).

Should also consider conducting pre-control tests in field conditions, to assess the effectiveness of extracts of *Narcissus broussonetii (Larg.)* the protection of vegetable crops, as plant extracts lose their biological activity under solar radiation (Scott et al., 2003), modes of action, application and impact of physical factors on the degradation of botanical compounds tested must also be studied.

Acknowledgments

Our thanks go to Professor J. BASTIDA from Department of Natural Products, Plant Biology, Faculty of Pharmacy, University of Barcelona Spain, for the identification of plant material and the National Institute of Agronomic Research for providing strains of Bemisia tobacco and tomato plants.

References

Applebaum SW, Marco S and Birk Y 1969. Saponins as possible factors of resistance of legume seeds to the attack of insects. *J. Agr. Food Chem.* 17: 618–622.

Brown JK 1994. Current status of Bemisia tabaci as a plant pest virus vector in agro-ecosystems worldwide. *FAO Plant Prot. Bull.* 42: 3–32.

Butler GD, Coudriet DL and Henneberry TJ 1988. Toxicity and repellency of soybean and cotton seed oil to the sweetpotato whitefly and the cotton aphid on cotton in greenhouse studies. *Southwest. Entomol.* 13: 81–86.

De Souza AP and Vendramim JD 2000. Efeito de axtratos aquosos de Meliaceas sobre Bemisia tabaci biotipo B em tomateiro. *Bragantia*. 59: 173–179.

EL-Lakwah F, Khaled OM, Kattab MM and Abdel-Rahman TA 1997. Effectiveness of some plant extracts and powders against the lesser grain borer Ryzopertha dominica (F.). *Ann. Agric. Sci.* 35: 567–578.

Gerling D and Howoritz A.R 1986. Autoecology of Bemisia tabaci. *Agriculture, Ecosystems & Environment* 17:5-19.

Gomez P, Cubillo D, Mora GA and Hilje L1997. Evalucion de posiblies repelentes de Bemisia tabaci. II. Extractos vegetales. *Manejo Integrado de Plagas* (Costa Rica). 29: 17–25.

Harborne JB 1998. Phytochemical methods: a guide to modern techniques of plant analysis. (3rd ed.) Chapman & Hall London.

Jabbour F, Abaha L, Kaak H and Moryane R 2001. La maladie des feuilles jaunes en cuillère de la tomate (TYLCV) au Maroc. Proceedings AMPP. Rabat: 41–46.

Lopez-Avila A 1986. Natural enemies. In: Bemisia tabaci a literature survey on the cotton whitefly with an annotated bibliography. Cock M.J.W., C.A.B. *International Institute of Biological Control.*

Magalhaes AF, Goulant AM, Tozzi de A, Santos CC, Serrano DR, Zanotti-Magalhaes EM, Magalhaes EG and Magalhaes AL 2003. Saponins from Swartizia langsdorffii: Biological activities. *Mem. Inst. Oswaldo Cruz, Rio de Janeiro* 98: 713–718.

Mbata GN and Ekpendu OT 1992. The insecticidal action of four botanicals against three storage beetles. Med. Fac. Landbouww. *Univ. Gent*. 57: 723–733.

Messaien C-M, Blancard D, Rouxel F and Lafon R 1991. Les maladies des plantes maraîchères, INRA, Paris: 183-194.

Moretti MDL, Bazzoni E, Giovanni SP and Romolo P 1998. Antifeedant effects of some essential oils on ceratitis capitataWied. (Diptera, Tephritidae). *J. Essent. Oil. Res.* 10: 405–412.

Moussaid M, Elamrani A, Codina C, Bastida J, Bourhim N and Benaissa M. (2010) Contribution à l'étude phytochimique et biologique de l'huile essentielle de Narcissus broussonetii (Lag.) Intern*ational Congress on Medicinal and Aromatic Plants, Fes. Morocco.*

Nardo EAD, De Costa AS and Lorencao AL 1997. Melia azadirach extract as an antifeedent to Bemisia tabaci (Homoptera: Aleyrodidae). *Florida Entomol.* 80: 92–94.

Nozzolillo C, Arnason JT, Campos F, Donskov N and Jurzysta M 1997. Alfalfa leaf saponins and insect resistance. *J. Chem. Ecol.* 23: 995–1002.

Rubinstein G, Morin S and Czosnek H 1999. Transmission of tomato yellow leaf curl geminivirus to imidacloprid treated tomato plants by the whitefly *Bemisia tabaci* (*Homoptera*: *Aleyrodidae*). *J. Econ. Entomol.* 92: 658–662.

Sanon A, Ouedraogo AP, Tricault Y, Credland PF and Huignard J 1998. Biological control of bruchids in cowpea stores by release of *Dinarmus basalis* adults. *Environ. Entomol.* 27: 717-725.

Schulten GGM 1997. Overview of the whitefly - Virus problem: objectives of the workshop. *FAO Plant Prod. Prot. Paper.* 143: 7–10.

Scott IM, Jensen H, Scott JG, Isman MB, Arnason JT and Philogène BJR 2003. Botanical insecticides for controlling agricultural pests: Piperamides and the Colorado potato beetle Leptinotarsa decemlineata Say (Coleoptera: Chrysomelidae). *Arch. Insect Bioch. Physio.* 54: 212–225.

Tsai JH and Wang K 1996. Development and reproduction of Bemisia argentifolii (Homoptera: Aleyrodidae) on five host plants. *Environmental Entomology* 25: 810-816.

Van der Werf HMG 1997. Evaluer l'impact des pesticides sur l'environnement. *Courr. Environ.* 31: 5–22.

In: Biotechnology and Conservation … (2 Volume Set) ISBN: 978-1-63117-869-6
Editors: S. A. Farooq, R. Abed and S. Baqir

Chapter 36

Existence for Two Ploidy Levels in a Relict Olive Population of Hoggar (Algeria)

D. Baali-Cherif[1,*], G. Besnard[2] and MS. Abdellaoui[3]
[1]Laboratoire de Recherche sur les Zones Arides LRZA/
USTHB BP 44 16000 Algiers-gare, Algeria
[2]CNRS-UPS-ENFA, EDB, UMR 5174, Toulouse, France
[3]Searchers INRF station, Hoggar, Tamanrasset, Algeria

Abstract

Polyploidy was recently reported in two endangered olive subspecies from North-West Africa. The origin of this phenomenon remains unclear. In the present study, 107 genotypes of a relic Laperrine's olive population from Hoggar (Algeria) were analysed using eight nuclear microsatellites. Diploid and triploid genetic profiles were observed. A flow cytometry analysis confirmed that trees displaying three alleles at several loci are effectively triploid. This report constitutes the first evidence for the coexistence of two ploidy types in an olive population. The triploid genotypes, probably in mean more vigorous than diploid trees, may be positively selected in the absence of sexual regeneration since a very long time, explaining their relative high frequency (2.8%) in the investigated population.

Keywords: Flow cytometry, Hoggar, Laperrine's olive, Microsatellite, Triploidy

1. Introduction

Polyploidy is a major evolutionary process, which is considered to have promoted speciation, contributed in the adaptation of organisms to new ecological niches, or helped the acquisition of novelties to face environmental changes [1-5]. In plants, numerous examples of

* Corresponding author: D. Baali-Cherif. Teatcher, ENSA (ex INA) El-Harrach, E-mail: bacherdj@yahoo.fr.

polyploids have been reported and it is clearly demonstrated that ancestors of all present species have experienced several rings of polyploidisation.

Variable polyploidy states have also been reported in some species, even at the population level. In trees, spontaneous intraspecific ploidy variation has been observed in numerous species, for instance, in poplar [6], oak [7], elm [8], ash [9], or acacia [10], but in such cases, frequency of polyploids remains usually low in natural populations (e.g., less than 1% in *Quercus robur* [7]).

Polyploidy was recently evidenced in two relict wild subspecies of the olive tree (*Olea europaea* L., Oleaceae) in South Morocco (subsp. *maroccana* [Greut. and Burd.] P. Vargas et al.; [11]) and Madeira (subsp. *cerasiformis* G. Kunkel and Sunding; [11, 12]). It was proposed that such polyploids were advantaged in very small endangered populations by limiting effects of inbreeding depression. However, the origins of these polyploids remain poorly understood and deserve some investigations [13]. Particularly, the mechanisms involved in the formation of such polyploids are unclear.

Previous studies also demonstrated that the polyploidy level of olive trees can be inferred from nuclear microsatellite data offering an alternative to the fastidious chromosome counting; subspp. *maroccana* and *cerasiformis* were thus suspected to be respectively hexaploid and tetraploid from genetic analyses, and this was then confirmed by flow cytometry analyses [11, 12].

Some intriguing genetic profiles were also reported in the Laperrine's olive [*O. europaea* subsp. *laperrinei* (Batt. and Trab.) Ciferri] populations, since a few trees with three different alleles at several microsatellite loci were observed in the Hoggar and Air mountains [14-17]. The diploid status of these trees was thus questioned but not more investigated, since no living material was available in collection for a flow cytometry analysis.

In the present study, we report the characterization of a population of Laperrine's olive with nuclear microsatellite loci coupled to a flow cytometry analysis to determine the ploidy of a few trees displaying a complex genetic profile.

Results clearly demonstrate that two polyploidy levels (e.g., *2x* and *3x*) coexist in the Laperrine's olive populations. The evolutionary significance of this finding for the olive polyploidy formation is discussed.

2. MATERIAL AND METHODS

A population of Laperrine's olive from the Adrar Hegguehene massif in the Hoggar was considered in the present study. In this massif, several hundreds of olive individuals are distributed in a fragmented habitat on cliffs and borders of temporary rivers. Leaf samples of 107 genets were collected on five distinct sites during several field works (Table 1). Most of these individuals were partially characterized in previous studies with different sets of nuclear microsatellite loci [14, 17] and a few new samples from Hadriane (15 individuals) were added in the present study. Genotyping of all these trees was completed for eight microsatellite loci (DCA01, DCA03, DCA05, DCA08, DCA09, DCA15, DCA18, GAPU71A [18, 19]) according to the methodology described by Baali-Cherif and Besnard [14]. The complete microsatellite genetic dataset is available in the online supplementary material.

The number of alleles (*Na*), allelic frequencies, and Nei's gene diversity (*HT*) were computed for each locus with FSTAT [20]. From allele frequencies in the 104 diploid individuals (see below), the probability a triploid genotype displays three distinct alleles was computed at each locus:

$$p3x(i) = \sum_{j=n}[6\ (fa\ fb\ fc)j]$$

where "6 (*fa fb fc*)*j*" is the probability of the triploid genotype *j* at a locus *i* with three distinct alleles (*a*, *b* and *c*), and *fa*, *fb* and *fc* are allelic frequencies. In other words, *p3x* is the sum of probabilities of all possible genetic profiles at a locus with three distinct alleles. Using our seven loci with more than three alleles (excluding DCA15; see below), the probability a triploid genotype should be detected (by the presence of three alleles on at least one locus) was then estimated by the following formula:

$$P3x = 1 - \prod_{i=7}(1 - p3x)i$$

where (1 - *p3x*)*i* is the probability a triploid genotype is not detected at a locus *i*.

Lastly, cuttings of two genets ("I-n-Toûnine 1-2-3", "Hadriane 19"; see below) displaying three different alleles at three or four loci, and therefore suspected to be triploid, were collected for a flow cytometry analysis. The genet "Hadriane 19" is presently maintained in a living collection at the INRA Montpellier. Two additional Laperrine's olive trees from Ti-n-Hâmor ("Ti-n-Hâmor 1" and "6") with one or two different alleles on all loci were also analysed as diploid references. The 2C-value (nuclear DNA content) of these trees was estimated using flow cytometry as reported by Besnard et al. [11]. The measures were done three times independently for each individual.

3. Results and Discussion

3.1. Evidence of Triploids in the Laperrine's Olive Population

As expected from previous studies [14-17], the eight loci used to characterize the Laperrine's olive population are variable, and five of them display a relatively high number of alleles *Na* (> 12) and/or gene diversity *HT* (> 0.7; Table 2).

Table 1. Origin of the plant material characterized in the present study. The number of diploid (2*x*) and triploid (3*x*) genets

Locality	Altitude	Longitude	Latitude	2*x**	3*x**
Tahâssa	1480 m	22°47 N	5°39 E	14	0
I-n-Toûnine	1390 m	22°45 N	5°38 E	6	1
I-n-Ezzebib	1470 m	22°48 N	5°37 E	22	0
Hadriane	1600 m	22°47 N	5°36 E	23	1
Ti-n-Hâmor	1600 m	22°51 N	5°37 E	39	1
Total				104	3

*Ploidy level as deduced from microsatellite profiles.

Table 2. Characteristics of the eight microsatellites used for the Laperrine's olive genotyping

Locus	Allele range	*Na*	*HT**	*p3x**
DCA1	226–284	26	0.92	0.782
DCA3	229–253	7	0.70	0.318
DCA5	199–249	22	0.90	0.688
DCA8	119–147	12	0.84	0.580
DCA9	169–193	12	0.56	0.196
DCA15	251–258	2	0.23	0.000
DCA18	152–183	12	0.81	0.561
GAPU71A	217–225	4	0.63	0.215
Total	–	97	0.70	*P3x*= 0.995

*Measured on the 104 diploid profiles.

Na = Number of alleles, *HT* = Nei's gene diversity, *p3x* = Probability of a triploid genotype with three different alleles at a locus, *P3x* = Probability of triploid detection on at least one microsatellite locus with the seven more variable loci (excluding DCA15).

Among the 107 genets analysed, we found three individuals (2.8%) with more than three different alleles on three or four loci (i.e. "I-n- Toûnine 1-2-3", "Ti-n-Hâmor 9-10" and "Hadriane 19"; Figure 1).

These individuals were sampled on three relatively distant sites. The remaining 104 trees display one or two different alleles at each locus. The diploid status of genets "I-n-Toûnine 1-2-3", "Ti-n-Hâmor 9-10" and "Hadriane 19" can thus be questioned. The 2C-value for two of them ("I-n-Toûnine 1-2-3" and "Hadriane 19") was therefore measured by flow cytometry and was respectively 4.97 pg (± 0.05) and 4.91 pg (± 0.02). A 2C-value of 3.34 pg (± 0.17) was obtained for two diploid Laperrine's olive trees.

This latter estimate is congruent with previous 2C-values reported in diploid olives [11-12]. These observations clearly support that "I-n-Toûnine 1-2-3" and "Hadriane 19" have a genome about 1.5 bigger than diploid trees, and we can thus contend that they are triploid as also suggested by their microsatellite profiles [6].

The efficiency of our eight microsatellite loci to detect triploid genotypes is very variable ($0 < p3x < 0.78$; Table 2). The two more efficient ones, which also display the highest gene diversity and number of alleles, are DCA01 and DCA05. Each of these loci should allow triploid detection in more than 70% of cases (Table 2). The combined use of the seven loci with more than three alleles (excluding DCA15) is expected to allow the detection of 99.4% (*P3x*) of triploids, and we can thus consider that all of them have been detected in the studied population.

However, we should also mention that the presence of only one locus with three 30 different alleles should not be enough to conclude on the polyploid status of a tree. Indeed, it is not excluded that mutations can sometime lead to chimerism at some loci as demonstrated for instance in grapevine [21].

But in this case, frequency of loci with more than two different alleles should be very low, and in order to detect efficiently triploids with nuclear microsatellites, we recommend accumulating evidences on several loci (as in the present study).

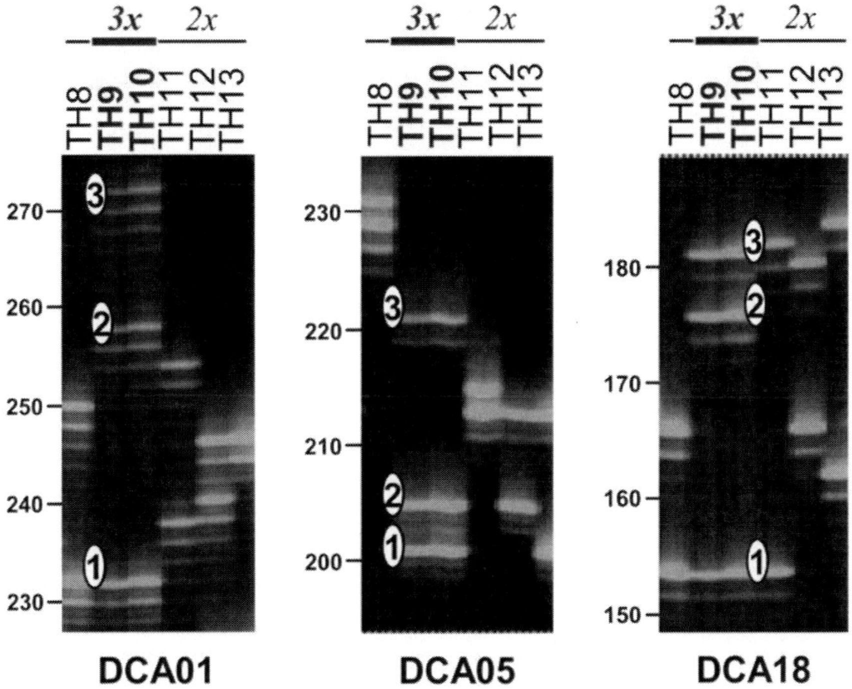

Figure 1. Example of microsatellite profiles for diploid (TH8, TH11, TH12 and TH13) and triploid (TH9-10) individuals in the population Ti-n-Hâmor on loci DCA01, DCA05 and DCA 18. The 3 alleles observed in the genet TH9-10 ("Ti-n-Hâmor 9-10") are numbered 1 to 3. Fragment-size (bp) on left of pictures.

3.2. On the Origin of Triploidy in the Laperrine's Olive

Variable ploidy levels in a population of the same subspecies were never reported in *O. europaea*, and all previous microsatellite studies of the "diploid" Mediterranean and African olives did not reveal complex genetic profiles in cultivars [e.g., 22] as well as in wild populations [e.g., 13, 23-25]. This suggests that the pattern observed in the Laperrine's olive is quite unusual in the *O. europaea* complex, and this leads us to wonder on the significance of the triploidy in the Laperrine's olive. Some molecular evidences suggest that recurrent migrations from both Mediterranean and Tropical Africa contributed to establishment of the present Laperrine's olive populations [26]. Spontaneous polyploids have been shown to be more common in hybrid systems [1], and the intraspecific hybrid status of the Laperrine's olive could have thus been an important factor in the polyploidy formation. However, additional experiments based on single-copy genes are needed to test more accurately such hypotheses on autopolyploidy vs. segmental allopoyploidy. Several explanations can also be proposed to explain triploid formation. Some tetraploid individuals could have crossed with diploids to lead to triploids, but this remains unlikely because tetraploids have never been detected yet in the Laperrine's populations. The triploids could have also resulted of the zygote development from the endosperm, or more likely, should be from one gamete that fails to undergo meiotic reduction [1, 7].

A relatively high proportion of triploids (2.8%) is observed in our olive population if we compare it to frequencies observed in different species of oaks (from 0.25 to 0.57% in natural populations [7]) or acacias (from 0 to 1.3% in seedling populations [9]). Nevertheless, we did not observe triploids in 211 seeds/seedlings recently collected in the Hoggar Mountains and characterized with the same set of microsatellite loci [17]. This suggests that triploid individuals could be positively selected during tree development and their frequency should thus increase on a long period of vegetative growth (without sexual regeneration). Indeed, it has been reported that triploids are unusually vigorous [e.g., 7, 27], probably because it reduces inbreeding depression effects, and this should favour their establishment in extreme conditions. During stages of sexual regeneration, inbreeding is expected to be quite high in the Laperrine's olive populations due to the distribution of trees in very small patches and limited gene dispersal [17]. In such conditions, triploids should be more successful than diploids [28], and this may explain their relatively high frequency in the Hoggar Mountains. In some particular conditions (e.g., strong isolation leading to high inbreeding), we can also speculate that this process could have led to the establishment of fully polyploid populations (e.g., *4x* or *6x*) as observed in Madeira and South Morocco. The emergence of triploids in a diploid population could be interpreted as the first step towards such populations. Indeed, even if the fertility of triploids is considerably reduced when compared to diploids, several authors proposed that such genotypes should be involved in tetraploid or hexaploid formation (via a triploid-bridge [1]). This hypothesis should be experimentally tested by crossing genotypes with different ploidy levels (i.e. *2x* and *3x*) maintained in collection.

ACKNOWLEDGMENTS

We are grateful for the kind collaboration of Boudjemâa Azzouz, who helped us to collect the plant material in Hoggar for this study. We also thank Amélie Fragnière (Fribourg University) for her help in flow cytometry analyses. G.B. was funded by the Intra-European Fellowship PIEF-GA-2008-220813.

REFERENCES

[1] J. Ramsey, D. W. Schemske, Pathways, mechanisms, and rates of polyploid formation in flowering plants, *Annu. Rev. Ecol. Syst.* 29 (1998) 467–501.

[2] D. E. Soltis, P. S. Soltis, J. A. Tate, Advances in the study of polyploidy since Plant speciation, *New Phytol.* 161 (2004) 173–191.

[3] M. J. Hegarty, S. J. Hiscock, Genomic clues to the evolutionary success of polyploid plants, *Curr. Biol.* 18 (2008) R435–R444.

[4] L. E. Flagel, J. F. Wendel, Gene duplication and evolutionary novelty in plants, *New Phytol.* 183 (2009) 557–564.

[5] U. A. Treier, O. Broennimann, S. Normand, A. Guisan, U. Schaffner, T. Steinger, H. Müller-Schärer, Shift in cytotype frequency and niche space in the invasive plant *Centaurea maculosa*, *Ecology* 90 (2009) 1366–1377.

[6] H. D. Jr. Bradshaw, R. F. Stettler, Molecular genetics of growth and development in *Populus*. I. Triploidy in hybrid poplars, *Theor. Appl. Genet.* 86 (1993) 301–307.
[7] A. Dzialuk, I. Chybicki, M. Welc, E. Sliwinska, J. Burczyk, Presence of triploids among oak species, *Ann. Bot.* 99 (2007) 959–964.
[8] J. E. Armstrong, Polyploidy and wood anatomy of mature white ash, *Fraxinus americana*, *Wood Fiber Sci.* 14 (1982) 331–339.
[9] D. Blakesley, A. Allen, T. K. Pellny, A. V. Roberts, Natural and induced polyploidy in *Acacia dealbata* Link. and *Acacia mangium* Wild., *Ann. Bot.* 90 (2002) 391–398.
[10] J. L. Sherald, F. S. Santamour, R. K. Jr. Hajela, N. Hajela, M. B. Sticklen, A Dutch elm disease resistant triploid elm. *Can. J. Forest Res.* 24 (1994) 647–653.
[11] G. Besnard, C. García-Verdugo, R. Rubio de Casas, U. A. Treier, N. Galland, P. Vargas, Polyploidy in the olive complex (*Olea europaea* L.): evidence from flow cytometry and nuclear microsatellite analyses, *Ann. Bot.* 101 (2008) 25–30.
[12] G. Brito, J. Loureiro, T. Lopes, E. Rodriguez, C. Santos, Genetic characterisation of olive trees from Madeira Archipelago using flow cytometry and microsatellite markers, *Genet. Res. Crop Evol.* 55 (2008) 657–664.
[13] C. García-Verdugo, M. F. Fay, C. Granado-Yela, R. Rubio de Casas, L. Balaguer, G. Besnard, P. Vargas, Genetic differentiation under geographical isolation in a ploidy series of *Olea europaea*: A multiscale approach from subspecies to islands, *Mol. Ecol.* 18 (2009) 454–467.
[14] D. Baali-Cherif, G. Besnard, High genetic diversity and clonal growth in relict populations of *Olea europaea* subsp. *laperrinei* (Oleaceae) from Hoggar, Algeria, *Ann. Bot.* 96 (2005) 823–830.
[15] G. Besnard, P. A. Christin, D. Baali-Cherif, N. Bouguedoura, F. Anthelme, Spatial genetic structure in the Laperrine's olive (*Olea europaea* subsp. *laperrinei*), a long-living tree from the central Saharan mountains, *Heredity* 99 (2007) 649–657.
[16] F. Anthelme, A. Abdoulkader, G. Besnard, Distribution, shape and clonal growth of the rare endemic tree *Olea europaea* subsp. *laperrinei* (Oleaceae) in the Saharan mountains of Niger, *Plant Ecol.* 198 (2008) 73–87.
[17] G. Besnard, D. Baali-Cherif, S. Bettinelli-Riccardi, D. Parietti, N. Bouguedoura, Pollen mediated gene flow in a highly fragmented landscape: consequences for defining a conservation strategy of the relict Laperrine's olive, *C. R. Biol.* 332 (2009) 662–672.
[18] K. M. Sefc, M. S. Lopes, D. Mendonça, M. Rodrigues Dos Santos, M. Laimer Da Câmara Machado, A. Da Câmara Machado, Identification of microsatellite loci in olive (*Olea europaea*) and their characterization in Italian and Iberian olive trees, *Mol. Ecol.* 9 (2000) 1171–1173.
[19] F. Carriero, G. Fontanazza, F. Cellini, G. Giorio, Identification of simple sequence repeats (SSRs) in olive (*Olea europaea* L.), *Theor. Appl. Genet.* 104 (2002) 301–307.
[20] J. Goudet, FSTAT (vesion 1.2): a computer programme to calculate *F*-statistics, *J. Hered.* 86 (1995) 485–486.
[21] T. Franks, R. Botta, M. R. Thomas, J. Franks, Chimerism in grapevines: implications for cultivar identity, ancestry and genetic improvement, *Theor. Appl. Genet.* 104 (2002) 192–199.
[22] L. Baldoni, N. G. Cultrera, R. Mariotti, C. Ricciolini, S. Arcioni, G. G. Vendramin, A. Buonamici, A. Porceddu, V. Sarri, M. A. Ojeda, I. Trujillo, L. Rallo, A. Belaj, E. Perri, A. Salimonti, I. Muzzalupo, A. Casagrande, O. Lain, R. Messina, R. Testolin, A

consensus list of microsatellite markers for olive genotyping, *Mol. Breed.* 24 (2009) doi:10/1007/s11032- 009-9285-8.

[23] C. Breton, M. Tersac, A. Bervillé, Genetic diversity and gene flow between the wild olive (oleaster, *Olea europaea* L.) and the olive, *J. Biogeogr.* 34 (2006) 1916–1928.

[24] A. Belaj, C. Muñoz-Diez, L. Baldoni, A. Porceddu, D. Barranco, Z. Satovic, Genetic diversity and population structure of wild olives from the north-western Mediterranean assessed by SSR markers, *Ann. Bot.* 100 (2007) 449–458.

[25] G. Besnard, P. Henry, L. Wille, D. Cooke, E. Chapuis, On the origin of the invasive olives (*Olea europaea* L., Oleaceae), *Heredity* 99 (2007) 608–619.

[26] G. Besnard, R. Rubio de Casas, P. Vargas, Plastid and nuclear DNA polymorphism reveals historical processes of isolation and reticulation in the olive tree complex (*Olea europaea* L.), *J. Biogeogr.* 34 (2007) 736–752.

[27] F. Lefort, G. C. Douglas, D. Thompson, Microsatellite DNA profiling of phenotypically selected clones of Irish oak (*Quercus* spp.) and ash (*Fraxinus excelsior* L.), *Silvae Genet.* 49 (2000) 21–28.

[28] J. H. Raush, M. T. Morgan, The effect of self-fertilization, inbreeding depression, and population size on autopolyploid establishment, *Evolution* 59 (2005) 1867–1875.

In: Biotechnology and Conservation … (2 Volume Set)
Editors: S. A. Farooq, R. Abed and S. Baqir
ISBN: 978-1-63117-869-6

Chapter 37

KINETIC STUDY OF ENZYMATIC HYDROLYSIS OF STARCH ISOLATED FROM SORGHUM GRAIN CULTIVARS BY VARIOUS METHODS

***Rachid Souilah*[1], *Nadia Boudries*[1], *Djaffar Djabali*[2], *Badreddine Belhadi*[1] and *Boubekeur Nadjemi*[1,2]**

[1]Laboratoire de Recherche sur Les Produits Bioactifs et Valorisation de la Biomasse, Département de Chimie, Ecole Normale Supérieure Kouba, Alger, Algeria
[2]Laboratoire d'Etudes et Développement des Techniques d'Epuration et de Traitement des Eaux et Gestion Environnementale, Département de Chimie, Ecole Normale Supérieure Kouba, Alger, Algeria

ABSTRACT

Sorghum (*Sorghum bicolor (L.) Moench*) crops as a source of starch was successfully cultivated in the Algerian Sahara. Starches from white and red sorghum grains were isolated under three steeping conditions for wet-milling process. The isolate starches susceptibility towards glucoamylase hydrolysis to glucose and the kinetic parameters (K_m, V_{max}, K_{cat} and K_{cat}/K_m) are studied. The yield of starch isolation ranged from 40.37 to 54.79%. The starch purity, amylose content, swelling power and water solubility index ranged from 90.00 to 99.08%, 23.78 to 25.89%, 10.37 to 12.47 g/g and 10.65 to 18.80 %, respectively. The values of V_{max} in $g.l^{-1}.min^{-1}$ for starch isolated by three methods are 0.20, 0.09 and 0.058 for white grains and 0.10, 0.08 and 0.20 for red grains. The glucose concentration values in $g.l^{-1}$ after 120 min of hydrolysis at the initial concentration of 20 $g.l^{-1}$ of starch isolated by three methods are 13, 10.20 and 7.80 for white grains and 10, 13.80 and 11 for red ones. The results of kinetic study show that the hydrolysis of starch by glucoamylase from *Aspergillus niger* are affected by the source of starch and the conditions of starch isolation from grains.

Keywords: Starch, Wet milling, Hydrolysis, Glucoamylase, Kinetic

INTRODUCTION

Sorghum ranks fifth among the cereal crops grown worldwide (Taylor et al., 2006). Many studies show that it will become the alternative of wheat in the future (Beta et al., 2001; Sang et al., 2008; Taylor et al., 2006). The grain chemical composition of sorghum genotypes from the world collection of the International Crops Research Institute for the Semi-Arid-Tropics (ICRISAT) showed that starch content was between 55.6 % and 75.2 %, with a mean value of 69.5% (FAO. 1995). According to a previous study, *Sorghum bicolor (L.) Moench* cultivars are cultivated with considerable magnitude in Algerian Sahara (Boudries et al., 2009).

Starch is today the most important biorenewable material because of its relative abundance and its relative ease of isolation in a highly pure form, which is relatively easily solubilized and enzymatically hydrolyzed to glucose and /or different maltodextrin product (Robyt 2008). It therefore, is a natural product that find many industrial uses, such as the formation of glucose syrups, high fructose syrups, maltodextrins, cyclomaltodextrins and the formation of ethanol, acetic acid, D-lactic acid, and other organic compounds by fermentation (Li et al., 2004; Robyt 2008; Uthumporn et al., 2010).

Glucoamylases (1,4-α-D-glucan glucohydrolase, EC 3.2.1.3), also known as amylo-glucosidase or γ-amylase, catalyse the hydrolysis of α-1,4 and α-1,6 glucosidic linkages to release β-D-glucose from the non-reducing ends of starch (Sauer et al., 2000; Robyt 2008). But because of the hydrolysis of the α-1,6 branch linkages, glucoamylases can completely hydrolyze all the starch to β-D-glucose (Robyt 2008).

Starch susceptibility to hydrolysis by amylolytic enzymes is affected by many factors notably the botanical origin, the nature of the granule surface and amylose content (Kimura and Robyt. 1995; Uthumporn et al., 2010). A prior study of the reaction of starch granules from seven botanical sources (waxy maize, maize, barley, tapioca, amylomaize-7, shoti, and potato) with glucoamylases from *Rhizopus niveus*, showed a wide degree of variance in their susceptibility to enzymatic hydrolysis, in one hand (Kimura and Robyt 1995). On the other hand, an investigation of the action pattern of various amylases on granular starches showed an influence of the amylase source (Planchot et al., 1995).

Because of the water-insolubility of starch granules, they are relatively easy to isolate from their plant sources. Wet-milling conditions, meanwhile, have an effect on physical and chemical properties; lactic acid and steeping time reduce the viscosity of starch suspensions (Haros et al., 2004). Steeping temperature, lactic acid and sulfur dioxide concentrations affect starch functional and thermal properties as well as swelling power, amylose content and pasting properties of maize starch (Brandemarte et al., 2004; Pérez et al., 2001; Shandera and Jackson 1996). Studies on enzymatic hydrolysis of Hull-Less barley starch by α-amylases and amyloglucosidase suggest that both of its morphological and ultrastructural features influence the hydrolysis. However, the rate and extent of α-amylolysis of black bean and lentil starches were mainly influenced by starch chain, amylose chain, and crystalline structure (Li et al., 2004; Zhou et al., 2004).

The objectives of the present study were: (1) to isolate starches from sorghum grains under steeping conditions for wet milling process, (2) to assess the hydrolysis of isolate starch by glucoamylase from *Aspergillus niger* using a Michaelis-Menten kinetic model, (3) and to evaluate the effect of cultivars differences and starch isolation methods on the kinetic parameters of hydrolysis.

Samples are white and red grains from two sorghum cultivars *(Sorghum bicolor (L.) Moench)*, cultivated in Tidikelt, a hyper arid region situated in the south of Algeria and known to have temperatures ranging from 7.8 to 45.2 °C with very low annual rainfall rate (16.9 mm). All chemical products were obtained from Sigma Chemical Co. (St. Louis, MO) and Merck certified grade. The commercial enzyme (glucoamylase from *Aspergillus niger*) used in the present study was purchased from Sigma-Aldrich, A-7255 Co. The density of the enzyme is 1.2 g/ml at 25 °C, the minimum activity 300 U/ml.

Starch Isolation

Sorghum grains were cleaned in order to remove foreign materials and broken grains. Representative 100 g sorghum samples were prepared and stored at 4°C until used. The moisture content of the sorghum was determined in triplicates by using approved procedure method 44-15A (AACC 2000).

Starch was isolated from white and red grains using procedures involving: steeping, wet milling, filtration, centrifugation and drying.

Steeping Treatments: Sorghum grain was steeped using three different treatments in steeping aqueous solutions: Treatment (1): grains (20g) were steeped in a 20 ml sodium hydroxide (NaOH) solution 0.25 % (w/v) for 24hr at 8°C (Beta et al., 2001). Treatment (2): grains (50g) were steeped in a 75 ml sulfur dioxide (SO_2) solution 0.25 % for 48hr at 58° C (Pérez et al., 2001). Treatment (3): grains (50g) were steeped in a 75 ml Lactic acid 0.5% and sulfur dioxide (SO_2) 0.25 % at 58°C (Pérez et al., 2001).

Wet milling procedures: samples from white and red sorghum grains pretreated by the three steeping treatments were wet milled using the procedure described by Wang et al., (2000). The wetted grains were ground by IKA labotechnik A10. The homogenate obtained was screened through a 125, 80 and 63-mesh sieves. The material remaining on sieve was rinsed with water. The filtrate was centrifuged at 5500 g, for 10 min by refrigerated centrifuge (Jouan E96). The mucilaginous top layer was removed. The rinsing and centrifugation are repeated until the starch slurry was near to neutrality and top starch layer was white and free from top protein layer. Finally, the starch isolate was dried overnight in an oven at 40 °C.

Total Starch (TS) in Starch Isolate and Yield of Extraction (Y %)

The total starch in starch isolate was carried out by methods given by Goni et al. (1997); Hu et al. (2004) and Rosin et al. (2002). 50 ± 0.1 mg of weighed sample was dispersed in 6 mL of KOH (2M) solution, shaken in a vortex, left for 1-2 h at ambient temperature until total dissolution of the starch, and then neutralized with acetic acid. Three ml of a buffer solution of sodium acetate 0.4 M (pH 4.75) and 1 ml of the enzyme glucoamylase (300 U/ml., sigma, A-7255) were added, and the mixture was introduced in a thermostatic bath (60 °C) for 45 min with occasional shaking. The reaction mixture was diluted in a flask of 50 ml with distilled water.

One mL of the obtained solution was diluted to 10 ml with distilled water in order to obtain a glucose concentration lower than 100 μg/ml. The glucose produced was analyzed by the glucose oxidase-peroxidase method, and the concentration of starch was obtained by multiplying the concentration of glucose by 0.9.

The yield Y(%) of extraction by dry weight of sorghum is calculated using Equation (1):

$$Y(\%)=m_s(100-H_s(\%))/m_g(100-H_g(\%))\ 100 \quad (1)$$

where: m_s: starch isolate mass in (g). H_s: starch isolate moisture. m_g: sorghum grain mass in (g). H_g: sorghum grain moisture.

Amylose Content

Amylose content was determined by iodine colorimetric procedure of Juliano described in Beta et al. (2001). Starch samples (100 mg) were weighed in 100 ml volumetric flasks. Ethanol (1 ml, 95% (v/v)) was used to wash the sample down the flask. NaOH (9 ml, 1 mol/l) was added to the starch sample before heating the flasks in a boiling water bath for 10 min. The samples were cooled and the volume made up to 100 ml with distilled water, and the contents were mixed vigorously to disperse the starch. For the iodine color development, a 5 ml aliquot of each solution was taken, to which 1 ml of acetic acid (1 mol/l) was added. The contents were mixed. The solutions were left for 20 min after the addition of 2 ml of iodine solution (0.2 % (w/w) I_2 in 2 % (w/w) KI) by mixing them occasionally in that time. The absorbance of the formed blue complex was then measured at 620 nm. Amylose and amylopectin standard mixtures were prepared from 0 to 60 % (w/w) of amylose content using high amylose starch potato (70% (w/w)) and pure maize amylopectin from sigma.

Swelling Power and Water Solubility Index

The Swelling power (SP) and water solubility index (WSI) of starch samples were determined at 85 °C by the method described by Li and Yeh (2001). A centrifuge tube containing 0.1 g of starch and 10 ml of distilled water was heated at 85 °C in a water bath for one hour. The suspension was then cooled at the room temperature and centrifuged (6000 rpm, 20 min). Only the materials adhered to the wall of the centrifuge tube were considered as the sediment and was weighed (W_s). The supernatant was dried to constant weight (W_1) in an air oven at 100 °C. The water soluble index WSI and SP were calculated as follows:

$$WSI= (W_1/0.1) \times 100\%,\ SP=W_S/\ (0.1\ (100\%-WSI))\ g/g.$$

Starch Hydrolysis

The substrates were prepared with different concentrations ranging from 0.65 to 20 g/l, by dissolving starch in 25 ml of acetate buffer solution with a pH 4.5.

Then, it was gelatinized in a water bath at 90 °C. The obtained suspension was shaken continuously at a speed of 1000 t/min and 55 °C temperature. After that, 25 ml of glucoamylase enzyme with a concentration of 0.01U/ml was added. This amount of enzyme is chosen after many experiments to arise the hydrolysis kinetic conditions to obey a Mechaelis-Menten model. The samples were taken out after different times with micropipette. Each time an aliquot was taken, the reactions were stopped by adding 0.1 ml of Trichloroacetic acid 50% (w/v), and the pH was adjusted to 7.0 by adding sodium bicarbonate powder. The extent of hydrolysis was measured from the amount of liberated glucose, using glucose oxidase-peroxidase (Biomaghreb, Tunisia) Kit. The values of the Michaels constant K_m and the maximum velocity V_{max} were determined from Lineweaver-Burk plots.

STARCH ISOLATION

The starch yield of extraction by dry weight of sorghum grain ranged from 40.37 to 54.79% (Table 1). The starch purity evaluated by total starch analysis ranged from 90 to 99.55% of the sorghum starch isolate. The results revealed a high purity expressed as total starch for all isolation methods. The above value for yield was less than that reported (58.5 to 60.1%) by Wang et al. (2000). But higher than that reported (27.73 to 30%) by Pérez Sira and Amaiz (2004), for two cultivars of sorghum starches. The treatment in lactic acid and sulfur dioxide yielded greater amount of starch.

SWELLING POWER AND SOLUBILITY, AMYLOSE CONTENT

Swelling power and water solubility index ranged from 10.37 to 12.47 g/g and 10.65 to 18.8 %, respectively (Table 1). All steeping factors (steeping solutions concentrations from sodium hydroxide, sulfur dioxide and lactic acid with sulfur dioxide, time and temperature) affected both properties. The amylose content of starch isolated from two sorghum cultivars ranged between 23.78 and 25.89%. These percentages indicate that cultivars had non waxy starch with normal amylose content (Sang et al., 2008).

HYDROLYSIS OF STARCH AND KINETIC PARAMETERS

The kinetic curves show that the sorghum starches from white and red grains were hydrolyzed by glucoamylase from *Aspergillus niger.* (Figure 1). The extents of reaction indicate that these starches have high susceptibilities for hydrolysis to glucose. The graphic of enzyme kinetic data using Lineweaver-Burk. plots (Figure 2) demonstrate that the hydrolysis of starches can be described by Michaelis-Menten Kinetic model. The values of the kinetic parameters (K_m, V_{max}, K_{cat} and K_{cat}/K_m), (table 2), calculated from the curves of Lineweaver-Burk, show that the steeping conditions for wet-;milling process and the source of starch affect the kinetic of starch hydrolysis to glucose. The V_{max} values for white and red sorghum starches, from the three different tested treatments, ranged from 0.058 to 0.2 g/l.min.

Table 1. Steeping treatments conditions, yield of isolation and some physico-chemical characteristics of starch isolate from white and red sorghum

Sorghum cultivars	white			red		
Steeping treatments	(1)[c]	(2)[c]	(3)[c]	(1)[c]	(2)[c]	(3)[c]
Aqueous solution	NaOH	SO_2	Lac.ac/SO_2	NaOH	SO_2	Lac.ac/SO_2
Solution conc, w/v (%)	0.25	0.25	0.5/0.25	0.25	0.25	0.5/0.25
Temperature, °C	8	58	58	8	58	58
Time, hr	24	48	48	24	48	48
Starches characteristics						
Yield[a], %	40.37	49.93	54.79	45.39	52.65	54.22
Moisture, %	11.25 ±0.60	11.56±0.08	10.96±0.04	12.58±0.03	11.22±0.47	11.95±0.18
Purity[b], %	96.06±0.06	99.59±0.34	90.00±1.05	99.08±1.84	96.36±1.55	94.51±0.51
Amylose[b], %	24.67 ±0.10	25.89±1.01	25.21±0.30	24.70±1.80	23.78±1.39	25.68±0.51
Swelling Power[a], g/g	10.37	10.55	11.33	12.47	10.67	10.44
Water solubility index[a], %	18.80	10.65	11.35	12.30	11.00	10.90

[a]Each value represents the mean of three replicates for each sample.
[b](Mean ±S.D, n=3).
[c]Numerous of treatment.

According to these values, the relative order of hydrolysis rates for white sorghum was as follows: treatment (1) > treatment (2) > treatment (3). While for red sorghum: treatment (3) > treatment (1)> treatment (2). The differences in measured V_{max} values reflect true differences in K_{cat}, which provided that E_0 is constant. The K_m values for white and red sorghum starches, from the three different tested treatments, ranged from 1.52 to 10.9 g/l. K_m is low in the starches extracted by treatments in sulfur dioxide, and in sulfur dioxide and lactic acid for white grains and treatment in sulfur dioxide for red grains; thus, this shows greater affinity for glucoamylase. The treatment in sodium hydroxide recovered the starch with low affinity to enzyme. The ratio of K_{cat}/K_m is defined as the catalytic efficiency or specifity constant. The high values of K_{cat}/K_m in sulfur dioxide solution are 1.88 and 2.86 l/g.min for white and red grains respectively; these values indicated a high specificity.

The kinetics curves of reaction in Figure 3 showed some disparity of starches for different steeping treatments. The greatest yield of glucose is obtained from white sorghum starch, treated by sodium hydroxide, and red sorghum starch treated by sulfur dioxide. These differences show the effects of steeping conditions, and botanic origin of the starch granule, mentioned by Tester and Morrison (1990).

Our results indicate that starches from sorghum grains cultivated in a hyper arid region situated in south of Algeria was isolated with high purity and yield of extraction. The amylose content of starch indicate that cultivars had non waxy starch with normal amylose. The steeping procedures for isolation treatments affect the yield of extraction from each cultivar grains. The starches have high susceptibility for enzymatic hydrolysis by glucoamylase from *Aspergillus niger*.

And the kinetic of hydrolysis can be assessed by Michaelis-Menten equation. The kinetic study shows that the parameters values K_m, V_{max}, K_{cat} and K_{cat}/K_m were affected by the source of starch and the steeping conditions of starch isolation.

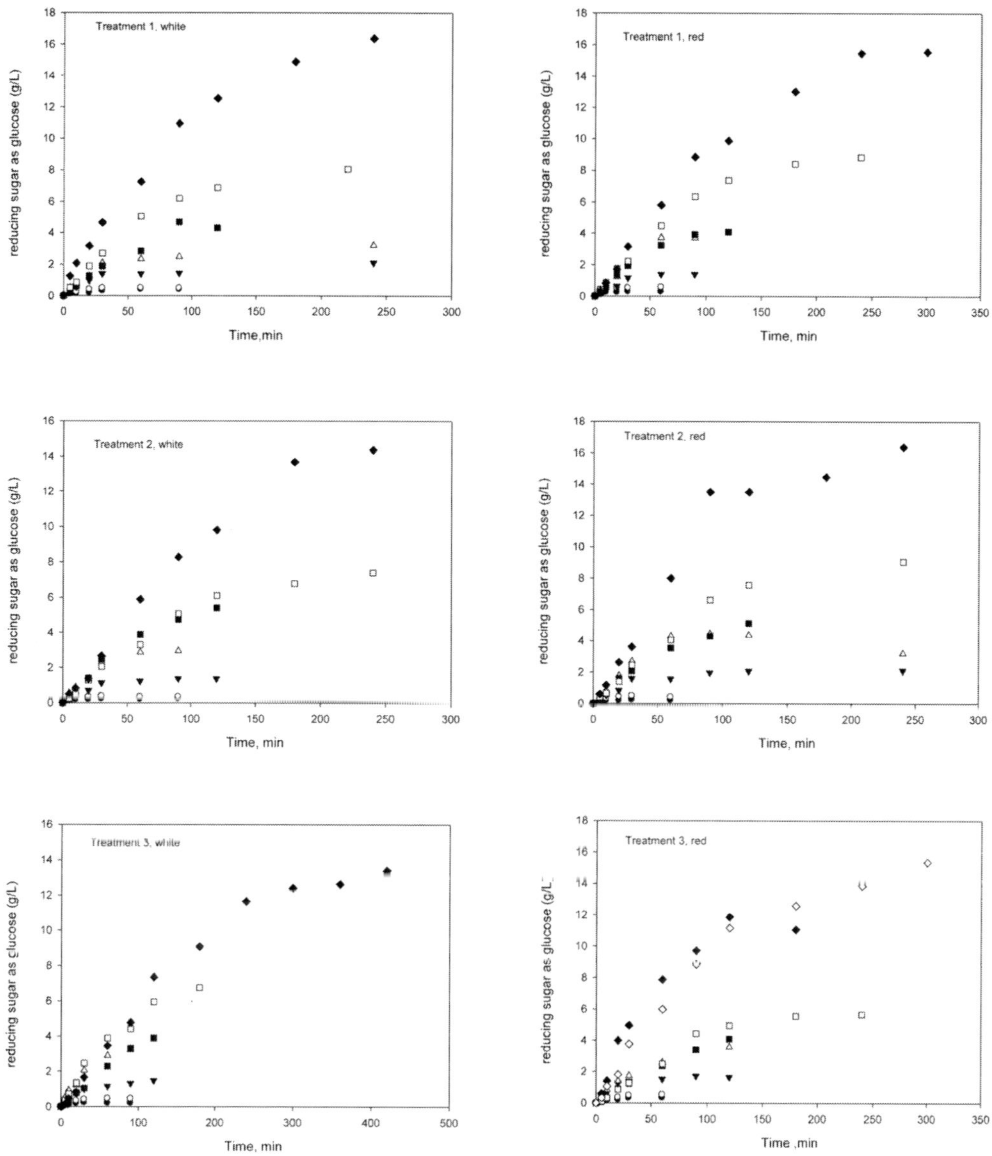

SO-0,65 g/L.
SO-1,25 g/L.
SO-2,5 g/L.
SO-5 g/L.
SO-7,5 g/L.
SO-10 g/L.
SO-20 g/L.

Figure 1. The progress curves of amyloglucosidase hydrolysis of white and red sorghum starch extracted by treatments 1-3. Effect of initial starch concentration ranging from 0.05 to 20 $g.L^{-1}$, at a 0.01U/mL concentration of the enzyme, pH 4.5 and 55 °C. Steeping treatment (1) in NaOH, treatment (2) in SO_2, treatment (3) in Lactic acid and SO_2.

The parameters values K_m, V_{max}, K_{cat} and K_{cat}/K_m show that the source of starch and the steeping conditions for wet-milling process affect the kinetic of starch hydrolysis to glucose.

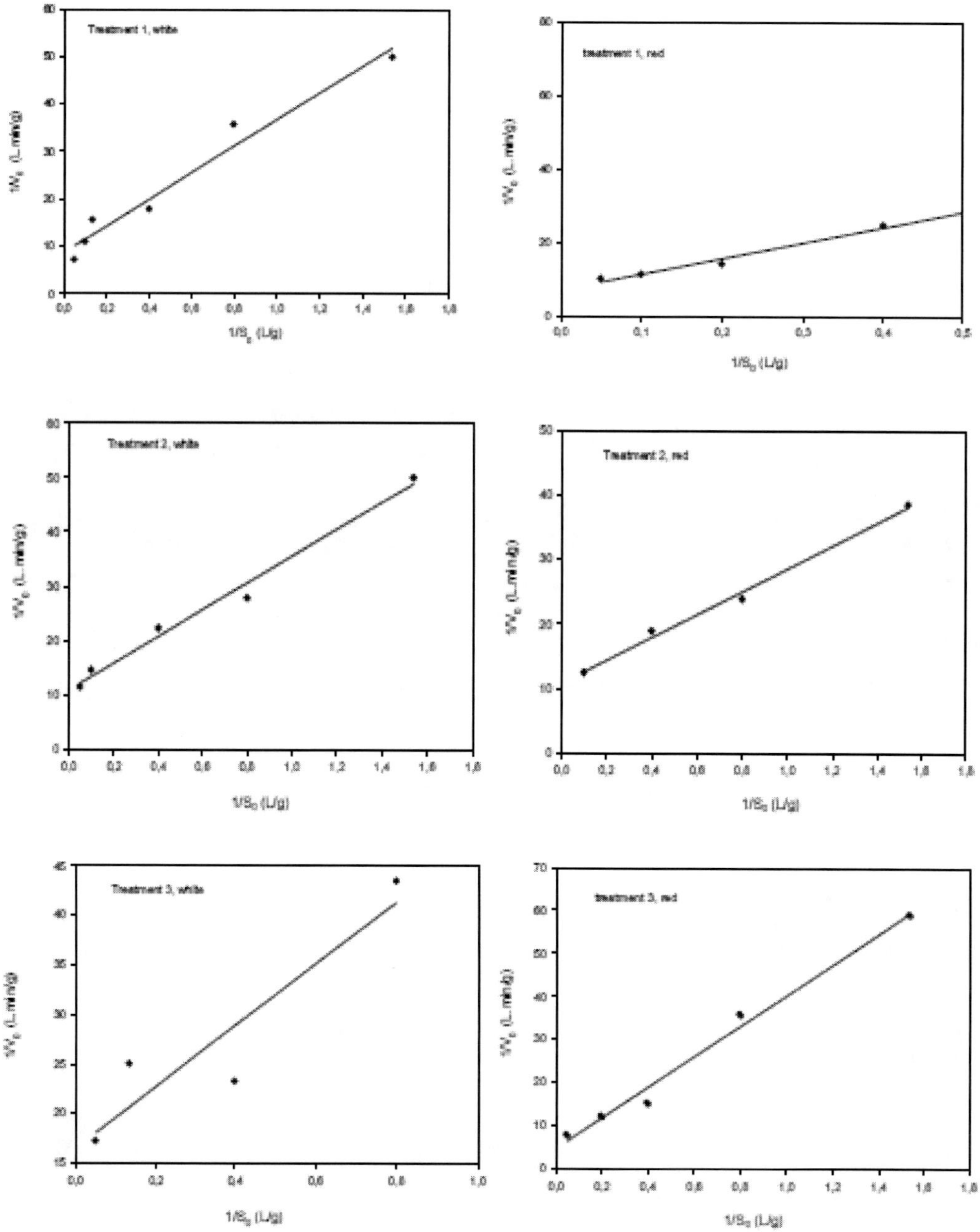

Figure 2. Lineweaver-Burk plot of amyloglucosidase hydrolysis of white and red sorghum starch extracted by treatments 1-3. Steeping treatment (1) in NaOH, treatment (2) in SO_2, treatment (3) in Lactic acid and SO_2.

Table 2. Kinetic parameters of sorghum starches hydrolysis to glucose by glucoamylase from *Aspergillus niger*

Parameters	K_m (g/L)	V_{max} (g/L.min)	K_{cat} (min^{-1})	K_{cat}/K_m (L/g.min)
White sorghum				
Treatment: 1[a]	10.90	0.20	10	0.92
Treatment: 2[b]	2.40	0.09	4.50	1.88
Treatment: 3[c]	1.52	0.058	1.90	1.25
Red sorghum				
Treatment: 1[a]	4.50	0.10	5	1.11
Treatment: 2[b]	1.40	0.08	4	2.86
Treatment: 3[c]	4.70	0.20	10	2.13

[a]Treatment (1) steeping in NaOH.
[b]Treatment (2) steeping in SO_2.
[c]Treatment (3) steeping in lactic acid and SO_2.

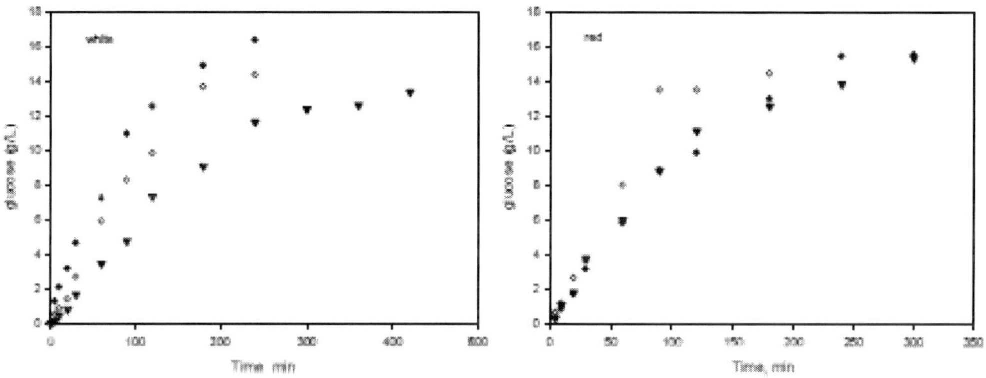

*Treatment (1).
°Treatment (2).
▾Treatment (3).

Figure 3. The progress curves of amyloglucosidase hydrolysis of white and red sorghum starch at S_0=20 g/L, extracted by all treatments, at a 0.01U/mL concentration of the enzyme, pH 4.5 and 55 °C.

REFERENCES

American Association of cereal chemists. 2000. *Approved Methods of the AACC*, 10th ed. Method 44-15A, The Association St. Paul, MN.

Beta, T., Obilana, A. B., Corke, H. 2001. Genetic diversity in properties of starch from Zimbabwean sorghum landraces. *Cereal Sci.* 78: 583-589.

Boudries, N., Belhaneche, N., Nadjemi, B., Deroanne, C., Mathlouthi, M., Roger, B., Sindic, M. 2009. Physicochemical and functional properties of starches from sorghum cultivated in the Sahara of Algeria. *Carbohydr. Polym.* 78: 475-480.

Brandemarte, E. A., Franco, C. M. L., Lopes-Filho, J. F. 2004. Physicochemical properties of maize starch obtained from intermittent milling and dynamic steeping (IMDS) under various steeping conditions. *Cereal Chem.* 81: 369-376.

FAO. 1995. Sorghum and millets in human nutrition. collection FAO. *Food and Nutrition*. Rome. Italy.

Goni, I., Garcia-Alonsa, A., Saura-Calixto, F. 1997. A starch hydrolysis procedure to estimate glycemic index. *Nutr. Res.* 17: 427-437.

Haros, M., Perez, O. A., Rosell, C. M. 2004. Effect of steeping corn with lactic acid on starch properties. *Cereal Chem.* 81: 10-14.

Hu, P., Zhao, H., Duan, Z., Linlin, Z., Wu, D. 2004. Starch digestibility and the estimated glycemic score of different type of rice differing in amylose contents. *J. Cereal Sci.* 40: 231-237.

Kimura, A., Robyt, J. F., 1995. Reaction of enzymes with starch granules: kinetics and products of the reaction with glucoamylase. *Carbohydr. Res.* 277: 87-107.

Li Jeng-Yune and Yeh An-I. 2001. Relationship between thermal, Rheological charectristics and swelling power for various starches. *J. Food Eng.* 50: 141-148.

Li, J. H., Vasanthan, T., Hoover, R., Rossnagel, B. G. 2004. Starch from hull-less barely: V. In-vitro susceptibility of waxy, normal, and high-amylose starches towards hydrolysis by alpha-amylases and amyloglucosidase. *Food Chem.* 84: 621-632.

Pérez, O. E., Haros, M., Suarez, C. 2001. Corn steeping: influence of time and lactic acid on isolation and thermal properties of starch. *J. Food Eng.* 48: 251-256.

Pérez Sira, E. E. and Amaiz, M. L. 2004. A laboratory scale method for isolation of starch from pigmented sorghum. *J. Food Eng.* 64: 515-519.

Planchot, V., Colonna, P., Gallant, D. J., Bouchet, B. 1995. Extensive Degradation of native starch granules by alpha-amylase from Aspergillus fumigates. *J. Cereal Sci.* 21: 163-171.

Sauer, J., Sigrurskjold, B. W., Christensen, U., Frandsen, T. B., Mirgorodskaya, E., Harrison, M., Roepstorff, P., Svensson, B. 2000. Glucoamylase: structure/function relationships, and protein engineering. *Biochim. Biophys. Acta* 1543: 275-293.

Robyt, J. F. 2008. Starch, Structure, Properties, Chemistry, and Enzymology. In: Fraser-Reid, B., Tatsuta, K., Thien, J. (eds) *Glycoscience*, 2nd edn. Springer, New York, pp. 1437-1473.

Rosin, P. M., Lajolo, F. M., Menezes, E. W. 2002. Measurement and characterization of dietary starches. *J. Food Compost. Anal.* 15: 367-377.

Sang, Y., Bean, S., Seib, P. A., Pedersen, J., Shi, Y. C. 2008. Structure and functional properties of sorghum starches differing in amylose content. *J. Agric. Food Chem.* 65: 6680-6685.

Shandera, D. L. and Jackson, D. S. 1996. Effect of corn wet-milling conditions (sulfur dioxide, lactic acid, and steeping temperature) on starch functionality. *Cereal Chem.* 73: 632-637.

Taylor, J. R. N., Schober, T. J., Bean, S. R. 2006. Novel food and non-food uses for sorghum and millets. *J. Cereal Sci.* 44: 252-271.

Tester, R. F. and Morrison, W. R. 1990. Swelling and gelatinization of cereal starches. I. effects of amylopectin, amylose and lipids. *Cereal Chem.* 67: 551-557.

Uthumporn, U., Zaidul, I. S. M., Karim, A. A. 2010. Hydrolysis of granular starch at sub-gelatinization temperature using a mixture of amylolytic enzymes. *Food Bioprod. Process* 88: 47-54.

Wang, F. C., Chung, D. S., Seib, P. A., Kim, Y. S. 2000. Optimum steeping process for wet milling of sorghum. *Cereal Sci.* 77: 478-483.

Zhou, Y., Hoover, R., Liu, Q. 2004. Relationship between α-amylase degradation and the structure and physicochemical properties of legume starches. *Carbohydr. Polym.* 57: 299-317.

In: Biotechnology and Conservation … (2 Volume Set) ISBN: 978-1-63117-869-6
Editors: S. A. Farooq, R. Abed and S. Baqir

Chapter 38

EFFECT OF STARCH ISOLATION METHODS ON STARCH RECOVERY AND PHYSICOCHEMICAL PROPERTIES FOR ALGERIAN SORGHUM (*SORGHUM BICOLOR* L. MOENCH)

***B. Belhadi*[1,*]*, D. Djabali*[2]*, R. Souilah*[1]*, M. Yousfi*[3] *and B. Nadjemi*[1,2]**

[1]Laboratoire de recherche sur les produits bioactifs et valorisation de la biomasse, Ecole Normale Supérieure, vieux kouba, Algiers, Algeria
[2]Laboratoire d'étude et de développement des techniques de traitement et d'épuration des eaux et de gestion environnementale, Ecole Normale Supérieure, vieux kouba, Algiers, Algeria
[3]Laboratoire des Sciences Fondamentales, Université Amar Telidji, Laghouat, Algeria

ABSTRACT

Sorghum (*Sorghum bicolor* (L.) Moench) is an important cereal that can be grown in the semi-arid regions of Asia, Africa and North America, and also under cool climatic conditions. A little research has been done on improving sorghum grains for nutritional quality or on their uses as industrial raw material.

Starch was isolated from two white sorghum using procedures involving: steeping, wet milling procedures, filtration, centrifugation, separation and drying without washing with organic or inorganic solutions for further purification.

The starch yield, starch purity, moisture, amylase and protein content ranged from 43.0% to 58.6%, 94.95 ± 0.35% to 97.29 ± 1.04%, 11.1 ± 0.18% to 11.76 ± 0.05%, 24.43 ± 0.56% to 26.01 ± 0.85% and 0.55 ± 0.20% to 2.34 ± 0.06%, respectively. These results demonstrate that our cultivars could have many industrial applications.

Keywords: Sorghum; starch; isolation; wet-milling

* Corresponding author: Belhadi Badreddine. Laboratoire de recherche sur les produits bioactifs et valorisation de la biomasse, Ecole Normale Supérieure, Bp 92, vieux kouba, Algiers, Algeria. E-mail: badrbel2004@yahoo.fr.

INTRODUCTION

Sorghum (*Sorghum bicolor* (L.) Moench) is an important cereal that can be grown in the semi-arid regions of Asia, Africa and North America, and under cool climatic conditions (Beta et al., 2001). (Beta et al., 2000; De wet, 1995; Liu et al., 2011; Mokrane et al., 2010) reported that little research has been done on improving sorghum grains for nutritional quality or on their uses as industrial raw material.

Starch is the major storage form of carbohydrate in sorghum grains. The starch content ranged from 60% to 77% and 55.6% to 75.2% according to (Watson, 1967) and Jambunathan (FAO and ICRISAT, 1997) respectively. Satin (2002) reported that there is tremendous potential for the commercial use of new starches, but considerable research and product development is necessary to thoroughly exploit the specific functional properties of these materials. Varied and numerous applications for native and modified sorghum starch in food industry were published: bakery, snacks, flavors and beverage clouds, canning, batters and breading, dressings, soups and sauces, confectionery and dairy products (FAO, 2000, Munck, 1995; Satin, 2002).

In general, corn and sorghum starches have similar properties (Watson, 1967). Starch production from sorghum has not been attractive for a number of reasons. Sorghum grain available for wet milling tends to be less consistent than corn in terms of size, color, and defects. Sorghum also yields less starch, as some starch occurs in the thick pericarp and some peripheral cells are not opened during the grinding of steeped grain (Beta et al., 2001). Sorghum starch is also associated with more highly cross-linked kafirin proteins (Munck, 1995).

In a conventional wet milling process, the sorghum grain was steeped in different aqueous solution, containing SO_2: a reducing and antimicrobial agent (Shandera et al., 1995). Lactic acid is often added to the steeping water used in laboratory batch steeping (Shandera et al., 1995). The role of lactic acid in steeping is not completely understood. Sodium hydroxide solution is used during the steeping procedure in order to prevent microbial growth.

Moreover, besides helping dehulling of the grain, alkaline medium could improve the brightness of the starch (Yang and Seib, 1995).

At the present time sorghum production in these areas was used as animal feed generally and less of human consumption. In the past, a wide range of traditional food products have been made from sorghum including kisra, porridges, and couscous.

The aim of the present work was to isolate starch from grain sorghum cultivated in a hyper arid region of Algeria and to prepare samples with high purity grade for physico-chemical and functional characterization. Our approach in this work was to apply and evaluate improved procedures for extracting starch from sorghum cultivars. Three laboratory steeping and wet milling procedures were used. The physicochemical properties of starch isolate were also evaluated.

White grain sorghum (*Sorghum bicolor* (L.) Moench) from two major cultivars grown under uniform field conditions at Tidikelt were selected for this study. This region is situated in the Algerian Sahara bordering the Sahel countries such as Niger and Mali, known as important sorghum producer countries (FAO and ICRISAT, 1997). The agro-climatic characteristics of the region include low annual rainfall rate (16.9 mm) and high temperature reaching a monthly mean of 45.2 °C.

In the International Reference Base (IRB) for soil classification, the soil group in the Tidikelt is solontchaks (salt-affected soils). The soil texture is sandy with clay ranging from 7.50% to 23.75 %. The irrigation used is saline underground water.

Starch Isolation

Starch was isolated from white sorghum grain using procedures involving steeping, wet milling, filtration, centrifugation, separation and drying (Figure 1).

Steeping Procedures

Three laboratory steeping procedures for starch isolation from sorghum grain were used:

- Procedure 1: Sorghum grains (100 g) were steeped in a 200 ml of 0.25 % (w/v) NaOH aqueous solution at 7 °C for 24 hr (Beta et al., 2001).
- Procedure 2: Sorghum grains (100 g) were steeped in a 200 ml of solution prepared from distilled water and sodium metabisulfite (0.25 % sulfur dioxide) at 58 °C for 48 hr (Pérez, Haros, and Suarez, 2001).
- Procedure 3: Sorghum grains (100 g) were steeped in a 200ml SO2 aqueous solution with the addition of 0.50% (v/v) lactic acid at 58 °C, for 48 hr (Pérez, Haros, and Suarez, 2001).

Laboratory Wet Milling, Filtration, Centrifugation and Drying Procedures

Grains from two sorghum samples pretreated by the three steeping procedures were washed with distilled water and wet milled using the procedure described by Wu et al. (1995), Perez et al. (1993) and Zhao and Whistler (1994) (Beta et al., 2001) with some minor modification.

The sorghum grains were washed and ground with an equal volume of water using a Waring blender for 5 min. The slurry was filtered through a 125 μm, 80 μm and 50 μm mesh sieve grains. The material remaining on the sieve was rinsed with water (3×25 ml). The grinding and filtering processes were repeated on this material. After rinsing, the material still remaining on the sieve was discarded.

The collected filtrate was allowed to stand for 1 hr. The filtrate was centrifuged at 760×g for steeping Procedures 1 and 5500×g for steeping Procedures 2 and 3, for 10 min (Yang and Seib, 1996). The gray-colored top protein layer was removed using a spatula.

Excess water was added to suspend the sediment and was centrifuged for 3 min. Washing and centrifugation were repeated several times until the top starch layer became white. The starch isolate was dried for 24 hr at 40 °C (Pérez-Sira and Amaiz, 2004; Beta et al., 2000).

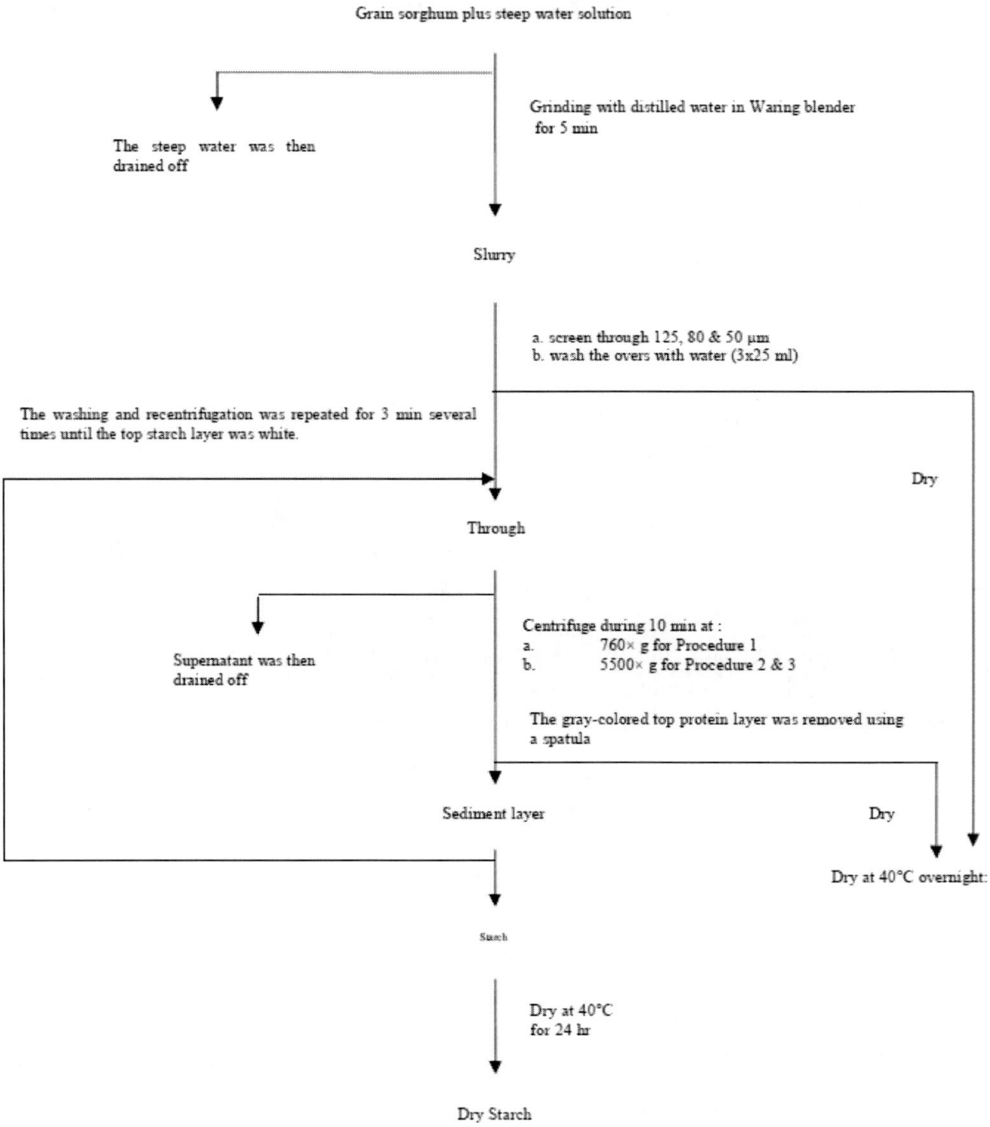

Figure 1. Laboratory scale schema to isolate starch from two sorghum grains.

The Starch Yield (Y%)

The starch yield (Y %) of extraction by dry weight of sorghum grain is calculated by equation 1.

$$Y\% = \{[m_s\,(100 - H_s\%)] / [m_g\,(100 - H_g\%)]\}.100 \quad (1)$$

where:

$H_g\%$: sorghum grain moisture;

$H_s\%$: starch isolate moisture;
m_g: sorghum grain mass in (g);
m_s: starch isolate mass in (g).

CHEMICAL COMPOSITION OF STARCH ISOLATE

Standard AACC methods were used to determine moisture (AACC, 2000). The protein content was determined using Kjeldahl method using nitrogen conversion factor of 6.25.

Total starch was determined by the enzymatic method (Goni et al., 1997). fifty mg of starch isolate was dispersed in 6 ml of KOH (2 mol/l) and energically shaken at room temperature for 30 min. three ml of sodium acetate buffer (0.4 mol/l, pH 4.75) and 1 ml of amyloglucosidase (300U/ml, Sigma A-7255) were added to this suspension and incubated for 45 min at 60 °C in a controlled shaking water bath. After centrifugation (3000 xg, 10 min), the glucose concentration in the supernatant was determined by using a glucose oxidase-peroxidase kit (ELITCH, France). Absorbance was measured at 500 nm and the glucose concentration was converted into starch content using a 0.9 factor. Each cultivar grain was analyzed in triplicates.

Amylose content was determined by iodine colorimetric procedure of Juliano (Beta et al., 2001). Starch samples (100 mg) were weighed in 100 ml volumetric flasks. Ethanol (1 ml, 95 % (v/v)) was used to wash the sample down the flask. NaOH (9 ml, 1 mol/l) was added to the starch sample before heating the flasks in a boiling water bath for 10 min. The samples were cooled and the volume made up to 100 ml with distilled water, and the contents were mixed vigorously to disperse the starch. For iodine colour development, a 5 ml aliquot of each solution was taken to which 1 ml of acetic acid (1 mol/l) was added. The contents were mixed. The solutions were allowed to stand for 20 min after the addition of 2 ml of iodine solution (0.2% (w/w) I_2 in 2.0% (w/w) KI) and mixing. The absorbance of the formed blue complex was then measured at 620 nm. Amylose and amylopectin standard mixtures were prepared to 0–60% (w/w) amylose content using high amylose starch potato (70% (w/w)) and pure maize amylopectin from Sigma. The starch samples from the white and red grain sorghum using procedure 1 were visually white, while the starch samples from both procedures 2 and 3 were visually white with very light pink color. Yield of starch (Y %) obtained from white sorghum samples ranged between 43.0% and 58.6%. The yield of starch (Y %) extracted from white sorghum by procedure 3 were higher than those obtained by procedure 2 and procedure 1 (Table 1, 2 and 3).

Table 1. Starch Yield (Y%), Starch purity (TSs%) and Moisture (Hs%) of Starch isolate after steeping in NaOH Solution at 7 0C, (db). (Procedure 1)

An example of a column heading	SBO8F1AS	SBO8F$_2$AS
Starch Yield (%)	49.9	43.0
Moisture (%)	11.90±0.10	11.31±0.13
Total starch (%)	95.35±1.91	97.63±1.04
Amylose (%)	26.01±0.85	24.70±0.72
Protein content (%)	0.55±0.20	0.69±0.11

Table 2. Starch Yield (Y%), Starch purity (TSs%) and Moisture (Hs%) of Starch isolate after steeping in SO2 Solution at 58 °C, (db). (Procedure 2)

An example of a column heading	SBO8F1AS	SBO8F_2AS
Starch Yield (%)	55.0	51.8
Moisture (%)	11.38±0.03	11.1±0.18
Total starch (%)	94.95±0.35	96.21±1.00
Amylose (%)	24.43±0.56	25.32±0.63
Protein content (%)	1.34±0.20	1.94±0.06

Table 3. Starch Yield (Y%), Starch purity (TSs%) and Moisture (Hs%) of Starch isolate after steeping in SO2 containing Lactic Acid solution at 58 °C, (db). (Procedure 3)

An example of a column heading	SBO8F1AS	SBO8F_2AS
Starch Yield (%)	58.6	52.3
Moisture (%)	11.76±0.05	11.33±0.06
Total starch (%)	96.67±1.63	95.53±1.29
Amylose (%)	24.49±0.91	25.06±1.13
Protein content (%)	1.75±0.17	2.34±0.06

This result indicates that the yield of starch obtained by procedure 2 was similar than the results found by Yang and Seib. (1996) (50.5) from two commercial yellow sorghum with SO2 procedure, whereas, they were lower and higher for procedure 1 and procedure 3, respectively. The addition of lactic acid to the SO_2 steepwater increased starch yield from sorghum with a percentage ranged between 0.5% and 3.6%. Similar results are observed on starch yield from corn (Eckhoff and Tso, 1991).

The purity of starch isolate evaluated by total starch analysis (TS_s) ranged between 94.95 ± 0.35% and 97.29 ± 1.04%. Moisture of starch isolate ranged from 11.1 ± 0.18% to 11.76 ± 0.05% (Table 1, 2 and 3).

The amylose content was ranged between 24.43 ± 0.56% and 26.01±0.85%, which indicate that sorghum cultivars were non waxy starch with normal amylose content.

Protein content in starch isolate gave an indication of the degree of separation and purity, while it's mean was from 0.61% (db) for Procedure 1, from 1.64% (db) for Procedure 2 and from 2.05% (db) for Procedure 3. The addition of NaOH in steep water solution decreased the protein content in starch isolate. Similar protein contents were obtained by Wang et al. (2000) and Xie, and Sieb. (2002) in starch isolated using Procedure 1, whereas, they were higher for Procedure 2 and Procedure 3 from white and red sorghum, ranging from 0.3% to 0.5%. In general, the steeping procedures affected protein isolation from starch.

ACKNOWLEDGMENTS

This work was financially supported by "Laboratoire de recherche sur les produits bioactifs et valorisation de la biomasse. Ecole Normale Supérieure. Vieux Kouba. Alger. Algérie: projet, valorisation des bio polymères des agro ressources".

REFERENCES

AACC, 2000. *Approved Methods*, 10th ed. American Association of Cereal Chemists, St. Paul.

Beta, T., Obilana, A. B., Corke, H., 2001. Genetic diversity in properties of starch from zimbabwean sorghum landraces. *Cereal Chem*., 78 (5), 583-589.

Beta, T., Rooney, L. W., Marovatsanga, L. T., Taylor, J. R. N., 2000. Effect of chemical treatments on polyphenols and malt quality in sorghum. *J. Cereal Sci*., 31, 295–302.

De Wet, J. M. J., 1995. Foreword. In: Dendy, A. V. (Ed.), *Sorghum and Millets: Chemistry and Technology.,* American Association of Cereal Chemists, St. Paul, Mn.

Eckhoff, S. R., Tso, C. C., 1991. Wet milling of corn using gaseous SO_2 addition before steeping and the effect of lactic acid on steeping. *Cereal Chem*., 68 (3), 248-251.

FAO, 1995. Le sorgho et le mil dans la nutrition humaine, *collection FAO: Alimentation et Nutrition,* n° 27, Rome, Italy, pp. 198.

FAO., ICRISAT, 1997. L'économie mondiale du sorgho et du mil: faits, tendances et perspectives. <http://www.fao.org/docrep/W1808F/W1808F00.htm> consulted 15.06.06.

Goni, I., Garcia-Aloso, A., Saura-Calixto, F., 1997. A starch hydrolysis procedure to estimate glycemic index. *Nutrition Research*, 17 (3), 427-437.

Liu, L., Herald, T. J., Wang, D., Wilson, J. D., Bean, S. R., Aramouni, F. M., 2012. Characterization of sorghum grain and evaluation of sorghum flour in a Chinese egg noodle system. *J. Cereal Sci*., 55, 31-36.

Mokrane, H., Amoura, H., Belhaneche-Bensemra, N., Courtin, C. M., Delcour, J. A., Nadjemi, B., 2010. Assessment of Algerian sorghum protein quality [*Sorghum bicolor* (L.) Moench] using amino acid analysis and in-vitro pepsin digestibility. *Food Chem*., 121,719–723.

Munck, L., 1995. New Milling Technologies and Products: Whole Plant Utilization by Milling and Separation of the Botanical and Chemical Components. In: Dendy, A. V. (Ed.), *Sorghum and Millets: Chemistry and Technology*., American Association of Cereal Chemists, St. Paul, Mn, pp. 223–281.

Pérez, O. E., Haros, M., Suarez, C., 2001. Corn steeping: influence of time and lactic acid on isolation and thermal properties of starch. *J. Food Eng*., 48, 251-256.

Pérez-Sira, E. E., Amaiz, M. L., 2004. A Laboratory scale method for isolation of starch from pigmented sorghum. *J. Food Eng*., 64, 515-519.

Satin, M., Functional properties of Starches in <http://www.fao.org/ag/ags/agsi/starchh 41.htm> consulted 22.10.08.

Shandera, D. L., Parkhurst, A. M., Jackson, D. S., 1995. Interactions of sulfur dioxide, lactic acid, and temperature during simulated corn wet milling. *Cereal Chem*., 72 (4), 371-378.

Wang, F. C., Chung, D. S., Seib, P. A., Kim, Y. S., 2000. Optimum steeping process for wet milling of sorghum. *Cereal Chem*., 77 (4), 478-483.

Watson, 1967. Manufacture of corn and milo starches. In: R. L. Whistler, Paschall, E. F. (V II), *Starch: Chemistry and technology*. New York and London: Academic Press, pp. 1-51.

Xie, J. X., Sieb, P. A., 2002. Laboratory wet-milling of grain sorghum with abbreviated steeping to give two products. *Starch/Stärke*, 54, 169-178.

Yang, P., Seib, P. A., 1995. Low-input wet-milling of grain sorghum for readily accessible starch and animal feed. *Cereal Chem.*, 72 (5), 498-503.

Yang, P., Seib, P. A., 1996. Wet milling of grain sorghum using a short steeping period. *Cereal chem.*, 73 (6), 751-755.

In: Biotechnology and Conservation … (2 Volume Set) ISBN: 978-1-63117-869-6
Editors: S. A. Farooq, R. Abed and S. Baqir

Chapter 39

GROWING TOMATO IN SALTY SOIL: SCREENING

Ahmed Al-Busaidi, Salim Al-Rawahy and Mushtaque Ahmed*
College of Agricultural & Marine Sciences,
Sultan Qaboos University, Al-Khod, Oman

ABSTRACT

Pot experiments were carried out to evaluate the effects of saline irrigations on five varieties of tomato (4, 22, 38, 46 & 54). Plants were irrigated with diluted seawater adjusted to three levels of electrical conductivity; freshwater (control), 3, and 6 dS m^{-1}. The results of the experiment showed that saline waters remarkably affected the evapotranspiration rate, soil moisture, salts accumulation and plant biomass production. Saline irrigation had the ability to keep much water in the soil with higher value of salt content. Low salinity treatment exhibited highest plant growth and lowest soil moisture and salts deposition. Varieties number 38 and 46 gave the highest values for fruits number and weight. Whereas, variety number 22 got the lowest values. However, variety number 4 was the tallest and had the highest value for green matter even under high salinity treatment. Overall, under saline condition it was observed that all plant parameters of different varieties were reduced compared to the control except for the number of fruits of some varieties such as 38, 46 and 54. However fruit fresh weight for variety number 38 was enhanced by saline irrigation which could be a good sign for salt tolerance in saline conditions.

Keywords: Saline irrigation, evapotranspiration, plant growth, stress coefficient

INTRODUCTION

Conventional water resources of good quality are scarce especially in arid and semiarid regions. The salinization of soils and water at these places is a substantial constraint of crop productivity. It is well documented that the amount and quality of irrigation water available in

* Corresponding author: College of Agricultural & Marine Sciences, Sultan Qaboos University, P.O. Box 34, Al-Khod 123, Oman, Tel: +968 24143736, Fax: +968 24413418, e.mail: ahmed99@squ.edu.om.

many of the arid and semiarid regions of the world are the main limiting factors to the extension of agriculture (Munns, 2002). Saline-sodic irrigation water, coupled with the low annual rainfall and high evaporation and transpiration in the arid and semi-arid regions, have resulted in accumulation of soluble salts in the soil solution, which can alter the structure and, consequently, affect the soil hydraulic conductivity (Sameni and Morshedi, 2000). The build-up of salts in irrigated regions is of particular concern since 14% of cultivated land that is irrigated supplies approximately half of the world's food (Ben-Hur et al., 2001), This has prompted researchers to study the impact of salinity on plant. Several studies showed external signs of salt toxicity due to irrigation with saline water such as sclerosis, leaf burning and poor vegetative growth (Munns, 2002).

Tomato (*Lycopersicon esculentum*) is a major food plant, and it is moderately sensitive to salinity (Peralta et al., 2005). Extensive research is necessary to develop growing conditions in moderate salinity to produce good vegetative growth. The effect of salinity concentration on plant growth has been studied in different tomato cultivars. Adler and Wilcor (1987) found that salinity adversely affected the vegetative growth of tomato, and it reduced plant length and dry weight. Salinity also reduced the fresh and dry shoot and root weight of tomato (Shannon et al., 1987). Increased salinity over 4000 ppm led to reduction in dry weight, leaf area, plant stem, and roots of tomatoes (Li et al., 2001). The reduction of dry weights due to increased salinity may be a result of a combination of osmotic and specific ion effects of Cl and Na (Al-Rwahy, 1989). The leaf and stem dry weights of tomato were also reduced significantly in plants irrigated with saline nutrient solution in contrast with control plants (Satti and Al-Yahyai, 1995). Byari and Almaghrabi (1991) found that tomato cultivars varied greatly in their response to different salinity levels. Increasing NaCl concentration in nutrient solution adversely affected tomato shoot and roots, plant height, K concentration, and K/Na ratio (Al- Karaki, 2000).

Tomato is a major vegetable crop that has achieved tremendous popularity over the last century. It is grown in practically every country in the world, in outdoor fields, greenhouses and net houses (Peralta et al., 2005). The main purpose of this study was to evaluate the ability of five different varieties of tomato for growing under different saline condition and screen them for salt tolerance. Moreover, this study will help Omani farmers in selecting the right variety for growing in salt affected soils.

The pots experiments were carried out in Agricultural Experiment Station at the College of Agricultural & Marine Sciences, Sultan Qaboos University, Oman. The study was conducted in a glasshouse. Air temperature (^{o}C) and relative humidity (%) were measured continuously by relative humidity and temperature meter (HOBO, Pro Series, onset, Japan) and Pyranometer (EKO, MS-601F) for solar radiation ($mV/Kw.m^2$). In all treatments, seeds of five varieties of tomato (4, 22, 38, 46 & 54) were sown in plastic trays and later transferred to plastic pots (depth=30cm, diameter=25cm) filled with a low salinity sandy soil. Irrigation with diluted seawater in different concentrations was started after 21 days from sowing. Treatments were made of three levels of water salinities (control, 3 and 6 dS m^{-1}). The three saline water treatments were factorially combined with the five varieties of tomato and arranged into a completely randomized design with four replications. Irrigation frequency was every two days and in amount depending on crop evapotranspiration (ETc). The ETc was measured gravimetrically in the pots by weighing some treatments before and after irrigation and the required quantity of water was added plus extra water for leaching. Saline water for each treatment was prepared by mixing tap water with seawater at appropriate ratios to obtain the desired electrical conductivity (EC). In addition, solid fertilizer was added to the irrigation water as recommended. Plant growth (height and leaf area) was monitored and at the end of the experiment, plant height, leaf area (using a portable area meter LI-3000A), number and weight of fruits and plant fresh weight were recorded. Soil samples at a depth of 15 cm were taken from each treatment.

Statistical analysis of the data (ANOVA) was performed and the means compared at 5% probability level.

Growth Conditions

Tomato plants are widely grown in sub-tropical regions where they often experience high temperatures during fruit setting. It has been reported that heat stress can occur at temperatures just a few degrees above the optimal mean daily temperatures of 27 and 29 °C (Peet et al., 1998). Climatic conditions in the glasshouse were varied during the study with average temperature of 29 °C and relative humidity of 60%. As can be seen from Figure 1, air temperature remained within the suitable conditions (21-32 °C) for tomato growth for the most of the days. Moreover, the relative humidity indicates that plant was growing without any environmental stress conditions. Berry and Uddin (1988) demonstrated that the periods of high temperature during the reproductive stage cause interruption in fruit set and spilt of fruit, resulting in losses of yield in the temperate growing areas of the world. They also indicated a highly significant negative correlation between the number of flowers pollinated and the percentage of flowers that had set fruit when plants were exposed to 35/26 °C day/night temperature.

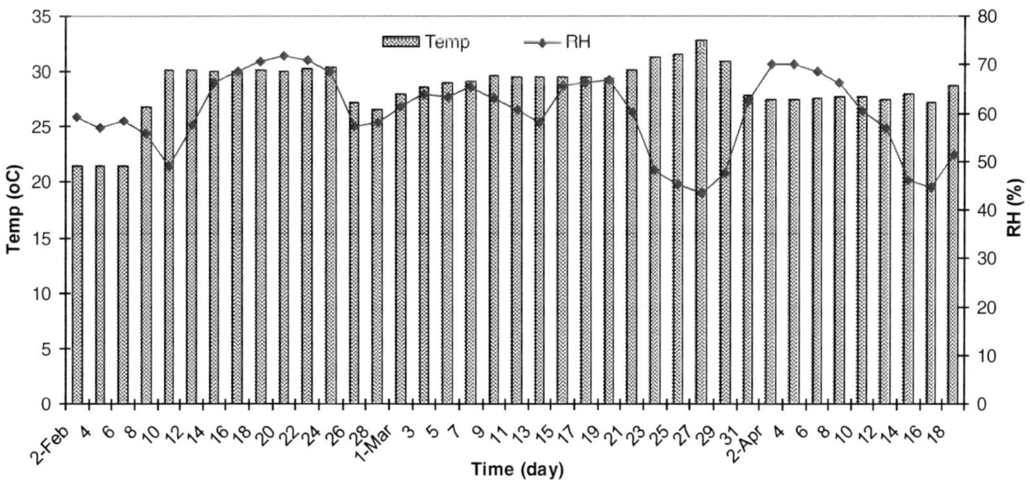

Figure 1. Mean value of the meteorological data during study.

Evapotranspiration (ET) remained as the function of growing conditions and salinity treatments. Water lost by evapotranspiration was directly related to the prevailing environmental conditions in the glasshouse. Figure 2 shows evapotranspiration variation with time (2-12 mm/day) with an average value of 6 mm/day. At the beginning of the study when the plant was small, the value of ET between different varieties was almost same but due to salt tolerance of some varieties and other physical properties of the plants such as leaf area and plant height the variation started to increase. It could be seen that variety number 4 gave the highest ET value whereas 22 was assessed as the lowest. Air temperature and salt stress led to several changes in the plant growth parameters. Plant substantially enhanced the evapotranspiration at the peak growth stage. Environmental condition along with the growing stage of plants tremendously affected the evapotranspiration and salt accumulation in the soils irrespective of the salt treatments (Al-Busaidi et al., 2007).

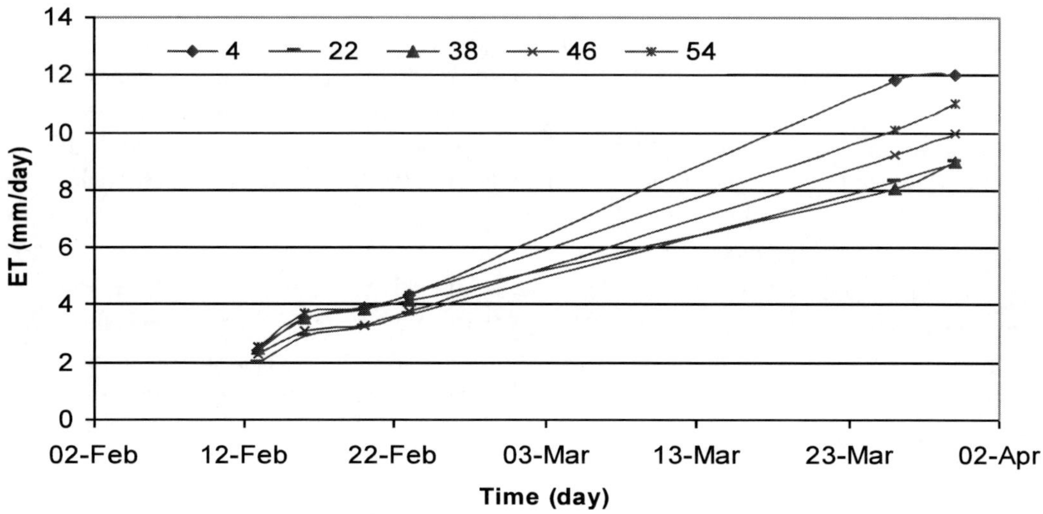

Figure 2. Evapotranspiration of different varieties under saline irrigation.

The evapotranspiration values were generally higher under low salt treatment regardless of the weather conditions (Fig. 3). In other words evapotranspiration was negatively related to the quality of irrigation water. A reduced water loss under high saline treatment was measured as compared to low saline water. Reduced bioavailability of water and retarded plant growth under saline irrigation produced poor evapotranspiration in the system. The depressing effects of salinity on plant growth have been reported by various researchers (Heakal et al., 1990; Abdul et al., 1988). Saline soils inhibit plant growth through reduced water absorption, reduced metabolic activities due to salt toxicity and nutrient deficiency caused by ionic interferences (Yeo, 1983). Salt concentrations in irrigation water inhibited evaporation from the soil surface (Fig. 3). This phenomenon could be related to the enhanced water density, viscosity and chemical bonds in the soil-salt system. High concentrations of salts also form salt crusts, which could reduce soil evaporation. Richards et al. (1998) reported that density, temperature and salinity affected several water characteristics e.g., evaporation etc. Al-Busaidi and Cookson (2005) reported salt crust formation on the soil surface due to saline irrigation, which inhibited evaporation and reduced leaching efficiency.

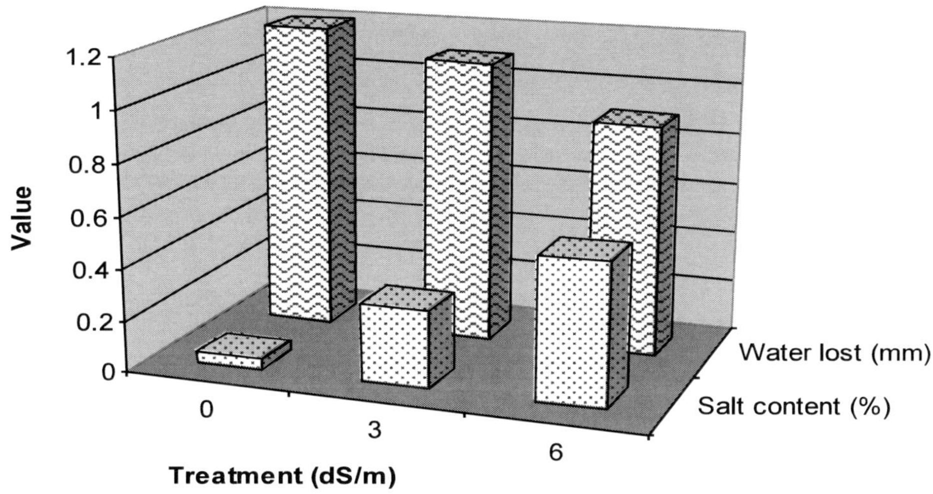

Figure 3. Salts concentration and water loss as affected by saline water treatments.

Salts accumulation in soil was highly affected by the saline irrigation water. Moreover, the data showed that soil water was apparently affected by the quality of irrigation water and salt accumulation (Fig 4). The amount of water in each treatment was following same sequence as indicated in Figure 3. When salinity of soil and water increase, the ability of plant to absorb more water will decrease and the rate of evapotranspiration will decrease. This phenomenon can be seen clearly in Figure 4 with the positive relationship between soil water content and salinity. The trend shown between lowest (0 dS/m) and highest (6 dS/m) salinity treatments is the best example for that.

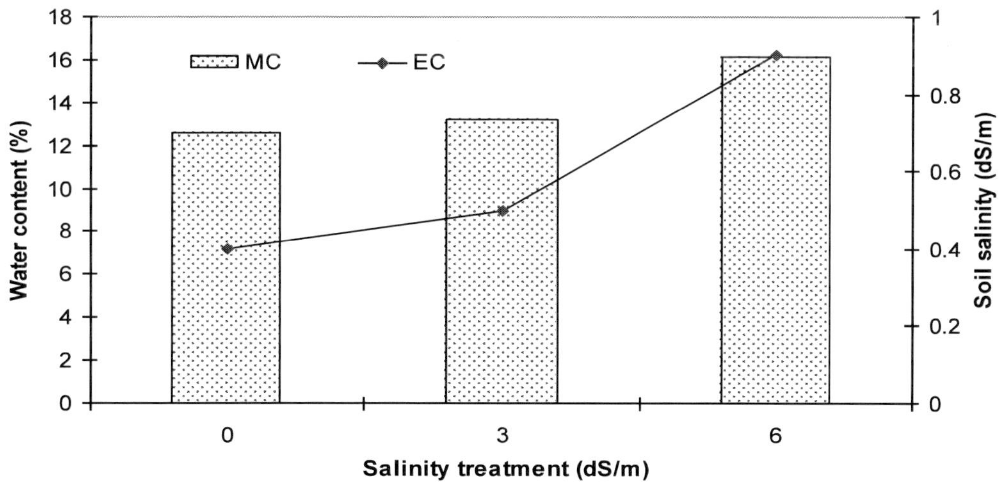

Figure 4. Soil water content and salt accumulation as affected by saline irrigation.

Water uptake by plants and evaporation from the soil surface were reported as the main factors for salts accumulation in the root zone (Ben-Hur et al., 2001; Bresler et al., 1982). This phenomenon can be seen clearly even with fresh water. Moreover, Blanco and Folegatti

(2002) found linear values of soil salinity through application of saline water down the soil profile with higher salts contents near the surface and this is why treatment of highest salinity (6 dS/m) got the highest value for salt accumulation. However, leaching soil is one of the ways to reduce salt accumulation and salinity stress problem. Petersen (1996) reported low soil salinity with increased volume of irrigation water due to salt transportation below the root zone.

PLANT GROWTH

Soil salinity is one of the principal abiotic factors affecting crop yields in the arid and semi-arid irrigated areas. Plant growth was significantly affected by different varieties as well as saline irrigation (Table 1). Treatment with lower salinity gave the higher values of most plant parameters as compared to the high salinity (Table 2).

Table 1. Analysis of variance (ANOVA) for plant parameters

Parameter	Sum of Squares	DF	Mean Square	F	Sig.*
Height	44712.98	14	3193.78	7.83	0.00
Leaf area	3769.91	14	269.28	4.22	0.00
Fruit No.	3866.53	14	276.18	7.96	0.00
Fruit fresh weight	774626.68	14	55330.48	50752.35	0.00
Plant fresh weight	3680804.75	14	262914.63	8.57	0.00

* Level of significance at $P < 0.05$.

Table 2. Plant growth parameters as affected by saline irrigation

Treatment		Height	Leaf area	Fruit	Fruit fresh weight	Plant fresh weight
Saline water	Variety	Cm	cm^2	No.	g	g
Control	4	161.33b*	67.48a	7f	323.28j	1377.20a
	22	108.67d	65.77a	15f	687.62a	635.63c
	38	101.67d	47.26d	29c	525.47d	638.07c
	46	144.67b	53.29ead	13f	453.79h	686.30c
	54	110.33d	47.33de	9f	485.95f	511.53c
3 dS m^{-1}	4	169.33a	48.39c	6d	234.33o	967.37b
	22	88.67f	50.38c	10d	513.25e	456.90e
	38	87.33f	38.57c	36a	566.58b	532.47e
	46	169.33a	43.16c	12d	274.63m	1007.87b
	54	105.00f	50.44c	19d	469.47g	590.43e
6 dS m^{-1}	4	151.00c	41.88f	7e	315.91k	610.10d
	22	84.67fe	42.71f	9e	309.65l	228.37f
	38	74.33f	36.68g	33b	555.26c	318.63f
	46	107.00e	54.90b	17e	344.34i	555.10d
	54	97.00ef	35.04g	12e	260.56n	330.97f

*Means in the column with same letter indicate no difference at Duncan's Multiple Range Test at $P < 0.05$.

Among different varieties and with higher salinity treatment, varieties number 38 and 46 got the highest values for fruits number and weight (33, 17 and 555.23g, 344.34g respectively). Whereas, variety number 22 got the lowest values. The biomass yield was reduced typically due to higher amount of salt depositions in the rhizosphere. However, variety number 4 was the tallest and got the highest value for green matter even under high salinity treatment and this is why the fruits production in this variety was low. Generally, the incorporation of salinity stress and weakness to tolerate salinity could lead to higher loss of plant production (Daoud et al., 2001).

Comparing the response of different varieties to saline irrigation, it could be seen that all plant parameters of different varieties were reduced compared to control except number of fruits of some varieties such as 38, 46 and 54 (Fig. 5). Moreover, the fruit fresh weight for variety number 38 was enhanced by saline irrigation. This evidence could be a good sign for positive response of plants to saline irrigation.

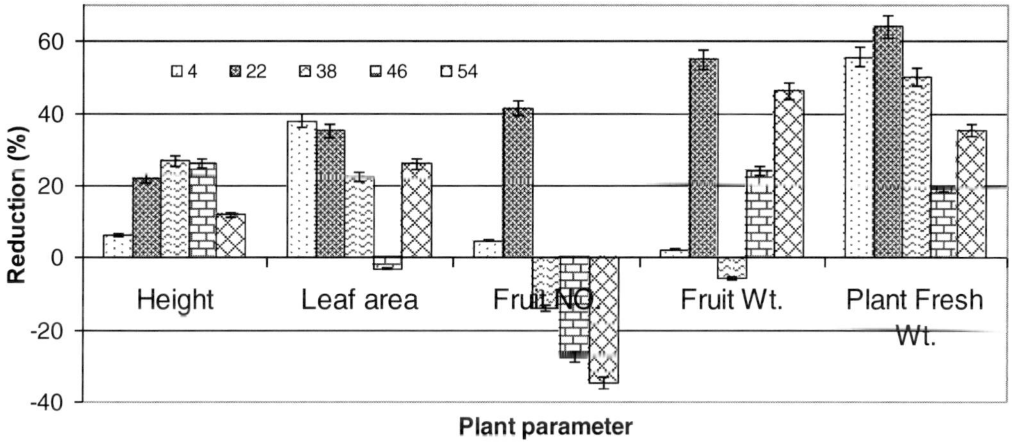

Figure 5. Reduction in plant growth parameters due to salinity treatments.

Generally soil salinity affects the plant growth by producing an ionic imbalance or water deficit state in the expanded leaves. Shani et al. (2001) related the yield loss to reduced photosynthesis, high energy and carbohydrate expenses in osmoregulation, and interference with cell functions under saline conditions. Heakal et al. (1990) reported that dry matter yield of plant shoots decreased with increasing salinity of water. The incorporation of some salts with high temperature could lead to higher loss of plant production (Daoud et al., 2001).

The effect of NaCl stress on the growth of tomato plants is reflected in lower dry weights. The reduction of the dry weights due to increased salinity may be a result of a combination of osmotic and specific ion effects of Cl and Na (Al-Rwahy, 1989). The results indicated that the stem leaves and root dry weights decreased in saline condition, due to the exposure to seawater stress. Similar outcome were obtained earlier by Mohammad et al. (1998) in other tomato cultivars. Saline stress leads to changes in growth, morphology and physiology of the roots that will in turn change water and ion uptake. The whole plants are then affected when roots are growing in saline medium. The results also indicate that salt tolerance of tomato plants tends to increase with age. The same trend was observed on the leaves and roots as also documented by other workers (Al- Rawahy, 1989; Pessarakli and Tucker 1988; Munns,

2002). Finally, in this study, salinity stress resulted in a clear stunting of plant growth, which results in a considerable decrease in the fresh weight of leaves and stems. Increasing salinity is accompanied also by significant reductions in shoot weight and plant height.

There are inconsistencies in the literature regarding the contribution of fruit number to EC-induced reductions in tomato fruit yield. Li et al. (2001) and Eltez et al. (2002) reported that the number of fruits was unaffected by moderate salinity, and that reduced yield was entirely due to smaller fruit. Results of this study are consistent with Adams and Ho (1989) and Van-Ieperen (1996) who observed that the number of harvested fruits per plant decreased with salinity, and was a contributing factor to reduced fruit yield. The decrease of fruit number in the present study was affected by EC and the duration of the harvesting period. The differences in fruit number were larger with increasing duration of the harvesting period as reported in other studies (Adams and Ho, 1989; Van Ieperen, 1996). The reduction in fruit number observed in the present study appeared to be related to a reduction in the average number of flowers per truss and per plant observed with increasing salinity (Maga´ n, 2005). This is consistent with the hypothesis of Cuartero and Ferna´ndez- Mun˜ oz (1999) that stress restricts the number of flowers per truss.

Stress factor (K_s) is an additional parameter to determine crop evapotranspiration. It is an indicator of unusual plants stress such as salinity, deficit water, disease or nutrient imbalance. It implies when its value decreases by less than 1 and smaller K_s value means higher stress. The stress co-efficient was found in the order of highest saline treatment > medium > control (Fig. 6). The K_s values greatly decreased under high level of salinity and heat conditions. Control plants irrigated with fresh water produced more biomass which did not decline K_s values and as salinity increased the K_s values decreased. It was reported that increased evaporation from the soil surface can counteract the reductions in crop coefficient factor (K_c) and K_s caused by high EC_e of the root zone (FAO, 1998). Letey et al. (1985) and Shalhevet (1994) reported that the effects of soil salinity and water stress were interactive to crop evapotranspiration.

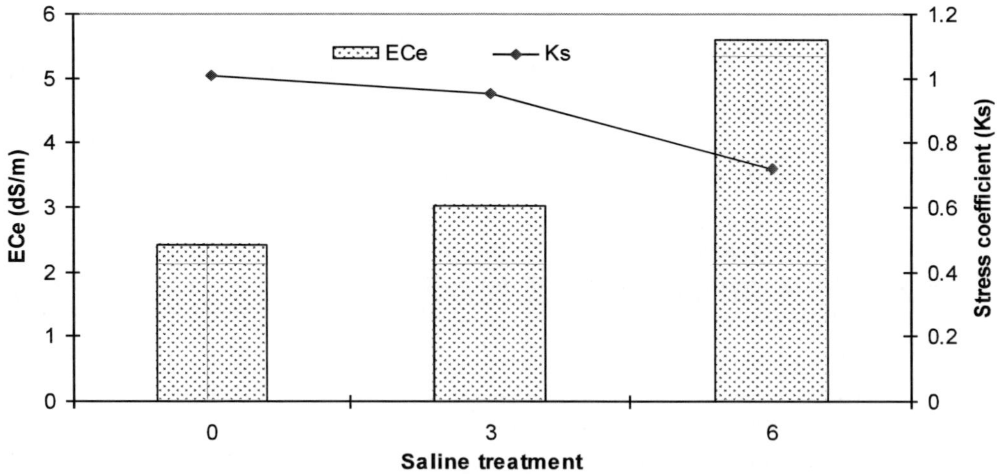

Figuer 6. Stress coefficient as affected by saline treatments.

The water deficit conditions under high salinity treatments could be directly attributed to the impaired water flow from soil to plant. Yeo (1999) reported that root selectivity and transpirational water flow provide the net uptake of salts whereas the salt concentration develops with the growth rate. The greater mass flow of solution through the soil-root interface or higher magnitude of evapotranspiration would increase the salt transport in plants. Thus there is a potential risk of higher salt damages in hot climate. Ghadiri et al. (2005) reported restricted water uptake by salinity due to the high osmotic potential in the soil and high concentrations of specific ions that may cause physiological disorders in the plant tissues and reduce yields. Whereas, Hajer et al. (2006) and Reina-Sa´nchez et al. (2005) reported that plants irrigated with saline water reached maximum daily water uptake earlier than control plants because salinity enhanced plant senescence.

Management of soil and water salinity in Oman is direly needed if agriculture of the country has to be kept alive and the ever increasing desertification is to be mitigated at all. A comprehensive research project in the management of saline soils is required to generate data under agro-climatic conditions of Oman, preparation of economically useful techniques and formulation of recommendations to the farmers.

CONCLUSION

Soil salinity is a major constraint to economic use of land for agriculture especially in the arid and semiarid regions. The results of the present pot study showed that the fresh fruit yield of tomato grown in glasshouse was reduced by increasing salinity. Saline irrigation added much salts to the soil and inhibited plant growth. Treatment of less salinity gave higher values for most plant parameters and as salinity increased there was a reduction in plant growth and final yield. Some varieties such as No. 38 showed an optimistic response to saline agriculture by producing more yield under saline conditions. An extended field study should be done before final recommendation can be given.

REFERENCES

Abdul KS, Alkam FM, Jamal MA. 1988. Effects of different salinity levels on vegetative growth, yield and its components in barley. *ZANCO*. 1: 21-32.

Adams P, Ho LC. 1989. Effect of constant and fluctuating salinity on the yield, quality and calcium status of tomatoes. *J. Hort. Sci*. 64: 725-732.

Adler PR, Wilcor GE. 1987. Salt stress, mechanical stress, or Chloromequat chloride effects on morphology and growth recovery of hydroponic tomato transplants. *J. Amer. Hort. Sci*. 112: 22-25.

Al-Busaidi A, Cookson P. 2005. Leaching potential of sea water. *Journal for Scientific Research: Agricultural and Marine Sciences* 9: 27-30.

Al-Busaidi A, Yamamoto T, Inoue M, Irshad M, Mori Y, Satoshi T. 2007. Effects of seawater salinity on salt accumulation and barley (Hordeum vulgare L.) growth under different metrological conditions. *Journal of Food, Agriculture and Environment* 5: 270-279.

Al-Karaki GN. 2000. Growth, water use efficiency, and sodium and potassium acquisition by tomato cultivars grown under salt stress. *J. Plant Nutr.* 23: 1-8.

Al-Rwahy SA. 1989. *Nitrogen uptake, growth rate and yield of tomatoes under saline condition. PhD. Dissertation*, University of Arizona, Tucson, p: 118.

Ben-Hur M, Li FH, Keren R, Ravina I, Shalit G. 2001. Water and salt distribution in a field irrigated with marginal water under high water table conditions. *Soil Sci. Soc. Am. J.* 65: 191-198.

Berry S, Uddin MR. 1988. Effect of high temperature on fruit set in tomato cultivars and selected germplasm. *Hort. Science* 23: 606-608.

Blanco FF, Folegatti MV. 2002. Salt accumulation and distribution in a greenhouse soil as affected by salinity of irrigation water and leaching management. *Revista brasileira de engenharia agricola e ambiental* 6: 414–419.

Bresler E, McNeal BL, Carter DL. 1982. *Saline and sodic soils: Principles-dynamics-modeling*. Berlin: Springer Verlag. p: 236.

Byari SH, Al-Maghrabi AA. 1991. Effect of salt concentration on morphological and physiological traits of tomato cultivar. *Al-Azhar J. Agric. Res.* 14: 9-11.

Cuartero J, Ferna´ndez-Mun˜ oz R. 1999. Tomato and salinity. *Sci. Hort.* 78: 83-125.

Daoud S, Harrouni MC, Bengueddour R. 2001. Biomass production and ion composition of some halophytes irrigated with different seawater dilutions. *First International Conference on Saltwater Intrusion and Coastal Aquifers; Monitoring, Modeling, and Management. Essaouira, Morocco.*

Eltez RZ, Tu¨zel Y, Gu¨ lA, Tu¨ zel IH, Duyar H. 2002. Effects of different EC levels of nutrient solution on greenhouse tomato growing. *Acta Hortic.* 573: 443-448.

FAO. 1998. Crop evapo-transpiration. *FAO Irrigation and Drainage Papers No. 56*. Rome.

Ghadiri H, Dordipour I, Bybordi M, Malakourti MJ. 2005. Potential use of Caspian Sea for supplementary irrigation in North Iran. *Agricultural Water Management* 79: 209-224.

Hajer AS, Malibari AA, Al-Zahrani HS, Almaghrabi OA. 2006. Responses of three tomato cultivars to seawater salinity 1. Effect of salinity on the seedling growth. *African Journal of Biotechnology* 5: 855-861.

Heakal MS, Modaihsh AS, Mashhady AS, Metwally AI. 1990. Combined effects of leaching fraction salinity and potassium content of waters on growth and water use efficiency of wheat and barley. *Plant and Soil* 125: 177-184.

Letey J, Dinar A, Knapp KC. 1985. Crop – water production function model for saline irrigation waters. *Soil Sci. Soc. Am. J.* 49: 1005-1009.

Li YL, Stanghellini C, Challa H. 2001. Effect of electrical conductivity and transpiration on production of greenhouse tomato (Lycopersicon esculentum L.). *Sci. Hort.* 88: 11–29.

Maga´ n JJ. 2005. *Respuesta a la salinidad del tomate larga vida en cultivo sin suelo recirculante en el sureste espan˜ ol.* Ph.D. Dissertation. Almeria University, Almeria, Spain, p: 171.

Mohammad M, Shibli R, Ajouni M, Nimri L, 1998. Tomato root and shoot responses to salt stress under different levels of phosphorus nutrition. *J. Plant Nutr.* 21: 1667-1680.

Munns R. 2002. Comparative physiology of salt and water stress. *Plant Cell and Environ.* 25: 239-250.

Peet MM, Sato S, Gardner R.G. 1998. Comparing heat stress effects on male-fertile and male-sterile tomatoes. Plant, *Cell and Environment* 21: 225-231.

Peralta E, Knapp S, Spooner DM. 2005. New species of wild tomato (*Solanum* section *Lycoperisicon*: Solanaceae) from northern Peru. Systematic Botany 30: 424-434.

Pessarakli, M. 1999. *Handbook of plant and crop stress*. CRC, USA.

Pessarakli M, Tucker TC. 1988. Dry matter yield and nitrogen-15 uptake by tomatoes under sodium chloride stress. *Soil Sci. Soc. Am. J.* 52: 698-700.

Petersen FH. 1996. Water testing and interpretation. In: Reed, D.W. (ed.) Water, media, and nutrition for greenhouse crops. *Batavia: Ball.*, pp: 31-49.

Reina-Sa´nchez A, Romero-Aranda R, Cuartero J. 2005. Plant water uptake and water use efficiency of greenhouse tomato cultivars irrigated with saline water. Estacion Experimental La Mayora, CSIC, 29750 Algarrobo-Costa, Malaga, Spain, *AWM*. 78: 54-66.

Richards GA, Pereira LS, Raes D, Smith M. 1998. Crop evapo-transpiration: Guidelines for computer crop water requirements. *FAO Irrigation and Drainage Paper* 65, p: 5.

Sameni AM, Morshedi A. 2000. Hydraulic conductivity of calcareous soils as affected by salinity and sodicity. II. Effect of gypsum application and flow rate of leaching solution carbohydrate pol. *Soil Sci. Plant Anal.* 31: 69-80.

Satti SM, Al-Yahyai RA. 1995. Salinity tolerance in tomato: Implications of potassium, calcium and phosphorus. *Soil Sci. Plant Anal.* 26: 2749-2760.

Shalhevert J. 1994. Using water of marginal quality for crop production: major issues. *Agricultural Water Management* 25: 233-269.

Shani U, Dudley LM. 2001. Field studies of crop response to water and salt stress. *Soil Sci. Soc. Am. J.* 65: 1522-1528.

Shannon MC, Gronwald JW, TAL M. 1987. Effect of salinity on growth and accumulation of organic and inorganic ions in cultivated and wild tomato species. *J. Am. Hort. Sci.* 112: 516-523.

Van Ieperen W. 1996. Effects of different day and night salinity levels on vegetative growth, yield and quality of tomato. *J. Hort. Sci.* 71: 99-111.

Yeo A. 1999. Prediction of the interaction between the effects of salinity and climate change on crop plants. *Scientia Hort.* 78: 159-174.

Yeo AR. 1983. Salinity resistance: physiologies and prices. *Physiol. Plant* 58: 214-222.

In: Biotechnology and Conservation … (2 Volume Set)   ISBN: 978-1-63117-869-6
Editors: S. A. Farooq, R. Abed and S. Baqir

Chapter 40

VINBLASTINE PRODUCTION INDUCED BY TRYPTOPHAN IN CALLUS CULTURE OF *CATHARANTHUS ROSEUS* L.

Hana S. Al-Salih*
Biology Department /College of Science /Mosul University, Iraq

ABSTRACT

The Study included callus initiation from seedling explants of *Catharanthus roseus* L. cultured on MS medium supplemented with 0.5 mgl^{-1} of each Benzyl adenine (BA) and Naphthalene acetic acid (NAA). The results revealed that the addition of tryptophan enhanced callus growth and differentiation of shoots. Mass spectrometry showed that the addition of both 6.0 and 8.0 $mg.l^{-1}$ of tryptophan encouraged the production level of the indole alkaloids. Vinblastine was clearly detected clearly in callus cultures as compared with the control treatments.

Keywords: Vinblastine, Catharanthus, Callus cultures, Tryptophan, BA, NAA

INTRODUCTION

Secondary plant metabolites encompass a huge number of natural compounds with a wide diversity chemical structure. They provide human beings with unique resources for medicines, food additives, fragrances, and fine chemicals, Therefore production of secondary plant metabolites, by cultivation of plants and chemical synthesis, are important agronomic and industrial objectives (Zhao and Verpoorte, 2007). Secondary metabolites are plant products, which occur as terpenoids, glycosides and alkaloids. These compounds are commonly obtained commercially by extraction from intact plants. But problems related to the production and marketing of such compounds, as well as the supply of raw materials, can

* h.alsalih59@yahoo.co.uk.

be erratic for several reasons (Smith, 1995). The need for methods of increasing the production of plant-derived pharmaceuticals cost-effectively and with environmental consideration is becoming more important (Karuppusamy, 2009). So, it may become critical to develop an alternative source of important natural therapeutic products. Plant cell cultures are a potential source of medicinally important substances (Misawa, 1994; Vijaya Sree et al., 2010). *Madagascar periwinkle,* also known as (*Catharanthus roseus*), is a plant belonging to the family *Apocynaceae* and is considered to be extremely interesting. It has received particular attention because of the indole alkaloids they are expected to produce (Carew, 1975; Van der Heijden et al., 2004). It was mentioned that this plant produces several commercially valuable alkaloids including the anti-cancer compounds vincristine, vinblastine and the anti-hypertensive compound ajmalicine (Moreno, etal.1995), and sedative compounds ajmalicine and serpentine (Gaines, 2004). The anti-cancer agents vinblastine and vincristine have yet to be produced in Catharanthus cell cultures (Zhao and Verpoorte, 2007; Shams et al., 2009).

Considering the continuous decrease in arable lands and increased considerations of environmental problems, the production of secondary plant metabolites by traditional plant cultivation and chemical synthesis may become extremely limited in the future (Zhao and Verpoorte, 2007).

Although most plant cell culture processes are not yet competitive for commercial application due to the high-cost caused by low productivity, to date there are already some successful examples of commercial production of valuable secondary metabolites by plant cell cultures (Alfermann and Petersen, 1995; Smith, 1995).

The present study was carried out to scale up methods for the in Vitro production of the valuable secondary metabolite, Vinblastine, by using the callus cultures of *Catharanthus roseus* and trying to enhance the growth of callus and levels of vinblastine by adding tryptophan to the medium used for callus growth.

CALLUS INITIATION

Catharanthus roseus L. seeds were surface sterilized under aseptic conditions in the laminar flow hood using 96% ethanol for 2 min. then 2% sodium hypochlorite for 10 min. (Al-Salih and Al-Katib, 2011). The sterilized seeds were germinated on Arnon and Hoagland medium (Arnon and Hoagland, 1940). Seedlings aged 1 month were used as a source of explants.

Callus initiation was achieved using seedling stem explants cultured on Murashige and Skoog (MS) medium (Murashige and Skoog, 1962) supplemented with 0.5 $mg.l^{-1}$ of Benzyl adenine (BA) and 1.0 $mg.l^{-1}$ of Naphthalene acetic acid (NAA) (Al-Salih and Al-Katib, 2011). The callus was recultured on the same medium for growth maintenance.

For Tryptophan elicitation, callus cultures aged three months were used to establish the experiments to evaluate the role of Tryptophan in enhancing the level of indole alkaloids in the callus cultures of *Catharanthus roseus.*

For this purpose the callus was divided into pieces of 0.5 g for each, then recultured on the same MS medium as in the initiation stage but with modification by adding 0.5, 1.0, 2.0, 4.0, 6.0, and 8.0 $mg.l^{-1}$ of tryptophan.

Callus fresh weights were recorded after 6 weeks of growth. All experiments were of completely randomized design and the data obtained was statistically analyzed using standard error (SE) according to SPSS.

Determination of Indole Alkaloid

Extraction was carried out by methanol (Verpotee et al., 2002), then analysis was done by TLC using Silica gel plates and solvent composed of (MeOH: NH_4OH) (3:1 V:V). 254 nm was used for detecting samples of the Ultra violet (UV) source.

Samples analyzed by TLC were compared with Standard Vinblastine, which is dissolved in methanol and analyzed by the same conditions. Results were compared according to RF values.

Mass spectrometry was used to determine vinblastine, (Bruker (Daltonics) Reflex IV, Maldi Source (Scout)/in the UK by injection of 2µg/min, using 20Kv.

Results showed that actively grown callus was obtained when tryptophan was added to the MS medium, All concentrations of tryptophan used in this study enhanced callus growth but the best fresh weight was obtained with the addition of 8.0 mg.l^{-}. Callus fresh weight reached 1.163 g after 6 weeks of culture. The addition of 0.5,4.0 and 6.0 mg.l^{-1} of tryptophan also encouraged callus growth to reach 0.957, 0.965 and 0.909 in fresh weight respectively, in comparison with 0.883 g for the callus grown on the standard medium or the control (0.5 mg.l^{-1} of BA and 1.0 mg.l^{-1} of NAA), whereas callus grown on the medium with the addition of 1.0 and 2.0 mg.l^{-1} of tryptophan reached to 0.889 and 0.806 g in fresh weight, which is almost the same as that of the standard medium (Figure 1). The addition of tryptophan helps to obtain the callus, which is more active than that in the control treatment (Figure 2-a, b). The application of Tryptophan also induced shoot regeneration from the callus (Figure 2-c).

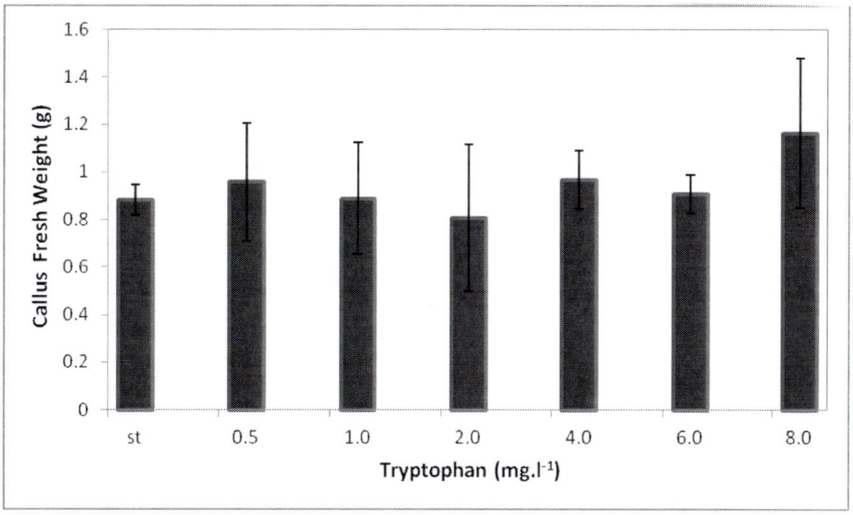

Figure 1. Callus fresh weight of *Catharanthus roseus* grown for 6 weeks on MS medium with addition of different concentrations of Tryptophan.

TLC results for the standard Vinblastine showed that (RF= 0.820) whereas callus extract grown on MS standard with the addition of $8.0 mg.l^{-1}$ of tryptophan showed separation of three bands of RF= 0.71, 0.794 and 0.811.

Results in Figure 3 showed that standard Vinblastine gave an absorption peak of 805.338, whereas callus grown on MS standard with the addition of $8.0 mg.l^{-1}$ of tryptophan showed separation of three peaks of molecular weights 805.396, 819.420 and 833.428. Although there were many other peaks separated the three mentioned above were shown to have higher levels.

When compared with the standard Vinblastine, Figure-3(a), results confirmed that vinblastine was identified in all the callus cultures treated with tryptophan, but with different levels (Figure 4). The best and higher level of Vinblastine was identified in the callus sample treated with $8.0 mg.l^{-1}$ of tryptophan.

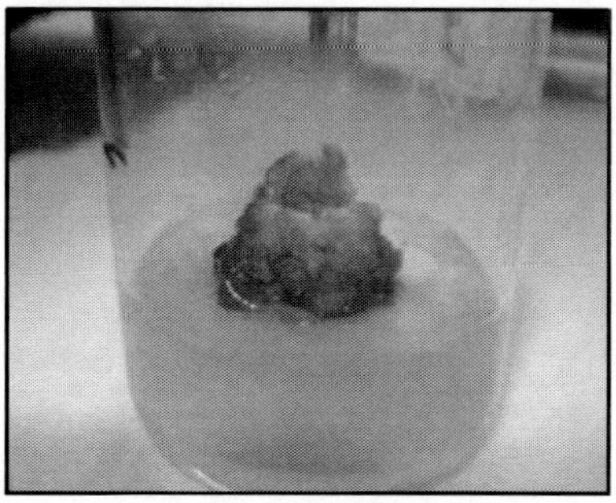

a

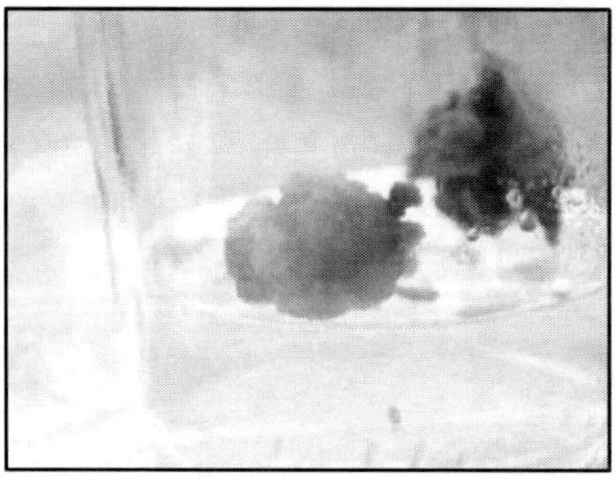

b

c

Figure 2. Callus cultures of *Catharanthus roseus* grown on MS medium (Standard) with the addition of $0.5 mg.l^{-1}$ of BA and $1.0 mg.l^{-1}$ of NAA (a), the addition of $8.0 mg.l^{-1}$tryptophan (b) and regeneration of shoots by the addition of $8.0 mg.l^{-1}$ tryptophan(c).

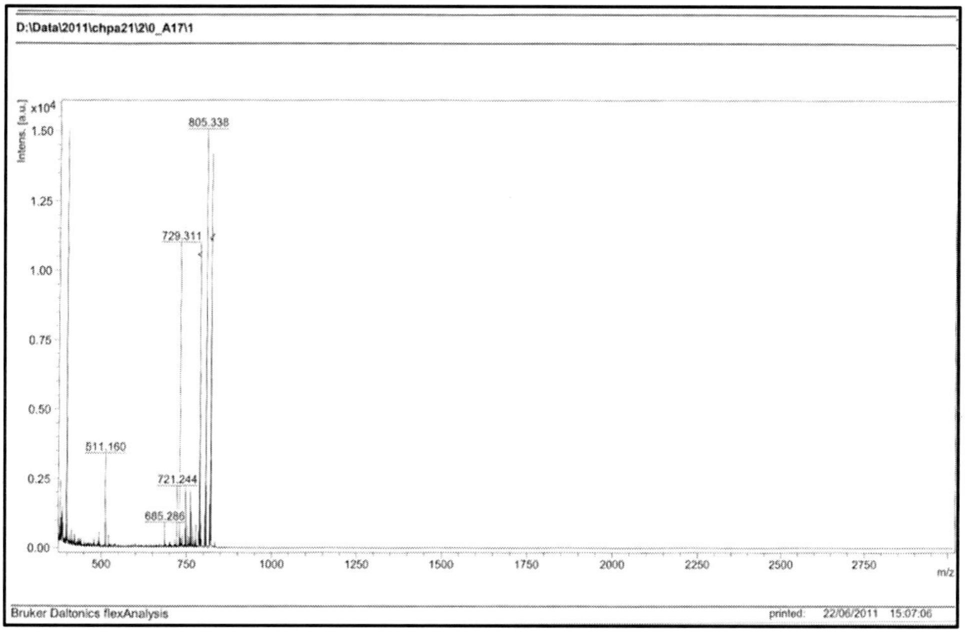

a

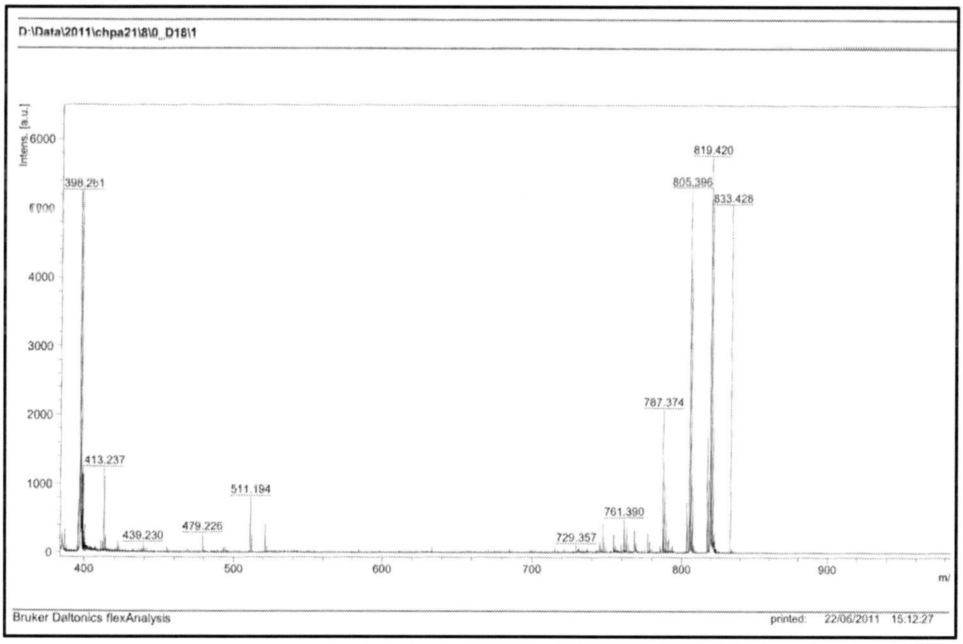

b

Figure 3. Mass Spectrometry chromatogram of standard Vinblastine (a) and callus grown on MS standard with the addition of 8.0mg.l^{-1} of tryptophan (b).

This study allowed us to obtain increasing fresh weight callus with the addition of tryptophan. Activation of fresh weight callus by tryptophan is considered to be an initial step for raising the level of secondary metabolite production in callus cultures of *Catharanthus roseus*. As mentioned in the results, all concentrations of tryptophan used encouraged growth of callus with different levels but the best was 8.0mg.l^{-}. Both auxins and cytokinins were considered to be the best plant growth regulators responsible for directing the growth of callus (Razdan, 2003). The addition of BA and NAA to MS medium encouraged callus initiation and growth (Al-Salih and Al-Katib, 2011). Enhancement of callus growth by tryptophan application may be because this amino acid is shared in many biosynthetic pathways in plants (Talaat et al., 2005).

Identification of Vinblastin in the callus cultures of *Catharanthus roseus* refer to those secondary metabolites that could accumulate in callus cultures. It was reported by (Sarin, 2005) that plant tissue cultures provided tools for manipulating many metabolic pathways. This study provided evidence that application of tryptophan enhanced Vinblastine production. It was reported that one of the main problems encountered is the lack of basic knowledge of the biosynthetic routes and the mechanisms responsible for the production of plant metabolites. Where the productivity of the desired metabolites is limited by the lack of particular precursors, biotransformation using an exogenous supply of biosynthetic precursors and genetic manipulation may improve the accumulation of compounds. Elicitors, compounds triggering the formation of secondary metabolites, can be abiotic or biotic (Verpotee et al., 2002). There have been reports that the addition of the amino acid tryptophan enhanced the vegetative growth in many plants, among them was *Catharanthus roseus* L. (Talaat et al., 2005). Results of the TLC analysis showed separation of vinblastine with RF 0.811 compared with 0.820 for the standard. This result, ensured with the analysis using Mass spectrometry, showed a peak of Vinblastine (805.396) in callus cultures grown in the presence of tryptophan compared with (805.338) for the standard. *Catharanthus roseus* was indicated to produce many terpenoid indole alkaloids. Over 130 compounds have been isolated and identified; tryptophan was one of the precursors of these indole alkaloids via its conversion to tryptamine (Wang et al., 2012). So this study succeeded in describing the steps for production of secondary metabolite vinblastine using levels of tryptophan, which are considered to be precursors of the biosynthesis pathway.

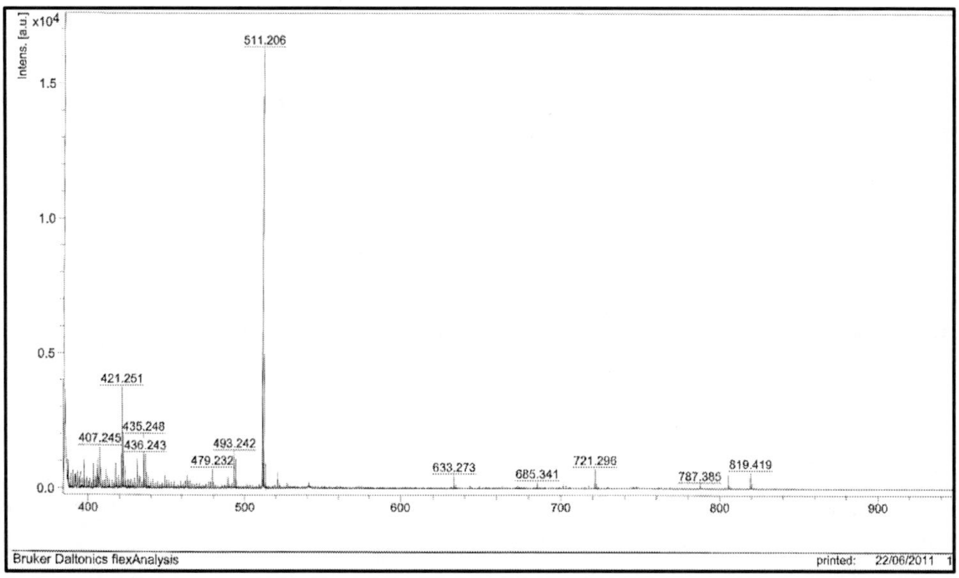

a

Figure 4. (Continued).

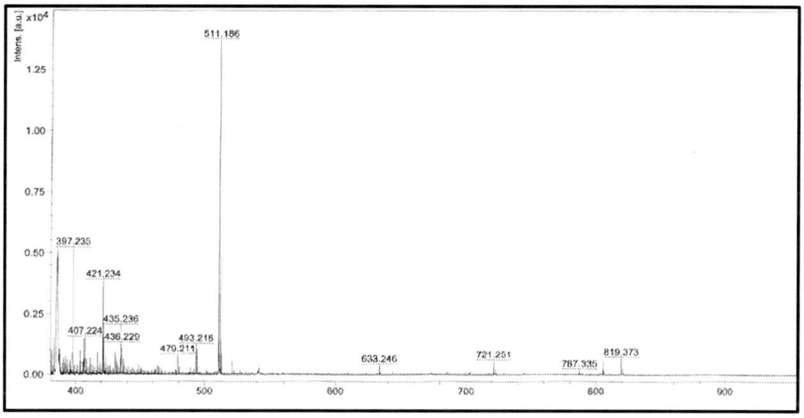

b

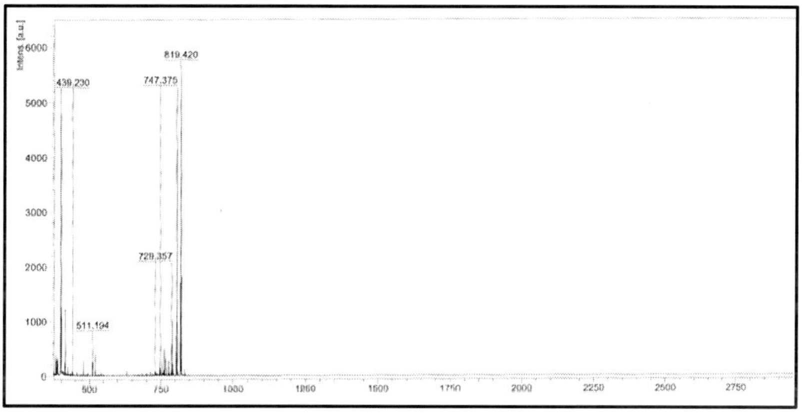

c

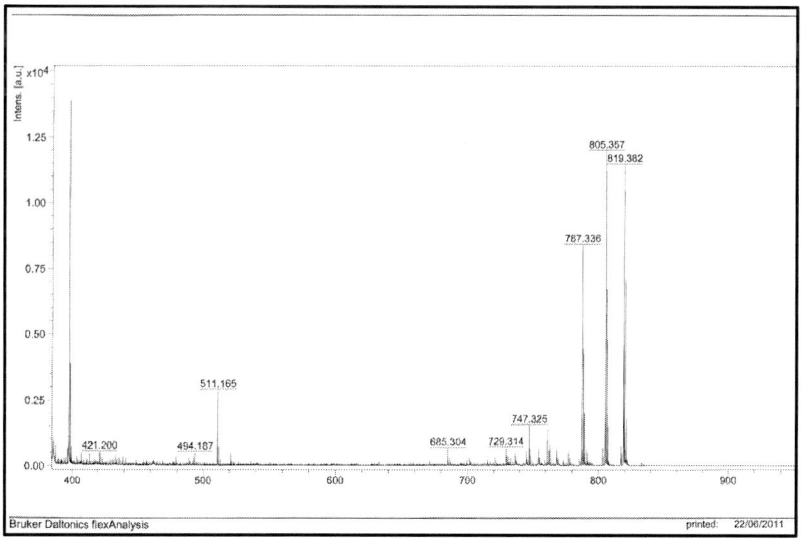

d

Figure 4. Mass Spectrometry chromatogram of *Catharanthus roseus* callus grown on MS standard medium with 0.5 $mg.l^{-1}$ of BA and 1.0 $mg.l^{-1}$ of NAA (a) and callus grown on with addition of 2.0 $mg.l^{-1}$ of tryptophan (b), addition of 4.0 $mg.l^{-1}$ of tryptophan (c), addition of 6. 0 $mg.l^{-1}$ of tryptophan (d).

ACKNOWLEDGMENTS

My thanks to Prof. Jumaa Aldulaimy from Bangor University/Wales/United Kingdom for his efforts that helped me to accomplish Mass Spectrometry analysis.

REFERENCES

Alfermann, A. W. and Petersen, M. 1995. Natural product formation by plant cell biotechnology: results and perspectives. *Plant Cell Tissue Organ Cult* 43:199–205.

Al-Salih, H. S. and Al-Katib, M.U., 2011. Callus cultures Formation of *Vinca rosea* L. and Extraction of Phenolic compounds from it. The Second Scientific conference of Biological Sciences-Mosul/IRAQ (in Arabic).

Arnon, D. I. and Hoagland, D. R., 1940. Crop production in artificial solutions and soils with special reference to factors influencing yield and absorption in of organic nutrients. *Soil Sci.*50: 463-471.

Carew, D. P. 1975. Tissue culture studies of *Catharanthus roseus*. In: Taylor WI, Farnsworth NR (eds) The Catharanthus Alkaloids. New York: Marcel Dekker Inc., pp 193–208.

Gaines, J.L. 2004. Increasing alkaloid production from *Catharanthus roseus* suspensions through Methyl Jasmonate elicitation. *Pharmaceutical Engineering*, 24, 4.

Karuppusamy, S. 2009. A review on trends in the production of secondary metabolites from higher plants by *in vitro* tissue, organ and cell cultures. *Journal of Medicinal Plants Research* Vol. 3(13): 1222-1239.

Misawa, M. 1994. Plant Tissue Culture: an alternative for production of useful metabolites, chapter 6. FAO Agricultural Services Bulletin No. 108, Misawa, M., ed., Bio International, Toronto, Canada.

Moreno, P. R. H.; Ven der Heijden, R.; Verpoorte,R. 1995. Cell and tissue cultures of *Catharanthus roseus*: A literature survey: II. Updating from 1988 to 1993. *Plant Cell Tiss. Org. Cult.* 2 (1), 1-25

Murashige, T. and Skoog, F. 1962. A revised medium for rapid growth and bioassays with tobacco tissue culture. *Physiol. Plant.* 15: 473-497.

Razdan, M. K. 2003. Introduction to Plant Tissue Culture, Scientific Publishers, Inc., Enfield, NH, USA, Printed in India.

Sarin, R. 2005. Useful metabolites from plant tissue cultures. *Biotech.* 4 (2): 79-93.

Shams, K. A.; Nazif,N.M.; Abdel Azim,N.S.; Shafeek,K.A.; El-Missiry,M.M.; Ismail,S.I. and Seif El Nasr,M.M. 2009. Isolation and Characterization of Antineoplastic Alkaloids from *Catharanthus Roseus* L. Don. Cultivated in Egypt. *Afr. J. Tradit. Complement Altern. Med.* 6(2): 118–122.

Smith, MAL, 1995. Large scale production of secondary metabolites. In: Terzi M, Cella R, Falavigna A (eds) Current Issues in Plant Molecular and Cellular Biology. Kluwer Academic Publishers, Dordrecht, the Netherlands, pp 669–674.

Talaat, I. M.; Bekheta, M. A. and Mahgoub, M. H. 2003. Physiological Response of Periwinkle Plants (*Catharanthus roseus* L.) to Tryptophan and Putrescine. *International J. of Agricul. and Biol.*2, 210–213.

Van der Heijden R, Jacobs DI, Snoeijer W, Hallard D, Verpoorte R. 2004. The Catharanthus alkaloids: pharmacognosy and biotechnology. *Curr. Med. Chem.* 11:607–628.

Verpoorte, R.; Contin, A. and Memelink, J. 2002. Biotechnology for the production of plant secondary metabolites. *Phytochem. Rev.* 1, 13-25.

Vijaya Sree N, Udayasri P, V. V. Aswani kumar Y, Ravi Babu B, Phani kumar Y, Vijay Varma M, 2010. Advancements in the Production of Secondary Metabolites. *Journal of Natural Products*, Vol. 3: 112-123.

Wang, Q.; Xing, S.; Qifang Pan, Q.; Yuan, F.; Zhao, J.; Tian, Y. Chen, Y.; Wang, G. and Kexuan Tang, K. 2012. Development of efficient catharanthus roseus regeneration and transformation system using agrobacterium tumefaciens and hypocotyls as explants. *Biotechnology*, 12:34.

Zhao, J. and Verpoorte R 2007. Manipulating indole alkaloid production by *Catharanthus roseus* cell cultures in bioreactors: from biochemical processing to metabolic engineering. *Phytochem. Rev.* 6:435–457.

In: Biotechnology and Conservation … (2 Volume Set) ISBN: 978-1-63117-869-6
Editors: S. A. Farooq, R. Abed and S. Baqir

Chapter 41

ANTIMUTAGENIC POTENTIAL OF OLIVE OIL ON CHROMOSOMAL ABERRATIONS IN *VICIA FABA* ROOT TIP

Galawezh O. Othman** *and Kazhal M. Sluiman
Department of Biology, Collage of Education, University of Salahaddin Erbil,
Kurdistan region, Iraq

ABSTRACT

The antimutagenic potential of olive oil was evaluated in *Vicia faba* root meristem cells. So far there is no report on the biological properties of Olive oil in plant test systems. The root tip cells were treated with Ethyl glycol (EG) at, 200 and 300 µg/ml for 3 h followed by the treatment with olive oil for 16 h. The tips were squashed after colchicine treatment and the cells were analysed for chromosome aberrations and mitotic index. The ability of EG to increase chromosomal aberrations in root tip cells of *Vicia faba* such as chromatid bridge, chromatid break, ring chromosome, and dicentric chromosome is recorded. The use of mutagen (EG) treatments show high level of Chromatid Bridge and low level of ring chromosome. The use of Olive oil was effective in decreasing the chromosome aberrations revealing the potential of antimutagen activity of olive oil.

Keywords: Chromosomal aberrations, olive oil antimutagenic, Ethyl glycol (EG), Vicia faba

INTRODUCTION

The use of antimutagens and anticarcinogens in everyday life is the most effective procedure for preventing human cancer and genetic diseases. There are several ways in which the action of mutagens can be reduced or prevented. Natural chemicals which act with DNA repair or with mutagen metabolism can be effective antimutagens (Ferguson, 1994).

---

* Corresponding author: email: galawezhbio@gmail.com.

Group of chemicals are reported to induce chromosomal aberration commonly known as chemical mutagens having specific and limited action, and found to induce specific mutation or aberration in organisms. Chemical mutagens are being used in inducing variability in plant breeding programmers. Geneticists are using chemical mutagens as potential tools and it is reported that a number of chemicals influence the sensitivity as well as increase the frequency and spectrum of mutations (zaman and salih, 2005). The studies of a number of compounds and the development of chemical mutagenesis have been reported by different workers (Raghuvanshi et al.,1978 and Bose and Dutta, 1973). In view of these research aspects, many mutant varieties have been developed through mutagenesis (Sandra, et al., 1990).

Olive oil with its strong antimutagenic and antioxidative properties has also been described as a valuable nutrient. Olive oil contains phenolic compounds that play important roles in treating diseases. A number of investigations have reported that phenolic compounds, such as hydroxytyrosol, present predominantly in olive and virgin olive oils, are strong antioxidants and could provide protective effects by inhibiting oxidative damage (Tuck and Hayball, 2002). Cell culture experiments have demonstrated that olive oil phenolic compounds have antiatherogenic and antioxidant effects, and could exert cardio protective affects in vivo (Turner et al., 2005). The beneficial health effects of olive oil are due to both its high content of mono-unsaturated fatty acids and its high content of anti-oxidative substances. Although the antimutagenic potential of Olive oil has been extensively studied and well documented, yet there is no report on the biological effects of Olive oil in plant test systems. This study is designed to evaluate the potential antimutagenic effect of Olive oil against ethyl glycol induced chromosomal aberrations in *Vicia faba* root meristem cells.

Healthy seeds of *V. faba* were selected washed and soaked in water for 12 h and were germinated in the dark. Seedlings with roots about 2-5 cm in length were treated with two doses of the mutagen 200, 300 μg/ml of EG for 3 h. Later, for preparation of metaphase chromosomes, root tips were treated with 0.5% colchicine for 3 hours at 20±1°C. Slides are prepared from the fixed root tips following acetocarmine squash technique. The frequency and types of traditional chromosome aberrations were determined. Mitotic index was determined for cytotoxicity evaluation. To study the antimutagenic potential of olive oil, the seedlings were treated with 200μg/ml or 300μg/ml of EG for 3hrs were put in 2ml olive oil for 16 h in dark.

The EG induced chromosomal aberrations included chromosome bridge, chromatid break, ring chromosome and dicentric chromosome. The number of chromosomal aberrations increased significantly by EG with increasing the dosage of mutagen (Table 1). The highest aberrations were found to be the chromosome bridges and the lowest were dicentric chromosomes. In the case of MI the highest value was in experiment with olive oil treatment that lowered the chromosomal aberrations. So olive oil antimutagenity leads to lowering the chromosome aberrations that happened by EG for. In all the concentrations tested, olive oil indicative of its antimutagenic potential in *Vicia faba* root cell the effect of olive oil on the reduction of total number of aberrations induced by EG was statistically significant when compared with EG control.

Plant test system is widely used for monitoring genotoxicity of chemicals because of many advantages such as low cost, easily available throughout the year, ease to handle, good chromosome condition for the study of chromosome damage and above all good correlation with other test systems. Root tip cells of *Vicia faba* constitute an excellent system for such cytogenetic tests (Abraham and John 1989; Gowrisankar et al., 1993; John and Abraham

1991; Upadhya et al., 1996). Historically, plants have been used as indicator organisms, in studies on mutagenesis in higher eukaryotes. Plant systems have a variety of well-defined genetic endpoints including alterations in ploidy, chromosomal aberrations and sister chromatid exchanges (Grant, 1994).

DNA damage is often measured as single-strand breaks, double-strand breaks or chromosomal aberrations, and increase in their frequencies is frequently associated with mutagenesis and carcinogenesis (Surh and Ferguson, 2003). In the present study, we investigated the cytogenetic effects of one dose of dietary olive oil, and its antimutagenic effect against EG in vicia faba root tip cells using the chromosomal aberrations assay induced damage in a dose dependent manner. Our result is agree with a study Showed that chromosomal abnormalities increase with increasing the doses of EG. Similar results were also reported (zaman and salih, 2005). In a study by(Othman, 2002)on albino mice showed that a significant effect were found between concentration and chromosomal aberration, the highest con, was most effective to cause all type of chromosomal aberration. Black cumin oil has anti-mutagen effect in leading to lowering the effect of EMS and decrease the level of chromosome aberrations in all concentrations (Othman and Suleiman, 2012).

The protective effect of olive oil is due to its antioxidant action, trapping of free radicals, formation of complex with mutagens, modulation of mutagen metabolism by absorbing the xenobiotic (Premkumar et al., 2004). And its mechanism for the protective effects of vegetable oils is that their phenolic compounds have antioxidant and antimutagenic properties in vivo (Evangelista, et al., 2006).

The cytogenetic effect of diuron and linuron, sister chromatid exchanges (SCE) in cells of Vicia faba root tips and in human lymphocytes in culture were used as test systems. Both herbicides did not induce SCE in these biological materials except at the lowest concentration of linuron (10 ppm) which produced a significant response only in Vicia faba (Mezarroy et al., 1990)

Table 1. Chromosomal aberrations in *Vicia faba* root cells analyzed and mitotic index after treatment with200, 300 μg/ml of Ethyl glycol and 2ml olive oil

Treatment	Number of cells analyzed	Chrom. bridge	Chro. break	ring Chrom.	Dicen.Chro.	Total	Percentage %	MI
Untreated control	250	3	1	–	1	5	2%	6.06
Olive 2ml	250	2	1	2	2	7	2.8%	4.53
EG 200 μg/ml	250	24	6	3	2	35	14%	3.3
Olive oil 2ml+E.G	250	20	2	3	3	28*	11.2%	2.73
EG 300 μg/ml	250	33	4	5	4	46	18%	3.76
Olive oil2ml+E.G	250	12	3	2	3	20*	8%	2.80

On the basis of experiments, it is concluded that olive oil has antimutagenic potential against EG induced clastogenic damage in *vicia faba* in a dose dependent manner. At the same time it exhibits a mild cytotoxic action; similar to the earlier reports in an in vitro and in vivo test systems. However the mechanism by which it acts remains to be investigated in plant test system and further studies are necessary to clarify these point. The antioxidant effects of olive oil are probably ascribable to a combination of its high oleic acid content and its content of a variety of plant antioxidants, (Visioli and Galli, 1998; Wahle et al., 2004).

REFERENCES

Abraham, S. & John, A. T. (1989). Clastogenic effects produced by black pepper in mitotic cells of *Vicia faba*. *Mut. Res.*, *224*, 281-285.

Bose, S. & Dutta, G. C. (1973). Effects of treatments of colchicine, dimethyl sulphonate, ethylene glycol, hydroxyl amine and triethanol amine in Jute (Corchoruscapsularis L.) *Bangladesh J Bot*, *2*, 1-9.

Evangelista, C. M. W., Antunes, L. M. G., Francescato, H. D. C. & Bianchi, M. L. P. (2004). Effects of the olive, extra virgin olive and canola oils on cisplatin-induced clastogenesis in Wistar rats. *Food Chem Toxicol*, *42*, 1291-1297.

Ferguson, L. R. (1994). Antimutagens as cancer chemopreventive agents in the diet. *Mutat. Res.*, *307*(1), 395-410.

Grant, W. F. (1994). The present status of higher plant assays for the detection of environmental mutagens. *Mutat. Res.*, *310*(2), 175-185.

Gowrishanker, B. & Vivekananda, O. S. (1993).Cytotoxic effects of whisky on Vicia faba in vivo. *Nucleus*, *36*(1,2), 62-65.

John, A. T. & Abraham, S. (1991). Cytological changes produced by red pepper in mitotic cells of Vicia faba L. *Caryologia*, *44*(3,4), 325-331.

Mezarroy, S. G., et al. (1990).Cytogenetic Effect Produced by the diuron and linuron Herbicide in Vicia faba and human lymphocytes culture. *Int. Contam. Ambient.*, *6*, 69-74.

Othman, G. O. (2002). Mutagenic effect of nicotine and test some antioxidants on Albino laboratory mice Mus musculus. MSC. Thesis- College of Educat/ University of Salahaddin- Erbil-Iraq.

Othman, G. O. & Suleiman, K. M. (2012). Preventive effective of Black cumin and Caster plant oils on chromosomal aberration in Onion Allium cepa root. *Tikret J Agric.*, *12* (1), 197-200.

Premkumar, K., Kavitha, S., Santhiya, S. T., Ramesh, A. R. & Suwanteerangkul, J. (2004). Interactive effects of saffron with garlic and curcumin against cyclophosphamide induced genotoxicity in mice. Asia Pac. *J. Clin. Nutr.*, *13*(3), 292-294.

Raghuvanshi, S. S., Pathak, C. S. & Singh, A. K. (1978). Effect of preirradiation colchicine treatment on mutation spectrum of phaseolusaureusRoxb. *Cytologia*, *43*, 143-151.

Sandra, G., RROYO, M. E., Josefina, -Z, A., CORTES, ESLAVA., Armando, RAM~REZDOM NGUEZA., lfredo, GUTIERREZ- ABAD. & Rafael, V. (1990). Cytogenetic Effects Produced by the uriceHerbiceidsdiourn and Linorin in Viciafaba and human lymphocytes.

Surh, Y. –J. & Ferguson, L. R. (2003). Dietary and medicinal antimutagens and anticarcinogens: Molecular mechanisms and chemo preventive potential - Highlights of a symposium. *Mutat Res*, *523*, 1-8.

Turner, R., Etienne, N., Alonso, M. G., de Pascual-Teresa, S., Minihane, A. M., Weinberg, P. D. & Rimbach, G. (2005). Antioxidant and anti-atherogenic activities of olive oil phenolics. *Int J Vitam Nutri Res*, *75*, 1-70.

Tuck, K. L. & Hayball, P. J. (2002). Major phenolic compounds in olive oil: Metabolism and health effects. *J Nutr Biochem*, *13*, 636-644.

Upadhya, M. D., Tiwari, S. P. & Chandra, Ramesh. (1986). Localised chromosome breakage induced by sodium metabisulphide in Vicia faba. In: GK Manna and U Sinha (eds), *"Perspectives in Cytology and Genetics"*, n. *5*, 565-569.

Visioli, F. & Galli, C. (1998). Olive oil phenols and their potential effects on human health. *J Agric Food Chem*, *46*, 4292-4296.

Wahle, K. W., Caruso, D., Ochoa, J. J. & Quiles, J. L. (2004). *Olive oil and modulation of cell signaling in disease prevention Lipids*, *39*, 1223-1231.

Zaman, S. & Saleh, M. A. (2005). Mutagenic Effect of Ethylene Glycol on somatic cells of wheat (TRITICUM AESTIVUM L.) *J. Life Earth Science*, Vol. *1*, 43-49.

In: Biotechnology and Conservation … (2 Volume Set)
Editors: S. A. Farooq, R. Abed and S. Baqir
ISBN: 978-1-63117-869-6

Chapter 42

REDUCTION OF WATER STRESS EFFECT BY SEAWEED LIQUID EXTRACT USED AS FOLIAR SPRAY OF *TRITICUM DURUM* VARIETY KARIM

Salma Latique**, *Mounir Mansori, Halima Chernane and Mimoun El Kaoua

Department of Biology, Laboratory of Biotechnology, Valorization and Protection of Agro-Resources, Marrakech, Morocco

ABSTRACT

The present study was undertaken to evaluate the effectiveness of two species of Moroccan marine algae *Ulva rigida* and *Fucus spiralis* on plants development cultivated under water deficit conditions.

Seaweed liquid extracts were used as foliar spray at different concentration (0%, 12.5%, 25% and 50%) and under water stress conditions (moderated and severe water stress). Seaweed extracts of the two algae species influenced leave growth and photosynthetic pigments of wheat plant when compared to the control plants. Indeed a maximum value was recorded for 50% and 25% of *Ulva rigida* and *Fucus spiralis* extracts respectively.

The development and the growth of one economic interest plant species, Durum wheat (*Triticum durum* Desf.v. Karim), cultivated under water deficit, was investigated with or without extract foliar spray.

The results have shown that water stress leads to a leaf area and chlorophyll content reduction and a high proline accumulation. However, seaweed liquid extracts (25% Fucus and 50% Ulva) decreased the water deficit effect on leaf area of plants subjected to stress conditions and especially under moderate stress.

The reduction of water stress effect by seaweed extract was studied with the decrease of proline accumulation under water deficit. Thus, under severe water stress, the proline accumulation decreased from 16 μmol/100mg fwt in untreated plants to 11 and 7.4 μmol/100mg fwt in plants treated with 25% Fucus extracts and 50% Ulva respectively.

---

* l.salma86@gmail.com.

In conclusion we can note a seaweed extract importance in water stress tolerance. Thus, it would be very interesting to study the biochemical and physiological origin of this stress tolerance. The seaweed extract application is probably rich on osmolyte which improves water stress tolerance. We will envisage to study the accumulation of osmolytes in plants treated with seaweed extracts, in particular the glycinebetaine and some others amino acids.

Keywords: Chlorophylls, Fucus spiralis, Leaf area, Proline, Triticum durum, Ulva rigida, Water deficit

INTRODUCTION

Seaweeds from an integral part of marine coastal ecosystems are classified into three main groups based on their pigmentation (for example phaeophyta, rhodophyta and chlorophyta; or the brown, red and green algae, respectively).

Brown algae such as *Fucus spp., laminaria spp., Sagrassum spp.,* and *Turbinaria spp.* are used as biofertilizers in agriculture (Hong and others 2007).

One approach to reduce the negative environmental impact of agriculture is the use of biostimulants, which can enhance the effectiveness of conventional mineral fertilizers (Rathore et al., 2009).

Liquid extracts obtained from seaweeds (LSE/SLE) have recently gained importance as foliar sprays for several crops (Thivy,1961; Metha et al.,1967;Bokil et al.,1974) because the LSE components such as macro- and microelement nutrients, aminoacids, vitamins, cytokinins, auxins and abscisic acid (ABA)-like growth substances affect cellular metabolism in treated plants leading to enhanced growth and crop yield (Crouch and others 1992; crouch and van staden 1993a; Reitz and trumble 1996; Durand and others 2003; Stirk and others 2003; Ördög and others 2004).

The present study was undertaken to evaluate the effectiveness of two species of Moroccan marine algae *Ulva rigida* and *Fucus spiralis* on one of economic interest plant species, Durum wheat (*Triticum durum Desf v. Karim*) cultivated under deficit water conditions and treated or not by LSE foliar spray.

The seaweeds used in the present study were *Fucus spiralis* and *Ulva rigida* belonging to the classes Phaeophyceae and chlorophyceae, respectively. Fresh thalli of Fucus and Ulva were handpicked from the coastal area of sidi Bouzid near El jadida city Morocco, in March-April 2012.

The algal material was washed thoroughly with sea water to remove all the unwanted impurities, adhering sand particals and epiphytes (S. Sivasankari et al., 2006). Morphologically distinct thalli of algae were placed separately in polythene bags and were kept in an ice box and brought to the laboratory. Samples were washed thoroughly using tap water to remove the surface salt and then blotted to remove excess water (N. M. Pise and A. B. Sabale, 2010), then stored in a refrigerator at -20°C.

One Kg of seaweed was cut into small pieces and boiled separately with 1l of distilled water for an hour and filtered. The filtrate was taken as 100% concentration of the seaweed extract and from this different concentrations (25%, 50%) were prepared using distilled water (Bholse et al., 1975). As seaweed liquid fertilizers contained organic matter, the seaweed liquid fertilizers were refrigerated between 0 and 4°C until use. We tested the effect of SLE on the strength of plants under water stress (moderate stress and sever stress). This experiment lasted 45 days:

Without stress:

Lot 1: Plants watered with distilled water for 45 days with a field capacity maintained at 82%.

Lot 2: Plants watered by 25% *Fucus spiralis* for 45 days with a capacity field maintained at 82%.

Lot 3: Plants watered by 50% *Ulva lactuca* for 45 days with a field capacity maintained at 82%.

Moderate stress:

Lot 1: Plants watered with distilled water for 1 month with a field capacity maintained at 82% and for 15 days with a field capacity maintained at 18%.

Lot 2: Plants watered with 25% Fucus spiralis for 1 month with a field capacity maintained at 82% and for 15 days with a field capacity maintained at 18%.

Lot 3: Plants watered with 50% Ulva rigida for 1 month with a field capacity maintained at 82% and for 15 days with a field capacity maintained at 18%.

Severe stress:

Lot 1: Plants watered with distilled water for 1 month with a field capacity maintained at 82% and for 15 days with a field capacity maintained at 7%.

Lot 2: Plants watered with 25% Fucus spiralis for 1 month with a field capacity maintained at 82% and for 15 days with a field capacity maintained at 7%.

Lot 3: Plants watered with 50% Ulva rigida for 1 month with a field capacity maintained at 82% and for 15 days with a field capacity maintained at 7%.

Chlorophyll Assay

After pigments extraction from 0.5 g of leaf fragments (collected from 20 plants from at least 3 pots) with 15 ml of acetone/ distilled water (90:10 v/v), the chlorophyll content was determined in three independent aliquots and expressed on a fresh weight basis (mg. g^{-1}FW) following Arnon (1949). The pigments content was determined in three independent aliquots.

Chlorophyll reduction rate (in %) was calculated by applying the following formula:

$$(ChCPC - ChCPT)*100/ ChCPC$$

ChCPC = Chlorophyll content of plant control

ChCPT = Chlorophyll content of plant treated by SWE

Proline Assay

Proline was determined five times in 0.2 g of leaf fragments collected from five different control well-watered plants and water deprived ones. Proline was assayed according to the method of Sing et al., (1973) modified by Paquin and Lechasseur, (1979). The assay is based on the quantification of the reaction proline-ninhydrin by spectrophoteometric measurement. Proline reacts with ninhydrin by giving a colored complex. The intensity of the color is proportional to the quantity of proline.

The effect of seaweed liquid extract of *Fucus spiralis* and *Ulva rigida* on leaves area and pigment content is presented in Table 1. The use of seaweed extract increased the two growth parameters measured for wheat plants. The 25% of Fucus and 50% of Ulva were the better concentrations showing the significant effect of SLE. Indeed, seaweed extract application of 25F% Fucus and 50% Ulva of seaweed extract significantly increased the leaves area by 16,9% and 16,3% and Chlorophyll content by 38,1% and 45,4% respectively over the control.

Figure 1 illustrates the effect of the application of algal extracts on the leaf area of plants subjected or not to water stress conditions. On the two cases, we can note a significant difference in the leaf area of plants treated with algal extracts.

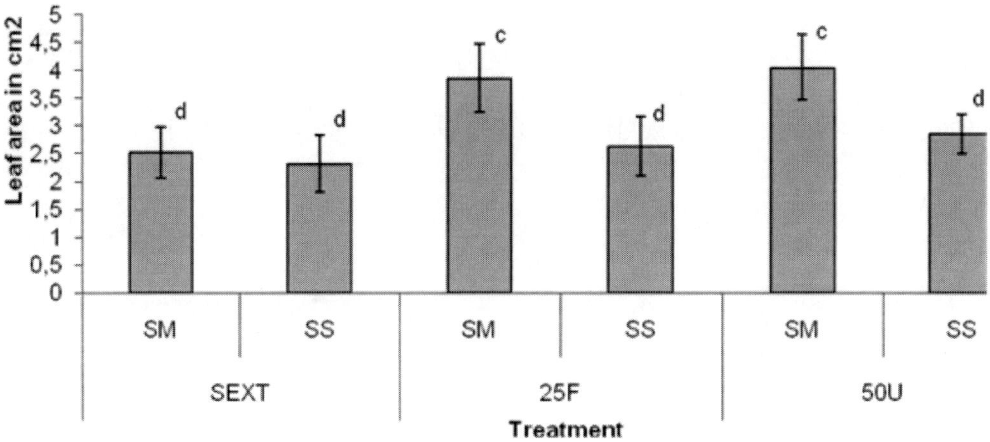

Figure 1. Effect of seaweed extract, *Fucus spiralis*, and *Ulva rigida* on the fourth leaf of *Triticum Durum* plants cultivated, during 15 days, under different conditions of water stress. T: Control, SM: moderate stress, SS: severe stress, SEXT: No extract; 25F: 25% *Fucus spiralis*, 50U: 50% *Ulva rigida*, the results represent the average of 10 measurements ± SD.

Table 1. Effect of seaweed liquid extract (SLE) on leaves area and chlorophyll content

Treatment	Leaves area (mm^2)	Chl a mg/g. fwt.	Chl b mg/g. fwt.
Control	511,00	1,96570	1,06070
25% *Fucus spiralis*	597,42	2,71522	1,04775
50% *Ulva rigida*	594,54	2,85930	1,15894

Rate reduction of the area recorded in stressed plants and untreated with the extracts algal is 50% compared to control plants. The application of the extract from 25% Fucus and 50% Ulva lowered the rate of reduction of leaf area (33%) for plants submitted especially under moderate stress. While the application algal extracts to plants under conditions of severe stress has no significant effect on the leaf area of wheat plants. Under optimal conditions, the application of algal extracts improves the surface leaf plants. Treatment with extracts of 25%

Fucus and 50% Ulva increases leaf area respectively 34.74% and 37.80% compared to control plants unstressed.

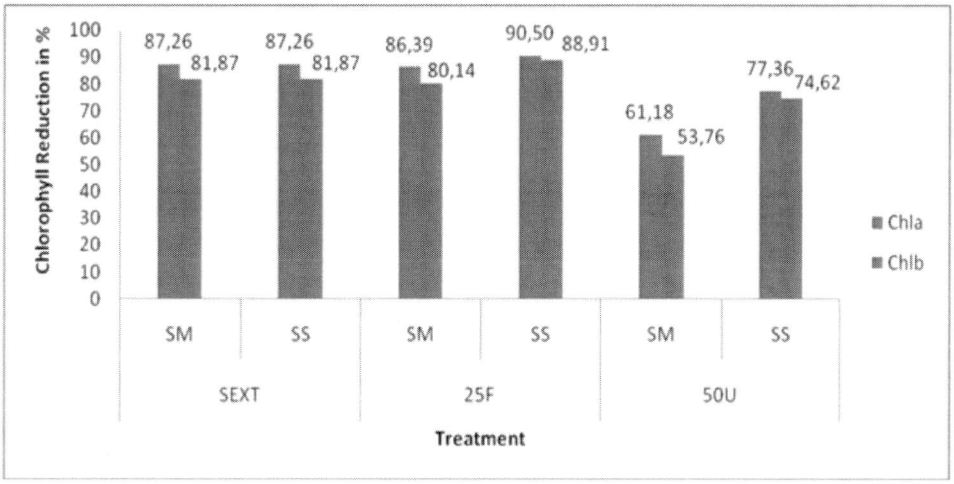

Chla: chlorophyll a; Chlb: chlorophyll b. SEXT: No extract; 25F: 25% *Fucus spiralis*, 50U: 50% *Ulva rigida,* SM: moderate stress, SS: severe stress, T: Control without stress. The results represent the mean ± SD of 5 replicates.

Figure 2. Chlorophyll content of durum wheat (Triticum durum Desf.) Variety Karim after 15 days of treatment under different water stress conditions with or without seaweed extracts treatment.

Figure 2 shows a decrease in chlorophyll content (chla and Chlb) in leaves of wheat under water stress. However, treating plants by 50% of *Ulva rigida* extracts reduces the effect of water stress levels on the chlorophyll (a and b) content. These results are consistent with those reported by other studies (Booth, 1966; Stephenson, 1966; Senn and Kingman, 1978), which shown that plant tolerance to abiotic stress and particularly water stress is significantly improved by applying extracts algal.

Proline Content

The treatment of plants with the algal extracts helps to reduce the rate of accumulation of proline.

In the case of wheat (Figure 3), treating plants with extracts of 25% Fucus and 50% Ulva favourably reduces the severity of both types of stress (moderate and severe). Thus, under conditions of severe stress rate of proline decreased by 16 µmol/100mg fwt in untreated control plants at 11 and 7.4 µmol/100mg fwt when the plants were treated respectively with 25% Fucus and 50 % Ulva extracts. We also noticed a significant decrease in the rate of accumulation of proline in plants treated with algal extracts and subjected under conditions of moderate stress.

Our findings coincide with those of earlier studies carried out on soybean (Rathore, Chaudhary et al., 2009) where there was an increase in vegetative growth by the application of seaweed extract. Similar results were also observed in Cajanus cajan (L.) Millsp. (Mohan et al., 1994) and Vigna sinensis L. (Sivasankari et al., 2006). The increased growth of these crops may be due to the presence of some growth promoting substances present in the seaweed extract (Mooney and Van Staden, 1986; Blunden, 1991). In addition, the growth enhancing potential of the seaweed extract might be attributed to the presence of macro and micronutrients. The test crop under present investigation was grown under optimal conditions, and then faced water stress conditions.

According to Debaeke et al., (1996), reducing the leaf area is caused by the effect water stress and how it affects the normal growth and development of plants leaves. The reduction of leaf area among the plant strategies allows to avoid water stress. The leaf curlling in some varieties of wheat can be considered an indicator of loss of turgor together with a character of the avoidance of dehydration. It leads to a reduction of 40 to 60% perspiration (Amokrane et al. 2002).

The decrease in leaf area of leaves and number of tillers are considered a response or adaptation to water scarcity (Blum, 1996). So our results are similar to those obtained by these authors. They are also consistent with those reported by other authors under stress conditions. Taiz and Zeiger, 1991 showed that the reduction leaf area of treated plants is due to algal extracts and phytohormones particularly cytokinins, a large amount of cytokinin actively involved in cell division and an increased fresh weight and leaf area. Other work (Featonby-Smith and van Staden, 1983; Mooney et al., 1985) reported that plants treated with algal extracts drought resistant.

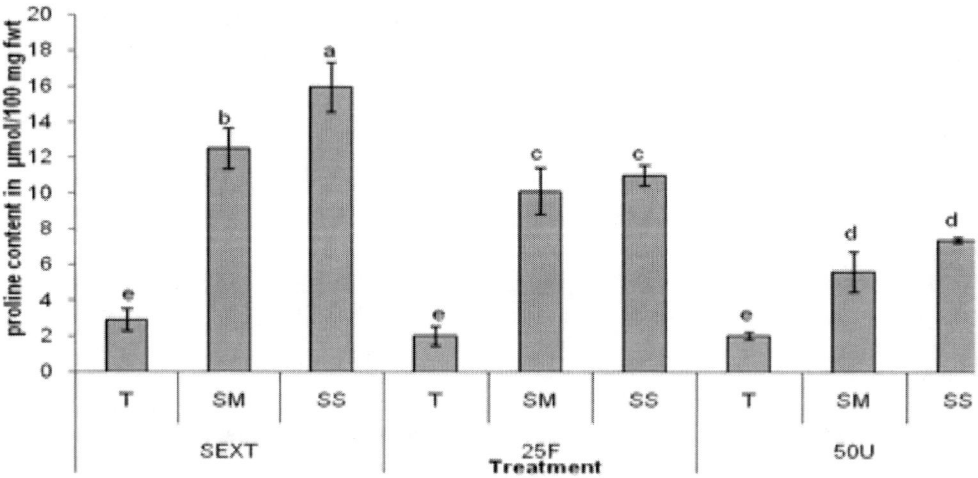

Figure 3. Proline content of wheat plants (Triticum durum Desf.) Variety Karim after 15 days under different water stress conditions with or without seaweed extract treatment
T: Control without stress SM: moderate stress, SS: severe stress SEXT: No extract; 25F: Fucus spiralis 25% 50U: Ulva rigida 50%. The results represent the average of 3 replicates ± SD.

According to Mourey-Bringuier, 1986, algae extracts increase the content through the phytohormones chlorophyll present in these extracts, and more specifically cytokinin which provides protection against chlorophyll degradation under the condition of water stress and the increase in nitrogen assimilation. While Blunden et al. (1997) reported that betaine present in extracts of algae are responsible for reduced degradation of chlorophyll.

Our results are in agreement with those obtained by some studies (Vendruscole et al., 2007) in plants subjected to stress conditions. The accumulation of proline is an indicator of stress. According to El Mourid, 1988 and Casals, 1996, maintaining the balance of the osmotic force is provided by the accumulation of proline, betaine and soluble sugars. These are among the mechanisms of adaptation to water stress in plants, after the fall of the water potential.

The application of algal extracts on wheat plants has reduced significantly the content of proline, this means a reduction of the severity of stress using seaweed extracts. According to Mooney et al., 1985, plants treated with algae resist to water stress through the phytohormones content.

CONCLUSION

In conclusion, we have observed in the present study that lower concentration seaweed extract Ulva rigida and Fucus spiralis, applied at wheat plants showed higher vegetative growth. The same result showed that seaweed extract treatment reduced the water deficit effect and enhanced the antioxidant activities which protected the plant against peroxidation imposed by water stress. Undeniably, present findings encourage the application of such seaweeds as natural fertilizer in agricultural sector. As perspective, we will look into opportunities for deepening the study of the biochemical and physiological effects that are the origin of the interest of liquid seaweed extracts on growth and plants tolerance to water stress.

ACKNOWLEDGMENTS

Part of this work was supported by the Project RS/2011/22 of CNRST. The CNRST project is funded by Ministry of Education in Higher Education, Staff Training and Scientific Research under thematic: Agriculture in difficult conditions.

REFERENCES

Arnon, D. I. (1949). Copper enzymes in isolated chloroplasts. Polyphenoloxidase in Beta Vulgaris. *Plant Physiol*, *24*, 1-15.

Bhosle, N. B., Untawale, A. G. & Dhargalker, V. K. (1975). Effect of seaweed extract on growth of Phaseolus vulgaris. *Indian J. Mar. Sci.*, *4*, 208-210.

Blunden, G. (1991). Agricultural uses of seaweeds and seaweed products in guiry, M.D, Blunden, G.(Eds.), Seaweed Resources in Europe: Uses and Potential. John wiley and sons, *Chichester*, 65-81.

Bokil, K. K, Metha, V. C. & Datar, D. S. (1974). Seaweeds as manure: II pot culture manorial experiments on wheat. *Phykos.*, *13* (1), 1-5.

Crouch, I. J. & Van Staden, J. (1992). Effect of seaweed concentrate on the establishment and yield of greenhouse tomato plants. *J Appl Phycol*, *4*, 291- 296.

Crouch, I. J. & Van staden, J. (1993a). Evidence for the presence of plant growth regulators in commercial seaweed products. *Plant Growth Regul*, *13*, 21-29.

Durand, N., Briand, X. & Meyer, C. (2003). The effect of marine bioactive substances (NPRO) and exogenous cytokinins on nitrate reductase activity in Arabidopsis thaliana. *Physiol Plant*, *119*, 489-493.

Hong, D. D., Hien, H. M. & Son, P. N. (2007). Seaweeds from Vietnam used for functional food, medicine and biofertilizer. *JAppl Phycol*, *19*, 817–826

Metha, V. C., Trivedi, B. S., Bokilk, K. k. & Narayama, M. R. (1967). Seaweed as manure, studies on nitrification. *Porc. semi. Sea. salt and Plants* (CSMCR). Bhavnager, 357-365.

Mooney, P. A. & Van Staden, J. (1986). Algae and cytokinins. *Journal of Plant Physiology*, *123*, 1-2.

Ordog, V., Stirk, W. A., van Staden, J., Novak, O. & Strnad, M. (2004). Endogenous cytokinins in the three genera of microalgae from the Chlorophyta. *J Phycol*, *40*, 88–95

Paquin, R. & Lechasseur, P. (1979). Observation sur une methode de dosage de la proline libre dans les extraits de plantes, can. *J. Bot.*, *57*, 1515-1854.

Pise, N. M. & Sabale, A. B. (2010). Effect of seaweed concentrates on the growth and biochemical constituents of trigonella foenum-graecum *l. Journal of Phytology*, 2010, *2*(4), 50–56.

Rathore, S. S., Chaudhary, D. R., Boricha, G. N., Ghosh, A. & Bhatt, B. P. (2009). Effect of seaweed extract on the growth, yield and nutrient uptake of soybean (Glycine max) under rainfed conditions. *South African Journal of botany*, *75* (2009), 351-355.

Reitz, S. R. & Trumble, J. T. (1996). Effects of cytokinin-containing seaweed extract on Phaseolus lunatus L.: Influence of nutrient availability and apex removal. *Bot Mar*, *39*, 33–38.

Sing, T. N., Paleg, L. G. & Aspinall, D. (1973). Stress metabolism. I. Nitrogen metabolism and growth in the barley plant. *J Biol. Sci.*, *26*, 45-56.

Sivasankari S., Venkatesalu, V., Anantharaj, M. & Chandrasekaran, M. (2005). Effect of seaweed extracts on the growth and biochemical constituents of vigna sinensis. *Bioresources Technolgy*, *97* (2006), 1745-1751.

Stirk, W. A., Novák, O., Strnad, M. & van Staden, J. (2003). Cytokinins in macroalgae. *Plant Growth Regul*, *41*, 13–24

Thivy, F. (1961). Seaweed manure for perfect soil and soiling fields. *Salt Res. Indust.*, *1*, 1-4

Vendruscolo, A. C. G., Schuster, I., Pileggi, M., Scapim, C. A., Molinari, H. B. C., Marur ad, C. J. & Vieira, L. G. C. (2007). Stress- induced synthesis of proline confers tolerance to water deficit in transgenic wheat. *J. Plant Physiol.*, *164* (10), 1367-1376.

In: Biotechnology and Conservation … (2 Volume Set) ISBN: 978-1-63117-869-6
Editors: S. A. Farooq, R. Abed and S. Baqir

Chapter 43

INDIGENOUS BACTERIAL CONSORTIA FROM OIL WELLS AND THEIR ROLE IN *IN-SITU* MICROBIAL ENHANCED OIL RECOVERY

***Ratiba A. Al-Maaini*[1], *Saif N. Al-Bahry*[1,*], *Abdulkader E. Elshafie*[1], *Yahya M. Al-Wahaibi*[2], *Ali S. Al-Bimani*[2], *Sanket J. Joshi*[1,3] *and Wafa J. Al-Alawi*[1]**

[1]College of Science, Department of Biology, Sultan Qaboos University, Oman
[2]College of Engineering, Petroleum and Chemical Engineering Department, Sultan Qaboos University, Oman
[3]Central Analytical and Applied Research Unit, College of Science, Department of Biology, Sultan Qaboos University, Oman

ABSTRACT

Worldwide decline in oil production rates and ever increasing demand for petroleum is gaining worldwide attention of the oil producing companies. Different enhanced oil recovery techniques like chemical and thermal recovery processes are employed worldwide to solve the problem. Those techniques currently in use are expensive and have certain environmental drawbacks. Microbial enhanced oil recovery (MEOR) is one of the most economical and efficient methods for extending the life of production wells in a declining reservoir. MEOR can be applied to the oil fields as either *in-situ* or *ex-situ* process, of which *in-situ* MEOR processes are comparatively economical. For *in-situ* applications, one has to identify the indigenous microbial populations of that oil field. After identifying the microbes, specifically designed nutrients can be injected to the oil-wells to improve oil recovery. Those nutrients will lead to growth of indigenous microbes, thus producing different bioproducts (like biosurfactants) and thus enhancing oil recovery. We have isolated and identified 58 species belonging to 33 different bacterial genera from Wafra oil wells and Suwaihat production water in Oman. All of the identified microbial genera were first records in Oman and most of them were found to be anaerobic, thermophilic and halophilic, which might be good candidates for MEOR.

* Corresponding author email address: snbahry@squ.edu.om.

Caminicella sporogenes is the second world record and first world record from an oil field. Those bacterial consortia were studied at laboratory scale for different bioproduct production and found to be quite promising for future applications.

Keywords: *In-situ* microbial enhanced oil recovery, Omani oil fields, bacterial consortia, *Caminicella sporogenes*

Oil Formation

Crude oil was produced from the remains of plants and animals that lived on earth millions of years ago. When organisms died and sank to the bottom, they were covered in mud, sand, and other mineral deposits. This rapid burial prevented immediate decay, which normally occurs if organisms remain exposed on the sea floor. The lack of oxygen in the sedimentary layers caused organisms to slowly decay into carbon-rich compounds. These compounds mixed with surrounding sediments, formed source rock; which is a type of fine-grained shale. As more layers were deposited on top of one another, pressure and heat acting on the source rock compressed the organic materials into crude oil. As time passed, oil traveled into rocks that have larger pores, such as limestone and sandstone which are called porous rocks (Freudenrich, 2005).

Oil Reservoir Features

Oil reservoirs are underground natural porous structures that may extend over many thousands of square meters and may be several hundred meters thick. However, the oil itself is found within natural pores of the permeable sandstone (silica, 'SiO_2'), limestone rocks (calcium carbonate, '$CaCO_3$') or double carbonate of calcium or magnesium, which make up the reservoir, in association with water, and sometimes gas. The flow of fluids within these reservoir rock pores is very complex because of the multiphase interactions of the oil, water, gas and the heterogeneous nature of the reservoir (Dawe and Grattoni, 1998). Planckaert (2005) reported that, any petroleum system is composed of three elements which are: a source rock, a reservoir rock and a cap rock. The source rock is a rock in which the crude oil is formed by thermal cracking of organic fossils buried in the sediment. The reservoir rock is porous and permeable rock that can store and eject fluid, like a sponge. It has a large capacity for storing hydrocarbon fluid. This capacity depends on the porosity (the ability of the rock to store fluid), and the permeability (the ability of rock to flow fluid). The cap rock is impermeable rock that has a plastic behaviour in which it can deform without breaking. It stops hydrocarbon migration since hydrocarbon and gas are less dense than water so they tend to migrate upwards. The petroleum system will not form if it misses one of these elements. Dawe and Grattoni (1998) reported that, a reservoir rock is an intricate network of solid matrix and void space with channels that are irregular in cross section with irregular pore wall. Pan et al., (2007) suggested that, the fluid in the reservoir is distributed into three layers according to their density in which, water forms the lower layer followed by oil layer and the top layer of gases. These layers vary in their size between reservoirs. Röling et al., (2003) reported that the oil saturated part of the reservoir (the "oil leg"), consists of oil that occupies

80% of the pore space with discontinuous water filling the rest. Below the oil leg is the water saturated part of the reservoir (the "water leg"), where 100% of the pore space would be continuously saturated with waters of variable salinity. Petroleum reservoirs are generally characterized by high temperatures, as temperature typically increases 2 to 3°C per 100 meters with depth of burial. Ramsay (1987) proposed that reservoir fluids are held under considerable pressure due to the combined weight of the formation rock and the reservoir fluids. In most cases the pressure increases with depth with a typical pressure gradient (2.5atm/m).

Compositions of Petroleum Fluids

Petroleum is a complex mixture of hydrocarbons and other inorganic compounds, such as vanadium and nickel. Petroleum recovered from different reservoirs varies widely in compositional and physical properties. Long chain hydrocarbons are substrates that support microbial growth; which could be both a target and a product of microbial metabolism (Van Hammeet al., 2003).

Petroleum oil composed of the following chemical constituents:

- Alkanes or paraffins: can be found in two structures, straight or normal alkancs and branched alkanes.
- Cyloalkanes or naphthalene: includes all saturated cyclic hydrocarbons and their derivatives such as cyclopentane and cyclohexane.
- Aromatic hydrocarbons: contains at least one benzene ring such as toluene and xylcnc (Planckaert, 2005).

The Mechanisms of Oil Productions

Moses (1987) reported that, there are 2 important factors in oil production. Firstly, there is a great variety among reservoirs with respect to parameters such as: rock porosity, permeability, homogeneity, temperature, pressure, salinity of brine water, viscosity and composition of the oil. Secondly, the longer a particular reservoir is operated the better it is understood. On the other hand, Westlake (1991) found that the porosity and permeability of the rock matrix and the temperature of the reservoir are the primary parameters that determine the susceptibility of a reservoir to microbial penetration. Other parameters such as pressure and salinity, which influence the types of bacteria found, can be considered as secondary controlling physio-chemical factors. Moreover, Planckaert (2005) suggested other parameters such as: wettability (the tendency of a fluid to spread on or adhere to a solid surface in the presence of another immiscible fluid), fluid saturation (the amount of fluid that is trapped in the reservoir rock pores) and capillary pressure (the pressure that controls the ability of a fluid to pass through a pore).

Conventional oil production occurs in three stages which are: primary, secondary and tertiary. During the stage of primary production the initial reservoir pressure is usually sufficient to drive oil from the rock pores to the surface of the well and mechanical pumps or lifting systems are used to supplement the energy within a reservoir. As the oil is withdrawn, the pressure tends to fall because the dissolved gas expands and the underlying aquifer doses not occupy the volume vacated by the oil quickly enough. This leads to a decrease in oil production by pumping processes and the amount of water increases in the production oil. Additional production becomes uneconomical, because only 15-20% of the original oil has been produced (Moses, 1987; Sheehy, 1991; Khire and Khan, 1994; Planckaert, 2005).

The next stage is secondary production in which it is essential to maintain reservoir pressure artificially to force the oil to the production wells. Typically, it involves the direct displacement of oil by injection of gas or water into the reservoir. An external fluid can be injected either into the oil bearing horizon or below into the aquifer. The most common fluid is water, which could be fresh water or sea water. Both contain adequate amount of inorganic materials and a natural mixed microbial population which may have harmful effects in the reservoir. The harmful effect may take place due to the stimulation of the growth of microorganisms such as sulfate reducing bacteria (SRB) and other sulfidogenic or fermentative bacteria which can be either introduced into the reservoir or stimulated by nutrients supplementary with injected fluid.

Vance and Thrasher (2005) suggested that, SRBs are nutritionally diverse and able to use electron donors and carbon sources present in petroleum reservoirs such as acetate, propionate, naphthenic acids, n-alkanes (C6-C20), hexadecane, benzoate, benzene, toluene, xylene and phenol. SRBs reduce sulfate in the injection water to sulfide, while oxidizing degradable organic (lactate) electron donor present in the oil reservoir. Hubert and Voordouw (2007) reported that the concentration of sulfate introduced depends on the source of the injection water, and it was found to be high with concentration of 30 mM when seawater is injected during offshore operations. Because large volumes of water are injected (typically 10, 000m^3/day), large amounts of biogenic sulfide can be co-produced with the oil and gas, up to 1,100Kg/day. Vance and Thrasher (2005) suggested that, one of the most significant economic impacts of hydrogen sulfide is reservoir souring which is characterized by an increasein the hydrogen sulfide (H_2S) concentration in production gas and water.

McInerney and Sublette (1997) documented that the consequences of souring are numerous which includes (i) corrosion in production and injection wells and other water handling equipments, (ii) aquatic toxicity and odour associated with high H_2S levels in the produced water, (iii) safety hazards due to H_2S in retention ponds or other open vessels, and (iv) increased costs to remove H_2S from natural gas before it enters the pipeline. Researchers have reported that, biofilms and precipitation of inorganic sulfide leads to the loss of the reservoir permeability due to plugging of the pores (Moses 1987; McInerney and Sublette, 1997; Gevertz et al., 2000).

McInerney et al. (2003) reported several alternative biological processes that can be used to control sulfide production. They reported that addition of nitrate to the brines helps denitrifying bacteria out-compete the SRBs for organic electron donors and the flow of electrons will shift from sulfate reduction to nitrate reduction. In addition, when nitrate levels are high, then denitrifiers will produce nitrous oxide, which will increase the oxidation/reduction potential of the environment and inhibit or wipe out the growth of the strictly anaerobic sulfate reducing bacteria. The second process involves the use of a sulfide resistant strain of *Thiobacillus denitrificans* that can oxidize the sulfide to sulfate by reducing nitrate to ammonium. The third process involves the addition of nitrate along with inhibitors of sulfate reducers (molybdate and nitrite) to not only inhibit sulfide production, but also to stimulate oil recovery. McInerney et al. (2007) reported that not all sulfides found in oil and gas facilities are from biological sources, but sulfides can also be formed thermogenically in the presence of sulfate and organic compounds.

Gevertz (1991) reported that conventional water flooding techniques used in secondary oil production recovers only 30-40% of oil while a large percentage of the remaining oil is still tapped in the reservoir pores. Khire and Khan (1994) reported that, the water flooding is practised until the ratio of oil produced to water injected yields more water than oil. They explained that water prefers to sweep in high permeability pore of the reservoir, leaving oil behind in the low permeability pore. At this point the processes become uneconomical to separate oil from the mixture (oil and water). In addition, Moses (1987) documented that at the end of secondary production, the distribution of the residual oil is very different from what it was in the virgin reservoir. It will be more unevenly distributed and the reservoir pressure will also fall. Some of the oil will be present as droplets trapped in pore and to force them out of the constructions requires the enlargement of contact surface area between oil and water. This in turn, requires more pressure than was available to overcome the high interfacial tension between the two phases (oil-water).

Zheng et al. (1998) suggested that the amount of residual oil after secondary recovery depends on three factors: oil in place at the start of water flooding; reservoir sweep efficiency; and microscopic displacement efficiency. Reservoir sweep efficiency is a key factor in determining the final residual oil after water flooding within economic limit. It is also dependent on the mobility ratio and reservoir heterogeneity. All of these issues require going for several tertiary recovery methods that are also known as Enhanced Oil Recovery (EOR) techniques. Sheehy (1991) reported that the objectives of EOR are: to increase the pressure differential between the reservoir and production well and to increase the oil mobility by reducing its viscosity, or decrease the interfacial tension between the displacing fluids and the oil. Current EOR methods can be grouped as thermal, chemical, miscible displacement and microbial.

Thermal Methods

Thermal methods are the most commonly used techniques and it includes steam injection and *in-situ* combustion. The principle of thermal methods is to decrease the oil viscosity by high temperature with the help of saturated steam injection into the oil wells.

Chemical Methods

Gabitto (2006) suggested that addition of chemicals to water will alter fluid properties or interfacial conditions that are more favourable for oil displacement. Planckaert (2005) reported that chemical processes improve the flood of water by injecting polymer, surfactant and alkaline agents. Moses (1987) and Ramsay (1987) reported that the purely injected water does not have the required pressure to push the oil from the reservoir pores, because the injected water is less viscous than the crude oil and therefore, it tends to form channels and by-pass much of oil. When polymer is injected with water, it increases the viscosity of water and therefore decreases the mobility ratio of water to oil (i. e., the water flows at the same velocity as the oil). Gabitto (2006) claimed that, polymer flooding using polyacrylamides or polysaccharides, is theoretically simple and inexpensive, and its commercial use is increasing despite the fact that it raises potential production by only small increments. On the other hand, Ramsay (1987) reported that polymer flooding does not displace the oil trapped by capillary forces. In addition, it is sensitive to shear forces, microbial attack, heat and the presence of electrolytes as they flow through the porous reservoir rock. It also adsorbs onto the rock surface and this eventually leads to reduced sweep efficiency. Water flooding with a surfactant has been found to displace the oil trapped by capillary forces. Brown et al. (1985) and Moses (1987) reported that surfactants reduces the interfacial tension between (oil- water and oil-rock) and displaces oil. Ramsay (1987) suggested that surfactants must reduce the interfacial tension between oil and water to a level as low as 10^{-4}mN/m. This reduces the capillary pressure sufficiently to allow the available pressure gradient to push the oil out of rock-pores, and the displaced oil droplets form an oil bank. This oil bank is pushed by a polymer which in turn is pushed by water-flooding to the production well. However, surfactants act as losing factors since they are easily adsorbed onto the reservoir rock. Furthermore, it is sensitive to high temperature and salinity greater than 0.3 mole/l. Brown et al. (1985) studied the effects of alkalinity by injection of sodium hydroxide in the flooded water to increase the pH values of the flood from pH 12 to 13. However, the mechanism that was responsible for improving oil recovery under these conditions was not well understood. Twenty years later, Planckaert (2005) reported that alkaline injection cause change in the wettability combined with lowering of the oil-water interfacial tension and allow the mobility of oil. McInerney et al. (2005) claimed that, although chemical flooding technologies (EOR) such as polymer, alkaline, surfactant flooding displaces tertiary oil efficiently, it is considered a high risk process due to:

- The chemical solutions containing surfactants or polymers are expensive.
- Under poor mobility control severe chemical losses are observed, due to adsorption.
- The performance of these processes is complicated.

Miscible Displacement Methods

Miscible displacement involves the injection of solvents such as: alcohol, refined hydrocarbons, condensed hydrocarbon gases, liquefied petroleum gases and carbon dioxide which dissolves in the reservoir oil. The injected solvent reduces the capillary forces which causes oil retention, whereas carbon dioxide reduces the oil viscosity (Ramsay, 1987; Sheehy, 1991). Miscible gas injection method allows the reduction or suppression of interfacial tension between the oil and the injected fluid and thus increases the oil mobility (Planckaert, 2005).

Microbial Enhanced Oil Recovery (MEOR)

Oil is a critical energy source that drives industrialization and sustained economic development of the world (Youssef et al., 2007), but when this is threatened by potentially high crude oil prices due to increased demand, the need for new oil production technologies are required (Caneba and Axland, 2002). McInerney et al. (2005) documented that current oil production technologies recover only about one-third to one-half of the oil originally present in an oil reservoir. It is widely known that there are still about 377 billion barrels of oil trapped within mineral rock matrices globally (Caneba and Axland, 2002). This huge oil potential is the target of Microbial Enhanced Oil Recovery (Al-Sulaimani et al., 2011).

MEOR is a process that utilizes the ability of microorganisms (either indigenous or injected in the reservoir), to produce useful products such as gases (carbon dioxide, methane and hydrogen), biosurfactants and biopolymers to improve oil production (Bhupathiraju et al., 1991; Youssef et al., 2007; Singh et al., 2007). The first attempt of MEOR field application was more than 60 years ago. Since then several techniques of using microbes and their products for EOR processes have been reported (Hitzman, 1991).

Microbial byproducts are produced from inexpensive fermentable sugar feedstock, whereas many of the chemicals proposed for EOR are derived from crude oil itself and therefore are expensive. In addition, the yield of microbial by-products can be increased by manipulating growth conditions and employing genetic techniques for strain improvement, thus making the products environment friendly, cheap and economic (Brown et al., 1985; CSIRO, 2007).

The concept of MEOR was established in 1926 by Beckman when he reported the action of bacteria on mineral oil. He proposed that bacterial enzymes could be used in oil recovery (Gabitto, 2006). The actual work started with methods developed by ZoBell in 1940, when he used solid state mineral media (sand, crushed limestone, glass beads etc.) saturated with crude oil and inoculated with bacteria suspended in suitable aqueous medium. The oil was gradually released from the solid medium and floated to the surface of aqueous medium. Later modifications in the study of MEOR involved actual cores of petroleum bearing reservoir rocks (Khire and Khan, 1994).

Al Blehed et al. (1996) reported that amongst different microorganisms, only bacteria are proposed as promising candidates for MEOR due to their small size, exponential growth and metabolic products such as acids, gases, biosurfactants and biopolymers. In addition, they can tolerate harsh environments such as high formation water salinity, high pressure and high temperature. On the other hand, molds, yeasts, algae and protozoa are not suitable either due to their morphological characteristics or due to the growth conditions present in the reservoirs (Singh et al., 2007).

Sheehy (1991) found that, ZoBell's research highlighted the compounds used in chemical and miscible EOR processes to improve waterflood efficiency, and the microbial fermentation products are all regular and predictable products of microbial metabolism. Gevertz, (1991) documented that microorganisms can be grown in bioreactors and their products such as emulsifiers, solvents, and biopolymers can be injected into reservoir to improve oil recovery. Indigenous and/or injected cells can be stimulated by the injection of nutrients (generally carbohydrates), to grow and produce products *in situ*. Alternately, bacteria capable of using components present in crude oil can also be injected to limit the need for nutrient supplementation.

Furthermore, the ability of bacteria to survive, grow and produce useful metabolites under reservoir conditions is a prerequisite for successful MEOR process. A multitude of extreme environments may be present in oil reservoirs with conditions such as anaerobiosis, high pressure, high temperature and high salinity, a broad spectrum of pH values, low nutrient concentrations, and the presence of potentially toxic compounds. The criterion for selection of suitable conditions includes temperature below 75 °C, near neutral pH, and salt concentrations of less than 10 percent. Fujiwara et al. (2004) and Al-Sulaimani et al. (2011) also suggested other reservoir conditions which are essential for MEOR processes such as: permeability (above 50-75 mD), porosity (above 20%) and low oil viscosity (above 5-50 mPa.s). These environmental conditions have led to the suggestion that halotolerant-thermophilic bacteria capable of anaerobic growth are good candidates for MEOR processes.

Brown et al. (1985) suggested that selection or recombinant techniques might be useful for improving the resistance of microorganisms to such adverse reservoir conditions. Illias et al. (2001) proposed that microorganisms associated with hypersaline ecosystems could be assigned into four specific groups: fermentative bacteria, homoacetogenic bacteria, sulfate reducing bacteria and methanogenic bacteria. Moreover, Bhupathiraju et al. (1991) reported the key requirement to initiate MEOR process is to understand and study the factors that affect the microbial activities in the oil reservoir which helps in selection of compatible organisms and nutrients. Jang et al. (1983) also reported that the success of MEOR processes depends on the selection of the candidate reservoir; proper choice of potential bacterial species; the viability of bacteria under reservoir conditions; the amount of metabolites generated and their effects on releasing residual oil; the spreading of bacteria and nutrient in the reservoir; and other economic factors. In addition, Brown et al. (1985) reported that two extreme physical environments affect the living organisms: they have a direct influence on equilibrium and rate of reaction which affects the cell density and integrity and activities of the enzymes. Second, they disrupt macromolecules, cellular structures such as membranes, nuclear bodies and ribosomes.

Bhupathiraju et al. (1991) stated that, the determination of reservoir fluid composition is the most critical challenge for MEOR process because it represents the most important part of the reservoir in which nutrients essential for growth and metabolism will be transported. The composition of fluid in the reservoir can influence the microorganisms' ability to metabolize certain carbon sources, the displacement of oil by water flooding, and property of clay. Furthermore, McInerney et al. (2007) reported that oil and brine often contain sufficient amounts of nutrients like carbon, phosphorus, nitrogen and sometimes sulphur, which are enough to support microbial growth and metabolism. Some microbial processes involve the use of hydrocarbons present in the formation.

However, nutrient limitations can easily be determined by supplementing the brine with different combinations of nutrients to verify which are required to support growth and the useful activities of the microbes. Gevertz (1991) reported that the indigenous bacteria present in an oil reservoir are a significant element in MEOR process for at least two reasons. First, they can be stimulated directly by the addition of nutrients. Second, they may interact with exogenous cells injected into formation. Therefore, it is essential to understand the microflora of the reservoir so as to predict how this population will respond to perturbation by nutrients and other bacteria.

Sepahy et al. (2005) reported that the investigation of indigenous microorganisms in the reservoir was initiated in 1920s, since then different types of microorganism have been detected in reservoirs. Microbes live in a large variety conditions, aerobic or anaerobic, acidic or basic (pH 2 to 9), low temperature or high temperature (from 0 °C to nearly100 °C), and pressure up to 20000 psi. Various studies have shown that microbes are found in many oil well environments where they adapt, grow and propagate.

MECHANISMS OF *IN-SITU* MEOR

There are several approaches by which microorganism or nutrients injected into oil reservoirs can improve oil recovery efficiency. Some methods depend on the growth of bacteria, and some rely on the by-products of bacterial growth to improve oil recovery (Ag-West-Biotech Inc., 1999; McInerney et al., 2005). Brown et al. (1985) suggested two major and distinct modes in which microbes may be employed in EOR processes. First, products can be synthesized in conventional fermentors by aerobic or anaerobic microorganisms. The byproducts can be extracted, purified and injected into a reservoir as a chemical flood. Second, anaerobic microorganisms may be injected into a reservoir with or without an organic nutrient source to stimulate the production of compounds useful for EOR within the reservoir rock.

Fujiwara et al. (2004) suggested three variations of the MEOR processes. The first process involves the injection of both microbes and nutrients. The microbes to be used should be selected for their ability to make products such as: gases, biosurfactants and biopolymers. The second variation involves the injection of nutrients that can be utilized by indigenous microbes within the reservoir. These nutrients may consist of molasses and/or plant fertilizer. The third process uses microbes which can utilize hydrocarbons. It has been reported that the mechanisms by which microbial consortia work can be quite complex and may involve multiple biochemical processes (Almeida et al., 2004). During microbial growth, several metabolites and microbial by-products are being formed in the oil well as a result of complex biochemical pathways, which are involved in the process of oil recovery. For example, acid, gas and solvents are produced by the process of catabolism, where bacteria utilize nutrients to obtain energy in the form of ATP. The produced ATP is used in anabolism to make new cells (biomass), biopolymers and biosurfactant. Some bacteria produce enzymes that degrade oil, such as heavy oil, into simpler compounds (Sepahy et al., 2005).

The type of by-products depends on the type and composition of the nutrients consumed by microbes from their environment. The basic components of nutrients consist of water, energy source, carbon source, electron acceptor, essential minerals, nitrogen source, and growth factor (Almeida et al., 2004). Some abiotic factors must be taken into account very carefully in case of MEOR: salinity, temperature, pH and macronutrients (such as sulfur, magnesium, iron, phosphate, nitrate and oxygen). These factors can influence microbial growth and the efficiency of MEOR (Sepahy et al., 2005). A better understanding of the mechanisms of oil recovery is extremely useful for selection of appropriate bacterial strains and design optimal operation procedures (Sarkar et al., 1991). Once bacteria are established in the well and reservoir areas, they start to metabolize nutrients, hydrocarbons, and excrete by-products (gases, biopolymers, solvent and biosurfactant) *in situ,* which changes the chemical and physical properties of the oil as well as the reservoir environment, thus improving oil production (Almeida et al., 2004). Furthermore, Khire and Khan (1994) proposed various MEOR mechanisms which could acts individually or interacting with each other to improve oil production: biosurfactants (normally lipid compounds), biopolymers (polysaccharides), gas generation and solvents production.

Biosurfactants

Poor oil recovery in oil-producing wells may be due to either low permeability of some reservoirs or high viscosity of the crude oil resulting in poor mobility (Singh et al., 2007). One of the major factors that limit oil production is the entrapment of oil in small pores by capillary force (McInerney and Sublette, 1997). The reduction in the interfacial forces between the oil and brine by biosurfactant can release the entrapped oil. Biosurfactants are amphipathic molecules (with both hydrophilic and hydrophobic parts) produced by a variety of microorganisms which reduces the surface and interfacial tension by accumulating at the interface of immiscible fluids and increase the solubility and mobility of hydrophobic or insoluble organic compounds (Amiriyan et al., 2004; Singh et al., 2007). Biosurfactants either adheres to cell surface or are excreted extracellular in the growth medium (Tabatabaee et al., 2005).

Al Araji et al. (2007) found that extracellular surfactants are involved in cell adhesion, emulsification, diffusion, flocculation and cell aggregation. It has been reported that, biosurfactants play an importantrole in increasing bioavailability of hydrophobic molecules ; swarming motility of microorganisms; participate in cellular physiological processes of cell-signalling and differentiation; and biofilm formation (Singh et al., 2007). Accordingly, Gautam and Tyagi (2006) categorized biosurfactants into the following groups: (i) glycolipids which includes rhamnolipids, trehalose lipids and sophorolipids; (ii) lipopeptide and lipoprotein; (iii) fatty acid, phospholipids and neutral lipids; (iv) polymeric biosurfactants; and (v) particulate biosurfactants.

Kosaric (1992) and Tabatabaee et al., (2005), reported that chemically-synthesized surfactants have been used in the oil industry to aid the clean-up of oil spills, as well as to enhance oil recovery from oil reservoirs. These compounds are not biodegradable and can be toxic to the environment. Brown et al.,(1985) studied the effect of oil-synthesized surfactant sulfonates and they found that these compounds have an optimal activity over a narrow range of temperatures and salinities. Thus in the reservoir, effectiveness of these compounds are questionable. Joshi et al. (2008) documented that the unique properties of biosurfactants allow their use and possible replacement of chemically synthesized surfactants in a great number of industrial operations.

Accordingly, Kosaric (1992) and Tabatabaee et al. (2005) stated that biosurfactants have special advantages over their commercially manufactured counterparts because of their: (a) biodegradability; (b) generally low toxicity; (c) effectiveness at extreme temperature, pH and salinity; (d) biocompatability and digestibility, which allows their application in cosmetics, pharmaceuticals and as functional food additives; (e) availability of raw materials as biosurfactants can be produced from cheap raw materials that are available in large quantities.

Kosaric (1992) reported that biosurfactants are proposed as an alternative to chemical surfactants foe EOR applications in petroleum industries. For production of the biosurfactant (*ex situ*), the selected culture is grown in bioreactors, the biosurfactant recovered or simply concentrated in the broth and then pumped into the reservoir. This external production of the biosurfactant must be in large quantity and produced under optimum fermentation conditions. The use of cheap and waste substrates for this production would be particularly advantageous. Agro-industrial wastes with a high content of carbohydrates, or lipids meet the requirement for use as substrate for biosurfactants production (Joshi et al., 2008). Several renewable substrates from various sources especially from industrial wastes have been extensively studied for microbial production at an experimental scale such as: frying oil waste, oil refinery wastes, molasses and cassava water. Sugarcane molasses is commonly used as a substrate, date molasses is also reported as a suitable carbon source for biosurfactant production (Al-Bahry et al., 2013a).

Kosaric (1992) suggested that a logical and more attractive approach is to produce the biosurfactant in the reservoir itself (*in situ*) by providing nutrients to indigenous microbial population in the reservoir or to a seed culture which produces surfactant mixed with the medium and pumped into the well. Youssef et al. (2005) suggested that, candidate microorganisms for enhanced oil recovery should produce biosurfactants at low oxygen concentrations, slightly elevated temperatures, and high salt concentrations, since these are the conditions encountered in majority of oil reservoirs. Javaheri et al.,(1985) reported that various microorganisms produce biosurfactants which have been found to significantly reduce the interfacial tension between oil and brine. Al Araji et al. (2007) found that, the type, quality and quantity of biosurfactant produced are influenced by the nature of the carbon substrate, the concentration of nitrogen, phosphorous, magnesium, ferric, and manganese ions in the medium and the culture conditions (such as pH, temperature, agitation and dilution rate in continues culture).

Some biosurfactants are reported to be active and functional at high temperature even after autoclaving at 120 °C for 15 min (Al-Sulaimani et al., 2011). Furthermore, salt concentration has been found to affect biosurfactant production due to its effect on the cellular activity of the bacteria. However, some biosurfactants are still efficient even at 10% salt concentrations (Gautam and Tyagi 2006; Al Araji et al., 2007).

Biopolymers

One of the most important factors that affect the sweep efficiency of waterflood or an enhanced oil recovery process is the variation in the permeability of the reservoir regions (Pfiffner et al., 1986). Brown et al., (1985) reported that because of the least resistance to flow, injected water sweeps preferentially through the high permeability zones and fractures in heterogeneous reservoirs. Thus, oil in the low permeability zones is bypassed resulting in a progressively increasing water to oil ratio at the producing well. Biopolymers are exopolysaccharide produced by microorganisms outside the cell as slime layer or capsules. These compounds are useful components in MEOR process (Illias et al., 2001). McInerney and Sublette (1997) reported that *in situ* growth and biopolymer production by microorganisms reduce the permeability variation and block water channels. Brown et al., (1985) suggested that, both of these factors improve the sweep efficiency of a recovery process and thus increases the oil production. Biopolymers can be added to a waterflood to increase the viscosity of the water which results in a reduction of the mobility ratio between oil and water which lead to a decrease in permeability to the injected fluid. McInerney et al., (2005) documented that numerous microorganisms produces biopolymers which can be used as a mobility control agents and emulsifiers. A variety of aerobic *Bacillus* species were found to produce extracellular polysaccharide at temperature up to 50 °C and 10% NaCl salinity, which reduced the permeability of Berea sandstone core and lead to recovery from 8% to 35% of residual oil.

Acid, Gas and Solvent

McInerney et al. (2005) reported that several fermentative anaerobes produce large amounts of acids (such as acetic acid, latic acid, and butyric acids), solvents (such as ethanol, acetone, butanol, and 2,3- butanediol) and gases (such as CO_2, N_2, CH_4 and H_2). The production of organic acids by microorganisms leads to the dissolution of the rock matrix which alters and increases the pore size and enhance the permeability (McInerney et al., 1999). The carbon dioxide produced from carbohydrate feed stocks (such as molasses), swells the oil and reduces its viscosity and increase oil mobility. On the other hand, H_2 produced by an aerobic bacterium is utilized by different bacteria such as: methanogens to reduce CO_2 to CH_4, homoacetogens to reduce CO_2 to acetic acid; sulfate reducing bacteria to reduce sulfate to sulfide; and denitrifiers to reduce nitrate to nitrogen gas.

It has been reported that, the most important gas-producing bacteria are *Clostridium*, *Desulfovibrio*, *Pseudomonas* and certain methanogenes (Behlülgil and Mehmetoglu, 2002). Such microorganisms were isolated and enumerated by media containing utilizable organic carbon source (McInerney and Sublette, 1997). Bryant (1987) isolated an aerobic *Clostridium* species which produce gases, acids, alcohol and biosurfactant. McInerney et al. (2005) found that the injection of cell-free culture fluids containing the acids and solvents did not recover residual oil but *in situ* CO_2 production was the main factor for oil displacement mechanism. *In situ* CO_2 lead to formation of calcite ($CaCO_3$) saturated fluid, and reduced the reservoir permeability which eventually altered the flow of the displacement fluid.

Biomass and Selective Plugging

Many of the carbonate reservoirs with excessive natural fractures have early water cut and becomes matured quite fast. These fractures cause early breakthrough in water flooding operations leaving majority of residual oil un-swept. Microbial biomass is one of the proposed solutions to seal these fractures and hence improve oil recovery. The mechanism of the microbial biomass in MEOR involves selective plugging of high permeability zones where the microbial cells will grow at the larger pore throats restricting the undesirable water flow through them (Al-Hattali et al., 2012). This will force the water-flood to divert its path through the smaller pores and hence displacing the un-swept oil and increasing the oil recovery. Several lab and field tests are reported to test the feasibility of this mechanism. Jenneman (1984) reported that the addition of nutrients (carbon, nitrogen and phosphate) into Berea sandstone resulted in permeability reduction of 60-80%. The advantage of this process is that it does not interfere with the normal waterflood operation. It is also environmentally friendly and is considered as the cheapest MEOR mechanism.

Laboratory Scale *In Situ* MEOR Work at Sultan Qaboos University, Oman

Fifty eight species belonging to 33different bacterial genera were isolated from Wafra oil wells and Suwaihat production water in Oman (Al-Bahry et al., 2013b). All of the identified microbial genera were first records in Oman and most of them were found to be anaerobic, thermophilic and halophilic, which might be good candidates for MEOR. *Caminicella sporogenes* is the second world record and first report from an oil field. Those bacterial consortia were studied at laboratory scale for different bioproduct production and found to be quite promising for future applications. No work has been carried out on the microflora of the Omani oil fields and no information is available on either indigenous or exogenous microorganisms of Oman oil wells. In addition, no research work has been conducted on MEOR in Oman. Therefore, isolation and identification of potentially useful microbes in oil biotechnology is so important for conservation of such microbes in this part of the world.

ACKNOWLEDGMENTS

Authors acknowledge the funding by Petroleum Development of Oman and His Majesty research fund, SQU during the course of this research.

REFERENCES

Al Araji L., Abd. Rahman, RNZ R., Basri M, Salleh A. 2007. Minireview: Microbial surfactant. *Asia Pacific Journal of molecular Biology and biotechnology.*15: 99-105.

Al-Bahry SN, Al-Wahaibi Y, Elshafie AK, Al-Bemani AS, Joshi SJ, Al-Makhmari H, Al-Sulaimani H. 2013a. Biosurfactant production by *Bacillus subtilis* B20 using date molasses and its application in enhanced oil recovery. *International Biodeterioration & Biodegradation*. 81: 141-146.

Al-Bahry S, Elshafie A, Al-Wahaibi Y, Al-Bemani A, Joshi SJ, Al-Maaini RA, Al-Alawi W, Sugai Y, Al-Mandhari M. 2013b. Microbial consortia in Oman oil fields: A possible use in enhanced oil recovery. *Journal of Microbiology and Biotechnology*. 23: 106-117.

Al Blehed MS, Sayyouh MH, Shoeb HA, Awwad AM, Desouky SM, Hemeida AM. 1996. Laboratory investigation of Microbial Enhanced oil Recovery. *Journal of King Soud University*. 8: 165-186.

Al-Hattali R, Al-Sulaimani H, Al-Wahaibi Y, Al-Bahry SN, Elshafie A, Al-Bemani A, Joshi S. 2012. Improving sweep efficiency in fractured carbonate reservoirs by microbial biomass, *SPE* 154679.

Almeida PF, Moreira RS, Almeida RCC, Guimarães AK, Carvahlo AS, Quintella C, Esperidia MCA, Taft CA. 2004. Selection and application of microorganisims to improve oil recovery. *Engineering in Life Sciences*. 4:319-25.

Al-Sulaimani H, Joshi S, Al-Wahaibi Y, Al-Bahry SN, Elshafie A, Al-Bemani A. 2011. Microbial biotechnology for enhancing oil recovery: current developments and future prospects. *Biotechnology, Bioinformatics and Bioengineering Journal*. 1:147-158.

Al-Sulaimani H, Al-Wahaibi Y, Al-Bahry SN, Elshafie A, Al-Bemani A, Joshi S, Zaragi S. (2011). Optimization and partial characterization of biosurfactants produced by Bacillus species and their potential for ex-situ enhanced oil recovery. *Society of Petroleum Engineering Journal*. 16, 672-682.

Amiriyan A, Assadi MM, Saggadian VA, Noohi A. 2004. Bioemulsan production by Iranian oil reservoirs microorganisms. *Iranian Journal of Environmental Health Science & Engineering*. 1: 28-35.

Ag-West Biotech Inc. 1999. Bacteria biotech: Developing environmentalalternatives in the oil industry (Part II). *Agbiotechnology Information Publication for Schools*. 43: 1-2.

Behlülgil K, Mehmetoglu MT. 2002. Bacteria for improvement of oil recovery: A laboratory study. *Energy Sources*. 24:413-421.

Bento FM, Camargo FO, Okeke BC, Frankenberger WT. 2005. Diversity of biosurfactant producing microorganisms isolated from soils contaminated with diesel oil. *Microbiological Research*. 160: 249-255.

Bhupathiraju VK, Sharma PK, McInerney MJ, Knapp RM, Fowler K, Jenkins W. 1991. Isolation and characterization of novel halophilic anaerobic bacteria from oil field brines. *Developments in Petroleum Science*. 31:131-143.

Brown MJ, Moses V, Robinson JP, Sapringham DG. 1985. Microbial enhanced oil recovery progress and prospect. *Critical Reviews in Biotechnology*. 3:159-179.

Bryant RS. 1987. Potential uses of microorganisms in petroleum recovery technology. *Proceedings of the Oklahoma Academy of Science*. 67:97-104.

Caneba GT, Axland J. 2002. Vinyl acetate-acrylic acid compolymer for enhanced oil recovery. *Journal of Minerals & Materials Characterization & Engineering*. 1: 97-109.

CSIRO. 2007. An integrated solution to the key challenges in microbial enhanced oil recovery (MEOR). *Australian Science*. 1-9.

Dawe RA, Grattoni CA. 1998. The visualization of the pore-scale physics of the hydrocarbon recovery from reservoirs. *First Break*.16:371-386.

Freudenrich C. 2005. How oil drilling work. Cald Energine.com. http://www.caldenergine.com/How%20Oil%20Drilling%20Works.pdf. Acecced March 2013.

Fujiwara K, Sugai Y, Yazawa N, Ohno K, Hong CX, Enomoto H. 2004. Biotechnological approach for development of microbial enhanced oil recovery technique. 405-445. *In*: Vazquez-Duhalt R, Quintero-Ramirez R. *Petroleum biotechnology developments and perspectives*. ELSEVIER. Mexico.

Gabitto JF. 2006. Combined microbial surfactant-polymer system for improved oil mobility and conformance control. *SPE* 102328.

Gautam KK, Tyagi VK. 2006. Microbial surfactants: A review. *Journal of Oleo Science*. 55: 155-166.

Gevertz D, Telang AJ, Voordouw G, Jenneman G. 2000. Isolation and characterization of strains CVO and FWKOB, two novel nitrate-reducing, sulfide-oxidizing bacteria isolated from oil field brine. *Applied and Environmental Microbiology*. 66:2491-2501.

Gevertz D. 1991. Isolation and characterization of anaerobic halophilic bacteria from oil reservoir brine.115-129. *In*: DonaldsonEC. Microbial Enhancement of oil recovery. *Recent advances*. New York. USA

Hitzman DO. 1991. Microbial enhanced oil recovery-the time is now. *Developments in Petroleum Science*. 31:11–20.

Hubert C, Voordouw G. 2007. Oil field souring control by nitrate-reducing *Sulfurospirillum* spp. That outcompete sulfate-reducing bacteria for organic electron donors. 73:2644-2652.

Illias RM, Ooi SW, Idris AK, Wan Abdul Rahman WA. 2001. Isolation and characterization of halotolerant aerobic bacteria from oil reservoir. *Jurnal Teknologi*. 35: 1-10.

Jang LK, Chang PW, Findley JE, Yen TF. 1983. Selection of bacteria with favorable transport properties through porous rock for the application of microbial enhanced oil recovery. *Applied and Environmental Microbiology*. 46: 1066-1072.

Javaheri M, Jenneman GE, McInerny MJ, Knapp RM. 1985. Anaerobic production of a biosurfactant by *Bacillus licheniformis* JF-2. *Applied and Environmental Microbiology*. 50: 698-700.

Jenneman GE, Knapp RM, McInerney MJ, Menzie DE, Revus DE. 1984. Experimental studies of in-situ microbial enhanced oil recovery. *Society of Petroleum Engineering Journal*. 24: 33-37.

Joshi S, Bharucha C, Jha S, Yadav S, Nerurkar A, Desai AJ. 2008. Biosurfactant production using molasses and whey under thermophilic conditions. *Bioresource Technology*. 99: 195-199.

Khire JM, Khan MI. 1994. Microbially enhanced oil recovery (MEOR). Part 1. Importantce and mechanism of MEOR. *Enzyme and Microbial Technology*.16:170-172.

Kosaric N.1992. Biosurfactants in industry. *Pure and Applied Chemistry*. 64:1731-1737.

McInerney MJ, Sublette KL. 1997. Petroleum microbiology: biofouling, souring, and improved oil recovery, p. 600-607. *In*: HurstCJ, KnudsenGR, McInerneyMJ, Stetzenbach LD, and Walter MV. (ed.). *Manual of environmental microbiology. American Society forMicrobiology*. Washington DC. USA.

McInerney MJ, Nagle DP, Knapp RM. 2005. Microbially enhanced oil recovery: past, present and future. 215-237. *In:* Ollivier B, Magot M. *Petroleum Microbiology*. ASM Press, Washington DC. USA.

McInerney MJ, Voordouw G, Jenneman GE, Sublette KL. 2007. Oil field microbiology. 898-911. *In*: Hurst JC, Garland JL, Mills AL, Crawford RL, Lipson DA, and Stetzenbach LD. *Mannual of Environmental Microbiology*. ASM Press, Washington, USA.

McInerney MJ, Knapp RM, Chisholm JL, Bhupathiraju VK, Coates JD. 1999. Use of indigenous or injected microorganisms for enhanced oil recovery. *Microbial Ecology of Oil Fields*. University of Oklahoma, Norman, Oklahoma, 73019, USA.

McInerney MJ, Madugalya SO, Mouttaki H, Folmsbee M, Knapp R, Nagle D, Jackson BE, Staudt M, Frye W. 2003. Development of more effective biosurfactant for enhanced oil recovery. National Energy Technology Laboratory National Petroleum Technology Office, Oklahoma.

Moses V. 1987. Microbes and oil recovery. *Microbiological Sciences*. 4: 306-309.

Pan C, TanY, Feng J, Jin G, Zhang Y, Sheng G, Fu J. 2007. Mixing and biodegradation of hydrocarbons in the Daerqi oilfield, Baiyinchagan Depression, northern China. *Organic Geochemistry*.38:1479-1500.

Pfiffner SM, McInerny MJ, Jenneman GE, Knapp RM. 1986. Isolation of halotolerant, thermotolerant, facultative polymer producing bacteria and characterization of exopolymer. *Applied and Environmental Microbiology*. 51: 1224-1229.

Planckaert M. 2005. *Oil reservoirs and oil production*. 3-19. *In:* Ollivier B, Magot M. ASM Press, Washington, USA.

Ramsay JA. 1987. Microbial products in enhances oil recovery. *PhD. Thesis.* McGill University Montreal, Canada.

Röling WFM, Head IM, Larter SR. 2003. The microbiology of hydrocarbon degradation in subsurface petroleum reservoirs: perspectives and prospects. *Research in Microbiology*. 154:321-328.

Sarkar K. 1991. Cmpositional numerical simulation of MEOR processes. 331-343. *In*: Donaldson EC. Microbial Enhancement of oil recovery. *Recent advances*. New York.

Sepahy AA, Assadi MM, Saggadian V, Noohi A. 2005. Production of biosurfactant from Iranian oil fields by isolated Bacilli. *International Journal of Environmental Science and Technology*. 1: 287-293.

Sheey AJ. 1991. Microbial physiology and enhanced oil recovery. 37-44. *In*: Donaldson EC. Microbial Enhancement of oil recovery. *Recent advances*. New York.

Singh A, Van Hamme JD, Ward OP. 2007. Surfactants in microbiology and biotechnology: Part 2. Application aspects. *Biotechnology Advances*. 25: 99-121.

Tabatabaee A, Assadi MM, Noohi AA, Sajadian VA. 2005. Isolation of biosurfactant producing bacteria from oil reservoirs. *Iranian Journal of Environmental Health Science & Engineering*.2: 6-12.

Vance I, Thrasher DR. 2005. Resevoir souring: mechanisms and prevention. 123-142. *In:* Ollivier B, Magot M. *Petroleum Microbiology*. ASM Press, Washington, USA.

Van Hamme JD, Singh A, Ward OP. 2003. Recent advances in petroleum microbiology. *American Society for Microbiology*. 67:503-549.

Westlake DWS. 1991. Microbialecology of corrosion and reservoir souring. 257-276. In: Donaldson EC. Microbial Enhancement of oil recovery. *Recent advances*. New York.

Youssef NH, Duncan KE, McInerney MJ. 2005. Importance of 3-Hydroxy fatty acid composition of lipopeptides for biosurfactant activity. *Applied and Environmental* Microbiology. 71: 7690-7695.

Youssef N, Simpson DR, Duncan KE, McInerney MJ, Folmsbee M, Fincher T, Knapp RM. 2007. In situ biosurfactant production by *Bacillus* strains injected into a limestone petroleum reservoir. *Applied and Environmental Microbiology*. 73: 1239-1247.

Zheng C, Gall BL, French TR, Bryant RS. 1998. Modeling and simulation of polymer-enhanced oil production. *System and Information Technology Review Journal*. 6:55-67.

In: Biotechnology and Conservation … (2 Volume Set) ISBN: 978-1-63117-869-6
Editors: S. A. Farooq, R. Abed and S. Baqir

Chapter 44

BIOSURFACTANT MEDIATED MICROBIAL ENHANCED OIL RECOVERY AND POTENTIAL FOR APPLICATION IN SOME OMANI OIL FIELDS

***Yahya M. Al-Wahaibi*[1]*, Saif N. Al-Bahry*[2]*, Abdulkader E. Elshafie*[2]*, Ali S. Al-Bemani*[1]*and Sanket J. Joshi*[2]**

[1]College of Engineering, Petroleum and chemical Engineering Department, Sultan Qaboos University, Oman

[2]College of Science, Department of Biology, Sultan Qaboos University, Oman

ABSTRACT

Surfactants are one of the most versatile compounds widely used in industries, household, and pharmaceutical applications. But recent awareness for environmental pollution and recalcitrant nature of chemical surfactants encouraged using surfactants of biological origin – biosurfactants. Various types of biosurfactants are produced by different microorganisms, of which lipopeptides (surfactins and lichenysins) produced by spore-forming *Bacilli* species are quite industrially promising. These lipopeptide biosurfactants can be used for bioremediation, oil recovery, textile industry, medicinal use as well as in food industries. Petroleum industries have the largest applications for biosurfactants, where it can be used for microbial enhanced oil recovery. We have isolated and identified microorganisms indigenous as wells as exogenous to oil wells from Oman, which produce biosurfactants and have the potential to enhance oil recovery in declining oil reservoirs across the sultanate. The identification and preservation of those microbes is very important for the biodiversity conservation of the species belonging to this region.

Keywords: Biosurfactants, Spore-forming bacilli, Surfactins, Lichenysins, Microbial enhanced oil recovery

SURFACTANTS

Characteristic Features of Surfactants

A surfactant (*Surface active* agent) when present in low concentration in a system, has the property of adsorbing onto the surfaces or interfaces of the system and altering to a marked degree the surface or interfacial free energies of those surfaces (or interfaces). It consists of both hydrophilic and hydrophobic groups. The hydrophobic group is usually a long-chain hydrocarbon residue, and less often a halogenated or oxygenated hydrocarbon or siloxane chain; the hydrophilic group is an anionic or highly polar group. Depending on the nature of the hydrophilic group, surfactants are classified as: Anionic (The surface active portion of the molecule bears a negative charge, for e.g., $RC{=}O{-}O{-}Na^+$ (soap), $RC_6H_4SO_3^- Na^+$ (alkylbenzene sulphonate); Cationic (The surface active portion bears a positive charge, for e.g., $RNH_3^+Cl^-$ (salt of a long chain amine), $RN\ (CH_3)_3^+Cl^-$ (quaternary ammonium chloride); Zwitterionic (Both positive and negative charges may be present in the surface active portion, for e.g., $R^+NH_2CH_2COO^-$ (long chain amino acid), $R^+N\ (CH_3)_2CH_2CH_2SO_3^-$ (sulphobetaine) or Nonionic (The surface active portion bears no apparent ionic charge, for e.g., $RCOOCH_2CHOHCH_2OH$ (monoglyceride of long-chain fatty acid), $RC_6H_4\ (OC_2H_4)_xOH$ (polyoxyethylenated alkylphenol) (Schramm et al., 2003).

Industrial Uses of Chemical Surfactants

Surfactants are among the most versatile products of the chemical industry appearing in diverse products such as the motor oils, the pharmaceuticals, the detergents, the drilling muds used in processing for petroleum, and the flotation agents used in benefication of ores. Last decade has seen the extension of surfactant application to high-technology areas as electronic printing, magnetic recording, biotechnology, microelectronics and viral research. World surfactant production in 1999 was 9 million tonnes, with a value at around 10 billion Euros (Clapes and Infante, 2002; Kosaric, 2001). The applications are very wide in a variety of industries with the largest industrial sectors that uses surfactants is household (61%) followed by industrial processes (25%), personal care (8%) and specialty cleaning (6%) (Greek, 1991).

BIOSURFACTANTS

Biosurfactants can be defined as biologically produced surfactants, produced by microorganisms, plants and animals. Microbial compounds that exhibit pronounced surface and emulsifying activities are classified as biosurfactants. Biosurfactants comprise a wide range of chemical structures, such as glycolipids, lipopeptides, polysaccharide–protein complexes, phospholipids, fatty acids and neutral lipids (Desai and Banat, 1997).

Mechanisms of Interaction

Biosurfactants are microbial amphiphilic and polyphilic polymers that tend to interact with the phase boundary between two phases in a heterogeneous system, defined as the interface. Biosurfactants are amphipathic molecules with both hydrophilic and hydrophobic moieties that partition preferentially at the interface between fluid phases that have different degrees of polarity and hydrogen bonding, such as oil and water or air and water interfaces. These biomolecules may also decrease interfacial tension (Banat et al., 2000). Such characteristics confer excellent detergency, emulsifying, foaming and dispersing traits, which makes biosurfactants versatile biological-process chemicals.

Classification and Microbial Origin

Unlike chemically synthesized surfactants, which are classified according to the nature of their polar grouping, biosurfactants are categorized mainly by their chemical composition and their microbial origin. In general, their structure includes a hydrophilic moiety consisting of amino acids or peptides, anions or cations; mono-, di-, or polysaccharides; and a hydrophobic moiety consisting of unsaturated or saturated, fatty acids. Accordingly, the major classes of biosurfactants include glycolipids, lipopeptides/lipoproteins, phospholipids and fatty acids, polymeric surfactants, and particulate surfactants. The biosurfactant-producing microbes are distributed among a wide range of genera. The detailed description can be found in Desai and Desai (1993), Rosenberg (1986), Kosaric et al. (1983), and Banat (1995).

Chemical Surfactants v/s Biosurfactants

Current worldwide surfactant markets are around \$9.4 billion per annum, and their demand is expected to increase at a rate of 35% toward the end of the century. Almost all surfactants currently in use, are chemically derived from petroleum; however, interest in microbial surfactants has been steadily increasing in recent years due to their diversity, environment friendly nature, the possibility of their production through fermentation, and their potential applications in the environmental protection, crude oil recovery, health care, and food-processing industries. Hester (2001) estimated that biosurfactants could capture 10% of the surfactant market by the year 2010 with sales of \$US200 million. According to Transparency Market Research report, New York (2011), the global biosurfactants market was worth USD 1,735.5 million in 2011 and is expected to reach USD 2,210.5 million in 2018, growing at a CAGR of 3.5% from 2011 to 2018 (http://www. transparency marketresearch.com/biosurfactants-market.html). Besides these future business potentials, there are several advantages of biosurfactants if compared to their chemically synthesized counterparts (Nitschke, and Costa, 2007): Surface and interfacial activity; temperature, pH and ionic strength tolerance, biodegradability, low toxicity, biocompatibility and specificity, and ability to be synthesized from renewable feedstock.

Current Scenario of Biosurfactants in Production Economy

Despite possessing many commercially attractive properties and clear advantages compared with their synthetic counterparts, the production of microbial surfactants on a commercial scale has not been realized because of their low yields and high production costs (at least 50 times more expensive, depending on the biosurfactant and its purity) – for such purposes, it is necessary that they are produced and recovered profitably on a large scale. The production economy of every microbial metabolite is governed by three basic factors: (i) initial raw material costs; (ii) availability of suitable economic production and recovery procedures; and (iii) the product yield of the producer microorganisms. Thus, in light of the economic constraints associated with biosurfactant production, three basic strategies are adopted worldwide to make this process cost-competitive: (i) the use of cheaper waste substrates to lower the initial raw material costs involved in the process (Al-Bahry et al., 2012; Joshi et al., 2008a; Makkar and Cameotra, 2011); (ii) development of efficient bioprocesses, including optimization of the culture conditions and cost-effective separation processes for maximum biosurfactant production and recovery (Joshi et al., 2007; 2008b; Joshi and Desai, 2013; Sen and Swaminathan, 2004); and (iii) development and use of overproducing mutant or recombinant strains for enhanced biosurfactant yields (Mukherjee et al., 2006). The first two strategies have been explored to a greater extent and have been reported to be effective in substantially increasing the production of biosurfactants. However, the third approach, using recombinant hyper-producing strains, has still not been properly tested, despite the fact that the hyper-producers have been reported to increase yields several fold. This area of biosurfactant research is still in its infancy (Mukherjee et al., 2006; Deleu and Paquot, 2004).

Applications of Biosurfactants

The largest possible market for unpurified crude biosurfactant is the oil industry, both for petroleum production and for incorporation into oil formulations (Van Dyke et al., 1991). An area of considerable potential for biosurfactant application is in the field of microbial enhanced oil recovery (MEOR). Other applications related to the oil industries include oil spill bioremediation/dispersion, both inland and at sea, removal/mobilization of oil sludge from storage tanks and enhanced oil recovery (Georgiou et al., 1992; Khire and Khan 1994a, b; Banat et al., 1991). The second largest market for biosurfactants is emulsion polymerization for paints, paper coatings and industrial coatings. Layman (1985) described other uses of surfactants including asphalt, cement, textile and fiber manufacturing, metal treatment, mining, water treatment, coal slurry defoamers and as wood preservatives. Biosurfactants are also used for metal-contaminated soil remediation. Frazer (2000) has shown the use of rhamnolipids for the removal of lead and cadmium from soil. Surfactants are also used in food and cosmetic industries, industrial cleaning of products and in agricultural chemicals to dilute and disperse fertilizers and pesticides and to enhance penetration of active compounds into plants (Kosaric et al., 1987). Ishigami (1997) has speculated various potential applications of biosurfactants in bioengineering including their use as cryopreservatives, protein solubilizers, enzyme stabilizers, DNA isolating agents, preservatives for cut flowers, growth enhancers for plants, recovery enhancers for wounds and swelling and for the control of biomembranous functions (Banat et al., 2000). The use and potential commercial application of biosurfactants in the medical field has increased during the past decade. Their antibacterial, antifungal and antiviral activities make them relevant molecules for applications in combating many diseases and as therapeutic agents. Surfactants are used in drug formulation and ~ 40 % of drugs are lipophilic hence they are poorly water-soluble which result in slow onset of action, high variability in bioavailability, inconsistent performance and undesirable side effects. Presently chemical surfactants are utilized for this purpose but biosurfactants can very well replace this synthetic surfactant in the near future. Recently biosurfactant are also used in genetic engineering. A major problem of gene therapy in treatment of diseases is the lack of efficient introduction of DNA into the cell nucleus. Biosurfactant can be used successfully to transport a piece of DNA to the target site e.g., gene delivery into mammalian cells using cationic liposomes including mannosylerythritol lipids (Kitamoto et al., 2002).

WHY SPORE-FORMING BACILLI ISOLATES FOR BIOSURFACTANT PRODUCTION?

Spore Forming *Bacillus* Species (Especially *B. licheniformis* and *B. subtilis*)

Bacillus species have been major workhorse industrial microorganisms with roles in applied microbiology, which date back more than a thousand years, since the production of natto by solid-state fermentation of soybeans using *Bacillus subtilis* (natto) was first practiced in Japan. These roles have continually expanded and evolved over the past century. The development of strains and production strategies has recently been influenced or facilitated by the application of molecular biology techniques to strain development. *Bacillus* species are attractive industrial organisms for a variety of reasons, including their high growth rates leading to short fermentation cycle times, their capacity to secrete proteins into the extracellular medium, and the GRAS (generally regarded as safe) status with the Food and Drug Administration (USFDA) for species, such as *B. subtilis* and *B. licheniformis*. In addition, much is now known about the biochemistry, physiology, and genetics of *B. subtilis* and other species, which facilitates further development and greater exploitation of these organisms in industrial processes (Schallmey et al., 2004). *Bacillus licheniformis* is a Gram-positive, spore-forming bacterium widely distributed as a saprophytic organism in the environment. This species is a close relative of *Bacillus subtilis*, an organism that is second only to *Escherichia coli* in the level of detail at which it has been studied. Unlike most other bacilli, which are predominantly aerobic, *B. licheniformis* is a facultative anaerobe, which may allow it to grow in additional ecological niches. Certain *B. licheniformis* isolates are capable of denitrification; the relevance of this characteristic to environmental denitrification may be small, however, as the species generally persists in soil as endospores (Alexander, 1977). Joshi et al., (2013) have reported occurrence of biosurfactant producing *Bacillus* spp., from diverse habitats, like garages, hot-water springs, oil wells etc. Recent taxonomic studies indicate that *B. licheniformis* is closely related to *B. subtilis* and *Bacillus amyloliquefaciens* on the basis of comparisons of 16S rDNA and 16S-23S internal transcribed spacer (ITS) nucleotide sequences (Xu et al., 2003). Lapidus et al. (2002), recently constructed a physical map of the *B. licheniformis* chromosome using a PCR approach, and established a number of regions of co-linearity where gene content and organization were conserved with the *B. subtilis* genome.

Economic Importance of *B. licheniformis* and *B. subtilis*

There are numerous commercial and agricultural uses for *B. licheniformis* and its extracellular products. The species has been used for decades in the manufacture of industrial enzymes including several proteases, α-amylase, penicillinase, pentosanase, cycloglucosyltransferase, β-mannanase and several pectinolytic enzymes. The alkaline serine proteases (subtilisins) that are manufactured with *B. licheniformis* and also with *B. pumilus* and *B. subtilis* have a primary application as additives to household detergents. Their annual output has been estimated to about 500 metric tons of pure enzyme protein (Schallmey et al., 2004). Amylases from *B. licheniformis* are deployed for the hydrolysis of starch, desizing of textiles and sizing of paper (Erickson, 1976). Specific *B. licheniformis* strains are also used to produce peptide antibiotics such as bacitracin and proticin in addition to a number of specialty chemicals such as citric acid, inosine, inosinic acid and poly- γ-glutamic acid (Gherna et al., 1989). Some *B. licheniformis* isolates can mitigate the effects of fungal pathogens on maize, grasses and vegetable crops (Neyra et al., 1996). As an endospore-forming bacterium, the ability of the organism to survive under unfavorable environmental conditions may enhance its potential as a natural biocontrol agent.

Biosurfactants Produced by *B. subtilis* and *B. licheniformis*

Many strains of *Bacillus* are known to produce lipopeptides with remarkable surface-active properties (Fiechter, 1992). The most prominent of these powerful lipopeptide biosurfactant is surfactin from *Bacillus subtilis*. Surfactin is an acylated cyclic heptapeptide that reduces the surface tension of water from 72 to 27mNm^{-1} even at concentration below 0.05% and showed some antibacterial and antifungal activities (Arima et al., 1968). Some *B. subtilis* strains are also known to produce other structurally related lipoheptapeptides like iturin and bacillomycin, or the lipodecapeptides fengycin and plipastatin (Konz et al., 1999).

In addition to *B. subtilis*, several strains of *B. licheniformis* have been described as producing the lipopeptide lichenysin (Horowitz et al., 1990; Jenny et al., 1991; Lin et al., 1994; McInerney et al., 1990; Yakimov et al., 1995). Lichenysins can be grouped under the general sequence L-Glx–L-Leu–D-Leu–L-Val–L-Asx–D-Leu–LIle/ Leu/Val. The first amino acid is connected to a β-hydroxy fatty acid, and the carboxy-terminal amino acid forms a lactone ring to the β-OH group of the lipophilic part of the molecule. Lichenysins were also reported to be synthesized during growth under aerobic and anaerobic conditions. The structural elucidation of the compounds revealed slight differences, depending on the producer strain. Various distributions of branched and linear fatty acid moieties of diverse lengths and amino acid variations in three defined positions have been identified (Lichenysins A, B, C, D; Surfactant 86; Surfactin; Iturins A, C; Bacillomycin L, D, F) (Konz et al., 1999).

World Oil Reserves and Scenario in Oman

Crude oil, natural gas and coal form the main constituent of fossil fuel energy. These energy resources were formed millions of years ago as the consequence of microbial decomposition of organic remains trapped in the sediments subjected to harsh environmental conditions under the subsurface of the earth. Amongst those energy sources, crude oil accounts for almost 33.2% of the world's total energy demand, being widely used for routine day to day activities in life (Hughes and Rudolph, 2011). It has been noticed that the potential oil resources are unevenly distributed over the globe, of which most of the favorable locations of oil deposits are confined to former Soviet Union, Middle East and North Africa and Asia Pacific and some other regions of the world. The OPEC members hold around 75% of world crude oil reserves. The countries with the largest oil reserves are from the Middle East (Saudi Arabia, Iran, Iraq, Kuwait, United Arab Emirates -UAE), accounting for almost 60%, and others in order, Venezuela, Russia, Libya, Kazakhstan and Nigeria (Thakur and Rajput, 2011). Like most of the Middle-Eastern countries, Oman is also dependent upon its oil sector for the majority of its export revenues and government spending and annual budget. Oman possesses the largest oil reserves of any non-OPEC country in the Middle East and significant reserves of natural gas, of which it is a leading exporter regionally. Oman has total proven reserves of 5.5 billion barrels of oil as of January 2012, which is quite low as compared to neighboring countries (Oman - Analysis – U.S. Energy Information Administration (EIA), www.eia.gov) like Saudi Arabia (264.5 billion barrels), Iran (151 billion barrels), Iraq (143 billion barrels), Kuwait (101.5 billion barrels), UAE (97.8 billion barrels), and Qatar (25.4 billion barrels).

Microbial Enhanced Oil Recovery (MEOR)

Crude Oil Recovery and Enhanced Oil Recovery (EOR)

There is an enormous importance of energy in determining the social stability and economic viability of a nation. Crude oil, the world's major source of energy, is essentially a mixture of compounds formed from hydrogen and carbon, although it may contain traces of nitrogen, sulfur, nickel and vanadium (Paul and Ladd, 1981). The oil recovery process starts with locating the oil field and drilling oil from the rock formation (reservoir). Petroleum is initially forced out or recovered, naturally, as a result of pressure in the reservoir and this is referred to as primary recovering phase (Berger and Anderson, 1992). This natural recovery is followed by enhanced oil recovery (EOR) process also known as Improved Oil Recovery (IOR). When the EOR process involves the injection of water (or gas) it is called secondary recovery, and if it involves the use of heat, chemical, miscible displacement and microbial products, etc., it is known as tertiary recovery (Moses, 1987; Donaldson et al., 1985). Enhanced oil recovery (EOR) processes rely upon the use of chemical or thermal energy to recover crude oil that is trapped in pores of reservoir rock after primary and secondary (water flood) crude oil production has ceased. The residual crude oil in reservoirs makes up about 67% of the total petroleum reserves, indicating the relative inefficiency of primary and secondary production (Bryant, 1987). The concept of microbial enhanced oil recovery

(MEOR) was first proposed nearly 80 years ago but received only limited attention until the early 1980's (Stosur, 1991). MEOR technology has advanced from laboratory- based studies in the early 1980's to field applications in the 1990's. The first suggestion for using MEOR was made as early as 1926, when Beckman reported the action of bacteria on mineral oil (Hitzman, 1983). As a result of his investigation, he proposed that bacterial enzymes could be used in oil recovery. However, little work was done until ZoBell (1947) started a series of systematic laboratory investigations in the 1940's in connection with the American Petroleum Institute. MEOR offers major advantages over conventional EOR in that lower capital and chemical/energy costs are required (Sarkar et al., 1989).

Microbial Enhanced Oil Recovery (MEOR)

MEOR is an important tertiary recovery technology utilizing microorganisms and/or their metabolic end products for recovery of residual oil. It is generally accepted that approximately 30% of the oil present in a reservoir can be recovered using current EOR technology (Singer & Finnerty, 1984). Poor oil recovery in existing producing wells may be due to several factors. The main factor is the low permeability of some reservoirs or the high viscosity of the oil which results in poor mobility. High interfacial tensions between the water and oil may also result in high capillary forces retaining the oil in the reservoir rock (Bubela, 1987). Since most of the oil remains in the reservoir following primary and secondary recovery techniques, interest has evolved in tertiary recovery techniques. Techniques involving the use of chemical or physical processes such as pressurization, water flooding or steaming, however, are generally not applicable to most oil reservoirs. The use of chemical surfactants for cleaning-up oil reservoirs is an unfavorable practice that is hazardous, costly and will leave undesirable residues which are difficult to dispose of without adversely affecting the environment (Banat, 1995). An area of considerable potential for biosurfactant application is MEOR. Enhanced oil recovery methods were devised to recover oil remaining in reservoirs after primary and secondary recovery procedures. In MEOR, microorganisms in reservoirs are stimulated to produce polymers and surfactants, Biosurfactants aid MEOR by lowering interfacial tension at the oil-rock interface. This reduces capillary forces that prevent oil from moving through rock pores. Biosurfactants can also aid in oil emulsification, and assist in the detachment of oil films from rocks (Van Dyke et al., 1991).

Why MEOR?

Microbial processes have several unique advantages that may result in the development of economically attractive technologies. Microbial processes neither consume large amounts of energy as the thermal processes, nor depend on the price of crude oil as many chemical processes do. Because microbial growth can occur at exponential rates, it should be possible to produce large amounts of useful products rapidly from inexpensive and renewable resources. Microbial processes have the potential to be more cost effective than conventional IOR and EOR processes. Portwood (1995) reviewed 322 MEOR projects and estimated that incremental oil can be produced for less than one USD per barrel. Even though more advantageous, microbial IOR and EOR technologies are not widely implemented because of the inconsistent technical performance, the lack of understanding of the mechanism of oil recovery, and lack of quantitative measures of microbial performance. The majority of the MEOR work leading to field trials has been completed in the past few years. The technology has advanced from a laboratory-based evaluation of microbial processes to field applications internationally. MEOR has been recognized as a potentially cost-effective method particularly for stripper wells.

Strategies and Factors Affecting MEOR Process

The appropriate remedy for any given oil reservoir will vary and be based on present conditions. Temperature, pressure, pH, porosity, salinity, geologic make-up of the reservoir, available nutrients and the presence of indigenous flora must all be taken into consideration. It is estimated, based on criteria developed by the National Institute for Petroleum & Energy Research, that 27% of the oil reservoirs in the major oil-producing states in the USA may be suitable for MEOR (Bryant, 1991). It has also been estimated that 40% of the oil-producing carbonate reservoirs in the USA may also be suitable for MEOR (Tanner et al., 1991). Poor oil recovery in oil-producing wells may be due to either low permeability of some reservoirs or high viscosity of the crude oil, resulting in poor mobility.

Mechanisms of MEOR

Microorganisms produce a variety of products that are potentially useful for oil recovery (biosurfactants, biopolymers, bio-solvents, bio-acids etc.). The mechanisms for MEOR remain essentially unchanged from those originally proposed by Zobell (1947). Although these mechanisms are often discussed independently, it is likely that microbial processes act synergistically to effect oil recovery. Their relative importance depends on conditions that limit oil production within a given reservoir, the strains of microorganism involved, and the protocols used for injection-volume of nutrients and microbial inoculum (McInerney and Sublette, 1997). The different mechanisms by which microorganisms could enhance oil recovery are:

1. Biotransformation: Microbial conversion of crude oil to remove paraffinic and asphaltic fractions; reduce the carbon-chain length of heavy oil and breaking down long chain compounds into less viscous, smaller molecular weight hydrocarbons or by the production of emulsifiers, thus changing the viscosity of hydrocarbons (McInerney et al., 2004).
2. Bio-Acid production: Production of acid because of microbial metabolic activity leads to the dissolution of the rock matrix and increase in pore size and permeability. This will be particularly important in limestone or carbonaceous sandstone formations. Acid reacts with minerals, especially carbonates and loosens clay particles and other organic deposits, and thus oil can be removed by this method from shale and tar sands also.

3. Bio-Gas production: Production gases by bacteria (e.g., CO_2) leads to decreased oil viscosity, repressurization of the reservoir, and swelling of the oil phase as gas is dissolved in the liquid.
4. Bio-Solvent production: Bacteria are known to produce a wide variety of solvents (aldehydes, ketones and alcohols), which act directly to reduce oil viscosity or reduce wettability (of rock surfaces), or indirectly as co-surfactant to decrease interfacial tension. These mechanisms help to improve oil recovery as well as help high viscosity oil flow through rocks by reducing the viscosity.
5. Bio-Polymer production: Many bacteria are known to produce different biopolymers (Xanthan, scleroglucan, γ-Polyglutamic acid), which could be beneficial for IOR. Biopolymers increase the viscosity of the aqueous phase and improve the mobility ratio and sweep efficiency of the crude oil from reservoirs.
6. Microbial Permeability Profile Modifications (MPPM) and Selective plugging: Selective plugging and MPPM by microbial growth helps to block the dominant flow channels or reduce permeability variation within a reservoir thus improving sweep efficiency. Different microorganisms form biofilms on solid-surfaces and cause selective or non-selective plugging, emulsification through adherence to hydrocarbons, modification of solid surface (e.g., wetting and alteration of rock surfaces), degradation and alteration of crude oil, reduction of oil viscosity, and various other effects.
7. Bio-Surfactant production: Biosurfactants reduces ST and IFT and improve microscopic displacement efficiency. This ability is helpful in desorbing oil from rock surfaces, placing it in to water phase surrounding the rock (by inducing oil in water or water in oil emulsions) and mobilizing it so that it may be removed along with water component.

Biosurfactants and EOR

The presence of different types of microorganisms with varying growth properties and metabolite production will have different effects on the reservoir environment. Thus, it is important to consider all aspects of MEOR when trying to influence oil production by one mechanism, such as the use of biosurfactants. Among microorganisms, only bacteria are considered promising candidates for MEOR. Molds, yeasts, algae and protozoa are not suitable either due to their morphological characteristics and/or to the growth conditions present in reservoirs. There are several strategies involving the use of biosurfactants in MEOR (Shennan & Levi, 1987):

1. The first involves injection of biosurfactant-producing microorganisms into a reservoir through the well, with subsequent propagation *in situ* through the reservoir rock (Bubela, 1985).
2. The second involves the injection of selected nutrients into a reservoir, thus stimulating the growth of indigenous biosurfactant-producing microorganisms.
3. The third mechanism involves the production of biosurfactants in bioreactors *ex situ* and subsequent injection into the reservoir.

Technologies being Practiced for MEOR

Various technologies being practiced for MEOR are as follows:

(A) **Cyclic microbial recovery/ single well treatment:** Injection of bacteria and nutrient solution into the injector well, a shut in period and production from the same injector well.

(B) **Microbial flooding recovery/ multi-well treatment:** Full field-flooding by injection of nutrient solution and bacteria into one or several injector wells, a shut-in period and production from the neighbouring producing wells.

(C) **Activation of natural microflora indigenous to oil wells:** Full field-flooding with nutrient mineral salts and some carbon/nitrogen sources to support indigenous bacteria.

(D) **Selective plugging recovery:** Injection of biopolymer producing bacteria and nutrients to block high permeability zones/fractures, together with other flooding technologies.

(E) **Microbial fracturing fields:** Microbial decomposition of long chain hydrocarbon compounds within the oil formation, into smaller fractions.

The injection of the nutrient solution supports the growth of microorganisms prevailing at the site. In MEOR, microorganisms in the reservoir are stimulated to produce polymers and surfactants which aid MEOR by lowering interfacial tension at the oil–rock interface. To produce microbial surfactants *in situ,* microorganisms in the reservoir are usually provided with low-cost substrates, such as molasses and inorganic nutrients, to promote growth and surfactant production. To be useful for MEOR *in situ,* bacteria must be able to grow under extreme conditions encountered in oil reservoirs such as high temperature, pressure, salinity, and low oxygen level. The fermentation of the substrate leads to metabolic products like organic acids, alcohols and hydrogen which aid the final anaerobic microbial decomposition. The final decomposition aims at the sulfate reduction and methane formation which is influenced by the water composition and reservoir rock, the microbial population of the rock and the biological technology. Injection of biosurfactants and bacteria such as *Pseudomonas aeruginosa*, *Xanthomonas campestris*, *B. licheniformis* and *Desulfovibrio desulfuricans* along with nutrients showed increase in oil recovery by 30-200%. More than 400 MEOR tests have been conducted in the United States alone and the results have indicated that reservoir heterogeneity significantly affects oil recovery efficiency (Singh et al., 2007). He et al., (2000) showed increase in the rate of production in single-well stimulation treatment, and observed the sustained increase in rate for 2–6 months without additional treatments.

EOR Programmes in Oman

Unlike neighboring Middle-Eastern countries, Oman has less oil reserves and due to complexity of reservoir rock and fluid properties, recovery factors are very low compared to other countries worldwide (Al-Sulaimani et al., 2011). Oman has thus far implemented a successful program to reverse the decline in production, deploying some of the most sophisticated methods of oil extraction.

In recent years, Petroleum Development Oman (PDO) and other major oil companies in Oman has embarked on a major drive to sustain oil production. As primary and secondary recovery techniques come to the end of their natural life, oil companies are increasingly turning to Enhanced Oil Recovery (EOR) methods to extend Oman's production plateau into the coming decades. In 2006, the PDO itself spent more than US$2 billion in oilfield development projects and is likely to see continued high capital expenditure in the coming years. The average recovery factor of the Omani portfolio is around 21 per cent from primary and secondary methods and they hypothesized that another 6 percent can be added by the application of further secondary and EOR methods. Gas injection is the most commonly used EOR technique, worldwide. Where, gases such as carbon dioxide (CO_2), natural gas, or nitrogen is injected into the reservoir, which it expands and pushes additional oil to a production wellbore, thereby dissolving into the oil to lower its viscosity and to improve the flow rate of the oil. The prospects of using CO_2 has gathered much interest, as this would allow storing it away from the atmosphere and hence a tool to combat global warming. Oil displacement by CO_2 injection relies on the phase behaviour of CO_2 and crude oil mixtures that are strongly dependent on reservoir temperature, pressure and crude oil composition. These mechanisms range from oil swelling and viscosity reduction for injection of immiscible fluids (at low pressures) to completely miscible displacement in high-pressure applications. In these applications, more than half and up to two-thirds of the injected CO_2 returns with the produced oil and is usually re-injected into the reservoir to minimize operating costs. The remainder is trapped in the oil reservoir by various means. Other techniques include thermal recovery by steam injection and, chemical injection, where polymers are injected to increase the effectiveness of water-floods, or the use of detergent-like surfactants to help lower the surface tension that often prevents oil droplets from moving through a reservoir. At present, PDO has three other EOR projects under way using two other main EOR technologies. At Harweel, (http://main.omanobserver.om/node/25568) they are working towards the completion of a major project using miscible gas injection, while at Qarn Alam and Amal, they are using steam injection technology. They reported that once these projects have been completed, PDO will be the only oil company in the world using each of the three proven EOR technologies. PDO is also producing over a million barrels of oil equivalent (BOE) per day of oil, gas and condensate from more than 120 active fields.

However, matching the right EOR technology to individual fields is a complex task. The available processes are not universally applicable as each field has its own unique characteristics. Indeed, for some fields, an appropriate EOR method might not exist at this time. More importantly, EOR technology is costly, requiring the careful appraisal of each project to ensure that it is economically viable. Unless oil prices are very high, many EOR technologies such as the use of chemical surfactants or direct heat injection are not economically viable even though they might be technologically appropriate. Another environmentally friendly approach for EOR is MEOR. MEOR researchers at Sultan Qaboos University, Oman, have reported isolation and characterization of biosurfactant producing bacteria indigenous to Omani oil wells and also from other oil contaminated sites (Al-Bahry et al., 2013). They reported 33 genera and 58 species for the first time from Omani oil fields. Most of the identified microorganisms were found to be anaerobic, thermophilic, and halophilic, and produced biogases, bio-solvents, and biosurfactants as by-products, which may be good candidates for MEOR. Amongst those identified bacteria, *Caminicella sporogenes* is the first report from an oil field, and rest of the identified microbes like *Thermotoga* spp., *Halomonas* spp., and *Petrotoga* spp., followed by *Clostridium* spp., *Geobacter* spp., and Desulfotomaculum spp., were predominant and were reported to be present in oil fields and have role in enhancing oil recovery. Al-Sulaimani et al. (2011) have reported biosurfactant producing *Bacillus subtilis* W19, which produced a lipopeptide biosurfactant, quite stable in harsh conditions of temperature, salinity and pH and giving 16-20% extra oil recovery from core-flood experiments. Al-Hattali et al. (2012) also reported enhanced oil recovery from fractured core-plugs by microbial biomass due to microbial permeability profile modification mechanism. These studies reported for MEOR research in the Sultanate of Oman, might be implemented in the field scale studies in near future.

IMPORTANCE OF ISOLATION, IDENTIFICATION AND CONSERVATION OF THE MICROBIAL SPECIES FROM OMAN

It is highly important to isolate and identify these microbial species from Oman, which are capable of producing biosurfactants and other microbial products, and to preserve it for further studies. Oman has quite unique and diverse environmental landscape, like deserts, sea beaches and mountains as well. It provides opportunity to find out many novel microorganisms, never been reported before, as not much information available for microbial species from this region. Our study was the first attempt to identify the microbes from harsh environments of oil reservoirs and to study their metabolites. We have identified those microbes and submitted their gene sequences in the databases like NCBI GenBank, USA.

CONCLUSION

Energy sector forms one of the major parameters in assessing any nation's economic growth, and crude oil is one of the world's major sources of energy. The world is not running out of oil itself, but rather its ability to produce high-quality cheap and economically extractable oil on demand. Several EOR techniques are currently employed in petroleum industries. Enhanced oil recovery or tertiary oil recovery follows water-flood operations termed primary and secondary oil recovery. It aims at recovering previously unrecoverable oil amounting to 30-40% of the total. MEOR offers an environmentally friendly alternative to chemical EOR processes. Microbial enhanced oil recovery (MEOR) involves the application of microbes and/or the exploitation of microbial metabolic processes and products to increase the production of residual oil reservoir. The mechanism of MEOR *in situ* involves most probably multiple effects of the microorganisms on the environment and oil. The microorganisms play various roles in the reservoir. The most important ones are: production of surfactants, solvents, organic and other acid production, gases, selective plugging of the reservoir and polymer production. A more focused MEOR strategy would include production of biosurfactant by microbial species from inexpensive raw materials and injection of this biosystem into the oil well.

ACKNOWLEDGMENTS

We acknowledge the funding by Petroleum Development of Oman and HM research fund, SQU during the course of this research.

REFERENCES

Al Hattali, R., Al-Sulaimani, H., Al-Wahaibi, Y., Al-Bahry, S. N., Elshafie, A., Al-Bemani, A. & Joshi, S. (2012). Improving sweep efficiency in fractured carbonate reservoirs by microbial biomass. SPE 154679.

Al-Bahry, S., Elshafie, A., Al-Wahaibi, Y., Al-Bemani, A., Joshi, S. J., Al-Maaini, R. A., Al-Alawi, W., Sugai, Y. & Al-Mandhari, M. (2013). Microbial consortia in Oman oil fields: A possible use in enhanced oil recovery. *J. Microbiol. Biotechnol.*, *23*, 106-117.

Al-Bahry, S. N., Al-Wahaibi, Y., Elshafie, A., Al-Bemani, A., Joshi, S., Al-Makhmari, H. & Al-Sulaimani, H. (2012). Biosurfactant Production by *Bacillus subtilis* B20 using Date Molasses and its Application in Enhanced Oil Recovery. *Int. Biodet. Biod.*, 81, 141-146.

Alexander, M. (1977). Introduction to Soil Microbiology. John Wiley, N.Y.

Al-Sulaimani, H., Al-Wahaibi, Y., Al-Bahry, S. N., Elshafie, A., Al-Bemani, A., Joshi, S. & Zaragari, S. (2011). Optimization and partial characterization of biosurfactant produced by *Bacillus* species and their potential for enhanced oil recovery. *SPEJ.*, *16*, 672-682.

Al-Sulaimani, H., Joshi, S., Al-Wahaibi, Y., Al-Bahry, S. N., Elshafie, A. & Al-Bemani, A. (2011). Microbial biotechnology for enhancing oil recovery: current developments and future prospects. Invited Review, *Biotechnol. Bioinf. Bioeng. J.*, *1*, 147-158.

Arima, K., Kakinuma, A. & Tamura, G. (1968). Surfactin, a crystalline peptidelipid surfactant produced by *Bacillus subtilis*: Isolation, characterization and its inhibition of fibrin clot formation. *Biochem. Biophys. Res. Commun.*, *31*, 488-494.

Banat, I. M., Makkar, R. S. & Cameotra, S. S. (2000). Potential commercial applications of microbial surfactants. *Appl. Microbiol. Biotechnol.*, *53*, 495-508.

Banat, I. M., Samarah, N., Murad, M., Horne, R. & Banerjee, S. (1991). Biosurfactant production and use in oil tank clean-up. *World J. Microbiol. Biotechnol.*, *7*, 80-88.

Banat, I. M. (1995). Biosurfactants production and possible uses in microbial enhanced oil recovery and oil pollution remediation: A review. *Bioresour. Technol.*, *51*, 1-12.

Berger, B. D. & Anderson, K. E. (1992). Modern Petroleum: A Basic Primer of the Industry. Penn Well Books, PennWell Publishing Company, Tulsa, Oklahoma.

Bryant, R. S. (1987). Potential Uses of Microorganisms in Petroleum Recovery Technology. *Proc. Okla. Acad. Sci.*, *67*, 97-104.

Bryant, R. S. (1991). MEOR screening criteria fit 27% of U.S. oil reservoirs. *Oil & Gas J.*, *89*, 56-9.

Bubela, B. (1985). Effect of biological activity on the movement of fluids through porous rocks and sediments and its application to enhanced oil recovery. *Geomicrobiol. J.*, *4*, 313-27.

Bubela, B. (1987). A comparison of strategies for enhanced oil recovery using *in situ* and *ex situ* produced biosurfactants. *Surf. Sci. Series.*, *25*, 143-61.

Clapes, P. & Infante, M. R. (2002). Amino Acid-based Surfactants: Enzymatic Synthesis, Properties and Potential Applications. *Biocat. Biotransform.*, *20*, 215–233.

Deleu, M. & Paquot, M. (2004). From renewable vegetables resources to microorganisms: new trends in surfactants. *Comptes. Rendus. Chimie.*, *7*, 641-646.

Desai, J. D. & Banat, I. (1997). Microbial production of surfactants and their commercial potential. *Mol. Microbiol. Rev.*, *61*, 47-64.

Desai, J. D. & Desai, A. J. (1993). Production of biosurfactants, p. 65-97. In N. Kosaric (ed.), *Biosurfactants: production, properties, applications.* Marcel Dekker, Inc., New York.

Donaldson, E. C., Yen, T. F. & Chilingarian, G. V. (1985). Enhanced Oil Recovery, Vol. 1 Fundamentals and Analysis. Elsevier, Amsterdam.

Erickson, R. J. (1976). Industrial applications of the bacilli: a review and prospectus, 406-419, *In*: *Microbiology* (*Ed.*) Schlesinger, D., Washington: American Society for Microbiology.

Fiechter, A. (1992). Biosurfactants: moving towards industrial application. *Trends Biotechnol.*, *10*, 208-217.

Frazer, L. (2000). Lipid Lather Removes Metals, Environmental Health Perspectives., *108*, A320-A323.

Georgiou, G., Lin, S. C. & Sharma, M. M. (1992). Surface-active compounds from microorganisms. *Biotech.*, *10*, 60-65.

Gherna, R., Pienta, P. & Cote, R. (1989). *American Type Culture Collection Catalogue of Bacteria and Phages,* Rockville: American Type Culture Collection.

Greek, B. F. (1991). Sales of detergents growing despite recession. *Chem. Eng. News.*, *28*, 25–52.

He, Z., She, Y., Xiang, T., Xue, F., Mei, B. & Li, Y. et al. (2000). MEOR pilot sees encouraging results in Chinese oil field. *Oil Gas J.*, *98*, 4.

Hester, A. (2001). I.B. Market Forecast. *Ind. Bioprocess.*, *23*(5), 3.

Hitzman, D. O. (1983). Petroleum microbiology and the history of its role in enhanced oil recovery. 162-218, *In*: Proceedings of the International Conference on Microbial Enhancement of Oil Recovery. (E. C. Donaldson and J. B. Clark, *eds.*), Technology Transfer Branch, U. S. Department of Energy, Bartlesville, OK.

Horowitz, S., Gilbert, J. N. & Griffin, W. M. (1990). Isolation and characterization of a surfactant produced by *Bacillus licheniformis* 86. *J. Ind. Microbiol.*, *6*, 243-248.

Hughes, L. & Rudolph, J. (2011). Future world oil production: growth, plateau, or peak? *Curr. Opin. Env. Sust.*, *3*, 225–234

Ishigami, Y. (1997). Characterization of biosurfactants, 197-226, *In*: Esumi K, Ueno M (eds) Structure-performance relationships in surfactants. Dekker, N.Y.

Jenny, K., Käppeli, O. & Fiechter, A. (1991). Biosurfactants from *Bacillus licheniformis*: Structural analysis and characterization. *Appl. Microbiol. & Biotechnol.*, *36*, 5-13.

Joshi, S., Bharucha, C., Jha, S., Yadav, S., Nerurkar, A. & Desai, A. J. (2008a). Biosurfactant production using substrates from renewable-resources, molasses and whey under thermophilic conditions. *Bioresour. Technol.*, *99*, 195-199.

Joshi, S., Suthar, H., Yadav, A., Hingurao, K. & Nerurkar, A. (2013). Occurrence of biosurfactant producing *Bacillus* spp. in diverse habitats and possible uses in oil recovery. *ISRN Biotechnol.* vol. 2013, Article ID 652340, 6 pages. doi: 10.5402/2013/652340.

Joshi, S., Yadav, S. & Desai, A. J. (2008b). Application of response surface methodology to evaluate the optimum medium components for the enhanced production of lichenysin by *B. licheniformis* R2. *Biochem. Eng. J.*, *41*, 122-127.

Joshi, S., Yadav, S., Nerurkar, A. & Desai, A. J. (2007). Statistical Optimization of Medium Components for the Production of Biosurfactant by *Bacillus licheniformis* K51. *J. Microbiol. Biotechnol.*, *17*, 313-319.

Joshi, S. J. & Desai, A. J. (2013). Bench-scale production of biosurfactants and its potential in *ex-situ* MEOR application. *Soil Sedi. Contam. Int. J.*, 22, 705-715.

Khire, J. M. & Khan, M. I. (1994a). Microbial enhanced oil recovery (MEOR). Part I - Importance and mechanism of MEOR. *Enzyme Microb. Technol.*, *16*, 170-172.

Khire, J. M. & Khan, M. I. (1994b). Microbial enhanced oil recovery (MEOR). Part II - Microbes and the subsurface environment for MEOR. *Enzyme Microb. Technol.*, *16*, 258-259.

Kitamoto, D., Isoda, H. & Nakahara, T. (2002). Functions and potential applications of glycolipids biosurfactants – from energy saving materials to gene delivery carriers. *J. Biosci. Bioeng.*, *94*, 187-201.

Konz, D., Doekel, S. & Marahiel, M. A. (1999). Molecular and biochemical characterization of the protein template controlling biosynthesis of the lipopeptide lichenysin. *J. Bacteriol.*, *181*, 133-140.

Kosaric, N., Cairns, W. L. & Gray, N. C. C. (1987). Microbial emulsifiers and deemulsifiers. 247-331, *In*: Kosaric N, Cairns WL, Gray NCC (*eds*) Biosurfactants and biotechnology. Surfactant science series, vol 25, Dekker, N.Y.

Kosaric, N., Gray, N. C. C. & Cairns, W. L. (1983). Microbial emulsifiers and de-emulsifiers, 575-592. *In* H. J. Rehm and G. Reed (*ed.*), Biotechnology, vol. 3. Verlag Chemie, Deerfield Beach, Fla.

Kosaric, N. (2001). Biosurfactants and their application for soil bioremediation. *Food Technol. Biotechnol.*, *39*, 295-304.

Lapidus, A., Galleron, N., Andersen, J. T., Jørgensen, P. L., Ehrlich, S. D. & Sorokin, A. (2002). Co-linear scaffold of the *Bacillus licheniformis* and *Bacillus subtilis* genomes and its use to compare their competence genes. *FEMS Microbiol. Lett.*, *209*, 23-30.

Layman, P. L. (1985). Industrial surfactants set for strong growth. *Chem. Eng. News.*, *23*, 23-48.

Lee, B. S. & Kim, E. K. (2004). Lipopeptide production from *Bacillus* sp. GB16 using a novel oxygenation method. *Enzyme Microb. Technol.*, *35*, 639-647.

Lin, S. C., Minton, M. A., Sharma, M. K. & Georgiou, G. (1994). Structural and immunological characterization of a biosurfactant produced by *Bacillus licheniformis* JF-2. *Appl. & Environ. Microbiol.*, *60*, 31-38.

Makkar, R. S., Cameotra, S. S. & Banat, I. M. (2011). Advances in utilization of renewable substrates for biosurfactant production. *AMB Express.*, *1*, 5.

McInerney, M. J., Javaheri, M. & Nagle, Jr. D. P. (1990). Properties of biosurfactants produced by *Bacillus licheniformis* strain JF-2. *J. Ind. Microbiol.*, *5*, 95-102.

McInerney, M. J., Maudgalya, S. K., Nagle, D. P. & Knapp, R. M. (2004). Critical assessment of the use of microorganisms for oil recovery. *Recent Res. Devel. Microbiol.*, *6*, 269-284.

McInerney, M. J. & Sublette, K. J. (1997). Petroleum microbiology: Biofouling, souring, and Improved oil recovery. 600-607, *In*: *Manual of Environmental Microbiology*, (Ed.) C.J. Hurst, ASM press, Washington, DC.

Moses, V. (1987). Microbiological methods in oil recovery. *Microbiol. Sci.*, *4*, 304 -309.

Mukherjee, S., Das, P. & Sen, R. (2006). Towards commercial production of microbial surfactants. *Trends Biotechnol.*, *24*, 509-515.

Neyra, C., Atkinson, L. A., Olubayi, O., Sadasivan, L., Zaurov, D. & Zappi, E. (1996). Novel microbial technologies for the enhancement of plant growth and biocontrol of fungal diseases in crops. *Cahiers Opt. Méd.*, *31*, 447-456.

Nitschke, M. & Costa, S. G. V. A. O. (2007). Biosurfactants in food industry. *Trends in Food Sci. Technol.*, *18*, 252-259.

Paul, E. A. & Ladd, J. N. (1981). Soil Biochemistry, Vol. 5 Marcel Dekker, Inc. New York.

Portwood, J. T. (1995). A Commercial Microbial Enhanced Oil Recovery Technology: Evaluation of 322 Projects. SPE 29518-MS.

Rosenberg, E. (1986). Microbial surfactants. *CRC Crit. Rev. Biotechnol.*, *3*, 109-132.

Sarkar, A. K., Goursand, J. C., Sharma, M. M. & Georgiou, G. (1989). A critical evaluation of MEOR process. *In Situ*, *13*, 207-238.

Schallmey, M., Singh, A. & Ward, O. P. (2004). Developments in the use of *Bacillus* species for industrial production. *Can. J. Microbiol.*, *50*, 1-17.

Schramm, L. L., Stasiuk, E. N. & Marangoni, D. G. (2003). Surfactants and their applications. *Annu. Rep. Prog. Chem., Sect. C.*, *99*, 3-48.

Sen, R. & Swaminathan, T. (2004). Response surface modeling and optimization to elucidate the effects of inoculum age and size on surfactin production. *Biochem. Eng. J.*, *21*, 141-148.

Shennan, J. L. & Levi, J. D. (1987) In-situ microbial enhanced oil recovery. 163-81, *In* Biosurfactants and Biotechnology, (*Eds.*) N. Kosaric, W. L. Cairns and N. C. C. Gray. Marcel Dekker, New York.

Singer, M., Vogt, E. & Finnerty, W. R. (1984). Microbial metabolism of straight and branched alkanes, 1-59, *In:* Petroleum Microbiology, (*ed.*) R. Atlas., Collier and MacMillan, N.Y.

Singh, A., Van Hamme, J. D. & Ward, O. P. (2007). Surfactants in microbiology and biotechnology: Part 2. Application aspects. *Biotechnol. Adv.*, *25*, 99–121.

Stosur, G. J. (1991). Unconventional EOR concepts. *Crit. Rev. Appl. Chem.*, *33*, 341-73.

Tanner, R. S., Udegbunam, E. O., McInerney, M. J. & Knapp, R. M. (1991). Microbially enhanced oil recovery from carbonate reservoirs. *Geomicrobiol. J.*, *9*, 169-95.

Thakur, N. K. & Rajput, S. (2011). Exploration of Gas Hydrates, World's Oil and Natural Gas Scenario, Springer-Verlag Berlin Heidelberg.

Van Dyke, M. I., Lee, H. & Trevors, J. T. (1991). Application of microbial surfactants. *Biotechnol. Adv.*, *9*, 241-252.

Xu, D. & Côté, J. C. (2003). Phylogenetic relationships between *Bacillus* species and related genera inferred from comparison of 3' end 16S rDNA and 5' end 16S-23S ITS nucleotide sequences. *Int. J. Syst. Evol. Microbiol.*, *53*, 695-704.

Yakimov, M. M., Timmis, K. N., Wray, V. & Fredrickson, H. L. (1995). Characterization of a new lipopeptide surfactant produced by thermotolerant and halotolerant subsurface *Bacillus licheniformis* BAS50. *Appl. Environ. Microbiol.*, *61*, 1706-1713.

Zobell, C. E. (1947). Bacterial release of oil from oil bearing materials. *Oil Gas J.*, *46*, 36-37.

In: Biotechnology and Conservation … (2 Volume Set)
Editors: S. A. Farooq, R. Abed and S. Baqir
ISBN: 978-1-63117-869-6

Chapter 45

IMPROVING THE CATALYTIC PROPERTIES OF PECTINASE FROM *BACILLUS LICHENIFORMIS* KIBGE-IB 21 BY ENTRAPMENT IN CALCIUM ALGINATE BEADS

Haneef Ur Rehman*, Shah Ali Ul Qader and Afsheen Aman
The Karachi Institute of Biotechnology and Genetic Engineering (KIBGE),
University of Karachi, Karachi, Pakistan

ABSTRACT

Pectinase is a heterogeneous group of enzymes that catalyze the hydrolysis of pectin substances which are responsible for the turbidity and undesirable cloudiness in fruit juices. Partially purified pectinase produced by *Bacillus licheniformis* KIBGE-IB21 was entrapped in calcium alginate beads. It was found that the immobilization increased the optimal reaction time from 5 to 10 minutes for pectin degradation and temperature from 45 °C to 55 °C, whereas the optimal pH remained same with reference to free enzyme. The immobilized enzyme also exhibited good operational stability and retained 65 % of its initial activity even after repeating the reaction three times. The results demonstrated that this calcium alginate immobilized pectinase has potential applications for the enzymatic degradation of complex carbohydrates like pectin which is commonly used in the food industry.

Keywords: Pectin, Pectinase, Immobilization, Calcium alginate beads

---

* Corresponding author email address: haneef.rehman@kibge.edu.pk.

INTRODUCTION

Pectinase is widely used in food industries for the clarification of fruit and vegetable juices as well as in wine industries due to its catalytic activity of degrading pectin. In spite of having high catalytic properties, the free enzyme as a biocatalyst always faces some problems such as low stability under operational conditions, difficulty of product recovery and impossibility of continuous uses in industrial processes (Sheldon 2007).

Enzyme immobilization is the technique which not only permits easy recovery of the product and the reusability of costly enzymes but also increases the operational stability of enzymes against different industrial processes (Xu et al. 2006). Currently, different methods of immobilization are used which are divided into three main categories: (1) binding to a supports (carrier), (2) cross-linking of enzyme molecules and (3) entrapment or encapsulation in organic and inorganic polymers. Among these techniques, enzyme entrapment within insoluble calcium alginate has been shown to be the most effective approach because it is a rapid, nontoxic, inexpensive and versatile method for the immobilization of enzymes as well as cells (Fraser et al., 1997).

To date, pectinase has been immobilized in various supports including agar gel (Tuoping et al., 2008), nylon (Lozano et al., 1987), silk (Zhu et al., 1998) and chitin (Iwasaki et al., 1998). Due to limited data on the immobilization of pectinase within calcium alginate beads, it is very important to perform this kind of experiment to improve the operational stability and reusability of pectinase for industrial processes. This study was based on the immobilization of pectinase from *Bacillus licheniformis* KIBGE-IB21 within calcium alginate beads.

METHODOLOGY FOR PECTINASE PRODUCTION, IMMOBILIZATION AND ENZYME ACTIVITY

Pectinase was produced from *B. licheniformis* KIBGE-IB21 by growing the bacterial strain in a medium containing pectin (1.0%), yeast extract (0.3%), K_2HPO_4 (0.2%), KH_2PO_4 (0.2%) and KNO_3 (0.2%) at pH 7.0 and 37°C of incubation for 48 hours of fermentation period (Rehman et al., 2012). Cells were harvested by centrifugation at 15000 rpm for 15 minutes at 4°C and supernatants were used for further studies.

The partially purified pectinase was immobilized using 3.0% sodium alginate. Ca-alginate beads were formed when the enzyme alginate mixed solution was dripped into 0.2M calcium chloride solution at 4°C.

The enzyme activity of free and immobilized pectinase was determined by the 3' 5' dinitrosalicylic acid method (Miller, 1959) using citrus pectin as a substrate and galacturonic acid monohydrate as the standard. One unit of pectinase was defined as the amount of enzyme producing 1 μmole of galacturonic per minute at pH 10.0 and 45°C.

The effect of optimum temperature and pH on the activity of free and immobilized pectinase was determined by performing the reaction in different ranges of temperature and pH levels (temperature from 30-60°C and pH from 5-10).

The operational stability of pectinase entrapped in Ca-alginate beads was determined by repeatedly reusing the beads in a hydrolysis reaction.

Pectinase Production, Immobilization and Its Effect on Enzyme Activity

Temperature and pH Profile

The impact of temperature on the activities of free and immobilized pectinase were investigated in a range of 30-60 °C and it was found that after immobilization, the optimal temperature of enzyme activity was increased from 45 °C to 55 °C with reference to free enzymes (Figure 1).

The optimal pH of free and immobilized pectinase were analyzed in the range of pH from 5 to 12 and the maximum pectinase activity was taken as 100% (Figure 2). It was observed that the activities of both the free and immobilized pectinase were the same as the pH shifted from 5 to 10. However, the influence of pH below 7 on free enzymes seemed relatively higher as compared to immobilized enzymes.

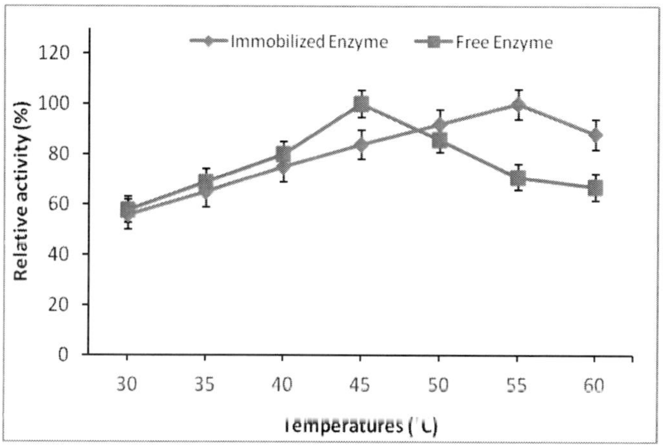

Figure 1. Effect of different temperatures (from 20 to 60 °C) on the relative activities of immobilized and free enzyme at constant pH (10.0) and substrate (1.0 %) concentration.

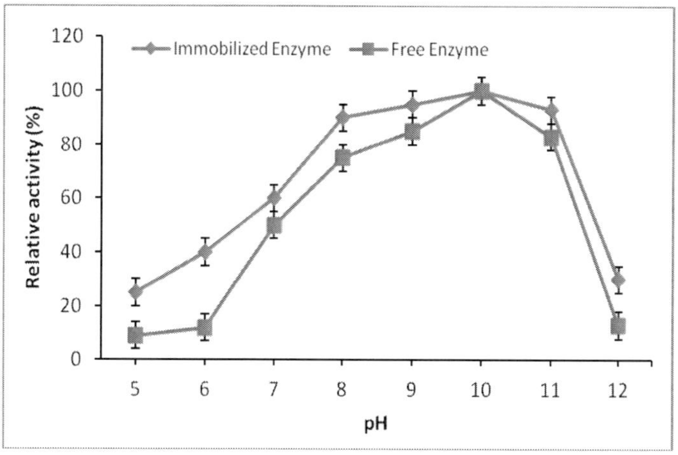

Figure 2. Effect of various levels of pH (5 to 12) on the relative activities of immobilized and free enzymes keeping the temperature (55 °C for immobilized pectinase and 45 °C for free pectinase) and substrate concentration (1.0 %) constant.

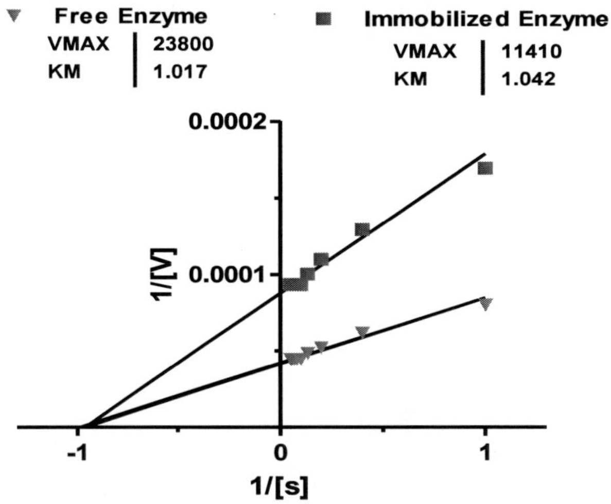

Figure 3. Linewaver-Burk plot for the calculation of Km and Vmax values of free and immobilized pectinase.

Table 1. Reusability of immobilized pectinase

Number of cycles	Residual Activity (%)
1	100
2	80
3	65
4	38
5	30
6	20
7	9

Kinetic Parameters of Immobilized Enzyme

The kinetic parameters were determined from Line Weaver – Burk plot (Figure 3). The Vmax and Km value of immobilized enzymes were 1342 U/ml/min and 2.526 mg/ml, respectively (Figure 3). The increase in the Km value of pectinase after immobilization suggested that the immobilization decreased the apparent affinity to the substrate, most likely as a result of diffusion limitation.

Reusability of Immobilized Pectinase

The reusability of immobilized pectinase was carried out at 55 °C in batch reactions (Table 1). The immobilized pectinase retained 80 % of its initial activity in the second batch and 65 % during the third cycle. However, the immobilized pectinase retained 9 % of its initial activity even after repeating the reaction 7 times.

Pectinase is an enzyme that catalyzes the degradation of pectin and has been widely used in food and textile industries. For enhancing the catalytic properties of pectinase produced from *Bacillus licheniformis* KIBGE IB-21, it has been immobilized within calcium alginate beads through entrapment. The effect of immobilization on the catalytic properties including temperature, pH and kinetic parameters were studied to evaluate the success of the immobilization method. It was observed from the results that the immobilization increased the optimum temperature for maximum enzymatic activity from 45 to 55°C. The shift of temperature toward a higher level may be due to the increase in the conformational rigidity of enzymes within calcium alginate gel. Calcium alginate protects the pectinase against higher temperatures for denaturation. Next to the temperature, pH is also an important parameter for the maximum enzymatic activity of soluble and insoluble enzymes. The optimum pH for the maximum activity of pectinase didn't change after immobilization and both the free as well as immobilized pectinase showed maximum activity at the same pH. However, the effect of acidic pH (below 7.0) on the activity of free enzymes was higher as compared to immobilized enzymes and immobilized pectinase showed a higher relative activity at an acidic pH range. The immobilized pectinase may be less affected by the acidity of the reaction environment due the conformational stability of an enzyme within the polymeric network of calcium alginate beads.

The kinetic parameters such as the Michealis-Menten constant (Km) and maximum reaction rate (Vmax) are very important parameters for the evaluation of the immobilization method. It has been observed that the immobilization increased the Km and decreased Vmax for pectin degradation. After immobilization the affinity of an enzyme toward its substrate lowered which ultimately reduced the maximum reaction rate for pectin hydrolysis. The increased Km and decreased Vmax might be due to the substrate diffusion limitation effect caused by the support.

The reusability of pectinase was demonstrated by repeatedly reusing pectinase in batch reaction under standard assay conditions and the same conditions were used in all reaction batches. It was observed from the results that the immobilized pectinase showed high reusability for pectin hydrolysis and it retained more than 60% of its original activity after being reused three times

In summary, pectinase produced from *B. licheniformis* KIBGE IB-21 was immobilized on Ca-alginate beads. After immobilization, the optimal pH and temperature for enzymatic activity was slightly broadened as compared to free enzymes. In addition, the immobilized pectinase exhibited excellent reusability for the pectin hydrolysis and retained 65 % activity after its third cycle.

REFERENCES

Fraser, F. and Bickerstaff,G. F.(1997). Entrapment of enzymes and cells in calcium alginate. In immobilization of enzymes and cells, *Humana Press*, 61-66.

Iwaski, K. Inoue, M. and Matsubara, Y. (1998). Continuous hydrolysis of pectate by immobilized endo-polygalacturonase in a continuously stirred tank reactor. *Bioscience Biotechnology Biochemistry,* 62: 262-267.

Lozano, P., Manjio, A., Eomojaro, F. Canovas, M. and Iborraet, J. L.(1987). A cross-flow reactor with immobilized pectolytic enzymes for juice clarification. *Biotechnology Lett*er, 9: 875-880.

Miller GL (1959). Use of dinitrosalicylic acid reagent for determination reducing sugar. *Analytical Chemistry*, 31: 426-428.

Rehman HU, Qader SAU and Aman A. 2012. Polygalacturonase: Production of pectin depolymersing enzyme from *Bacillus licheniformis*KIBGE IB-21. *Carbohydrate Polymer,* 90: 387-391.

Sheldon RA. 2007. Enzyme immobilization: the quest for optimum performance. *Advance Synthesis and Catalysis*, 3489: 1289-1307.

Tuoping L, Suhong L, Na W and Lirui T. 2008. Immobilization and stabilization of pectinase by multipoint attachment onto an activated agar-gel support. *Food Chemistry,*109: 703-708.

Xu SW, Lu Y, Li J, Jiang ZY and Wu H. 2006. Efficient conversion of CO_2 to methanol catalyzed by three dehydrogenases co-encapsulated in an alginate-silica (ALG-SiO_3). Industrial *and Engineering Chemistry Research,* 45: 4567-4573.

Zhu, R. X, Lin R and Wang JOJ. 1998. Studies on immobilization of pectatelyase with silk fibroin. *Zhejiang Agric. Univ.* 24: 74-78.

In: Biotechnology and Conservation … (2 Volume Set)
Editors: S. A. Farooq, R. Abed and S. Baqir
ISBN: 978-1-63117-869-6

Chapter 46

Potential of Microbial Biotechnology in Heavy Oil Enhanced Recovery

Balqees S. Al-Hinai[1], Saif N. Al-Bahry[2], Abdulkader E. Elshafie[2], Yahya M. Al-Wahaibi[3], Ali S. Al-Bemani[3] and Sanket J. Joshi[2]

[1]The Research Council, Muscat, Oman

[2]Department of Biology, College of Science, Sultan Qaboos University, Muscat, Oman

[3]Department of Petroleum and Chemical Engineering, College of Engineering, Sultan Qaboos University, Muscat, Oman

Abstract

Oil will remain as a dominant factor of energy sectors throughout the world for many years. The International Energy Agency (IEA) expected 70% growth in worldwide demand for energy by 2030 and 88% of this increased demand will be covered by oil, natural gas, and coal. In Oman, oil and gas are the backbone of the economy. Their contribution to national economy reached 80.71%. Oman oil discoveries are largely considered difficult oil or heavy oil. Heavy crude oil reserves will meet high world demands on energy. Heavy oil recovery requires more advanced and costly techniques compared to light oil. Conventional oil production technologies recover only about one-third of the oil originally in place in a reservoir. The ability of microorganisms to live in petroleum environment opens a window for many biotechnology based applications in oil industries and its application to enhance heavy oil recovery. Microbial enhanced oil recovery (MEOR) involves use of bacteria or their products, such as biosurfactants, biopolymers, gases, enzymes and others. MEOR processes do not consume large amounts of energy as thermal processes; do not depend on crude-oil-price fluctuations; and unlike chemical processes which are not environmental friendly. Therefore, isolation, identification and understanding biotechnological properties of the microbes is important in oil industry and it becomes crucial for setting up conservation strategies of such microbes.

Keywords: Heavy oil, enhanced oil recovery, microbial biotechnology

INTRODUCTION

Importance of Oil in World Economy

Crude oil is the major source of energy in the world. International Energy Agency (EIA, 2011) expects that in 2030, the oil production in the world will increase to 107 million barrels per day (Montiel et al., 2009). Continued industrialization and economic growth will raise the demand for oil. By 2030 there will be 70% growth in worldwide demand for primary sources of energy. About 88% of this bigger demand will be met by oil, natural gas, and coal (Flint, 2004). The production peak is estimated to be stable between 2010 and 2020, and then the oil resources will be significantly reduced at the end of this century (Campbell, 1980). Technologies, which improve recovery efficiency and reduce environmental impacts, will be promising alternates to traditional oil recovery and refinery. Biotechnology is a new field presented to the oil refining industry (Gieg et al., 2008; Duhalt et al., 2002).

Oil Components and Classification

Petroleum is a complex mixture of hydrocarbons, organometallo elements, vanadium and nickel (Hamme et al., 2003). Petroleum hydrocarbons are divided into four classes: Saturates, Aromatics, Asphaltenes (phenols, fatty acids, ketones, esters, and porphyrins), and Resins (pyridines, quinolines, carbazoles, sulfoxides, and amides) (Leahy and Colwell, 1990). Oils are classified for economic value according to American Petroleum Institute (API) gravity, based on a surface measurement of the specific gravity of degassed oil. Typical light oil has an API gravity around 36-38°, whereas heavy oils have API gravities of 20° or less, super-heavy oils have API gravities of 10 or less due to the presence of a high proportion of heavy hydrocarbon fractions (Head et al., 2003).

Oil in Oman

The Sultanate of Oman's economy is heavily dependent on oil revenues, which accounts for about 81% of the country income (Hasan, 2011). Oman's petroleum deposits were discovered in 1962. Confirmed oil reserves in Oman stand at 5.5 billion as of January 2010 data (Central Intelligence Agency World Fact book, 2011). The major oil fields in Oman are Yibal, Qarn Alam, Athel-Marmul, Bahja, Rima, Jalmud, Nimr, Karim, Harweel, Mukhaizna and Safah. Omani fields are smaller, more widely scattered, less productive, and pose higher production costs compared to other gulf countries. In April 2006, Oman's Minister of Oil and Gas announced, that the country planned to invest 10 billion US Dollars in upstream oil and natural gas projects especially in enhanced oil recovery applications (Oil & Gas Directory Middle East - 2009). Presently, Oman is one of the countries which use the Enhanced Oil Recovery (EOR) technique for its oil production.

Heavy Oil Recovery and Components

Heavy oil is a thick, black, sticky fluid characterized by high viscosities and densities compared to conventional oil. The World Petroleum Congress defines heavy oil as oil whose gas-free viscosity is between 100 cp and 10,000 cp at reservoir temperature (Sette et al., 2007). Heavy oil is deficient in hydrogen and has high carbon, sulfur, heavy metal contents. It is formed in deep formations as conventional oil, but migrated to the surface where they were degraded by bacteria and weathering, and the lightest hydrocarbons escaped (Sette et al., 2007; Head et al., 2003).

The increasing demand for energy has increased the interest in heavy oil and bitumen. Heavy oil represents the largest potentially recoverable petroleum energy resource known (Bryant, 1998). The IEA estimates that there are 6 trillion barrels of heavy oil in place worldwide (Clark et al., 2007). Heavy oil requires additional processing to become a suitable feedstock for normal refinery (Sette et al., 2007). Technologies that upgrade value, decrease costs, and diminish environmental impacts will have the greatest effect on increasing heavy oil production. Heavy oil production methods are divided into four main categories: cold-depletion production, water flood production, thermal production, and solvent-flood production processes (Memon et al., 2010). Heavy oil upgrading is achieved by decreasing viscosity using heat, solvents and catalysts. But these techniques faced technical, economic and environmental challenges that have limited their application (Clark et al., 2007).

In recent years, there is an increased interest within oil companies, research institutions and universities to develop new technologies. Recovery, separation and processing problems of heavy oils are linked to high concentrations of asphaltenes. Asphaltene, is the highest-molecular-weight fraction of petroleum. It is dark solid rich in heteroatoms (S, O, N), and metals (Fe, Ni, V), and represent as 50% of the total, which is responsible for high viscosity (Duhalt et al., 2002). In relation to heavy oil recovery challenges, EOR will target large amount of heavy oil compared to light oil.

Heavy Oil in Oman

The oil fields of south Oman contain heavier and more viscous oil than fields in north and central Oman and in most of the major fields in Arabian Peninsula. Significant discoveries of heavy oil accumulation, around 7+ billion barrels Stock Tank Oil Initially In Place (STOIIP) in south Oman have led to a greater emphasis on their development (Nadyal, 1993). Mukhazina is the third largest oil field in south Oman. It was discovered in 1975 and was found to contain heavy viscous oil (14-16 API) (Al-Azkawi, 2002). The average oil viscosity at reservoir temperature of 50° C varies from 1200 to 3600 cp (Malik, 2010). Engineers at the Petroleum Development Oman (PDO), inject steam into the oil wells to heat heavy oil and to make it thinner and easy to recover. This method can recover 25% to 30%, but they are very expensive. As oil recovery techniques are developed and spread across countries, it may change the economics of the oil industry. In the future, a greater emphasis will be on maximizing recovery, by using enhanced recovery techniques (BBC News, Published: 2006/02/15).

Oil Recovery Techniques

Enhanced oil recovery (EOR) technologies, which kept the oil industry strong and competitive, are needed to overcome stagnant oil production and unimpressive recovery (Sette, 2007). Crude petroleum exists in the complex capillary network of oil reservoirs. Traditional recovery technologies can recover a maximum of 40-50% of the oil (Sen, 2008). According to world reserve statistics, about 2 trillion barrels of conventional oil and 5 trillion barrels of heavy oil remain in reservoirs worldwide after using conventional recovery methods (Al-Sulaimani et al., 2011). About 5-10% of oil is recovered by primary techniques using natural pressure drive of the reservoir. Secondary recovery techniques like, fluid injection, fracture the hydrocarbon-bearing formation and improve the flow of oil to the wellhead by 10%to 40% (Mclnerney et al., 2005; Sen, 2008). Chemical (polymer, alkaline and surfactants) and thermal energy (steam, combustion and hot water) are used to recover trapped crude oil after primary and secondary techniques (Bryant, 1987; Ayirala, 2002). These tertiary recovery techniques are economically unattractive, so scientists look for cost effective alternative which is MEOR (Sen, 2008). These days, biotechnology open new window by applying microbes in oil industry. Biotechnology is production of beneficial materials by means of biological systems or their products. The implementation of biological processes in the oil industry to explore, produce, refine, transform petroleum into valuable derivatives and clean the pollution is known as petroleum biotechnology (Montiel et al., 2009). Biotechnology can improve and expand fuel upgrading worldwide, as it does not require hydrogen and produce far less carbon dioxide than thermochemical processes (Borgne and Quintero, 2003).

Microbial Enhanced Oil Recovery

MEOR refers to petroleum recovery methods, which involve the use of microbes and their metabolic products. The bio-products include biosurfactants, biopolymers, biomass, acids, solvents, gases and enzymes to increase recovery of oil from depleted and marginal reservoirs (Sen, 2008; Behlülgil, 2002; Bryant, 1987; McInerney, 1999). MEOR is considered as an environmentally compatible tertiary oil recovery method with high performance and cost-effective EOR method (Sen 2008; Behlülgil, 2002). Microbial cell need little input of energy to produce the MEOR agents and the application of microbial processes does not directly depend on the global crude oil price (Sen, 2008; McInerney, 2007).

The Role of Bacteria in Oil Industry

Bacteria are abundant in oil reservoirs and can cause desired changes in the physicochemical properties of the oil (Sen 2008; Sette, 2007). A large number of microorganisms belonging to the phyla of eubacteria, which are yeast, and fungi, like strictly anaerobic thermophilic and hyperthermophilic, thermotolerant have been detected *ex-situ* or *in-situ*. They are able to use long-chain n-alkanes as carbon and energy source (Beilen et al., 2003). Bacteria were isolated and identified by cultivation-independent and culture-based enrichment methods (Al-Bahry et al., 2013a; Sette, 2007). A high rate of hydrocarbons utilization was carried by bacterial enrichment consortium isolated from oil exposed environments. Bacterial consortia display a wide array of metabolic mechanisms in the breakdown of hydrocarbon (Menezes, 2005; Kebria, 2009).

A microbial response to the hydrocarbon substrates depends on accession and transformation of hydrocarbons, cell membrane architecture, cell-surface adhesion, hydrophobicity, chemotaxis, and quality of petroleum feedstock (Sen, 2008). The major leading factors in MEOR are the bottom-hole conditions, formations nature, bacterial composition, inoculation concentration, their adaptation time, optimum slug size (Zhang, 1999).

Microbes and their metabolites mobilize residual oil and improve the areal sweep. Surfactants lead to reduction of interfacial tension and alteration of wettability (Al-Bahry et al., 2013b; Al-Sulaimani et al., 2012, Al-Hattali et al., 2013, Nielsen, 2010). Microbial growth and polymer production change fluid components of oil. Viscosity reduction is achieved by degradation oil components or gas production (Nielsen, 2010).

Adapted microorganisms (extremophiles) to "harsh" environments with very high temperatures to 85°C, pressure over 17.23MPa, extreme pH ($pH < 3$ or $pH > 10$) and high salt concentrations (from 5% to saturated sodium chloride) are attractive catalysts for oil industry (Duhalt, 2002; Sen, 2008). Difficulty in isolating and/or engineering microbial strains, which can survive in the extreme environment of the oil reservoirs led to enzyme enhanced oil recovery (EEOR) as an advanced MEOR methods (Sen, 2008).

Bacterial Contribution in Heavy Oil Industry

Large production of heavy oil in the near future forces scientists to transform it into light oil in reservoirs or during storage. The viscous and capillary forces that hold this oil in place are expressed as a ratio called the capillary number (Nca): Nca = (μw vw)/(ow) where μw is the viscosity, vw is the flux of fluid, and ow is the oil-water interfacial tension. Large changes in the capillary number (about a factor of 1000) are needed for substantial oil recovery. Large changes in viscous forces are only possible for the recovery of heavy oil. Interfacial tension reduction by surfactants is one way to reach large change in capillary number (McInerney, 2007). Reduction of oil viscosity after growth of some strains in the crude oil in the lab can lead to effective strategy for recovering residual oil from oil field (Zhang, 1999). Oxidization of asphaltenes, responsible for viscosity on heavy crude, is the success way to enhance fluidity and pumping of oil (Montiel et al., 2009).

Bio-upgrading of long carbon chain, can reduce viscosity or remove the more difficult hydrocarbon species (Flint, 2004). Bacteria can degrade crude oil and consume its heavy fractions. So, oil becomes lighter as a result of a decrease in viscosity (Bryant and Burchfield, 1989). *Pseudomonas*, *Arthrobacter*, and other aerobic bacteria are especially effective in the degradation of crude oil. *Clostridium*, *Desulfovibrio*, *Pseudomonas*, and certain methanogens produce CO_2, N_2, H_2, and CH_4 which dissolves in the crude oil and reduces its viscosity or increases the pressure in the reservoir (Behlülgil, 2002). Microbial systems have not been developed into a commercial service for heavy oil recovery today because there is little data or research specifically with heavy oil to support commercialization. To date, Microbial technologies have been designed and optimized for medium weight oil only. Use of solvent-, gas-, and surfactant-producing bacteria that reduce oil viscosity and improve mobility will be helpeful in oil recovery (Bryant, 1998).

Asphaltenes fraction is the most resistant fraction of oil. There are microorganisms associated with bitumens that contain high amounts of asphaltenes (Duhalt, 2002). Chloroperoxidase enzyme from the fungus *Caldariomyces fumago* was able to degrade petroporphyrins and asphaltenes. The destruction of petroporphyrins by chloroperoxidase in the presence of hydrogen peroxide leads to removal of Ni and V from asphaltene. The low energy requirements for operation and easy process control are two of the advantages that make the enzymatic treatment an interesting and attractive technology over conventional treatments (Duhalt, 2002). Enzymes from some bacteria have proved to be capable of minimal upgrading in controlled surface environments (Clark et al., 2007). Specific enzymes such as peroxidases and lactases can be used to treat heavy oil (Montiel et al., 2009).

Enzyme Enhanced Oil Recovery

Enzymes are polymers of amino acids with a high molecular weight. They act as catalysts which speed up the chemical reactions, under controlled conditions of pH and temperature. Enzymes activities are specific; certain enzymes break down or synthesize certain compounds (Harris, 1998). Enzymes are water soluble and environment friendly molecules (Meintanis, 2006; Harris, 1998). Hydrolase and oxidoreductase enzymes produced by bacteria, are novel effective enzymes in oil field. These enzymes break a covalent bond in their substrate and adding a water molecule in the process (Harris, 1998). Temperature is an important factor in controlling microbial hydrocarbon degradation nature and efficiency. Thermophilic hydrocarbon degraders of *Bacillus, Thermus, Thermococcus* and *Thermotoga* species occurring in natural high-temperature or sulfur-rich environments will be efficient in oil recovery (Meintanis, 2006).

Jumpstart Energy Services in USA developed Greenzyme® enhanced oil recovery. This hydrolase is able to reduce interfacial tension (IFT), paraffin waxy components and viscosity. Enzyme active site attaches to the oil and catalyzes it, so larger oil droplets are broken into smaller ones allowing more oil release and improve mobility permeability. Greenzyme improved recovery of medium and heavy oil up to 50-100%. Greenzyme is extracted from oil loving microbes where the oil digesting property is neutralized, and the ability to seek hydrocarbons is left as such. After the release of hydrocarbons the enzyme will remain in the water phase in the reservoir and be available to release oil from new oil wetted surfaces (Lau and Gray, 2008).

Alkanes Utilization by Bacterial Enzymes

Hydrocarbons need to undergo cellular uptake or attachment to become available for cell's catabolic machinery. Substrate is first broken down by extracellular hydrolytic enzymes before taking them as reaction products (Wentzel et al., 2007).

Alkanes are saturated insoluble hydrocarbons representing the main components of mineral oil. Microorganisms use specialized enzyme systems and metabolic pathways to access n-alkanes as a carbon and energy source (Wentzel et al., 2007). Hydrophobic nature of the bacterial cell surface is necessary for *n*-alkanes metabolism. The initial step in aliphatic and aromatic hydrocarbon degradation is oxidation reactions catalyzed by cell-surface-associated oxygenases. Interfacial accession (by direct contact of the cell with the hydrocarbon) and biosurfactant-mediated accession (by cell contact with emulsified hydrocarbons) are two mechanisms for long chain n-alkanes accession to the cell (Wentzel et al., 2007). Fimbriae on the cell surface enable *Acinobacter* sp. to attach n-alkanes (Rosenberg, 1982).

The *n*-alkane is oxidized to primary alcohol by terminal monooxygenases/hydroxylases. There are two classes of alkane hydroxylases involved in bacterial aerobic long-chain *n*-alkane metabolism; cytochrome-P450-related enzymes in both yeasts and bacteria, e.g., bacterial CYP153 enzymes, and bacterial particulate alkane hydroxylases (pAHs) (Beilen and Funhoff, 2007). After initial oxidation of the *n*-alkane, the resultant alcohol is oxidized by alcohol dehydrogenase and aldehyde dehydrogenase to the aldehyde and carboxylic acid. The carboxylic acid then serves as a substrate for acyl-CoA synthetase, and the resulting acyl-CoA enters the β- oxidation pathway. The most extensively studied bacterial alkane degradation pathway by enzymes (AlkB) is that of *Pseudomonas putida* GPo1 (Wentzel et al., 2007) which is encoded by the OCT plasmid (Hamme, 2003).

Depending on the chain-length of the alkane substrate, different enzyme systems are required. C1–C4 (methane to butane, oxidized by methane monooxygenase-like enzymes), C5–C16 (pentane to hexadecane, oxidized by integral membrane non-heme iron or cytochrome P450 enzymes), and C17+ (longer alkanes, oxidized by essentially unknown enzyme systems). Alkanes longer than C16 support growth of many microorganisms. For example, many *Rhodococcus* isolates grow well on purified alkanes up to C32 by virtue of unknown enzyme systems whereas uncharacterized alkane oxygenases allow *P. fluorescens* to grow on C18–C28 alkanes (Beilen and Funhoff, 2007). N-alkane dioxygenase in *Acinetobacter sp.* M-1, has been found to be long chain (C10-C30) utilizer *alm*A gene that, encodes a putative monooxygenase, found to be involved in the utilization of *n*-alkanes with a chain length of C32 or longer. Genes homologous to *alm*A have been identified in *Acinetobacter sp.* RAG-1, *Acinetobacter sp.* M-1 and *Acinetobacter baylyi* ADP1 (Wenzel et al., 2007).

Biodegradation rates are highest for saturates, followed by the light aromatics, and polar compounds. Higher temperatures increase the rates of hydrocarbon metabolism to a maximum, typically in the range of 30 to 40 °C. Oxidation is the initial step in aliphatic, cyclic and aromatic hydrocarbons catabolism by bacteria and fungi. Aerobic conditions are therefore necessary for this route of microbial oxidation of hydrocarbons in the environment. Based on the number of published reports, the most important hydrocarbon-degrading bacteria in both marine and soil environments are *Achromobacter, Acinetobacter, Alcaligenes, Arthrobacter, Bacillus, Flavobacterium, Nocardia,* and *Pseudomonas* spp. and the coryneforms (Leahy And Colwell, 1990).

Mixed microbial communities have the most powerful bio-degradative potential because the genetic information of more than one organism is necessary to degrade the complex mixtures of organic compounds present in contaminated areas (Fritsche and Hofrichter).

Laboratory Scale Enhanced Heavy Oil Recovery (EHOR) Experimental Studies at Sultan Qaboos University, Oman

Bacteria are able to live with oil and cause desired changes in its physiochemical properties. In Oman, thermophillic spore forming bacteria were isolated from contaminated soil (Al-Bahry et al. 2013a, 2013b; Al-Sulaimani et al., 2012). In another study, the spore forming isolates were found to degrade heavy oil forming clearing zones on oil plate agars after twenty one days of incubation. The filtrate from microbial culture formed similar clearing zone suggesting that the isolates secretes components, probably enzymes, which break heavy oil. Heavy oil (C50-C70) was converted to light chain hydrocarbons ranging from C11-C27. Microbial density was found to increase as viscosity of liquid medium. Sequencing of 16 sRNA revealed that the isolates belong to alpha-proteobacteria, gamma-proteobacteria and bacilli.

This study suggests that there is a potential use of thermophilic spore-forming bacteria or their by-products to recover heavy oil in Omani oil fields. If this technology is up-scaled, it can play a significant role to recover heavy oil and reduce the expensive and none-environmental friendly chemical and physical based technologies. Isolation and identification of such microbes is very important for their conservation. Different bacteria isolates from this study may be contributed in may contribute to heavy oil recovery by breaking long C-chain of hydrocarbons.

REFERENCES

Al-Azkawi, A., Taylor, G., Chadwick, R. & Mcgarian, B. (2002). Multilateral wells improve development of heavy crude production in the Mukhaizna field. *International Thermal Operation and Heavy Oil Symposium and International Horizontal Well Technology Conference*. Society of Petroleum Engineer. Calgary, Alberta, Canada.

Al-Bahry, S. N., Elshafie, A. E., Al-Wahaibi, Y. M., Al- Bemani, A. S., Joshi, S. J., Al-Maaini, R. A., Al-Alawi, W. J., Sugai, Y. & Al- Mandhari, M. (2013a). Microbial consortia in Oman oil fields: a possible use in enhanced oil recovery. *Journal of Microbiology and Biotechnology.*, *23*, 106–117.

Al-Bahry, S. N., Al-Wahaibi, Y., Elshafie, A. K., Al-Bemani, A. S., Joshi, S. J., Al-Makhmari, H. & Al- Sulaimani, H. (2013b). Biosurfactant production by *Bacillus subtilis* B20 using date molasses and its application in enhanced oil recovery. *International Biodeterioration & Biodegradation*, *81*, 141-146.

Al-Hattali, R., Al-Sulaimani, H., Al-Wahaibi, Y., Al-Bahry, S., Elshafie, A., Al-Bemani, A. & Joshi, S. (2012). Improving sweep efficiency in fractured carbonate reservoirs by microbial biomass. International Enhancement Oil Recovery Conference at Oil and Gas West Asia. *Society of Petroleum Engineering* (*SPE*). 154679-MS.

Al-Sulaimani, H., Joshi, S., Al-Wahaibi, Y., Al-Bahry, S. N., Elshafie, A. & Al-Bemani, A. (2011). Microbial biotechnology for enhancing oil recovery: Current developments and future prospects. *Biotechnology, Bioinformatics and Bioengineering*, *1*, 147-158.

Al-Sulaimani, H., Al-Wahaibi, Y., Al-Bahry, S. N., Elshafie, A., Al-Bemani, A., Joshi, S. & Ayatollahi, S. (2012). Residual oil recovery via injection of biosurfactant, chemical surfactant and mixtures of both under reservoir condition: Induced Wettability and Interfacial Tension Effects. *SPE Reservoir Evaluation & Engineering, 15*, 210-217.

Ayirala, S. C. (2002). *Surfactant-induced relative permeability modifications for oil recovery enhancement.* Master of Science in Petroleum Engineering, Louisiana State University.

Behlülgil, K. & Lu, M. T. M. (2002). Bacteria for Improvement of Oil Recovery: A Laboratory Study. *Energy Sources, 24*, 413–421.

Beilen, J. B., Li, Z., Duetz, W. A., Smits, T. H. M. & Witholt, B. (2003). Diversity of Alkane Hydroxylase Systems in the Environment. *Oil Gas Science and Technology., 58*, 427-440.

Beilen, J. V. & Funhoff, E. (2007). Alkane hydroxylases involved in microbial alkane degradation. *Applied Microbiology and Biotechnology., 74*, 13-21.

Borgne, S. L. & Quintero, R. (2003). Biotechnological processes for the refining of petroleum. *Fuel Processing Technology, 81*, 151-169.

Bryant, R. S. (1987). Potential uses of microorganisms in petroleum recovery technology. *Proceedings of the Oklahoma Academy of Science, 67*, 97-104.

Bryant, R. S., Bailey, S. A., Stepp, A. K., Evans, D. B., Parli, J. A. & Kolhatkar, A. R. (1998). *Biotechnology for Heavy Oil Recovery,110.*

Bryant, R. S. & Burchfield, T. E. (1989). Review of Microbial Technology for Improving Oil Recovery. *SPE Reservoir Engineering, 4*, 151-154.

Campbell, C. J. & Laherrère, J. H. (1998). The End of Cheap Oil. *Scientific American*, 78-83.

Central Intelligence Agency. (2013). Oman Economy. http://www.theodora.com/wfbcurrent/oman/oman_economy.html. Accessed, March 30.

Clark, B. (2007). *Heavy Oil.* Working Document of the NPC Global Oil and Gas Study. National Petroleum Council. Heavy Oil Subgroup of The Technology Task Group of The NPC Committee on Global Oil And Gas.

Duhalt, V. R., Torres, E., Valderrama, B. & Borgne, S. L. (2002). Will biochemical catalysis impact the petroleum refining industry? *Energy & Fuels., 16*, 1239-1250.

Flint, L. (2004). Bitumen and very heavy crude upgrading technology - a review of long term r&d opportunities. LENEF Consulting (1994) Limited http://www.ptac.org/links/dl/osdfnlreport.pdf.

Fritsche, W. & Hofrichter, M. Aerobic Degradation by Microorganisms. *In: Hans-Jördening J, Winter J.* Environmental Biotechnology: Concepts and Applications. 146-155. John Wiley & Sons. Weinheim, Germany.

Gieg, L. M., Duncan, K. E. & Suflita, J. M. (2008). Bioenergy Production via Microbial Conversion of Residual Oil to Natural Gas. *Applied and Environmental Microbiology, 7*, 3022-3029.

Hamme, J. D. V., Singh, A. & Ward, O. P. (2003). Recent advances in petroleum microbiology. *Microbiology and Molecular Biology Reviews., 67*, 503-549.

Harris, R. E. & McKay, I. D. (1998). New Applications for enzymes in oil and gas production. *Society of Petroleum Engineers., 50621*, 467-475.

Hasan, F. (2011). *Oman Economic Overview.* Global Research. Global Investment House. Safat, Kuwait.

Head, I. M., Jones, D. M. & Larter, S. R. (2003). Biological activity in the deep subsurface and the origin of heavy oil. *Nature., 426*, 344-352.

IEA (International Energy Agency). (2011). Oil market report.

Kebria, D. Y., Khodadadi, A., Ganjidoust, H., Badkoubi, A. & Amoozegar, M. A. (2009). Isolation and characterization of a novel native *Bacillus* strain capable of degrading diesel fuel. *International Journal of Environmental Science and Technology.*, *6*, 435-442.

Lau, Y. P. & Gray, L. J. (2008). Enzyme enhanced oil recovery (EEOR) for waterflooding operations. *Patent application.* US20080196892A1. Houston. Texas. USA.

Leahy, J. G. & Colwell, R. R. (1990). Microbial Degradation of Hydrocarbons in the Environment. *Microbiological Reviews*, *54*, 305-315.

Malik, S., Zhang, Y. M., Asimi, M. A. & Gould, T. L. (2010). Modified SADG in multizones at Mukhazina. Eor Conference at Oil and Gasm West Asia. Society of Petroleum Engineer. Muscat, Oman.

McInerney, M. J., Knapp, R. M., Chisholm, J. L., Bhupathiraju, V. K. & Coates, J. D. (1999). *Use of indigenous or injected microorganisms for enhanced oil recovery.* 8th International Symposium on Microbial Ecology, Halifax, Canada.

McInerney, M. J., Knapp, R. M., Duncan, K., Simpson, D. R., Youssef, N., Ravi, N. & Weiland, S. (2007). Development of an *in situ* biosurfactant production technology for enhanced oil recovery (1-161), Technical Report. University of Oklahoma Department of Botany and Microbiology.

Memon, A. I., Taylor, S. D., Davies, T. L. & Jia, N. (2010). A Systematic Workflow process for heavy oil characterization: experimental techniques and challenges. Canadian Unconventional Resources and International Petroleum Conference. Society of Petroleum Engineer. Calgary, Canada.

Meintanis, C., Chalkou, K. I., Kormas, K. A. & Karagoun, A. D. (2006). Biodegradation of crude oil by thermophilic bacteria isolated from a volcano island. *Biodegradation*, *17*, 3-9.

Menezes, B. F., Camargoa, F. A., Okekeb, B. C. & Jr,W. T. (2005). Diversity of biosurfactant producing microorganisms isolated from soils contaminated with diesel oil. *Microbioogical Research*, *160*, 249-255.

Montiel, C., Quintero, R. & Aburto, J. (2009). Petroleum biotechnology: technology trends for the future. *African Journal of Biotechnology*, *8*, 2653-2666.

Nadyal, M., Thurber, S. S., Al-Hinai, K. M. & Jaffer, N. A. (1983). Development of heavy oil reserves in south Oman. The Middle East Oil Technical Conference of the Society of Petroleum Engineers. Manama, Bahrain.

Nielsen, S. M. (2010). *microbial enhanced oil recovery-advanced reservoir simulation.* PHD dissertation, Technical University of Denmark, Denmark.

Oil & Gas Directory Middle East. (2009). *Oman research profile*, 963-968. Musct, Oman.

Rosenberg, A. & Alexander, M. (1979). Microbial cleavage of various organophosphorus insecticides. *Applied Environmental Microbiology*, *37*, 886-891.

Rosenberg, M., Bayer, E. A., Delarea, J. & Rosenberg, E. (1982). Role of thin fimbriae in adherence and growth of *Acinetobacter calcoaceticus* rag-1 on hexadecane. *Applied and Environmental Microbiology*, *44*, 929-937.

Sen, R. (2008). Biotechnology in petroleum recovery: the microbial EOR. *Progress in Energy and Combustion Science*, *34*, 714-724.

Sette, L. D., Simioni, K. C. M., Vasconcellos, S. P., Dussan, L. J., Neto, E. V. & Oliveira, V. M. (2007). Analysis of the composition of bacterial communities in oil reservoirs from a southern offshore Brazilian basin. *Antonie Van Leeuwenhoek*, *91*, 253-266.

Zhang, Y., Xu, Z., Ji, P. & Hou, W. (1999). Microbial EOR Laboratory studies and application results in Daqing oil field. *In*: Proceedings of the 1999 SPE Asia Pacific Oil and Gas Conference and Exhibition, Jakarta, Indonesia, April 20–22 1999, pages 1–8. *Society of Petroleum Engineers* (*SPE*). 54332.

In: Biotechnology and Conservation … (2 Volume Set)
Editors: S. A. Farooq, R. Abed and S. Baqir
ISBN: 978-1-63117-869-6

Chapter 47

SCREENING AND ANTIMICROBIAL ACTIVITY OF ACTINOMYCETES STRAINS ISOLATED FROM EXTREME ECOSYSTEMS OF ALGERIA

M. Kitouni[*] and L. Oulmi
Laboratoire de Biologie appliquée et santé. Faculté des Sciences de la Nature et de la vie. Université de Constantine 1, Constantine, Algerie

ABSTRACT

The constant development of bacterial resistance to antibiotics and the appearance of new infectious diseases justify the emergency of searching new antimicrobial molecules.

Actinomycetes are the main source of secondary metabolites with anti-cellular activity. In Algeria, few scientific works has been done on their presence and diversity in different ecosystems.

In this study, forty five Actinomycetes strains were isolated from various extreme ecosystems of Algeria characterized by high salinity, the aridity region and polluted water. The isolates were purified and identified.

Antibacterial activity was screened against three Gram-positive bacteria (*Staphylococcus aureus* ATCC 25923, *Staphylococcus aureus* Mu 50 resistant to vancomycin and *Streptococcus faecalis* ATCC 19433) and three Gram-negative bacteria (*Escherichia coli* ATCC 25922, *Proteus vulgaris* ATCC 13315 and *Pseudomonas aeruginosa* ATCC 10145).

Antifungal activity of actinomycetes isolates was estimated by the agar cylinders method against filamentous fungi (*Fusarium oxysporum* UMIP 625.72, *Aspergillus fumigatus* UMIP 1082.74 and *A. niger* ATCC 16404) and the yeasts (*Candida albicans* UMIP 884.65, *C. albicans* UMIP 48.72 and *C. tropicalis* R2 UMIP 1275.81 (resistant to amphotericin B and nystatin) in the presence of *Streptomyces noursei* (NRRL B-1714) and *Streptomyces nodosus* (NRRL B-2371) nystatin and amphotericin A and B, producers respectively.

Among the 45 tested isolates, 17 (37.78%) showed activity vis-à-vis at least one of the studied bacteria-tests and 2 (4.44%) showed antifungal activity. The molecular

[*] Corresponding author: mahmoudkitouni@yahoo.fr.

identification of isolates revealed the predominance of the *Streptomyces* genus with 93%, followed by *Nocardia* genus with 4% then *Actinomadura* 2%.

Keywords: Actinomycetes, antimibacterial activity, antifungal activity, extreme ecosystems

INTRODUCTION

If the advent of antibiotics represented a new stage in the improvement of the quality and the lasting life, the constant evolution of bacteria antibioresistance represents a major limitation of their use. Indeed, the increasing emergence of multiresistant bacteria throughout the world and the lack of antibiotics to combat such pathogenic agents continue to be the major concern of the medical community (Kuti et al., 2002). In the United States of America (USA), community-acquired pneumonia infects each year more than 4 million people, of which 20% require hospitalization (Kuti et al., 2002). The species most frequently isolated during its infections are *Streptococcus pneumoniae*, *Hemophilus influenzae* and *Moraxella catarrhalis* (Sutcliffe, 2003). Thirty-four percent of *S. pneumonia* isolated in the USA are resistant to penicillin, 32% of *H. influenzae* are resistant to ampicillin and 91.5% of *M. catarrhalis* are resistant to penicillin and 15% to erythromycin.

In addition, the vancomycin was the antibiotic of choice for the treatment of the infections caused by methicillin-resistant *Staphylococcus aureus*, until the appearance of the first resistant strains, initially in Japan (Hiramatsua, 1997), then in the USA (CDC, 1997) and in France (Poly et al., 1998). To all that is added the resistance of 50% of *Enterococcus* to vancomycin. These bacteria which until 1990 were regarded as nonpathogenic and in particular *Enterococcus faecium* which develops among patients under treatment a resistance to several antibiotics (Auckland et al. 2002).

In addition to the bacterial infections, the fungal infections which constituted a problem exclusively relevant to dermatology have been recognized during the 1990s as a major cause of mortality (Odds, 1995).

According to Pfaller and Wenzel (1992) *Candida* spp. became the third most usually present isolates found in hemocultures in the USA. Moreover, the death rate due to the infections with *Candida*, *Aspergillus* and *Fusarium* among patients under immunosuppressive chemotherapy or transplant patients is relatively high (Odds, 1995).

Amphotericine B was the most appropriate antifungal agent for the treatment of the majority of the mycoses, nevertheless undesirable side effects are observed during the use of this antifungal agent in particular renal toxicity (Odds, 1995). Beside the amphotericine B, the antifungals belonging to the derivatives azoles class are not very toxic, but their efficacies is limited in the treatments of the mycoses infections. In addition, it has been reported that the frequency of multiazole-resistant strains belonging to *Candida* species other than *Candida albicans* is increasing (Hitchcok et al., 1993). From this report the search for new antibiotics enters a new phase of development. The actinomycetes represent the principal sources of secondary metabolites with anticellular activities, and among-them rare actinomycetes constitutes a potentially important source of novel antibiotics (Iwai and Takahashi, 1992). For that, many laboratories endeavored to diversify the sources of actinomycetes by the use of samples coming from unexplored habitats (kitouni et al., 2005).

The aims of this study is based on the isolation and the selection of antibiotics producing actinomycetales strains isolated from various areas of the north–east of Algeria and the molecular identification of these strains.

COLLECTION OF SAMPLES

The soil samples of collected according to the technique of Pochon and Tardieux (1962) from a semi-arid soil surrounding the Sebkha of Ain M'lila located at the North-East of Algeria. Water samples, were taken from the tables (basins) of the salt company (ENA Sel Algeria) located at Ain M'lila and from the lakes closed and strongly polluted of Djbel Ouahche mountain located at Constantine, city of the Algerian North-East according to the technique of Rodier (1984). Trees barks were collected from Oak (Quercus) and Cedars (Cedrus) of Djbel Ouahche Mountain, by cutting out pieces of barks using a sterile blade. The taken barks are deposited on a sterile aluminum sheet, then recovered in sterile flasks and transported to the laboratory.

BACTERIAL ISOLATION AND IDENTIFICATION

Three culture media were used for actinomycetes isolation - GLM: [yeast extract 3 g/L, malt extract 3 g/L, peptone 5 g/L, glucose 10 g/L, agar 20 g/L, (pH 7.2)], Bennett: [yeast extract 1 g/L, meat extract 1 g/L, casaminoacides 2 g/L, glucose 10 g/L, agar 15 g/L, (pH 7.3)] and Glucose-Asparagine: [glucose 10 g/L, K_2HPO_4 0,5 g/L, asparagine 0,5 g/L, agar 15 g/L, (pH 7,0)]. These media are supplemented with 75 µg/mL of amphotericin B to inhibit the development of filamentous fungi and 10 µg/mL of polymixin to inhibit Gram-negative bacteria. Inocula consisting of 0.1 mL of dilutions 10^{-3}, 10^{-4} and 10^{-5} in the physiological water (NaCl 9 g/L) of different samples were spread over the surface of the three culture media. The plates were then incubated at 28°C. All the plates were observed after two, three and four weeks of incubation at 28°C.

ADN was extracted according to (Provost et al., 1997) Fragments of The 16S rRNA genes were amplified with PCRs using primers couple 91E (3'-TCAAAK* GAATTGACGGGGGC-5') and $16S_2$ (5'-CCCGGGAACGTATTCAC-3'). Amplified reactions were carried out in a thermal cycler (MinicyclerTM Mj. Research), after initial denaturation at 94°C for 10 minutes followed for 45 cycles of PCR at 94, 55 and 72°C for 30 seconds each and finally for 10 minutes extension period at 72°C. Sequences were determined using an ABI PRISM Taq DyeDeoxy DNA sequencer (Applied Biosystems). Phylogenetic analysis was carried out using MEGA, version 3.1 (Kumar et al., 2004), after multiple alignment of the data by CLUSTAL W, version 1.8 (Thompson et al., 1994). Distances and clustering were determined using the neighbor-joining method with bootstrap analyses based on 1000 replications.

BIOACTIVITY OF ISOLATES

Antibacterial Activity

Actinomycetes isolates were grown on Bennett's and GLM agar plates for 7 days at 28°C. Agar cylinders (3 mm in diameter) were then taken with hollow punch and deposited on the surface of the Mueller-Hinton media (Merck), which had previously been seeded with each test bacteria obtained from the American Type Culture Collection (ATCC) and from Juntendo Hospital, Tokyo Japan; they are three Gram-positive bacteria (*Staphylococcus aureus* ATCC 25923, *Staphylococcus aureus* Mu 50 resistant to vancomycin and *Streptococcus faecalis* ATCC 19433) and three Gram-negative bacteria (*Escherichia coli* ATCC 25922, *Proteus vulgaris* ATCC 13315 and *Pseudomonas aeruginosa* ATCC 10145). Plates were kept at 4°C for 4 hours, and then incubated at 37°C for 18 to 24 hours. Inhibition diameters were then measured.

Antifungal Activity

Antifungal activity of actinomycetes isolates was estimated by the agar cylinders method against filamentous fungi obtained from the ATCC and the fungi collection of the Mycology Unit of the Pasteur Institute (UMIP) [*F. oxysporum* UMIP 625.72, *A. fumigatus* UMIP 1082.74 and *A. niger* ATCC 16404) and counters the yeasts (*C. albicans* UMIP 884.65, *C. albicans* UMIP 48.72 and *C. tropicalis* R2 UMIP 1275.81 (resistant to amphotericin B and nystatin)] in the presence of *S. noursei* (NRRL B-1714) and *S. nodosus* (NRRL B-2371) producing of nystatin and amphotericin A and B, respectively, obtained from Northern Regional Research Laboratory (NRRL). The activity was tested in casitone medium, except for *C. tropicalis* R2, for which the activity is carried out in YMA medium. Inhibition diameters were measured after 24 h of incubation at 28 °C for yeasts and after 48 h for filamentous fungi.

Diversity and Characterictis of the Isolates

Results presented in table 1 reveal an important variation of the number of actinomycetes strains isolated from each sample on the whole of the three selective media. Among the 45 isolated strains of actinomycetes 24 were isolated from telluric samples (soil of sebkha) i.e. 53.33 %, 10 from trees barks thus 22.22 % and 11 from the water samples thus 24.44 %. The presence of high number of actinomycetes in the soil and tree barks are in agreement with the bibliographical data which let appear the soil as the principal reservoir of actinomycetes.

The phylogenetic diversity of actinomycetales isolates was determined by the universal PCR. Fragments of 479pb of 16S rDNA were obtained using primers 91E and 16S2. Sequences analysis of these fragments enabled us to classify the isolates in the *Actinobacteria* phylum, class V of *Actinobacteridae* and the order I of *Actinomycetales*. They are spread over three different genera; *Streptomyces* and *Actinomadura* who belonged to the *Streptomycineae* order and *Nocardia* genus belong to the *Corynebacterineae* order. Ninety-three percent of actinomycetes identified by the universal PCR primers are belonging to the *Streptomyces* genus, 4 % to the *Nocardia* genus and 2 % to the *Actinomadura* genus. These results confirm the abundance of *Streptomyces* in telluric ecosystems compared to the other genera of actinomycetes (Figure 1).

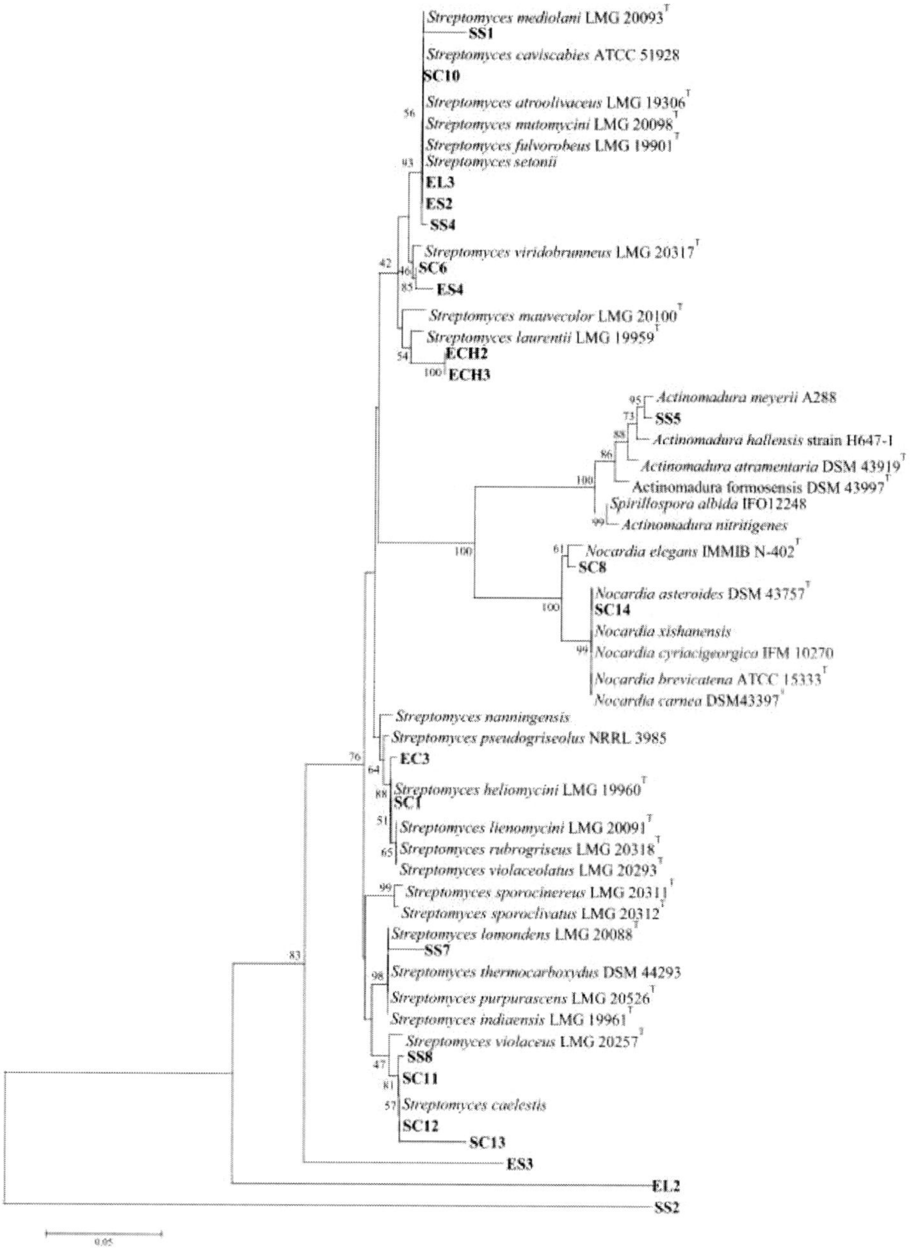

Figure 1. Neighbour-joining phylogenetic tree, based on fragments of 479pb 16S rRNA gene sequences, showing the relationship of the 222 actinomycetes isolates and species of the genus Streptomyces, genus Nocardia and genus Actinomadura Bootstrap percentages (bases on 1000 replications) are shown if greater than 40%. Bar, 0.05 substitutions per site.

Table 1. Origin of isolated actinomycetals strains

Samples	Number of actinomycetes isolates
Soil of sebkha (Ain M'lila)	08
Water of sebkha de (Ain M'lila)	04
Lake water (Constantine)	03
Treesde barks (Constantine)	10
Thermal spring water (Guelma)	00
Thermal spring water (Bniharoun)	04
Mountain soil (Constantine)	16
Total	**45**

The results of of the antibacterial and antifungal activity of the isolated actinomycetals strains appear in table 2 and table 3. The diffusion method was used for the screening of antibacterial activity of the 45 isolated actinomycetales (Figure 2). Among the 45 tested isolates, only 17 strains (37.78 %) were active against at least one of the test-bacteria (table 2) and two isolates (4.44 %) presents antifungal activity (table 3).

Isolates SS2, SS4 and SC10 showed activity against all of the tested bacteria. Only EC5 isolate presents activity only against *S. aureus* Mu 50 (Juntendo Hospital, Tokyo, Japan) and ES3 against *E. coli* ATCC 25922, EC1, and EC3 inhibits all the Gram-positive bacteria, SS6 and EC6 isolates were the unique actinomycetes which inhibited only *Proteus vulgaris* ATCC 13315.

Isolates SS2 showed activity against all of the tested fungi and EL3 presents activity only against *C. albicans* UMIP 48.72, *C. albicans* UMIP 884.65, *C. tropicalis* R_2 UMIP 1275.81 and A. *niger* ATCC 16404.

Table 2. Antibacterial activity of the isolated actinomycetes strains

Isolate	**Origin**	***E. coli* ATCC 25922**	***P. vulgaris* ATCC 13315**	***P. aeruginosa ATCC 10145***	***S. faecalis ATCC 19433***	***S. aureus Mu 50***	***S. aureus* ATCC 25923**
SS2	Soil of sebkha	+	+	+	+	+	+
SS3	Soil of sebkha	-	+	-	+	+	-
SS4	Soil of sebkha	+	+	+	+	+	+
SS5	Soil of sebkha	+	-	-	-	+	+
SS6	Soil of sebkha	-	+	-	-	-	-
ES2	Sebkha water	-	-	-	+	-	+
ES3	Sebkha water	+	-	-	-	-	-
ES4	Sebkha water	-	+	-	-	+	+
EL2	Lake water	+	-	-	+	-	-
EL3	Lake water	+	-	-	+	-	+
EC1	*Cedrus* bark	-	-	-	+	+	+
EC3	*Cedrus* bark	-	-	-	+	+	+
EC5	*Cedrus* bark	-	-	-	-	+	-
EC6	*Cedrus* bark	-	+	-	-	-	-
SC1	Soil of Constantine	-	+	-	+	-	+
SC6	Soil of Constantine	+	+	-	+	+	+
SC10	Soil of Constantine	+	+	+	+	+	+

Table 3. Antifungal activity of the isolated actinomycetes strains

Isolate	Origin	*C. albicans* UMIP 48.72	*C. albicans* UMIP 884.65	*C. tropicalis* R_2 UMIP 1275.81	*A. fumigatus* UMIP 1082.74	*A. niger* ATCC 16404	*F. oxysporum* UMIP 625.72
SS2	Sol de sebkha	+	+	+	+	+	+
EL2	Eau de lac	+	+	+	-	+	-
Streptomyces noursei NRRL B-1714	-	+	+	-	+	+	+
Streptomyces nodosus NRRL B-2371	-	+	+	-	+	+	+

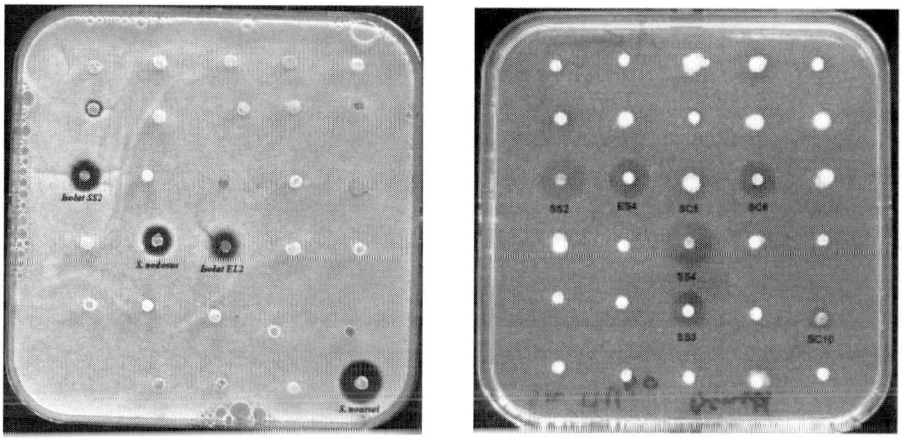

Figure 2. The activity of actinomycetes isolates against *S. aureus* Mu 50 and *Aspergillus niger* ATCC 16404.

References

Auckland, C., Teare, L., Cooke, F., Kaufmann, M. E., Warner, M., Jones, G., et al. (2002). Linezolid-resistant enterococci: report of the first isolates in the United Kingdom. *J Antimicrob Chemother., 50*, 743–6.

Centers for Disease Control and Prevention. (1997). *Staphylococcus aureus* with reduced susceptibility to vancomycine-United States, 1997. *Morbid Mortal Wkly Rep., 46,* 765–76.

Hiramatsua, K., Hanaki, H., Ino, T., Yabuta, K., Oguri, T. & Ten-over, F. C. (1997). Methicillin-resistant *Staphylococcus aureus* clinical strain with reduced vancomycin susceptibility. *J Anti-microb Chemotherapy., 40*, 135–46.

Hitchcok, C. A., Pye, G. W., Tork, D. F., Johnson, E. M. & Warnok, D. W. (1993). Fluconazole resistance in *Candida glabrata. Antimicrob Agents Chemother., 37*, 1962–5.

Iwai, Y. & Takahashi, Y. (1992). Selection of Microbial Sources of Bioactive Compounds. In the Search for Bioactive Compounds from Microorganisms. In: Omura S, editor. New York: Springer-Verlag; 281–301.

Kitouni, M., Boudemagh, A., Oulmi, L., Reghioua, S., Boughachiche, F., Zerizer, H., Hamdiken, H., Couble, A., Mouniee, D., Boulahrouf, A. & Boiron, P. (2005). Isolation of actinomycetes producing bioactive substances from water, soil and tree bark samples of the north–east of Algeria. *Journal de Mycologie Médicale*, *15*, 45–51

Kumar, S., Tamura, K. & Nei, M. (2004). MEGA3: integrated software for Molecular Evolutionary Genetics Analysis and sequence alignment. *Brief Bioinform.*, *5,* 150–163.

Kuti, J. L., Capitano, B. & Nicolau, D. P. (2002). Cost-effective approaches to the treatment of community-acquired pneumonia in the era of resistance. Pharmacoeconomics., *20*, 513–28.

Odds, F. C. (1995). Les agents antifongiques, leur passé, leur présent et leur avenir. *Bull Soc Fr Microbiol.*, *10*, 285–93.

Pochon, J. & Tardieux, P. (1962). Techniques d'analyse en microbiologie du sol. Saint Mandé: Edition de la tourtourelle.

Poly, M. C., Greland, C., Martin, D., Lumley, L. & Denis, F. (1998). First clinical isolate of vancomycin intermediated *Staphylococcus aureus* in French hospital. *Lancet*, *351*, 1212.

Provost, F., Laurent, F., Camachouz categeri, L. R. & Boiron, P. (1997). Molecular study of persistence of *Nocardia asteroides* and *Nocardia otitidiscaviarum* strains in patients with longterm nocardiosis. *J Clin Microbiol.*, *35*, 1157–60.

Rodier, J. (1984). L'analyse de l'eau. Paris: Dunod; 7ème édition.

Sutcliffe, J. A. (2003). Antibacterial agents: solutions for the evolving problems of resistance. *Bioorg Med Chem Letter.*, *13*, 4159–61.

Thompson, J. D., Higgins, D. G. & Gibson, T. J. (1994). CLUSTAL W: improving the sensitivity of progressive multiple sequence alignment through sequence weighting, position-specific gap penalties and weight matrix choice. *Nucleic Acids Res*, *11*, 4673-80.

In: Biotechnology and Conservation … (2 Volume Set)
Editors: S. A. Farooq, R. Abed and S. Baqir
ISBN: 978-1-63117-869-6

Chapter 48

MICROBIAL CELLULASES AND ITS POTENTIAL APPLICATIONS IN OMAN

***Amira A. Al Zadjali*[1], *Saif N. Al Bahry*[2] *and Nallusamy Sivakumar*[2]**
[1]Higher College of Technology, Alkuwair, Muscat, Oman
[2]Department of Biology, College of Science, Sultan Qaboos University, Muscat, Oman

ABSTRACT

Cellulase is a microbial enzyme, which degrades the cellulose into glucose. There is an increasing interest in the cellulase studies because the produced glucose from cellulose could be used as a substrate to produce biofuels and other important products. The screening and preservation of cellulase producing microorganisms from arid region has a significance industrial importance. Because of the diverse environmental conditions exist in the arid region of Oman, the chances of getting extremophilic cellulase producing microorganisms are more. In this chapter structure of cellulosic biomass, the mode of action of cellulase enzymes, microbial cellulases and their applications are discussed.

Keywords: Cellulose, cellulase, arid region, paper recycling

INTRODUCTION

The increase of solid waste is becoming an important issue worldwide (Wyk and Mohulatsi, 2003). Improper management of solid wastes contaminated air, soil and water is considered as a global problem (Fujita et al., 2004). Cellulosic wastes such as paper, wood, agriculture residues and cardboard materials are considered as the major components of solid wastes (Bayer et al., 2007). Different ways such as burial, incineration and recycling are used to get rid of solid wastes. Disposal of solid wastes in landfills contaminates the ground water. Moreover, landfills cause emission of greenhouse gases such as carbon dioxide and methane (Connett, 1998). This causes an increase in temperature of the earth's atmosphere and change climate. Most of the carbon dioxide and methane are produced from biodegradable compounds such as wood, leaves and waste papers (Barlaz et al., 1997). In addition, incineration of materials containing carbon, especially cellulosic materials, generates the same problem.

Therefore, recycling of cellulosic wastes such as wood, agriculture residues and waste papers will decrease the greenhouse effect and will protect the environment (Kazaragis, 2005). There is an increasing interest in the use of cellulosic biomass as a renewable source of energy by breaking down cellulose into sugars using different cellulase enzymes. Then, the sugar can be converted to some useful products such as liquid fuel (Demain et al., 2005). In this review, we discuss the mode of action of cellulase, cellulase producing microorganisms, cellulase fermentation conditions and application of cellulase enzyme.

Structure of Cellulosic Biomass

Cellulose is synthesized by all higher plants and other organisms such as bacteria, fungi, protists and invertebrates (Haigher and Wemier, 1991). It is considered as the most abundant renewable natural biological resource (Zhang et al., 2006). Cellulose is a linear polysaccharide composed of β-1, 4 linked D-glucopyranosly units (Figure 1). It is synthesized as individual molecules, which undergo self-assembly to give protofibrils. These protofibrils are assembled together to form crystalline cellulose. Fibrous structure and strong glucosidic bond of cellulose make it highly resistant to be hydrolyzed by enzymes and water. In most cases, cellulose does not occur alone in nature but is associated with other biopolymers. This association affects its degradation rate. Cellulose fibrils are embedded in a matrix of other biopolymers such as hemicelluloses, pectins, lignins and proteins (Lynd et al., 2002).

Hemicelluloses are heteropolysaccharides of pentoses, hexoses and sugar acids. Pentoses are the dominant in the hemicelluloses with relatively small amounts of the L-sugars. They include xylose and L-arabinose. Hexoses, which include mannose are found too in the hemicelluloses as well as sugar acids. Xylans are often the most abundant hemicelluloses. Its chemical structure is made of a β-1,4-linked xylopyranose backbone with attached side groups of acetate, arabinofuranose, and O-methyl glucuronic acid (Saha and Bothast, 1997).

Lignin is hetergenous polymers composed of three subunits: p-coumaryl alcohol (p-hydroxyphenyl propanol), coniferyl alcohol (guaiacyl propanol), and sinapyl alcohol (syringyl propanol). These building blocks are connected by ether bond and carbon to carbon linkage. Lignin found in plant in plant cell wall. It gives the plant strength and protects it from fungal infection (Tuomela et al., 2000).

Mechanism of Cellulosic Biomass Degradation

Cellulose is degraded by cellulase enzymes which are secreted by different microorganisms such as bacteria and fungi (Lynd et al., 2002; and Zhang et al., 2006). There are three types of celluases, which are involved in cellulose hydrolysis: endoglucanases, exoglucanases and β-glucosidases (Saqib and Whitney, 2006).

Endoglucanases consist of different classes produced by bacteria and fungi. They cleave randomly intramolecular β-1, 4-glucosidic bonds of cellulose chain and produce oligosaccharides (Zhang et al., 2006). Some substrates such as carboxymethylcellulose, acid –

swollen cellulose and cellulose azure can be hydrolysed by endoglucanases only (Han et al., 1995).

Exoglucanases hydrolyse cellulose chain at the reducing and non-reducing ends to produce cellobiose or glucose (Lynd et al., 2002). β-glucosidases convert cellobiose to glucose and cleave oligosaccharide (Han et al., 1995). Complete hydrolysis of cellulose occurs by synergistic action of the three types of cellulases.

Hemicelluloses are heterogeneous polymers of pentoses (xylose and L-arabinose), hexoses (mannose), and sugar acids. Xylans, major hemicelluloses of many plant materials, contain xylose, L-arabinose, and D-glucoronic acid, among others. Two major enzyme are needed to hydrolysis xylan; endo β-1, 4 xylanase, β-xylosidase and several accessory enzyme activities such as α- L- arabinosidase and α– glucoronidase. Endo-xylanase randomly attacks the main chains of xylans. β-xylosidase hydrolyzes xylooligosacharides to xylose. The α-L-arabinosidase and α-glucoronidase remove the arabinose and 4-O-methyl glucoronic acid substituents, respectively, from the xylan backbone (Jeffries, 1994).

Lignin is biodegraded by bacteria and fungi. Some fungi and bacteria have ability to secret liginases enzymes. Lignin peroxidases, manganese peroxidases, and laccases are three families of enzymes that degraded the lignin (Saha and Bothast, 1997).

Microbial Cellulases

Celluloytic enzymes are produced by both bacteria and fungi. Cellulytic fungi include *Aspergillus* spp., *Fusarium* spp., *Chaetomium* spp. and *Trichoderma* spp. (Lynd et al., 2002). Celluloytic bacteria include aerobic, anaerobic and facultative anaerobic bacteria. The aerobic bacteria include *Bacillus* spp., *Pseudomonas* spp. and *Actinomycetes*. The anaerobic bacteria include the group of *Clostridium* spp., while the facultative anaerobic bacteria include *Cellulomonas* spp. (Heck et al., 2002).

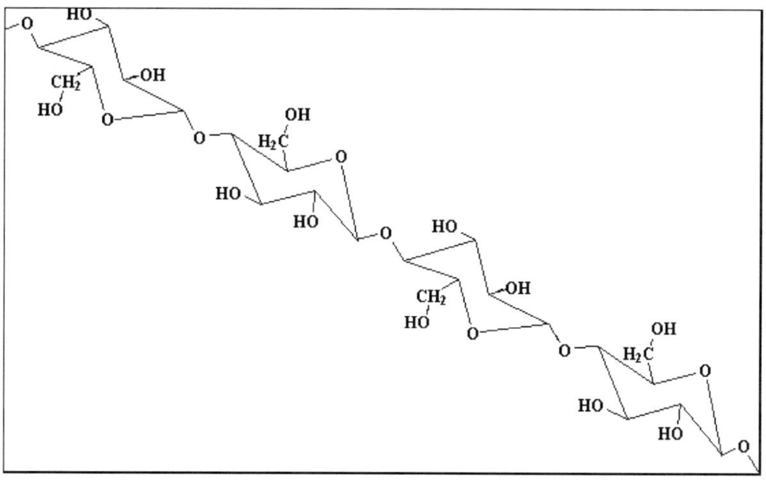

Figure 1. Structure of cellulose polymer.

Fungal Cellulases

Recently *Aspergillus* spp. and *Trichoderma* spp. were used for commercial production of cellulases (Immanuel et al., 2007). Cellulases produced by *Trichoderma* spp. were extensively investigated. *T. reesi* has three cellulase enzymes: endoglucanases, exoglucanases and β-glucosidases. Zhou, et al., (2008), reported that mutant strain of *T. viride* T100-14 secretes cellobiohydrolases, endoglucanses and high amount of β-glucosidases, which are able to break cellobiose to glucose. However, *Trichoderma* cellulases have several disadvantages such as low cellulases yields, low specific activities, end-product inhibition of the cellulases (Zaldívar et al., 2001).

Bacterial Cellulases

Bacterial cellulases have been studied more extensively from *Bacillus* spp., *Clostridium* spp., *Cellulomonas* spp. and *Thermomonaspora* spp. Bacteria have the ability to produce thermostable and alkalinestable cellulases are considered as industrial cellulases. Thermostable cellulases have potential application in industries because higher processing temperatures can speed up reaction rates, increasing the solubility of reactants and minimize the contamination (Ng et al., 2009).

Han and Kim (1987) reported that the optimum temperature for cellulases activity produced by *Cellulomonas* spp. ATCC was 55°C at optimum pH 5.5. Moreover, they reported that the cellulase was stable for 2.5 h at 60°C. Barman et al., 2011 isolated three celluloytic bacteria such as *Moraxella* sp., *Cellulomonas* sp. and *Planococcus* sp from soil. Among the three isolates, *Moraxella* sp. showed the highest ability for degradation of solid organic waste such as kitchen and agriculture waste. Acharya and Chaudhary, 2012 isolated *Bacillus thermoaerophilus* WBS2, which can secret thermostable cellulase from hot spring in India.

On the other hand, Huang et al., 2012 reported novel cellulolytic microorganisms such as *Siphonobacter aquaeclarae*, *Cellulosimicrobium funkei*, *Paracoccus sulfuroxidans*, *Ochrobactrum cytisi*, *Ochrobactrum haematophilum*, *Kaistia adipata*, *Devosia riboflavina*, *Labrys neptuniae*, *Ensifer adhaerens*, *Shinella zoogloeoides*, *Citrobacter freundii* and *Pseudomonas nitroreducens*.

Secretion of Cellulase by *Bacillus*

Bacillus species have been extensively used as rich sources of industrial enzymes such as proteases (Kotchoni et al., 2003). Furthermore, different *Bacillus* species secrete extracellular enzymes such as polysaccharide hydrolysing enzymes (Mawadza et al., 2000). These enzymes can degrade and hydrolyse several polysaccharides (Ranoa et al., 2004). Most celluloytic *Bacillus* secrete endoglucanases, which are capable of degrading carboxymethyl cellulose (Robson and Chambliss, 1984). Examples of *Bacillus* spp. secreting cellulases are *Bacillus licheniformis*, *Bacillus subtilis* and *Bacillus polymyxa* (Han et al., 1995).

Thermophilic *Bacillus* spp. isolated from compost was reported to produce highly active thermostable cellulases (Mayende et al., 2006). Moreover, another study showed that

Anoxybacillus flarithermus EHP1 isolated from hot spring has optimum temperature for cellulases production at 75°C and optimum pH 7.5. Furthermore, it showed high degradation rate towards CMC. Thermostable alkaline cellulase from *Bacillus* sp.KSM-S237 had been isolated from soil samples collected from Okinawa, Japan (Hakamada et al., 1997). Aygan and Arikan (2008) have reported that the *Bacillus* sp.C14 produced halo-alkalophilic thermostable cellulase.

Usually *Bacillus* spp. are widely used in industries because they grow and reproduce easily. Most of *Bacillus* spp. are not pathogenic, do not produce foul odours or gases, some species can survive at alkaline condition, some species can survive at high temperature, secrete proteins extracellularly and are considered relatively safe to use with regard to health and environmental aspects (Beukes and Pletschke, 2006). The optimum temperature and pH of cellulases produced by some *Bacillus* strains are listed below (Table 1).

Cellulases Research

The major goal of the cellulase research is to optimize the conditions of cellulase production and improve the performance of cellulases to make them more effective. This task can be done by measure the optimum growth conditions for cellulase production. Cellulase yields appear to depend on a complex relationship involving a variety of factors like inoculum size, pH value, temperature, presence of inducers, medium additives, aeration and growth time.

In addition, genetic engineering could improve the performance of cellulase properties (Sukumaran et al., 2005). Genetic techniques were used to clone the cellulase coding sequences into bacteria, yeasts, fungi, plants and animals to create new cellulase production systems with possible improvement of enzyme production and activity. Cellulase production in microorganisms can be regulated by promoter. Promoters are molecular biological tools crucial for regulation of the expression of genes of interest. Functional analyses of the promoter regions of several genes encoding cellulases genes have been reported. For example, *Trichoderma* spp. cbhI promoter is a strong promoter for expression of cellulase genes. This promoter introduce in bacteria to enhance cellulase production (Rahman et al., 2009).

Many individual enzymes acting synergistically form an effective cellulase mix for conversion of lignocellulosics to sugars. Production of cellulase components individually is not economically possible; instead all the proteins necessary should be expressed and secreted by a single host and this can be obtained by application of recombinant-DNA-technology and associated techniques such as protein engineering. Moreover, expression of cellulases in fermenting organisms, or transfer of the biofuel-synthesizing pathway into a cellulase-producing organism, is being pursued in a process termed the consolidated bioprocess (CBP). CBP, however, is presently hampered by the relative inability of yeast to process recombinant cellulases at high rates through their endoplasmic reticulum and secretory pathways, and the relative lag in development of molecular biological methods to manipulate organisms (such as *Bacillus* spp.) that secrete cellulases naturally (Stephanopoulos, 2007).

Table 1. Optimum temperature and pH for cellulases of some *Bacillus* strains

Bacillus sp.	Optimum temperature(°C)	pH	References
Bacillus sp.C14	50	11	(Aygan and Arikan, 2008)
Bacillus strain CH43	65	5-6.5	(Mawadza et al., 2000)
Bacillus strain HR68	70	5-6.5	(Bajaj et al., 2009)
Bacillus strain M-9	60	5	(Bajaj et al., 2009)
Bacillus licheniformis-1	55	6.1	(Dhillon et al., 1985)
Bacillus licheniformis	65	6	(Bischoff et al., 2006)
Bacillus sp. NZ	50	9-10	(Nizamudeen & Bajaj, 2009)
Bacillus circulans	50	4.5	(Kim, 1995)
Bacillus circulans	55	8.5	(Hakamada et al., 2002)
Bacillus sphaericus JS1	65	7–10.5	(Singh et al., 2004)
Bacillus sp.	50	6	(Vijayaraghavan & Vincent 2012)

The synthesis of polymer intermediates, pharmaceuticals and agrochemicals is often hindered by expensive processes that suffer from low selectivity and undesirable byproducts. Because of the lack of enzyme stability, mesophilic enzymes are often not well suited for the harsh reaction conditions required in several industrial processes. For this reason, the use of biocatalysts in organic reactions represented only a small fraction of the potential industrial market in the past. The discovery of new extremophilic microorganisms, organisms that have evolved to exist in a variety of extreme environments and their enzymes has had a great impact on the industrial processes.

Running biotechnological processes at elevated temperature has many advantages. The increase of temperature has a significant influence on the bioavailability and solubility of organic compounds. The elevation of temperature is accompanied by a decrease in viscosity and an increase in the diffusion coefficient of organic compounds. Therefore, higher reaction rates due to smaller boundary layers are expected (Becker et al., 1997). Furthermore, the risk of contamination is reduced by performing biological processes at temperatures above 60 °C and controlled processes under strict conditions can be carried out.

Application

Cellulases achieved significant importance in several industries like laundry, food industry, textile, pulp and paper and ethanol production (Bhat, 2000). For recycling, manufacturing paper from waste paper generally requires less energy than manufacturing paper from wood and pulp. Consequently, paper recycling results in lower greenhouse gases (GHG) emissions than manufacturing them from virgin inputs such as wood (Pelach et al., 2003). However, paper recycling is still not an easy task and can cause harm to the environment. Biotechnologists are trying to make paper recycling meet environmental goals by using different techniques such as novel enzyme technologies in the de-inking process. In current paper recycling, the de-inking process of impact ink, which is generally used for newsprint, is easy because impact ink does not fuse with the paper (Jeffries et al., 1996). On the other hand, non-impact inks, used in photocopying, ink-jet and laser printing result in the ink fusing with the paper and making the de-inking process more difficult (Kinsella and Gleason 2003). As a result, during paper de-inking alkalis such as sodium hydroxide, chlorine and sodium silicate, oxidative bleaching agents such as hydrogen peroxide, chelating agents, and agents such as surfactants are added to the waste paper to accelerate the release of ink (Gleisner et al., 2005). Chlorine can combine with organic matter under certain conditions to produce organo-carbons, including dioxins, which are toxic pollutants. Although the de-inking process uses water and chemicals, it is still less harmful to the environment than the manufacturing process of new paper (Oanh et al., 1999). Therefore, it is crucial to develop an environmentally friendly de-inking method that improves the current technology.

In paper recycling, de-inking the waste paper using cellulase enzyme will avoid detrimental effect of the chemical de-inking on environment (Pelach et al., 2003; Jeffries et al., 1996). Also, cellulase enzymes prevent the alkaline yellow tint of paper therefore improving paper brightness (Bhat, 2000).

In laundry detergent cellulases enriched with endoglucanase are used in household powder to enhance the removal of small fuzzy fibrils from fabric surface. Cellulases, such as endoglucanase, enhance the fabric softness and brightness (Schulein, 2000). Endoglucanases have high activity in defibrillation and therefore are used in textile to remove microfibril from the surface of lyocell (Shimonaka, 2006). In addition, cellulases enriched with endoglucanase are used in bipolishing of fabrics and generating a stonewashed looking fabrics by removing excess dye from them, especially denim fabrics (Mawadza et al., 2000).

In food industry, cellulases are used in extraction and clarification of fruit and vegetable juices. Moreover, cellulases are used in carotenoid extraction in production of coloring agents. Cellulases, hemicellulases and pectinases are used for extraction of olive oil. In animal food industry, cellulase and hemicellulases are used to improve the nutritional quality of animal feed (Bhat, 2000).

Agricultural and agroindustrial activities produce large amount of dry materials contains cellulose. However, the problem of feeding agricultural residues to farm animals are low protein content, high crude fiber, low digestibility and some antinutrients factors such as tannins and alkaloids. Thus, to increase digestibility of agricultural residues, it is important to breakdown the linkage between cellulose, hemicelluloses and lignin. Supplementing diets of ruminants with cellulases and xylanases enzymes can improve feed utilization, milk production, body weight and hydrolyze the anti-nutritional factors (Karmakar and Ray, 2011).

The other important application of cellulase enzyme is in biofuel production. The main steps of cellulosic biomass conversion to ethanol are pre-treatment, cellulose hydrolysis, glucose fermentation, pentose fermentation and ethanol recovery. In pre-treatment step, cellulosic biomass is breakdown to make it more accessible to enzmaytic hydrolysis. Then, cellulose is hydolysed by a mixture of cellulolytic enzymes such as endoglucanases, exoglucanases, β-glucosidases and cellobiohydrolases into simple sugar. Glucose produced from cellulose hydrolysis was fermented to ethanol by bacteria and yeast. Pentose (five-carbon sugars) which obtained from hemicellulose hydrolysis was fermented to ethanol by *Zymomonas mobilis* (Hossain et al., 2008).

Ethanol produced from cellulosic wastes used as transporting fuel either pure or mixed with gasoline in different quantities (Fujita et al., 2004). Ethanol is an excellent transportation fuel when compared with gasoline (Demain et al., 2005). It can be burned with greater efficiency because it produces smaller amounts of ozone precursors (Fujita et al., 2004), therefore decreasing air pollution and minimizing the amount of CO_2 released into the atmosphere. It is also less toxic to humans than gasoline (Demain et al., 2005). The production of fuels and chemicals from cellulosic substrates using microbial enzymes such as cellulases reduce the use of fossil fuels and save the environment (Zhou et al., 2001).

The arid environment present in the Oman would be a good resource of potential cellulose-producing microorganisms. The chances of getting extremophilic cellulase degrading microbes are high in this unexplored arid region. Some recent laboratory studies showed that some cellulase degrading bacteria isolated in Oman were able to produce high quantities of cellulase at 70°C. It was reported that some *B. licheniformis* isolated from compost produced a maximum activity of cellulase at 4 h of incubation (Sivakumar et al., 2012). Till date detergent, leather, food and other industries in Oman use the imported cellulase enzyme to mix in their range of products.

Conclusion

In Oman, very few studies were conducted related to the cellulose-degrading microorganisms. The active research in the cellulase producing microorganisms would lead to the production and utilization of indigenous cellulase enzyme. Further the native cellulase producers would play a significant role in the utilization of solid wastes accumulating in this country. The preservation of the cellulase producing microorganisms from this arid region is also of significant importance in terms of microbial biodiversity.

References

Acharya, S. & Chaudhary, A. (2012). Alkaline cellulase produced by a newly isolated thermophilic *Aneurinibacillus thermoaerophilus* WBS2 from hot spring, India. *African Journal of Microbiology Research.*, *26*, 5453-5458.

Aygan, A. & Arikan, B. (2008). A new halo-alkaliphilic, thermostable endoglucanase from moderately halophilic *Bacillus* sp. C14 isolated from Van Soda Lake. *International Journal of Agriculture Biology.*, *10*, 369-374.

Bajaj, B. K., Himani, P., Wani, W. M. A., Priyanka, S. & Ajay, S. (2009. Partial purification and characterization of a highly thermostable and pH stable endoglucanase from a newly isolated *Bacillus* strain M-9. *Indian Journal for chemical technology.*, *16*, 382-387.

Barlaz, M. A., Eleazer, W. E., Odle, W. S., Qian, X. & Wang, Y. S. (1997). Biodegradative analysis of municipal solid waste in laboratory-scale landfills. *Environment Protection Agency*, 1-5.

Bayer, A. E., Lamed, R. & Himme, M. E. (2007). The potential of cellulases and cellulosomes for cellulosic waste management. *Current Opinion in Biotechnology.*, *18*, 237–245.

Becker, P., Abu-Reesh, I., Markossian, S., Antranikian, G. & Markl, H. (1997). Determination of the kinetic parameters during continuous cultivation of the lipase producing thermophile Bacillus sp. IHI-91 on olive oil. *Applied Microbiology and Biotechnology.*, *48*, 184–190

Beukes, N. & Pletschke, B. I. (2006). Effect of sulfur-containing compounds on Bacillus cellulosome associated'CMCase'and'Avicelase'activities. *FEMS Microbiology Letters., 264*, 226–231.

Bhat, M. K. (2000). Cellulase and related enzymes in biotechnology. *Biotechnology Advances., 18*, 335-383.

Bischoff, K. M., Rooney, A. P., Li, X. L., Liu, S. & Hughes, S. R. (2006). Purification and characterization of a family 5 endoglucanase from a moderately thermophilic strain of *Bacillus licheniformis. Biotechnology Letters., 28*, 1761–1765.

Connett, P. (1998). *Municipal waste incineration a poor solution for the twenty first century.* 4th Annual International Management Conference. Amsterdam.

Demain, A. L., Newcomb, M., David, Wu J. H. (2005). Cellulase, clostridia, and ethanol. *American society for Microbilogy., 69*, 124-154.

Dhillon, N., Chhibber, S., Saxena, M., Dajani, S. & Vadehra, D. V. (1985). A constitutive endoglucanase (CMCase) from *Bacillus licheniformis*-1. *Biotechnology Letters., 7*, 695-697.

Fujita, Y., Ito, J., Ueda, M., Fukuda, H. & Kondo A. (2004). Synergistic Saccharification, and Direct Fermentation to Ethanol, of amorphous cellulose by use of an engineered yeast strain co-displaying three types of cellulolytic Enzyme. *Applied and Environmental Microbiology., 70*, 1207-1212.

Gleisner, R., Scallon, K., Fleischmann, M., Blankenburg, J. & Sykes, M. (2005). Annual Patents Review. *Progress in Paper Recycling., 14*, 24- 39.

Haigher, C. H. & Weimer, P. J. (1991). *Biosynthesis and biodegradation of cellulose.* Marcel Dekker. New York.

Hakamada, Y., Endo, K., Takizawa, S., Kobayashi, T., Shirai, T., Yamane, T. & Ito, S. (2002(. Enzymatic properties, crystallization, and deduced amino acid sequence of an alkaline endoglucanase from *Bacillus circulans. Biochimica et Biophysica Acta., 1570*, 174– 180.

Hakamada, Y., Koike, K., Yoshimatsu, T., Mori, H., Kobayashi, T. & Ito, S. (1997). Thermostable alkaline cellulase from an alkaliphilic isolate, *Bacillus* sp. KSM-S237. *Extremophiles. 1*, 151-156.

Han, D. P. & Kim, T. Y. (1987). Isolation and characterization of cellulase from cellulomonas sp. ATCC 21399. *Koren Biochemical Journal., 20*, 273-278.

Han, S. J., Yoo, Y. J. & Kang, H. S. (1995). Characterization of bifunctional cellulase and its structural gene. The American Society for *Biochemistry and Molecular Biology., 270*, 26012-26019.

Heck, J. X., Hertz, P. F. & Ayub, M. Z. (2002). Cellulaes and xylanase production by isolated Amazon Bacillus strains using soybean industrial residue based solid- state cultivation. *Brazilian Journal of Microbiology, 33*, 213-218.

Hossain, A. B. M. S., Saleh, A. A., Aishah, S. A., Boyce, N., Chowdhury, P. P. & Naqiuddin, M. (2008). Bioethanol production from agricultural waste biomass as a renewable bioenergy resource in biomaterials. *Biomed. Proceedings, 21*, 300–305. Springer-Verlag Berlin.

Huang, S., Sheng, P. & Zhang, H. (2012). Isolation and Identification of Cellulolytic Bacteria from the Gut of *Holotrichia parallela* Larvae (Coleoptera: Scarabaeidae). *International Journal of Molecular Sciences., 13*, 2563-2577.

Immanuel, G., Bhagavath, C., Raj, P. I., Esakkiraj, P. & Palavesam, A. (2007). Production and Partial Purification of Cellulase by *Aspergillus niger* and *A. fumigatus* Fermented in Coir waste and Sawdust. The *Internet Journal of Microbiology*., *3*, 1937-8289.

Jeffries, T. W. (1994). Biodegradation of lignin and hemicelluloses. In: Ratledge, C. (ed.), *Biochemistry of Microbial Degradation*. 233–277. Kluwer Academic Publishers. Netherlands.

Jeffries, T. W., Sykes, M. S., Rutledge-Cropsey, K., Klungness, J. H. & Abubakr, S. (1996). Enhanced removal of tones from offices waste paper by microbial cellulases. Proceedings of 6th international conference on biotechnology in the pulp paper industry: Advance in Applied and fundamental research. 141-144. Facultas-Universitatsverlag. Vienna, Austria. *Journal of Engineering Technology*., *3*, 47-53.

Kazaragis, A. (2005). Minimization of atmosphere pollution by utilization of cellulose waste. *Journal of Environment and Landscape Management*., *2*, 81-90.

Karmakar, M. & Ray, R. (2011). Current trends in research and application of microbial cellulases. *Research Journal of Microbiology*., *6*, 41-53.

Kim, C. H. (1995). Characterization and substrate specificity of an endo-beta-1,4-D-glucanase I (Avicelase I) from an extracellular multienzyme complex of *Bacillus circulans*. *Applied Environmental Microbiology*., *61*, 959-965.

Kinsella, S. & Gleason, G. (2003). Recycled Content Paper: Use In Office Machines. 1-50. The Environmental Paper Listening Study.

Kotchoni., O. S., Shonukan, O. O. & Gachomo, W. E. (2003). *Bacillus pumilus* BpCRI 6, a promising candidate for cellulase production under conditions of catabolite repression. *African Journal of Biotechnology*., *2*, 140-146.

Lynd, L. R., Weimer, P. J., Van Zyl, W. H. & Pretorius, I. S. (2002). Microbial cellulose utilization, Fundementals and biotechnology. *Microbiology and Molecular Biology Reviews*., *60*, 506-577.

Mawadza, C., Hatti-Kaul, R., Zvauya, R. & Mattiasson, B. (2000). Purification and characterization of cellulases produced by two *Bacillus* strains. *Journal of Biotechnology*. *83*, 177-187.

Mayende, L., Wilhelmi, B. S. & Pletschke, B. I. (2006). Cellulases (CMCases) and polyphenol oxidases from thermophilic *Bacillus* spp. isolated from compost. *Soil Biology and Biochemistry*., *38*, 2963-2966.

Ng, Son I., Li, C. W., Yeh, Y. F., Chen, P. T. & Chir, J. L. (2009). A novel endo-glucanase from the thermophilic bacterium *Geobacillus* sp. 70PC53 with high activity and stability over a broad range of temperatures. *Extremophiles*., *13*, 425–435.

Nizamudeen, S. & Bajaj, B. K. (2009). A Novel Thermo-Alkalitolerant Endoglucanase Production Using Cost-Effective Agricultural Residues as Substrates by a Newly Isolated *Bacillus* sp. *NZ. Food Technology and Biotechnology*., *47*, 435–440.

Oanh, K. N. T., Bengtsson, B. E., Reutergardh, L. B., Bergqvist, P. A., Hoa, D. T., Broman, D. & Zebuhr, Y. (1999). Dioxins in the Effluents from a Chlorine Bleached Kraft Integrated Pulp and Paper Mill in Southeast Asia. The R´99 Congress (Recovery, Recycling, Re-integration). 1-6. Environmental Expert.

Pelach, M. A., Pastor, F. J., Puig, J., Vilaseca, F. & Mutje, P. (2003). Enzymic deinking of old newspapers with cellulase. *Process Biochemistry*., *38*, 1063-1067.

Rahman, Z., Shida, Y., Furukawa, T., Suzuki, Y., Okada, H., Ogasawara, W., Morikawa, Y. (2009. Evaluation and characterization of *Trichoderma reesei* cellulase and xylanase promoters. *Applied Microbiology and Biotechnology, 82*, 899–908

Ranoa, D. R. E., Lumibao, C. Y., Roxas, J. L., San Luis, B. B. & Hedreyda, C. T. (2004). Comparison of Conventional Plate Assays with DNA-based screening protocols for protease and cellulase production from putative *Bacillus* isolates. *Science Diliman A journal of Pure Science., 17*, 31-36.

Robson, L. M. & Chambliss, G. H. (1984). Characterization of the celluloytic activity of *Bacillus* isolate. *Applied and environmental Microbiology., 47*, 1039-1046.

Saha, B. C. & Bothast, R. J. (1997). Enzymes in lignocellulosic biomass conversion. In: Saha, B.C., and Woodward, J. Fuels and Chemicals from Biomass. ACS Symposium Series 666. American Chemical Society, Washington, DC.

Saqib, A. A. N. & Whitney, P. J. (2006). Role of fragmentation activity in cellulose hydrolysis. *International Biodeterioration & Biodegradation., 58*, 180-185.

Schulein., M. (2000). Protein engineering of cellulases. *Biochimica et Biophysica Acta., 1543*, 239-252.

Singh, J., Batra, N. & Sobti, RC. (2004). Purification and characterization of alkaline cellulase produced by a novel isolate *Bacillus sphaericus* JS1. *Journal of Industrial Microbiology and Biotechnology., 31*, 51–56.

Sivakumar, N., Al Zadjali, A. A. & Al Bahry, S. N. (2012). Production of cellulase by *Bacillus licheniformis*. The 2nd Korea - Indonesia Workshop & International Symposium on Bioenergy from Biomass. Indonesia.

Stephanopoulos, G. (2007). Challenges in Engineering Microbes for Biofuels Production. *Science, 9*, 801-804.

Sukumaran, R. K., Singhania, R. R. & Pandey, A. (2005). Microbial cellulases—production, applications and challenges. *Journal of Scientific and Industrial Research, 64*, 832–844.

Tuomela, M., Vikman, M., Hatakka, A. & Itavaara, M. (2000). Biodegradation of lignin in a compost environment: a review. *Bioresource Technology., 72*, 169-183.

Vijayaraghavan, P. & Vincent, S. G. (2012). Purification and characterization of carboxymethyl cellulase from *Bacillus* sp. isolated from a paddy field. *Polish Journal of Microbiology., 6*, 51-55

Wyk, J. P. H. & Mohulatsi, M.. (2003). Biodegradation of wastepaper by cellulase from *Trichoderma viride*. *Bioresource Technology., 86*, 21–23.

Zaldívar, M., Velásquez, J. C., Contreras, I. & Pérez, L. M. (2001). *Trichoderma aureoviride* 7-121, a mutant with enhanced production of lytic enzymes, its potential use in waste cellulose degradation and/or biocontrol. *Electronic Journal of Biotechnology, 4*, 0717-3458.

Zhang, Y. H. P., Himmel, M. E. & Mielenz, J. R. (2006). Outlook of cellulase improvement, screening and selection strategies. *Biotechnology Advances., 24*, 452–481.

Zhou, J., Wang, Y. H., Chu, J., Zhuang, Y. P., Zhang, S. L. & Yin, P. (2008). Identification and purification of the main components of cellulases from a mutant strain of *Trichoderma viride* T 100-14. *Bioresource Technology., 99*, 6826–6833.

Zhou, S., Davis, F. C. & Ingram, L. O. (2001). Gene integration and expression and extracellular secretion of Erwinia chrysanthemi endoglucanase celY (celY) and celZ (celZ) in ethanologenic *Klebsiella oxytoca* P2. *Applied Environmental Microbiology., 67*, 6–14.

In: Biotechnology and Conservation … (2 Volume Set)
Editors: S. A. Farooq, R. Abed and S. Baqir
ISBN: 978-1-63117-869-6

Chapter 49

Application of Molecular Biology Methods in Identifying Biosurfactant Producers from Oil Contaminated Sites

***Asma K. Al-Bahri*[3], *Saif N. Al-Bahry*[1]*, *Abdulkadir E. Elshafie*[1], *Yahya M. Al-Wahaibi*[2], *Ali S. Al-Bemani*[2] *and Sanket J. Joshi*[3]**

[1]College of Science, Department of Biology, Sultan Qaboos University, Oman
[2]College of Engineering, Petroleum and Chemical Engineering Department, Sultan Qaboos University, Oman
[3]Central Analytical and Applied Research Unit, College of Science, Department of Biology, Sultan Qaboos University, Oman

Abstract

Crude oil is a major source of energy around the world. Recently the petroleum industry is facing many problems in extracting oil from mature wells. Many enhanced oil recovery (EOR) techniques are used, including microbial enhanced oil recovery (MEOR). Microbes produce different metabolites: biosurfactant, biopolymer, acids, solvents, and gases which can be used to increase oil recovery from reservoirs. Biosurfactants are amphiphilic molecules which can be used in MEOR; it reduces surface tension (ST) and interfacial tension (IFT) between oil and water, consequently, releasing oil trapped by capillary forces in rock pores. There are various microorganisms producing different types of biosurfactants (low or high molecular mass). As reported previously, surfactin and lichenysin are the best lipopeptidal-biosurfactants produced by *Bacillus* species which reduces the surface tension of water from 72-27 mN/m. Molecular biology methods are well adapted for identification and conservation of such novel microorganisms. Recent progress in molecular biology tools has played an important role in petroleum biotechnology. In this present review, genetic regulation of surfactin and lichenysin, factors affecting biosurfactant production and different techniques used to screen biosurfactant producers, are discussed briefly.

Keywords: Microbial enhanced oil recovery (MEOR), biosurfactant, genetic regulation, *Bacillus* species

Oil and Oil Reservoirs

Oil is an essential source of energy which is one of the main factors that drives the economic development around the world. The exploitation of oil resources in existing mature reservoirs is essential for meeting future energy demands. Crude oil is a natural resource which was formed over millions of years ago through the decomposition of plants and organisms (Phrommanich et al., 2007).

Petroleum oil consists of a complex array of gases, liquid, and solid n-alkanes, branched paraffins, cyclic paraffins and substituted cycloparaffins, aromatic compounds, sulfur compounds including benzo (b) thiophene and dibenzothiophene, along with many other organic compounds (Glazer and Nikaido, 1998). Petroleum reservoir rock is divided into three types: (1) source rock, where crude oil is formed by thermal cracking of organic and fossil remains; (2) cap rock (impermeable), where petroleum oil migrates to and is trapped there; (3) reservoir porosity rock which has a large capacity for oil storage. Plankaert (2005) reported that petroleum fluid cannot be produced if any of these forms of rock is not present.

Stages of Crude Oil Recovery from Reservoirs

Primary Recovery

During the primary recovery stage, reservoir drive comes from a number of natural mechanisms which include: natural water displacing oil upward into the well, expansion of the natural gas at the top of the reservoir and expansion of gas initially dissolved in the crude oil.

Secondary Recovery

Pressure falls gradually over the lifetime of the oil well. After the natural reservoir driving force diminishes, secondary recovery methods, using artificial lifts or beam pumps, are applied to inject water into wells to push trapped oil,.

Tertiary Recovery

Primary and secondary stages recover no more than 50% of the oil in the reservoir (Youssef, 2007). As a result, tertiary recovery is needed to extract the remaining oil from the wells. Tertiary recovery involves injecting of: gases (such as CO_2 and N_2), heat (steam or hot water), or chemicals (surfactants). Gas injection increases reservoir pressure and also reduces the viscosity and swelling of individual trapped droplets of crude oil (Almeida et al., 2004). Injection of surfactants alters the surface tension between water and oil in the reservoir, while heat injection decreases the viscosity of the oil and makes it easier to extract from the oil well. Haghighat et al., (2008) reported that 30% of additional oil present in a reservoir can be recovered using enhanced oil recovery technology such as CO_2 injection and steam injection.

MICROBIAL ENHANCED OIL RECOVERY

MEOR is an EOR method that uses microorganisms and their metabolic products to enhance and improve oil production. Gabitto and Barrufet (2004) reported that MEOR is a potentially cost-effective method especially for stripper wells that produce less than10 bb/day.

The concept of using microorganisms and their metabolites in EOR is not new. In 1926, Beckman was the first one who reported the action of bacteria on mineral oil and suggested its use in MEOR. He suggested that bacterial enzymes could be used in oil recovery. However, there was little work done until the 1940s when ZoBell initiated a systematic laboratory investigation in connection with the American Petroleum Institute. After that, several other researchers began the MEOR work (Khire and Khan, 1994).

In conventional MEOR methods special cultures of microorganisms are injected into the oil reservoir. This bacterial culture (pure or mixed) is selected for each individual reservoir according to different reservoir conditions (e.g., temperature and salt concentrations) with high metabolic activities and production of displacement agents (Gabitto and Barrufet, 2004).

There are two different types of MEOR processes: *in-situ* and *ex-situ*. In *in-situ* recovery, microbes are injected into oil wells where they secrete metabolites. Nutrients are supplied to the microbes to ensure effectiveness. Moreover, the microbes used *in-situ* MEOR should be able to survive in a harsh environment with high temperature, high salinity, and an anaerobic condition (Al-Wahibi et al., 2009). On the other hand, in *ex-situ* MEOR microbial metabolites are injected into oil wells. Bacteria are grown externally under optimum conditions to produce metabolites to be injected into the wells. In both types of MEOR methods (*in-situ* and *ex-situ)* indigenous and exogenous bacteria can be used (Al-Wahibi et al., 2009).

A study done by Almeida et al. (2004) identified 10 different bacterial strains having potential applications in oil recovery. These bacteria were: *Psedomonas aeruginosa*, *Bacillus licheniformis*, *Bacillus brevis*, *Bacillus polymyxa*, *Micrococcus varians*, *Micrococcus* sp., and two *Vibrio* species. They reported that strains of *B. licheniformis* and *B. polymyxa* produced the most active surfactants and proved to be the most themophilic and anaerobic among other isolated bacteria. They also reported that *Micrococcus varians* and *B. brevis* are highly salt-tolerant and polymer producing bacteria, respectively, and *Vibrio* sp. and *B. polymyxa* strains are the most potent gas producing bacteria. In addition, Almeida et al. (2004) concluded that the consortia of *B. brevis, B. licheniformis,* and *B. polymyxa* exhibited the best oil mobilization rate under anaerobic conditions and could easily be controlled through the administration of phosphate and inorganic electron acceptors.

The microbial processes for MEOR research involves the use of thermotolerant aerobic and facultative anaerobic species of *Pseudomonas, Bacillus, Acinetobacter, Nocardia, Rhodcoccus,* sulfide oxidizing bacteria, certain strains of fermenting methanognic thermophilic, as well as extremophilic bacteria as hyperthermophiles (Almeida et al., 2004). However, the environmental parameters of the reservoir (temperature, acidity, salinity, and permeability) limit the type of microorganisms that can be used for *in-situ* processes Compared to other EOR technologies, MEOR is an environmentally friendly, more economical technology. Different metabolic products are produced by different microorganisms in various media, and some of these metabolites have potential applications in EOR. The most extensively studied is biosurfactants. The other useful microbial metabolites are: biopolymer, biomass, acids, solvents, and gases. Microbial metabolites change the physical and chemical properties of the oil as well as the reservoir environment, thus improving the sweep efficiency of oil from reservoirs (Gabitto and Barrufet, 2004; and Almeida et al., 2004). In this chapter, biosurfactants are discussed in further detail.

Biosurfactants

Biosurfactants are microbial produced surface-active compounds which are amphiphilic molecules with both hydrophilic and hydrophobic regions that help them to aggregate at interface between fluids with different polarities (e.g., water and hydrocarbon). Biosurfactants are secondary metabolites which enhance nutrient transport across membrane, act in various host-microbe interactions and provide biocidal and fungicidal protection to the producing organism (Jenning and Tanner, 2000). It was reported that surfactants are widely used as an emulsifier in petroleum, pharmaceuticals, cosmetics and food industries (Al-Bahry et al., 2013a; Al-Bahry et al., 2013b; Al-Bahry et al., 2014; Al-Suleimani et al., 2011; Al-Sulaimani et al., 2012; Al-Wahaibi et al., 2014a; Al-Wahaibi et al., 2014b; Kim et al., 1997; Rashedi et al., 2005; Shibulal et al., 2014). Most of these compounds are chemically synthesized and only in the past few decades have surface-active compounds of biological origin (biodegradable and cheaper) been advocated. Many studies reported the usage of biosurfactants in environmental applications such as bioremediation, dispersion of oil spills, and treatment of waste (Al-Bahry et al., 2014; Al-Wahaibi 2014b).

Surfactants are widely used in enhanced oil recovery, as they decrease the surface and interfacial tension between oil and water, and as a consequence, decrease the pressure required to release oil trapped by capillary forces in the rock pores which disperse oil from the pores into the liquid mobile phase (Al-Suleimani et al., 2011).

Javaheri et al. (1985) reported that under a strictly anaerobic condition *B. licheniformis* JF-2 produced a biosurfactant that significantly lowered the surface tension of medium to <30 mN/m. Some studies also reported that biosurfactant production was stimulated when certain nutrients (especially nitrogen and phosphorus) were limited. Certain strains of *B. licheniformis* were also reported to degrade and consume hydrocarbons in addition to production of biosurfactants, polymers, and solvents. Al-Bahry et al., 2013b characterized biosurfactant producing spore forming microbes from diverse oil contaminated soil using several production media. Li et al. (2008) reported that different types of media led to production of different biosurfactants. They used nine different substrates to cultivate nine strains of *B. licheniformis* under the same incubation conditions. The strains produced four [Leu] surfactin homologues when cultivated in glucose, yeast extract and ammonium chloride medium, but it produced five lichenysin homologues when cultivated in eight other different media.

Rosenberg and Ron (1999), and Desai and Banat (1997) reported that microorganisms synthesize a wide variety of high- and low-molecular-mass biosurfactants. Glycolipids and lipopeptides are considered to be low molecular-mass emulsifiers, whereas polysaccharides, proteins, lipopolysaccharides, lipoproteins and fatty acids are considered to be high molecular-mass emulsifiers. However, different reviews divided biosurfactants depending on differences in size, biochemical nature, nature of hydrophilic part, type of carbon source, and location (intracellular, cell surface and extracellular pool) (Neu, 1996). Moreover, Arima et al. (1968) claimed that lipopeptide biosurfactant was first reported by *Bacillus subtilis* IAM1213.

Recent Advances in Molecular Biology Techniques in the Petroleum Industry

Further consideration is given to the study of microorganisms in the oil field to evaluate the beneficial or harmful effects that could be raised with specific microorganisms in the oil reservoir (enhanced oil production or metal corrosion). Use of traditional microbial cultivation technology faced many difficulties, in correctly understanding the microbial communities, as only 1% of microbes could be cultured (Al-Bahry 2013a). The rest (99%) of microbial ecosystem is uncultivable, and even if cultured, its role and contribution to the ecosystem is not always clear due to the manipulated artificial environment in laboratories (Van Hamme et al., 2003).

Recently, advances in molecular biology and related biotechnological techniques such as DNA extraction, sequencing of 16S rRNA, denaturant gradient gel electrophoresis (DGGE), terminal restriction fragment length polymorphism (T-RFLP), fluorescent *in-situ* hybridization (FISH), quantitative dot blot, quantitative polymerase chain reaction (qPCR), micro auto radiography (MAR) and biochip (Wen and Hong-Bo, 2011), have a lot of potential applications in the petroleum biotechnology industry. The most used techniques are discussed briefly in the following paragraphs.

After DNA extraction from soil samples, amplified sequences of 16S rRNA gene by PCR are used to study bacterial phylogeny, taxonomy and evolution relationship of different bacteria present in the sample by comparing the 16S rRNA sequence result with known databases like National Centre for Biotechnology Information (NCBI) GenBank. Also, from 16S rRNA sequence results, a phylogenetic tree could be constructed to reveal the evolutionary distance between different microbes present in different samples (Al-Bahry et al., 2013a; Al-Bahry et al., 2013b).

DGGE was invented by Leonard Lerman (Fischer and Lerman, 1983). It is an electrophoresis technique, where gradient acrylamide gel is used to separate DNA samples through the electrical field. The principle of DGGE is as follows: the PCR amplified regions of 16S rRNA gene run in a gradient of acrylamide gel, the sequence will stop at specific concentration in the gel depending on which base the DNA contains (Ferris et al., 1996). Because the sample contains many different strains of bacteria, many bands are observed at different positions (Al-Bahry et al., 2013a). Band numbers and positions give sufficient information about microbial composition of different samples (Ovreås et al., 1997). In addition, Wen and Hong-Bo (2011) reported that the number and intensity of bands in DGGE results gave an estimation of species number and abundance of specific strain because each band was expected to be derived from one (or few) phylogenetically distinct population. DNA band can be excised from the gel and sequence to infer the members of microbial communities (Ovreås et al., 1997). Wang et al. (2008) successfully used DGGE with PCR-amplified 16S rRNA fragments to investigate the diversity of indigenous bacteria during a MEOR field trial and monitored the changes in bacterial population after introducing exogenous bacteria into the oil reservoirs.

Fluorescent *in-situ* hybridization (FISH) technique is based on designing a probe that will hybridize with specific sequence of interest (e.g., 16S rRNA) to identify specific groups, genera or strains of microbes (e.g., bacteria). Fluorophore (fluorescent dyes) is added to the probe to enable the detection of hybridization between the target sequence and the probe. Moreover, it is possible to use different fluorophores to identify different groups of microorganisms (Hesham and Alamri (2012). Kleikemper and his group (2002) used FISH technique to characterize the sulfur reducing bacteria (SRB) population in petroleum hydrocarbons contaminated aquifer samples. Hesham and Alamri (2012) also used FISH technique to study the SRB community and its dynamics in the activated sludge from oil contaminated wastewater treatment system.

Quantitative polymerase chain reaction (qPCR) is used to quantify the concentration of genetic material of specific gene in the sample. PCR facilitates exponential multiplication of the DNA and with few adjustments and adding fluorescent probes into the PCR reaction, it can be quantitative. The technique is called real time quantitative PCR (qPCR). Emission of fluorescent signal after each PCR cycle reflects the initial number of specific bacterial cells in the tested samples. This method is very sensitive and can detect and amplify very few copies in the initial sample (Skovhus et al., 2004).

On a global scale, some research and studies showed that using molecular and biotechnological techniques made the analysis and identification of complex microbial community structure fast and accurate irrespective of some microorganisms which could not be cultured in the laboratory environment (Van Hamme et al., 2003).

BIOSURFACTANT FROM *BACILLUS* SPECIES

Bacillus species are spore forming, Gram positive and thermophilic bacteria. They can survive under reservoir conditions such as high temperature, high salinity and anaerobic conditions. It was reported that *Bacillus* spp. isolated from oil wells lowered the water surface tension (Singh et al., 2006). Several biosurfactants produced by *Bacillus* spp were reported to lower the interfacial tension between hydrocarbon and aqueous phases required to mobilize the oil (Al-Bahry et al., 2013b; Al-Bahry et al., 2014; Al-Sulaimani et al., 2012; Al-Wahaibi et al., 2014a; Youssef et al., 2007). Also, lipopeptide biosurfactants (LPBSs) exhibited various biological activities, including surface activity as well as anti-cellular and anti-enzymatic activities (Roongsawang et al., 2011). Members of *Bacillus* are known to inhabit soil and can easily be cultivated and used for commercial production of biosurfactants. A well-known biosurfactant produced by this group of microorganism is lipopeptides; a group of biosurfactants whose structure consists of a fatty acyl chain and peptide. Surfactins, lichenysins and iturins are examples of highly studied lipopeptides biosurfactant produced by *Bacillus* species.

GENETIC REGULATION OF BIOSURFACTANT PRODUCTION

The genetic regulation of rhamnolipid (glycolipid biosurfactant produced by *Pseudomonas aeruginosa*) and surfactin (lipopeptide biosurfactant produced by *Bacillus subtilis*) were the first to be studied in detail. In recent years, the molecular genetics of other biosurfactants comprised of lichenysin and iturin from *Bacillus* species, arthrofactin from

Pseudomonas species, emulsan from *Acinetobacter* and mannosylerythritol lipids (MEL) from *Candida* species (Das et al., 2008a). Little information is available about the molecular regulation of other biosurfactants, such as serrawettin, alasan, tensin, amphisin, lokisin, hydrophobin, putisolvin and viscosin.

Quorum sensing is a process which depends on cell density that allows bacterial cells to express some specific genes under high cell-density. It was reported that quorum sensing is involved in biosurfactant production from different bacteria (Daniels et al., 2004). On the other hand, Das et al., (2008a) claimed that it was not known if the quorum sensing is the cause of biosurfactant production.

Surfactin

Surfactin is an effective biosurfactant which has many pharmacological applications, including antiviral properties, inhibition of fibrin clotting formation and lysis of erythrocytes and bacterial spheroplasts (Sullivan, 1998). Arima et al. (1968) and Grangemard et al. (2001) reported that surfactin reduces surface tension from 72mN/m to 27mN/m at 25–220mg/L critical micelle concentration (CMC). It consists of a heptapeptide moiety attached to a fatty acid chain. Sullivan (1998) also reported that surfactin is composed of a cyclic heptapeptide (Glu Leu DLeu Val Asp D Leu Leu) linked to a fatty acid moiety of 13-15 carbons (3-hydroxy-13-methyl tetradecanoic acid is the major component).

Surfactin biosynthesis is non-ribosomally catalyzed by a large multienzyme peptide synthetase complex 'surfactin synthetase', consisting of three protein subunits-SrfA, ComA (earlier known as SrfB) and SrfC (Das et al., 2008b and Sullivan, 1998). The peptide synthetase required for amino acid moiety of surfactin is encoded by four open reading frames (ORF) in the srfA operon (27 kb fragment) namely SrfAA, SrfAB, SrfAC and SrfAD or SrfA-TE (Fabret et al., 1995). This operon also contains comS gene lying within and out-of-frame with the *srf*B. It had been indicated by deletion analysis that the SrfAD is not essential for surfactin biosynthesis; however, the other three ORFs are essential for this process. Another important gene is *Sfp* which is an essential gene for surfactin production and it encodes for phosphopantetheinyl transferase required for activation of surfactin synthetase by post-translational modification (Das et al., 2008b). This has been supported by the findings of a few mutants that have all the genes required for surfactin biosynthesis except *sfp*. Moreover, an acyl transferase gene is required to transfer a hydroxy fatty acid moiety to SrfAA (Peypoux et al., 1999).

Das et al. (2008a) claimed that *B. subtilis* regulates surfactin production by a cell density-responsive mechanism utilizing a peptide pheromone, ComX not based on homoserine lactone (Menkhaus et al., 1993). ComX is known as a signal peptide which accumulates in the growth medium when the cell density is high (Griffith and Grossman, 2008). It becomes a signal peptide after being modified by the gene product of comQ. By ComX, quorum sensing controls srfA expression which activates the signal transduction system when there is a interaction with ComP and ComA (Sullivan, 1998). Donation of phosphate from histidine protein kinase ComP to the regulator ComA activates and stimulates the transcription of the srf operon. Also, transcription of srf is activated by the competence-inducing pheromone (CSF) by means of inhibiting RapC, the phosphatase enzyme ComA-phosphate (Das et al., 2008b). CSF is an extracellular peptide factor and should be imported into the cell by the oligopeptide permease Spo0K. Moreover, it was reported that ComR and SinR are other factors that influence srfA expression as follows: ComR post-transcriptionally enhances srfA expression and SinR (perhaps by regulating comR) negatively controls srfA (Cosby et al., 1998; Luttinger et al., 1996; Liu et al., 1996).

Lychenisin

Lichenysin is a lipopeptide produced by several *B. licheniforrnis* strains under both aerobic and anaerobic condition (Yakimov et al., 1995 and Sullivan, 1998) which lowers the water surface tension from 72mN/m to 28mN/m (Das et al., 2008b). Lichenysin A, B, C, D, and G are different types of lichenysin biosurfactant based on species specific variations (Nerurkar, 2010). Lichenysin is a surfactin-like biosurfactant, but it is two times more efficient than surfactin, possibly as a result of substitution of Glu1 by Gln1(Grangemard et al., 1999; 2001). It was found that structural genes required for lichenysin synthesis had high sequence homology with those of surfactin (Das et al., 2008b; Nerurkar, 2010). Consequently, it can be expected that surfacin and lichenysin follow similar pathways in their biosyntheses. Moreover, Sullivan (1998) reported that structurally lichenysin differs from surfactin in the substitution of a leucine to an isoleucine in the last amino acid of the peptide and in the lipophilic moiety, which contains a mixture of straight and branched β-hydroxy fatty acids. Lichenysin is also non-ribosomally synthesized by a multienzyme peptide synthetase complex. Yakimov et al. (1998) claimed that lichenysin-A synthetase operon contains seven amino acid activation-thiolation, two epimerization and one thioesterase domain similar to that of surfactin (Das et al., 2008a). The sequenced lichenysin operon (*lic* operon: 26.6 kb long) from *B. licheniformis* consisted of three peptide synthetase genes which were transcribed in the same direction: *lic*A (three modules), *lic*B (three modules) and *lic*C (one module) (Marahiel et al., 1999; Nerurkar, 2010). The domain structures of these seven modules and the organization of lichenysin synthetases LicA to LicC was similar to surfactin synthetases SrfA-C (Das et al., 2008a). The analysis of the gene sequences encoding the lichenysin synthetase (lchA) revealed strong structural similarity to the srfA operon of surfactin synthetase (Sullivan, 1998).

Screening for Biosurfactant Producers

The different techniques used to detect biosurfactant production included: blood agar lysis, drop-collapse method, oil spreading techniques and surface tension (ST) measurements (Pyaza et al., 2005). Youssef et al., (2007) found that blood lysis technique was a poor technique to detect biosurfactant production. They reported that 16% of the strains that lysed blood agar tested were negative for biosurfactant production compared with other methods used in their experiment. However, 38% of the strains that did not lyse blood agar tested positive for biosurfactant production with the other methods and had surface tension values as low as 35mN/m. Simpson et al. (2011) designed degenerate primers to detect the presence of surfactin/lichenysin (*srf*A3/*lic*A3) gene which was involved in production of lipopeptide biosurfactant in *Bacillus subtilis* and *Bacillus licheniformis* respectively. Also, they used primers for *rhl*R gene which is responsible for the regulation of rhamnolipids production. The experiment was successful in amplifying the correct size of PCR product. Using oil spreading technique, the result was one-to-one relation between the presence of the PCR product and biosurfactant production. These results are promising, easier and consume less time in screening for the production of biosurfactant compared to other known available methods. Also, Al-Bahry et al. (2013) successfully screened and identified microbial consortia in Wafra oil wells and Suwaihat production water in Oman by molecular techniques, using denaturing gradient gel electrophoresis (DGEE) and 16S rRNA gene sequencing. Recently, molecular techniques are being used more frequently to identify microbial species as well as their genetic determinants, due to its precision and advantages of consuming low materials and less time.

FACTORS AFFECTING BIOSURFACTANT PRODUCTION

A number of factors affect the production of microbial surfactant: environmental factors as well as the composition of the production media (Rahman and Gakpe, 2008). The growth conditions (environmental factors) such as pH, temperature, agitation and oxygen availability affect the production of biosurfactant through their effects on cellular growth or activities (Gautam and Tyagi, 2006). Moreover, Desai and Banat (1997) reported that carbon and nitrogen sources are the main composition that causes the fluctuations in biosurfactant production. Joshi et al. (2008) identified in their experiment that the concentration of NH_4NO_3, glucose, Na_2HPO_4 and $MnSO_4{\cdot}4H_2O$ were the most significant variables affecting the fermentation process during biosurfactant production. Gautam and Tyagi (2006) reported the effect of the nature of fatty acid on lipase catalyzed ester formation and the effect of support material and relationship with water as other factors. Other researchers reported that Mg^{2+}, K^+, Mn^{2+}, and Fe^{2+} were the five trace elements that are more significant factors affecting surfactin production by the *B. subtilis* strain (Wei et al., 2006). Wei et al. (2006) observed a decrease of 25% in the surfactin production in the absence of Mg^{2+} or K^+. Also, they found that the production yield of surfactin was reduced about one-third compared to the control value when Fe^{2+} and Mn^{2+} were both absent. In the optimization process of the medium used for biosurfactant production, it was observed that the JF-2 biosurfactants (from *B. licheniformis* JF-2) was produced in the mid-linear growth phase and then deactivated by the stationary phase (Lin et al., 1998). It was found that this deactivation was independent of metabolic energy, affected by the concentration of phosphate and magnesium and could be overcome by continuous fermentation at low dilution rates (Lin et al., 1994).

For the first time, Sekhon and his colleagues (2011) have shown a possible association between biosurfactant production and esterase activity in any *Bacillus* species. They found that esterase is a protein which is associated with the release of the biosurfactant and they observed that after the successful cloning of the biosurfactant genes from *Bacillus subtilis* SK320 into *E. coli*, biosurfactant production and esterase activities doubled. Moreover, the symbiotic relationship between these two genes was confirmed by multiple sequence alignment which showed regions of similarity and conserved sequences.

Conclusion

Microbial surfactant is a degradable and low toxic compound that makes it environmentally friendly. It has enormous applications in medicine, industrial and environmental processes. Like any other biotechnological products, the commercialization of biosurfactant depends on the final yield and its production economics. Previous methods known in screening for biosurfactant producers are time consuming and not very accurate, sometimes giving false positive results, as reported by some studies. Recently, use of molecular techniques such as DNA extraction, polymerase chain reaction (PCR), sequencing and DGGE were very useful in identifying indigenous microorganisms from the oil fields and screening for the biosurfactant (or any other metabolites) producers. These molecular techniques will help by saving material and time, which are the most critical issues in petroleum research. Additionally, molecular techniques deal with genome which gives more precise results than other classical techniques. Detection and identification of biosurfactant determinants and their regulation mechanisms are crucial for petroleum biotechnology and conservation of the isolates where molecular biology techniques have many potential applications.

References

Al-Bahry SN, Elsahfie AE, Al-Wahaibi YM., Al-Bimani AS, Joshi SJ, Al-Maaini RA, Al-Alawai WJ, Sugai Y, Al-Mandhari M. 2013a. Microbial consortia in Oman oil fields: a possible use in enhanced oil recovery. *Journal of Microbiology and Biotechnology*. 23: 106–117.

Al-Bahry SN., Elashafie AE., Al-Wahaibi YM, Al-Bemani AS., Joshi SJ. Al-Lawati A. 2013b. Isolation and characterization of biosurfactant/biopolymer producing spore forming bacteria from oil contaminated sites and oil field of Oman. *Procedia APCBEES*. 5: 242 – 246.

Al-Bahry SN, Al-Wahaibi YM, Elshafie AE, Al-Bemani AS, Joshi SJ. 2014. Isolation of biosurfactant producing spore-forming bacteria from Oman: potential applications in bioremediation. *Advances in Environmental Science and Development.* (In press).

Al-Sulaimani H, Joshi S, Al-Wahaibi Y, Al-Bahry SN, Elshafie A, Al- Bemani A. 2011. Microbial biotechnology for enhancing oil recovery: current developments and future prospects. *Biotechnology, Bioinformatic and Bioengineering.* 1: 147-158.

Al-Sulaimani H, Al-Wahaibi Y, Al-Bahry S, Elshafie A, Al-Bemani A, Joshi S, Ayatollahi S. 2012. Residual oil recovery via injection of biosurfactant, chemical surfactant and mixtures of both under reservoir condition: induced wettability and interfacial tension effects. *SPE Reservoir Evaluation & Engineering*. 15: 210-217.

Al-Wahibi Y, Al-Bahry S, Al-Sulaimani H, Al-Manadhari M, Ghosh B. 2009. Microbial technology applications in wellbore stimulation and oil recovery enhancement: a review. *International Journal of Oil, Gas and Coal Technology*. 2: 315-33.

Al-Wahaibi Y, Joshi S, Al-Bahry S, Elshafie A, Al-Bemani A, Shibulal B. 2014a. Biosurfactant production by *Bacillus subtilis* B30 and its application in enhancing oil recovery. *Colloids and Surfaces B: Biointerfaces*. 114: 324-333.

Al-Wahaibi YM, Al-Bahry SN, Elshafie AE, Al-Bemani AS, Joshi SJ, Al-Bahri AK. 2014b. Screening of minimal salt media for biosurfactant production by *Bacillus* spp. *Advances in Environmental Science and Development.* (In press).

Almeida PF, Moreira RS, Almeida RCC, Guimaraes AK, Carvalho AS, Quintella C, Esperidia MCA Tafi CA. 2004. Selection and application of microorganisma to improve oil recovery. *Engineering in Life Sciences*. 4: 319-324.

Arima K, Kakinuma A, Tamura G. 1968. Surfactin, a crystalline peptidelipid surfactant produced by *Bacillus subtilis*: isolation, characterization and its inhibition of fibrin clot formation. *Biochemical and Biophysical Research Communications.* 31: 488–494.

Cosby WM, Vollenbroich D, Lee OH, Zuber P.1998. Altered srf expression in *Bacillus subtilis* resulting from changes in culture pH is dependent on the Spo0K oligopeptide permease and the ComQX system of extracellular control. *Journal of Bacteriology.* 180:1438-1445.

Daniels R, Vanderleyden J, Michiels J. 2004. Quorum sensing and swarming migration in bacteria. *FEMS Microbiology Reviews*. 28: 261-289.

Das P, Mukherjee S, Sen R. 2008a. Genetic regulations of the biosynthesis of microbial surfactants: an overview. *Biotechnology and Genetic Engineering Reviews*. 25: 165-186.

Das P, Mukherjee S, Sen R. 2008b. Antimicrobial potential of a lipopeptide biosurfactants derived from a marine *Bacillus circulans. Journal of Applied Microbiology.* 104: 1675–1684.

Desai JD, Banat IM. 1997. Microbial production of surfactant and their commercial potential. *Microbiology and Molecular Biology Reviews*. 61: 47-59.

Fabret C, Quentin Y, Guiseppi A, Busuttil J, Haiech J, Denizot F.1995. Analysis of errors in finished DNA sequences: the surfactin operon of *Bacillus subtilis* as an example. *Microbiology*. 141:345-350.

Ferris M, Muyzer G, Ward D. 1996. Denaturing gradient gel electrophoresis profiles of 16S rRNA-defined populations inhabiting a hot spring microbial mat community. *Applied and Environmental Microbiology*. 62: 340-346.

Fischer SG, Lerman LS. 1983. DNA fragments differing by single base-pair substitutions are separated in denaturing gradient gels: correspondence with melting theory. *Proceedings of the National Academy of Sciences of the United States of America.* 80: 1579-1583.

Gabitto J and Barrufet M. 2004. Combined microbial surfactant-polymer system for improved oil mobility and conformance control. 3rd Annual Progress Report.

Gautam KK, Tyagi VK. 2006. Microbial surfactants: a review. *Japan Oil Chemists's Society.* 55: 155-166.

Glazer AN, Nikadido H. 1998. Microbial Biotechnology: Fundamentals of Applied Microbiology. Second Edition. New York.

Grangemard I, Wallach J, Maget-Dana R, Peypoux F. 2001. Lichenysin: a more efficient cation chelator than surfactin. *Applied Biochemistry and Biotechnology*. 90: 199–210.

Griffith KL, Grossma AD.2008. A degenerate tripartite DNA-binding site required for activation of coma-dependent quorum response gene expression in *Bacillus subtilis. Journal of Molecular Biology*. 381: 261–275.

Haghighat S, Sepahy AA, Assadi MM, Pasdar H. 2008. Ability of indigenous *Bacillus licheniformis* and *Bacillus subtilis* in microbial enhanced oil recovery. *International journal of Environmental Science and Technology*. 5: 385-390.

Hesham AEL, Alamri SA. 2012. Application of fluorescence *in situ* hybridization (FISH) to the analysis of sulfate reducing bacterial community in an oily bench scale reactor. *African Journal of Biotechnology*.11: 10221-10226

Javaheri M, Jenneman GE, McInerney MJ and Knapp RM. 1985. anaerobic production of biosrfactant by *Bacillus licheniformis* JF-2. *Applied and Environmental Microbiology*. 50: 698-700.

Jenning EM, Tanner RS. 2000. Biosurfactant-producing bacteria found in contaminated and uncontaminated soils. *Proceedings of 2000 Conference on Hazardous Waste Research*.

Joshi S, Yadav S, Desai AJ. 2008. Application of response-surface methodology to evaluate the optimum medium components for the enhanced production of lichenysin by *Bacillus licheniformis* R2. *Biochemical Engineering Journal*. 41: 122–127.

Khire JM., Khan MI. 1994. Microbially enhanced oil recovery (MEOR). Part 1. Importance and mechanism of MEOR. *Enzyme Microbial Technology*. 16:17-172.

Kim HS, Yoon BD, Lee CH, Suh HH, Oh HM, Katsuragi T, Tani Y. 1997. Production and properties of lipopeptide biosurfactant from *Bacillus subtills* C9. *Journal of Fermentation and Bioengineering*. 84: 41-46.

Kleikemper J, Pelz O, Schroth MH, Zeyer J. 2002. Sulfate-reducing bacterial community response to carbon source amendments in contaminated aquifer microcosms. *FEMS Microbiology Ecology*. 42:109-18

Li YM., Haddad NIA, Yang SZ, Mu BZ. 2008. Variants of lipopeptides produced by *Bacillus licheniformis* HSN221 in different medium components evaluated by a rapid method ESI-MS. *International Journal of Peptide Research and Therapeutics*. 14: 229-235.

Lin SC, Lin KG, Lo CC, Lin YM. 1998. Enhanced biosurfactant production by a *Bacillus licheniformis* mutant. *Enzyme and Microbial Technology*. 23:267–273.

Lin SC, Carswell KS, Sharma MM., Georgiou G. 1994. Continuous production of the lipopeptide biosurfactant of *Bacillus licheniformis* JF-2. *Applied Microbiology and Biotechnology*. 41: 281–285.

Liu L, Nakano M, Lee OH, Zuber P. 1996. Plasmid-amplified comS enhances genetic competence and suppresses sinR in *Bacillus subtilis*. *Journal of Bacteriology*. 178: 5144-5152.

Luttinger A, Hahn J, Dubnau D. 1996. Polynucleotide phosphorylase is necessary for competence development in *Bacillus subtilis*. *Molecular Microbiology*. 19: 343-356.

Marahiel MA, Konz D, Dokel S. 1999. Molecular and biochemical characterization of the protein template controlling biosynthesis of the lipopeptide lichenysin of *Bacillus licheniformis*. *Journal of Bacteriology*. 181: 133-140.

Menkhaus M, Ullrich C, Kluge B, Vater J, Vollenbroich D. 1993. Structural and functional organization of the surfactin synthetase multienzyme system. *Journal of Biological Chemistry*. 268: 7678-7684.

Nerurkar AS. 2010. Structural and molecular characteristics of lichenysin and its relationship with surface activity. *Advances in Experimental Medicine and Biology*. 672:304-15.

Neu TR.1996. Significance of bacterial surface-active compounds in interaction of bacteria with interfaces. *Microbiology Reviews*. 60: 151-166.

Ovreås L, Forney L, Frida LD. 1997. Distribution of bacterioplankton in meromictic lake sælenvannet, as determined by denaturing gradient gel electrophoresis of PCR-amplified gene fragments coding for 16S rRNA. *Applied and Environmental Microbiology*. 63: 3367-3373.

Peypoux F, Bonmatin JM, Wallach J. 1999. Recent trends in the biochemistry of surfactin. *Applied Microbiology and Biotechnology*. 51: 553-563.

Phrommanich S, Suanjit S, Upatham S, Grams SV, Kruatrachue M, Pokethitiyook P, Korge G, Hofman A. 2009. Quantitative detection of the oil-degading bacterium *Acinetobacter* sp. strain MUB 1 by hybridization probe based real-time PCR. *Microbiological Research*. 164: 486-92.

Plankaert, M. 2005. Oil reservoir and oil production. *In*: B. Ollivier and M. Magot. *Petroleum Microbiology*. ASM Press, Washington, DC. pp. 3-19.

Pyaza GA, Zjawiony I, Banat IM. 2005. Use of different methods for detection of thermophilic biosurfactantproducing bacteria from hydrocarbon-contaminated and bioremediated soils. *Journal of Petroleum Science and Engineering*. 50: 71- 77.

Rahman PKSM and Gakpe E. 2008. Production, characteristation and application of biosurfactant-review. *Biotechnology*. 7: 360-370.

Rashedi H, Jamshidi E, Assadi MM, Bonakdarpour B. 2005. Isolation and production of biosurfactant from *Pseudomonas aeruginosa* isolated from Iranian southern wells oil. *International Journal of Environmental Science Technology*. 2: 121-127.

Roongsawang N, Washio K and Morikawa M. 2011. Diversity of nonribosomal peptide synthetases involved in the biosynthesis of lipopeptide biosurfactants. *International Journal of Molecular Sciences*. 12: 141-172.

Rosenberg E and Ron EZ. 1999. High- and low-molecular mass microbial surfactant. *Applied Biochemistry and Biotechnology*. 52: 154-162.

Sekhon KK, Khanna S, Cameotra SS. 2011. Enhanced biosurfactant production through cloning of three genes and role of esterase in biosurfactant release. *Microbial Cell Factories*. 10: 49.

Skovhus TL, Ramsing NB, Holmstrom C, Kjelleberg S, Dahllof I. 2004. Real-time quantitative PCR for assessment of abundance of *Pseudoalteromonas* species in marine samples. *Applied and Environmental Microbiology*. 70: 2273.

Simpson DR, Natraj NR, McInerney MJ, Duncan KE. 2011. Biosurfactant-producing *Bacillus* are present in produced brines from Oklahoma oil reservoirs with a wide range of salinities. *Applied Microbiology and Biotechnology*. 91:1083–1093.

Singh A, Hamme JDV, Ward OP. 2006. Surfactants in microbiology and biotechnology: part 2. application aspects. *Biotechnology Advances*. 25: 99–121.

Shibulal B, Al-Bahry S, Al-Wahaibi Y, Elshafie A, Al-Bemani A, Joshi S. 2014. Microbial enhanced heavy oil recovery by the aid of inhabitant spore-forming bacteria: an insight review. *The Scientific World Journal*. 2014: 1-12.

Sullivan ER. 1998. Molecular genetics of biosurfactant production. *Environmental biotechnology*. 9: 263-269.

Van Hamme JD, Singh A, Ward OP. 2003. Recent advances in petroleum microbiology. *Microbiology and Molecular Biology Reviews*. 67:503-49.

Wang J, Ma T, Zhao L, Lv J, Li G, Zhang H, Zhao B, Liang F, Liu R. 2008. Monitoring exogenous and indigenous bacteria by PCR-DGGE technology during the process of microbial enhanced oil recovery. *Journal of Industrial Microbiology and Biotechnology*. 35: 619-628.

Wei YH, Lai CC, Chang JS. 2006. Using Taguchi experimental design methods to optimize trace element composition for enhanced surfactin production by *Bacillus subtilis* ATCC 21332. *Process Biochemistry*. 42: 40-45.

Wen Z, Hong-Bo S. 2011. Applications of molecular biology and biotechnology in oil field microbial biodiversity research. *African Journal of Microbiology Research.* 5: 3103-3112.

Yakimov MM, KroÈger A, Slepak TN, Giuliano L, Timmis, KN, Golyshin PN. 1998. A putative lichenysin A synthetase operon in *Bacillus licheniformis*: initial characterization. *Biochimica et Biophysica Acta*. 1399: 141-53.

Youssef N, Simpson DR, Duncan KE, McInerney MJ, Folmsbee M. 2007. *In situ* biosurfactant production by *Bacillus* strains injected into a limestone petroleum reservoir. *Applied and Environmental Microbiology.* 73: 1239-1247.

In: Biotechnology and Conservation … (2 Volume Set)
Editors: S. A. Farooq, R. Abed and S. Baqir
ISBN: 978-1-63117-869-6

Chapter 50

The Influence of Bacterial Solution on the Mineral Surface Characterization

***Mahvash Karimi*[1,3], *Maziyar Mahmoodi*[1], *Ali Niazi*[2] *and Shahab Ayatollahi*[1, 3*]**

[1]School of Chemical and Petroleum Engineering, Shiraz University, Shiraz, Iran
[2]Institute of Biotechnology, Shiraz University, Shiraz, Iran
[3]EOR Research Centre, Shiraz University, Shiraz, Iran

Abstract

Reservoir wettability plays a profound role in influencing flow through the porous media, results in more recovery of residual oil during microbial enhanced oil recovery (MEOR). Understanding the mechanisms of wettability changes is the main key to understand the oil recovery mechanism during microbial treatments of the oil reservoirs.

In an attempt for better discerning of the active mechanisms of wettability alteration, this study examines the effects of bacterial metabolites, bacterial adhesion and bacterial solution on the surface characteristics using a variant of *Bacillus thermodenitrificans* strain with two different carbon sources. Using n-*dodecane* and sucrose as carbon sources, this strain has the ability to produce biosurfactant only when sucrose is presented as the carbon source. In order to have various wetting conditions, the experiments were performed on both fresh substrates and surfaces aged in crude oil.

Contact angle measurements have been used as measurement technique to assess the wetting condition of surfaces after treatments. Atomic force microscopy (AFM) tests were also performed to asses microscopic changes of the surface pointed to the macroscopic contact angles measurement.

The results indicated that the treatments using bacterial products in the presence of biosurfactant would change the wettability from water-wet to less water-wet and from oil-wet to water-wet conditions. Results showed that surface wettability changes due to bacterial adhesion to the surface were insignificant. Biosurfactant adsorption to the surface imparted wettability changes to the surface.

---

* Corresponding author: email address: shahab@shirazu.ac.ir.

Keywords: Microbial enhanced oil recovery, Wettability alteration, Biosurfactant, AFM

INTRODUCTION

As world oil demands keeps rising and not all oil can be promptly withdrawn from the reservoir, oil recovery efficiency is a crucial issue in petroleum production. About two-thirds of the oil is still trapped in reservoirs after primary and secondary oil production. Recovering such residual oil in the reservoirs, is the target of enhanced oil recovery (EOR) technologies. Microbial enhanced oil recovery (MEOR) is an environmentally compatible process that may provide an economic advantage among other EOR processes. By the use of genetic engineering techniques, it is possible to develop more effective microorganisms that produce large amounts of useful products rapidly from inexpensive, abundant and renewable resources.

Some of the MEOR key mechanisms leading to more oil recovery are: interfacial tension (IFT) reduction due to biosurfactant production, wettability alteration, biodegredation, bioclogging due to biomass formation and gas production (Youssef et al., 2009; Armstrong and Wildenschild, 2012, Kowalewski et al., 2005; Crescente et al., 2006; Soudmand-asli et al., 2007; Sen, 2008).

Wettability describes the preference of a solid surface to be in contact with one fluid in the presence of other immiscible fluids (Anderson, 1986). Reservoir wettability plays a profound role in influencing flow through the porous media, results in more recovery of residual oil during MEOR. More oil is recovered, if rock wettability changes toward neutral-wet and water-wet state in non-fractured and fractured reservoirs, respectively (Anderson, 1987; Tiab and Donaldson, 2004; Donaldson and Alam, 2008). Hence, understanding the mechanisms of wettability changes is the main key to understand the oil recovery mechanism during microbial treatment of the oil reservoir.

Although wettability is an important parameter, few studies have focused on the wetting effects of the bacteria. And understanding of wettability alteration mechanisms during MEOR process is still far from being clear. Proposed mechanisms of wettability changes are biofilm formation, bacterial adhesion to the surface, adsorption of ingredients from bacterial metabolites and biosurfactants (Crescente et al., 2006; Shabani Afrapoli et al., 2009; Gandler et al., 2006; Hiorth et al., 2007; Zargari et al., 2010). However, more than one mechanism can act at the same time.

Initial wetting state, surface characteristics, types of microorganism and the metabolites affect surface wettability. Depending on the situation, microbial treatment can alter the initial wettability both toward more water-wet (Mu et al., 2002; Zekri et al., 2003; Shabani Afrapoli et al., 2008; Karimi et al., 2012) and more oil-wet conditions (Kowalewski et al., 2005; Crescente et al., 2006; Polson et al., 2010; Zargari et al., 2010).

The goal of this study was to evaluate wetting changes and find active mechanisms responsible for these changes in order to determine the best conditions for enhanced oil recovery. Accordingly contact angle measurements were used to assess wetting changes induced by bacterial activity on glass and glass surfaces aged in crude oil. Atomic force microscopy (AFM) experiments were also used to gain a better understanding of how the microbial solution interact with the surfaces.

Bacillus thermodenitrificans isolate was performed in this research, as a strain that produce different amount of biosurfactant by the use of sucrose and n-*dodecane* as carbon source in the growth medium. In an attempt to find the active mechanism of wettability alteration, the effects of bacterial solution, bacterial adhesion and extracellular biosurfactant solution on the wettability of treated surfaces were characterized macroscopically using contact angle measurements and microscopically using atomic force microscopy.

Bacteria and Growth Condition

A facultative anaerobic, gram positive biosurfactant producing bacteria, *Bacillus thermodenitrificans* strain, was employed in this study, that was obtained from Shiraz University Collection of Petroleum Microorganisms (SUCPM). This microorganism was isolated from an operating oil reservoir in the south-western part of Iran. The isolation, screening and selection program of the bacteria were performed in the Shiraz University Institute of Biotechnology. Experimental results showed this bacterium produce different amount of biosurfactant utilizing sucrose and n-*dodecane* in growth culture. Investigations demonstrated, this strain is capable of produce remarkably more biosurfactant using sucrose as carbon sources in culture medium.

Bacillus thermodenitrificans strain was cultured on a mineral salt solution. The composition of mineral salt solution is presented in Table 1. The mineral salt solution was supplemented with 10 ml stock solutions of 0.25% (w/v) $MgSO_4$, 1% $(NH_4)_2SO_4$ and trace salt solution, which had been sterilized separately by filtration through 0.20-μm filters (Millipore Corp., Bedford, MA, USA). The trace salt solution contained the following components (g/l): EDTA, 0.5; $MnSO_4.H_2O$, 3; NaCl, 1; $CaCl_2.2H_2O$, 0.1; $ZnSO_4.7H_2O$, 0.1; $FeSO_4.7H_2O$, 0.1; $CuSO_4.5H_2O$, 0.01; $AlK(SO_4)_2$, 0.01; $Na_2MoO_4.2H_2O$, 0.01; H_3BO_3, 0.01; Na_2SeO_4, 0.005; $NiCl_2.6H_2O$, 0.003. The strain was incubated for specified time intervals (2, 5 or 7 days) in a shaker incubator at 160 rpm and 40 °C.

Table 1. Composition of the mineral salt solution

Component	Concentration (g/l)
KH_2PO_4	2.7
K_2HPO_4	13.9
Carbon source (sucrose or *n*-dodecane)	10
NaCl	1.0
Yeast extract	0.5
$NaNO_3$	1.0

Surfaces

Microscope slide was used as a glass surface for this study. Glass does not represent all reservoir pore surfaces, but it is similar in surface properties to the silica, the main component in sandstone. Glass surface is a mineralogical surface with constant mineralogy and chemistry. In this regard, the surface heterogeneity would not affect the adsorption and desorption of the microbial ingredients. All samples had the size of 38×12.5 mm.

When a water-wet surface aged in crude oil, it can be converted to oil-wet surface due to adsorption of polar compounds and asphaltenes (Buckley et al., 1997). Aging the glass surface in crude oil that was obtained from Gachsaran oil field, located in southwest Iran; 32 °API, 18 cp viscosity; increase the contact angle from a water wet behaviour (angles below 10°) to an oil wet situation, with angles larger than 110° (Karimi et al., 2012). The crude oil composition and its properties are given elsewhere (Roosta et al., 2009; Seiedi et al., 2011). In order to change wetting behaviour of glass surfaces, they were soaked in crude oil for 21 days at 80°C. Then the aged surfaces were washed with toluene, to remove any oil residues.

For all tests, prior to treatment surfaces were sterilized in an autoclave at 120 °C for 20 minutes.

Contact Angle and Surface Tension Measurements

Surfaces were soaked in different solutions for 2, 5 and 7 days and the contact angle values, measured in the water phase, were determined using a drop shape analysis system (Krüss, Germany) by the captive drop method. Hence, as contact angle increases, water-wet behaviour of the surface descends and the oil-wet behaviour ascends. Contact angles, were measured by carefully depositing a 25 μl n-*decane* droplet beneath the solid surface, which was already immersed in the distilled water. n-*decane* has a density of 0.731 g/ml and 0.92 cp viscosity. At least each case repeated three times and the reported values were the average of several independent measurements. Surface tension of solutions were determined, utilizing the pendant drop method (Krüss, Germany).

To evaluate the effect of bacterial cells on surface wettability and adhesion of bacteria to the surface, *Bacillus thermodenitrificans* strain was grown in the nutrient broth for 24 hours. The suspension was washed three times with the washing phosphate buffer solution having the following composition (g/l): NaCl 8, K_2HPO_4 5.6 and KH_2PO_4 2.2 by centrifugation at 3000 rpm for 10 minutes. Subsequently, the pellet was resuspended in the washing buffer solution, at a concentration of 10^6 cells/ml. This solution prevents bacterial growth because it has no nutrients. Each surface was placed in the pellet suspension for different time periods (2, 5 and 7days).

Contact angle and surface tension measurements were performed for the following cases:

1) Bacterial solution (mineral salt solution which contained bacteria and their products),
2) Cell free solution (supernatant of centrifuged bacterial solution which contained microbial products and produced biosurfactant) and;

3) Buffer solution (pellets of bacterial cells that were re-suspended in the washing buffer solution, according to the instructions mentioned above, contained only bacterial cells).

In order to have a comparison base, control solution was used in each case as a reference. For the adhesion test, a sterile washing phosphate buffer was the control solution. While in other cases, a sterile mineral salt solution was utilized as the control solution.

AFM Surface Analysis

Measuring the force between the tip (imaging source) and a surface, AFM is a very useful technique in understanding changes in surface characteristics and surficial deposits sorbed onto a surface. This force is measured by the instrument, which then can be used to determine the height and phase signal by the instrument.

In this study, fresh and aged glass surfaces were soaked in bacterial, cell free and control solutions that was described earlier. The effect of pure biosurfactant solution with equal surface tension to the cell free solution on the fresh glass surface is also investigated. After two days of exposure, surfaces were removed from the solution and allowed to dry thoroughly in air. To evaluate an image of the bacterial cells, 50 μl of the bacterial suspension was put onto a fresh glass surface. This surface also left to dry at room temperature and then imaged with AFM.

The AFM samples were imaged at ambient temperature using Dual Scope^{TM} DS 95-200/50 in the tapping mode. Using this mode, the tip tapped the surface as it moves along and the height and phase signals were taken. Imaging was conducted with a silicone tip. The height signal shows the topography of the surface. The phase signal is able to distinguish different materials on the surface due to varying elasticity. The phase signals unveil more details of the morphology than the height image. Phase images could reveal significantly different morphologies. The data was processed with DME software (Danish Micro Engineering A/S (DME), version 1.6.0, Copenhagen, Denmark).

To determine and compare roughness of different surfaces, the most commonly specified parameters namely, arithmetic average roughness (S_a), root mean square roughness (S_q), and Z value range (S_y) were utilized in this study. the vertical distance between the highest and the lowest point in the height image is given by S_y. If some compounds adsorb on the solid surface peaks, the Z value range amplified. S_a and S_q are calculated as follows:

$$S_a = \iint_a |Z(x,y)dxdy| \quad (1)$$

$$S_q = \sqrt{\iint_a (Z(x,y)^2)dxdy} \quad (2)$$

Observations for Influence of Bacterial Solutions on the Mineral Surfaces

Contact Angle and Surface Tension Measurements

The results of surface tension measurements for two different carbon sources used in the solutions, are presented in Table 2. Reference values represent the surface tension of the control solutions in each case, which described earlier (Section 2.3). Surface tension measurements reveal that, utilizing sucrose as carbon source *B. thermodenitrificans* could produce high amount of biosurfactant and lowered the surface tension of solution remarkably. In general, surface active compounds in solutions are stable, whereas variations of surface tension values were small over time.

Figure 1 represents the results of contact angle measurements for glass surfaces when sucrose was the carbon source of mineral salt solution. These data indicate that the contact angle of the control surface was remained constant during the process at 7.65° and significant differences are observed for bacterial and cell free solutions compared to control surface, with the average values of 29.9° and 36.5°, respectively. The measured contact angle for these two surfaces are much higher than control surface and wettability change toward less water-wetness.

On the other hand, when bacteria used n-*dodecane* as carbon source, bacterial and cell free solutions did not alter wettability of glass surface considerably (data are not shown here). Additionally, values of contact angle of surfaces that immersed in the buffer solution did not change markedly (data are not shown here). Having no nutrients the bacteria did not grow on the buffer solution and this solution just contained bacterial cells.

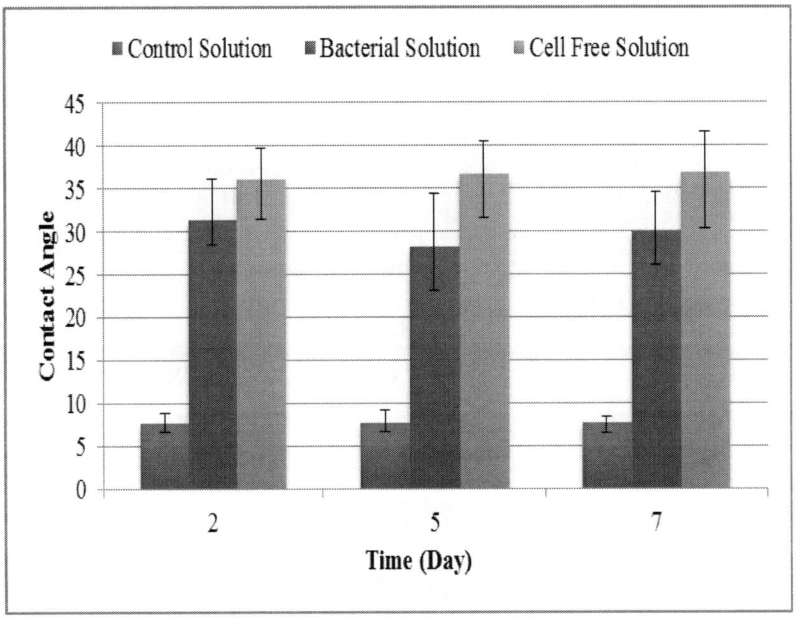

Figure 1. Wettability alteration caused by bacterial and cell free solutions for glass surface. The error bars represent minimum and maximum contact angles obtained in the repeated experiments.

Table 2. Surface tension measurements (mN/m). Data are the mean value of triplicate samples

Sucrose as carbon source			
Time(day)	**Reference**	**Bacterial solution**	**Cell free solution**
2	59.9 ± 0.6	29.2±0.4	26.4±0.6
5	64.3 ± 0.8	34.1±0.7	25.8±0.3
7	65.6 ± 0.7	31.5±0.5	26.6±0.5
***n*-dodecane as carbon source**			
2	54.0 ± 0.6	46.1±0.5	48.2±0.7
5	56.4 ± 0.7	56.5±0.1	50.1±0.6
7	59.4 ± 0.4	57.4±0.8	49.6±0.3

Bacterial metabolites produced the greatest wetting alteration. The presence of surface active compounds, in the case of utilizing sucrose as carbon source in bacterial solution, draw less water-wet behaviour and contact angle increase for glass surface, while in the absence of these compounds wetting behaviour changed insignificantly. Results suggest that surface active compounds in the solution are responsible for wettability alteration of glass surface.

Based on surface tension and contact angle measurements, adsorption of surface active agents play an important role in the glass surface wettability alteration.

Figure 2 shows the contact angle measurements of aged glass surfaces. As discussed previously, hence changes in contact angle were noticeable only when sucrose was used as carbon source in mineral salt solution, this case is mentioned here.

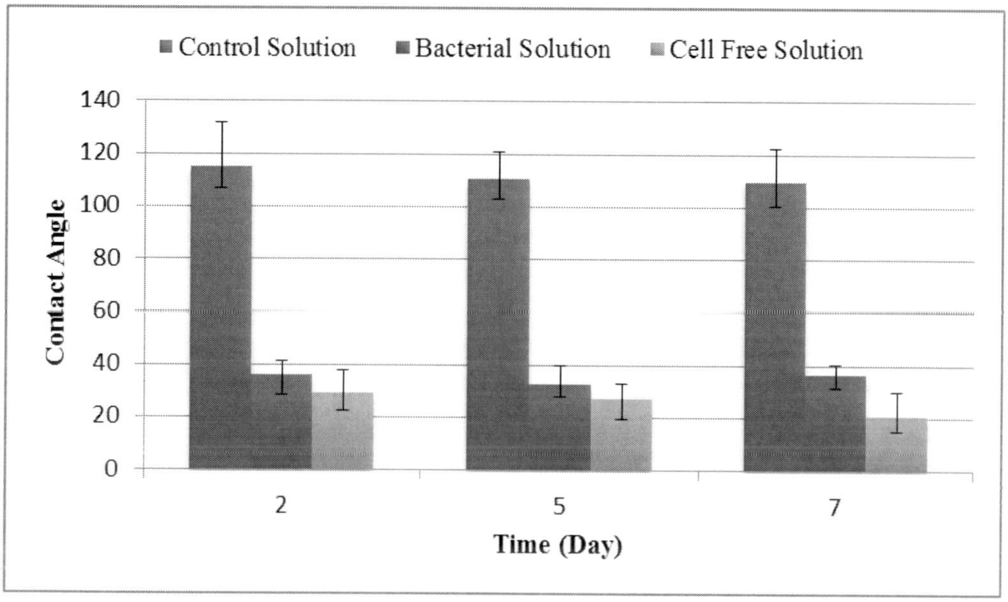

Figure 2. Wettability alteration caused by bacterial and cell free solutions for aged glass surface. The error bars represent minimum and maximum contact angles obtained in the repeated experiments.

After 21 days aging of glass in crude oil, the surface became oil-wet and contact angle increased from 7.65 ° to 115°, which is indicated the adsorption of polar and asphaltenic compounds of crude oil on the glass surface. Wettability of aged surfaces exposed to bacterial solutions and cell free solutions changed toward water-wet situation and contact angles descended considerably. Although, lower contact angles could be demonstrated when aged glass was soaked in bacterial products.

Significant differences are observed upon the presence of bacterial metabolite, which strongly favour the transition from an oil-wet to water-wet behaviour. Results display a more extensive change of wettability behaviour with respect to glass surface. Effects of bacterial metabolite produced water-wet conditions. In comparison with the values assessed for glass surface (Figure 1), a similar conclusion could be obtained, that confirmed the previous observations. Surface active compounds altered the wettability of both fresh and aged glass surfaces.

AFM Surface Analysis

AFM give a greatly detailed insight into the nanoscopic morphology revealing very good description of the surface. AFM is used for assessment of surface characteristics before and after microbial treatments. Height and the corresponding phase images were obtained for 3×3 μm scan sizes for each surface.

Figure 3(a) shows a topographic and phase image of glass surface without any treatment, called as fresh glass. An image of fresh glass was required to compare a clean surface to that of a treated one. There appears to be nothing on the surface of the fresh glass. This surface is a very smooth surface with average roughness (S_a) of 0.164 nm.

When glass exposed to bacterial and cell free solutions (Figure 3(b) and (c)), there appeared to be some small features on the surface, of what is presumed to be biosurfactant due to contact angle measurements. These two surfaces have rougher surface compared to fresh glass. And adsorption of particles on the surface had an impact on surface roughness. The roughness measurements for the topography images over the scanned area are presented in Table 3. Figure 3(b) shows very small sharp features on the surface. A similar morphology can be observed in Figure 3(c). Soaking in bacterial and cell free solutions have a pronounced effect on the surface characteristics, hence different height and phase images observed.

In order to find the nature of adsorbed species on the glass surfaces, height and phase images of bacterial cell on the surface and glass surface after adsorption of pure biosurfactant were investigated. Figure 4(a) demonstrates bacterial cells on the glass surface. While, Figure 4(b) shows the glass surface after it had been treated with pure biosurfactant solution. It is clear that the size of this bacterium is about 670 nm. Hence, adsorbed particles on the surface are much smaller than the bacterial cells. Figure 4(b) has the same appearance as Figure 3(b) and 3(c), suggesting that the produced biosurfactant adsorbed on the surface and change surface wetting behaviour.

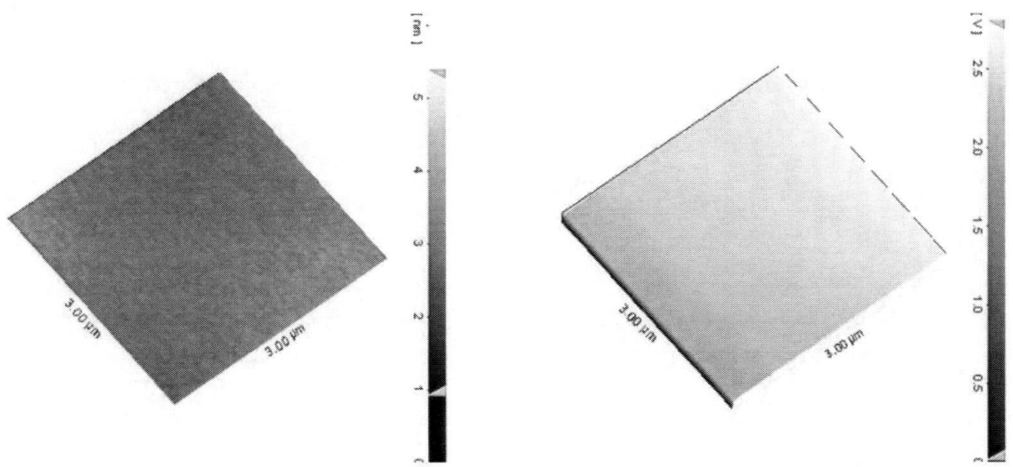

(a)

Figure 3. (Continued).

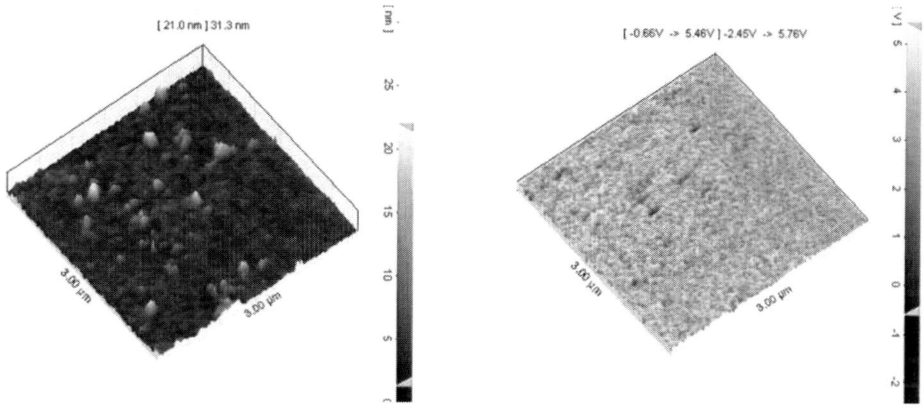

(b)

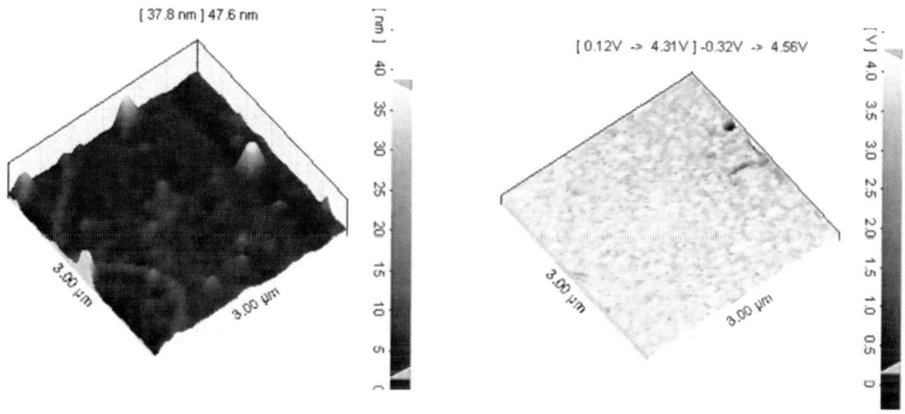

(c)

Figure 3. From left to right three dimensional height and phase images of 3×3 μm scan area of (a) fresh galss surface, (b) glass surface after exposure to bacterial solution and (c) glass surface after exposure to cell free solution.

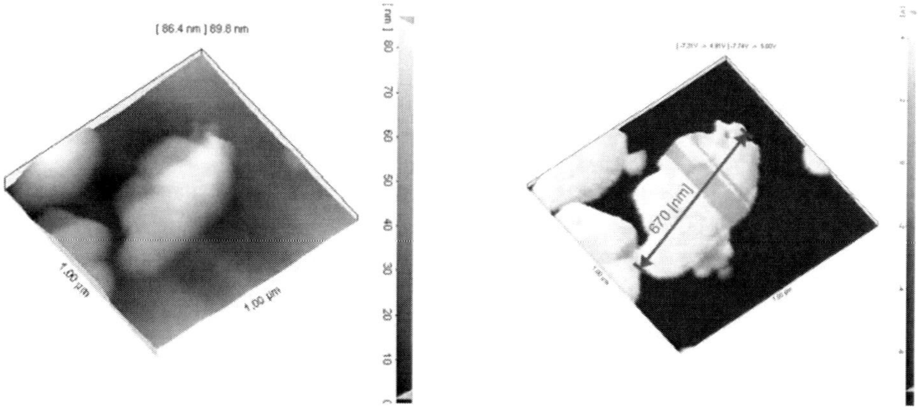

(a)

Figure 4. (Continued).

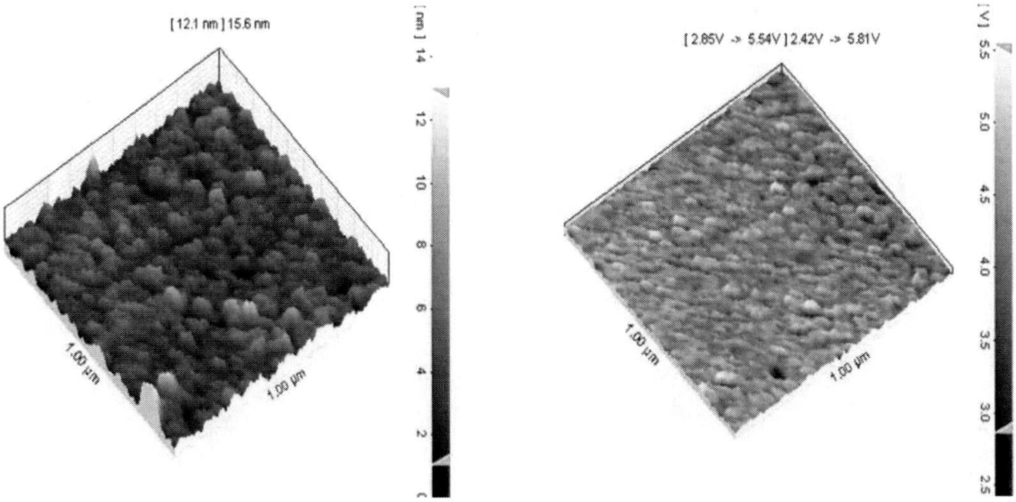

(b)

Figure 4. From left to right three dimensional height and phase images of 1×1 µm scan area of (a) bacterial cells on the glass surface and (b) glass surface after exposure to pure biosurfactant solution.

As shown in Figure 5(a), when glass exposed to crude oil, that organic compounds of the crude completely cover the surface. Adsorption of organic compound cause increase in surface roughness.

The mean roughness of the aged surface is 22 nm compared to the significantly different value of 0.164 nm for fresh glass. The fresh glass has the smallest surface roughness, and the aged glass without any treatments has the highest one. As a reference we used an aged glass surface without any treatments. In this way we could discriminate the original polar compounds and adsorbed compounds after microbial activity.

Figure 5(b) and (c) present the topography and phase images of the aged glass surface treated by bacterial and cell free solutions, respectively. These two images look fairly similar. It can be conceived that, biosurfactant adsorbed onto the surface.

The surface roughness of the glass and also aged glass, exposed to bacterial and cell free solutions are very close to each other. As the height difference between peaks and valleys (S_y) is reduced, it is possible that the bacterial products adsorbed on the organic compounds on the surface, but that is probably not the case since the contact angle was approximately 30° (Figure 2).

A contact angle of 30° and similarity of topographic and phase images and surface roughness of glass surfaces and also aged glass, exposed to bacterial and cell free solutions, suggest that biosurfactant remove the organic compound and then adsorbed on the surface. It can be concluded that, organic compounds on the surface washed off from the surface during soaking aged glass in bacterial and cell free solutions; and biosurfactants wrapped all over the surface.

AFM generally confirms the interpretation derived from contact angle measurements. There is complete coverage of the glass and aged glass surface by what is presumed to be biosurfactant which could alter surface wettability.

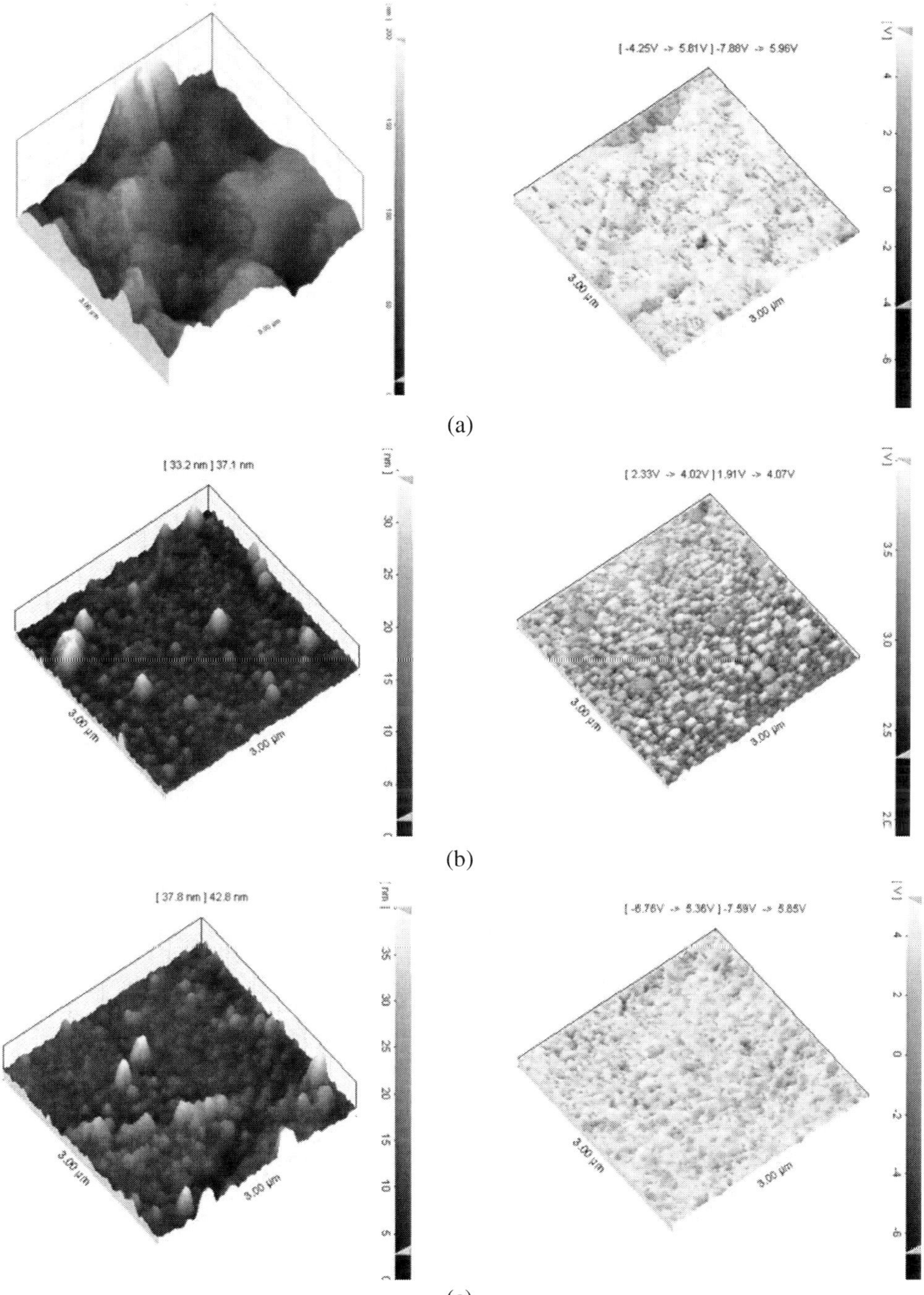

Figure 5. From left to right three dimensional height and phase images of 3×3 μm scan area of aged glass surface. (a) without treatments (b)after exposure to bacterial solution and (c) after exposure to cell free solution.

FUTURE PERSPECTIVE

Based on this work, the following conclusions are made:

Bacillus thermodenitrificans strain changed the wettability of water-wet surface towards less water-wet state and alter the wettability of oil-wet surface to a water-wet condition. AFM experiments were consistent with interpretations of macroscopic contact angle measurements, revealing that, among the principal mechanisms responsible for surface wettability changes namely, biosurfactant and other microbial product adsorption, bacterial adhesion and biofilm formation, biosurfactant adsorption considered to be the dominant mechanism for wettability alteration. Biosurfactant could remove the adsorbed oil film on the surface and adsorbed to the surface, cause significant changes in surface wettability. Only in the presence of extracellular biosurfactant, wettability alteration could be seen.

REFERENCES

Anderson, W. G. (1986). Wettability Literature Survey Part 2: Wettability Measurement. *Journal of Petroleum Technology*, *38*(11), 1246-1262.

Anderson, W. G. (1987). Wettability Literature Survey Part 4: Effects of Wettability on Capillary Pressure. *Journal of Petroleum Technology*, *39*, 1283–1300.

Armstrong, R. T. & Wildenschild, D. (2012). Investigating the pore scale mechanisms of microbial enhanced oil recovery. *Journal of Petroleum Science and Engineering*, 94-95: 155–163

Buckley, J. S., Liu Y. & Montsterleet S. (1997). Mechanisms of wetting alteration by crude oils. *Proceedings SPE International Symposium on Oilfield Chemistry*, Richardson, TX, USA, 191-202.

Crescente, C. Totsaeter, O. Hultmann, L., Stroem, M., Rasmussen, K. & Kowalewski, E. (2006). An experimental study of driving mechanisms in MIOR processes by using Rhodococcus sp. 094. *Presented at the International Symposium on Improved Oil Recovery. SPE Paper100033*, 22-26 April, Tulsa, Oklahoma.

Donaldson E. C. & Alam W. (2008). *Wettability*. Gulf Publishing Company, Houston, Texas.

Gandler, G. L., Gbosi, A., Bryant, S. L. & Britton, L. N. (2006). Mechanistic understanding of microbial plugging for improved sweep efficiency. *Proceedings SPE/DOE Symposium on Improved Oil Recovery*, SPE 10048. Tulsa, Oklahoma, USA, 22-26 April.

Hiorth, A., Kaster, K., Lohne, A., Siqveland, O. K., Berland, H., Giske, N. H. & Stavland, A. 2007. Microbial enhanced oil recovery- Mechanism. *Paper presented at the International Symposium of the Society of Core Analysts*, Calgary, Canada, 10-13 September.

Karimi M., Mahmoodi M., Niazi A., Al-Wahaibi Y. & Ayatollahi Sh. (2012). Investigating wettability alteration during MEOR process, a micro/macro scale analysis. *Colloids and Surfaces B: Biointerfaces.*, *95*, 129–136.

Kowalewski, E., Rueslatten, I., Boassen, T., Sunde, E., Stensen, J. A., Lillebo, B. L. P., Bodtker, G. & Torsvik, T. (2005). Analyzing microbial improved oil recovery processes from core floods. *Proceedings International Symposium of the Society of Core Analysis*, Toronto, Canada, 21-25 August.

Mu, B., Wu, Z., Chen, Z., Wang, X., Ni, F. & Zhou, J. (2002). Wetting behavior on quartz surfaces by the microbial metabolism and metabolic products. *Paper presented at the 7th international symposium on wettability and its effect on oil recovery*, Tasmania, Australia, 12-14 March.

Polson, E. J., Buckman, J. O., Bowen, D., Todd, A. C., Gow, M. M. & Cuthbert, S. J. (2010). An Environmental Scanning Electron Microscope Investigation Into the Effect of Biofilm on the Wettability of Quartz. *SPE Journal.*, *15* (1), 223-227.

Roosta, A., Escrochi, M., Varzandeh, F., Khatibi, J. & Ayatollahi, Sh. (2009). Investigating the mechanisms of thermally induced wettability alteration. SPE 120354. *Proceedings 16th Middle East Oil & Gas Show and Conference*, Bahrain, Kingdom of Bahrain, 15-18 March.

Seiedi, O., Rahbar, M., Nabipour, M., Emadi, M. A., Ghatee, M. H. & Ayatollahi, Sh. (2011). Atomic Force Microscopy (AFM) Investigation on the Surfactant Wettability Alteration Mechanism of Aged Mica Mineral Surfaces. *Energy Fuels*, *25*, 183–188.

Sen R. (2008). Biotechnology in petroleum recovery: The microbial EOR. *Progress in Energy and Combustion Science*, *34*, 714-724.

Shabani Afrapoli, M., Crescente, C. & Torsaeter, O. (2008). The effect of bacterial solution on the wettability index and residual oil saturation in sandstone. *Proceedings 10th International Symposium on Evaluation of Wettability and Its Effects on Oil Recovery*, Abu Dhabi, UAE, 26–28 October.

Shabani Afrapoli, M., Crescente, C., Alipour, S. & Torsaeter, O. (2009). The effect of bacterial solution on the wettability index and residual oil saturation in sandstone. *Journal of Petroleum Science and Engineering.*, *69*, 255-260.

Soudmand-asli A., Ayatollahi S. S., Mohabatkar H., Zareie M. & Shariatpanahi S. F. (2007). The in-situ microbial enhanced oil recovery in fractured porous media. *Journal of Petroleum Science and Engineering.*, *58*(1–2), 161–172.

Tiab, D. & Donaldson, E. C. (2004). *Petrophysics*. Elsevier, Gulf Professional Publishing.

Youssef, N., Elshahed, M. S. & McInerney, M. J. (2009). Microbial processes in oil fields: culprits, problems, and opportunities. *Advances in Applied Microbiology*, vol. *66*. Elsevier Academic Press Inc., SanDiego (edited, 141–251).

Zargari, S., Ostvar, S., Niazi, A. & Ayatollahi, Sh. (2010). Atomic force microscopic and wettability study of the alteration of mica and sandstone by a biosurfactant-producing bacterium Bacillus *Thermodenitrificans. Journal of Advanced Microscopy Research, 5.* 1-6.

Zekri, A. Y., Ghannam, M. T. & Almehaideb, R. A. (2003). Carbonate rocks wettability changes induced by microbial solution. *SPE 80527*, *Proceedings SPE Asia Pacific Oil and Gas Conference and Exhibition*, Jakarta, Indonesia, 15–17 April.

EDITOR'S CONTACT INFORMATION

Dr. Sardar A. Farooq,
Associate Professor
Department of Biology, College of Science
Sultan Qaboos University
P.O. No: 36, Al-Khoud,
Muscat, 123, Sultanate of Oman
Tel: 00968 2414 1487
Fax: 00968 24141437
Emai: sfarook@squ.edu.om

Dr. Raeid M. M. Abed,
Assistant Professor
Department of Biology, College of Science
Sultan Qaboos University
P.O. No: 36, Al-Khoud,
Muscat, 123, Sultanate of Oman
Tel: 00968 24142406
Fax: 00968 24141437
E-mail: rabed@squ.edu.om

Dr. Senan Baqir,
Assistant Professor
Department of Biology, College of Science
Sultan Qaboos University
P.O. No: 36, Al-Khoud,
Muscat, 123, Sultanate of Oman
Tel: 00968 24141450
Fax: 00968 24141437
E-mail: baqirs@squ.edu.om

INDEX

A

B

C

D

E

F

G

H

I

J

K

L

M

N

O

P

Q

R

S

T

U

V

W

Y

Z